STUDENT'S SOLUTIONS MANUAL

TO ACCOMPANY
MICHAEL SULLIVAN'S

PRECALCULUS

THIRD EDITION

KATY MURPHY
and
MICHAEL SULLIVAN

DELLEN
an imprint of
MACMILLAN PUBLISHING COMPANY
NEW YORK

MAXWELL MACMILLAN CANADA
TORONTO

MAXWELL MACMILLAN INTERNATIONAL
NEW YORK OXFORD SINGAPORE SYDNEY

Macmillan Publishing Company
866 Third Avenue
New York, NY 10022

Macmillan Publishing Company is part of the Maxwell Communication
Group of Companies.

Maxwell Macmillan Canada, Inc.
1200 Eglinton Avenue East, Suite 200
Don Mills, Ontario M3C 3N1

ISBN: 0-02-418425-X

Printing: 1 2 3   Year: 3 4 5

≡ PREFACE

This manual contains detailed solutions to all the odd-numbered problems in *Precalculus*, Third Edition, by Michael Sullivan. Preceding the solutions to some of the groups of problems we have listed step-by-step procedures which may be applied to each exercise in the group. Hopefully, these will enable you to develop a systematic approach to solving certain types of exercises. Our desire is that after seeing several examples, you will be able to solve problems without referring to this manual.

We wish to thank everyone who has helped with this project. Special credit goes to our publisher, Don Dellen, for his many efforts and encouragements. The enormous job of typing was done by Brenda Dobson, and the art work was prepared by Kelly Evans. A special thank you to our families for their support during this project.

Finally, we would be grateful to hear from readers who discover any errors in this solutions manual.

Katy Murphy
and
Michael Sullivan

≡ CONTENTS

CONTENTS

PRELIMINARIES

≡ **EXERCISE 1.1 REVIEW TOPICS FROM ALGEBRA AND GEOMETRY**

1. $\frac{1}{2} > 0$ since $\frac{1}{2} - 0 = \frac{1}{2}$ is positive.

3. $-1 > -2$ since $-1 - (-2) = -1 + 2 = 1$ is positive.

5. $\pi > 3.14$ since $\pi \approx 3.1416$ and $3.1416 - 3.1400 = .0016$ is positive, so then $\pi - 3.14$ is positive.

7. $\frac{1}{2} = 0.5$

9. $\frac{2}{3} < 0.67$ since $\frac{2}{3} = 0.666 \ldots \approx 0.667$ and $0.670 - 0.667 = 0.003$ is positive, so that $0.67 - \frac{2}{3}$ is positive.

11.

13. $x > 0$

15. $x < 2$

17. $x \leq 1$

19. $2 < x < 5$

21. $[0, 4]$

23. $[4, 6)$

25. $2 \leq x \leq 5$

27. $x \geq 4$

29. $|x + y| = |2 + (-3)|$
         $= |2 - 3| = |-1|$
         $= 1$

31. $|x| + |y| = |2| + |-3|$
         $= 2 + 3 = 5$

33. $3^0 = 1$

35. $4^{-2} = \dfrac{1}{4^2} = \dfrac{1}{16}$

37. $\left(\dfrac{2}{3}\right)^2 = \dfrac{2^2}{3^2} = \dfrac{4}{9}$

39. $3^{-6} \cdot 3^4 = 3^{-6+4} = 3^{-2}$
$$= \dfrac{1}{3^2} = \dfrac{1}{9}$$

41. $\left(\dfrac{2}{3}\right)^{-2} = \dfrac{1}{\left(\dfrac{2}{3}\right)^2} = \dfrac{1}{\dfrac{2^2}{3^2}} = \dfrac{3^2}{2^2} = \dfrac{9}{4}$

43. $\dfrac{2^3 \cdot 3^2}{2 \cdot 3^{-2}} = \dfrac{2^3}{2} \cdot \dfrac{3^2}{3^{-2}}$
$$= 2^{3-1} \cdot 3^{2-(-2)}$$
$$= 2^2 \cdot 3^4 = 4 \cdot 81$$
$$= 324$$

45. $9^{3/2} = \sqrt{9^3} = \left(\sqrt{9}\right)^3 = 3^3 = 27$

47. $(-8)^{4/3} = \left(\sqrt[3]{-8}\right)^4 = (-2)^4 = 16$

49. $\sqrt{32} = \sqrt{16 \cdot 2} = \sqrt{16} \cdot \sqrt{2} = 4\sqrt{2}$

51. $\sqrt[3]{\dfrac{-8}{27}} = \dfrac{-2}{3}$

53. $x^0 y^2 = 1 \cdot y^2 = y^2$

55. $x^{-2}y = \dfrac{1}{x^2}y = \dfrac{y}{x^2}$

57. $\dfrac{x^{-2}y^3}{xy^4} = \dfrac{x^{-2}}{x} \cdot \dfrac{y^3}{y^4} = x^{-2-1}y^{3-4}$
$$= x^{-3}y^{-1} = \dfrac{1}{x^3}\dfrac{1}{y} = \dfrac{1}{x^3 y}$$

59. $\left(\dfrac{4x}{5y}\right)^{-2} = \dfrac{1}{\left(\dfrac{4x}{5y}\right)^2} = \dfrac{1}{\dfrac{(4x)^2}{(5y)^2}} = \dfrac{(5y)^2}{(4x)^2} = \dfrac{5^2 \cdot y^2}{4^2 \cdot x^2} = \dfrac{25y^2}{16x^2}$

61. $\dfrac{x^{-1}y^{-2}z}{x^2 y z^3} = \dfrac{x^{-1}}{x^2}\dfrac{y^{-2}}{y}\dfrac{z}{z^3} = x^{-1-2}y^{-2-1}z^{1-3} = x^{-3}y^{-3}z^{-2} = \dfrac{1}{x^3}\dfrac{1}{y^3}\dfrac{1}{z^2} = \dfrac{1}{x^3 y^3 z^2}$

63. $\dfrac{(-2)^3 x^4 (yz)^2}{3^2 x y^3 z^4} = \dfrac{-8}{9}\dfrac{x^4}{x}\dfrac{y^2}{y^3}\dfrac{z^2}{z^4} = \dfrac{-8}{9}x^{4-1}y^{2-3}z^{2-4} = \dfrac{-8}{9}x^3 y^{-1} z^{-2} = \dfrac{-8}{9}x^3 \dfrac{1}{y}\dfrac{1}{z^2}$
$$= \dfrac{-8x^3}{9yz^2}$$

65. $\dfrac{x^{-2}}{x^{-2} + y^{-2}} = \dfrac{\dfrac{1}{x^2}}{\dfrac{1}{x^2} + \dfrac{1}{y^2}} = \dfrac{\dfrac{1}{x^2}}{\dfrac{y^2 + x^2}{x^2 y^2}} = \dfrac{x^2 y^2}{x^2(y^2 + x^2)} = \dfrac{x^2}{x^2} \cdot \dfrac{y^2}{y^2 + x^2}$
$$= x^{2-2}\dfrac{y^2}{y^2 + x^2} = x^0 \dfrac{y^2}{y^2 + x^2} = \dfrac{y^2}{x^2 + y^2}$$

67. $\left(\dfrac{3x^{-1}}{4y^{-1}}\right)^{-2} = \dfrac{1}{\left(\dfrac{3x^{-1}}{4y^{-1}}\right)^2} = \dfrac{1}{\dfrac{(3x^{-1})^2}{(4y^{-1})^2}} = \dfrac{(4y^{-1})^2}{(3x^{-1})^2} = \dfrac{4^2y^{-2}}{3^2x^{-2}} = \dfrac{16}{9}\,\dfrac{\dfrac{1}{y^2}}{\dfrac{1}{x^2}} = \dfrac{16}{9}\,\dfrac{x^2}{y^2}$

$$= \dfrac{16x^2}{9y^2}$$

For legs a and b of a right triangle in Problems 69 - 78, we use $c^2 = a^2 + b^2$ to find the hypotenuse c:

69. For $a = 5$ and $b = 12$, $\begin{aligned} c^2 &= a^2 + b^2 \\ &= 5^2 + 12^2 \\ &= 25 + 144 \\ c^2 &= 169 \\ \text{then } c &= 13 \end{aligned}$

71. For $a = 10$ and $b = 24$, $\begin{aligned} c^2 &= 10^2 + 24^2 \\ &= 100 + 576 \\ c^2 &= 676 \\ \text{then } c &= 26 \end{aligned}$

73. For $a = 7$ and $b = 24$, $\begin{aligned} c^2 &= 7^2 + 24^2 \\ &= 49 + 576 \\ c^2 &= 625 \\ \text{then } c &= 25 \end{aligned}$

75. For $a = 3$ and $c = 5$, $\begin{aligned} c^2 &= a^2 + b^2 \\ 5^2 &= 3^2 + b^2 \\ b^2 &= 5^2 - 3^2 \\ b^2 &= 25 - 9 \\ b^2 &= 16 \\ \text{then } b &= 4 \end{aligned}$

77. For $b = 7$ and $c = 25$, $\begin{aligned} c^2 &= a^2 + b^2 \\ a^2 &= c^2 - b^2 \\ a^2 &= 25^2 - 7^2 \\ &= 625 - 49 \\ a^2 &= 576 \\ \text{then } a &= 24 \end{aligned}$

In Problems 79 - 83, we will test whether the given triangles are right triangles by using $c^2 = a^2 + b^2$. The hypotenuse must be the longest side in any case:

79. For sides 3, 4, 5, let $c = 5$: $\begin{aligned} c^2 &= a^2 + b^2 \\ 5^2 &= 3^2 + 4^2 \\ 25 &= 9 + 16 \\ 25 &= 25 \end{aligned}$

The given triangle is a right triangle with hypotenuse of length 5.

1.1 REVIEW TOPICS FROM ALGEBRA AND GEOMETRY   3

81. For sides 4, 5, and 6, let $c = 6$: but $6^2 \neq 4^2 + 5^2$

$$\text{since } 36 \neq 16 + 25 = 41$$

The triangle is not a right triangle.

83. For sides 7, 24, 25, let $c = 25$:
$$25^2 = 7^2 + 24^2$$
$$625 = 49 + 576$$
$$625 = 625$$

The triangle is a right triangle with hypotenuse of length 25.

85. Enter $\boxed{8.51}$ Press $\boxed{x^2}$

Display: 8.51   72.4201

Thus, $(8.51)^2 \approx 72.42$

87. Enter $\boxed{4.1}$ Press $\boxed{+}$ Enter $\boxed{3.2}$ Press $\boxed{\times}$

Display   4.1   3.2

Enter $\boxed{8.3}$ Press $\boxed{=}$

Display: 8.3   30.66

Thus, $4.1 + (3.2)(8.3) = 30.66$

89. Enter $\boxed{8.6}$ Press $\boxed{x^2}$ Press $\boxed{+}$ Enter $\boxed{6.1}$

Display: 8.6   73.96   6.1

Press $\boxed{x^2}$ Press $\boxed{=}$

Display: 37.21   111.17

Thus, $(8.6)^2 + (6.1)^2$ 111.17

91. $8.6 + 10.2/4.2 \approx 11.02857143 \approx 11.03$

93. $\dfrac{2.3 - 9.25}{8.91 + 5.4} = \dfrac{-6.95}{14.31} \approx -0.4856744 \approx -0.49$

95. $\dfrac{\pi + 8}{10.2 + 8.6} \approx 0.592637907 \approx 0.59$

97. The diagonal of the rectangle forms two right triangles. The length, 8 inches, is the base of the triangle, and the width, 5 inches, is the altitude or height of the triangle. The diagonal is the hypotenuse, so

$$c^2 = a^2 + b^2$$
$$c^2 = 5^2 + 8^2$$
$$= 89$$
$$c = \sqrt{89} \approx 9.4 \text{ inches}$$

1 PRELIMINARIES

99.

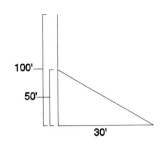

$$c^2 = a^2 + b^2$$
$$c^2 = 50^2 + 30^2$$
$$c^2 = 3400$$
$$c \approx 58.3 \text{ feet}$$

The guy wire needs to be 58.3 feet.

101. We have the triangle with side 3960 miles and hypotenuse $3960 + \dfrac{1454}{5280} = 3960.275$ miles. Let side a be the distance in miles one can see from the Sears Tower: Here $c = 3960.3$ and $b = 3960$:

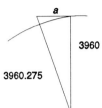

$$(3960.275)^2 = a^2 + (3960)^2$$
$$15683778 = a^2 + 15681600$$
$$2178 = a^2$$
$$46.7 \text{ miles} = a$$

103. We have the triangle with hypotenuse c representing the distance $3960 + \dfrac{6}{5280} = 3960.0011$ miles. Let a be the distance in miles of the ship from shore that a 6-foot-tall person can see.

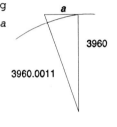

$$c^2 = a^2 + b^2$$
$$(3960.0011)^2 = a^2 + (3960)^2$$
$$15681609 = a^2 + 15681600$$
$$9 = a^2$$
$$3 \text{ miles} = a$$

105. Since $a \le b$, then $a - b \le 0$
$$(a - b)c \le 0 \cdot c$$
$$ca - cb \le 0$$
$$ca \le cb \text{ or } ac \le bc$$
Thus, if $a \le b$ and $c > 0$, then $ac \le bc$

107. Since $a < b$, $a \cdot \dfrac{1}{2} < b \cdot \dfrac{1}{2}$
$$\frac{a}{2} < \frac{b}{2}$$
$$\frac{a}{2} + \frac{a}{2} < \frac{b}{2} + \frac{a}{2}$$
$$a < \frac{a + b}{2}$$

Also,
$$\frac{a}{2} < \frac{b}{2}$$
$$\frac{a}{2} + \frac{b}{2} < \frac{b}{2} + \frac{b}{2}$$
$$\frac{a + b}{2} < b$$

Thus,
$$a < \frac{a + b}{2} < b$$

1.1 REVIEW TOPICS FROM ALGEBRA AND GEOMETRY

5

≡ EXERCISE 1.2 EQUATIONS

1.
$$6 - x = 2x + 9$$
$$(6 - x) - 6 = (2x + 9) - 6$$
$$-x = 2x + 3$$
$$-x - 2x = 2x + 3 - 2x$$
$$-3x = 3$$
$$\frac{-3x}{-3} = \frac{3}{-3}$$
$$x = -1$$

Check:
$$6 - x = 2x + 9$$
$$6 - (-1) \stackrel{?}{=} 2(-1) + 9$$
$$6 + 1 \stackrel{?}{=} -2 + 9$$
$$7 = 7$$

3.
$$2(3 + 2x) = 3(x - 4)$$
$$6 + 4x = 3x - 12$$
$$(6 + 4x) - 6 = (3x - 12) - 6$$
$$4x = 3x - 18$$
$$4x - 3x = 3x - 18 - 3x$$
$$x = -18$$

Check:
$$2(3 + 2x) = 3(x - 4)$$
$$2(3 + 2)(-18)) \stackrel{?}{=} 3(-18 - 4)$$
$$2(3 - 36 \stackrel{?}{=} 3(-22)$$
$$2(-33) \stackrel{?}{=} -66$$
$$-66 = -66$$

5.
$$8x - (2x + 1) = 3x - 10$$
$$8x - 2x - 1 = 3x - 10$$
$$6x - 1 = 3x - 10$$
$$(6x - 1) + 1 = (3x - 10 + 1$$
$$6x = 3x - 9$$
$$6x - 3x = (3x - 9) - 3x$$
$$3x = -9$$
$$\frac{3x}{3} = \frac{-9}{3}$$
$$x = -3$$

Check:
$$8x - (2x + 1) = 3x - 10$$
$$8(-3) - (2(-3) + 1) \stackrel{?}{=} 3(-3) - 10$$
$$-24 - (-6 + 1) \stackrel{?}{=} -9 - 10$$
$$-24 - (-5) \stackrel{?}{=} -19$$
$$-24 + 5 \stackrel{?}{=} -19$$
$$-19 = -19$$

7.
$$\frac{1}{2}x - 4 = \frac{3}{4}x$$
$$4\left[\frac{1}{2}x - 4\right] = 4\left[\frac{3}{4}x\right]$$
$$2x - 16 = 3x$$
$$(2x - 16) + 16 = 3x + 16$$
$$2x = 3x + 16$$
$$2x - 3x = 3x + 16 - 3x$$
$$-x = 16$$
$$\frac{-x}{-1} = \frac{16}{-1}$$
$$x = -16$$

Check:
$$\frac{1}{2}x - 4 = \frac{3}{4}x$$
$$\frac{1}{2}(-16) - 4 \stackrel{?}{=} \frac{3}{4}(-16)$$
$$-8 - 4 \stackrel{?}{=} -12$$
$$-12 = -12$$

9.
$$0.9t = 0.4 + 0.1t$$
$$0.9t - 0.1t = (0.4 + 0.1t) - 0.1t$$
$$0.8t = 0.4$$
$$\frac{0.8t}{0.8} = \frac{0.4}{0.8}$$
$$t = 0.5$$

Check:
$$0.9t = 0.4 + 0.1t$$
$$0.9(0.5) \stackrel{?}{=} 0.4 + 0.01(0.5)$$
$$0.45 \stackrel{?}{=} 0.4 + 0.05$$
$$0.45 = 0.45$$

11. $\dfrac{2}{y} + \dfrac{4}{y} = 3$

$y\left[\dfrac{2}{y} + \dfrac{4}{y}\right] = [3]$

$2 + 4 = 3y$

$6 = 3y$

$\dfrac{6}{3} = \dfrac{3y}{3}$

$2 = y$

$y = 2$

Check:

$\dfrac{2}{y} + \dfrac{4}{y} = 3$

$\dfrac{2}{2} + \dfrac{4}{2} \overset{?}{=} 3$

$1 + 2 \overset{?}{=} 3$

$3 = 3$

13. $(x + 7)(x - 1) = (x + 1)^2$

$x^2 + 6x - 7 = x^2 + 2x + 1$

$(x^2 + 6x - 7) - x^2 = (x^2 + 2x + 1) - x^2$

$6x - 7 = 2x + 1$

$(6x - 7) + 7 = (2x + 1) + 7$

$6x = 2x + 8$

$6x - 2x = (2x + 8) - 2x$

$4x = 8$

$\dfrac{4x}{4} = \dfrac{8}{4}$

$x = 2$

Check:

$x + 7)(x - 1) = (x + 1)^2$

$(2 + 7)(2 - 1) \overset{?}{=} (2 + 1)^2$

$(9)(1) \overset{?}{=} 3^2$

$9 = 9$

15. $x(2x - 3) = (2x + 1)(x - 4)$

$2x^2 - 3x = 2x^2 - 7x - 4$

$(2x - 3x) - 2x^2 = (2x^2 - 7x - 4) - 2x^2$

$-3x = -7x - 4$

$-3x + 7x = (-7x - 4) + 7x$

$4x = -4$

$\dfrac{4x}{4} = \dfrac{-4}{4}$

$x = -1$

Check:

$x(2x - 3) = (2x + 1)(x - 4)$

$-1(2(-1) - 3) \overset{?}{=} (2(-1) + 1)(-1 - 4)$

$-1(-2 - 3) \overset{?}{=} (-2 + 1)(-5)$

$-1(-5) \overset{?}{=} (-1)(-5)$

$5 = 5$

17. $z(z^2 + 1) = 3 + z^3$

$z^3 + z = 3 + z^3$

$(z^3 + z) - z^3 = (3 + z^3) - z^3$

$z = 3$

Check: $z(z^2 + 1) = 3 + z^3$

$3(3^2 + 1) \overset{?}{=} 3 + 3^3$

$3(9 + 1) \overset{?}{=} 30$

$3(10) \overset{?}{=} 30$

$30 = 30$

19. $\dfrac{x}{x - 3} + 3 = \dfrac{3}{x - 3}$

Note that $x - 3$ cannot equal zero so $x = 3$ is <u>NOT</u> in the domain of the variable.

$(x - 3)\left(\dfrac{x}{x - 3} + 3\right) = \dfrac{3}{x - 3}(x - 3)$

$x + 3(x - 3) = 3$

$x + 3x - 9 = 3$

$4x = 12$

$x = 3$

But $x = 3$ is not in the domain of the variable. Hence, the equation has no solution.

21.

$$x^2 = 9x$$
$$x^2 - 9x = 0$$
$$x(x - 9) = 0$$
$$x = 0, \quad x = 9$$

The solution set is $\{0, 9\}$

23.

$$t^3 - 9t^2 = 0$$
$$t^2(t - 9) = 0$$
$$t^2 = 0, \quad t - 9 = 0$$
$$t = 0, \quad t = 9$$

The solution set is $\{0, 9\}$

25.

$$\frac{2x}{x^2 - 4} = \frac{4}{x^2 - 4} - \frac{1}{x + 2}$$

Note that $x^2 - 4$ cannot equal zero and $x + 2$ cannot equal zero. Therefore, $x = -2$ and $x = 2$ are not in the domain of the variable.

$$(x^2 - 4)\frac{2x}{x^2 - 4} = (x^2 - 4)\left[\frac{4}{x^2 - 4} - \frac{1}{x + 2}\right]$$

$$2x = 4 - \frac{x^2 - 4}{x + 2}$$

$$2x = 4 - \frac{(x - 2)(x + 2)}{x + 2}$$

$$2x = 4 - (x - 2)$$
$$2x = 4 - x + 2$$
$$2x = 6 - x$$
$$3x = 6$$
$$x = 2$$

But $x = 2$ is not in the domain of the variable. The equation has no solution.

27.

$$\frac{x}{x + 2} = \frac{1}{2}$$

$$2(x + 2)\left[\frac{x}{x + 2}\right] = 2(x + 2)\left[\frac{1}{2}\right]$$
$$2x = x + 2$$
$$2x - x = (x + 2) - x$$
$$x = 2$$

Check:

$$\frac{x}{x + 2} = \frac{1}{2}$$
$$\frac{2}{2 + 2} \stackrel{?}{=} \frac{1}{2}$$
$$\frac{2}{4} \stackrel{?}{=} \frac{1}{2}$$
$$\frac{1}{2} = \frac{1}{2}$$

29.

$$\frac{3}{2x - 3} = \frac{2}{x + 5}$$

$$(2x - 3)(x + 5)\left[\frac{3}{2x - 3}\right] = (2x - 3)(x + 5)\left[\frac{2}{x + 5}\right]$$
$$3(x + 5) = 2(2x - 3)$$
$$(3x + 15) = 4x - 6$$
$$(3x + 15) - 15 = (4x - 6) - 15$$
$$3x = 4x - 21$$
$$3x - 4x = (4x - 21) - 4x$$
$$-x = -21$$
$$\frac{-x}{-1} = \frac{-21}{-1}$$
$$x = 21$$

8   1 PRELIMINARIES

Check:
$$\frac{3}{2x - 3} = \frac{2}{x + 5}$$

$$\frac{3}{2(21) - 3} \overset{?}{=} \frac{2}{21 + 5}$$

$$\frac{3}{42 - 3} = \frac{2}{26}$$

$$\frac{3}{39} \overset{?}{=} \frac{1}{13}$$

$$\frac{1}{13} = \frac{1}{13}$$

31.
$$(x + 2)(3x) = (x + 2)(6)$$
$$3x^2 + 6x = 6x + 12$$
$$(3x^2 + 6x) - 6x = (6x + 12) - 6x$$
$$3x^2 = 12$$

$$\frac{3x^2}{3} = \frac{12}{3}$$
$$x^2 = 4$$
$$x = 2 \quad \text{or} \quad x = -2$$
$$\{-2, \ 2\}$$

Alternate Solution:

$$(x + 2)(3x) = (x + 2)(6)$$
$$\frac{(x + 2)(3x)}{x + 2} = \frac{(x + 2)(6)}{x + 2}$$
$$3x = 6$$
$$x = 2$$

Check:
$$(x + 2)(3x) \overset{?}{=} (x + 2)(6)$$
$$(2 + 2)(3(2)) \overset{?}{=} (2 + 2)(6)$$
$$(4)(6) \overset{?}{=} (4)(6)$$
$$24 = 24$$

Check:
$$(x + 2)(3x) \overset{?}{=} (x + 2)(6)$$
$$(-2 + 2)(3(-2)) \overset{?}{=} (-2 + 2)(6)$$
$$0 = 0$$

33.
$$\frac{6t + 7}{4t - 1} = \frac{3t + 8}{2t - 4}$$

$$(4t - 1)(2t - 4)\left[\frac{6t + 7}{4t - 1}\right] = (4t - 1)(2t - 4)\left[\frac{3t + 8}{2t - 4}\right]$$
$$(2t - 4)(6t + 7) = (4t - 1)(3t + 8)$$
$$12t^2 - 10t - 28 = 12t^2 + 29t - 8$$
$$(12t^2 - 10t - 28) - 12t^2 = (12t^2 + 29t - 8) - 12t^2$$
$$-10t - 28 = 29t - 8$$
$$(-10t - 28) + 28 = (29t - 8) + 28$$
$$-10t = 29t + 20$$
$$-10t - 29t = 29t + 20 - 29t$$
$$-39t = 20$$
$$\frac{-39t}{-39} = \frac{20}{-39}$$
$$t = \frac{-20}{39}$$

Check:
$$\frac{6t + 7}{4t - 1} = \frac{3t + 8}{2t - 4}$$

$$\frac{6\left(\frac{-20}{39}\right) + 7}{4\left(\frac{-20}{39}\right) - 1} \overset{?}{=} \frac{3\left(\frac{-20}{39}\right) + 8}{2\left(\frac{-20}{39}\right) - 4}$$

$$\frac{\frac{-120}{39} + 7}{\frac{-80}{39} - 1} \overset{?}{=} \frac{\frac{-60}{39} + 8}{\frac{-40}{39} - 4}$$

$$\frac{\frac{-120 + 273}{39}}{\frac{-80 - 39}{39}} \overset{?}{=} \frac{\frac{-60 + 312}{39}}{\frac{-40 - 156}{39}}$$

$$\frac{\frac{153}{39}}{\frac{-119}{39}} \overset{?}{=} \frac{\frac{252}{39}}{\frac{-196}{39}}$$

$$\frac{153}{39} \cdot \frac{39}{-119} = \frac{252}{39} \cdot \frac{39}{-196}$$

$$\frac{153}{119} - \frac{252}{196} = -\frac{252}{39} \cdot \frac{39}{-196}$$

$$-\frac{153}{119} = -\frac{252}{196}$$

$$-\frac{9 \cdot 17}{7 \cdot 17} = -\frac{4 \cdot 7 \cdot 9}{4 \cdot 7 \cdot 9}$$

$$-\frac{9}{7} = -\frac{9}{7}$$

35.

$$\frac{2}{x-2} = \frac{3}{x+5} + \frac{10}{(x+5)(x-2)}$$

$$(x+5)(x-2)\left[\frac{2}{x-2}\right] = (x+5)(x-2)\left[\frac{3}{x+5} + \frac{10}{(x+5)(x-2)}\right]$$

$$2(x+5) = 3(x-2) + 10$$
$$2x + 10 = 3x - 6 + 10$$
$$2x + 10 = 3x + 4$$
$$(2x+10) - 10 = (3x+4) - 10$$
$$2x = 3x - 6$$
$$2x - 3x = (3x-6) - 3x$$
$$-x = -6$$
$$\frac{-x}{-1} = \frac{-6}{-1}$$
$$x = 6$$

Check:

$$\frac{2}{x-2} = \frac{3}{x+5} + \frac{10}{(x+5)(x-2)}$$

$$\frac{2}{6-2} \overset{?}{=} \frac{3}{6+5} + \frac{10}{(6+5)(6-2)}$$

$$\frac{2}{4} \overset{?}{=} \frac{3}{11} + \frac{10}{(11)(4)}$$

$$\frac{1}{2} \overset{?}{=} \frac{12}{44} + \frac{10}{44}$$

$$\frac{1}{2} \overset{?}{=} \frac{22}{44}$$

$$\frac{1}{2} = \frac{1}{2}$$

37.
$$|2x| = 6$$
$$2x = 6 \quad \text{or} \quad 2x = -6$$
$$x = 3 \quad \text{or} \quad x = -3$$
The solution set is {-3, 3}.

39.
$$|2x+3| = 5$$
$$2x + 3 = 5 \quad \text{or} \quad 2x + 3 = -5$$
$$2x = 2 \quad \text{or} \quad 2x = -8$$
$$x = 1 \qquad\qquad x = -4$$
The solution set is (-4, 1}.

41.
$$|1 - 4t| = 5$$
$$1 - 4t = 5 \quad \text{or} \quad 1 - 4t = -5$$
$$-4t = 4 \quad \text{or} \quad -4t = -6$$
$$t = -1 \text{ or } \quad t = \frac{3}{2}$$
The solution set is $\left\{-1, \frac{3}{2}\right\}$.

43.
$$|-2x| = 8$$
$$-2x = 8 \quad \text{or} \quad -2x = -8$$
$$x = -4 \qquad\qquad x = 4$$
The solution set is {-4, 4}.

45.
$$|-2|x = 4$$
$$2x = 4$$
$$x = 2$$
The solution is 2.

47.
$$\frac{2}{3}|x| = 8$$
$$|x| = 12$$
$$x = 12 \quad \text{or } x = -12$$
The solution set is {-12, 12}.

49.
$$\left|\frac{x}{3} + \frac{2}{5}\right| = 2$$
$$\frac{x}{3} + \frac{2}{5} = 2 \quad \text{or} \quad \frac{x}{3} + \frac{2}{5} = -2$$
$$5x + 6 = 30 \quad \text{or} \quad 5x + 6 = -30$$
$$5x = 24 \quad \text{or} \quad 5x = -36$$
$$x = \frac{24}{5} \quad \text{or} \quad x = \frac{-36}{5}$$

The solution set is $\left\{\frac{-36}{5}, \frac{24}{5}\right\}$.

51.
$$|x - 2| = -\frac{1}{2}$$

It is not possible that $|x - 2| < 0$. Therefore, there is no real solution.

1 PRELIMINARIES

53.
$$\left| x^2 - 4 \right| = 0$$
$$x^2 - 4 = 0$$
$$(x + 2)(x - 2) = 0$$
$$x = -2 \quad \text{or} \quad x = 2$$

The solution set is {-2, 2}.

55.
$$\left| x^2 - 2x \right| = 3$$
$$x^2 - 2x = 3 \quad \text{or} \quad x^2 - 2x = -3$$
$$x^2 - 2x - 3 = 0 \quad \text{or} \quad x^2 - 2x + 3 = 0$$
$$(x - 3)(x + 1) = 0 \quad \text{or} \quad \text{No real solution.}$$
$$(\text{Note that } b^2 - 4ac = -8)$$
$$x = -1 \quad \text{or} \quad x = 3$$

The solution set is {-1, 3}.

57.
$$\left| x^2 + x - 1 \right| = 1$$
$$x^2 + x - 1 = 1 \quad \text{or} \quad x^2 + x - 1 = -1$$
$$x^2 + x - 2 = 0 \quad \text{or} \quad x^2 + x = 0$$
$$(x + 2)(x - 1) = 0 \quad \text{or} \quad x(x + 1) = 0$$
$$x = -2 \quad \text{or} \quad x = 1 \quad \text{or} \quad x = -1 \quad \text{or} \quad x = 0$$

The solution set is {-2, -1, 0, 1}.

59.
$$x^2 = 4x$$
$$x^2 - 4x = 0$$
$$x(x - 4) = 0$$
$$x = 0 \quad \text{or} \quad x - 4 = 0$$
$$x = 4$$

The solution set is {0, 4}.

61.
$$z^2 + 4z - 12 = 0$$
$$(z + 6)(z - 2) = 0$$
$$z + 6 = 0 \quad z - 2 = 0$$
$$z = -6 \quad \quad z = 2$$
$$\{-6, 2)$$

63.
$$2x^2 - 5x - 3 = 0$$
$$(2x + 1)(x - 3) = 0$$
$$2x + 1 = 0 \quad \text{or} \quad x - 3 = 0$$
$$x = -\frac{1}{2} \quad \quad x = 3$$

The solution set is $\left\{ -\frac{1}{2}, 3 \right\}$.

65.
$$x(x - 7) + 12 = 0$$
$$x^2 - 7x + 12 = 0$$
$$(x - 3)(x - 4) = 0$$
$$x - 3 = 0 \quad \text{or} \quad x - 4 = 0$$
$$x = 3 \quad \quad x = 4$$

The solution set is {3, 4}.

67.
$$4x^2 + 9 = 12x$$
$$4x^2 - 12x + 9 = 0$$
$$(2x - 3)(2x - 3) = 0$$
$$2x - 3 = 0$$
$$x = \frac{3}{2}$$

The solution is $\frac{3}{2}$.

69.
$$6x - 5 = \frac{6}{x}$$
$$6x^2 - 5x = 6$$
$$6x^2 - 5x - 6 = 0$$
$$(2x - 3)(3x + 2) = 0$$
$$2x - 3 = 0 \quad \text{or} \quad 3x + 2 = 0$$
$$x = \frac{3}{2} \quad \quad x = -\frac{2}{3}$$

The solution set is $\left\{ -\frac{2}{3}, \frac{3}{2} \right\}$.

71.
$$\frac{4(x-2)}{x-3} + \frac{3}{x} = \frac{-3}{x(x-3)}$$
$$4x(x-2) + 3(x-3) = -3$$
$$4x^2 - 8x + 3x - 9 = -3$$
$$4x^2 - 5x - 6 = 0$$
$$(4x+3)(x-2) = 0$$
$$4x + 3 = 0 \quad \text{or} \quad x - 2 = 0$$
$$x = \frac{-3}{4} \qquad\qquad x = 2$$

The solution set is $\left\{\frac{-3}{4}, 2\right\}$.

For Problems 73-85, we get each equation into standard form

$ax^2 + bx + c = 0$ *and use the quadratic formula* $x = \dfrac{-b \pm \sqrt{b^2 - 4ac}}{2a}$

73.
$$x^2 - 4x + 2 = 0$$
Here, $a = 1$, $b = -4$, $c = 2$.
$$x = \frac{-(-4) \pm \sqrt{(-4)^2 - 4(1)(2)}}{2(1)}$$
$$= \frac{4 \pm \sqrt{16 - 8}}{2}$$
$$= \frac{4 \pm 2\sqrt{2}}{2}$$
$$= 2 \pm \sqrt{2}$$
$$\left\{2 - \sqrt{2}, \; 2 + \sqrt{2}\right\}$$

75.
$$x^2 - 5x - 1 = 0$$
Here, $a = 1$, $b = -5$, $c = -1$.
$$x = \frac{-(-5) \pm \sqrt{(-5)^2 - 4(1)(-1)}}{2(1)}$$
$$= \frac{5 \pm \sqrt{25 + 4}}{2}$$
$$= \frac{5 \pm \sqrt{29}}{2}$$
$$\left\{\frac{5 - \sqrt{29}}{2}, \; \frac{5 + \sqrt{29}}{2}\right\}$$

77.
$$2x^2 - 5x + 3 = 0$$
Here, $a = 2$, $b = -5$, $c = 3$.
$$x = \frac{-(-5) \pm \sqrt{(-5)^2 - 4(2)(3)}}{2(2)}$$
$$= \frac{5 \pm \sqrt{25 - 24}}{4}$$
$$= \frac{5 \pm 1}{4}$$
$$\left\{1, \; \frac{3}{2}\right\}$$

79.
$$4y^2 - y + 2 = 0$$
Here, $a = 4$, $b = -1$, $c = 2$.
$$x = \frac{-(-1) \pm \sqrt{(-1)^2 - 4(4)(2)}}{2(4)}$$
$$= \frac{1 \pm \sqrt{1 - 32}}{8}$$
$$= \frac{1 \pm \sqrt{-31}}{8}$$
No real solution.

81.
$$4x^2 = 1 - 2x$$
$$4x^2 + 2x - 1 = 0$$
Here, $a = 4$, $b = 2$, $c = -1$.
$$x = \frac{-2 \pm \sqrt{2^2 - 4(4)(-1)}}{2(4)}$$
$$= \frac{-2 \pm \sqrt{4 + 16}}{8}$$
$$= \frac{-2 \pm 2\sqrt{5}}{8}$$
$$= \frac{-1 \pm \sqrt{5}}{4}$$
$$\left\{\frac{-1 - \sqrt{5}}{4}, \; \frac{-1 + \sqrt{5}}{4}\right\}$$

83.
$$x^2 - 4x + 2 = 0$$
Here, $a = 1$, $b = -4$, $c = 2$.
$$x = \frac{-(-4) \pm \sqrt{(-4)^2 - 4(1)(2)}}{2(1)}$$
$$= \frac{4 \pm \sqrt{16 - 8}}{2}$$
$$= \frac{4 \pm 2\sqrt{2}}{2} = 2 \pm \sqrt{2}$$
$$= 2 \pm 1.41$$
$$\{0.59, \; 3.41\}$$

85. $x^2 + \sqrt{3}\,x - 3 = 0$

    Here $a = 1$, $b = \sqrt{3}$, $c = -3$

$$x = \frac{-\sqrt{3} \pm \sqrt{\left(\sqrt{3}\right)^2 - 4(1)(-3)}}{2(1)}$$

$$= \frac{-\sqrt{3} \pm \sqrt{3 + 12}}{2}$$

$$= \frac{-\sqrt{3} \pm \sqrt{15}}{2} = \frac{-1.73 \pm 3.87}{2}$$

    $\{-2.80,\ 1.07\}$

In Problems 87 - 91, we use the discriminant $b^2 - 4ac$:

87. $x^2 - 5x + 7 = 0$
    Here $a = 1$, $b = -5$, $c = 7$.
    $b^2 - 4ac = (-5)^2 - 4(1)(7)$
               $= -3$

    No real solution.

89. $9x^2 - 30x + 25 = 0$
    Here $a = 9$, $b = -30$, $c = 25$.
    $b^2 - 4ac = (-30)^2 - 4(9)(25)$
               $= 0$

    Repeated real solution.

91. $3x^2 + 5x - 2 = 0$
    Here $a = 3$, $b = 5$, $c = -2$.
    $b^2 - 4ac = 5^2 - 4(3)(-2) = 1$

    Two unequal real solutions.

93.   (a) For $s = 96 + 80t - 16t^2$,
        $s = 0$ when ball strikes
        ground:

$$0 = 96 + 80t - 16t^2$$
$$0 = 16(6 + 5t - t^2)$$
$$0 = 6 + 5t - t^2$$
$$0 = t^2 - 5t - 6$$
$$0 = (t - 6)(t + 1)$$

    $t - 6 = 0$ or $t + 1 = 0$
        $t = 6$ $t = -1$

    Thus, after 6 seconds, the
    ball strikes the ground.

    (b) $s = 96$ when the ball is
        at the level of the top
        of the building:

$$96 = 96 + 80t - 16t^2$$
$$0 = 80t - 16t^2$$
$$0 = -16t(t - 5)$$

    $-16t = 0$ or $t - 5 = 0$
         $t = 0$ $t = 5$

    Thus, the ball passes the
    top of the building after
    5 seconds.

95.   The roots of a quadratic equation are

$$\frac{-b - \sqrt{b^2 - 4ac}}{2a} \quad \text{and} \quad \frac{-b + \sqrt{b^2 - 4ac}}{2a}$$

    The sum $\dfrac{-b - \sqrt{b^2 - 4ac}}{2a} + \dfrac{-b + \sqrt{b^2 - 4ac}}{2a}$

$$= \frac{-b}{2a} - \frac{\sqrt{b^2 - 4ac}}{2a} + \frac{-b}{2a} + \frac{\sqrt{b^2 - 4ac}}{2a}$$

$$= \frac{-2b}{2a} = \frac{-b}{a}$$

1.2 EQUATIONS

97. $kx^2 + x + k = 0$

Here, $a = k$, $b = 1$, $c = k$ and $b^2 - 4ac = 1 - 4(k)(k) = 1 - 4k^2$. For repeated real roots, we want

$$1 - 4k^2 = 0$$
$$1 = 4k^2$$
$$\frac{1}{4} = k^2$$
$$\pm \frac{1}{2} = k$$

$k = \frac{-1}{2}$ or $k = \frac{1}{2}$ will give a repeated real solution.

99. For $ax^2 + bx + c = 0$, the real solutions are:

$$x = \frac{-b \pm \sqrt{b^2 - 4ac}}{2a}, \quad \text{or} \quad \frac{-b - \sqrt{b^2 - 4ac}}{2a}, \quad \text{or} \quad \frac{-b + \sqrt{b^2 - 4ac}}{2a}$$

For $ax^2 - bx + c = 0$, the real solutions are:

$$x = \frac{-(-b) \pm \sqrt{b^2 - 4ac}}{2a} = \frac{b \pm \sqrt{b^2 - 4ac}}{2a}, \quad \text{or} \quad \frac{b - \sqrt{b^2 - 4ac}}{2a},$$

$$\text{or} \quad \frac{b + \sqrt{b^2 - 4ac}}{2a}$$

Since $\sqrt{b^2 - 4ac}$ is the same in both answers, we have

$$-\frac{-b - \sqrt{b^2 - 4ac}}{2a} = \frac{b + \sqrt{b^2 - 4ac}}{2a} \quad \text{and}$$

$$-\frac{-b + \sqrt{b^2 - 4ac}}{2a} = \frac{b - \sqrt{b^2 - 4ac}}{2a}$$

so the real solutions to $ax^2 - bx + c = 0$ are the negatives of those of $ax^2 + bx + c = 0$.

101. For $\frac{1}{2}n(n + 1) = 666$, we solve for n to get the number of integers:

$$2\left(\frac{1}{2}n(n + 1)\right) = (666)(2)$$
$$n(n + 1) = 1332$$
$$n^2 + n - 1332 = 0$$
$$(n + 37)(n - 36) = 0$$

$n = -37$ or $n = 36$ so that 36 consecutive integers, starting with 1, add up to 666.

≡ EXERCISE 1.3 SETTING UP EQUATIONS: APPLICATIONS

1. Let A represent area of the circle and r the radius:

   Area of a circle is the product of π times the square of the radius.

   | A | = | π | · | r^2 |

   $$A = \pi r^2$$

3. Let A represent the area of the square and s the length of a side:

   Area of a square is the square of the length of a side.

   | A | = | s^2 |

   $$A = s^2$$

5. Let F represent the force, m the mass, and a the acceleration:

   Force equals the product of mass times acceleration.

   | F | = | m | · | a |

   $$F = ma$$

7. Let W represent the work, F the force, and d the distance:

   Work equals force times distance.

   | W | = | F | · | d |

   $$W = Fd$$

9. **Step 1:** We are being asked for two integers
   Step 2: Let x represent the smaller integer. The other, the next consecutive integer, is $x + 1$.
   Step 3: We can make a table:

One integer	Other integer	Reason
x	$x + 1$	The integers are "consecutive."

   Since the **sum** of these integers is 83, then

   $$x + (x + 1) = 83$$

   Step 4:
   $$2x + 1 = 83$$
   $$2x = 82$$
   $$x = 41$$

   then $x + 1 = 42$

   Step 5: Checking, the sum of 41 and 42, both integers, is 83.

11. **Step 1:** We are being asked for two consecutive integers.

    Step 2: Let x represent the smaller integer. The other, the next consecutive integer, is $x + 1$.

    Step 3: We can make a table:

One integer	Other integer	Reason
x	$x + 1$	The integers are "consecutive."

Since the **difference** of their **squares** is 27, then

$$(x + 1)^2 - (x)^2 = 27$$

Step 4:
$$x^2 + 2x + 1 - x^2 = 27$$
$$2x + 1 = 27$$
$$2x = 26$$
$$x = 13$$

Then the other integer $x + 1 = 14$

Step 5: Checking, the squares are $13^2 = 169$ and $14^2 = 196$. The difference, $196 - 169 = 27$

13. We want two odd consecutive integers. Let x be the smaller. Then $x + 2$ is the larger. Then:

$$(x)(x + 2) = 143$$
$$x^2 + 2x = 143$$
$$x^2 + 2x - 143 = 0$$
$$(x + 13)(x - 11) = 0$$
$$x + 13 = 0 \quad \text{or} \quad x - 11 = 0$$
$$x = -13 \qquad\qquad x = 11$$

If $x = 11$, then $x + 2 = 13$ and $(11)(13) = 143$.

15. **Step 1:** We are being asked for an hourly wage in dollars per hour.

Step 2: Let x represent the hourly wage.

Step 3: We set up a table:

	Hourly wage	Salary
Regular hours, 40	x	$40x$
Overtime hours, 8	$1.5x$	$8(1.5x) = 12x$

The total of regular salary plus overtime is $442.00, then

$$40x + 12x = 442$$

Step 4:
$$52x = 442$$
$$x = 8.50$$

The hourly wage is $8.50 per hour.

Step 5: Forty hours yields a salary of $40(8.50) = \$340$, and 8 hours of overtime yields a salary of $8(1.5)(8.50) = \$102$, for a total of $442.

17. **Step 1:** We are being asked to find the number of touchdowns scored.

Step 2: Let x represent the number of touchdowns scored.

Step 3: We set up a table:

	Point value	Points earned
Safeties, 1	2	$(1)(2) = 2$
Field goals, 2	3	$(2)(3) = 6$
Touchdowns without extra points, 2	6	$(2)(6) = 12$
Touchdowns with extra points, $x - 2$	7	$(x - 2)(7) = 7x - 14$

The total points scores is 41; thus

$$2 + 6 + 12 + 7x - 14 = 41$$

Step 4:
$$7x + 6 = 41$$
$$7x = 35$$
$$x = 5$$

There were two touchdowns without extra points and three touchdowns with extra points for a total of 5 touchdowns.

Step 5: Two safeties (for 2 points) and two field goals (for 6 points) and two touchdowns without extra points (for 12 points) and three touchdowns with extra points (for 21 points) give a total of $2 + 6 + 12 + 21 = 41$ points.

19. ℓ = length, w = width

$$2\ell + 2w = 40 \qquad \text{Perimeter} = 2\ell + 2w$$
$$\ell = w + 8 \quad \text{The length is 8 more than the width.}$$
$$2(w + 8) + 2w = 40$$
$$2w + 16 + 2w = 40$$
$$4w + 16 = 40$$
$$4w = 24$$
$$w = 6 \text{ feet}$$
$$\ell = 14 \text{ feet}$$

21. ℓ = length of garden
     w = width of garden

(a) If the length of garden is to be twice its width, then the width is to be half the length. Thus,

$$w = \frac{1}{2}\ell$$

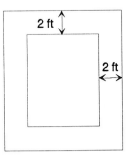

The dimensions of the fence are $\ell + 4$ and $w + 4$, which means the dimensions are $\ell + 4$ and $\frac{1}{2}\ell + 4$.

Its perimeter is 46 feet, so

$$2[(\ell + 4) + \frac{1}{2}\ell + 4] = 46$$
$$2\ell + 8 + \ell + 8 = 46$$
$$3\ell = 30$$
$$\ell = 10$$

The dimensions of the garden are 10 feet by 5 feet.

(b) Area = $\ell \cdot w = 10 \cdot 5 = 50$ square feet

(c) If the dimensions of the fence are the same, then the dimensions are $\ell + 4$ and $\ell + 4$. Its perimeter is 46 feet, so

$$2[(\ell + 4) + (\ell + 4)] = 46$$
$$2\ell + 8 + 2\ell + 8 = 46$$
$$4\ell = 30$$
$$\ell = 7.5$$

The dimensions of the garden are 7.5 feet by 7.5 feet.

(d) The area of this square garden is $\ell \cdot w$ (7.5)(7.5) = 56.25 square feet.

23. **Step 1:** We want to find two dollar amounts, the principle to invest in B-rated bonds pay 15% per year and the principle to invest in a certificate paying 7% per year.

Step 2: Let x represent the amount invested in bonds at 15%. Then $50,000 - x$ is the amount that will be invested in a certificate at 7%.

Step 3: We make a table:

	Principle	Rate	Time (yr)	Interest
Bonds at 15%	x	0.15	1	0.15
Certificate at 7%	$50,000 - x$	0.07	1	$0.07(50,000 - x)$

Since the total interest is to be $6000, we have

$$0.15x + 0.07(50,000 - x) = 6000$$

Step 4:
$$15x + 7(50,000 - x) = 600,000$$
$$15x + 350,000 - 7x = 600,000$$
$$350,000 + 8x = 600,000$$
$$8x = 250,000$$
$$x = 31,250$$

Thus, $31,250 will be invested in bonds at 15% and $18,750 in a certificate at 7%.

Step 5: The interest on the bond after one year is $0.15(31,250) = \$4687.50$, and the interest on the certificate is $0.07(18,750) = \$1312.50$, for a total interest of $\$4687.50 + \$1312.50 = \$6000.00$.

25. **Step 1:** We want to find the dollar amount loaned at 8%.

Step 2: Let x represent the amount invested at 8%. Then the amount invested at 18% is $10,000 - x$.

Step 3: We make a table:

	Principle	Rate	Time (yr)	Interest
Loan at 8%	x	0.08	1	$0.08x$
Loan at 18%	$10,000 - x$	0.18	1	$0.18(10,000 - x)$

Since the total interest is to be $1000, we have

$$0.8x + 0.18(10,000 - x) = 1000$$

Step 4:
$$8x + 18(10,000 - x) = 100,000$$
$$8x + 180,000 - 18x = 100,000$$
$$-10x = -80,000$$
$$x = 8,000$$

Thus, $8,000 will be loaned at 8% and $2,000 at 18%.

Step 5: The interest on the loan at 8% is $0.08(8,000) = \$640$, and the interest on the loan at 18% is $0.18(2,000) = \$360$; thus, the total interest is $1000.

27. Let ℓ = length, w = width, Perimeter = $2\ell + 2w$.
$$(1) \quad 2\ell + 2w = 30$$
$$\ell w = 56$$
Simplifying (1), we get,
$$\ell + w = 15$$
$$w = 15 - \ell$$
$$\ell(15 - \ell) = 56$$
$$15\ell - \ell^2 = 56$$
$$\ell^2 - 15\ell + 56 = 0$$
$$(\ell - 8)(\ell - 7) = 0$$

$\ell = 8$ meters	$\ell = 7$ meters
$w = 7$ meters	$w = 8$ meters

The dimensions of the rectangle are 7 ft by 8 ft.

29. We want the dimensions of a box with square base and volume of 4 cubic feet. Let x be the side of the sheet of metal. Then $v = \ell w$ gives

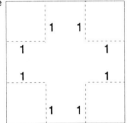

$$4 = (x - 2)(x - 2)$$
$$4 = x^2 - 4x + 4$$
$$0 = x^2 - 4x$$
$$0 = x(x - 4)$$
$$x = 0 \quad \text{or} \quad x - 4 = 0$$
$$x = 4$$

Thus, the sheet should be 4 ft. × 4 ft. to give a box of $2 \times 2 = 4$ cu. ft.

31. (a) For $s = 160 + 48t - 16t^2$, $s = 0$ when ball strikes ground:

$$0 = 160 + 48t - 16t^2$$
$$0 = 16(10 + 3t - t^2)$$
$$0 = 10 + 3t - t^2$$
$$0 = t^2 - 3t - 10$$
$$0 = (t - 5)(t + 2)$$
$$t - 5 = 0 \text{ or } t + 2 = 0$$
$$t = 5 \qquad t = 2$$

Thus, after 5 seconds, the ball strikes the ground.

(b) $s = 160$ when the ball is at the level of the top of the building:

$$160 = 160 + 48t - 16t^2$$
$$0 = 48t - 16t^2$$
$$0 = -16t(t - 3)$$

$$-16t = 0 \quad \text{or} \quad t - 3 = 0$$
$$t = 0 \qquad\qquad t = 3$$

Thus, the ball passes the top of the building after 3 seconds.

33. **Step 1:** We are asked to find the original price and **difference** between the original price and the new price.

Step 2: Let x be the **original price** of the house.

Step 3: Note that

$$\text{original price} = x$$
$$\text{amount reduced} = 0.15x$$
$$\text{new price} = \$93{,}500$$

Original price − amount reduced = new price
$$x \qquad - \qquad 0.15x \qquad = \$93{,}500$$

Step 4:
$$.85x = \$93{,}500$$
$$x = \$110{,}000$$

Thus, the original price is \$110,000, and the amount of the reduction is 0.15(\$110,000) = \$16,500.

Step 5: The original price \$110,000 less then 15% reduction 0.15(110,000) gives the new price \$93,500.

35. **Step 1:** We are asked to find the amount the bookstore paid for the book.

Step 2: Let x be the bookstore's price for the book.

Step 3: Note that

$$\text{bookstore price} = x$$
$$\text{percent increase} = 0.25x$$
$$\text{selling price} = \$35.00$$

bookstore price + percent increase = selling price
$$x \qquad + \qquad 0.25x \qquad = \$35.00$$

Step 4:
$$1.25x = \$35.00$$
$$x = \$28.00$$

Step 5: The amount that the bookstore paid for the book was \$28.00. The 25% markup of the price increases the cost of the book \$7.00, making the selling price of the book \$35.00.

37. We want to find the average speed from Chicago to Miami.

	Velocity	Time	Distance
Chicago to Atlanta	45	t_1	$45t_1$
Atlanta to Miami	55	t_2	$55t_2$

If Atlanta is halfway between Chicago and Miami, then the distances from Chicago to Atlanta and Atlanta to Miami are equal.

$$45t_1 = 55t_2 \;\rightarrow\; t_1 = \frac{55}{45}t_2$$

$$
\begin{aligned}
\text{Average speed} &= \frac{\text{Distance}}{\text{Time}} \\[4pt]
&= \frac{45t_1 + 55t_2}{t_1 + t_2} \\[4pt]
&= \frac{45\left(\frac{55}{45}t_2\right) + 55t_2}{\frac{55}{45}t_2 + t_2} \\[4pt]
&= \frac{55t_2 + 55t_2}{\frac{55t_2 + 45t_2}{45}} \\[4pt]
&= \frac{110t_2}{\frac{100}{45}t_2} = \frac{110}{\frac{100}{45}} \\[4pt]
&= \frac{(45)(110)}{100} = 49.5 \text{ miles per hour}
\end{aligned}
$$

The average speed from Chicago to Miami is 49.5 mph.

1 PRELIMINARIES

39.   **Step 1:** We want to find a final exam score.

    Step 2: Let x be the final exam score to yield an average of 80 when counted twice and combined with the other scores.

    Step 3: The total points scored with the exam counted twice is

$$80 + 83 + 71 + 61 + 95 + 2x = 390 + 2x$$

        Averaging, we get $\dfrac{390 + 2x}{7} = 80$

    Step 4:
$$390 + 2x = 560$$
$$2x = 170$$
$$x = 85$$

        Thus, a score of 85 is needed to get an average of 80 if the final counts as two tests.

    Step 5: $\dfrac{80 + 83 + 71 + 61 + 95 + 2(85)}{7} = 80$

41.   **Step 1:** We want to find a position on a football field.

    Step 2: Let s be the distance the tight end runs after catching the ball.

    Step 3: We can make a table (using $v = \dfrac{s}{t}$ and $t = \dfrac{s}{v}$):

	Velocity	Time (seconds)	Distance (yards)
Tight end	$\dfrac{100}{12} = \dfrac{25}{3}$	$\dfrac{s}{\frac{25}{3}} = \dfrac{3s}{25}$	s
Defensive back	$\dfrac{100}{10} = 10$	$\dfrac{s + 5}{10}$	$s + 5$

        Since the time is the same for both runners:

$$\frac{3s}{25} = \frac{s + 5}{10}$$

    Step 4:
$$30s - 25s = 125$$
$$5s = 125$$
$$s = 25$$

        Thus, the tight end goes 25 yards from his 20 yard line, or to his 45 yard line.

    Step 5: Checking the time for the appropriate distances:

        Tight end: $\dfrac{25 \text{ yds}}{\frac{25}{3} \text{ yds/sec}} = 3 \text{ sec}$

        Defensive back: $\dfrac{30 \text{ yds}}{10 \text{ yds/sec}} = 3 \text{ sec}$

43. We are looking for a width. Let x represent the width of the border in feet. It is best to convert all units to feet now:

$$1 \text{ cubic yard} = 27 \text{ cubic feet}$$
$$3 \text{ inches} = \frac{3}{12} = \frac{1}{4} \text{ ft.}$$

From the figure: The total area is $A_T = (6 + 2x)(10 + 2x)$
The area of the garden is $A_G = 6 \times 10 = 60$ sq. ft.

Then, the area of the border, A_B, is
$$A_B = (6 + 2x)(10 + 2x) - 60$$

The volume of the border is 27 cubic feet or $\left(\frac{1}{4} \text{ ft}\right)(A_B)$ so that

$$27 = \frac{1}{4}A_B$$
$$108 = A_B$$

Thus, $108 = (6 + 2x)(10 + 2x) - 60$
$108 = 60 + 12x + 20x + 4x^2 - 60$
$0 = 4x^2 + 32x - 108$
$0 = x^2 + 8x - 27$

Here, $a = 1$, $b = 8$, $c = -27$,
and $b^2 - 4ac = 8^2 - 4(1)(-27) = 172$

Then, $x = \dfrac{-8 \pm \sqrt{172}}{2(1)} = \dfrac{-8 \pm 2\sqrt{43}}{2} = -4 \pm \sqrt{43}$

so that $x = -4 - \sqrt{43}$ or $x = -4 + \sqrt{43}$
cannot be negative $\qquad\qquad x = -4 + 6.56$
$\qquad\qquad\qquad\qquad\qquad\qquad x = 2.56$ ft.

The width of the border is 2.56 ft.

Check: Thus, the area of the border is:

$$A_B = (6 + 2(2.56))(10 + 2(2.56)) - (6)(10)$$
$$= 108.13 \text{ which is close to } 108 \text{ sq. ft.}$$

The given volume of cement is then $(108)\left(\frac{1}{4}\right) = 27$ cubic ft.

45. We want to find the new dimensions of length and width in centimeters. Let x be the amount of reduction of the length and width measured in centimeters.

The current bar has volume $V_c = (12)(7)(3) = 252$ cubic centimeters.

The new volume is to be $V_N = .90V_c = .9(252) = 226.8$ cubic centimeters.

Then $V_N = 226.8 = (3)(12 - x)(7 - x)$
$$226.8 = 3(84 - 19x + x^2)$$
$$75.6 = 84 - 19x + x^2$$
$$0 = 8.4 - 19x + x^2$$

Here, $a = 1$, $b = -19$, $c = 8.4$ and
$b^2 - 4ac = (-19)^2 - 4(1)(8.4) = 327.4$

Then, $x = \dfrac{-(-19) \pm \sqrt{327.4}}{2(1)} = \dfrac{19 \pm 18.09}{2}$

$x = 0.455$ or $x = 18.55$; but since 18.55 exceeds the measurements it would be subtracted from, it is not a practical solution.

The new dimensions are:

7 - 0.455, and 12 - 0.455, and 3 centimeters

or   width = 6.55, length = 11.55, and thickness = 3 centimeters.

Check: $(6.55)(11.55)(3) = 226.9 \cong 226.8$.

47.   **Step 1:** We want to find a speed in miles per hour.

Step 2: Let v be the speed of the current in miles per hour.

Step 3: We make a table (using $s = vt$):

	Velocity of boat	Time (hr.)	Distance (mi.)
Upstream	$15 - v$	$\dfrac{20}{60} = \dfrac{1}{3}$	$\dfrac{15 - v}{3}$
Downstream	$15 + v$	$\dfrac{15}{60} = \dfrac{1}{4}$	$\dfrac{15 + v}{4}$

Since the distance is the same in each direction:

$$\frac{15 - v}{3} = \frac{15 + v}{4}$$

Step 4: $4(15 - v) = 3(15 + v)$
$$60 - 4v = 45 + 3v$$
$$15 = 7v$$
$$v = \frac{15}{7} = 2.14$$

Thus, the speed of the current is 2.14 miles per hour.

Step 5: The distance is the same in each direction:

$$\frac{15 - v}{3} = \frac{15 - 2.14}{3} = 4.29 \text{ miles}$$
$$\frac{15 + v}{4} = \frac{15 + 2.14}{4} = 4.29 \text{ miles}$$

1.3 SETTING UP EQUATIONS: APPLICATIONS   23

49. We want to find the width of a concrete pool border. Let x represent the width in feet of the border. It is best to convert all units to feet now:

> 1 cubic yard = 27 cubic feet
>
> 3 inches = $\dfrac{3}{12}$ = $\dfrac{1}{4}$ foot

We will use $A = \pi r^2$.

We are given that the distance across the pool is 10 feet, which means the radius is $\dfrac{10}{2}$ = 5 feet.

The total area, pool and border, is $A_T = \pi(5 + x)^2$.

The area of the pool alone is $A_p = \pi(5)^2 = 25\pi$.

The area of the border is $A_B = A_T = A_p = \pi(5 + x)^2 - 25\pi$.

The volume of the border, $\dfrac{1}{4}A_B = \dfrac{1}{4}\left(\pi(5 + x)^2 - 25\pi\right)$. Then,

$$\frac{1}{4}\left(\pi(5 + x)^2 - 25\pi\right) = 27$$
$$\pi(25 + 10x + x^2 - 25) = 108$$
$$x^2 + 10x - \frac{108}{\pi} = 0$$
$$x^2 + 10x - 34.38 = 0$$

Here $a = 1$, $b = 10$, $c = -34.38$, and
$b^2 - 4ac = 10^2 - 4(1)(-34.38) = 237.52$

Then,

$$x = \frac{-10 \pm \sqrt{237.52}}{2(1)}$$
$$= \frac{-10 \pm 15.41}{2}$$

Thus, $x = 2.71$ or $x = -12.71$

We ignore -12.71 since a measurement must be positive.

The width of the border is 2.71 feet.

Check: Area of pool and border = $\pi(10 + 1.59)^2 = 422.00$
Area of border = 422 - area of pool
= 422 - 100π
= 107.84
Volume of border = $\dfrac{1}{4}(107.84) = 26.96$

51. We want to find a speed in miles per hour. Let x represent the speed of the current in miles per hour. We can make a table using $s = vt$:

	Velocity	Time	Distance
Downstream	$15 + x$	$\dfrac{10}{15 + x}$	10
Upstream	$15 - x$	$\dfrac{10}{15 - x}$	10

Then the total time

$$\frac{10}{15 + x} + \frac{10}{15 - x} = 1.5$$

$$10(15 - x) + 10(15 + x) = 1.5(15 + x)(15 - x)$$

$$150 - 10x + 150 + 10x = 1.5(225 - x^2)$$

$$-1.5(225 - x^2) + 300 = 0$$

$$-225 + x^2 + 200 = 0$$

$$x^2 - 25 = 0$$

$$(x + 5)(x - 5) = 0$$

$$x = -5 \quad \text{or} \quad x = 5$$

We choose the positive solution. The speed of the current is 5 miles per hour.

Check: Downstream $(15 + 5)\left(\dfrac{10}{15 + 5}\right) = 10$ miles

Upstream $(15 - 5)\left(\dfrac{10}{15 - 5}\right) = 10$ miles

53. **Step 1:** We want to find a time in minutes and a distance in miles.

Step 2: Let t be the time in minutes to run the mile.

Step 3: We can make a table:

	Minutes to run the mile	Time	Part of mile run in one minute	Distance
Mike	6	t	$\dfrac{1}{6}$	$\dfrac{1}{6}t$
Dan	9	$t + 1$	$\dfrac{1}{9}$	$\dfrac{1}{9}(t + 1)$

Step 4:

$$\frac{1}{6t} = \frac{1}{9}(t + 1)$$

$$\frac{1}{6}t = \frac{1}{9}t + \frac{1}{9}$$

$$\frac{3t - 2t}{18} = \frac{1}{9}$$

$$\frac{1}{18}t = \frac{1}{9}$$

$$t = 2$$

$$d = \frac{1}{6}t = \frac{1}{9}(t + 1)$$

$$= \frac{1}{6} \cdot 2 = \frac{1}{9}(2 + 1)$$

$$= \frac{1}{3}$$

Thus, after 2 minutes and a distance of $\dfrac{1}{3}$ mile, Mike will pass Dan.

1.3 SETTING UP EQUATIONS: APPLICATIONS

Step 5: If Mike gives Dan a headstart of 1 minute, Mike will pass Dan $\frac{1}{3}$ of a mile from the start in 2 minutes.

55. **Step 1:** We need to find speed in miles per hour and distance in miles.

Step 2: Let x be the average speed of the slower car and $x + 10$ be the average speed of the faster car.

Step 3: Let us set up a table.

	Velocity	Time	Distance
Slower car	x	3.5	$3.5x$
Faster car	$x + 10$	3	$3(x + 10)$

$$3.5x = 3(x + 10)$$

Step 4:
$$3.5x = 3x + 30$$
$$.5x = 30$$
$$x = 60$$

Thus, the slower car travels at a speed of 60 miles per hour, and the faster car travels at a speed of 70 miles per hour.

Step 5: After $3\frac{1}{2}$ hours, the slower car, travelling at an average speed of 60 mph, goes a distance of $(3.5)(60) = 210$ miles. The faster car travels $(70)(3)$ or 210 miles.

57.

	AMOUNT	RATE	MONTHLY INTEREST
CD	x	9% = 0.09	$.09x$
Bond	$100,000 - x$	12% = 0.12	$.12(100,000 - x)$
Equity Loan	$100,000$	10% = 0.10	$.10(100,000)$

$$.09x + .12(100,000 - x) = .10(100,000)$$
$$.09x + 12,000 - .12x = 10,000$$
$$-.03x = -2000$$
$$x = 66,667$$

You can invest no more than $66,667 in the CD to ensure the monthly home equity loan payment is made.

1 PRELIMINARIES

☰ EXERCISE 1.4 INEQUALITIES

1.
$$3x - 1 \geq 3 + x$$
$$3x - 1 + 1 \geq 3 + x + 1$$
$$3x \geq 4 + x$$
$$3x - x \geq 4 + x - x$$
$$2x \geq 4$$
$$\frac{2x}{2} \geq \frac{4}{2}$$
$\{x \mid 2 \leq x < \infty\}$ or $[2, \infty)$

3.
$$-2(x + 3) < 6$$
$$\frac{-2(x + 3)}{-2} > \frac{6}{-2}$$
$$x + 3 > -3$$
$$x + 3 - 3 > -3 - 3$$
$\{x \mid -6 < x < \infty\}$ or $(-6, \infty\}$

5.
$$4 - 3(1 - x) \leq 3$$
$$4 - 3 + 3x \leq 3$$
$$1 + 3x - 1 \leq 3 - 1$$
$$3x \leq 2$$
$$\frac{3x}{3} \leq \frac{2}{3}$$
$\left\{x \mid -\infty < x < \frac{2}{3}\right\}$ or $\left(-\infty, \frac{2}{3}\right]$

7.
$$\frac{1}{2}(x - 4) > x + 8$$
$$\frac{1}{2}x - 2 > x + 8$$
$$\frac{1}{2}x - 2 - 8 > x + 8 - 8$$
$$\frac{1}{2}x - 10 > x$$
$$\frac{1}{2}x - 10 - \frac{1}{2}x > x - \frac{1}{2}x$$
$$-10 > \frac{1}{2}x$$
$$2(-10) > 2\left(\frac{1}{2}x\right)$$
$$-20 > x$$
$\{x \mid -\infty < x < -20\}$ or
$(-\infty, -20)$

9.
$$\frac{x}{2} \geq 1 - \frac{x}{4}$$
$$\frac{x}{2} + \frac{x}{4} \geq 1 - \frac{x}{4} + \frac{x}{4}$$
$$\frac{3x}{4} \geq 1$$
$$\frac{4}{3}\left(\frac{3x}{4}\right) \geq \frac{4}{3}(1)$$
$\left\{x \mid \frac{4}{3} \leq x < \infty\right\}$ or $\left[\frac{4}{3}, \infty\right)$

11.
$$0 \leq 2x - 6 \leq 4$$
$$0 + 6 \leq 2x - 6 + 6 \leq 4 + 6$$
$$6 \leq 2x \leq 10$$
$$\frac{6}{2} \leq \frac{2x}{2} \leq \frac{10}{2}$$
$\{x \mid 3 \leq x \leq 5\}$ or $[3, 5]$

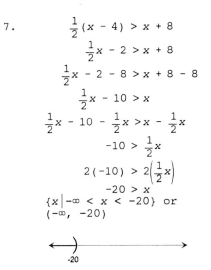

13.
$$-6 \le 1 - 3x \le 2$$
$$-6 - 1 \le 1 - 3x - 1 \le 2 - 1$$
$$-7 \le -3x \le 1$$
$$\frac{-7}{-3} \ge \frac{-3x}{-3} \ge \frac{1}{-3}$$
$$\frac{7}{3} \ge x \ge -\frac{1}{3}$$
$$\left\{x \mid -\frac{1}{3} \le x \le \frac{7}{3}\right\} \text{ or } \left[\frac{-1}{3}, \frac{7}{3}\right]$$

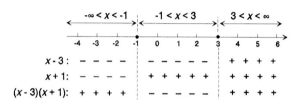

15. $(x - 3)(x + 1) < 0$

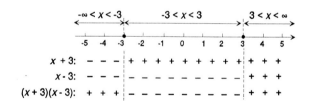

The solution set is $\{x \mid -1 < x < 3\}$ or $(-1, 3)$.

17.
$$-x^2 + 9 > 0$$
$$x^2 - 9 < 0$$
$$(x + 3)(x - 3) < 0$$

	$-\infty < x < -3$		$-3 < x < 3$		$3 < x < \infty$
$x + 3$:	$- - -$	$+ + + + + + + + +$			$+ + +$
$x - 3$:	$- - -$	$- - - - - - - - - -$			$+ + +$
$(x + 3)(x - 3)$:	$+ + +$	$- - - - - - - - - -$			$+ + +$

The solution set is $\{x \mid -3 < x < 3\}$ or $(-3, 3)$.

19.
$$x^2 + x > 6$$
$$x^2 + x - 6 > 0$$
$$(x + 3)(x - 2) > 0$$

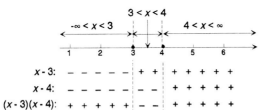

The solution set is $\{x | -\infty < x < -3 \text{ or } 2 < x < \infty\}$.

21.
$$x(x - 7) > -12$$
$$x^2 - 7x + 12 > 0$$
$$(x - 3)(x - 4) > 0$$

The solution set is $\{x | -\infty < x < 3 \text{ or } 4 < x < \infty\}$; $(-\infty, 3)$ or $(4, \infty)$.

23.
$$4x^2 + 9 < 6x$$
$$4x^2 - 6x + 9 < 0$$

Since for $a = 4$, $b = -6$, $c = 9$, $b^2 - 4ac = -108 < 0$, then $4x^2 - 6x + 9$ has no real roots. For $x = 0$, $4x^2 - 6x + 9 = 9 > 0$. Thus, there is no real solution.

25. $(x - 1)(x^2 + x + 1) > 0$ gives boundary point 1 only since $x^2 + x + 1 > 0$ for all x. [Note that for $a = 1$, $b = 1$, $c = 1$, $b^2 - 4ac = 1^2 - 4(1)(1) = -3 < 0$ implies $x^2 + x + 1$ has no real roots.]

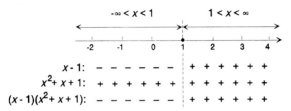

The solution set is $\{x | 1 < x < \infty\}$ or $(1, \infty)$.

27. $(x - 1)(x - 2)(x - 3) < 0$ gives boundary points 1, 2, 3:

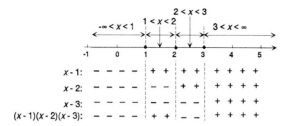

The solution set is $\{x\,|\,-\infty < x < 1 \text{ or } 2 < x < 3\}$; $(-\infty, 1)$ or $(2, 3)$.

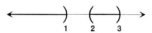

29. $-x^3 + 2x^2 + 8x < 0$
$x^3 - 2x^2 - 8x > 0$
$x(x^2 - 2x - 8) > 0$
$x(x + 2)(x - 4) > 0$ gives boundary points -2, 0, 4:

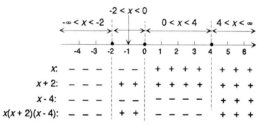

The solution set is $\{x\,|\,-2 < x < 0 \text{ or } 4 < x < \infty\}$; $(-2, 0)$ or $(4, \infty)$.

31. $\qquad\qquad x^3 > x$
$\qquad\quad x^3 - x > 0$
$\qquad x(x^2 - 1) > 0$
$x(x + 1)(x - 1) > 0$ gives boundary points -1, 0, 1:

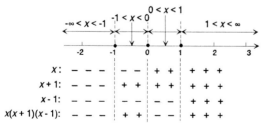

The solution set is $\{x\,|\,-1 < x < 0 \text{ or } 1 < x < \infty\}$; $(-1, 0)$ or $(1, \infty)$.

33. $$x^3 > x^2$$
$$x^3 - x^2 > 0$$
$x^2(x - 1) > 0$ gives boundary points 0, 1:

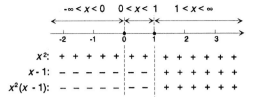

The solution set is $\{x \mid 1 < x < \infty\}$; $(1, \infty)$.

35. $$\frac{x + 1}{1 - x} < 0$$

$$\frac{x + 1}{x - 1} > 0 \text{ gives boundary points } -1, 1:$$

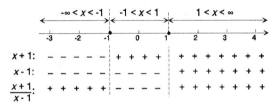

The solution set is $\{x \mid -\infty < x < -1 \text{ or } 1 < x < \infty\}$; $(-\infty, -1)$ or $(1, \infty)$.

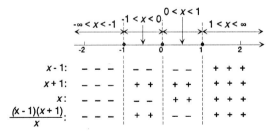

37. $$\frac{(x - 1)(x + 1)}{x} < 0 \text{ gives boundary points } -1, 0, 1:$$

The solution set is $\{x \mid -\infty < x < -1 \text{ or } 0 < x < 1\}$; $(-\infty, -1)$ or $(0, 1)$.

39. $\dfrac{x-2}{x^2-1} \geq 0$

$\dfrac{x-2}{(x+1)(x-1)} \geq 0$ gives boundary points -1, 1, 2:

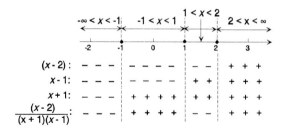

Note that $\dfrac{x-2}{(x+1)(x-1)} = 0$ only if $x = 2$.

The solution set is $\{x \mid -1 < x < 1 \text{ or } 2 \leq x < \infty\}$; $(-1, 1)$ or $[2, \infty)$.

41. $\dfrac{x+4}{x-2} \leq 1$

$\dfrac{x+4}{x-2} - 1 \leq 0$

$\dfrac{x+4-(x-2)}{x-2} \leq 0$

$\dfrac{6}{x-2} \leq 0$ gives boundary point 2:

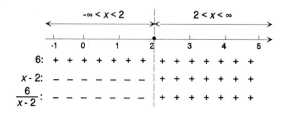

Note that $\dfrac{6}{x-2} \neq 0$ for any x.

The solution set is $\{x \mid -\infty < x < 2\}$; $(-\infty, 2)$.

43.

$$\frac{2x + 5}{x + 1} > \frac{x + 1}{x - 1}$$

$$\frac{2x + 5}{x + 1} - \frac{x + 1}{x - 1} > 0$$

$$\frac{2x^2 + 3x - 5 - (x^2 + 2x + 1)}{(x + 1)(x - 1)} > 0$$

$$\frac{x^2 + x - 6}{(x + 1)(x - 1)} > 0$$

$$\frac{(x + 3)(x - 2)}{(x + 1)(x - 1)} > \text{gives boundary } -3,\ -1,\ 1,\ 2:$$

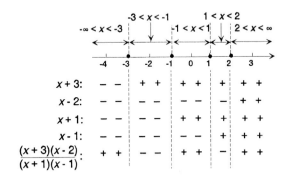

The solution set $\{x \mid -\infty < x < -3 \text{ or } -1 < x < 1 \text{ or } 2 < x < \infty\}$; $(-\infty,\ -3)$ or $(-1,\ 1)$ or $(2,\ \infty)$.

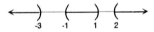

45. $|2x| < 6$

$$-6 < 2x < 6$$
$$\frac{-6}{2} < \frac{2x}{2} < \frac{6}{2}$$
$$-3 < x < 3$$

The solution set consists of $\{x \mid -3 < x < 3\}$ or $(-3,\ 3)$.

47. $|3x| > 12$

$$3x < -12 \quad \text{or} \quad 3x > 12$$
$$\frac{3x}{3} < \frac{-12}{3} \quad \text{or} \quad \frac{3x}{3} > \frac{12}{3}$$
$$x < -4 \quad \text{or} \quad x > 4$$

The solution set consists of $\{x \mid -\infty < x < -4 \text{ or } 4 < x < \infty\}$; $(-\infty,\ -4)$ or $(4,\ \infty)$.

49. $|x - 2| < 1$

$$-1 < x - 2 < 1$$
$$-1 + 2 < x - 2 + 2 < 1 + 2$$
$$1 < x < 3$$

The solution set consists of $\{x \mid 1 < x < 3\}$ or $(1, 3)$.

51. $|3t - 2| \leq 4$

$$-4 \leq 3t - 2 \leq 4$$
$$-4 + 2 \leq 3t - 2 + 2 \leq 4 + 2$$
$$-2 \leq 3t \leq 6$$
$$\frac{-2}{3} \leq \frac{3t}{3} \leq \frac{6}{3}$$
$$\frac{-2}{3} \leq t \leq 2$$

The solution set consists of $\left\{t \mid -\frac{2}{3} \leq t \leq 2\right\}$ or $\left[\frac{-2}{3}, 2\right]$.

53. $|x - 1| \geq 2$

$$x - 1 \leq -2 \quad \text{or} \quad x - 1 \geq 2$$
$$x \leq -1 \quad \text{or} \quad x \geq 3$$

The solution set consists of $\{x \mid -\infty < x \leq -1 \text{ or } 3 \leq x < \infty\}$; $(-\infty, -1]$ or $[3, \infty)$.

55. $|1 - 4x| < 5$

$$-5 < 1 - 4x < 5$$
$$-5 - 1 < 1 - 4x - 1 < 5 - 1$$
$$-6 < -4x < 5$$
$$\frac{-6}{-4} > \frac{-4x}{-4} > \frac{4}{-4}$$
$$\frac{3}{2} > x > -1$$
$$-1 < x < \frac{3}{2}$$

The solution set consists of $\left\{x \mid -1 < x < \frac{3}{2}\right\}$ or $\left(-1, \frac{3}{2}\right)$.

57. x differs from 2 by less than $\frac{1}{2}$

$$|x - 2| < \frac{1}{2}$$
$$|x - 2| < \frac{1}{2}$$
$$-\frac{1}{2} < x - 2 < \frac{1}{2}$$
$$-\frac{1}{2} + 2 < x - 2 + 2 < \frac{1}{2} + 2$$
$$\frac{3}{2} < x < \frac{5}{2}$$

The solution set consists of $\left\{x \mid \frac{3}{2} < x < \frac{5}{2}\right\}$.

59. x differs from -3 by more than 2

$$|x - (-3)| > 2$$
$$|x + 3| > 2$$
$$x + 3 < -2 \quad \text{or} \quad x + 3 > 2$$
$$x < -5 \quad \text{or} \quad x > -1$$

The solution set consists of $\{x \mid -\infty < x < -5 \text{ or } -1 < x < \infty\}$.

1 PRELIMINARIES

61. A temperature x that differs from 98.6°F by at least 1.5°.

$$|x - 98.6°| \geq 1.5°$$

$$x - 98.6° \leq -1.5° \quad \text{or} \quad x - 98.6° \geq 1.5°$$
$$x \leq 97.1° \quad \text{or} \quad x \geq 100.1°$$

The bodily temperatures that are considered unhealthy are those that are less than 97.1°F or greater than 100.1°F inclusive.

63. Let k = the monthly usage in kilowatt hours and C = the total monthly charges per customer. Then $C = 0.10819k = 9.06$ and

$$
\begin{array}{lll}
82.14 \leq & C & \leq 279.63 \\
82.14 \leq & 0.10819k + 9.06 & \leq 279.63 \\
73.08 \leq & 0.10819k & \leq 270.57 \\
675.48 \leq & k & \leq 2500.88
\end{array}
$$

65. Let C = the dealer cost and M = the markup over dealer's cost. If the price is $8800, then $8800 = C + MC = C(1 + M)$ which gives

$$C = \frac{8800}{1 + M}$$

Also,
$$
\begin{array}{lll}
0.12 \leq & M & \leq 0.18 \\
1.12 \leq & 1 + M & \leq 1.18 \\
\dfrac{1}{1.12} \geq & \dfrac{1}{1 + M} & \geq \dfrac{1}{1.18} \\[2mm]
\dfrac{8800}{1.12} \geq & \dfrac{8800}{1 + M} & \geq \dfrac{8800}{1.18} \\[2mm]
7857.14 \geq & C & \geq 7457.63 \\
7457.63 \leq & C & \leq 7857.14
\end{array}
$$

The cost ranged from $7457.63 to $7857.14, inclusive.

Check: For a cost of $7457.63 and markup of 18%,
$7457.63 + 0.18(7457.63) = \$8800$.
For a cost of $7857.14 and markup of 12%,
$7857.14 + 0.12(7857.14) = \$8800$

67. Let T = the score on the last test and G = the resulting grade. Then $G = (70 + 82 + 85 + 89 + T) + 5$ so that

$$T = 5G - 326 \text{ and}$$

$$
\begin{array}{lll}
80 \leq & G & \leq 90 \\
400 \leq & 5G & \leq 450 \\
74 \leq & 5G - 326 & \leq 124
\end{array}
$$

The fifth test score must be 74 or greater.

Check: $G = \dfrac{70 + 82 + 85 + 89 + 74}{5} = 80$

69. Let G = the amount (in gallons) of gasoline in the car at the start of the trip, and let D = the distance covered in the trip. Then $D = 25G$ or $G = \dfrac{D}{25}$, and

$$
\begin{array}{l}
300 \leq D \\
12 \leq \dfrac{D}{25} \\
12 \leq G
\end{array}
$$

Thus, there were between 12 and 20 gallons of gasoline at the beginning of the trip.

Check: For 12 gallons at 25 miles per gallon, the car will go $(12)(25) = 300$ miles.

71. $x^2 - a = \left(x - \sqrt{a}\right)\left(x + \sqrt{a}\right) < 0$

   $-\sqrt{a} < x < \sqrt{a}$

73. $x^2 < 1$

   $\{x \mid -1 < x < 1\}$ or $(-1, 1)$

75. $x^2 \geq 9$

   $\{x \mid -\infty < x \leq -3$ or $3 \leq x > \infty\}$; $(-\infty, -3]$ or $[3, \infty)$

77. $x^2 \leq 16$

   $\{x \mid -4 \leq x \leq 4\}$ or $[-4, 4]$

79. $x^2 > 4$

   $\{x \mid -\infty < x < -2$ or $2 < x < \infty\}$; $(-\infty, -2)$ or $(2, \infty)$

≡ EXERCISE 1.5 COMPLEX NUMBERS

1. $(2 - 3i) + (6 + 8i) = (2 + 6) + (-3 + 8)i = 8 + 5i$

3. $(-3 + 2i) - (4 - 4i) = (-3 - 4) + (2 + 4)i = -7 + 6i$

5. $(2 - 5i) - (8 + 6i) = (2 - 8) + (-5 - 6)i = -6 - 11i$

7. $3(2 - 6i) = 6 - 18i$

9. $2i(2 - 3i) = 4i - 6i^2 = 4i - 6(-1) = 6 + 4i$

11. $(3 - 4i)(2 + i) = 3(2 + i) - 4i(2 + i)$
    $= 6 + 3i - 8i - 4i^2$
    $= 6 - 5i - 4(-1)$
    $= 10 - 5i$

13. $(-6 + i)(-6 - i) = -6(-6 - i) + i(-6 - i)$
    $= 36 + 6i - 6i - i^2$
    $= 36 - (-1)$
    $= 37$

15. $\dfrac{10}{3 - 4i} = \dfrac{10}{3 - 4i} \cdot \dfrac{3 + 4i}{3 + 4i} = \dfrac{30 + 40i}{9 + 12i - 12i - 16i^2} = \dfrac{30 + 40i}{9 - 16(-1)}$
    $= \dfrac{30 + 40i}{25} = \dfrac{30}{25} + \dfrac{40}{25}i = \dfrac{6}{5} + \dfrac{8}{5}i$

17. $\dfrac{2 + i}{i} = \dfrac{2 + i}{i} \cdot \dfrac{-i}{-i} = \dfrac{-2i - i^2}{-i^2} = -2i + 1 = 1 - 2i$

19. $\dfrac{6 - i}{1 + i} = \dfrac{6 - i}{1 + i} \cdot \dfrac{1 - i}{1 - i} = \dfrac{6 - 6i - i + i^2}{1 - i^2} = \dfrac{6 - 7i + (-1)}{1 - (-1)}$
    $= \dfrac{5 - 7i}{2} = \dfrac{5}{2} - \dfrac{7}{2}i$

21. $\left(\frac{1}{2} + \frac{\sqrt{3}}{2}i\right)^2 = \frac{1}{4} + 2\left(\frac{1}{2}\right)\left(\frac{\sqrt{3}}{2}i\right) + \frac{3}{4}i^2 = \frac{1}{4} + \frac{\sqrt{3}}{2}i + \frac{3}{4}(-1) = -\frac{1}{2} + \frac{\sqrt{3}}{2}i$

23. $(1 + i)^2 = 1 + 2i + i^2 = 1 + 2i - 1 = 2i$

25. $2^{3} = i^{22+1} = i^{22} \cdot i = (^2)^{11} \cdot i = (-1)^{11}i = -i$

27. $i^{-15} = \dfrac{1}{i^{15}} = \dfrac{1}{i^{14+1}} = \dfrac{1}{i^{14}i} = \dfrac{1}{(i^2)^7 i} = \dfrac{1}{(-1)^7 i} = \dfrac{1}{-i} = \dfrac{1}{-i}\dfrac{i}{i} = \dfrac{i}{-i^2}$

$$= \dfrac{i}{-(-1)} = i$$

29. $i^6 - 5 = (^2)^3 - 5 = (-1)^3 - 5 = -1 - 5 = -6$

31. $6i^3 - 4i^5 = i^3(6 - 4i^2) = i^2 \cdot i(6 - 4(-1)) = -1 \cdot i(10) = -10i$

33. $(1 + i)^3 = 1^3 + 3i + 3i^2 + i^3 = 1 + 3i + 3(-1) + i^2 \cdot i$
$$= -2 + 3i + (-1)i$$
$$= -2 + 2i$$

35. $i^7(1 + i^2) = i^7(1 + (-1)) = i^7(0) = 0$

37. $i^6 + i^4 + i^2 + 1 = (i^2)^3 + (i^2)^2 + (-1) + 1 = (-1)^3 + (-1)^2$
$$= -1 + 1 = 0$$

39. $\sqrt{-4} = \sqrt{4}\,i = 2i$ 41. $\sqrt{-25} = \sqrt{25}\,i = 5i$

43. $\sqrt{(3 + 4i)(4i - 3)} = \sqrt{12i - 9 + 16i^2 - 12i} = \sqrt{-9 - 16} = \sqrt{-25}$
$$= \sqrt{25}\,i = 5i$$

For Problems 45 - 57 we use $x = \dfrac{-b \pm \sqrt{b^2 - 4ac}}{2a}$, *Equation (9).*

45. $x^2 + 4 = 0$
Here $a = 1$, $b = 0$, $c = 4$, and $b^2 - 4ac = 0 - 4(1)(4) = -16$.

Then $x = \dfrac{-0 \pm \sqrt{-16}}{2(1)} = \dfrac{\pm\sqrt{16}\,i}{2} = \dfrac{\pm 4i}{2} = \pm 2i$

The solution set is $\{-2i, 2i\}$.

47. $x^2 - 16 = 0$
Here $a = 1$, $b = 0$, $c = -16$, and $b^2 - 4ac = 0 - 4(1)(-16) = 64$.

Then $x = \dfrac{-0 \pm \sqrt{64}}{2(1)} = \dfrac{\pm 8}{2} = \pm 4$

The solution set is $\{-4, 4\}$.

49. $x^2 - 6x + 13 = 0$
Here $a = 1$, $b = -6$, $c = 13$, and $b^2 - 4ac = (-6)^2 - 4(1)(13) = -16$.

Then $x = \dfrac{-(-6) \pm \sqrt{-16}}{2(1)} = \dfrac{6 \pm \sqrt{16}\,i}{2} = \dfrac{6 \pm 4i}{2} = 3 \pm 2i$

The solution set is $\{3 - 2i,\ 3 + 2i\}$.

51. $x^2 - 6x + 10 = 0$

Here $a = 1$, $b = -6$, $c = 10$, and $b^2 - 4ac = (-6)^2 - 4(1)(10) = -4$.

Then $x = \dfrac{-(-6) \pm \sqrt{-4}}{2(1)} = \dfrac{6 \pm \sqrt{4}\,i}{2} = \dfrac{6 \pm 2i}{2} = 3 \pm i$

The solution set is $\{3 - i,\ 3 + i\}$.

53. $8x^2 - 4x + 1 = 0$

Here $a = 8$, $b = -4$, $c = 1$, and $b^2 - 4ac = (-4)^2 - 4(8)(1) = -16$.

Then $x = \dfrac{-(-4) \pm \sqrt{-16}}{2(8)} = \dfrac{4 \pm \sqrt{16}\,i}{16} = \dfrac{4 \pm 4i}{16} = \dfrac{1}{4} \pm \dfrac{1}{4}i$

The solution set is $\left\{\dfrac{1}{4} - \dfrac{1}{4}i,\ \dfrac{1}{4} + \dfrac{1}{4}i\right\}$.

55. $5x^2 + 2x + 1 = 0$

Here $a = 5$, $b = 2$, $c = 1$, and $b^2 - 4ac = 2^2 - 4(5)(1) = -16$.

Then $x = \dfrac{-2 \pm \sqrt{-16}}{2(5)} = \dfrac{-2 \pm \sqrt{16}\,i}{10} = \dfrac{-2 \pm 4i}{10} = -\dfrac{1}{5} \pm \dfrac{2}{5}i$

The solution set is $\left\{-\dfrac{1}{5} - \dfrac{2}{5}i,\ \dfrac{1}{5} + \dfrac{2}{5}i\right\}$.

57. $x^2 + x + 1 = 0$

Here $a = 1$, $b = 1$, $c = 1$, and $b^2 - 4ac = 1^2 - 4(1)(1) = -3$.

Then $x = \dfrac{-1 \pm \sqrt{-3}}{2(1)} = \dfrac{-1 \pm \sqrt{3}\,i}{2} = -\dfrac{1}{2} \pm \dfrac{\sqrt{3}}{2}i$

The solution set is $\left\{-\dfrac{1}{2} - \dfrac{\sqrt{3}}{2}i,\ -\dfrac{1}{2} + \dfrac{\sqrt{3}}{2}i\right\}$.

59. $x^3 - 8 = 0$

$x^3 - 8 = (x - 2)(x^2 + 2x + 4) = 0$

$x - 2 = 0 \quad x^2 + 2x + 4 = 0$

$\qquad x = 2 \quad$ Here $a = 1$, $b = 2$, $c = 4$, and

$\qquad\qquad\qquad b^2 - 4ac = 2^2 - 4(1)(4) = -12$

$\qquad\qquad$ Then $x = \dfrac{-2 \pm \sqrt{-12}}{2(1)} = \dfrac{-2 \pm \sqrt{12}\,i}{2}$

$\qquad\qquad\qquad = \dfrac{-2 \pm \sqrt{3}\,i}{2} = -1 \pm \sqrt{3}\,i$

The solution set is $\left\{2,\ -1 - \sqrt{3}\,i,\ -1 + \sqrt{3}\,i\right\}$.

61. $x^4 - 16 = 0$

$x^4 - 16 = (x^2 - 4)(x^2 + 4) = (x - 2)(x + 2)(x^2 + 4) = 0$

$x - 2 = 0 \quad x + 2 = 0 \quad x^2 + 4 = 0$

$\quad x = 2 \qquad x = -2 \quad$ Here $a = 1$, $b = 0$, $c = 4$, and

$\qquad\qquad\qquad\qquad\qquad b^2 - 4ac = 0^2 - 4(1)(4) = -16$

Then, $x = \dfrac{0 \pm \sqrt{-16}}{2(1)} = \dfrac{\sqrt{16}\,i}{2} = \dfrac{\pm 4i}{2} = \pm 2i$

The solution set is $\{-2,\ 2,\ -2i,\ 2i\}$.

63. $x^4 + 13x^2 + 36 = 0$

$(x^2 + 9)(x^2 + 4) = 0$

$x^2 = -9 \qquad x^2 = -4$

$x^2 = 9i^2 \qquad x^2 = 4i^2$

$x = \pm 3i \qquad x = \pm 2i$

The solution set is $\{-3i, -2i, 2i, 3i\}$.

65. $3x^2 - 3x + 4 = 0$

Here $a = 3$, $b = -3$, $c = 4$, and $b^2 - 4ac = (-3)^2 - 4(3)(4) = -39$.
Hence, this equation has two complex solutions.

67. $2x^2 + 3x - 4 = 0$

Here $a = 2$, $b = 3$, $c = -4$, and $b^2 - 4ac = 3^2 - 4(2)(-4) = 41$.
Hence, this equation has two unequal real solutions.

69. $9x^2 - 12x + 4 = 0$

Here $a = 9$, $b = -12$, $c = 4$, and $b^2 - 4ac = (-12)^2 - 4(9)(4) = 0$.
Hence, this equation has a repeated real solution.

71. The other solution must be the conjugate of $2 + 3i$, or $2 - 3i$.

In Problems 73 - 76, z - 3 - 4i and w = 8 + 3i.

73. $z + \bar{z} = 3 - 4i + (\overline{3 - 4i}) = 3 - 4i + (3 + 4i)$
$$= (3 + 3) + (-4 + 4)i$$
$$= 6$$

75. $z\bar{z} = (3 - 4i)(\overline{3 - 4i}) = (3 - 4i)(3 + 4i)$
$$= 9 + 12i - 12i - 16i^2$$
$$= 9 - 16(-1) = 25$$

77. For $z = a + bi$, $z + \bar{z} = a + bi + \overline{a + bi} = (a + bi) + (a - bi)$
$$= (a + a) + (b - b)i$$
$$= 2a + 0i = 2a$$

and $z - \bar{z} = a + bi - (\overline{a + bi})$
$$= a + bi - (a - bi)$$
$$= (a - a) + (b - (-b))i$$
$$= 0 + 2bi = 2bi$$

79. For $z = a + bi$ and $w = c + di$,

$$\overline{z + w} = \overline{(a + bi) + (c + di)}$$
$$= \overline{(a + c) + (b + d)i}$$
$$= (a + c) - (b + d)i$$
$$= (a - bi) + (c - di)$$
$$= \overline{a + bi} + \overline{c + di} = \bar{z} + \bar{w}$$

≡ EXERCISE 1.6 RECTANGULAR COORDINATES AND GRAPHS

1. (a) Quadrant II

    (b) Positive x-axis

    (c) Quadrant III

    (d) Quadrant I

    (e) Negative y-axis

    (f) Quadrant IV

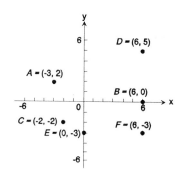

3. The points will be on a vertical line that is two units to the right of the y-axis.

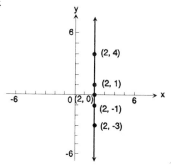

5. $P_1 = (3, -4)$; $P_2 = (3, 1)$

$$d(P_1, P_2) = \sqrt{(3 - 3)^2 + [1 - (-4)]^2}$$
$$= \sqrt{0 + 25}$$
$$= 5$$

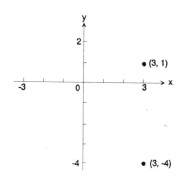

7. $P_1 = (-3, 2)$; $P_2 = (6, 0)$

$$d(P_1, P_2) = \sqrt{[6 - (-3)]^2 + (0 - 2)^2}$$
$$= \sqrt{(9)^2 + (-2)^2}$$
$$= \sqrt{81 + 4}$$
$$= \sqrt{85}$$

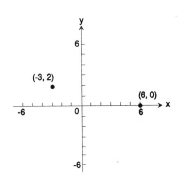

9. $P_1 = (4, -3)$; $P_2 = (6, 1)$

$$d(P_1, P_2) = \sqrt{(6 - 4)^2 + [1 - (-3)]^2}$$
$$= \sqrt{(2)^2 + (4)^2}$$
$$= \sqrt{4 + 16}$$
$$= \sqrt{20}$$
$$= 2\sqrt{5}$$

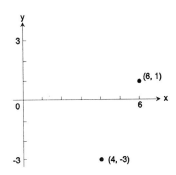

11. $P_1 = (-0.2, 0.3)$; $P_2 = (2.3, 1.1)$

$$d(P_1, P_2) = \sqrt{[2.3 - (-0.2)]^2 + (1.1 - 0.3)^2}$$
$$= \sqrt{(2.5)^2 + (.8)^2}$$
$$= \sqrt{6.25 + .64}$$
$$= \sqrt{6.89}$$
$$\approx 2.625$$

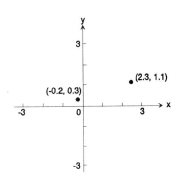

13. $P_1 = (a, b)$; $P_2 = (0, 0)$

$$d(P_1, P_2) = \sqrt{(0 - a)^2 + (0 - b)^2}$$
$$= \sqrt{a^2 + b^2}$$

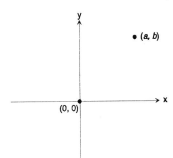

15. $A = (-2, 5)$, $B = (1, 3)$, $C = (-1, 0)$

$d(A, B) = \sqrt{[1 - (-2)^2 + (3 - 5)^2]}$

$d(A, B) = \sqrt{3^2 + (-2)^2}$

$d(A, B) = \sqrt{9 + 4}$

$d(A, B) = \sqrt{13}$

$d(B, C) = \sqrt{(-1 - 1)^2 + (0 - 3)^2}$

$d(B, C) = \sqrt{(-2)^2 + (-3)^2}$

$d(B, C) = \sqrt{4 + 9}$

$d(B, C) = \sqrt{13}$

$d(A, C) = \sqrt{[-1 - (-2)]^2 + (0 - 5)^2}$

$d(A, C) = \sqrt{(-1)^2 + (-5)^2}$

$d(A, C) = \sqrt{1 + 25}$

$d(A, C) = \sqrt{26}$

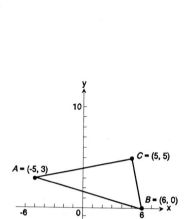

Verify that ABC is a right triangle by the Pythagorean Theorem:

$$[d(A, B)]^2 + [d(B, C)]^2 = [d(A, C)]^2$$
$$(\sqrt{13})^2 + (\sqrt{13})^2 = (\sqrt{26})^2$$
$$13 + 13 = 26$$
$$26 = 26$$

Area of a triangle is $A = \frac{1}{2}bh$. In this problem,

$A = \frac{1}{2}[d(B, C)][d(A, B)]$

$A = \frac{1}{2}(\sqrt{13})(\sqrt{13})$

$A = \frac{1}{2}(13)$

$A = \frac{13}{2}$ square units

17. $A = (-5, 3)$, $B = (6, 0)$, $C = (5, 5)$

$d(A, B) = \sqrt{[6 - (-5)^2 + (0 - 3)^2]}$

$d(A, B) = \sqrt{(11)^2 + (-3)^2}$

$d(A, B) = \sqrt{121 + 9}$

$d(A, B) = \sqrt{130}$

$d(B, C) = \sqrt{(5 - 6)^2 + (5 - 0)^2}$

$d(B, C) = \sqrt{(-1)^2 + (5)^2}$

$d(B, C) = \sqrt{1 + 25}$

$d(B, C) = \sqrt{26}$

1 PRELIMINARIES

$$d(A,\ C) = \sqrt{[5 - (-5)]^2 + (5 - 3)^2}$$

$$d(A,\ C) = \sqrt{(10)^2 + (2)^2}$$

$$d(A,\ C) = \sqrt{100 + 4}$$

$$d(A,\ C) = \sqrt{104}$$

Verify that ABC is a right triangle by the Pythagorean Theorem:

$$[d(A,\ C)]^2 + [d(B,\ C)]^2 = [d(A,\ B)]^2$$
$$\left(\sqrt{104}\right)^2 + \left(\sqrt{26}\right)^2 = \left(\sqrt{130}\right)^2$$
$$104 + 26 = 130$$
$$130 = 130$$

Area of a triangle is $A = \frac{1}{2}bh$. In this problem,

$$A = \frac{1}{2}[d(A,\ C)][d(B,\ C)]$$

$$A = \frac{1}{2}\left(\sqrt{104}\right)\left(\sqrt{26}\right)$$

$$A = \frac{1}{2}(2704)$$

$$A = \frac{1}{2}(52)$$

$$A = 26 \text{ square units}$$

19. All points having an x-coordinate of 2 would be of the form $(2, y)$. Those which are 5 units from $(-2, -1)$ would be:

$$\sqrt{[2 - (-2)]^2 + [y - (-1)]^2} = 5$$
$$\sqrt{(4)^2 + (y + 1)^2} = 5$$

Square both sides:

$$4^2 + (y + 1)^2 = 25$$
$$16 + y^2 + 2y + 1 = 25$$
$$y^2 + 2y - 8 = 0$$
$$(y + 4)(y - 2) = 0$$
$$y + 4 = 0 \qquad y - 2 = 0$$
$$y = -4 \qquad y = 2$$

Therefore, the points are $(2, -4)$, $(2, 2)$.

21. All points on the x-axis would be of the form $(x, 0)$. Those which are 5 units from $(2, -3)$ would be:

$$\sqrt{(x - 2)^2 + [0 - (-3)]^2} = 5$$
$$\sqrt{(x - 2)^2 + (3)^2} = 5$$

1.6 RECTANGULAR COORDINATES AND GRAPHS

Square both sides:

$$(x - 2)^2 + 9 = 25$$
$$x^2 - 4x + 4 + 9 = 25$$
$$x^2 - 4x + 13 = 25$$
$$x^2 - 4x - 12 = 0$$
$$(x - 6)(x + 2) = 0$$
$$x - 6 = 0 \qquad x + 2 = 0$$
$$x = 6 \qquad x = -2$$

Therefore, the points are $(-2, 0)$, $(6, 0)$.

23. $P_1 = (3, -4)$, $P_2 = (3, 1)$

Let $x_1 = 3 \quad y_1 = -4$
$x_2 = 3 \quad y_2 = 1$

Then, the coordinates (x, y) of the midpoint are:

$$x = \frac{x_1 + x_2}{2} = \frac{3 + 3}{2} = \frac{6}{2} = 3$$
$$y = \frac{y_1 + y_2}{2} = \frac{-4 + 1}{2} = -\frac{3}{2}$$

Midpoint $= \left(3, -\frac{3}{2}\right)$

25. $P_1 = (-3, 2)$, $P_2 = (6, 0)$

Let $x_1 = -3 \quad y_1 = 2$
$x_2 = 6 \quad y_2 = 0$

Then, the coordinates (x, y) of the midpoint are:

$$x = \frac{x_1 + x_2}{2} = \frac{-3 + 6}{2} = \frac{3}{2}$$
$$y = \frac{y_1 + y_2}{2} = \frac{2 + 0}{2} = 1$$

Midpoint $= \left(\frac{3}{2}, 1\right)$

27. $P_1 = (4, -3)$, $P_2 = (6, 1)$

Let $x_1 = 4 \quad y_1 = -3$
$x_2 = 6 \quad y_2 = 1$

Then, the coordinates (x, y) of the midpoint are:

$$x = \frac{x_1 + x_2}{2} = \frac{4 + 6}{2} = \frac{10}{2} = 5$$
$$y = \frac{y_1 + y_2}{2} = \frac{-3 + 1}{2} = \frac{-2}{2} = -1$$

Midpoint $= (5, -1)$

1 PRELIMINARIES

29.

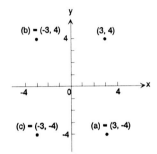

31.

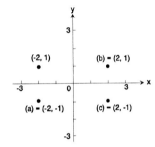

33.

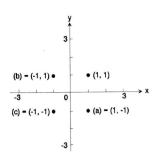

35.

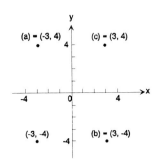

37. (a) $(-1, 0)$; $(1, 0)$

(b) x-axis, y-axis, origin

41. (a) $(0, 0)$

(b) x-axis

39. (a) $\left(-\dfrac{\pi}{2}, 0\right)$, $\left(\dfrac{\pi}{2}, 0\right)$, $(0, 1)$

(b) y-axis

43. $y = 3x + 2$

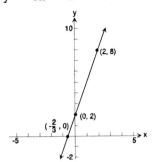

45. $3x - 2y + 6 = 0$

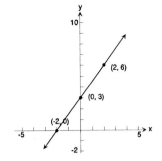

47. $y = -x^2$

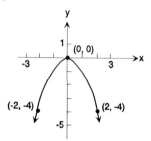

49. $y = x^2 + 3$

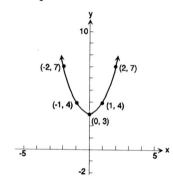

51. $y = x^3 - 1$

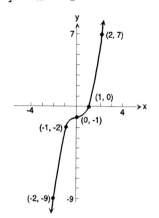

53. $x^2 = y + 1$

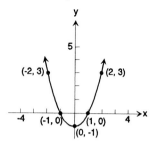

55. $y = \sqrt{x}$

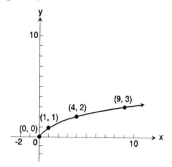

57. $y = \sqrt{x - 1}$

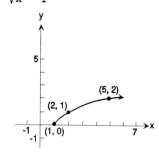

59. $y = \dfrac{1}{x - 2}$

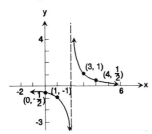

61. Use $(x - h)^2 + (y - k)^2 = r^2$, where $r = 2$; $(h,\ k) = (0,\ 2)$

$$(x - 0)^2 + (y - 2)^2 = 2^2$$
$$x^2 + (y - 2)^2 = 4$$
$$x^2 + y^2 - 4y + 4 = 4$$
$$x^2 + y^2 - 4y = 0$$

63. Use $(x - h)^2 + (y - k)^2 = r^2$, where $r = 5$; $(h,\ k) = (4,\ -3)$

$$(x - 4)^2 + [y - (-3)]^2 = (5)^2$$
$$(x - 4)^2 + (y + 3)^2 = 25$$

Square each part and gather up terms:

$$(x^2 - 8x + 16) + (y^2 + 6y + 9) = 25$$
$$x^2 + y^2 - 8x + 6y + 25 = 25$$
$$x^2 + y^2 - 8x + 6y = 0$$

65. Since $(h,\ k) = (0,\ 0)$, the center is at the origin, so use:

$$x^2 + y^2 = r^2 \text{ when } r = 2$$
$$x^2 + y^2 = 2^2$$
$$x^2 + y^2 = 4$$
$$x^2 + y^2 - 4 = 0$$

67. Center $= (2,\ 1)$
Radius $=$ Distance from $(0,\ 1)$ to $(2,\ 1)$

$$= \sqrt{(2 - 0)^2 + (1 - 1)^2} = \sqrt{4} = 2$$
$$(x - 2)^2 + (y - 1)^2 = 4$$

69. Center $=$ Midpoint of $(1,\ 2)$ and $(4,\ 2)$

$$= \left(\frac{1 + 4}{2},\ \frac{2 + 2}{2}\right) = \left(\frac{5}{2},\ 2\right)$$

Radius $=$ Distance from $\left(\frac{5}{2},\ 2\right)$ to $(4,\ 2)$

$$= \sqrt{\left(4 - \frac{5}{2}\right)^2 + (2 - 2)^2} = \sqrt{\frac{9}{4}} = \frac{3}{2}$$

$$\left(x - \frac{5}{2}\right)^2 + (y - 2)^2 = \frac{9}{4}$$

71. Since $x^2 + y^2 = 4$ is of the form $x^2 + y^2 = r^2$, it can be written as $x^2 + y^2 = (2)^2$, so $(h, k) = (0, 0)$ and $r = 2$.

73. $(x - 3)^2 + y^2 = 4$ can be written as:

$(x - 3)^2 + (y - 0)^2 = 2^2$, so compare to standard form
$(x - h)^2 + (y - k)^2 = r^2$ and get $(h, k) = (3, 0)$ and $r = 2$.

75. $x^2 + y^2 + 4x - 4y - 1 = 0$

Group together the x-terms and the y-terms and rearrange constant to right side:

$$(x^2 + 4x) + (y^2 - 4y) = 1$$

Complete the square of each expression in parentheses by taking $\frac{1}{2}$ (coefficient of variable to first degree) and then squaring it and adding the number to the left side and the right side.

Add $\left[\frac{1}{2}(4)\right]^2 = 4$ Add $\left[\frac{1}{2}(-4)\right]^2 = 4$

$$(x^2 + 4x + 4) + (y^2 - 4y + 4) = 1 + 4 + 4$$
$$(x + 2)^2 + (y - 2)^2 = 9$$
$$[x - (-2)]^2 + (y - 2)^2 = 3^2$$
$$(h, k) = (-2, 2)$$
$$r = 3$$

77.
$$x^2 + y^2 - x + 2y + 1 = 0$$
$$(x^2 - x) + (y^2 + 2y) = -1$$

Add $\left[\frac{1}{2}(-1)\right]^2 = \frac{1}{4}$, $\left[\frac{1}{2}(2)\right]^2 = 1$ to both sides)

$$\left(x^2 - x + \frac{1}{4}\right) + (y^2 + 2y + 1) = -1 + \frac{1}{4} + 1$$
$$\left(x - \frac{1}{2}\right)^2 + (y + 1)^2 = \frac{1}{4}$$
$$\left(x - \frac{1}{2}\right)^2 + [y - (-1)]^2 = \left(\frac{1}{2}\right)^2$$
$$(h, k) = \left(\frac{1}{2}, -1\right)$$
$$r = \frac{1}{2}$$

79.
$$2x^2 + 2y^2 - 12x + 8y - 24 = 0$$

The coefficients of x^2 and y^2 should be 1 in order to put the equation in standard form, so divide each term by 2.

$$x^2 + y^2 - 6x + 4y - 12 = 0$$
$$(x^2 - 6x) + (y^2 + 4y) = 12$$

Add $\left[\frac{1}{2}(-6)\right]^2 = 9$, $\left[\frac{1}{2}(4)\right]^2 = 4$ to both sides.

1 PRELIMINARIES

$$(x^2 - 6x + 9) + (y^2 + 4y + 4) = 12 + 9 + 4$$
$$(x - 3)^2 + (y + 2)^2 = 25$$
$$(x - 3)^2 + [y - (-2)]^2 = 5^2$$
$$(h, k) = (3, -2)$$
$$r = 5$$

81. Symmetry with respect to the x-axis means $(x, -y)$ is on the graph for every (x, y) on the graph. Therefore, given $(-4, 1)$, $(-2, 1)$, $(2, -1)$, $(4, 1)$; plot $(-4, -1)$, $(-2, -1)$, $(2, 1)$, $(4, -1)$.

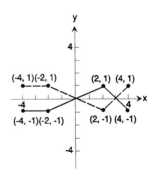

83. Symmetry with respect to the origin means $(-x, -y)$ is on the graph for every (x, y) on the graph. Therefore, given $(-4, 1)$, $(-2, 1)$, $(2, -1)$, $(4, 1)$; plot $(4, -1)$, $(2, -1)$, $(-2, 1)$, $(-4, -1)$.

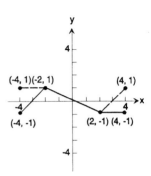

85. $$x^2 = y$$

   y-intercept: Let $x = 0$ so $y = 0$ $(0, 0)$
   x-intercept: Let $y = 0$ so $x = 0$ $(0, 0)$

   Test for symmetry:

   x-axis: Replace y by $-y$ so $x^2 = -y$, which is not equivalent to $x^2 = y$.

   y-axis: Replace x by $-x$ so $(-x)^2 = y$ or $x^2 = y$ is equivalent to $x^2 = y$.

   Origin: Replace x by $-x$ and y by $-y$ so $(-x)^2 = -y$ or $x^2 = -y$ is not equivalent to $x^2 = y$.

   Therefore, symmetric with respect to the y-axis.

87.
$$y = 3x$$

y-intercept: Let $x = 0$ so $y = 0$ $(0, 0)$
x-intercept: Let $y = 0$ so $x = 0$ $(0, 0)$

Test for symmetry:

    x-axis: Replace y by $-y$ so $-y = 3x$ is not equivalent to
        $y = 3x$.
    y-axis: Replace x by $-x$ so $y = -3x$ is not equivalent to
        $y = 3x$.
    Origin: Replace x by $-x$ and y by $-y$ so $-y = -3x$ is $y = 3x$
        which is not equivalent to $y = 3x$.

Therefore, symmetric with respect to the origin.

89.
$$x^2 + y - 9 = 0$$

y-intercept: Let $x = 0$ so $y = 9$ $(0, 9)$
x-intercept: Let $y = 0$ so $x^2 = 9$ $(-3, 0)$
                             $x = \pm 3$ $(3, 0)$

Test for symmetry:

    x-axis: Replace y by $-y$ so $x^2 - y - 9 = 0$ is not equivalent to
        $x^2 + y - 9 = 0$.
    y-axis: Replace x by $-x$ so $(-x)^2 + y - 9 = 0$ is $x^2 + y - 9$
        $= 0$, which is equivalent to $x^2 + y - 9 = 0$.
    Origin: Replace x by $-x$ and y by $-y$ so $(-x)^2 - y - 9 = 0$ is
        $x^2 - y - 9 = 0$, which is not equivalent to $x^2 + y - 9$
        $= 0$.

Therefore, symmetric with respect to the y-axis.

91.
$$4x^2 + 9y^2 = 36$$

y-intercept: Let $x = 0$ so $9y^2 = 36$ $(0, -2)$
                                $y^2 = 4$ $(0, 2)$
                                $y = \pm 2$
x-intercept: Let $y = 0$ so $4x^2 = 36$ $(-3, 0)$
                                $x^2 = 9$ $(3, 0)$
                                $x = \pm 3$

Test for symmetry:

    x-axis: Replace y by $-y$ so $4x^2 + 9(-y)^2 = 36$
                         $4x^2 + 9y^2 = 36$
        is equivalent to $4x^2 + 9y^2 = 36$
    y-axis: Replace x by $-x$ so $4(-x)^2 + 9y^2 = 36$
                         $4x^2 + 9y^2 = 36$
        is equivalent to $4x^2 + 9y^2 = 36$
    Origin: Replace x by $-x$ and y by $-y$ so $4(-x)^2 + 9(-y)^2 = 36$
                                $4x^2 + 9y^2 = 36$
        is equivalent to $4x^2 + 9y^2 = 36$

Therefore, symmetric with respect to the x-axis, y-axis, and origin.

1 PRELIMINARIES

93.
$$y = x^3 - 27$$

y-intercept: Let $x = 0$ so $y = -27$ (0, -27)
x-intercept: Let $y = 0$ so $0 = x^3 - 27$
$$27 = x^3$$
$$3 = x$$ (3, 0)

Test for symmetry:

x-axis: Replace y by $-y$ so $-y = x^3 - 27$ is not equivalent to
$y = x^3 - 27$.
y-axis: Replace x by $-x$ so $y = (-x)^3 - 27$
$y = -x^3 - 27$ is not equivalent to $y = x^3 - 27$
Origin: Replace x by $-x$ and y by $-y$ so $-y = (-x)^3 - 27$
$$-y = -x^3 - 27$$
$y = x^3 + 27$ is not equivalent to $y = x^3 - 27$.

Therefore, no symmetry.

95.
$$y = x^2 - 3x - 4$$

y-intercept: Let $x = 0$ so $y = -4$ (0, -4)
x-intercept: Let $y = 0$ so $0 = x^2 - 3x - 4$
$$0 = (x - 4)(x + 1)$$
$x - 4 = 0$ or $x + 1 = 0$
$x = 4$ or $x = -1$
(4, 0), (-1, 0)

Test for symmetry:

x-axis: Replace y by $-y$ so $-y = x^2 - 3x - 4$ is not equivalent
to $y = x^2 - 3x - 4$.
y-axis: Replace x by $-x$ so $y = (-x)^2 - 3(-x) - 4$
$y = x^2 + 3x - 4$ is not equivalent to $y = x^2 - 3x - 4$.
Origin: Replace x by $-x$ and y by $-y$ so $-y = (-x)^2 - 3(-x) - 4$
$$-y = -x^2 + 3x - 4$$
$y = -x^2 - 3x + 4$ is not equivalent to $y = x^2 - 3x - 4$.

Therefore, no symmetry.

97.
$$y = \frac{x}{x^2 + 9}$$

y-intercept: Let $x = 0$ so $y = 0$ (0, 0)

x-intercept: Let $y = 0$ so $0 = \frac{x}{x^2 + 9}$
$$0 = x$$ (0, 0)

Test for symmetry:

x-axis: Replace y by $-y$ so $-y = \dfrac{x}{x^2 + 9}$ is not equivalent to

$y = \dfrac{x}{x^2 + 9}$

y-axis: Replace x by $-x$ so $y = \dfrac{-x}{(-x)^2 + 9}$

$y = \dfrac{-x}{x^2 + 9}$ is not equivalent to $y = \dfrac{x}{x^2 + 9}$

Origin: Replace x by $-x$ and y by $-x$ so $-y = \dfrac{-x}{(-x)^2 + 9}$

$-y = \dfrac{-x}{x^2 + 9}$

$y = \dfrac{x}{x^2 + 9}$ is not equivalent to $y = \dfrac{x}{x^2 + 9}$

Therefore, symmetric with respect to the origin.

99. (a)

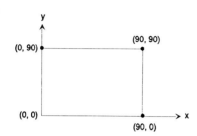

(b) $d = \sqrt{(310 - 90)^2 + (15 - 90)^2}$

$= \sqrt{(220)^2 + (75)^2}$

$= 232.4 \; feet$

(c) $d = \sqrt{(300 - 0)^2 + (300 - 90)^2}$

$= \sqrt{(300)^2 + (210)^2}$

$= 366.2 \; feet$

101. $d = rt$

Automobile's distance $= 40t$
Truck's distance $= 30t$
$d^2 = (40t)^2 + (30t)^2$
$d^2 = 1600t^2 + 900t^2$
$d^2 = 2500t^2$
$d = 50t \; miles$

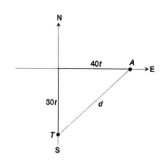

103. Earth:

$$x^2 + y^2 + 2x + 4y - 4091 = 0$$
$$(x^2 + 2x \qquad) + (y^2 + 4y \qquad) = 4091$$
$$(x^2 + 2x + 1) + (y^2 + 4y + 4) = 4091 + 1 + 4$$
$$(x + 1)^2 + (y + 2)^2 = 4096$$
$$(h, k) = (-1, -2)$$
$$r = \sqrt{4096} = 64$$

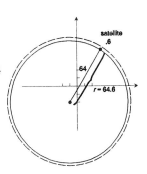

Orbit of satellite has center at $(-1, -2)$.

$$r = 0.6 + 64$$

$$[x - (-1)]^2 + [y - (-2)]^2 = (64.6)^2$$
$$(x + 1)^2 + (y + 2)^2 = 4173.16$$
$$x^2 + 2x + 1 + y^2 + 4y + 4 = 4173.16$$
$$x^2 + y^2 + 2x + 4y - 4168.16 = 0$$

≡ EXERCISE 1.7 THE STRAIGHT LINE

1. (x_1, y_1) (x_2, y_2)
    $(2, 3)$ $(1, 0)$

    slope $= \dfrac{y_2 - y_1}{x_2 - x_1} = \dfrac{0 - 3}{1 - 2} = \dfrac{-3}{-1} = 3$

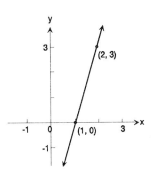

3. (x_1, y_1) (x_2, y_2)
    $(-2, 3)$ $(2, 1)$

    slope $= \dfrac{y_2 - y_1}{x_2 - x_1} = \dfrac{1 - 3}{2 - (-2)} = \dfrac{-2}{4} = -\dfrac{1}{2}$

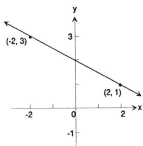

1.7 THE STRAIGHT LINE

5. $(x_1,\ y_1)\ (x_2,\ y_2)$
$\quad(-3,\ -1)\quad(2,\ 1)$

slope $= \dfrac{y_2 - y_1}{x_2 - x_1} = \dfrac{-1 - (-1)}{2 - (-3)} = \dfrac{0}{5} = 0$

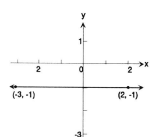

7. $(x_1,\ y_1)\ (x_2,\ y_2)$
$\quad(-1,\ 2)\quad(-1,\ -2)$

slope $= \dfrac{y_2 - y_1}{x_2 - x_1} = \dfrac{-2 - 2}{1 - (-1)} = \dfrac{-4}{0}$ (undefined)

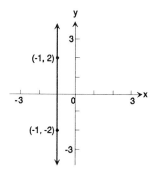

9. $(x_1,\ y_1)\ (x_2,\ y_2)$
$\quad\left(\sqrt{2},\ 3\right)\quad\left(1,\ \sqrt{3}\right)$

slope $= \dfrac{y_2 - y_1}{x_2 - x_1} = \dfrac{\sqrt{3} - 3}{1 - \sqrt{2}}$

$\approx \dfrac{1.732 - 3}{1 - 1.414} \approx \dfrac{-1.268}{-.414} \approx 3.06$

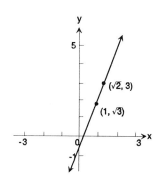

11.

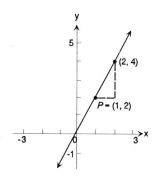

13.

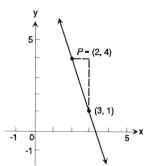

1 PRELIMINARIES

15.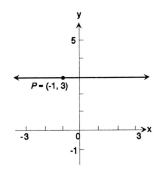

17.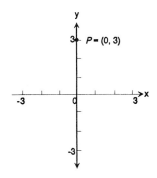

19. (0, 0) and (2, 1) are points on the line.

Slope = $\dfrac{1 - 0}{2 - 0} = \dfrac{1}{2}$

y-intercept is 0; so using $y = mx + b$, we get

$$y = \dfrac{1}{2}x + 0$$
$$2y = x$$
$$0 = x - 2y$$
$$x - 2y = 0$$

21. (-1, 3) and (1, 1) are points on the line.

Slope = $\dfrac{1 - 3}{1 - (-1)} = \dfrac{-2}{2} = -1$

Use $y - y_1 = m(x - x_1)$ with (x_1, y_1) being either point on the line.

$$y - 3 = -1[x - (-1)]$$
$$y - 3 = -(x + 1)$$
$$y - 3 = -x - 1$$
$$x + y - 2 = 0$$

23.
$$y - y_1 = m(x - x_1), \ m = 2$$
$$y - 3 = 2(x - 3)$$
$$y - 3 = 2x - 6$$
$$2x - y - 3 = 0$$

25.
$$y - y_1 = m(x - x_1), \ m = -\dfrac{1}{2}$$
$$y - 2 = -\dfrac{1}{2}(x - 1)$$
$$2y - 4 = -x + 1$$
$$x + 2y - 5 = 0$$

27. $(x_1, y_1) = (-2, 3)$

slope = 2

$$y - y_1 = m(x - x_1)$$
$$y - 3 = 2[x - (-2)]$$
$$y - 3 = 2(x + 2)$$
$$y - 3 = 2x + 4$$
$$0 = 2x - y + 7$$
$$2x - y + 7 = 0$$

1.7 THE STRAIGHT LINE

55

29. $(x_1, y_1) = (1, -1)$

slope $= -\dfrac{2}{3}$

$$y - y_1 = m(x - x_1)$$
$$y - (-1) = -\dfrac{2}{3}(x - 1)$$
$$y + 1 = -\dfrac{2}{3}x + \dfrac{2}{3}$$

Multiply both sides by 3: $\qquad 3y + 3 = -2x + 2$
$$2x + 3y + 1 = 0$$

31. Passing through $(1, 3)$ and $(-1, 2)$.

slope $= \dfrac{2 - 3}{-1 - 1} = \dfrac{-1}{-2} = \dfrac{1}{2}$

Use either point and slope.

$$y - 3 = \dfrac{1}{2}(x - 1)$$
$$y - 3 = \dfrac{1}{2}x - \dfrac{1}{2}$$

Multiply both sides by 2: $\qquad 2y - 6 = x - 1$
$$0 = x - 2y + 5$$
$$x - 2y + 5 = 0$$

33. slope $= -3$

$b = y$-intercept $= 3$

Use $y = mx + b$
$$y = (-3)x + 3$$
$$y = -3x + 3$$
$$3x + y - 3 = 0$$

35. x-intercept $= 2$

y-intercept $= -1$

$\dfrac{x}{a} + \dfrac{y}{b} = 1$ $\qquad$ or $\qquad$ slope $= \dfrac{-1 - 0}{0 - 2} = \dfrac{1}{2}$

$\dfrac{x}{2} + \dfrac{y}{-1} = 1$ $\qquad\qquad\qquad$ y-intercept $= -1$

Multiply both sides by 2: $\qquad\qquad\qquad$ $y = mx + b$

$\qquad x - 2y = 2$ $\qquad\qquad\qquad\qquad$ $y = \dfrac{1}{2}x - 1$

$\qquad x - 2y - 2 = 0$ $\qquad\qquad\qquad\qquad$ $2y = x - 2$

$\qquad\qquad\qquad\qquad\qquad\qquad\qquad$ $0 = x - 2y - 2$

$\qquad\qquad\qquad\qquad\qquad\qquad$ $x - 2y - 2 = 0$

37. Slope undefined; passing through $(1, 4)$

Vertical line $x = a$ has undefined slope.

Thus, $\qquad\qquad x = 1$
$$x - 1 = 0$$

39. Parallel to $y = 3x$; passing through $(-1, 2)$.

      slope = 3 and parallel line has same slope.
$$y - 2 = 3[x - (-1)]$$
$$y - 2 = 3(x + 1)$$
$$y - 2 = 3x + 3$$
$$0 = 3x - y + 5$$
$$3x - y + 5 = 0$$

41. Parallel to $2x - y + 2 = 0$; passing through $(0, 0)$.

$$2x + 2 = y$$
$$y = 2x + 2$$
      slope = 2 and parallel line has same slope.
$$y - 0 = 2(x - 0)$$
$$y = 2x$$
$$0 = 2x - y$$
$$2x - y = 0$$

43. Parallel to $x = 5$; passing through $(4, 2)$.

    $x = 5$ is a vertical line; slope is undefined.

Therefore, $x = 4$ is parallel to $x = 5$.

$$x - 4 = 0$$

45. Perpendicular to $y = \frac{1}{2}x + 4$; passing through $(1, -2)$.

$$\text{slope} = \frac{1}{2}$$

Perpendicular line has slope = -2

$$[y - (-2)] = -2(x - 1)$$
$$y + 2 = -2x + 2$$
$$2x + y = 0$$

47. Perpendicular to $2x + y - 2 = 0$; passing through $(-3, 0)$.

$$y = -2x + 2$$
$$\text{slope} = -2$$

Perpendicular line has slope = $\frac{1}{2}$

$$y - 0 = \frac{1}{2}[x - (-3)]$$
$$y = \frac{1}{2}(x + 3)$$
$$y = \frac{1}{2}x + \frac{3}{2}$$
$$2y = x + 3$$
$$0 = x - 2y + 3$$
$$x - 2y + 3 = 0$$

49. Perpendicular to $x = 5$; passing through $(3, 4)$.

    Since $x = 5$ is vertical, a line perpendicular would be horizontal of the form $y = b$.

Therefore, $y = 4$ is the line perpendicular to $x = 5$.

$$y - 4 = 0$$

51.
$$y = 2x + 3$$
$$y = mx + b$$
m: slope = 2
b: y-intercept = 3

Using intercepts, draw graph.
(0, 3)
$$\left(-\frac{3}{2}, 0\right)$$

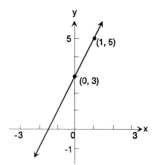

53.
$$\frac{1}{2}y = x - 1$$
$$y = 2x - 2$$
$$y = mx + b$$
m: slope = 2
b: y-intercept = -2

Using intercepts, draw graph.

(0, -2)
(1, 0)

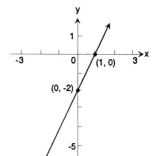

55.
$$2x - 3y = 6$$
$$-3y = -2x + 6$$
$$y = \frac{2}{3}x - 2$$
$$y = mx + b$$
m: slope = $\frac{2}{3}$
b: y-intercept = -2

Using intercepts, draw graph.
(0, -2)
(3, 0)

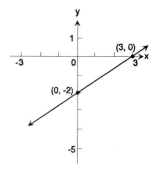

57.
$$x + y = 1$$
$$y = -x + 1$$
m: slope = -1
b: y-intercept = 1

Using intercepts, draw graph:

(0, 1)
(1, 0)

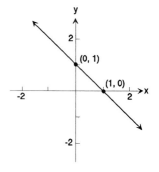

1 PRELIMINARIES

59.
$$x = -4$$
is of the form $x = a$, a vertical line. Slope is undefined; no y-intercept.

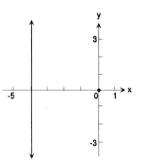

61.
$$2y - 3x = 0$$
$$2y = 3x$$
$$y = \frac{3}{2}x$$
$$y = mx + b$$
$$m: \text{slope} = \frac{3}{2}$$
$$b: y\text{-intercept} = 0$$

Using intercept and another point, draw graph.

$$(0, 0)$$
$$(2, 3)$$

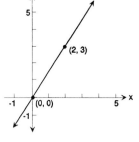

63. All points on the x-axis have a y-coordinate of 0, so the equation of the x-axis is $y = 0$.

65. Plot points $P_1(-2, 5)$, $P_2(1, 3)$, and $P_3(-1, 0)$.

$$\text{slope of } P_1P_2 = \frac{3 - 5}{1 + 2} = \frac{-2}{3} = m_1$$

$$\text{slope of } P_2P_3 = \frac{3 - 0}{1 - (-1)} = \frac{3}{2} = m_2$$

Since $m_1m_2 = \left(-\frac{2}{3}\right)\left(\frac{3}{2}\right) = -1$, the lines are perpendicular, so the points, P_1, P_2, P_3 form a right triangle.

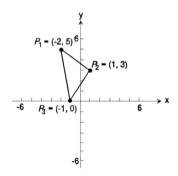

67. Plot $P_1(-1, 0)$, $P_2(2, 3)$, $P_3(1, -2)$, and $P_4(4, 1)$

$$\text{slope of } P_1P_2 = m_1 = \frac{3 - 0}{2 - (-1)} = \frac{3}{3} = 1$$

$$\text{slope of } P_3P_4 = m_2 = \frac{1 - (-2)}{4 - 1} = \frac{3}{3} = 1$$

$$\text{slope of } P_1P_3 = m_3 = \frac{-2 - 0}{1 - (-1)} = \frac{-2}{2} = -1$$

$$\text{slope of } P_2P_4 = m_4 = \frac{1 - 3}{4 - 2} = \frac{-2}{2} = -1$$

Since the opposite sides are parallel and the adjacent sides are perpendicular, the points form a rectangle.

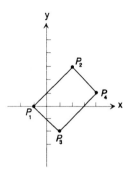

1.7 THE STRAIGHT LINE

69. (32°F, 0°C) and (212°F, 100°C)

$$m = \frac{100 - 0}{212 - 32} = \frac{100}{180} = \frac{5}{9}$$

$$°C - 0 = \frac{5}{9}(°F - 32)$$

$$°C = \frac{5}{9}(°F - 32)$$

When F = 70°, $°C = \frac{5}{9}(70 - 32)$

$$°C = \frac{5}{9}(38)$$

$$°C = \frac{190}{9} = 21\frac{1}{9} \approx 21°C$$

71. $2x - y + C = 0$

If $C = 4$, $2x - y + 4 = 0$

Plot points (0, 4)
             (1, -2)
             (2, 0)

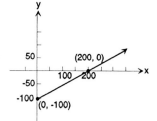

If $C = 0$, $2x - y + 0 = 0$ or $2x = y$

Plot points (0, 0)
             (1, 2)

If $C = 2$, $2x - y + 2 = 0$

Plot points (0, 2)
             (1, 4)

All have the same slope of 2. The lines are parallel.

73. Since there is only a profit of $0.50 per copy and the expense of $100 must be deducted, then the profit is
         $P = (0.50)x - 100$ or
         $P = 0.5x - 100$

Plot points: (0, -100)
                     (200, 0)

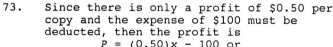

75. If $m_1 m_2 = -1$, then

$$d(A, B) = \sqrt{(1 - 1)^2 + (m_2 - m_1)^2} = \sqrt{(m_2 - m_1)^2}$$

$$d(O, A) = \sqrt{(1 - 0)^2 + (m_2 - 0)^2} = \sqrt{1 + m_2^2}$$

$$d(O, B) = \sqrt{(1 - 0)^2 + (m_1 - 0)^2} = \sqrt{1 + m_1^2}$$

Show: $[d(0, B)]^2 + [d(0, A)]^2 = [d(A, B)]^2$

$$\left(\sqrt{1 + m_1^2}\right)^2 + \left(\sqrt{1 + m_2^2}\right)^2 = \left(\sqrt{(m_2 - m_1)^2}\right)^2$$

$$1 + m_1^2 + 1 + m_2^2 = (m_2 - m_1)^2$$

$$m_1^2 + m_2^2 + 2 = m_2^2 - 2m_1m_2 + m_1^2$$

Since $m_1m_2 = -1$, $\quad m_1^2 + m_2^2 + 2 = m_2^2 - 2(-1) + m_1^2$

$$m_1^2 + m_2^2 + 2 = m_1^2 + m_2^2 + 2$$

77.   (a) $x^2 + (mx + b)^2 = r^2$

       $(1 + m^2)x^2 + 2mbx + b^2 - r^2 = 0$

       One solution if and only if discriminant = 0

$$(2mb)^2 - 4(1 + m^2)(b^2 - r^2) = 0$$

$$-4b^2 + 4r^2 + 4m^2r^2 = 0$$

$$r^2(1 + m^2) = b^2$$

     (b) $x = \dfrac{-2mb}{2(1 + m^2)} = \dfrac{-2mb}{2b^2/r^2} = \dfrac{-r^2m}{b}$

$$y = m\left(\dfrac{-r^2m}{b}\right) + b = \dfrac{-r^2m^2}{b} + b = \dfrac{-r^2m^2 + b^2}{b} = \dfrac{r^2}{b}$$

     (c) Slope of tangent line = m

        Slope of line joining center to point of tangency

$$= \dfrac{r^2/b}{-r^2m/b} = \dfrac{-1}{m}$$

79. $x^2 - 4x + y^2 + 6y = -4$

      $(x - 2)^2 + (y + 3)^2 = 9$

    Center $(2, -3)$

Slope from center to $\left(3, 2\sqrt{2} - 3\right)$ is $\dfrac{2\sqrt{2} - 3 + 3}{3 - 2} = 2\sqrt{2}$

Slope of tangent line is $\dfrac{-1}{2\sqrt{2}} = \dfrac{-\sqrt{2}}{4}$

$$y - \left(2\sqrt{2} - 3\right) = \dfrac{-\sqrt{2}}{4}(x - 3)$$

$$\sqrt{2}x + 4y - 11\sqrt{2} + 12 = 0$$

81. $x^2 + y^2 - 4x + 6y = 0$ $x^2 + y^2 + 6x + 4y + 9 = 0$

      $x^2 - 4x + y^2 + 6y = -4$ $x^2 + 6x + y^2 + 4y = -9$

      $(x - 2)^2 + (y + 3)^2 = 9$ $(x + 3)^2 + (y + 2)^2 = 4$

    Center $(2, -3)$ Center $(-3, -2)$

Slope of line joining centers is $\dfrac{-2 + 3}{-3 - 2} = \dfrac{1}{-5}$

$$y + 3 = \dfrac{-1}{5}(x - 2)$$

$$x + 5y + 13 = 0$$

1.7 THE STRAIGHT LINE

1.
$$2 - \frac{x}{3} = 5$$
$$2 - \frac{x}{3} - 2 = 5 - 2$$
$$-\frac{x}{3} = 3$$
$$(-3)\left(-\frac{x}{3}\right) = 3(-3)$$
$$x = -9$$

3.
$$-2(5 - 3x) + 8 = 4 + 5x$$
$$-10 + 6x + 8 = 4 + 5x$$
$$-2 + 6x - 4 = 4 + 5x - 4$$
$$6x - 6 = 5x$$
$$6x - 6 - 5x = 5x - 5x$$
$$x - 6 = 0$$
$$x = 6$$

5.
$$\frac{3x}{4} - \frac{x}{3} = \frac{1}{12}$$
$$12\left(\frac{3x}{4} - \frac{x}{3}\right) = \left(\frac{1}{12}\right)(12)$$
$$9x - 4x = 1$$
$$5x = 1$$
$$\frac{5x}{5} = \frac{1}{5}$$
$$x = \frac{1}{5}$$

7.
$$\frac{x}{x - 1} = \frac{5}{4}$$
$$(4(x - 1))\frac{x}{x - 1} = \frac{5}{4}(4(x - 1)) \quad x \neq 1$$
$$4x = 5(x - 1)$$
$$4x = 5x - 5$$
$$4x + 5 = 5x - 5 + 5$$
$$4x + 5 - 4x = 5x - 4x$$
$$5 = x$$

9.
$$x(1 - x) = 6$$
$$x - x^2 = 6$$
$$0 = x^2 - x + 6$$

Here, $a = 1$, $b = -1$, $c = 6$,
and $b^2 - 4ac = -23$.

No real solution.

11.
$$\frac{1}{2}\left(x - \frac{1}{3}\right) = \frac{3}{4} - \frac{x}{6}$$
$$12 \cdot \frac{1}{2}\left(x - \frac{1}{3}\right) = 12\left(\frac{3}{4} - \frac{x}{6}\right)$$
$$6x - 2 = 9 - 2x$$
$$6x - 2 + 2 = 9 - 2x + 2$$
$$6x = 11 - 2x$$
$$6x + 2x = 11 - 2x + 2x$$
$$8x = 11$$
$$\frac{8x}{8} = \frac{11}{8}$$
$$x = \frac{11}{8}$$

13.
$$(x - 1)(2x + 3) = 3$$
$$2x^2 + 3x - 2x - 3 = 3$$
$$2x^2 + x - 3 = 3$$
$$2x^2 + x - 3 - 3 = 3 - 3$$
$$2x^2 + x - 6 = 0$$
$$(2x - 3)(x + 2) = 0$$
$$2x - 3 = 0 \quad \text{or} \quad x + 2 = 0$$
$$x = \frac{3}{2} \quad \text{or} \quad x = -2$$
$$\left\{-2, \frac{3}{2}\right\}$$

15.
$$2x + 3 = 4x^2$$
$$2x + 3 - 2x = 4x^2 - 2x$$
$$3 = 4x^2 - 2x$$
$$3 - 3 = 4x^2 - 2x - 3$$
$$0 = 4x^2 - 2x - 3$$

Here, $a = 4$, $b = -2$, $c = -3$,
and $b^2 - 4ac = 52$.

Then,
$$x = \frac{-(-2) \pm \sqrt{52}}{2(4)}$$
$$= \frac{2 \pm 2\sqrt{13}}{8}$$
$$= \frac{1 \pm \sqrt{13}}{4}$$
$$\left\{\frac{1 - \sqrt{13}}{4}, \frac{1 + \sqrt{13}}{4}\right\}$$

17.
$$\sqrt[3]{x^2 - 1} = 2$$
$$\left(\sqrt[3]{x^2 - 1}\right)^3 = 2^3$$
$$x^2 - 1 = 8$$
$$x^2 = 9$$
$$x = \pm\sqrt{9} = \pm3$$

{-3, 3}

19.
$$x(x + 1) + 2 = 0$$
$$x^2 + x + 2 = 0$$

Here, $a = 1$, $b = 1$, $c = 2$, and $b^2 - 4ac = -7$.

No real solution.

21. $|2x - 3| = 5$

$2x - 3 = 5$ or $2x - 3 = -5$
$2x = 8$ or $2x = -2$
$x = 4$ or $x = -1$

The solution set is {-1, 4}.

23. $10a^2x^2 \ 0 \ 2abx - 36b^2 = 0$

Here $A = 10a^2$, $B = -2ab$, $C = -36b^2$ and
$$B^2 - 4AC = (-2ab)^2 - 4(10a^2)(-36b^2)$$
$$= 4a^2b^2 + 1440a^2b^2 = 1444a^2b^2$$

Then, $x = \dfrac{-(-2ab) \pm \sqrt{148a^2b^2}}{2(10a^2)}$

$$= \frac{2ab - 38ab}{20a^2}$$

or $x = \dfrac{2ab - 38ab}{20a^2} = \dfrac{-36ab}{20a^2} = \dfrac{-9b}{5a}$

and $x = \dfrac{2ab + 38ab}{20a^2} = \dfrac{40ab}{20a^2} = \dfrac{2b}{a}$

The solution set is $\left\{\dfrac{-9b}{5a}, \ \dfrac{2b}{a}\right\}$

25.
$$\frac{2x - 3}{5} + 1 \le \frac{x}{2}$$
$$(10)\left(\frac{2x - 3}{5} + 1\right) \le \left(\frac{x}{2}\right)(10)$$
$$4x - 6 + 10 \le 5x$$
$$4x + 4 - 4x \le 5x - 4x$$
$$4 \le x$$
$$\{x \mid 4 \le x < \infty\}$$

27.
$$-9 \le \frac{2x + 3}{-4} \le 7$$
$$(-4)(-9) \ge (-4)\frac{2x + 3}{-4} \ge 7(-4)$$
$$36 \ge 2x + 3 \ge -28$$
$$36 - 3 \ge 2x + 3 - 3 \ge -28 - 3$$
$$33 \ge 2x \ge -31$$
$$\frac{33}{2} \ge \frac{2x}{2} \ge \frac{-31}{2}$$
$$\left\{x \mid \frac{-31}{2} \le x \le \frac{33}{2}\right\}$$

29.
$$6 > \frac{3 - 2x}{12} > 2$$

$$12(6) > 12\left(\frac{3 - 3x}{12}\right) > (12)(2)$$

$$72 > 3 - 3x > 24$$

$$72 - 3 > 3 - 3x - 3 > 24 - 3$$

$$69 > -3x > 21$$

$$\frac{69}{-3} < \frac{-3x}{-3} < \frac{21}{-3}$$

$$\{x \mid -23 < x \; -7\}$$

31.
$$2x^2 + 5x - 12 < 0$$

$(2x - 3)(x + 4) < 0$ gives the boundary points $\frac{3}{2}$ and -4:

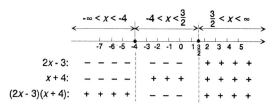

The solution set is $\left\{x \mid -4 < x < \frac{3}{2}\right\}$.

33.
$$\frac{6}{x + 2} \geq 1$$

$$\frac{6}{x + 2} - 1 \geq 0$$

$$\frac{6 - (x + 2)}{x + 2} \geq 0$$

$$\frac{4 - x}{x + 2} \geq 0 \text{ gives the boundary points } -2 \text{ and } 4:$$

	$-\infty < x < -2$	$-2 < x < 4$	$4 < x < \infty$

	-4 -3 -2 -1 0 1 2 3 4 5 6
4 - x:	+ + + | + + + + + + | - - -
x + 2:	- - - | + + + + + + | + + +
$\frac{4 - x}{x + 2}$:	- - - | + + + + + + | - - -

Note that $\frac{4 - x}{x + 2} = 0$ only at $x = 4$. The solution set is $\{x \mid -2 < x \leq 4\}$.

35.
$$\frac{2x - 3}{1 - x} < 2$$

$$\frac{2x - 3}{1 - x} - 2 < 0$$

$$\frac{2x - 3 - 2(1 - x)}{1 - x} < 0$$

$$\frac{2x - 3 - 2 + 2x}{1 - x} < 0$$

$$\frac{4x - 5}{1 - x} < 0 \text{ gives the boundary points 1 and } \frac{5}{4}:$$

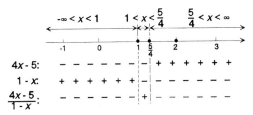

The solution set is $\left\{ x \mid -\infty < x < 1 \text{ or } \frac{5}{4} < x < \infty \right\}$.

37. $\dfrac{(x - 2)(x - 1)}{x - 3} > 0$ gives the boundary points 1, 2, and 3:

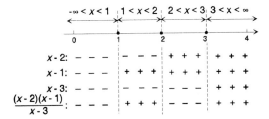

The solution set is $\{x \mid 1 < x < 2 \text{ or } 3 < x < \infty\}$.

39.
$$\frac{x^2 - 8x + 12}{x^2 - 16} > 0$$

$$\frac{(x - 6)(x - 2)}{(x + 4)(x - 4)} > 0 \text{ gives boundary points } -4, 2, 4, \text{ and } 6:$$

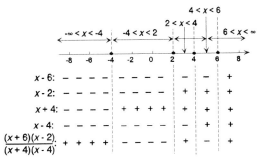

The solution set is $\{x \mid -\infty < x < -4 \text{ or } 2 < x < 4 \text{ or } 6 < x < \infty\}$.

41. $\left|3x + 4\right| < \dfrac{1}{2}$

$$-\dfrac{1}{2} < \quad 3x + 4 \quad < \dfrac{1}{2}$$

$$-\dfrac{1}{2} - 4 < 3x + 4 - 4 < \dfrac{1}{2} - 4$$

$$-\dfrac{9}{2} < \quad 3x \quad < -\dfrac{7}{2}$$

$$-\dfrac{9}{6} < \quad \dfrac{3x}{3} \quad < -\dfrac{7}{6}$$

$$\left\{x \,\middle|\, -\dfrac{3}{2} < x < -\dfrac{7}{6}\right\}$$

43.
$$\left|2x - 5\right| \geq 7$$
$$2x - 5 \leq -7 \quad \text{or} \quad 2x - 5 \geq 7$$
$$2x \leq -2 \quad \text{or} \quad 2x \geq 12$$
$$\{x \,|\, -\infty < x \leq -1 \text{ or } 6 \leq x < \infty\}$$

45. $x^2 + x + 1 \neq 0$

Here $a = 1$, $b = 1$, $c = 1$, and $b^2 - 4ac = -3$.

Then, $x = \dfrac{-1 \pm \sqrt{-3}}{2(1)}$

$\quad\quad = \dfrac{-1 \pm \sqrt{3}\,i}{2}$

The solution set is

$$\left\{\dfrac{-1 - \sqrt{3}\,i}{2}, \ \dfrac{-1 + \sqrt{3}\,i}{2}\right\}.$$

47. $2x^2 + x - 2 = 0$

Here $a = 2$, $b = 1$, $c = -2$ and $b^2 - 4ac = 17$.

Then, $x = \dfrac{-1 \pm \sqrt{17}}{2(2)}$

$\quad\quad = \dfrac{-1 \pm \sqrt{17}}{4}$

The solution set is

$$\left\{\dfrac{-1 - \sqrt{17}}{4}, \ \dfrac{-1 + \sqrt{17}}{4}\right\}.$$

49.
$$x^2 + 3 = x$$
$$x^2 - x + 3 = 0$$

Here, $a = 1$, $b = -1$, $c = 3$, and $b^2 - 4ac = -11$.

Then, $x = \dfrac{-(-1) \pm \sqrt{-11}}{2(1)}$

$\quad\quad = \dfrac{1 \pm \sqrt{11}\,i}{2}$

The solution set is

$$\left\{\dfrac{1 - \sqrt{11}\,i}{2}, \ \dfrac{1 + \sqrt{11}\,i}{2}\right\}.$$

51.
$$x(1 - x) = 6$$
$$x - x^2 = 6$$
$$-x^2 + x - 6 = 0$$
$$x^2 - x + 6 = 0$$

Here, $a = 1$, $b = -1$, $c = 6$, and $b^2 - 4ac = -23$.

Then, $x = \dfrac{-(-1) \pm \sqrt{-23}}{2(1)}$

$\quad\quad = \dfrac{1 \pm \sqrt{23}\,i}{2}$

The solution set is
$$\left\{\dfrac{1 - \sqrt{23}\,i}{2}, \ \dfrac{1 + \sqrt{23}\,i}{2}\right\}.$$

53. $\dfrac{x^{-2}}{y^{-2}} = \dfrac{\frac{1}{x^2}}{\frac{1}{y^2}} = \dfrac{1}{x^2} \cdot \dfrac{y^2}{1} = \dfrac{y^2}{x^2}$

55. $\dfrac{(x^2 y)^{-4}}{(xy)^{-3}} = \dfrac{\frac{1}{(x^2 y)^4}}{\frac{1}{(xy)^3}} = \dfrac{(xy)^3}{(x^2 y)^4}$

$\quad\quad = \dfrac{x^3 y^3}{x^8 y^4} = x^{3-8} y^{3-4}$

$\quad\quad = x^{-5} y^{-1} = \dfrac{1}{x^5} \dfrac{1}{y} = \dfrac{1}{x^5 y}$

57. $(25x^{-4/3}y^{-2/3})^{3/2} = 25^{3/2}x^{-12/6}y^{-6/6} = \left(\sqrt{25}\right)^3 x^{-2}y^{-1} = 5^3 \dfrac{1}{x^2}\dfrac{1}{y} = \dfrac{125}{x^2y}$

59. $(6 - 3i) - (2 + 4i) = (6 - 2) + (-3 - 4)i = 4 - 7i$

61. $4(3 - i) + 3(-5 + 2i) = 12 - 4i - 15 + 6i$
$= -3 + 2i$

63. $\dfrac{3}{3 + i} = \dfrac{3}{3 + i} \cdot \dfrac{3 - i}{3 - i} = \dfrac{3(3 - i)}{(3 + i)(3 - i)} = \dfrac{3(3 - i)}{9 - i^2} = \dfrac{3(3 - i)}{9 - (-1)}$
$= \dfrac{3(3 - i)}{10} = \dfrac{9}{10} - \dfrac{3}{10}i$

65. $i^{68} = (i^2)^{34} = (-1)^{34} = 1$

67. $(2 + 3i)^3 = 2^3 + 3 \cdot 2^2 \cdot 3i + 3 \cdot 2 \cdot (3i)^2 + (3i)^3$
$= 8 + 36i + 54i^2 + 27i^3$
$= 8 + 36i + 54(-1) + 27i^2 \cdot i$
$= -46 + 36i + 27(-1)i = -46 + 9i$

69. Slope = -2; passing through (2, -1)
Use $y - y_1 = m(x - x_1)$
$y - (-1) = -2(x - 2)$
$y + 1 = -2x + 4$
$2x + y - 3 = 0$

71. Slope undefined; passing through (-3, 4).
Use $x = a$ so $x = -3$ or $x + 3 = 0$

73. y-intercept = -2; passing through (5, -3)
y-intercept: (0, -2)
slope $= \dfrac{-3 - (-2)}{5 - 0} = \dfrac{-3 + 2}{5} = -\dfrac{1}{5}$
$y = mx + b$
$y = -\dfrac{1}{5}x - 2$
$5y = -x - 10$
$x + 5y + 10 = 0$

75. Parallel to $2x - 3y + 4 = 0$; passing through (-5, 3).
$-3y = -2x - 4$
$y = \dfrac{2}{3}x + \dfrac{4}{3}$
slope $= \dfrac{2}{3}$

Line parallel has same slope.
Use $y - y_1 = m(x - x_1)$
$y - 3 = \dfrac{2}{3}[x - (-5)]$
$y - 3 = \dfrac{2}{3}(x + 5)$
$y - 3 = \dfrac{2}{3}x + \dfrac{10}{3}$
$3y - 9 = 2x + 10$
$0 = 2x - 3y + 19$
$2x - 3y + 19 = 0$

77. Perpendicular to $x + y - 2 = 0$; passing through $(1, -3)$.

$$x + y - 2 = 0$$
$$y = -x + 2$$
$$\text{slope} = -1$$
$$\text{Perpendicular line has slope} = 1$$
$$y - (-3) = 1(x - 1)$$
$$y + 3 = x - 1$$
$$0 = x - y - 4$$
$$x - y - 4 = 0$$
$$\text{or} \quad -x + y + 4 = 0$$

79. $4x - 5y + 20 = 0$

x-intercept: $(-5, 0)$
y-intercept: $(0, 4)$

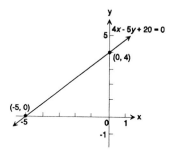

81. $\frac{1}{2}x - \frac{1}{3}y + \frac{1}{6} = 0$
$3x - 2y + 1 = 0$

Let $y = 0$ so x-intercept: $\left(-\frac{1}{3}, 0\right)$

Let $x = 0$ so y-intercept: $\left(0, \frac{1}{2}\right)$

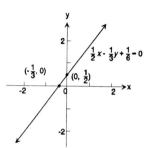

83. $\sqrt{2}x + \sqrt{3}y = \sqrt{6}$

Let $y = 0$, so $\sqrt{2}x = \sqrt{6}$

$$x = \frac{\sqrt{6}}{\sqrt{2}} = \sqrt{3}$$

x-intercept: $\left(\sqrt{3}, 0\right)$

Let $x = 0$, so $\sqrt{3}y = \sqrt{6}$

$$y = \frac{\sqrt{6}}{\sqrt{3}}$$

$$y = \sqrt{2}$$

y-intercept: $\left(0, \sqrt{2}\right)$

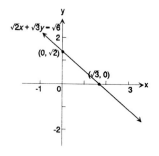

68   1 PRELIMINARIES

85. $$x^2 + y^2 - 2x + 4y - 4 = 0$$
        $$(x^2 - 2x \quad) + (y^2 + 4y \quad) = 4$$

Complete the square twice:

$$(x^2 - 2x + 1) + (y^2 + 4y + 4) = 4 + 1 + 4$$
$$(x - 1)^2 + (y + 2)^2 = 9$$

Center $(1, -2)$

Radius $= \sqrt{9} = 3$

87. $$3x^2 + 3y^2 - 6x + 12y = 0$$

Divide by 3:

$$x^2 + y^2 - 2x + 4y = 0$$
$$(x^2 - 2x \quad) + (y^2 + 4y \quad) = 0$$

Complete the square twice:

$$(x^2 - 2x + 1) + (y^2 + 4y + 4) = 1 + 4$$
$$(x - 1)^2 + (y + 2)^2 = 5$$
Center $(1, -2)$
Radius $= \sqrt{5}$

89. $2x = 3y^2$

x-intercept: $(0, 0)$
y-intercept: $(0, 0)$

Replace y by $(-y)$.

$2x = 3(-y^2)$
$2x = 3y^2$ so symmetric with respect to x-axis.

Replace x by $(-x)$.

$2(-x) = 3y^2$
$-2x = 3y^2$, not symmetric with respect to y-axis.

Replace x by $(-x)$ and y by $(-y)$.

$2(-x) = 3(-y)^2$
$-2x = 3y^2$, not symmetric with respect to origin.

91. $4x^2 + y^2 = 1$

To find x-intercepts, let $y = 0$.

$$4x^2 = 1$$
$$x^2 = \frac{1}{4}$$
$$x = \pm\frac{1}{2} \qquad \left(-\frac{1}{2},\ 0\right) \text{ and } \left(\frac{1}{2},\ 0\right)$$

To find y-intercepts, let $x = 0$.

$$y^2 = 1$$
$$y = \pm 1 \qquad (0,\ -1) \text{ and } (0,\ 1)$$

Replace y by $(-y)$.

$$4x^2 + (-y)^2 = 1$$
$$4x^2 + y^2 = 1 \text{ so symmetric with respect to } x\text{-axis.}$$

Replace x by $(-x)$.

$$4(-x)^2 + y^2 = 1$$
$$4x^2 + y^2 = 1 \text{ so symmetric with respect to } y\text{-axis.}$$

Replace x by $(-x)$ and y by $(-y)$.

$$4(-x)^2 + (-y)^2 = 1$$
$$4x^2 + y^2 = 1 \text{ so symmetric with respect to the origin.}$$

93. $y = x^4 + 2x + 1$

To find x-intercept, let $y = 0$.

$$x^4 + 2x^2 + 1 = 0$$
$$(x^2 + 1)^2 = 0$$
$$x^2 = -1 \qquad \text{No } x\text{-intercept}$$

To find y-intercept, let $x = 0$.

$$y = 0 + 0 + 1$$
$$y = 1 \qquad (0,\ 1)$$

Replace y by $-y$.

$$-y = x^4 + 2x^2 + 1 \text{ so \underline{not} symmetric with respect to } x\text{-axis.}$$

Replace x by $-x$.

$$y = (-x)^4 + 2(-x)^2 + 1$$
$$y = x^4 + 2x^2 + 1 \text{ so symmetric with respect to } y\text{-axis.}$$

Replace x by $(-x)$ and y by $(-y)$.

$$-y = (-x)^4 + 2(-x^2) + 1$$
$$-y = x^4 + 2x^2 + 1 \text{ so \underline{not} symmetric with respect to the origin.}$$

95. $x^2 + x + y^2 + 2y = 0$

To find x-intercept, let $y = 0$.

$x^2 + x = 0$
$x(x + 1) = 0$
$x = 0$ or $x + 1 = 0$
$\qquad\qquad\quad x = -1 \qquad\qquad (0, 0)$ and $(-1, 0)$

To find y-intercept, let $x = 0$.

$y^2 + 2y = 0$
$y(y + 2) = 0 \qquad\qquad (0,0)$ and $(0,-2)$
$y = 0$ or $y = -2$

$(-1,0)\,(0,0)\,(0, -2)$

Replace x by $-x$.

$(-x)^2 - x + y^2 + 2y = 0$
$\quad x^2 - x + y^2 + 2y = 0$ so <u>not</u> symmetric with respect to x-axis.

Replace y by $-y$.

$x^2 + x + (-y)^2 + 2(-y) = 0$
$\quad x^2 + x + y^2 - 2y = 0$ so <u>not</u> symmetric with respect to
$\qquad\qquad\qquad\qquad y$-axis.

Replace x by $-x$ and y by $-y$.

$(-x)^2 + (-x) + (-y)^2 + 2(-y) = 0$
$\qquad\quad x^2 - x + y^2 - 2y = 0$ so <u>not</u> symmetric with respect
$\qquad\qquad\qquad\qquad\qquad$ to the origin.

Therefore, no symmetry.

FUNCTIONS AND THEIR GRAPHS

≡ EXERCISE 2.1 FUNCTIONS

1. $f(x) = -3x^2 + 2x - 4$

(a) $f(0) = -3(0)^2 + 2(0) - 4$
$= -4$

(b) $f(1) = 3(1)^2 + 2(1) - 4$
$= -3 + 2 - 4$
$= -5$

(c) $f(-1) = -3(-1)^2 + 2(-1) - 4$
$= -3 - 2 - 4$
$= -9$

(d) $f(2) = 3(2)^2 + 2(2) - 4$
$= -12 + 4 - 4$
$= -12$

3. $f(x) = \dfrac{x}{x^2 + 1}$

(a) $f(0) = \dfrac{0}{0^2 + 1} = 0$

(b) $f(1) = \dfrac{1}{1^2 + 1} = \dfrac{1}{2}$

(c) $f(-1) = \dfrac{-1}{(-1)^2 + 1} = -\dfrac{1}{2}$

(d) $f(2) = \dfrac{2}{2^2 + 1} = \dfrac{2}{5}$

5. $f(x) = |x| + 4$

(a) $f(0) = |0| + 4 = 4$

(b) $f(1) = |1| + 4 = 5$

(c) $f(-1) = |-1| + 4 = 1 + 4$
$= 5$

(d) $f(2) = |2| + 4 = 2 + 4 = 6$

7. $f(x) = \dfrac{2x + 1}{3x - 5}$

(a) $f(0) = \dfrac{2(0) + 1}{3(0) - 5} = \dfrac{1}{-5} = -\dfrac{1}{5}$

(b) $f(1) = \dfrac{2(1) + 1}{3(1) - 5} = \dfrac{3}{-2} = -\dfrac{3}{2}$

(c) $f(-1) = \dfrac{2(-1) + 1}{3(-1) - 5} = \dfrac{-2 + 1}{-3 - 5} = \dfrac{-1}{-8} = \dfrac{1}{8}$

(d) $f(2) = \dfrac{2(2) + 1}{3(2) - 5} = \dfrac{5}{6 - 5} = -\dfrac{5}{1} = 5$

9. $f(0) = 3$ since $(0, 3)$
   is on graph

   $f(2) = 4$ since $(2, 4)$
   is on graph

11. $f(2) = $ is positive since
    $f(2) = 4$

13. $f(x) = 0$ when $x = -3$
    $x = 6$
    $x = 10$

15. Domain: $[-6, 11]$ or
    $\{x \mid -6 \leq x \leq 11\}$

17. The x-intercepts are -3, 6, 10. 19. 3 times

21. $f(x) = \dfrac{x + 2}{x - 6}$

   (a) $14 \stackrel{?}{=} \dfrac{3 + 2}{3 - 6}$

   $14 \stackrel{?}{=} \dfrac{5}{-3}$

   No $(3, 14)$ is not on
   the graph of f.

   (b) $f(4) = \dfrac{(4) + 2}{(4) - 6} = \dfrac{6}{-2} = -3$

   (c) $2 = \dfrac{x + 2}{x - 6}$

   $2x - 12 = x + 2$

   $x = 14$

   (d) Domain of $f = \{x \mid x \neq 6\}$

23. $f(x) = \dfrac{2x^2}{x^4 + 1}$

   (a) $1 \stackrel{?}{=} \dfrac{2(-1)^2}{(-1)^4 + 1}$

   $1 \stackrel{?}{=} \dfrac{2}{2}$

   $1 = 1$

   Yes, $(-1, 1)$ is on the
   graph of f.

   (b) $f(2) = \dfrac{2(2)^2}{(2)^4 + 1} = \dfrac{8}{17}$

   (c) $1 = \dfrac{2x^2}{x^4 + 1}$

   $x^4 + 1 = 2x^2$

   $x^4 - 2x^2 + 1 = 0$

   $(x^2 - 1)^2 = 0$

   $x = \pm 1$

   (d) Domain of
   $f = \{x \mid x \in \text{Real Numbers}\}$

25. Not a function since there are vertical lines that intersect the
    graph in more than one point.

2.1 FUNCTIONS

27. Function   (a) Domain: $-\pi \le x \le \pi$;   Range: $-1 \le y \le 1$
                       (b) $\left(-\frac{\pi}{2}, 0\right)$, $\left(\frac{\pi}{2}, 0\right)$, $(0, 1)$
                       (c) y-axis

29. Not a function since there are vertical lines that intersect the graph in more than one point.

31. Function   (a) Domain: $0 < x < \infty$; Range: all real numbers
                       (b) $(1, 0)$ There are no y-intercepts
                       (c) None

33. Function   (a) Domain: all real numbers; Range: $-\infty < y \le 2$
                       (b) $(-3, 0)$, $(3, 0)$, $(0, 2)$
                       (c) y-axis

35. Function   (a) Domain: $-4 \le x \le 4$; Range: $-2 \le y \le 2$
                       (b) $(-4, 0)$, $(0, 0)$, $(3, 0)$
                       (c) None

37. Function   (a) Domain: $\{x \mid x \ne 2\}$; Range: $\{y \mid y \ne 1\}$
                       (b) $(0, 0)$
                       (c) None

39. $f(x) = 2x + 1$
all real numbers

41. $f(x) = \dfrac{x}{x^2 + 1}$
all real numbers

43. $g(x) = \dfrac{x}{x^2 - 1}$ $\{x \mid x \ne -1, \; x \ne 1\}$
$x^2 - 1 \ne 0$
$x^2 \ne 1$
$x \ne \pm 1$

45. $F(x) = \dfrac{x - 2}{x^3 + x}$ $\{x \mid x \ne 0\}$
$x^3 + x \ne 0$
$x(x^2 + 1) \ne 0$
$x \ne 0 \quad x^2 \ne -1$

47. $h(x) = \sqrt{3x - 12}$ $\{x \mid x \ge 4\}$
$3x - 12 \ge 0$
$3x \ge 12$
$x \ge 4$

49. $f(x) = \sqrt{x^2 - 9}$
$x^2 - 9 \ge 0$
Solve this second-degree inequality.
$(x - 3)(x + 3) = 0$
$x = 3 \quad x = -3$

	Test Number	$(x + 3)$	$(x - 3)$	$(x + 3)(x - 3)$
$x < -3$	-4	$-$	$-$	$+$
$-3 < x < 3$	0	$+$	$-$	$-$
$x > 3$	4	$+$	$+$	$+$

Hence, $x \le 3$ or $x \ge 3$ or $(-\infty, -3]$ or $[3, \infty)$.

2 FUNCTIONS AND THEIR GRAPHS

51. $p(x) = \sqrt{\dfrac{x - 2}{x - 1}}$

$\dfrac{x - 2}{x - 1} \geq 0$

Solve this second-degree inequality.

$x - 2 = 0 \quad x - 1 = 0$

	Test Number	$(x + 1)$	$(x - 2)$	$\left(\dfrac{x - 2}{x - 1}\right)$
$x < 1$	0	$-$	$-$	$+$
$1 < x < 2$	$1\frac{1}{2}$	$+$	$-$	$-$
$x > 2$	3	$+$	$+$	$+$

Hence, $x < 1$ or $x \geq 2$ or $(-\infty, 1)$ or $[2, \infty)$.

53. $f(x) = 2x^3 + Ax^2 + 4x - 5$ and $f(2) = 3$

$f(2) = 2(2)^3 + A(2)^2 + 4(2) - 5 = 3$
$\quad\quad = 16 + 4A + 8 - 5 = 3$
$\quad\quad\quad\quad\quad 4A + 19 = 3$
$\quad\quad\quad\quad\quad\quad\quad 4A = -16$
$\quad\quad\quad\quad\quad\quad\quad\; A = -4$

55. $f(x) = \dfrac{3x + 8}{2x - A}$ and $f(0) = 2$

$f(0) = \dfrac{3(0) + 8}{2(0) - A} = 2$

$\quad\quad\quad \dfrac{8}{-A} = \dfrac{2}{1}$

$\quad\quad\quad\quad\quad\quad\quad$ (cross multiply)

$\quad\quad\quad -2A = 8$

$\quad\quad\quad\quad A = -4$

57. $f(x) = \dfrac{2x - A}{x - 3}$ and $f(4) = 0$

$f(4) = \dfrac{2(4) - A}{4 - 3} = 0$

$\quad\quad\quad \dfrac{8 - A}{1} = 0$

$\quad\quad\quad\quad 8 - A = 0$

$\quad\quad\quad\quad\quad\; 8 = A$

Since $x - 3 = 0$ when $x = 3$, then f is not defined at 3.

2.1 FUNCTIONS

59. $H(x) = 20 - 4.9x^2$

(a) $x = 1$
$$H(1) = 20 - 4.9(1)^2 = 20 - 4.9 = 15.1 \ m$$

$x = 1.1$
$$H(1.1) = 20 - 4.9(1.1)^2 = 20 - 4.9(1.21) = 20 - 5.929$$
$$= 14.071 = 14.07 \ m$$

$x = 1.2$
$$H(1.2) = 20 - 4.9(1.2)^2 = 20 - 4.9(1.44) = 20 - 7.056$$
$$= 12.944 = 12.94 \ m$$

$x = 1.3$
$$H(1.3) = 20 - 4.9(1.3)^2 = 20 - 4.9(1.69) = 20 - 8.281$$
$$= 11.719 = 11.72 \ m$$

(b) The rock strikes the ground when $H = 0$.
$$H(x) = 20 - 4.9x^2 = 0$$
$$20 = 4.9x^2$$
$$4.0816 = x^2$$
$$\sqrt{4.0816} = x$$
$$2.02 \ sec = x$$

61. $\ell = x$

$x = 2w$ or $\left(\dfrac{x}{2}\right) = w$

$A = \ell \cdot w$

$A(x) = x\left(\dfrac{x}{2}\right) = \dfrac{x^2}{2} = \dfrac{1}{2}x^2$

63. $G(x) = $ (amt. per hr.)(no. of hrs.)
$$G(x) = 5x$$

65. $A(x) = (7 - 2x)(11 - 2x)$

Domain: $0 \le x \le \dfrac{7}{2}$

Range: $0 \le A \le 77$

$A(1) = (7 - 2)(11 - 2)$
$\quad = 45$ square inches

$A(1.2) = (7 - 2.4)(11 - 2.4)$
$\quad = (4.6)(8.6)$
$\quad = 39.56$ square inches

$A(1.5) = (7 - 3)(11 - 3)$
$\quad = 32$ square inches

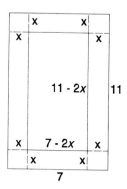

2 FUNCTIONS AND THEIR GRAPHS

67. $T(\ell) = 2\pi\sqrt{\dfrac{\ell}{g}}$, $g = 32.2$

$T(1) = 2\pi\sqrt{\dfrac{1}{32.2}} = 2\pi\sqrt{0.0310559} = 2\pi(0.1762)$

$= 1.1070973 \approx 1.11$ sec.

$T(2) = 2\pi\sqrt{\dfrac{2}{32.2}} = 2\pi\sqrt{0.06211} = 2\pi(0.2492)$

$= 1.5659 \approx 1.57$ sec.

The increase is 1.57 sec. - 1.11 sec. = 0.46 sec.

69. $y = x^2 + 2x$
Look at some ordered pairs (x, y) in this set:
$(-2, 0)$, $(-1, -1)$, $(0, 0)$, $(1, 3)$, $(2, 8)$
No two pairs have the same <u>first</u> element. <u>Function</u>.

71. $y = \dfrac{2}{x}$
Look at some ordered pairs (x, y) in this set:
$\left(-3, -\dfrac{2}{3}\right)$, $(-2, -1)$, $(-1, -2)$, $(1, 2)$, $(2, 1)$, $\left(3, \dfrac{2}{3}\right)$
No two pairs have the same <u>first</u> element. <u>Function</u>.

73. $y^2 = 1 - x^2$
Look at some ordered pairs (x, y) in this set:
$(-1, 0)$, $(0, -1)$, $(0, 1)$, $(1, 0)$

same first element
<u>Not</u> a function.

75. $x^2 + y = 1$
Look at some ordered pairs (x, y) in this set:
$(-1, 0)$, $(0, 1)$, $(1, 0)$, $(2, -3)$
No two pairs have the same <u>first</u> element. <u>Function</u>.

77. (a) $h(x) = 2x$
$h(a + b) \overset{?}{=} h(a) + h(b)$
$h(a + b) = 2(a + b) = 2a + 2b = h(a) + h(b)$

(b) $g(x) = x^2$
$g(a + b) \overset{?}{=} g(a) + g(b)$
$g(a + b) = (a + b)^2 + a^2 + 2ab + b^2 \overset{?}{=} a^2 + b^2 = g(a) + g(b)$

(c) $F(x) = 5x - 2$
$F(a + b) \overset{?}{=} F(a) + F(b)$
$F(a + b) = 5(a + b) - 2 = 5a + 5b - 2 = 5a - 2 + 5b - 2$
$= F(a) + F(b)$

(d) $G(x) = \dfrac{1}{x}$
$G(a + b) \overset{?}{=} G(a) + G(b)$
$G(a + b) = \dfrac{1}{a + b} \neq \dfrac{1}{a} + \dfrac{1}{b} = G(a) + G(b)$

2.1 FUNCTIONS

≡ EXERCISE 2.2 MORE ABOUT FUNCTIONS

1. C   3. E   5. B   7. F

9. (a) Domain: $\{x \mid -3 \leq x \leq 4\}$; Range: $\{y \mid 0 \leq y \leq 3\}$

    (b) In interval notation, increasing on $[-3, 0]$ and on $[2, 4]$; and decreasing on $[0, 2]$. In inequality notation, increasing on $-3 \leq x \leq 0$ and on $2 \leq x \leq 4$ and decreasing on $0 \leq x \leq 2$.

    (c) Since the graph is not symmetric with respect to the y-axis and is not symmetric with respect to the origin, it is NEITHER even nor odd.

    (d) The intercepts are $(-3, 0)$, $(0, 3)$, $(2, 0)$.

11. (a) Domain: all real numbers; Range: $\{y \mid 0 < y < \infty\}$

    (b) In inequality notation, increasing on $-\infty < x < \infty$. In interval notation, increasing on $(-\infty, \infty)$.

    (c) Since the graph is neither symmetric with respect to the y-axis nor the origin, it is NEITHER even nor odd.

    (d) The intercept is $(0, 1)$.

13. (a) Domain: $\{x \mid -\pi \leq x \leq \pi\}$; Range: $\{y \mid -1 \leq y \leq 1\}$

    (b) In interval notation, increasing on $\left[-\frac{\pi}{2}, \frac{\pi}{2}\right]$; and decreasing on $\left(-\pi, -\frac{\pi}{2}\right]$ and on $\left[\frac{\pi}{2}, \pi\right]$. In inequality notation, increasing on $-\frac{\pi}{2} \leq x \leq \frac{\pi}{2}$; and decreasing on $-\pi \leq x \leq -\frac{\pi}{2}$ and $\frac{\pi}{2} \leq x \leq \pi$.

    (c) Since the graph is symmetric with respect to the origin, the graph is ODD.

    (d) The intercept is $(0, 0)$.

15. (a) Domain: $\{x \mid x \neq 2\}$; Range: $\{y \mid y \neq 1\}$

    (b) In interval notation, decreasing on $(-\infty, 2)$ and on $(2, \infty)$. In inequality notation, decreasing on $-\infty < x < 2$ and on $2 < x < \infty$.

    (c) Since the graph is neither symmetric with respect to the y-axis nor symmetric with respect to the origin, it is NEITHER even nor odd.

    (d) The intercept is $(0, 0)$.

17. (a) Domain: $\{x \mid x \neq 0\}$; Range: all real numbers

    (b) In interval notation, increasing on $(-\infty, 0)$ and $(0, \infty)$. In inequality notation, increasing on $-\infty < x < 0$ and on $0 < x < \infty$.

(c) Since the graph is symmetric with respect to the origin, the graph is ODD.

(d) The intercepts are (-1, 0) and (1, 0).

19. (a) Domain: $\{x \mid x \neq -2, x \neq 2\}$; Range: $\{y \mid -\infty < y \leq 0$ and $1 < y < \infty\}$.

(b) In interval notation, increasing on $(-\infty, -2)$ and on $(-2, \infty]$, and decreasing on $[0, 2)$ and on $(2, \infty)$. In inequality notation, increasing on $-\infty < x < -2$ and on $-2 < x \leq 0$ and decreasing on $0 \leq x < 2$ and on $2 < x < \infty$.

(c) Since the graph is symmetric with respect to the y-axis, it is EVEN.

(d) The intercept is (0, 0).

21. $f(x) = 2x + 3$

(a) $f(-x) = 2(-x) + 3 = -2x + 3$
(b) $-f(x) = -(2x + 3) = -2x - 3$
(c) $f(2x) = 2(2x) + 3 = 4x + 3$
(d) $f(x - 3) = 2(x - 3) + 3 = 2x - 6 + 3 = 2x - 3$
(e) $f\left(\frac{1}{x}\right) = 2\left(\frac{1}{x}\right) + 3 = \frac{2}{x} + 3 = \frac{2}{x} + \frac{3x}{x} = \frac{3x + 2}{x}$

(f) $\dfrac{1}{f(x)} = \dfrac{1}{2x + 3}$

23. $f(x) = 2x^2 - 4$

(a) $f(-x) = 2(-x)^2 - 4 = 2x^2 - 4$
(b) $-f(x) = -(2x^2 - 4) = -2x^2 + 4$
(c) $f(2x) = 2(2x)^2 - 4 = 8x^2 - 4$
(d) $f(x - 3) = 2(x - 3)^2 - 4 = 2(x^2 - 6x + 9) - 4$
$= 2x^2 - 12x + 18 - 4 = 2x^2 - 12x + 14$

(e) $f\left(\frac{1}{x}\right) = 2\left(\frac{1}{x}\right)^2 - 4 = \frac{2}{x^2} - 4 = \frac{2}{x^2} - \frac{4x^2}{x^2} = \frac{2 - 4x^2}{x^2}$

(f) $\dfrac{1}{f(x)} = \dfrac{1}{2x^2 - 4}$

25. $f(x) = x^3 - 3x$

(a) $f(-x) = (-x)^3 - 3(-x) = -x^3 + 3x$
(b) $-f(x) = -(x^3 - 3x) = -x^3 + 3x$
(c) $f(2x) = (2x)^3 - 3(2x) = 8x^3 - 6x$
(d) $f(x - 3) = (x - 3)^3 - 3(x - 3) = x^3 - 9x^2 + 27x - 27 - 3x + 9$
$= x^3 - 9x^2 + 24x - 18$

(e) $f\left(\frac{1}{x}\right) = \left(\frac{1}{x}\right)^3 - 3\left(\frac{1}{x}\right) = \frac{1}{x^3} - \frac{3}{x} = \frac{1}{x^3} - \frac{3x^2}{x^3} = \frac{1 - 3x^2}{x^3}$

(f) $\dfrac{1}{f(x)} = \dfrac{1}{x^3 - 3x}$

2.2 MORE ABOUT FUNCTIONS

27. $f(x) = \dfrac{x}{x^2 + 1}$

(a) $f(-x) = \dfrac{-x}{(-x)^2 + 1} = -\dfrac{x}{x^2 + 1}$

(b) $-f(x) = -\left(\dfrac{x}{x^2 + 1}\right) = -\dfrac{x}{x^2 + 1}$

(c) $f(2x) = \dfrac{2x}{(2x)^2 + 1} = \dfrac{2x}{4x^2 + 1}$

(d) $f(x - 3) = \dfrac{x - 3}{(x - 3)^2 + 1} = \dfrac{x - 3}{x^2 - 6x + 9 + 1} = \dfrac{x - 3}{x^2 - 6x + 10}$

(e) $f\left(\dfrac{1}{x}\right) = \dfrac{\frac{1}{x}}{\left(\frac{1}{x}\right)^2 + 1} = \dfrac{\frac{1}{x}}{\frac{1}{x^2} + \frac{x^2}{x^2}} = \dfrac{\frac{1}{x}}{\frac{1 + x^2}{x^2}} = \dfrac{1}{x} \cdot \dfrac{x^2}{1 + x^2}$

$= \dfrac{x}{1 + x^2} = \dfrac{x}{x^2 + 1}$

(f) $\dfrac{1}{f(x)} = \dfrac{1}{\frac{x}{x^2 + 1}} = \dfrac{x^2 + 1}{x}$

29. $f(x) = |x|$

(a) $f(-x) = |-x| = |x|$
(b) $-f(x) = -|x|$
(c) $f(2x) = |2x| = 2|x|$
(d) $f(x - 3) = |x - 3|$

(e) $f\dfrac{1}{x} = \left|\dfrac{1}{x}\right| = \dfrac{1}{|x|}$

(f) $\dfrac{1}{f(x)} = \dfrac{1}{|x|}$

31. $f(x) = 1 + \dfrac{1}{x}$

(a) $f(-x) = 1 + -\dfrac{1}{x} = 1 - \dfrac{1}{x}$

(b) $-f(x) = -\left(1 + \dfrac{1}{x}\right) = -1 - \dfrac{1}{x}$

(c) $f(2x) = 1 + \dfrac{1}{2x}$

(d) $f(x - 3) = 1 + \dfrac{1}{x - 3}$

(e) $f\dfrac{1}{x} = 1 + \dfrac{1}{\frac{1}{x}} = 1 + x$

(f) $\dfrac{1}{f(x)} = \dfrac{1}{1 + \frac{1}{x}} = \dfrac{1}{\frac{x + 1}{x}} = \dfrac{x}{x + 1}$

33. $\dfrac{f(x) - f(1)}{x - 1} = \dfrac{3 - 3}{x - 1} = \dfrac{0}{x - 1} = 0$

35. $\dfrac{f(x) - f(1)}{x - 1} = \dfrac{(1 - 3x) - (-2)}{x - 1} = \dfrac{3 - 3x}{x - 1} = \dfrac{-3(x - 1)}{x - 1} = -3$

2 FUNCTIONS AND THEIR GRAPHS

37. $\dfrac{f(x) - f(1)}{x - 1} = \dfrac{(3x^2 - 2x) - (1)}{x - 1} = \dfrac{3x^2 - 2x - 1}{x - 1} = \dfrac{(3x + 1)(x - 1)}{x - 1}$

$$= 3x + 1$$

39. $\dfrac{f(x) - f(1)}{x - 1} = \dfrac{(x^3 - x) - (0)}{x - 1} = \dfrac{x^3 - x}{x - 1} = \dfrac{x(x - 1)(x + 1)}{x - 1} = x(x + 1)$

41. $\dfrac{f(x) - f(1)}{x - 1} = \dfrac{\dfrac{2}{x + 1} - 1}{x - 1} = \dfrac{\dfrac{2 - (x + 1)}{x + 1}}{x - 1} = \dfrac{1 - x}{(x + 1)(x - 1)} = \dfrac{-1}{x + 1}$

43. $\dfrac{f(x) - f(1)}{x - 1} = \dfrac{\sqrt{x} - 1}{x - 1} = \dfrac{\left(\sqrt{x} - 1\right)}{\left(\sqrt{x} - 1\right)\left(\sqrt{x} + 1\right)} = \dfrac{1}{\sqrt{x} + 1}$

45. $f(x) = 2x^3$

odd: $\quad\quad f(-x) = -f(x)$

$\quad\quad\quad\quad 2(-x)^3 = -(2x^3)$

$\quad\quad\quad\quad\quad -2x^3 = -2x^3$

47. $g(x) = 2x^2 - 5$

even: $\quad\quad f(-x) = f(x)$

$\quad\quad\quad 2(-x)^2 - 5 = 2x^2 - 5$

$\quad\quad\quad\quad 2x^2 - 5 = 2x^2 - 5$

49. $F(x) = \sqrt[3]{x}$

odd: $\quad f(-x) = -f(x)$

$\quad\quad\quad \sqrt[3]{-x} = -\sqrt[3]{x}$

$\quad\quad\quad -\sqrt[3]{x} = -\sqrt[3]{x}$

51. $f(x) = x + |x|$

even: $\quad\quad f(-x) = f(x)$

$\quad\quad -x + |-x| = x + |x|$

$\quad\quad -x + |x| \ne x + |x|$

odd: $\quad\quad f(-x) = -f(x)$

$\quad\quad -x + |-x| = (x + |x|)$

$\quad\quad -x + |x| \ne -x - |x|$

neither

53. $g(x) = \dfrac{1}{x^2}$

even: $\quad\quad f(-x) = f(x)$

$\quad\quad \dfrac{1}{(-x)^2} = \dfrac{1}{x^2}$

$\quad\quad\quad \dfrac{1}{x^2} = \dfrac{1}{x^2}$

55. $h(x) = \dfrac{x^3}{3x^2 - 9}$

odd: $\quad\quad f(-x) = -f(x)$

$\quad\quad \dfrac{(-x)^3}{3(-x)^2 - 9} = -\dfrac{x^3}{3x^2 - 9}$

$\quad\quad \dfrac{-x^3}{3x^2 - 9} = \dfrac{-x^3}{3x^2 - 9}$

57. One at most because if f is increasing it could only cross the x-axis at most one time. It could not "turn" and cross it again or it would start to decrease.

2.2 MORE ABOUT FUNCTIONS

59. $f(x) = 3x - 3$

(a) The domain is all real numbers.

(b) x-intercept(s): y-intercept:

$$0 = 3x - 3 \qquad y = 3(0) - 3$$
$$1 = x \qquad\qquad y = -3$$

The intercepts are $(0, -3)$, $(1, 0)$.

(c)

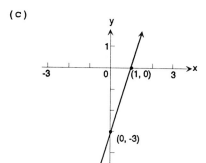

(d) The range is all real numbers.

61. $g(x) = x^2 - 4$

(a) The domain is all real numbers.

(b) x-intercept(s): y-intercept:

$$0 = x^2 - 4 \qquad y = (0)^2 - 4$$
$$4 = x^2 \qquad\qquad y = -4$$
$$x = -2, 2$$

The intercepts are $(-2, 0)$, $(2, 0)$, $(0, -4)$.

(c)

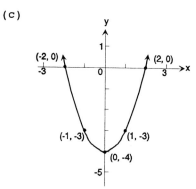

(d) The range is $\{y \,|\, -4 \le y < \infty\}$.

2 FUNCTIONS AND THEIR GRAPHS

63. $h(x) = -x^2$

   (a) The domain is all real numbers.

   (b) x-intercept(s):   y-intercept:
   $$0 = -x^2 \qquad\qquad y = -0^2$$
   $$0 = x \qquad\qquad\qquad y = 0$$

   The intercept is (0, 0).

   (c)

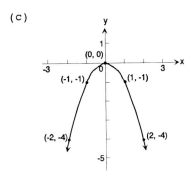

   (d) The range is $\{y \mid -\infty < y \le 0\}$.

65. $f(x) = \sqrt{x - 2}$

   (a) The domain is $[2, \infty)$.

   (b) x-intercept(s):   y-intercept:
   $$0 = \sqrt{x - 2} \qquad\qquad y = \sqrt{0 - 2}$$
   $$x = 2 \qquad\qquad\qquad y = \sqrt{-2}$$

   The intercept is (2, 0).

   (c)

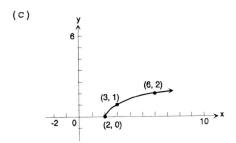

   (d) The range is $\{y \mid 0 \le y < \infty\}$.

2.2 MORE ABOUT FUNCTIONS

67. $h(x) = \sqrt{2 - x}$

(a) The domain is $\{x \mid -\infty < x \le 2\}$

(b) x-intercept(s): y-intercept:

$$0 = \sqrt{2 - x} \qquad\qquad y = \sqrt{2 - 0}$$

$$0 = 2 - x \qquad\qquad y = \sqrt{2}$$

$$x = 2$$

The intercepts are $(2, 0)$, $(0, \sqrt{2})$.

(c)

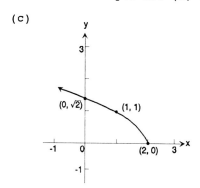

(d) The range is $\{y \mid 0 \le y < \infty\}$.

69. $f(x) = |x| + 3$

(a) The domain is all real numbers.

(b) x-intercept(s): y-intercept:

$$0 = |x| + 3 \qquad\qquad y = |0| + 3$$

$$-3 = |x| \qquad\qquad\quad y = 3$$

No solution.

The intercept is $(0, 3)$.

(c)

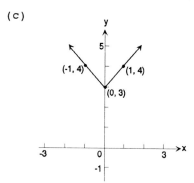

(d) The range is $\{y \mid 3 \le y < \infty\}$.

2 FUNCTIONS AND THEIR GRAPHS

71. $h(x) = -|x|$

(a) The domain is all real numbers.

(b) x-intercept(s): y-intercept:
          $0 = -|x|$ $y = -|0|$
          $x = 0$ $y = 0$

The intercept is $(0, 0)$.

(c)

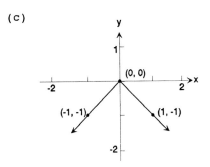

(d) The range is $\{y \,|\, -\infty < y \le 0\}$

73. $f(x) = \begin{cases} 2x & \text{if } x \ne 0 \\ 0 & \text{if } x = 0 \end{cases}$

(a) The domain is all real numbers.

(b) x-intercept(s): y-intercept:
          $0 = 2x$ $y = 0$
          $0 = x$

The intercept is $(0, 0)$.

(c)

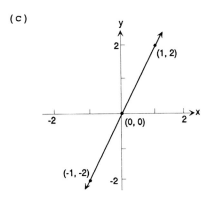

(d) The range is all real numbers.

2.2 MORE ABOUT FUNCTIONS

75. $f(x) = \begin{cases} 1 + x & \text{if } x < 0 \\ x^2 & \text{if } x \geq 0 \end{cases}$

(a) The domain is all real numbers.

(b) x-intercept(s):

$$0 = 1 + x \text{ or } 0 = x^2$$
$$-1 = x \text{ or } x = 0$$

y-intercept:

$$y = 0^2$$
$$y = 0$$

The intercept is $(-1, 0)$, $(0, 0)$.

(c)

(d) The range is all real numbers.

77. $f(x) = \begin{cases} |x| & \text{if } -2 \leq x < 0 \\ 1 & \text{if } \quad x = 0 \\ x^3 & \text{if } \quad x > 0 \end{cases}$

(a) The domain is $\{x \mid -2 \leq x < \infty\}$

(b) x-intercept(s):

None

y-intercept:

$y = 1$ if $x = 0$

The intercept is $(0, 1)$.

(c)

(d) The range is $\{y \mid 0 < y < \infty\}$.

2 FUNCTIONS AND THEIR GRAPHS

79. $g(x) = \begin{cases} 1 & \text{if } x \text{ is an integer} \\ 0 & \text{if } x \text{ is not an integer} \end{cases}$

(a) The domain is all real numbers.

(b) x-intercept(s): y-intercept:
           None $y = 1$ if $x = 0$

The intercept is $(0, 1)$.

(c)

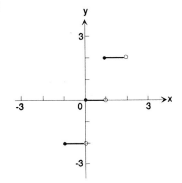

(d) The range is $\{-1, 1\}$.

81. $h(x) = 2[x]$

(a) The domain is all real numbers.

(b) x-intercept(s): y-intercept:
       $0 = 2[x]$ $y = 2[0]$
       $0 = [x]$ $y = 0$
       $0 \le x < 1$

The intercepts are all ordered pairs $(x, 0)$ when $0 \le x < 1$.

(c)

(d) The range is the set of even integers.

83. $f(x) = \begin{cases} -x & \text{if } -1 \le x \le 0 \\ \dfrac{1}{2}x & \text{if } 0 < x \le 2 \end{cases}$

Other answers are possible.

85. $f(x) = \begin{cases} -x & \text{if } x \le 0 \\ 2 - x & \text{if } 0 < x \le 2 \end{cases}$

Other answers are possible.

87. $f(x) = \begin{cases} x^2 + 4 & \text{if } x \ne 2 \\ 6 & \text{if } x = 2 \end{cases}$

To see if f is even, we need to show that $f(x) = f(-x)$ for all possible values of x.

$$f(-x) = (-x)^2 + 4 = x^2 + 4 = f(x)$$

However, when $x = -2$,

$$f(-2) = (-2)^2 + 4 = 8, \text{ but}$$
$$f(2) = 6$$

Because $f(-2) \ne f(2)$, the function is not even.

89. Each graph is that of $y = x^2$, but shifted vertically. If $y = x^2 + k$, $k > 0$, the shift is up k units; if $y = x^2 + k$, $k < 0$, the shift is down $|k|$ units. The graph of $y = x^2 - 4$ is the same as the graph of $y = x^2$ but shifted down 4. The graph of $y = x^2 + 5$ is the graph of $y = x^2$, but shifted up 5.

91. Each graph is that of $y = |x|$, but either compressed or stretched. If $y = k|x|$ and $k > 1$, the graph is stretched; if $y = k|x|$, $0 < k < 1$, the graph is compressed. The graph of $y = \dfrac{1}{4}|x|$ is the same as the graph of $y = |x|$, but compressed [for example, from $(2, 2)$ to $\left(2, \dfrac{1}{2}\right)$]. The graph of $y = 5|x|$ is the same as the graph of $y = |x|$, but stretched [for example, from $(2, 2)$ to $(2, 10)$].

93. The graph of $y = \sqrt{-x}$ is the reflection about the y-axis of the graph of $y = \sqrt{x}$. The same type of reflection occurs when graphing $y = 2x + 1$ and $y = 2(-x) + 1$. The conclusion is that the graph of $y = f(-x)$ is the reflection about the y-axis of the graph of $y = f(x)$.

95. (a) For 50 therms, the charge $C = 7.00 + 0.21054(50) + 0.26341(50)$
$$= \$30.70$$

(b) For 500 therms, the charge
$$C = 7.00 + 0.21054(90) + 0.11242(410) + 0.26341(500)$$
$$= \$203.75$$

2 FUNCTIONS AND THEIR GRAPHS

(c) If C is the monthly charge, then

$$C = \begin{cases} 7 + 0.21054x + 0.26341x & \text{if } 0 \le x \le 90 \\ 7 + 0.21054(90) + 0.11242(x - 90) + 0.26341x & \text{if } x > 90 \end{cases}$$

$$= \begin{cases} 7 + 0.47395x & \text{if } 0 \le x \le 90 \\ 25.95 + 0.11242(x - 90) + 0.26341x & \text{if } x > 90 \end{cases}$$

$$= \begin{cases} 7 + 0.47395x & \text{if } 0 \le x \le 90 \\ 15.83 + 0.37583x & \text{if } x > 90 \end{cases}$$

(d)

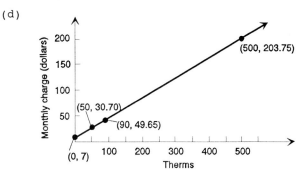

97. (a) $E(x)$ is even if $E(-x) = E(x)$

$$\frac{1}{2}\big[f(-x) + f(-(-x))\big] = \frac{1}{2}[f(x) + f(-x)]$$

$$\frac{1}{2}(f(-x) + f(x)] = \frac{1}{2}[f(x) + f(-x)]$$

$$\frac{1}{2}[f(x) + f(-x)] = \frac{1}{2}[f(x) + f(-x)]$$

(b) $0(x)$ is odd if $0(-x) = -0(x)$

$$\frac{1}{2}\big[f(-x) - f(-(-x))\big] = -\frac{1}{2}[f(x) - f(-x)]$$

$$\frac{1}{2}(f(-x) - f(x)] = -\frac{1}{2}[f(x) - f(-x)]$$

$$\frac{1}{2}f(-x) - \frac{1}{2}f(x) = -\frac{1}{2}f(x) + \frac{1}{2}f(-x)$$

$$\frac{1}{2}f(-x) - \frac{1}{2}f(x) = \frac{1}{2}f(-x) - \frac{1}{2}f(x)$$

(c) Show: $f(x) = E(x) + 0(x)$

$$f(x) = \frac{1}{2}[f(x) + f(-x)] + \frac{1}{2}[f(x) - f(-x)]$$

$$= \frac{1}{2}f(x) + \frac{1}{2}f(-x) + \frac{1}{2}f(x) - \frac{1}{2}f(-x)$$

$$f(x) = f(x)$$

(d) From parts (a), (b), and (c) we have shown that
$f(x) = E(x) + 0(x)$ and that $E(x)$ is even and $0(x)$ is odd.

2.2 MORE ABOUT FUNCTIONS

≡ EXERCISE 2.3 GRAPHING TECHNIQUES

1. B   3. H   5. I

7. L   9. F   11. G

13. $y = (x - 2)^3$ 15. $y = x^3 + 2$

17. $y = -x^3$ 19. $y = 2x^3$

21. $f(x) = x^2 - 1$

Using the graph of $y = x^2$, vertically
shift downward 1 unit.

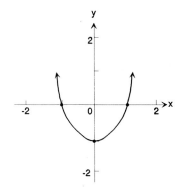

23. $g(x) = x^3 + 1$

Using the graph of $y = x^3$, vertically
shift upward 1 unit.

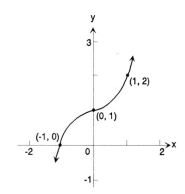

25. $h(x) = \sqrt{x - 2}$

Using the graph of $y = \sqrt{x}$,
horizontally shift to the right 2
units.

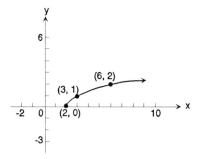

27. $f(x) = (x - 1)^3$

Using the graph of $y = x^3$, horizontally shift to the right 1 unit.

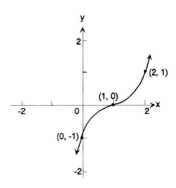

29. $g(x) = 4\sqrt{x}$

Using the graph of $y = \sqrt{x}$, vertically stretch so that $(1, 1)$ becomes $(1, 4)$.

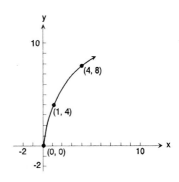

31. $h(x) = \dfrac{1}{2x}$

Using the graph of $y = \dfrac{1}{x}$, vertically compress so that $(1, 1)$ becomes $\left(1, \dfrac{1}{2}\right)$.

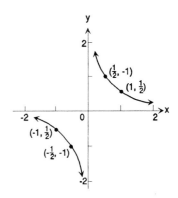

33. $f(x) = -|x|$

Reflect the graph of $y = |x|$ about the x-axis.

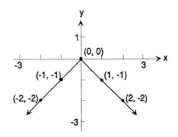

2.3 GRAPHING TECHNIQUES

35. $g(x) = \dfrac{-1}{x}$

Reflect the graph of $y = \dfrac{1}{x}$ about the x-axis.

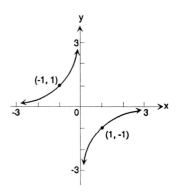

37. $h(x) = [\![x]\!]$

The greatest integer function $y = [\![x]\!]$ takes the integer values less than or equal to the given x. Thus, if we have the inequality, $0 \le x < 1$, taking the negative, we have $-(0 \le x < 1)$ $= 0 \ge x > -1$ or $-1 < x \le 0$. In other words, reflect the graph of $y = [\![x]\!]$ about the axis.

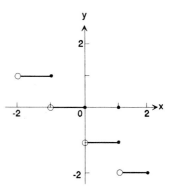

39. $f(x) = (x + 1)^2 - 3$

Using the graph of x^2, horizontally shift to the left 1 unit and vertically shift downward 3 units.

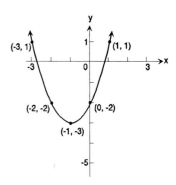

41. $g(x) = \sqrt{x - 2} + 1$

Using the graph of $y = \sqrt{x}$, horizontally shift to the right 2 units, and vertically shift upward 1 unit.

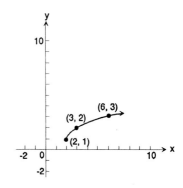

2 FUNCTIONS AND THEIR GRAPHS

43. $h(x) = \sqrt{-x} - 2$

Reflect the graph $y = \sqrt{x}$ about the y-axis, and vertically shift downward 2 units.

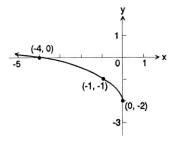

45. $f(x) = (x + 1)^3 - 1$

Using the graph of $y = x^3$, horizontally shift to the left 1 unit, and vertically shift downward 1 unit.

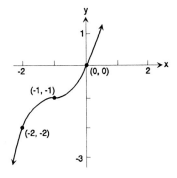

47. $g(x) = 2|1 - x|$

Using the graph of $y = |x|$, since $|1 - x| = |x - 1|$, horizontally shift to the right 1 unit and then vertically stretch so that $(0, 1)$ becomes $(0, 2)$.

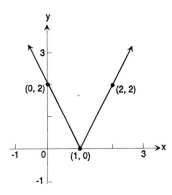

49. $h(x) = 2[\![x - 1]\!]$

Using the graph of $y = [\![x]\!]$, horizontally shift to the right 1 unit, then vertically stretch so that the range becomes even integers instead of all integers.

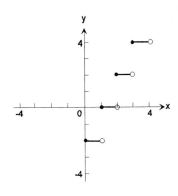

2.3 GRAPHING TECHNIQUES

51. (a) $F(x) = f(x) + 3$
Shift up 3 units.

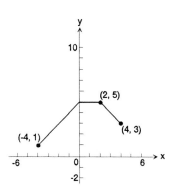

(b) $G(x) = f(x + 2)$
Shift left 2 units.

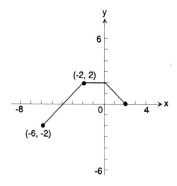

(c) $P(x) = -f(x)$
Reflect about the
x-axis.

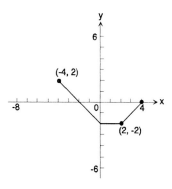

(d) $Q(x) = \frac{1}{2}f(x)$
Vertically compress so
(2, 2) becomes (2, 1).

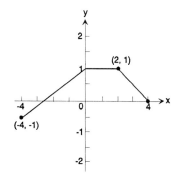

(e) $g(x) = f(-x)$
Reflect about the
y-axis.

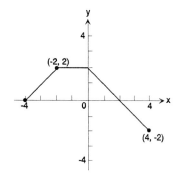

(f) $h(x) = 3f(x)$
Vertically stretch the graph
by a factor of 3.

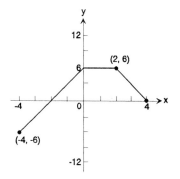

2 FUNCTIONS AND THEIR GRAPHS

53. (a) $F(x) = f(x) + 3$
Shift up 3 units.

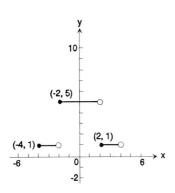

(b) $G(x) = f(x + 2)$
Shift left 2 units.

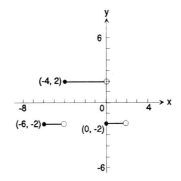

(c) $P(x) = -f(x)$
Reflect about the
x-axis.

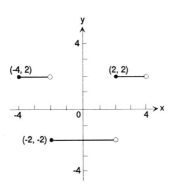

(d) $Q(x) = \frac{1}{2}f(x)$
Vertically compress by a
factor of $\frac{1}{2}$

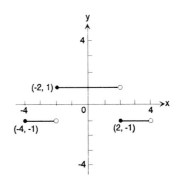

(e) $g(x) = f(-x)$
Reflect about the
y-axis.

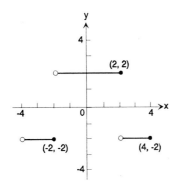

(f) $h(x) = 3f(x)$
Vertically stretch by a
factor of 3.

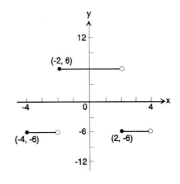

2.3 GRAPHING TECHNIQUES

55. (a) $F(x) = f(x) + 3$
Shift up 3 units.

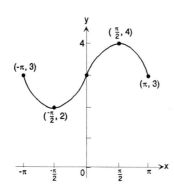

(b) $G(x) = f(x + 2)$
Shift left 2 units.

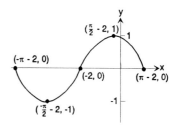

(c) $P(x) = -f(x)$
Reflect about the
x-axis.

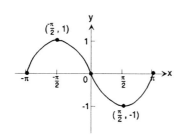

(d) $Q(x) = \frac{1}{2} f(x)$
Vertically compress by a
factor of $\frac{1}{2}$.

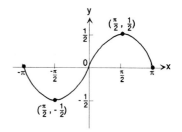

(e) $g(x) = f(-x)$
Reflect about the
y-axis.

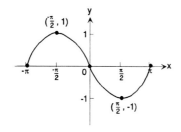

(f) $h(x) = 3f(x)$
Vertically stretch by a
factor of 3.

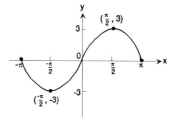

2 FUNCTIONS AND THEIR GRAPHS

57. $f(x) = x^2 + 2x$

$f(x) = (x^2 + 2x + 1) - 1$

$f(x) = (x + 1)^2 - 1$

Using $f(x) = x^2$, shift left 1 unit and shift down 1 unit.

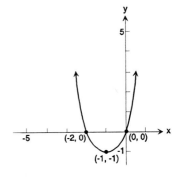

59. $f(x) = x^2 - 8x + 1$

$f(x) = (x^2 - 8x) + 1$

$f(x) = (x^2 - 8x + 16) + 1 - 16$

$f(x) = (x - 4)^2 - 15$

Using $f(x) = x^2$, shift right 4 units and shift down 15 units.

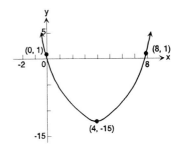

61. $f(x) = x^2 + x + 1$

$f(x) = (x^2 + x) + 1$

$f(x) = \left(x^2 + x + \dfrac{1}{4}\right) + 1 - \dfrac{1}{4}$

$f(x) = \left(x + \dfrac{1}{2}\right)^2 + \dfrac{3}{4}$

Using $f(x) = x^2$, shift left $\dfrac{1}{2}$ unit and shift up $\dfrac{3}{4}$ unit.

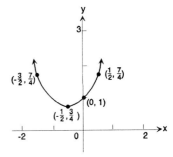

63. $y = (x - c)^2$

If $c = 0$, $y = x^2$.

If $c = 3$, $y = (x - 3)^2$, shift right 3 units.

If $c = -2$, $y = (x + 2)^2$, shift left 2 units.

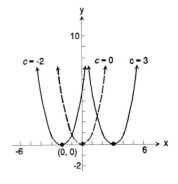

2.3 GRAPHING TECHNIQUES

65. $F = \dfrac{9}{5}C + 32$

Graph:

C	0	40	100
F	32	104	212

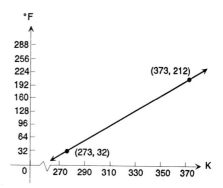

C	0	40	100
K = C + 273	273	313	373
F	32	104	212

or
$C = K - 273$

$F = \dfrac{9}{5}(K - 273) + 32$

Shift the graph of
$F = F = \dfrac{9}{5}C + 32$, 273 units to the
right.

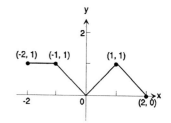

67. (a) Given the graph of $y = f(x)$, if
$y = |f(x)|$, then all negative
values for y become positive values
for y. So the portion of the graph
in quadrant III, where y
coordinates are negative, become
positive y coordinates in quadrant
II. In other words, reflect the
negative portion about the x-axis.

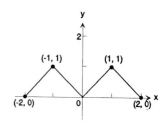

(b) Given the graph $y = f(x)$, if
$y = f(|x|)$, then we must reflect
about the y-axis, because $f(|+x|)$
$= f(|-x|)$.

2 FUNCTIONS AND THEIR GRAPHS

≡ EXERCISE 2.4 OPERATIONS ON FUNCTIONS; COMPOSITE FUNCTIONS

1. $f(x) = 3x - 4 \quad g(x) = 2x + 3$

    (a) $(f + g)(x) = 3x - 4 + 2x + 3 = 5x - 1$

    The domain is all real numbers.

    (b) $(f - g)(x) = (3x - 4) - (2x + 3) = 3x - 4 - 2x - 3 = x - 7$

    The domain is all real numbers.

    (c) $(f \cdot g)(x) = (3x - 4)(2x + 3) = 6x^2 + 9x - 8x - 12$
    $$= 6x^2 + x - 12$$

    The domain is all real numbers.

    (d) $\left(\dfrac{f}{g}\right)(x) = \dfrac{3x - 4}{2x + 3}$

    The domain is all real numbers except $-\dfrac{3}{2}$.

3. $f(x) = x - 1 \quad g(x) = 2x^2$

    (a) $(f + g)(x) = (x - 1) + 2x^2 = 2x^2 + x - 1$

    The domain is all real numbers.

    (b) $(f - g)(x) = (x - 1) - (2x^2) = -2x^2 + x - 1$

    The domain is all real numbers.

    (c) $(f \cdot g)(x) = (x - 1)(2x^2) = 2x^3 - 2x^2$

    The domain is all real numbers.

    (d) $\left(\dfrac{f}{g}\right)(x) = \dfrac{x - 1}{2x^2}$

    The domain is all real numbers except 0, $\{x \mid x \neq 0\}$.

5. $f(x) = \sqrt{x}, \ x \geq 0 \quad g(x) = 3x - 5$

    (a) $(f + g)(x) = \sqrt{x} + 3x - 5$

    The domain is $\{x \mid 0 \leq x < \infty\}$.

    (b) $(f - g)(x) = \sqrt{x} - (3x - 5) = \sqrt{x} - 3x + 5$

    The domain is $\{x \mid 0 \leq x < \infty\}$.

(c) $(f \cdot g)(x) = \sqrt{x}(3x - 5) = 3x\sqrt{x} - 5\sqrt{x}$,

The domain is $\{x \mid 0 \le x < \infty\}$.

(d) $\left(\dfrac{f}{g}\right)(x) = \dfrac{\sqrt{x}}{3x - 5}$

The domain is $\{x \mid 0 \le x < \infty \text{ and } x \ne \frac{5}{3}\}$.

7. $f(x) = 1 + \dfrac{1}{x}, \quad x \ne 0 \qquad g(x) = \dfrac{1}{x}, \quad x \ne 0$

(a) $(f + g)(x) = \left(1 + \dfrac{1}{x}\right) + \dfrac{1}{x} = 1 + \dfrac{2}{x}$

The domain is $\{x \mid x \ne 0\}$.

(b) $(f - g)(x) = \left(1 + \dfrac{1}{x}\right) - \dfrac{1}{x} = 1$

The domain is $\{x \mid x \ne 0\}$.

(c) $(f \cdot g)(x) = \left(1 + \dfrac{1}{x}\right)\left(\dfrac{1}{x}\right) = \dfrac{1}{x} + \dfrac{1}{x^2}$

The domain is $\{x \mid x \ne 0\}$.

(d) $\left(\dfrac{f}{g}\right)(x) = \dfrac{1 + \dfrac{1}{x}}{\dfrac{1}{x}} = \dfrac{\dfrac{x + 1}{x}}{\dfrac{1}{x}} = \dfrac{x + 1}{x} \cdot \dfrac{x}{1} = x + 1$

The domain is $\{x \mid x \ne 0\}$.

9. $f(x) = \dfrac{2x + 3}{3x - 2}, \quad x \ne \dfrac{2}{3} \qquad g(x) = \dfrac{x}{3x - 2}, \quad x \ne \dfrac{2}{3}$

(a) $(f + g)(x) = \dfrac{2x + 3}{3x - 2} + \dfrac{x}{3x - 2} = \dfrac{3x + 3}{3x - 2}$

The domain is $\left\{x \mid x \ne \dfrac{2}{3}\right\}$.

(b) $(f - g)(x) = \dfrac{2x + 3}{3x - 2} - \dfrac{x}{3x - 2} = \dfrac{x + 3}{3x - 2}$

The domain is $\left\{x \mid x \ne \dfrac{2}{3}\right\}$.

(c) $(f \cdot g)(x) = \left(\dfrac{2x + 3}{3x - 2}\right)\left(\dfrac{x}{3x - 2}\right) = \dfrac{2x^2 + 3x}{(3x - 2)^2}$

The domain is $\left\{x \mid x \ne \dfrac{2}{3}\right\}$.

2 FUNCTIONS AND THEIR GRAPHS

(d) $\left(\dfrac{f}{g}\right)(x) = \dfrac{\dfrac{2x + 3}{3x - 2}}{\dfrac{x}{3x - 2}} = \dfrac{2x + 3}{3x - 2} \cdot \dfrac{3x - 2}{x} = \dfrac{2x + 3}{x}$

The domain is $\left\{x \middle| x \ne \dfrac{2}{3} \text{ and } x \ne 0\right\}$.

11. $f(x) = 3x + 1$

$(f + g)(x) = 6 - \dfrac{1}{2}x$

$6 - \dfrac{1}{2}x = (3x + 1) + g(x)$

$-\dfrac{7}{2}x + 5 = g(x)$

$g(x) = 5 - \dfrac{7}{2}x$

13. Graph: $y = |x|$ on the interval $[0, 2]$
$y = x^2$ on the interval $[0, 2]$

Add the y-coordinates of the plotted points to get $y = |x| + x^2$.

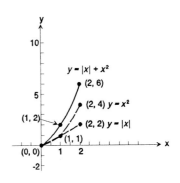

15. Graph: $y = |x|$ on the interval $[0, 2]$
$y = x^2$ on the interval $[0, 2]$

Add the y-coordinates of the plotted points to get $y = x^3 + x$.

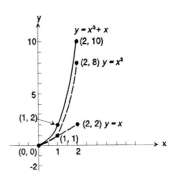

17. $f(x) = 2x \qquad g(x) = 3x^2 + 1$

(a) $(f \circ g)(4) = f(g(4)) = f(49) = 2(49) = 98$

(b) $(g \circ f)(2) = g(f(2)) = g(4) = 3(4)^2 + 1 = 48 + 1 = 49$

(c) $(f \circ f)(1) = f(f(1)) = f(2) = 2(2) = 4$

(d) $(g \circ g)(0) = g(g(0)) = g(1) = 3(1)^2 + 1 = 4$

2.4 OPERATIONS ON FUNCTIONS; COMPOSITE FUNCTIONS

19. $f(x) = 4x^2 - 3$ $g(x) = 3 - \frac{1}{2}x^2$

(a) $(f \circ g)(4) = f(g(4)) = f(-5) = 4(-5)^2 - 3 = 97$

(b) $(g \circ f)(2) = g(f(2)) = g(13) = 3 - \frac{1}{2}(13)^2 = 3 - \frac{169}{2}$

$$= \frac{6 - 169}{2} = -\frac{163}{2}$$

(c) $(f \circ f)(1) = f(f(1)) = f(1) = 1$

(d) $(g \circ g)(0) = g(g(0)) = g(3) = 3 - \frac{1}{2}(3)^2 = 3 - \frac{9}{2} = \frac{6-9}{2} = -\frac{3}{2}$

21. $f(x) = \sqrt{x}$ $g(x) = 2x$

(a) $(f \circ g)(4) = f(g(4)) = f(8) = \sqrt{8} = 2\sqrt{2}$

(b) $(g \circ f)(2) = g(f(2)) = g\left(\sqrt{2}\right) = 2\sqrt{2}$

(c) $(f \circ f)(1) = f(f(1)) = f(1) = \sqrt{1} = 1$

(d) $(g \circ g)(0) = g(g(0)) = g(0) = 2(0) = 0$

23. $f(x) = |x|$ $g(x) = \dfrac{1}{x^2 + 1}$

(a) $(f \circ g)(4) = f(g(4)) = f\left(\frac{1}{17}\right) = \left|\frac{1}{17}\right| = \frac{1}{17}$

(b) $(g \circ f)(2) = g(f(2)) = g(2) = \dfrac{1}{2^2 + 1} = \dfrac{1}{5}$

(c) $(f \circ f)(1) = f(f(1)) = f(1) = 1$

(d) $(g \circ g)(0) = g(g(0)) = g(1) = \dfrac{1}{1^2 + 1} = \dfrac{1}{2}$

25. $f(x) = \dfrac{3}{x^2 + 1}$ $g(x) = \sqrt{x}$

(a) $(f \circ g)(4) = f(g(4)) = f(2) = \dfrac{3}{2^2 + 1} = \dfrac{3}{5}$

(b) $(g \circ f)(2) = g(f(2)) = g\left(\frac{3}{5}\right) = \sqrt{\frac{3}{5}} = \dfrac{\sqrt{15}}{5}$

2 FUNCTIONS AND THEIR GRAPHS

(c) $(f \circ f)(1) = f(f(1)) = f\left(\dfrac{3}{2}\right) = \dfrac{3}{\left(\dfrac{3}{2}\right)^2 + 1} = \dfrac{3}{\dfrac{9}{4} + \dfrac{4}{4}}$

$= \dfrac{3}{\dfrac{13}{4}} = \dfrac{3}{1} \cdot \dfrac{4}{13} = \dfrac{12}{13}$

(d) $(g \circ g)(0) = g(g(0)) = g(0) = \sqrt{0} = 0$

27. $f(x) = 2x + 1 \quad g(x) = 3x$

(a) $f \circ g = f(g(x)) = f(3x) = 2(3x) + 1 = 6x + 1$

(b) $g \circ f = g(f(x)) = g(2x + 1) = 3(2x + 1) = 6x + 3$

(c) $f \circ f = f(f(x)) = f(2x + 1) = 2(2x + 1) + 1$
$= 4x + 2 + 1 = 4x + 3$

(d) $g \circ g = g(g(x)) = g(3x) = 3(3x) = 9x$

29. $f(x) = 3x + 1 \quad g(x) = x^2$

(a) $f \circ g = f(g(x)) = f(x^2) = 3x^2 + 1$

(b) $g \circ f = g(f(x)) = g(3x + 1) = (3x + 1)^2 = 9x^2 + 6x + 1$

(c) $f \circ f = f(f(x)) = f(3x + 1) = 3(3x + 1) + 1$
$= 9x + 3 + 1 = 9x + 4$

(d) $g \circ g = g(g(x)) = g(x)^2 = (x^2)^2 = x^4$

31. $f(x) = \sqrt{x} \quad g(x) = x^2 - 1$

(a) $f \circ g = f(g(x)) = f(x^2 - 1) = \sqrt{x^2 - 1}$

(b) $g \circ f = g(f(x)) = g\left(\sqrt{x}\right) - 1 = \left(\sqrt{x}\right)^2 - 1 = x - 1$

(c) $f \circ f = f(f(x)) = f\left(\sqrt{x}\right) = \sqrt{\sqrt{x}} = \sqrt[4]{x}$

(d) $g \circ g = g(g(x)) = g(x^2 - 1) = (x^2 - 1)^2 - 1$
$= x^4 - 2x^2 + 1 = x^4 - 2x^2$

33. $f(x) = \dfrac{x - 1}{x + 1} \quad g(x) = \dfrac{1}{x}$

(a) $f \circ g = f(g(x)) = f\left(\dfrac{1}{x}\right) = \dfrac{\dfrac{1}{x} - 1}{\dfrac{1}{x} + 1} = \dfrac{\dfrac{1 - x}{x}}{\dfrac{1 + x}{x}}$

$= \dfrac{1 - x}{x} \cdot \dfrac{x}{1 + x} = \dfrac{1 - x}{1 + x}$

(b) $g \circ f = g(f(x)) = g\left(\dfrac{x-1}{x+1}\right) = \dfrac{1}{\dfrac{x-1}{x+1}} = \dfrac{x+1}{x-1}$

(c) $f \circ f = f(f(x)) = f\left(\dfrac{x-1}{x+1}\right) = \dfrac{\dfrac{x-1}{x+1} - 1}{\dfrac{x-1}{x+1} + 1} = \dfrac{\dfrac{(x-1)-(x+1)}{x+1}}{\dfrac{(x-1)+(x+1)}{x+1}}$

$\qquad = \dfrac{\dfrac{-2}{x+1}}{\dfrac{2x}{x+1}} = \dfrac{-2}{(x+1)} \cdot \dfrac{(x+1)}{2x} = -\dfrac{1}{x}$

(d) $g \circ g = g(g(x)) = g\left(\dfrac{1}{x}\right) = \dfrac{1}{\dfrac{1}{x}} = x$

35. $f(x) = x^2 \qquad g(x) = \sqrt{x}$

(a) $f \circ g = f(g(x)) = f\left(\sqrt{x}\right) = \left(\sqrt{x}\right)^2 = x$

(b) $g \circ f = g(f(x)) = g(x^2) = \sqrt{x^2} = |x|$

(c) $f \circ f = f(f(x)) = f(x)^2 = (x^2)^2 = x^4$

(d) $g \circ g = g(g(x)) = g\left(\sqrt{x}\right) = \sqrt{\sqrt{x}} = \sqrt[4]{x}$

37. $f(x) = \dfrac{1}{2x+3} \qquad g(x) = 2x+3$

(a) $f \circ g = f(g(x)) = f(2x+3) = \dfrac{1}{2(2x+3)} = \dfrac{1}{4x+6+3}$

$\qquad\qquad = \dfrac{1}{4x+9}$

(b) $g \circ f = g(f(x)) = g\left(\dfrac{1}{2x+3}\right) = 2\left(\dfrac{1}{2x+3}\right) + 3$

$\qquad\qquad = \dfrac{2}{2x+3} + \dfrac{6x+9}{2x+3} = \dfrac{6x+11}{2x+3}$

(c) $f \circ f = f(f(x)) = f\left(\dfrac{1}{2x+3}\right) = \dfrac{1}{2\left(\dfrac{1}{2x+3}\right) + 3}$

$\qquad\qquad = \dfrac{1}{\dfrac{2}{2x+3} + \dfrac{6x+9}{2x+3}} = \dfrac{1}{\dfrac{6x+11}{2x+3}} = \dfrac{2x+3}{6x+11}$

(d) $g \circ g = g(g(x)) = g(2x+3) = 2(2x+3) + 3$
$\qquad\qquad\qquad = 4x+6+3 = 4x+9$

2 FUNCTIONS AND THEIR GRAPHS

39. $f(x) = ax + b$ $g(x) = cx + d$

    (a) $f \circ g = f(g(x)) = f(cx + d) = a(cx + d) + b = acx + ad + b$

    (b) $g \circ f = g(f(x)) = g(ax + b) = c(ax + b) + d = acx + bc + d$

    (c) $f \circ f = f(f(x)) = f(ax + b) = a(ax + b) + b = a^2x + ab + b$

    (d) $g \circ g = g(g(x)) = g(cx + d) = c(cx + d) + d = c^2x + cd + d$

41. $(f \circ g)(x) = f(g(x)) = f\left(\frac{1}{3}x\right) = 3\left(\frac{1}{3}x\right) = x$

    $(g \circ f)(x) = g(f(x)) = g(3x) = \frac{1}{3}(3x) = x$

43. $(f \circ g)(x) = f(g(x)) = f\left(\sqrt[3]{x}\right) = \left(\sqrt[3]{x}\right)^3 = x$

    $(g \circ f)(x) = g(f(x)) = g(x^3) = \sqrt[3]{x^3} = x$

45. $(f \circ g)(x) = f(g(x)) = f\left(\frac{1}{2}(x + 6)\right) = 2\left[\frac{1}{2}(x + 6)\right] - 6$
                                     $= x + 6 - 6$
                                       $= x$

    $(g \circ f)(x) = g(f(x)) = g(2x - 6) = \frac{1}{2}(2x - 6 + 6) = x$

47. $(f \circ g)(x) = f(g(x)) = f\left(\frac{1}{a}(x - b)\right) = a\left[\frac{1}{a}(x - b)\right] + b$
                                    $= a\left(\frac{x}{a} - \frac{b}{a}\right) + b$
                                    $= x - b + b = x$

    $(g \circ f)(x) = g(f(x)) = g(ax + b) = \frac{1}{a}(ax + b - b) = x$

49. $f(x) = 2x^3 - 3x^2 + 4x - 1$ $g(x) = 2$

    $(f \circ g)(x) = f(g(x)) = f(2)$
                                $= 2(2)^3 - 3(2)^2 + 4(2) - 1$
                                  $= 16 - 12 + 8 - 1 = 11$

    $(g \circ f)(x) = g(f(x)) = 2$

51. $f(x) = x^2$, $g(x) = \sqrt{x} + 2$, $h(x) = 1 - 3x$

    $[f \circ (g \circ h)](x) = f \circ (g(h(x))) = f\left(\sqrt{1 - 3x} + 2\right) = \left(\sqrt{1 - 3x} + 2\right)^2$
                        $= (1 - 3x) + 4\sqrt{1 - 3x} + 4 = 5 - 3x + 4\sqrt{1 - 3x}$

53. $[(f + g) \circ h](x) = \left[(f + g)(h(x))\right] = [h(x)]^2 + \sqrt{h(x)} + 2$

$$= (1 - 3x)^2 + \sqrt{1 - 3x} + 2$$

$$= 1 - 6x + 9x^2 + \sqrt{1 - 3x} + 2$$

$$= 9x^2 - 6x + 3 + \sqrt{1 - 3x}$$

55. $f(x) = x^2,\; g(x) = 3x,$

$h(x) = \sqrt{x} + 1$

$F(x) = 9x^2$
$F(x) = (3x)^2$
$F(x) = (g(x))^2$
$F(x) = f(g(x))$
$\quad F = f \circ g$

57. $H(x) = |x| + 1$

$H(x) = \sqrt{x^2} + 1$

$H(x) = \sqrt{f(x)} + 1$
$H(x) = h(f(x))$
$\quad H = h \circ f$

59. $q(x) = x + 2\sqrt{x} + 1$

$q(x) = \left(\sqrt{x} + 1\right)^2$
$q(x) = (h(x))^2$
$q(x) = f(h(x))$
$\quad q = f \circ h$

61. $P(x) = x^4$
$P(x) = (x^2)^2$
$P(x) = (f(x))^2$
$P(x) = f(f(x))$
$\quad P = f \circ f$

63. $H(x) = (2x + 5)^3 = f(g(x))$
$f(x) = x^3,\; g(x) = 2x + 5$

65. $H(x) = \sqrt{x^2 + x + 1} = f(g(x))$
$f(x) = \sqrt{x},\; g(x) = x^2 + x + 1$

67. $H(x) = \left(1 - \dfrac{1}{x^2}\right)^2 = f(g(x))$

$f(x) = x^2,\; g(x) = 1 - \dfrac{1}{x^2}$

69. $H(x) = [\![x^2 + 1]\!] = f(g(x))$

$f(x) = [\![x]\!],\; g(x) = x^2 + 1$

71. $(f \circ g)(x) = f(g(x)) = f(3x + a) = 2(3x + a)^2 + 5$

When $x = 0,\; (f \circ g)(x) = f \circ g)(0) = 23$

Then, $2(3 \cdot 0 + a)^2 + 5 = 23$
$$2a^2 + 5 = 23$$
$$2a^2 = 18$$
$$a^2 = 9$$
$$a = -3,\; 3$$

73. $S(r) = 4\pi r^2$

$r(t) = \dfrac{2}{3}t^3,\; t \geq 0$

$S(r(t)) = 4\pi\left(\dfrac{2}{3}t^3\right)^2$

$$= 4\pi\left(\dfrac{4}{9}t^6\right)$$

$$= \dfrac{16}{9}\pi t^6$$

75. $N(t) = 100\,t - 5t^2,\; 0 \leq t \leq 10$
$C(x) = 5000 + 6000x$
$C(N(t)) = 5000 + 6000(100t - 5t^2)$
$C(N(t)) = 5000 + 600{,}000t - 30{,}000t^2$

77. $p = -\frac{1}{4}x + 100 \qquad 0 \le x \le 400$

$\frac{1}{4}x = 100 - p$

$x = 4(100 - p)$

$C = \frac{\sqrt{x}}{25} + 600 = \frac{\sqrt{4(100 - p)}}{25} + 600$

$\qquad = \frac{2\sqrt{100 - p}}{25} + 600$

79. Since f and g are odd, $f(-x) = -f(x)$ and $g(-x) = -g(x)$.

$(f \circ g)(-x) = f(g(-x)) = f(-g(x)) = -f(g(x)) = -(f \circ g)(x)$.

Because $(f \circ g)(-x) = -(f \circ g)(x)$, by definition, $f \circ g$ is odd.

≡ EXERCISE 2.5 ONE-TO-ONE FUNCTIONS; INVERSE FUNCTIONS

1. Yes, any horizontal line intersects the graph of f at most in one point. <u>One-to-one</u>

3. No, there are horizontal lines which intersect the graph of f at more than one point. <u>Not one-to-one</u>

5. Yes, any horizontal line intersects the graph of f at most in one point. <u>One-to-one</u>

7. Reflect about the line $y = x$.

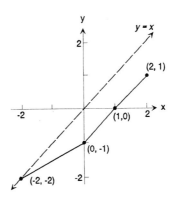

9. Reflect about the line $y = x$.

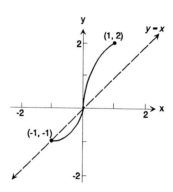

11. Reflect about the line $y = x$.

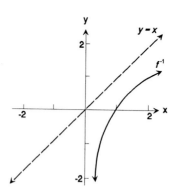

13. $f(x) = 3x - 4$, $g(x) = \frac{1}{3}(x + 4)$

$f(g(x)) = f\left(\frac{1}{3}(x + 4)\right) = 3\left[\frac{1}{3}(x + 4)\right] - 4 = (x + 4) - 4 = x$

$g(f(x)) = g(3x - 4) = \frac{1}{3}(3x - 4 + 4) = \frac{1}{3}(3x) = x$

15. $f(x) = 4x - 8$, $g(x) = \frac{x}{4} + 2$

$f(g(x)) = f\left(\frac{x}{4} + 2\right) = 4\left(\frac{x}{4} + 2\right) - 8 = (x + 8) - 8 = x$

$g(f(x)) = g(4x - 8) = \frac{4x - 8}{4} + 2 = x - 2 + 2 = x$

17. $f(x) = x^3 - 8$, $g(x) = \sqrt[3]{x + 8}$

$f(g(x)) = f\left(\sqrt[3]{x + 8}\right) = \left[\sqrt[3]{x + 8}\right]^3 - 8 = (x + 8) - 8 = x$

$g(f(x)) = g(x^3 - 8) = \sqrt[3]{x^3 - 8 + 8} = \sqrt[3]{x^3} = x$

2 FUNCTIONS AND THEIR GRAPHS

19. $f(x) = \dfrac{1}{x}, \quad g(x) = \dfrac{1}{x}$

$f(g(x)) = f\left(\dfrac{1}{x}\right) = \dfrac{1}{\dfrac{1}{x}} = x$

$g(f(x)) = g\left(\dfrac{1}{x}\right) = \dfrac{1}{\dfrac{1}{x}} = x$

21. $f(x) = \dfrac{2x + 3}{x + 4} \qquad g(x) = \dfrac{4x - 3}{2 - x}$

$f(g(x)) = f\left(\dfrac{4x - 3}{2 - x}\right) = \dfrac{2\left(\dfrac{4x - 3}{2 - x}\right) + 3}{\dfrac{4x - 3}{2 - x} + 4} = \dfrac{\dfrac{2(4x - 3) + 3(2 - x)}{2 - x}}{\dfrac{(4x - 3) + 4(2 - x)}{2 - x}}$

$= \dfrac{\dfrac{8x - 6 + 6 - 3x}{2 - x}}{\dfrac{4x - 3 + 8 - 4x}{2 - x}} = \dfrac{\dfrac{5x}{2 - x}}{\dfrac{5}{2 - x}} = \dfrac{5x}{2 - x} \cdot \dfrac{(2 - x)}{5} = x$

$g(f(x)) = g\left(\dfrac{2x + 3}{x + 4}\right) = \dfrac{4\left(\dfrac{2x + 3}{x + 4}\right) - 3}{2 - \left(\dfrac{2x + 3}{x + 4}\right)} = \dfrac{\dfrac{4(2x + 3) - 3(x + 4)}{x + 4}}{\dfrac{2(x + 4) - (2x + 3)}{x + 4}}$

$= \dfrac{\dfrac{8x + 12 - 3x - 12}{x + 4}}{\dfrac{2x + 8 - 2x - 3}{x + 4}} = \dfrac{\dfrac{5x}{x + 4}}{\dfrac{5}{x + 4}} = \dfrac{5x}{x + 4} \cdot \dfrac{(x + 4)}{5} = x$

23. $f(x) = 2x$
$y = 2x$
$x = 2y$
$y = \dfrac{x}{2}$
$f^{-1}(x) = \dfrac{x}{2}$

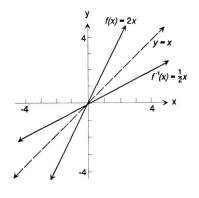

Verify: $f(f^{-1}(x)) = f\left(\dfrac{x}{2}\right) = 2\left(\dfrac{x}{2}\right) = x$

and $f^{-1}(f(x)) = f^{-1}(2x)$

$= \dfrac{2x}{2} = x$

Domain of f = range of f^{-1} = $(-\infty, \infty)$
Range of f = domain of f^{-1} = $(-\infty, \infty)$

25.

$$f(x) = 4x + 2$$
$$y = 4x + 2$$
$$x = 4y + 2$$
$$4y = x - 2$$
$$y = \frac{x}{4} - \frac{1}{2}$$
$$f^{-1}(x) = \frac{x}{4} - \frac{1}{2}$$

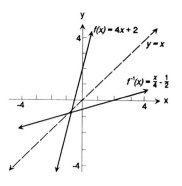

Verify: $f(f^{-1}(x)) = f\left(\frac{x}{4} - \frac{1}{2}\right)$

$$= 4\left(\frac{x}{4} - \frac{1}{2}\right) + 2$$

$$= x - 2 + 2 = x$$

$$f^{-1}(f(x)) = f^{-1}(4x + 2)$$

$$= \frac{4x + 2 - 2}{4} = x$$

Domain of f = range of f^{-1} = $(-\infty, \infty)$
Range of f = domain of f^{-1} = $(-\infty, \infty)$

27.

$$f(x) = x^3 - 1$$
$$y = x^3 - 1$$
$$x = y^3 - 1$$
$$y^3 = x + 1$$
$$y = \sqrt[3]{x + 1}$$
$$f^{-1}(x) = \sqrt[3]{x + 1}$$

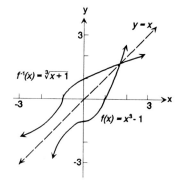

Verify: $f(f^{-1}(x))$

$$= f\left(\sqrt[3]{x + 1}\right)$$

$$= \left(\sqrt[3]{x + 1}\right)^3 - 1$$

$$= x + 1 - 1 = x$$

$$f^{-1}(f(x)) = f^{-1}(x^3 - 1)$$

$$= \sqrt[3]{x^3 - 1 + 1} = \sqrt[3]{x^3} = x$$

Domain of f = range of f^{-1} = $(-\infty, \infty)$
Range of f = domain of f^{-1} = $(-\infty, \infty)$

2 FUNCTIONS AND THEIR GRAPHS

29. $f(x) = x^2 + 4, \ x \geq 0$

$$y = x^2 + 4$$
$$x = y^2 + 4$$
$$y^2 = x - 4$$
$$y = \sqrt{x - 4}$$
$$f^{-1}(x) = \sqrt{x - 4}$$

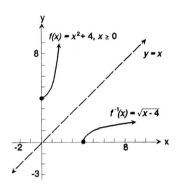

Verify: $f(f^{-1}(x)) = f\left(\sqrt{x - 4}\right)$

$$= \left(\sqrt{x - 4}\right)^2 + 4$$
$$= x - 4 + 4$$
$$= x$$

$$f^{-1}(f(x)) = f^{-1}(x^2 + 4)$$
$$= \sqrt{x^2 + 4 - 4}$$
$$= \sqrt{x^2}$$
$$= |x|$$
$$= x, \ x \geq 0$$

Domain of f = range of f^{-1} = $[0, \infty)$
Range of f = domain of f^{-1} = $[4, \infty)$

31. $f(x) = \dfrac{4}{x}$

$$y = \frac{4}{x}$$
$$x = \frac{4}{y}$$
$$yx = 4$$
$$y = \frac{4}{x}$$
$$f^{-1}(x) = \frac{4}{x}$$

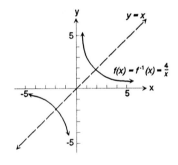

Verify: $f(f^{-1}(x)) = f\left(\dfrac{4}{x}\right) = \dfrac{4}{\frac{4}{x}} = 4 \cdot \dfrac{x}{4} = x$

$$f^{-1}(f(x)) = f^{-1}\left(\frac{4}{x}\right) = \frac{4}{\frac{4}{x}} = 4 \cdot \frac{x}{4} = x$$

Domain of f = range of f^{-1} = all real numbers except 0
Range of f = domain of f^{-1} = all real numbers except 0

33. $f(x) = \dfrac{1}{x - 2}$

$$y = \dfrac{1}{x - 2}$$

$$x = \dfrac{1}{y - 2}$$

$$xy - 2x = 1$$

$$y = \dfrac{2x + 1}{x}$$

$$f^{-1}(x) = \dfrac{2x + 1}{x}$$

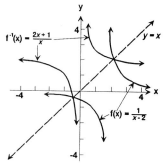

Verify: $f(f^{-1}(x)) = f\!\left(\dfrac{2x + 1}{x}\right) = \dfrac{1}{\dfrac{2x + 1}{x} - 2} = \dfrac{1}{\dfrac{2x + 1 - 2x}{x}} = \dfrac{x}{1} = x$

$$f^{-1}(f(x)) = f^{-1}\!\left(\dfrac{1}{x - 2}\right) = \dfrac{2\left(\dfrac{1}{x - 2}\right) + 1}{\dfrac{1}{x - 2}} = \dfrac{\dfrac{2 + (x - 2)}{x - 2}}{\dfrac{1}{x - 2}} = x$$

Domain of f = range of f^{-1} = all real numbers except 2
Range of f = domain of f^{-1} = all real numbers except 0

35. $f(x) = \dfrac{1}{3 + x}$

$$y = \dfrac{1}{3 + x}$$

$$x = \dfrac{1}{3 + y}$$

$$3x + xy = 1$$

$$y = \dfrac{1 - 3x}{x}$$

$$f^{-1}(x) = \dfrac{1 - 3x}{x}$$

Verify: $f(f^{-1}(x)) = f\!\left(\dfrac{1 - 3x}{x}\right) = \dfrac{1}{3 + \left(\dfrac{1 - 3x}{x}\right)} = \dfrac{1}{\dfrac{3x + 1 - 3x}{x}} = x$

$$f^{-1}(f(x)) = f^{-1}\!\left(\dfrac{1}{3 + x}\right) = \dfrac{1 - 3\left(\dfrac{1}{3 + x}\right)}{\dfrac{1}{3 + x}} = \dfrac{\dfrac{3 + x - 3}{3 + x}}{\dfrac{1}{3 + x}} = x$$

Domain of f = range of f^{-1} = all real numbers except -3
Range of f = domain of f^{-1} = all real numbers except 0

2 FUNCTIONS AND THEIR GRAPHS

37. $f(x) = (x + 2)^2, \ x \geq -2$

$\quad\quad y = (x + 2)^2, \ x \geq -2$

$\quad\quad x = (y + 2)^2$

$\quad\quad \sqrt{x} = y + 2$

$\quad\quad y = \sqrt{x} - 2$

$\quad f^{-1}(x) = \sqrt{x} - 2$

Verify: $f(f^{-1}(x)) = f\left(\sqrt{x} - 2\right) = \left(\sqrt{x} - 2 + 2\right)^2 = \sqrt{x^2} = x$

$\quad\quad\quad\quad f^{-1}(f(x)) = f^{-1}\left[(x + 2)^2\right] = \sqrt{(x + 2)^2} - 2 = x + 2 - 2 = x$

Domain of f = range of f^{-1} = $[-2, \ \infty)$
Range of f = domain of f^{-1} = $[0, \ \infty)$

39. $f(x) = \dfrac{2x}{x - 1}$

$\quad\quad y = \dfrac{2x}{x - 1}$

$\quad\quad x = \dfrac{2y}{y - 1}$

$\quad xy - x = 2y$

$\quad xy - 2y = x$

$\quad y(x - 2) = x$

$\quad\quad y = \dfrac{x}{x - 2}$

$\quad f^{-1}(x) = \dfrac{x}{x - 2}$

Verify: $f(f^{-1}(x)) = f\left(\dfrac{x}{x - 2}\right) = \dfrac{2\left(\dfrac{x}{x - 2}\right)}{\left(\dfrac{x}{x - 2}\right) - 1} = \dfrac{\dfrac{2x}{x - 2}}{\dfrac{x - (x - 2)}{x - 2}} = \dfrac{2x}{2} = x$

$\quad\quad\quad\quad f^{-1}(f(x)) = f^{-1}\left(\dfrac{2x}{x - 1}\right) = \dfrac{\dfrac{2x}{x - 1}}{\dfrac{2x}{x - 1} - 2} = \dfrac{\dfrac{2x}{x - 1}}{\dfrac{2x - 2x + 2}{x - 1}} = \dfrac{2x}{2} = x$

Domain of f = range of f^{-1} = all real numbers except 1
Range of f = domain of f^{-1} = all real numbers except 2

2.5 ONE-TO-ONE FUNCTIONS; INVERSE FUNCTIONS   113

41.

$$f(x) = \frac{3x + 4}{2x - 3}$$

$$y = \frac{3x + 4}{2x - 3}$$

$$x = \frac{3y + 4}{2y - 3}$$

$$2xy - 3x = 3y + 4$$

$$2xy - 3y = 3x + 4$$

$$y(2x - 3) = 3x + 4$$

$$y = \frac{3x + 4}{2x - 3}$$

$$f^{-1}(x) = \frac{3x + 4}{2x - 3}$$

Verify: $f(f^{-1}(x)) = f\left(\frac{3x + 4}{2x - 3}\right) = \dfrac{3\left(\frac{3x + 4}{2x - 3}\right) + 4}{2\left(\frac{3x + 4}{2x - 3}\right) - 3} = \dfrac{\frac{9x + 12 + 8x - 12}{2x - 3}}{\frac{6x + 8 - 6x + 9}{2x - 3}}$

$$= \frac{17x}{2x - 3} \cdot \frac{2x - 3}{17} = \frac{17x}{17} = x$$

$$f^{-1}(f(x)) = f^{-1}\left(\frac{3x + 4}{2x - 3}\right) = \dfrac{3\left(\frac{3x + 4}{2x - 3}\right) + 4}{2\left(\frac{3x + 4}{2x - 3}\right) - 3} = \dfrac{\frac{9x + 12 + 8x - 12}{2x - 3}}{\frac{6x + 8 - 6x + 9}{2x - 3}}$$

$$= \frac{17x}{2x - 3} \cdot \frac{2x - 3}{17} = \frac{17x}{17} = x$$

Domain of f = range of f^{-1} = all real numbers except $\frac{3}{2}$

Range of f = domain of f^{-1} = all real numbers except $\frac{3}{2}$

43.

$$f(x) = \frac{2x + 3}{x + 2}$$

$$y = \frac{2x + 3}{x + 2}$$

$$x = \frac{2y + 3}{y + 2}$$

$$xy + 2x = 2y + 3$$

$$xy - 2y = 3 - 2x$$

$$y(x - 2) = 3 - 2x$$

$$y = \frac{3 - 2x}{x - 2}$$

$$f^{-1}(x) = \frac{3 - 2x}{x - 2}$$

2 FUNCTIONS AND THEIR GRAPHS

Verify: $f(f^{-1}(x)) = f\left(\dfrac{3 - 2x}{x - 2}\right) = \dfrac{2\left(\dfrac{3 - 2x}{x - 2}\right) + 3}{\dfrac{3 - 2x}{x - 2} + 2} = \dfrac{\dfrac{6 - 4x + 3x - 6}{x - 2}}{\dfrac{3 - 2x + 2x - 4}{x - 2}}$

$= \dfrac{-x}{-1} = x$

$f^{-1}(f(x)) = f^{-1}\left(\dfrac{2x + 3}{x + 2}\right) = \dfrac{3 - 2\left(\dfrac{2x + 3}{x + 2}\right)}{\left(\dfrac{2x + 3}{x + 2}\right) - 2} = \dfrac{\dfrac{3x + 6 - 4x - 6}{x + 2}}{\dfrac{2x + 3 - 2x - 4}{x + 2}}$

$= \dfrac{-x}{-1} = x$

Domain of f = range of f^{-1} = all real numbers except -2
Range of f = domain of f^{-1} = all real numbers except 2

45. $f(x) = 2\sqrt[3]{x}$

$y = 2\sqrt[3]{x}$

$x = 2\sqrt[3]{y}$

$x^3 = 8y$

$y = \dfrac{x^3}{8}$

$f^{-1}(x) = \dfrac{x^3}{8}$

Verify: $f(f^{-1}(x)) = f\left(\dfrac{x^3}{8}\right) = 2\sqrt[3]{\dfrac{x^3}{8}} = 2\left(\dfrac{x}{2}\right) = x$

$f^{-1}(f(x)) = f^{-1}\left(2\sqrt[3]{x}\right) = \dfrac{\left(2\sqrt[3]{x}\right)^3}{8} = \dfrac{8x}{8} = x$

Domain of f = range of f^{-1} = $(-\infty, \infty)$
Range of f = domain of f^{-1} = $(-\infty, \infty)$

47. $f(x) = mx + b, \ m \neq 0$

$y = mx + b$

$x = my + b$

$my = x - b$

$y = \dfrac{x - b}{m}$

$f^{-1}(x) = \dfrac{x - b}{m}, \ m \neq 0$

49.   No. If a function is even, $f(-x) = f(x)$. Whenever x and $-x$ are in the domain of f, two equal y values, $f(x)$ and $f(-x)$ are present.

51. f^{-1} also lies in quadrant one because whenever (a, b) is on f, then (b, a) is on f^{-1}. In quadrant one (a, b) is $(+, +)$, so (b, a) is $(+, +)$, also in quadrant one.

53. $f(x) = |x|$, $x \geq 0$ is one to one. Thus,
$$f(x) = x$$
$$f^{-1}(x) = x$$

55. $f(x) = \dfrac{9}{5}x + 32$ $g(x) = \dfrac{5}{9}(x - 32)$

$$f(g(x)) = f\left(\frac{5}{9}(x - 32)\right) = \frac{9}{5}\left[\frac{5}{9}(x - 32)\right] + 32 = x - 32 + 32 = x$$

$$g(f(x)) = g\left(\frac{9}{5}x + 32\right) = \frac{5}{9}\left(\frac{9}{5}x + 32 - 32\right) = x$$

57. $T(\ell) = 2\pi\sqrt{\dfrac{\ell}{g}}$, $g \approx 32.2$

$$T = 2\pi\sqrt{\frac{\ell}{g}}$$

$$\frac{T}{2\pi} = \sqrt{\frac{\ell}{g}}$$

$$\frac{T^2}{4\pi^2} = \frac{\ell}{g}$$

$$\frac{gT^2}{4\pi^2} = \ell$$

$$\ell = \frac{gT^2}{4\pi^2}$$

$$\ell(T) = \frac{gT^2}{4\pi^2}$$

59. $$f(x) = \frac{ax + b}{cx + d}$$

$$y = \frac{ax + b}{cx + d}$$

$$x = \frac{ay + b}{cy + d}$$

$$cxy + dx = ay + b$$

$$cxy - ay = -dx + b$$

$$y(cx - a) = -dx + b$$

$$y = \frac{-dy + b}{cx - a}$$

$$f^{-1}(x) = \frac{-dx + b}{cx - a}$$

$f = f^{-1}$ if $\dfrac{ax + b}{cx + d} = \dfrac{-dx + b}{cx - a}$

This is true if $a = -d$.

2 FUNCTIONS AND THEIR GRAPHS

≡ EXERCISE 2.6 CONSTRUCTING FUNCTIONS: APPLICATIONS

1. If $V = \pi r^2 h$ and $h = 2r$, then $V(r) = \pi r^2 (2r) = 2\pi r^3$

3. If $p = \frac{-1}{5}x + 100$ and $R = xp$, then

$$R(x) = x\left(\frac{-1}{5}x + 100\right) = -\frac{1}{5}x^2 + 100x$$

5. If $x = -20p + 100$ and $R = xp$, then $p = \frac{100 - x}{20}$ and

$$R(x) = x\left(\frac{100 - x}{20}\right) = \frac{-1}{20}x^2 + 5x$$

7. We know that width $= x$. In order to enclose a rectangular area, we need the perimeter, P. Let ℓ = length. $P = 2\ell + 2x = 100$.

   Then $\ell = \frac{100 - 2x}{2} = 50 - x$. The area of the rectangle as a function of the width x, represented by $A(x) = \ell x = (50 - x)x = -x^2 + 50x$, $0 < x < 50$.

9. (a) Let C = circumference, r = radius. We know that $C = 2\pi r$ by definition. If a wire of length x is bent into a circle, then x is the circumference, so $C = x = 2\pi r$. Writing the circumference as a function of x we have $C(x) = x$.

   (b) We know that $C = x = 2\pi r$, so $r = \frac{x}{2\pi}$. By definition, the area of a circle is $A = \pi r^2$. Expressing the area of the circle as a function of x, we have $A(x) = \pi\left(\frac{x}{2\pi}\right)^2 = \frac{x^2}{4\pi}$

11. By definition, a triangle has area $A = \frac{1}{2}bh$, b = base, h = height. Because a vertex of the triangle is at the origin, we know that $b = x$ and $h = y$. Expressing the area of the triangle as a function of x, we have $A(x) = \frac{1}{2}xy = \frac{1}{2}x(x^3) = \frac{1}{2}x^4$.

13. (a) The distance d from P to the origin is $d = \sqrt{x^2 + y^2}$. Since P is a point on the graph of $y = x^2 - 4$, we have

$$d(x) = \sqrt{x^2 + (x^2 - 4)^2} = \sqrt{x^4 - 7x^2 + 16}$$

   (b) If $x = 0$, the distance d is $d(0) = \sqrt{16} = 4$

   (c) If $x = 1$, the distance d is $d(1) = \sqrt{1 - 7 + 16} = \sqrt{10}$

15. The distance d from P to the point $(1, 0)$ is $d = \sqrt{(x-1)^2 + y^2}$

Since P is a point on the graph of $y = \sqrt{x}$, we have

$$d(x) = \sqrt{(x-1)^2 + \left(\sqrt{x}\right)^2}$$

$$= \sqrt{x^2 - x + 1}$$

17. We know that distance = (velocity)(time), $d = vt$.
$d_1 = 25t$ and $d_2 = 40t$. By the Pythagorean Theorem,

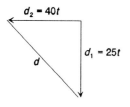

$$d^2 = d_1^2 + d_2^2$$

$$d^2 = (25t)^2 + (40t)^2$$

$$d(t) = \sqrt{625t^2 + 1600t^2}$$

$$d(t) = \sqrt{2225t^2} = 5\sqrt{89}\, t$$

19. By definition, Volume, V = (length)(width)(height)

length = $24 - 2x$, width = $24 - 2x$, height = x

Therefore, $V(x) = (24 - 2x)^2$

21. Volume (V) = (length)(width)(height) = 10

length = width = x, height = h,
so $10 = x^2 h$ and $h = \dfrac{10}{x^2}$

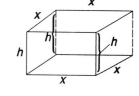

Area, $A = 2x^2 + 2xh + 2xh$

$$A(x) = 2x^2 + 4x\left(\frac{10}{x^2}\right) = 2x^2 + \frac{40}{x}$$

23. A = Area

$A(x) = xy = x(16 - x^2)$

Domain of $A = \{x \mid 0 \leq x \leq 4\}$ because $x \geq 0$ and $16 - x^2 \geq 0$

25. A = Area, p = perimeter

(a) $A(x)$ = (length)(width) = $(2x)(2y) = 4x(4 - x^2)^{1/2}$

(b) $p(x)$ = 2 length + 2 width = $2(2x) = 2(2y)$
$= 4x + 4(4 - x^2)^{1/2}$

2 FUNCTIONS AND THEIR GRAPHS

27. C = Cost of the material, r = radius, h = height

$$500 = \pi r^2 h, \quad h = \frac{500}{\pi r^2}$$

$$C(r) = 6(2\pi r^2) + 4(2\pi rh)$$

$$= 12\pi r^2 + 8\pi r\left(\frac{500}{\pi r^2}\right)$$

$$= 12\pi r^2 + \frac{4000}{r}$$

29. C = Circumference, A = Area, r = radius, x = side of square

$$C = 2\pi r = 10 - 4x; \quad r = \frac{5 - 2x}{\pi}$$

$$A(x) = x^2 + \pi r^2 = x^2 + \pi\left(\frac{5 - 2x}{\pi}\right)^2$$

$$= x^2 + \frac{25 - 20x + 4x^2}{\pi}$$

Since all lengths must be positive, we have $x > 0$ and
$$10 - 4 > 0$$
$$-4x > -10$$
$$x < 2.5$$

Thus, domain $A = \{x \mid 0 < x < 2.5\}$.

31. (a) Distance off road $= \sqrt{4 + (5 - x)^2} = \sqrt{x^2 - 10x + 29}$

C = Cost; x = distance from connection box to where the installation begins to be off the road.

$$C(x) = 10x + 14\sqrt{x^2 - 10x + 29}$$

(b) Since all distances must be positive,

we have $x \geq 0$ and $5 - x \geq 0$
$$-x \geq -5$$
$$x \leq 5$$

Thus, domain $C = \{x \mid 0 \leq x \leq 5\}$.

(c) $C(1) = 10 + 14\sqrt{20} = 10 + 28\sqrt{5} \approx \$72.61$

$C(2) = 20 + 14\sqrt{13} \approx \$70.48$

$C(3) = 30 + 14\sqrt{8} \approx \$69.60$

$C(4) = 40 + 14\sqrt{5} \approx \$71.30$

33. (a) A = Area, r = radius; diameter = $2r$
$$A(r) = (2r)r = 2r^2$$

(b) p = perimeter
$$p(r) = 2(2r) + 2r = 6r$$

35. Area of equilateral triangle = $\pi r^2 - \dfrac{\sqrt{3}}{4}x^2$

Area of equilateral triangle = $\dfrac{1}{2}x\sqrt{r^2 - \dfrac{x^2}{4}}$

$$= \frac{1}{3} \cdot \frac{\sqrt{3}}{4}x^2$$

$$\sqrt{r^2 - \frac{x^2}{4}} = \frac{2}{3}\frac{\sqrt{3}}{4}x = \frac{x}{2\sqrt{3}}$$

$$r^2 - \frac{x^2}{4} = \frac{x^2}{12}$$

$$r^2 = \frac{4x^2}{12} = \frac{x^2}{3}$$

Area $= \dfrac{\pi x^2}{3} - \dfrac{\sqrt{3}}{4}x^2$

$$= \left(\frac{\pi}{3} - \frac{\sqrt{3}}{4}\right)x^2$$

37. $C = \begin{cases} 95 & \text{if} \quad x = 7 \\ 119 & \text{if} \quad 7 < x \le 8 \\ 143 & \text{if} \quad 8 < x \le 9 \\ 167 & \text{if} \quad 9 < x \le 10 \\ 190 & \text{if} \quad 10 < x \le 14 \end{cases}$

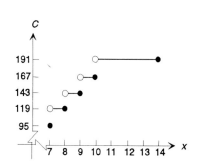

39. r = radius, h = height, V = volume of a cone

$$\frac{r}{h} = \frac{4}{16}$$

$$16r = 4h$$

$$r = \frac{1}{4}h$$

$$V = \frac{1}{3}\pi r^2 h$$

$$V(h) = \frac{1}{3}\pi\left(\frac{h}{4}\right)^2 h$$

$$= \frac{1}{48}\pi h^3$$

2 FUNCTIONS AND THEIR GRAPHS

41. Schedule X: $y = \begin{cases} .15x & \text{if } 0 < x \le 20{,}350 \\ 3052.50 + 0.28(x - 20{,}350) & \text{if } 20{,}350 < x \le 49{,}300 \\ 11{,}158.50 + 0.31(x - 49{,}300) & \text{if } 49{,}300 < x < \infty \end{cases}$

Schedule $Y - 1$: $y = \begin{cases} .15x & \text{if } 0 < x \le 34{,}000 \\ 5100 + 0.28(x - 34{,}000) & \text{if } 34{,}000 < x \le 82{,}150 \\ 18{,}582 + 0.31(x - 82{,}150) & \text{if } 82{,}150 < x < \infty \end{cases}$

≡ 2 - CHAPTER REVIEW

1. $f(4) = -2$

   $f(1) = 4$

   $f(x) = mx + b$

   $f(4): \quad 4m + b = -2$ $\left.\right\}$ Solve two equations
   $f(1): \quad m + b = 4$ $\left.\right\}$ in two unknowns.

   $\begin{array}{r} 4m + b = -2 \\ -m - b = -4 \\ \hline 3m = -6 \\ m = -2 \end{array}$

   $-2 + b = 4$
   $b = 6$

   Hence, $f(x) = -2x + 6$

3. $f(x) = \dfrac{Ax + 5}{6x - 2}$ and $f(1) = 4$

   $f(1): \quad \dfrac{A(1) + 5}{6(1) - 2} = 4$

   $\dfrac{A + 5}{4} = 4$

   $A + 5 = 16$

   $A = 11$

5. (a) B, C, and D pass the vertical line test.

   (b) Only D passes the horizontal line test.

7. $f(x) = \dfrac{x}{x^2 - 4}$

   (a) $f(-x) = \dfrac{-x}{(-x)^2 - 4} = \dfrac{-x}{x^2 - 4}$

   (b) $-f(x) = -\left(\dfrac{x}{x^2 - 4}\right) = -\dfrac{x}{x^2 - 4}$

(c) $f(x + 2) = \dfrac{x + 2}{(x + 2)^2 - 4} = \dfrac{x + 2}{x^2 + 4x + 4 - 4} = \dfrac{x + 2}{x^2 + 4x}$

(d) $f(x - 2) = \dfrac{x - 2}{(x + 2)^2 - 4} \quad \dfrac{x - 2}{x^2 - 4x + 4 - 4} = \dfrac{x - 2}{x^2 - 4x}$

9. $f(x) = \sqrt{x^2 - 4}$

(a) $f(-x) = \sqrt{(-x)^2 - 4} = \sqrt{x^2 - 4}$

(b) $-f(x) \; -\sqrt{x^2 - 4}$

(c) $f(x + 2) = \sqrt{(x + 2)^2 - 4}$

$\qquad = \sqrt{x^2 + 4x + 4 - 4}$

$\qquad = \sqrt{x^2 + 4x}$

(d) $f(x - 2) = \sqrt{(x - 2)^2 - 4}$

$\qquad = \sqrt{x^2 - 4x + 4 - 4}$

$\qquad = \sqrt{x^2 - 4x}$

11. $f(x) = \dfrac{x^2 - 4}{x^2}$

(a) $f(-x) = \dfrac{(-x)^2 - 4}{(-x)^2} = \dfrac{x^2 - 4}{x^2}$

(b) $-f(x) = -\left(\dfrac{x^2 - 4}{x^2} \right) = \dfrac{4 - x^2}{x^2}$

(c) $f(x + 2) = \dfrac{(x + 2)^2 - 4}{(x + 2)^2} = \dfrac{x^2 + 4x + 4 - 4}{x^2 + 4x + 4} = \dfrac{x^2 + 4x}{x^2 + 4x + 4}$

(d) $f(x - 2) = \dfrac{(x - 2)^2 - 4}{(x - 2)^2} = \dfrac{x^2 - 4x + 4 - 4}{x^2 - 4x + 4} = \dfrac{x^2 - 4x}{x^2 - 4x + 4}$

2 FUNCTIONS AND THEIR GRAPHS

13. $f(x) = x^3 - x$

if even:
$$f(-x) = f(x)$$
$$(-x)^3 - (-x) = x^3 - x$$
$$-x^3 + x \neq x^3 - x$$

if odd:
$$f(-x) = -f(x)$$
$$(-x)^3 - (-x) = -(x^3 - x)$$
$$-x^3 + x = -x^3 + x$$

Hence, function is odd.

15. $h(x) = \dfrac{1}{x^4} + \dfrac{1}{x^2} + 1$

if even:
$$h(-x) = h(x)$$
$$\frac{1}{(-x)^4} + \frac{1}{(-x)^2} + 1 = \frac{1}{x^4} + \frac{1}{x^2} + 1$$
$$\frac{1}{x^4} + \frac{1}{x^2} + 1 = \frac{1}{x^4} + \frac{1}{x^2} + 1$$

Hence, $h(x)$ is even.

17. $G(x) = 1 - x + x^3$

if even:
$$G(-x) = G(x)$$
$$1 - (-x) + (-x)^3 = 1 - x + x^3$$
$$1 + x - x^3 \neq 1 - x + x^3$$

if odd:
$$G(-x) = -G(x)$$
$$1 - (-x) + (-x)^3 = -(1 - x + x^3)$$
$$1 + x - x^3 \neq -1 + x - x^3$$

Hence, $G(x)$ is neither even nor odd since $G(-x) \neq G(x)$ and $G(-x) \neq -G(x)$.

19. $f(x) = \dfrac{x}{x^2 - 4}$

The domain is the set of all values x such that

$$x^2 - 4 \neq 0$$
$$(x - 2)(x + 2) \neq 0$$
$$x \neq 2, -2$$

The domain is $\{x \mid x \neq -2, x \neq 2\}$.

21. $f(x) = \sqrt{2 - x}$

The domain consists of all values such that $2 - x \geq 0$
$$x \leq 2$$

The domain is $\{x \mid x \leq 2\}$ or $(-\infty, 2]$.

23. $h(x) = \dfrac{\sqrt{x}}{|x|}$

The domain consists of all values x such that $x > 0$ or $(0, \infty)$.

25. $f(x) = \dfrac{x}{x^2 + 2x - 3}$

The domain consists of all values x such that
$$x^2 + 2x - 3 \ne 0$$
$$(x + 3)(x - 1) \ne 0$$
$$x \ne -3, 1$$

The domain is $\{x \mid x \ne -3, x \ne 1\}$.

27. $G(x) = \begin{cases} |x| & \text{if} \quad -1 \le x \le 1 \\ \dfrac{1}{x} & \text{if} \quad x > 1 \end{cases}$

The domain is $[-1, \infty)$.

29. $f(x) = \begin{cases} \dfrac{1}{x - 2} & \text{if} \quad x > 2 \\ 0 & \text{if} \quad x = 2 \\ 3 & \text{if} \quad 0 \le x \le 2 \end{cases}$

The domain is $[0, \infty)$.

31. $F(x) = |x| - 4$

(a) The domain is all real numbers.

(b) x-intercept(s): y-intercept:

$$0 = |x| - 4 \qquad\qquad y = |0| - 4$$
$$4 = |x| \qquad\qquad\qquad y = -4$$
$$x = -4, 4$$

The intercepts are $(-4, 0)$, $(4, 0)$, $(0, -4)$.

(c)

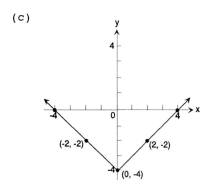

(d) The range is $\{y \mid -4 \le y < \infty\}$.

2 FUNCTIONS AND THEIR GRAPHS

33. $g(x) = -|x|$

    (a) The domain is all real numbers.

    (b) x-intercept(s): y-intercept:

$$0 = -|x|$$
$$x = 0$$

$$y = -|0|$$
$$y = 0$$

    The intercept is $(0, 0)$.

    (c)

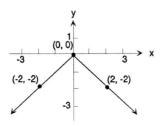

    (d) The range is $\{y \,|\, -\infty < y \leq 0\}$.

35. $h(x) = \sqrt{x - 1}$

    (a) The domain is $\{x \,|\, 1 \leq x < \infty\}$.

    (b) x-intercept(s): y-intercept:

$$0 = \sqrt{x - 1}$$
$$0 = x - 1$$
$$1 = x$$

$$y = \sqrt{0 - 1}$$
$$y = \sqrt{-1}$$
No solution.

    The intercept is $(1, 0)$.

    (c)

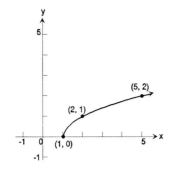

    (d) The range is $\{y \,|\, 0 \leq y < \infty\}$.

37. $f(x) = \sqrt{1 - x}$

(a) The domain is $\{x \mid -\infty < x \leq 1\}$.

(b) x-intercept(s): y-intercept:

$$0 = \sqrt{1 - x} \qquad\qquad y = \sqrt{1 - 0}$$
$$0 = 1 - x \qquad\qquad\quad y = 1$$
$$x = 1$$

The intercepts are $(1, 0)$, $(0, 1)$.

(c)

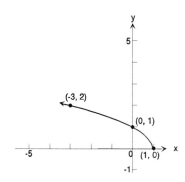

(d) The range is $\{y \mid 0 \leq y < \infty\}$.

39. $F(x) = \begin{cases} x^2 + 4 & \text{if } x < 0 \\ 4 - x^2 & \text{if } x \geq 0 \end{cases}$

(a) The domain is all real numbers.

(b) x-intercept(s): y-intercept:

$$x^2 + 4 = 0 \qquad \text{or} \qquad 0 = 4 - x^2 \qquad\qquad y = 4 - 0^2$$
$$\text{(No real solution)} \qquad x^2 = 4 \qquad\qquad\qquad y = 4$$
$$x = 2$$

The intercepts are $(2, 0)$, $(0, 4)$.

(c)

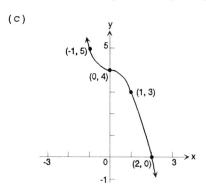

(d) The range is all real numbers.

2 FUNCTIONS AND THEIR GRAPHS

41. $h(x) = (x - 1)^2 + 2$

(a) The domain is all real numbers.

(b) x-intercept(s): y-intercept(s):

$0 = (x - 1)^2 + 2$ $y = (0 - 1)^2 + 2$
$0 = x^2 - 2x + 1 + 2$ $y = 1 + 2$
$0 = x^2 - 2x + 3$ $y = 3$
No real solution.

The intercept is $(0, 3)$.

(c)

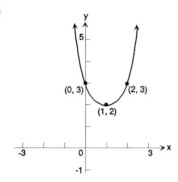

(d) The range is $\{y \mid 2 \le y < \infty\}$.

43. $g(x) = (x - 3)^3 + 1$

(a) The domain is all real numbers.

(b) x-intercept(s): y-intercept(s):

$0 = (x - 1)^3 + 1$ $y = (0 - 1)^3 + 1$
$0 = x^3 - x^2 + x - 1 + 1$ $y = -1 + 1$
$0 = x^3 - x^2 + x$ $y = 0$
$0 = x(x^2 - x + 1)$
$x = 0$

The intercept is $(0, 0)$.

(c)

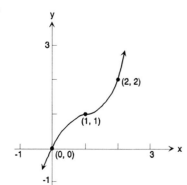

(d) The range is all real numbers.

45. $f(x) = \begin{cases} 2\sqrt{x} & \text{if} \quad x \geq 4 \\ x & \text{if} \quad 0 < x < 4 \end{cases}$

(a) The domain is $\{x \mid 0 < x < \infty\}$.

(b) x-intercept(s): y-intercept:

   None   None

   There are no intercepts.

(c)

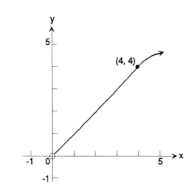

(d) The range is $(0, \infty)$.

47. $g(x) = \dfrac{1}{x - 1} + 1$

(a) The domain is $\{x \mid x \neq 1\}$.

(b) x-intercept(s): y-intercept(s):

$$0 = \frac{1}{x - 1} + 1 \qquad\qquad y = \frac{1}{0 - 1} + 1$$

$$-1 = \frac{1}{x - 1} \qquad\qquad\qquad y = -1 + 1$$

$$-x + 1 = 1 \qquad\qquad\qquad\qquad y = 0$$

$$x = 0$$

   The intercept is $(0, 0)$.

(c)

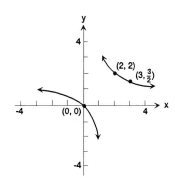

(d) Because $\dfrac{1}{x-1}$ can never be zero, the range is $\{y \,|\, y \neq 1\}$.

49. $h(x) = [\![-x]\!]$

(a) The domain is all real numbers.

(b) x-intercept(s): y-intercept:

$$0 = [\![-x]\!] \qquad\qquad y = [\![-x]\!]$$
$$-1 < x \leq 0 \qquad\qquad y = 0$$

(c)

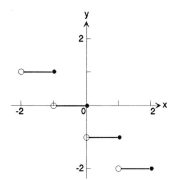

(d) The range is set of all integers.

51.
$$f(x) = \frac{2x + 3}{x - 2}$$

$$y = \frac{2x + 3}{x - 2}$$

$$x = \frac{2y + 3}{y - 2}$$

$$xy - 2x = 2y + 3$$

$$xy - 2y = 2x + 3$$

$$y(x - 2) = 2x + 3$$

$$y = \frac{2x + 3}{x - 2}$$

$$f^{-1}(x) = \frac{2x + 3}{x - 2}$$

Verify: $f(f^{-1}(x)) = f\left(\dfrac{2x + 3}{x - 2}\right) = \dfrac{2\left(\dfrac{2x + 3}{x - 2}\right) + 3}{\left(\dfrac{2x + 3}{x - 2}\right) - 2} = \dfrac{\dfrac{4x + 6 + 3x - 6}{x - 2}}{\dfrac{2x + 3 - 2x + 4}{x - 2}}$

$$= \frac{7x}{7} = x$$

$$f^{-1}(f(x)) = f^{-1}\left(\frac{x + 3}{x - 2}\right) = \frac{2\left(\dfrac{2x + 3}{x - 2}\right) + 3}{\left(\dfrac{2x + 3}{x - 2}\right) - 2} = x$$

Domain of f = range of f^{-1} = all real numbers except 2
Range of f = domain of f^{-1} = all real numbers except 2

53.
$$f(x) = \frac{1}{x - 1}$$

$$y = \frac{1}{x - 2}$$

$$x = \frac{1}{y + 1}$$

$$xy - x = 1$$

$$xy = x + 1$$

$$y = x + 1$$

$$f^{-1} = \frac{x + 1}{x}$$

Verify: $f(f^{-1}(x)) = \dfrac{1}{\dfrac{x + 1}{x} + 1} = \dfrac{1}{\dfrac{x + 1 - x}{x}} = x$

$$f^{-1}(f(x)) = \frac{\dfrac{1}{x - 1} + 1}{\dfrac{1}{x - 1}} = \frac{\dfrac{1 + x - 1}{x - 1}}{\dfrac{1}{x - 1}} = x$$

Domain of f = range of f^{-1} = all real numbers except 1
Range of f = domain of f^{-1} = all real numbers except 0

2 FUNCTIONS AND THEIR GRAPHS

55. $f(x) = \dfrac{3}{x^{1/3}}$

$y = \dfrac{3}{x^{1/3}}$

$x = \dfrac{3}{y^{1/3}}$

$y^{1/3}x = 3$

$y^{1/3} = \dfrac{3}{x}$

$y = \dfrac{27}{x^3}$

$f^{-1}(x) = \dfrac{27}{x^3}$

Verify: $f(f^{-1}(x)) = \dfrac{3}{\left(\dfrac{27}{x^3}\right)^{1/3}} = \dfrac{3}{\dfrac{3}{x}} = x$

$f^{-1}(f(x)) = \dfrac{27}{\left(\dfrac{3}{x^{1/3}}\right)^3} = \dfrac{27}{\dfrac{27}{x}} = x$

Domain of f = range of f^{-1} = all real numbers except 0
Range of f = domain of f^{-1} = all real numbers except 0

57. $f(x) = 3x - 5, \; g(x) = 1 - 2x^2$

(a) $(f \circ g)(2) = f(g(2)) = f(1 - 2(2)^2) = f(1 - 8) = f(-7)$
$= 3(-7) - 5 = -26$

(b) $(g \circ f)(-2) = g(f(-2)) = g(3(-2) - 5) = g(-11) = 1 - 2(-11)^2$
$= 1 - 242 = -241$

(c) $(f \circ f)(4) = f(f(4)) = f(3(4) - 5) = f(7) = 3(7) - 5 = 16$

(d) $(g \circ g)(-2) = g(g(-1)) = g(1 - 2(-1)^2) = g(1 - 2) = g(-1)$
$= 1 - 2(-1)^2 = 1 - 2 = -1$

59. $f(x) = \sqrt{x + 2}, \; g(x) = 2x^2 + 1$

(a) $(f \circ g)(2) = f(g(2)) = f(9) = \sqrt{11}$

(b) $(g \circ f)(-2) = g(f(-2)) = g(0) = 1$

(c) $(f \circ f)(4) = f(f(4)) = f(\sqrt{6}) = \sqrt{\sqrt{6} + 2}$

(d) $(g \circ g)(-2) = g(g(-1)) = g(3) = 19$

61. $f(x) = \dfrac{1}{x^2 + 4}$, $g(x) = 3x - 2$

(a) $(f \circ g)(2) = f(g(2)) = f(4) = \dfrac{1}{4^2 + 4} = \dfrac{1}{20}$

(b) $(g \circ f)(-2) = g(f(-2)) = g\left(\dfrac{1}{8}\right) = \dfrac{3}{8} - 2 = \dfrac{3}{8} - \dfrac{16}{8} = -\dfrac{13}{8}$

(c) $(f \circ f)(4) = f(f(4)) = f\left(\dfrac{1}{20}\right) = \dfrac{1}{\dfrac{1}{400} + 4} = \dfrac{1}{\dfrac{1601}{400}} = \dfrac{400}{1601}$

(d) $(g \circ g)(-1) = g(g(-1)) = g(-5) = -17$

63. $f(x) = \dfrac{2 - x}{x}$, $g(x) = 3x + 2$

$f \circ g = f(g(x)) = f(3x + 2) = \dfrac{2 - (3x + 2)}{3x + 2} = \dfrac{-3x}{3x + 2}$

$g \circ f = g(f(x)) = g\left(\dfrac{2 - x}{x}\right) = 3\left(\dfrac{2 - x}{x}\right) + 2 = \dfrac{6 - 3x}{x} + \dfrac{2x}{x} = \dfrac{6 - x}{x}$

$(f \circ f) = f(f(x)) = f\left(\dfrac{2 - x}{x}\right) = \dfrac{2 - \left(\dfrac{2 - x}{x}\right)}{\dfrac{2 - x}{x}} = \dfrac{\dfrac{2x - 2 + x}{x}}{\dfrac{2 - x}{x}}$

$= \dfrac{\dfrac{3x - 2}{x}}{\dfrac{2 - x}{x}} = \dfrac{3x - 2}{x} \cdot \dfrac{x}{2 - x} = \dfrac{3x - 2}{2 - x}$

$(g \circ g) = g(g(x)) = g(3x + 2) = 3(3x + 2) + 2 = 9x + 6 + 2$
$= 9x + 8$

65. $f(x) = 3x^3 + x + 1$, $g(x) = |3x|$

$f \circ g = f(g(x)) = f(|3x|) = 3(|3x|)^2 + |3x| + 1$
$= 27x^2 + 3|x| + 1$

$g \circ f = g(f(x)) = g(3x^3 + x + 1) = |3(3x^2 + x + 1)|$
$= |9x^2 + 3x + 3|$
$= 3|3x^2 + x + 1|$

$(f \circ f) = f(f(x)) = f(3x^2 + x + 1)$
$= 3(3x^2 + x + 1)^2 + (3x^2 + x + 1) + 1$
$= 3(3x^2 + x + 1)^2 + 3x^2 + x + 2$

$(g \circ g) = g(g(x)) = g(|3x|) = |3|3x|| = 9|x|$

2 FUNCTIONS AND THEIR GRAPHS

67.　$f(x) = \dfrac{x + 1}{x - 1}, \; g(x) = \dfrac{1}{x}$

$$f \circ g = f(g(x)) = f\!\left(\dfrac{1}{x}\right) = \dfrac{\dfrac{1}{x} + 1}{\dfrac{1}{x} - 1} = \dfrac{\dfrac{1 + x}{x}}{\dfrac{1 - x}{x}} = \dfrac{1 + x}{x} \cdot \dfrac{x}{1 - x} = \dfrac{1 + x}{1 - x}$$

$$g \circ f = g(f(x)) = g\!\left(\dfrac{x + 1}{x - 1}\right) = \dfrac{1}{\dfrac{x + 1}{x - 1}} = \dfrac{x - 1}{x + 1}$$

$$(f \circ f) = f(f(x)) = f\!\left(\dfrac{x + 1}{x - 1}\right) = \dfrac{\dfrac{x + 1}{x - 1} + 1}{\dfrac{x + 1}{x - 1} - 1} = \dfrac{\dfrac{x + 1 + x - 1}{x - 1}}{\dfrac{x + 1 - (x - 1)}{x - 1}}$$

$$= \dfrac{\dfrac{2x}{x - 1}}{\dfrac{2}{x - 1}} = \dfrac{2x}{x - 1} \cdot \dfrac{x - 1}{2} = x$$

$$(g \circ g) = g(g(x)) = g\!\left(\dfrac{1}{x}\right) = \dfrac{1}{\dfrac{1}{x}} = x$$

69.　(a) $y = f(-x)$

Reflect about the
y-axis.

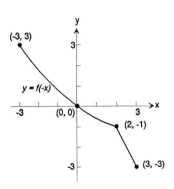

(b)　$y = -f(x)$

Reflect about the x-axis.

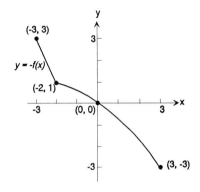

(c) $y = f(x + 2)$

    Shift left 2 units.

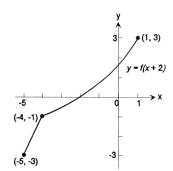

(d) $y = f(x) + 2$

    Shift up 2 units.

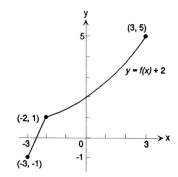

(e) $y = f(2 - x)$
    $y = f(-(x - 2))$

    Reflect about the
    y-axis and shift right
    2 units.

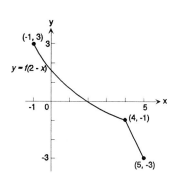

(f) $y = f^{-1}(x)$

    Reflect about the line
    $y = x$.

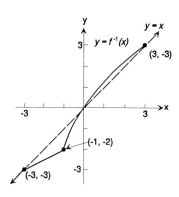

71. $T = T(h)$, $h = 0$ when $T = 30°$ and $h = 10,000$ when $T = 5°$

    $T(h) = mh + b$ (linear function)

$$T(0): \qquad\qquad m(0) + b = 30° \atop T(10,000): \quad m(10,000) + b = 5° \Bigg\}$$

$$b = 30°$$
$$10,000m + 30 = 5$$
$$10,000m = -25$$
$$m = \frac{-25}{10,000}$$
$$m = -0.0025$$

    Hence, $T(h) = -0.0025h + 30$

CHAPTER 3

POLYNOMIAL AND RATIONAL FUNCTIONS

≡ EXERCISE 3.1 QUADRATIC FUNCTIONS

1. D   3. A   5. B   7. E

In each of Problems 9-27, we start with the graph of $y = x^2$, and apply shifting, compressions and stretches and/or reflections.

9. The graph of $f(x) = \frac{1}{4}x^2$ is a vertically compressed version of the graph of $y = x^2$. For each x, the y-coordinate of a point on the graph of $f(x) = \frac{1}{4}x^2$ is $\frac{1}{4}$ times as large as the corresponding y-coordinate on the graph of $y = x^2$.

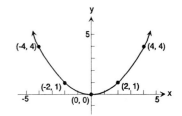

11. For $f(x) = \frac{1}{4}x^2 - 2$, we use the results of Problem 1, and then apply a vertical shift, moving the graph <u>down</u> 2 units.

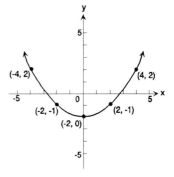

13. $f(x) = \frac{1}{4}x^2 + 2$. Use the results of Problem 1, followed by a vertical shift, 2 units <u>upward</u>.

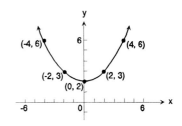

15. For $f(x) = x^2 + 1$, apply a vertical shift, one unit upward.

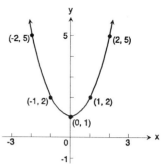

17. For $f(x) = -x^2 + 1$, first perform a reflection about the x-axis. This gives the graph of $f(x) = -x^2$. Then shift the graph vertically one unit upward.

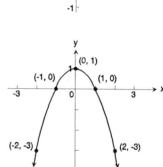

19. To graph $f(x) = \frac{1}{3}x^2 + 1$, apply a vertical compression of the graph of $y = x^2$, then shift upward one unit.

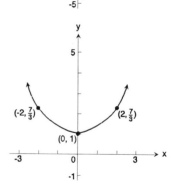

21. For $f(x) = x^2 + 4x + 2$, we must first complete the square in order to write the function in the form

$$f(x) = (x + h)^2 + k$$

We will then be able to perform horizontal and vertical shifts.

$$f(x) = x^2 + 4x + 2$$
$$= (x^2 + 4x + 4) + 2 - 4$$
$$= (x + 2)^2 - 2$$

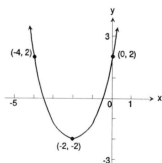

For $f(x) = (x + 2)^2$, shift the graph of $f(x) = x^2$ <u>horizontally</u> 2 units to the <u>left</u>. Then, to obtain $f(x) = (x + 2)^2 - 2$, shift the graph <u>down</u> 2 units.

136   3 POLYNOMIAL AND RATIONAL FUNCTIONS

23. By completing the square,
$f(x) = 2x^2 - 4x + 1$ can be put in the
form $f(x) = 2(x + h)^2 + k$:

$$f(x) = 2x^2 - 4x + 1$$
$$= 2(x^2 - 2x + \underline{\quad}) + 1 - \underline{\quad}$$
$$= 2(x^2 - 2x + \underline{1}) + 1 - 2$$

$$(2)(1) = 2$$
$$= 2(x - 1)^2 - 1$$

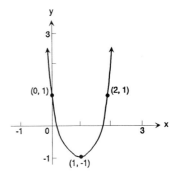

We can now perform the following steps:

(1) Starting graph: $f(x) = x^2$
(2) Vertical stretch of the graph of
$f(x) = x^2$: $f(x) = 2x^2$
(3) Shift to the <u>right</u> one unit: $f(x) = 2(x - 1)^2$
(4) Shift <u>down</u> one unit: $f(x) = 2(x - 1)^2 - 1$

The final graph is shown at right.

25. First complete the square:

$$f(x) = -x^2 - 2x$$
$$= -1(x^2 + 2x + \underline{\quad}) - \underline{\quad}$$
$$= -1(x^2 + 2x + \underline{1}) - (-\overline{1})$$

$$(-1)(1) = -1$$
$$= -1(x + 1)^2 + 1$$

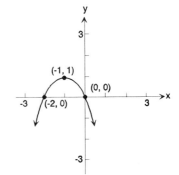

Now perform the following steps:
(1) Starting graph: $f(x) = x^2$
(2) Reflect about the x-axis: $f(x) = -x^2$
(3) Shift <u>left</u> one unit:
$f(x) = -(x + 1)^2$
(4) Shift <u>up</u> one unit:
$f(x) = -(x + 1)^2 + 1$

The final graph is shown at right.

27. Complete the square:

$$f(x) = \frac{1}{2}x^2 + x - 1$$
$$= \frac{1}{2}(x^2 + 2x) - 1$$
$$= \frac{1}{2}(x^2 + 2x + 1) - 1 - \frac{1}{2}$$

$$\left(\frac{1}{2}\right)(1) = \frac{1}{2}$$
$$= \frac{1}{2}(x + 1)^2 - \frac{3}{2}$$

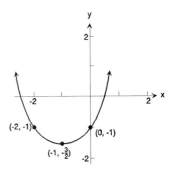

To graph $f(x) = \frac{1}{2}(x + 1)^2 - \frac{3}{2}$, apply a
vertical compression of the graph of $y = x^2$, shift to the <u>left</u> one
unit, and then shift down $\frac{3}{2}$ unit.

3.1 QUADRATIC FUNCTIONS

In Exercises 29-41, use the formulas in Display (2) to locate the vertex and axis. The vertex will be denoted (h, k).

29. $f(x) = x^2 + 2x - 3$. Here, $a = 1$,
$b = 2$, $c = -3$, so that
$h = \dfrac{-b}{2a} = \dfrac{-2}{2} = -1$, and $k = f(h) = f(-1)$
$= 1 - 2 - 3 = -4$. Therefore, the vertex
is $(h, k) = (-1, -4)$, the axis is the
line $x = h$ or $x = -1$, and the parabola
opens upward since $a = 1 > 0$.

To find the y-intercepts, set $x = 0$:
$y = f(0) = -3$. To find the
x-intercepts, set $y = 0$ and solve
for x:

$$y = x^2 + 2x - 3 = 0$$
$$(x + 3)(x - 1) = 0$$
$$x = -3 \quad \text{and} \quad x = 1$$

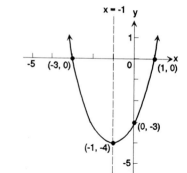

31. $f(x) = -x^2 - 3x + 4$
$(a = -1, \ b = -3, \ c = 4)$

$h = \dfrac{-b}{2a} = \dfrac{3}{-2} = \dfrac{-3}{2}$ and

$k = f(h) = f\left(-\dfrac{3}{2}\right) = -\left(\dfrac{9}{4}\right) - 3\left(-\dfrac{3}{2}\right) + 4 = \dfrac{25}{4}$

The parabola opens downward.

Set $x = 0$: $y = 4$ is the y-intercept.
Set $y = 0$: $y = -x^2 - 3x + 4 = 0$
$$x^2 + 3x - 4 = 0$$
$$(x + 4)(x - \ ) = 0$$
$$x = -4, \ x = 1 \ (x\text{-intercepts})$$

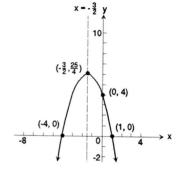

33. $f(x) = x^2 + 2x + 1$
$(a = 1, \ b = 2, \ c = 1)$

$h = \dfrac{-2}{2} = -1$ and $k = f(-1) = 0$.

Vertex at $(-1, 0)$

Set $x = 0$: $y = 1$
Set $y = 0$: $x^2 + 2x + 1 = 0$
$$(x + 1)(x + 1) = 0$$
$$x = -1$$

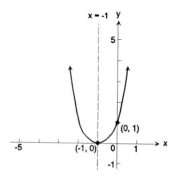

3 POLYNOMIAL AND RATIONAL FUNCTIONS

35. $f(x) = 2x^2 - x + 2$
$(a = 2, \ b = -1, \ c = 2)$

$h = \dfrac{-(-1)}{4} = \dfrac{1}{4}$ and $k = f\left(\dfrac{1}{4}\right) = \dfrac{15}{8}$

Vertex at $\left(\dfrac{1}{4}, \ \dfrac{15}{8}\right)$

Set $x = 0$: $y = 2$
Set $y = 0$: $2x - x + 2 = 0$

The discriminant of this quadratic is
$b^2 - 4ac = 1 - 16 < 0$, so there are no
solutions, i.e., no x-intercepts. The
parabola opens upward.

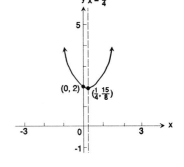

37. $f(x) = -2x^2 + 2x - 3$
Here, $a = -2, \ b = 2, \ c = -3$

$h = \dfrac{-2}{-4} = \dfrac{1}{2}$ and $k = f\left(\dfrac{1}{2}\right) = \dfrac{-5}{2}$,

so the vertex is at $\left(\dfrac{1}{2}, \ \dfrac{-5}{2}\right)$.

The vertex is below the x-axis, and the
parabola opens downward, since
$a = -2 < 0$. Therefore, there can be no
x-intercepts. To find the y-intercept,
set $x = 0$: $y = f(0) = -3$.

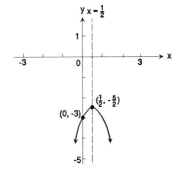

39. $f(x) = 3x^2 - 6x + 2$
$(a = 3, \ b = -6, \ c = 2)$

$h = \dfrac{6}{6} = 1$ and $k = f(1) = -1$, so the
vertex is at $(1, -1)$.

Set $x = 0$: $y = 2$ is the y-intercept.
Set $y = 0$: $3x^2 - 6x + 2 = 0$. Here we
must use the quadratic formula to solve
for x (the x-intercepts):

$x = \dfrac{-b \pm \sqrt{b^2 - 4ac}}{2a} = \dfrac{6 \pm \sqrt{36 - 4(6)}}{6}$

$= \dfrac{6 \pm \sqrt{12}}{6}$

or

$x = \dfrac{6 \pm 2\sqrt{3}}{6} = \dfrac{3 \pm \sqrt{3}}{3}, \ x \approx \dfrac{3 \pm 1.732}{3}$

Thus, the x-intercepts are at $x = 1.58$ and $x = 0.42$.

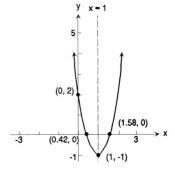

3.1 QUADRATIC FUNCTIONS   139

41. $f(x) = -4x^2 - 6x^2 + 2$
$(a = -4, b = -6, c = 2)$

Since $a = -4 < 0$, the parabola opens downward.

$h = \dfrac{6}{-8} = -\dfrac{3}{4}$ and $k = f\left(-\dfrac{3}{4}\right) = \dfrac{17}{4}$

so the vertex is at $\left(-\dfrac{3}{4}, \dfrac{17}{4}\right)$.

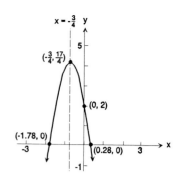

Set $x = 0$: $y = 2$
Set $y = 0$: $-4x^2 - 6x + 2 = 0$
or $-2x^2 - 3x + 1 = 0$.

By the quadratic formula, $x = \dfrac{3 \pm \sqrt{9 - 4(-2)}}{-4}$

$$x = \dfrac{3 \pm \sqrt{17}}{-4}$$

or $x = -1.78$ and $x = 0.28$

For Problems 43-47, if a parabola opens upward, $\smile$, then it has a minimum value, <u>at its vertex</u>. If it opens downward, $\frown$, then it has a maximum value, <u>at its vertex</u>. Thus, we must determine whether the parabola opens up or down (by looking at the coefficient of the x^2-term), and find the vertex of the parabola, using Display (2).

43. $f(x) = 6x^2 + 12x - 3$ $(a = 6, b = 12, c = -3)$

Since $a = 6 > 0$, the parabola opens <u>upward</u>, and therefore has a <u>minimum</u> at the vertex (h, k).

By (3), $h = \dfrac{-b}{2a} = \dfrac{-12}{12} = -1$, and $k = f(h) = f(-1)$ or $k = -9$.
Therefore the minimum value of $f(x)$ is $k = -9$.

45. $f(x) = -x^2 + 10x - 4$ $(a = -1, b = 10, c = -4)$

Here, $a = -1 < 0$, so the parabola opens downward and has a <u>maximum</u> at (h, k).

$h = \dfrac{-10}{-2} = 5$ and $k = f(5) = 21$. The maximum value of $f(x)$ is $k = 21$.

3 POLYNOMIAL AND RATIONAL FUNCTIONS

47. $f(x) = -3x^2 + 12x + 1$ $(a = -3, b = 12, c = 1)$

Here, $a = -3 < 0$, so the parabola opens downward and $f(x)$ has a maximum at (h, k).

$h = \dfrac{-12}{-6} = 2$ and $k = f(2) = 13$. So the maximum value of $f(x)$ is $k = 13$.

49. Graph $f(x) = x^2 + 2x + c$ for $c = -3, 0,$ and 1. In each case, $a = 1$, so the parabola opens upward.

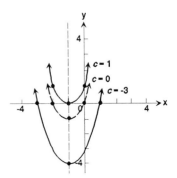

$h = \dfrac{-2}{2} = -1$, and
$k = f(-1) = 1 - 2 + c = c - 1$

Thus, the vertex is at the point $(-1, c - 1)$. To find the y-intercept, set $x = 0$: then $y = c$.

$c = -3$: Vertex at $(-1, c - 1)$
          $= (-1, -4)$
          y-intercept at $(0, -3)$
$c = 0$: Vertex at $(-1, -1)$
          y-intercept at $(0, 0)$
$c = 1$: Vertex at $(-1, 0)$
          y-intercept at $(0, 1)$

51. Each parabola opens up and passes through $(0, 1)$. Each one has the same shape.

53. We are given $R = -4p^2 + 400p$, which represents a parabola that opens downward (since $a = -4 < 0$). Thus, R will be a maximum at the vertex. Now, $h = \dfrac{-b}{2a} = \dfrac{-4000}{-8} = 500$, and $k = f(h) = f(500)$ $= -4(500)^2 + 4000(500) = 1,000,000$. The maximum revenue is $R = k = \$1,000,000$. It occurs when $p = h = 500$ dollars per dryer.

55. Let ℓ be the length and w be the width of the rectangle. Since the perimeter is 100 feet, we have $2\ell + 2w = 100$, so that $\ell = 50 - w$. We want to maximize the area, $A = \ell w = (50 - w)w = -w^2 + 50w$. Now $A = -w^2 + 50w$ is the equation of a parabola that opens downward, and hence has a maximum value w at $w = \dfrac{-b}{2a} = \dfrac{-50}{-2} = 25$, and the value is $f(25) = -(25)^2 + 50(25)$ $= 625$. The maximum area is 625 sq. ft., and this occurs when $w = 25$ and $\ell = 50 - w = 50 - 25 = 25$. Thus, the dimensions of the rectangle are 25 ft. by 25 ft.

57. The area of the rectangular plot is given by width times length, or $A = x(4000 - 2x)$. (Refer to the figure.) Thus, $A = -2x^2 + 4000x$; hence, A will be a maximum at the vertex.
$h = \dfrac{-4000}{-4} = 1000$, and $k = f(1000) = -2,000,000 + 4,000,000 = 2,000,000$. Thus, the largest area that can be enclosed is $A = k = 2,000,000$ square meters.

59. Consider the figure: The total length of
the fence is $x + x + x + y + y$, and this
must equal 30,000 meters: $3x + 2y =$
$30,000$; or $2y = 30,000 - 3x$ so that
$y = \dfrac{30,000 - 3x}{2}$. The total area enclosed

is $A = xy = x\left(\dfrac{30,000 - 3x}{2}\right)$, or $A = \dfrac{-3}{2}x^2$
$+ 15,000x$, which attains a <u>maximum</u> at the

vertex (since $a = \dfrac{-3}{2} < 0$). Now,

$h = \dfrac{-15,000}{-3} = 5,000$. Thus, the maximum

area is $A = k = f(5000) = 37,500,000$ square meters.

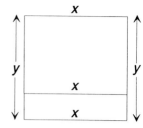

61. If the club has exactly 60 members, its revenue would be $60(200)$
$= 12,000$ dollars. Let x be the number in <u>excess</u> of 60. Then the
total membership is $60 + x$. Now the price for <u>every</u> member will
be <u>reduced</u> by $2 for each member in excess of 60, i.e., by $2x$.
Hence, the price per person will be $200 - 2x$. The total revenue
is equal to the number of members times the charge per member:
$R = (60 + x)(200 - 2x)$, or $R = -2x^2 + 80x + 12,000$. This
represents a parabola with a <u>maximum</u> at its vertex.
$h = \dfrac{-b}{2a} = \dfrac{-80}{-4} = 20$; $k = f(20) = -800 + 1600 + 12,000 = 12,800$.
The maximum possible revenue is $R = k = 12,800$, and this occurs
when $x = h = 20$, so that the number of members is $60 + x = 80$.

63. The situation at 4:00 P.M. is depicted below:

Let x denote the number of hours that
have elapsed since 4:00 P.M. In x
hours, the aircraft carrier (A) will
have travelled $10x$ nautical miles, and
the destroyer (D) will have gone $20x$
miles. The situation will then be as
indicated:

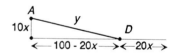

If we let y denote the distance from A
to D, then by the Pythagorean Theorem,

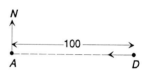

$y^2 = (10x)^2 + (100 - 20x)^2$
$\quad = 100x^2 + 10000 - 4000x + 400x^2$, or
$y^2 = 500x^2 - 4000x + 10000$

This equation is a parabola that opens upward, since $a = 500 > 0$,
so it will have a minimum value at its vertex, (h, k). Now
$h = \dfrac{-b}{2a} = \dfrac{4000}{1000} = 4$, and $k = f(h) = f(4) = 2000$. Thus, the
smallest possible value of y^2 is 2000, so the minimum value for

$y = \sqrt{2000} \approx 44.72$ nautical miles. But the question is to find
what time it is. The minimum distance y occurs when $x = h = 4$
hours, i.e., at 8:00 P.M.

65. For simplicity, locate the origin at the point where the cable touches the road:

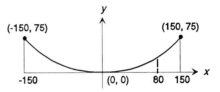

Then the equation of the parabola must be of the form:

$$y = ax^2, \text{ for some } a > 0$$

Since the point (150, 75) is on the parabola, we can determine the constant a:

$$75 = a(150)^2, \text{ or } a = \frac{75}{(150)^2} = .00333...$$

Then when $x = 80$, we have $y = ax^2$, or

$$y = (.00333)(80)^2 = 21.33... = 21\frac{1}{3} = \frac{64}{3} \text{ meters.}$$

67. The area A of the gutter of height x and length y is $A = xy$. Since $2x + y = 12$, we have $y = 12 - 2x$. We want to maximize the area $A = xy = x(12 - 2x) = -2x^2 + 12x$. This is the equation of a parabola that opens downward, and hence the maximum area occurs when $x = \frac{-b}{2a} = \frac{-12}{-4} = 3$, and the value is $f(3) = -2(3)^2 + 12(3) = 18$. Thus, a depth of 3 inches will provide maximum cross-sectional area.

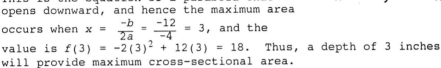

69. Since the diameter ($2r$) equals the width of the rectangle, then $w = 2r$. The perimeter is

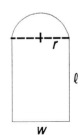

$$20 = 2\ell + w = \frac{1}{2}(2\pi r) \text{ or}$$
$$20 = 2\ell + 2r + \pi r \text{ so that}$$
$$2\ell = 20 - 2r - \pi r$$

$$A = \ell w + \frac{1}{2}(\pi r^2) = \ell(2r) + \frac{1}{2}(\pi r^2)$$

$$= r(20 - 2r - \pi r) + \frac{1}{2}\pi r^2$$

$$= \left(-2 - \frac{\pi}{2}\right)r^2 + 20r$$

The area is maximum at

$$r = \frac{-b}{2a} = \frac{-20}{-4 - \pi} \approx 2.80$$

Thus, $w = \frac{40}{\pi + 4} \approx 5.6$ ft., and

$$\ell = 10 - r - \frac{\pi}{2}r \approx 10 - 2.8 - \frac{\pi}{2}(2.8)$$
$$\approx 2.8 \text{ ft.}$$

3.1 QUADRATIC FUNCTIONS

143

71. If ℓ is the length of the rectangle, and x is the width, the perimeter of the window is $16 = 2\ell + 3x$ so that $\ell = 8 - \frac{3}{2}x$. The area of the window is

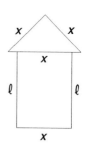

$$A = \ell x + \frac{\sqrt{3}}{4}x^2 \text{ or}$$

$$A = \left(8 - \frac{3}{2}x\right)x + \frac{\sqrt{3}}{4}x^2$$

$$A = \left(\frac{-3}{2} + \frac{\sqrt{3}}{4}\right)x^2 + 8x$$

The area is maximum at $x = \frac{-b}{2a} - \frac{-8}{-3 + \frac{\sqrt{3}}{2}} \approx \frac{16}{6 - \sqrt{3}} \approx 3.75$ ft. and

$$\ell = 8 - \frac{3}{2}x \approx 8 - \frac{3}{2}(3.75) \approx 2.38 \text{ ft.}$$

73. Since $f(x) = ax^2 + bx + c$ has vertex at $x = 0$, we have $\frac{-b}{2a}$ or $b = 0$. At $(0, 2)$, we have $2 = a(0)^2 + c$ or $c = 2$. At $(1, 8)$, we have $8 = a(1)^2 = 2$ or $a = 6$. Thus, $f(x) = 6x^2 + 2$.

75. We are given

$$V(x) = kx(a - x) = -kx^2 + kax$$

This is a maximum when

$$x = \frac{-ka}{-2k} = \frac{a}{2}$$

77. We have

$$ah^2 - bh + c = y_0$$

$$c = y_1$$

$$ah^2 + bh + c = y_2$$

so that

$$y_0 + y_2 = 2ah^2 + 2c$$

$$4y_1 = 4c$$

Hence,

$$\text{Area} = \frac{h}{3}(2ah^2 + 6c) = \frac{h}{3}(y_0 + 4y_1 + y_2)$$

≡ EXERCISE 3.2 POLYNOMIAL FUNCTIONS

1. $f(x) = 2x - x^3$ is a polynomial of degree 3.

3. $g(x) = \dfrac{1 - x^2}{2} = \dfrac{1}{2} - \dfrac{1}{2}x^2$ is a polynomial of degree 2.

5. $f(x) = 1 - \dfrac{1}{x} = 1 - x^{-1}$ is <u>not</u> a polynomial, since it contains x raised to a negative power.

7. $g(x) = x^{3/2} - x^2 + 2$ is not a polynomial since it contains x raised to a fractional power.

9. $F(x) = 5x^4 - \pi x^3 + \dfrac{1}{2}$ <u>is</u> a polynomial, of degree 4.

In Problems 11–17, start with the graph of $y = x^4$, and perform change of scale, shifting and reflection to obtain the graph of the given function.

11. For $f(x) = (x - 1)^4$, all that is needed is a horizontal shift, one unit to the right.

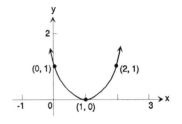

13. For $f(x) = \dfrac{1}{2}x^4$, vertically compress the graph of $y = x^4$. The graph will pass through $\left(1, \dfrac{1}{2}\right)$ instead of $(1, 1)$.

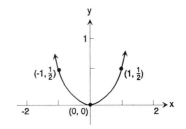

15. To graph $f(x) = 2(x + 1)^4 + 1$, vertically stretch, then shift to the <u>left</u> one unit, and finally shift <u>up</u> one unit.

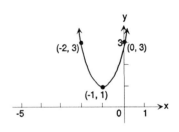

17. Perform the following steps:

(1) Starting graph: $y = x^4$

(2) Vertically compress: $y = \frac{1}{2}x^4$

(3) Reflect about the x-axis:
$$y = -\frac{1}{2}x^4$$

(4) Shift <u>right</u> 2 units:
$$y = -\frac{1}{2}(x - 2)^4$$

(5) Shift <u>down</u> one unit:
$$y = -\frac{1}{2}(x - 2)^4 - 1$$

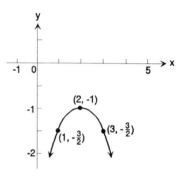

19. The zeros of $f(x) = 3(x - 4)(x + 5)^2$ are: 4, with multiplicity one; and −5, with multiplicity two. The graph touches the x-axis at −5 and crosses it at 4.

21. The zeros of $f(x) = 4(x^2 + 1)(x - 2)^3$ are: 2, with multiplicity three. Note: $x^2 + 1 = 0$ has no real solution. The graph crosses the x-axis at 2.

23. The zeros of $f(x) = -2\left(x + \frac{1}{2}\right)(x^2 + 4)^2$ are $\frac{-1}{2}$, with multiplicity two (and $x^2 + 4 = 0$ has no real solutions). The graph touches the x-axis at $\frac{-1}{2}$.

25. The zeros of $f(x) = (x - 5)^3(x + 4)^2$ are: 5, with multiplicity three; and −4, with multiplicity two. The graph touches the x-axis at −4 and crosses it at 5.

27. $f(x) = 3(x^2 + 4)(x^2 + 9)^2$ has no real zeros. The graph neither touches nor crosses the x-axis.

29. $f(x) = (x - 1)^2$

(a) x-intercept: 1; y-intercept: 1
(b) touches at 1
(c) $y = x^2$
(d) 1
(e)

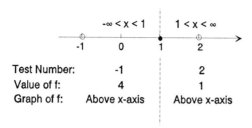

(f)

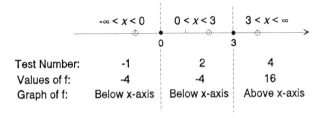

31. $f(x) = x^2(x - 1)$

   (a) x-intercepts: 0, 3; y-intercept: 0
   (b) touches at 0; crosses at 3
   (c) $y = x^3$
   (d) 2
   (e)

	$-\infty < x < 0$	$0 < x < 3$	$3 < x < \infty$
Test Number:	-1	2	4
Values of f:	-4	-4	16
Graph of f:	Below x-axis	Below x-axis	Above x-axis

   (f)

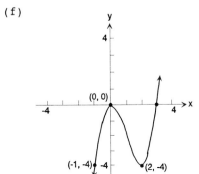

33. $f(x) = 6x^3(x + 4)$

   (a) x-intercepts: -4, 0; y-intercept: 0
   (b) crosses at -4 and 0
   (c) $y = 6x^4$
   (d) 3

(e)

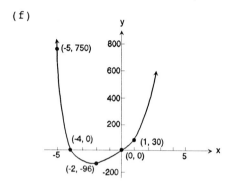

	$-\infty < x < -4$	$-4 < x < 0$	$0 < x < \infty$
Test Number:	-5	-2	1
Values of f:	750	-96	30
Graph of f:	Above x-axis	Below x-axis	Above x-axis

(f)

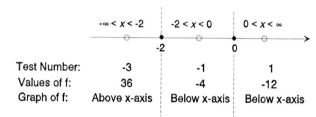

35. $f(x) = -4x^2(x + 2)$

(a) x-intercepts: -2, 0; y-intercept: 0
(b) crosses at -2; touches at 0
(c) $y = -4x^3$
(d) 2
(e)

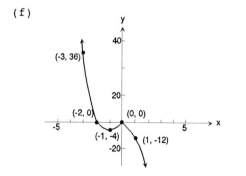

	$-\infty < x < -2$	$-2 < x < 0$	$0 < x < \infty$
Test Number:	-3	-1	1
Values of f:	36	-4	-12
Graph of f:	Above x-axis	Below x-axis	Below x-axis

(f)

3 POLYNOMIAL AND RATIONAL FUNCTIONS

37. $f(x) = x(x - 2)(x + 4)$

(a) x-intercepts: -4, 0, 2; y-intercept: 0
(b) crosses at -4, 0, and 2
(c) $y = x^3$
(d) 2
(e)

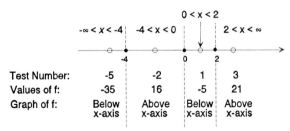

	$-\infty < x < -4$	$-4 < x < 0$	$0 < x < 2$	$2 < x < \infty$
Test Number:	-5	-2	1	3
Values of f:	-35	16	-5	21
Graph of f:	Below x-axis	Above x-axis	Below x-axis	Above x-axis

(f)

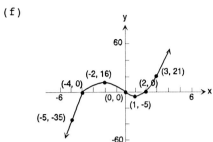

39. $f(x) = 4x - x^3 = -x(x^2 - 4) = -x(x + 2)(x - 2)$

(a) x-intercepts: -2, 0, 2; y-intercept: 0
(b) crosses at -2, 0, and 2
(c) $y = -x^3$
(d) 2
(e)

	$-\infty < x < -2$	$-2 < x < 0$	$0 < x < 2$	$2 < x < \infty$
Test Number:	-3	-1	1	3
Values of f:	15	-3	3	-15
Graph of f:	Above x-axis	Below x-axis	Above x-axis	Below x-axis

(f)

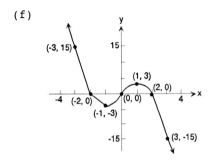

3.2 POLYNOMIAL FUNCTIONS   149

41. $f(x) = x^2(x - 2)(x + 2)$

(a) x-intercepts: $-2, 0, 2$; y-intercept: 0
(b) crosses at -2 and 2; touches at 0
(c) $y = x^4$
(d) 3
(e)

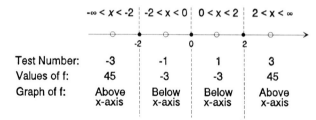

	$-\infty < x < -2$	$-2 < x < 0$	$0 < x < 2$	$2 < x < \infty$
Test Number:	-3	-1	1	3
Values of f:	45	-3	-3	45
Graph of f:	Above x-axis	Below x-axis	Below x-axis	Above x-axis

(f)

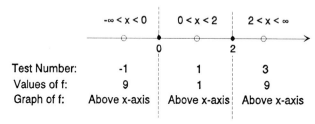

43. $f(x) = x^2(x - 2)^2$

(a) x-intercepts: $0, 2$; y-intercept: 0
(b) touches at 0 and 2
(c) $y = x^4$
(d) 3
(e)

	$-\infty < x < 0$	$0 < x < 2$	$2 < x < \infty$
Test Number:	-1	1	3
Values of f:	9	1	9
Graph of f:	Above x-axis	Above x-axis	Above x-axis

(f)

3 POLYNOMIAL AND RATIONAL FUNCTIONS

45. $f(x) = x^2(x - 3)(x + 1)$

(a) x-intercepts: -1, 0, 3; y-intercept: 0
(b) crosses at -1 and 3; touches at 0
(c) $y = x^4$
(d) 3
(e)

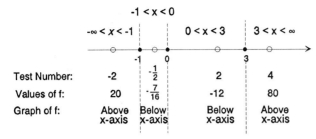

Test Number:	-2	$-\frac{1}{2}$	2	4
Values of f:	20	$-\frac{7}{16}$	-12	80
Graph of f:	Above x-axis	Below x-axis	Below x-axis	Above x-axis

(f)

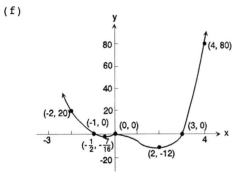

47. $f(x) = x(x + 2)(x - 4)(x - 6)$

(a) x-intercepts: -2, 0, 4, 6; y-intercept: 0
(b) crosses at -2, 0, 4, and 6
(c) $y = x^4$
(d) 3
(e)

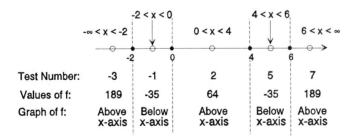

Test Number:	-3	-1	2	5	7
Values of f:	189	-35	64	-35	189
Graph of f:	Above x-axis	Below x-axis	Above x-axis	Below x-axis	Above x-axis

3.2 POLYNOMIAL FUNCTIONS

151

(f)

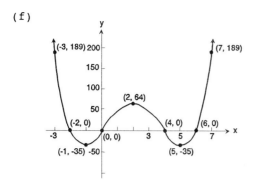

49. $f(x) = x^2(x - 2)(x^2 + 3)$

   (a) x-intercepts: 0, 2; y-intercept: 0
   (b) touches at 0; crosses at 2;
   (c) $y = x^5$
   (d) 4
   (e)

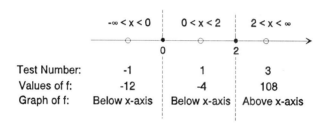

	$-\infty < x < 0$	$0 < x < 2$	$2 < x < \infty$
Test Number:	-1	1	3
Values of f:	-12	-4	108
Graph of f:	Below x-axis	Below x-axis	Above x-axis

   (f)

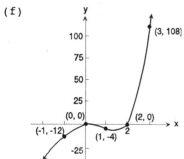

51. c, e, f

53. Each one resembles $y = x^3$ for large values of x. Each has two x-intercepts with each one touching the x-axis at 0 and crossing the x-axis at the other x-intercept.

55. Each one resembles $y = x^3$ for large values of x. Each has three x-intercepts, one being 0, and the graph crosses the x-axis at each one.

57. Each one resembles $y = x^4$ for large values of x. Each has two x-intercepts, one being 0. Each graph crosses the x-axis at the x-intercepts.

59. Each graph has one x-intercept, 0, at which the graph crosses the x-axis.

≡ EXERCISE 3.3 RATIONAL FUNCTIONS

In Problems 1-10, the domain consists of all real numbers except those for which the denominator, q(x), is zero.

1. $R(x) = \dfrac{x}{x - 2}$. Here the denominator is $q(x) = x - 2$, which has 2 as its only zero. Thus, the domain of $R(x)$ consists of all real numbers except 2.

3. For $H(x) = \dfrac{-4x^2}{(x - 2)(x + 1)}$, the denominator, $q(x) = (x - 2)(x + 1)$, has zeros at −1 and 2. Thus, the domain of $H(x)$ is all real numbers except −1 and 2.

5. In $F(x) = \dfrac{3x(x - 1)}{2x^2 - 5x - 3}$, the denominator is $q(x) = 2x^2 - 5x - 3 = (2x + 1)(x - 3)$, whose zeros are $\dfrac{-1}{2}$ and 3. Thus, the domain of $F(x)$ consists of all real numbers except $\dfrac{-1}{2}$ and 3.

7. For $R(x) = \dfrac{x}{x^3 - 8}$, the denominator is $q(x) = x^3 - 8$, which can be factored as a difference of cubes:

$$q(x) = x^3 - 8 = (x - 2)(x^2 + 2x + 4)$$

Now, $x^2 + 2x + 4$ has no real zeros since its discriminant is $b^2 - 4ac = 4 - 16 = -12 < 0$. Thus, the domain of $R(x)$ is all real numbers except 2.

9. In $H(x) = \dfrac{3x^2 + x}{x^2 + 4}$, the denominator has no real zeros, so the domain is <u>all</u> real numbers.

11. (a) Domain: $\{x \mid x \neq 2\}$; Range: $\{y \mid y \neq 1\}$
    (b) $(0, 0)$
    (c) $y = 1$
    (d) $x = 2$
    (e) none

13. (a) Domain: $\{x \mid x \neq 0\}$; Range: all real numbers
    (b) $(-1, 0)$, $(1, 0)$
    (c) none
    (d) none
    (e) $y = 2x$

15. (a) Domain: $\{x \mid x \neq -2, x \neq 2\}$;
        Range: $\{y \mid -\infty < y \leq 0 \text{ or } 1 < y < \infty\}$
    (b) $(0, 0)$
    (c) $y = 1$
    (d) $x = -2$, $x = 2$
    (e) none

Problems 17-25 are all based on the graphs of $R(x) = \frac{1}{x}$ and $H(x) = \frac{1}{x^2}$, shown below.

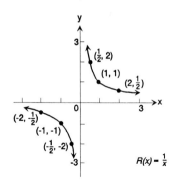

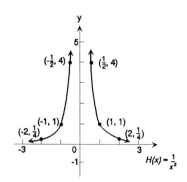

17. $R(x) = \dfrac{1}{(x - 1)^2}$. This graph can be obtained by shifting the graph of $H(x) = \dfrac{1}{x^2}$ horizontally, one unit to the right.

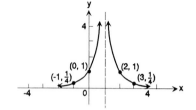

19. $H(x) = \dfrac{-2}{x + 1}$

This can be graphed in stages:

(1) Starting graph: $R(x) = \dfrac{1}{x}$

(2) Vertically stretch $y = \dfrac{2}{x}$

(3) Reflect about the x-axis: $y = \dfrac{-2}{x}$

(4) Shift one unit to the left: $y = \dfrac{-2}{x + 1}$

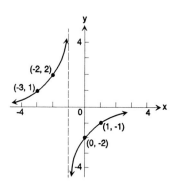

21. $R(x) = \dfrac{1}{x^2 + 4x + 4} = \dfrac{1}{(x + 2)^2}$

To obtain this graph, shift the graph of $H(x) = \dfrac{1}{x^2}$ two units to the left:

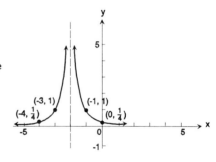

23. $F(x) = 1 - \dfrac{1}{x} = -\dfrac{1}{x} + 1$

Start with the graph of $R(x) = \dfrac{1}{x}$, do a reflection about the x-axis, and then shift <u>upward</u> one unit:

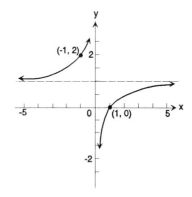

25. $R(x) = \dfrac{x^2 - 4}{x^2} = \dfrac{x^2}{x^2} - \dfrac{4}{x^2} = \dfrac{-4}{x^2} + 1$

(1) Start with $H(x) = \dfrac{1}{x^2}$

(2) Vertically stretch: $y = \dfrac{4}{x^2}$

(3) Reflect about the x-axis:
$y = \dfrac{-4}{x^2}$

(4) Shift up one unit: $y = \dfrac{-4}{x^2} + 1$

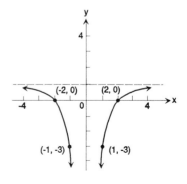

For Problems 27-35, refer to the summary concerning asymptotes.

27. $R(x) = \dfrac{x}{x + 1}$

The degree of the numerator, $p(x) = x$, is $n = 1$.

The degree of the denominator, $q(x) = x + 1$ is $m = 1$. Since $n = m$, the line $y = \dfrac{1}{1} = 1$ is a horizontal asymptote. The denominator is zero at $x = -1$, so $x = -1$ is a vertical asymptote.

3.3 RATIONAL FUNCTIONS   155

29. $H(x) = \dfrac{x^4 + 2x^2 + 1}{3x^2 - x + 1}$

$p(x) = x^4 + 2x^2 + 1$; degree $= n = 4$
$q(x) = 3x^2 - x + 1$; degree $= m = 2$

Since $n > m + 1$, $H(x)$ has no horizontal nor oblique asymptote. Since $q(x)$ has no real zeros, there is no vertical asymptote.

31. $T(x) = \dfrac{x^3}{x^4 - 1}$

$p(x) = x^3$; degree $= n = 3$
$q(x) = x^4 - 1$; degree $= m = 4$

Since $n < m$, the line $y = 0$ is a horizontal asymptote. We can factor $q(x) = x^4 - 1 = (x^2 - 1)(x^2 + 1) = (x - 1)(x + 1)(x^2 + 1)$, so the vertical asymptotes are the lines $x = -1$ and $x = 1$.

33. $Q(x) = \dfrac{5 - x^2}{3x^4}$

$p(x) = 5 - x^2$; $n = 2$
$q(x) = 3x^4$; $m = 4$

Since $n < m$, the line $y = 0$ is a horizontal asymptote. The vertical asymptote is $x = 0$.

35. $R(x) = \dfrac{3x^4 + 1}{x^3 + 5x}$

Since $n = m + 1$, we have an oblique asymptote, so it is necessary to perform long division:

$$
\begin{array}{r}
3x \\
x^3 + 5x \overline{)\, 3x^4 + 0x^3 + 0x^2 + 0x + 1} \\
\underline{3x^4 + 15x^2 } \\
-15x^2 + 0x + 1 \quad \leftarrow \text{Remainder}
\end{array}
$$

Thus, $R(x) = \dfrac{3x^4 + 1}{x^3 + 5x} = 3x + \dfrac{-15x^2 + 1}{x^3 + 5x}$

Therefore, $y = 3x$ is an oblique asymptote for $R(x)$.

We can factor the denominator: $q(x) = x^3 + 5x = x\underbrace{(x^2 + 5)}_{\text{no real zeros}}$.

The vertical asymptote is $x = 0$.

In Problems 37-55, we will use the following terminology: $R(x) = \dfrac{p(x)}{q(x)}$, where the degree of $p(x) = n$ and the degree of $q(x) = m$. In every problem, we will follow the steps listed below:

Step 0: *Preparation:*

　　　　(a) If $n = m + 1$, perform long division to determine the oblique asymptote.
　　　　(b) Factor both $p(x)$ and $q(x)$ over the reals, in order to determine their zeros.

Step 1: *Locate Intercepts:*

　　　　(a) For x-intercepts, find the zeros of $p(x)$.
　　　　(b) For the y-intercepts, evaluate $R(0)$.

Step 2: *Test for Symmetry.*

Step 3: *Locate Vertical Asymptotes:*
　　　　Find the zeros of $q(x)$.

Step 4: *Locate Horizontal and Oblique Asymptotes.*

Step 5: *Determine where the Graph is above or below the x-axis:*
　　　　Use all the zeros of $p(x)$ and $q(x)$.

Step 6: *Sketch the Graph.*

37.　　$R(x) = \dfrac{x + 1}{x(x + 4)}$　$p(x) = x + 1;\ q(x) = x(x + 4) = x^2 + 4x$
　　　　　　　　　　　　　$n = 1,\ m = 2$

　　Step 0: This is already done, with
　　　　　　$p(x) = x + 1$
　　　　　　$q(x) = x(x + 4)$

　　Step 1: (a) The x-intercept is the zero of $p(x)$: -1
　　　　　　(b) For y-intercept, $R(0)$ is not defined, since $q(0) = 0$, so there is no y-intercept.

　　Step 2: $R(-x) = \dfrac{-x + 1}{x^2 - 4x}$; this is neither $R(x)$ nor $-R(x)$, so there is no symmetry.

　　Step 3: The vertical asymptotes are the zeros of $q(x)$:　$x = -4$ and $x = 0$.

　　Step 4: Since $n < m$, the line $y = 0$ is the horizontal asymptote; intersected at $(-1, 0)$.

3.3 RATIONAL FUNCTIONS　　　　　　　　　　　　　　　157

Interval	Test Number	$R(x)$	Graph of R
$x < -4$	$x = -5$	$\dfrac{-4}{-5(-1)} = \dfrac{-4}{5}$	Below the x-axis
$-4 < x < -1$	$x = -2$	$\dfrac{-1}{-2(2)} = \dfrac{1}{4}$	Above the x-axis
$-1 < x < 0$	$x = -\dfrac{1}{2}$	$\dfrac{\frac{1}{2}}{-\frac{1}{2}\left(\frac{7}{2}\right)} = \dfrac{-2}{7}$	Below the x-axis
$x > 0$	$x = 1$	$\dfrac{2}{1(5)} = \dfrac{2}{5}$	Above the x-axis

Step 6:

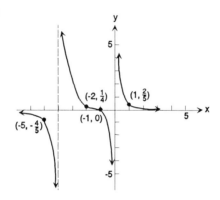

39. $R(x) = \dfrac{3x + 3}{2x + 4}$

Step 0: $p(x) = 3x + 3;\ n = 1$
$q(x) = 2x + 4;\ m = 1$

Step 1: (a) The x-intercept is the zero of $p(x)$: -1
(b) For y-intercept is $y = R(0) = \dfrac{3}{4}$

Step 2: No symmetry.

Step 3: The vertical asymptote is the zero of $q(x)$: $x = -2$.

Step 4: Since $n = m$, the horizontal asymptote is the line $y = \dfrac{3}{2}$; not intersected.

Step 5: We have a total of two zeros, -2, -1:

Interval	Test Number	$R(x)$	Graph of R
$x < -2$	$x = -3$	$\dfrac{-9 + 3}{-6 + 4} = 3$	Above the x-axis
$-2 < x < -1$	$x = \dfrac{-3}{2}$	$\dfrac{\frac{-9}{2} + 3}{-3 + 4} = \dfrac{-3}{2}$	Below the x-axis
$x > -1$	$x = 0$	$\dfrac{3}{4}$	Above the x-axis

Step 6:

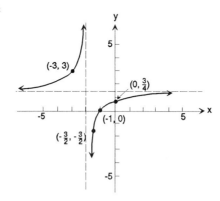

41. $R(x) = \dfrac{3}{x^2 - 4}$

Step 0: $p(x) = 3; \ n = 0$
$q(x) = x^2 - 4 = (x + 2)(x - 2); \ m = 2$

Step 1: (a) There are no x-intercepts, since $p(x)$ has no zeros.
(b) The y-intercept is $y = R(0) = \dfrac{-3}{4}$

Step 2: $R(-x) = \dfrac{3}{x^2 - 4} = R(x)$

Therefore, the graph of $R(x)$ is symmetric with respect to the y-axis.

Step 3: The vertical asymptotes are the zeros of $q(x)$: $x = -2$ and $x = 2$.

Step 4: Since $n < m$, the line $y = 0$ is a horizontal asymptote; not intersected.

We have a total of two zeros, -2, 2:

Interval	Test Number	$R(x)$	Graph of R
$x < -2$	$x = -3$	$\dfrac{3}{9 - 4} = \dfrac{3}{5}$	Above the x-axis
$-2 < x < 2$	$x = 0$	$\dfrac{-3}{4}$	Below the x-axis
$x > 2$	$x = 3$	$\dfrac{3}{9 - 4} = \dfrac{3}{5}$	Above the x-axis

Step 6:

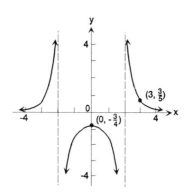

43. $P(x) = \dfrac{x^4 + x^2 + 1}{x^2 - 1}$

Step 0: $p(x) = x^4 + x^2 + 1$
$q(x) = x^2 - 1 = (x + 1)(x - 1)$

Step 1: (a) There are no x-intercepts, since $p(x)$ has no real zeros.
(b) The y-intercept is $y = P(0) = -1$.

Step 2: $P(-x) = \dfrac{x^4 + x^2 + 1}{x^2 - 1} = P(x)$, so we do have symmetry with respect to the y-axis.

Step 3: The vertical asymptotes are $x = -1$, $x = 1$.

Step 4: Since $n > m + 1$, we have no horizontal and no oblique asymptote.

Step 5: We have two zeros, -1 and 1; due to symmetry, we only need to check for $x > 0$:

Interval	Test Number	$P(x)$	Graph of P
$0 < x < 1$	$x = \frac{1}{2}$	$\dfrac{\frac{1}{16} + \frac{1}{4} + 1}{\frac{1}{4} - 1} = \dfrac{-7}{4}$	Below the x-axis
$x > 1$	$x = 2$	$\dfrac{16 + 4 + 1}{4 - 1} = 7$	Above the x-axis

<u>Step 6</u>:

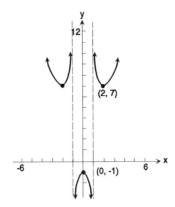

45. $H(x) = \dfrac{x^3 - 1}{x^2 - 9}$

<u>Step 0</u>: (a) $n = m + 1$, so we do long division:

$$x^2 - 9 \overline{\smash{\big)}\ \begin{array}{r} x \\ x^3 + 0x^2 + 0x - 1 \\ \underline{x^3 - 9x } \\ 9x - 1 \end{array}}$$

So $H(x) = x + \dfrac{9x - 1}{x^3 - 9}$, and we have an oblique asymptote, $y = x$.

(b) $p(x) = x^2 - 1 = (x - 1)\underbrace{(x^2 + x + 1)}$

 No real zeros

$q(x) = x^2 - 9 = (x + 3)(x - 3)$

<u>Step 1</u>: (a) The x-intercept is $x = 1$.

(b) The y-intercept is $y = H(0) = \dfrac{1}{9}$

<u>Step 2</u>: No symmetry.

<u>Step 3</u>: The vertical asymptotes are $x = -3$, $x = 3$.

Step 4: Since $n = m + 1$, we have only the oblique asymptote, $y = x$, found above; intersected at $\left(\dfrac{1}{9}, \dfrac{1}{9}\right)$.

Step 5: We have a total of three zeros, -3, 1, 3:

Interval	Test Number	$H(x)$	Graph of H
$x < -3$	$x = -5$	$\dfrac{-125 - 1}{25 - 9} = \dfrac{-63}{8}$	Below the x-axis
$-3 < x < 1$	$x = 0$	$\dfrac{1}{9}$	Above the x-axis
$1 < x < 3$	$x = 2$	$\dfrac{8 - 1}{4 - 9} = -\dfrac{7}{5}$	Below the x-axis
$x > 3$	$x = 5$	$\dfrac{125 - 1}{25 - 9} = \dfrac{31}{4}$	Above the x-axis

Step 6:

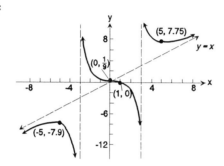

47. $R(x) = \dfrac{x^2}{x^2 + x - 6}$

Step 0: $p(x) = x^2$; $n = 2$
$q(x) = x^2 + x - 6 = (x + 3)(x - 2)$; $m = 2$

Step 1: (a) The x-intercept is $x = 0$
(b) The y-intercept is $y = R(0) = \dfrac{0}{-6} = 0$

The intercept is $(0, 0)$.

Step 2: No symmetry.

Step 3: Vertical asymptotes $x = -3$, $x = 2$.

Step 4: Since $n = m$, the line $\dfrac{1}{1} = 1$ is the horizontal asymptote; intersected at $(6, 1)$.

162   3 POLYNOMIAL AND RATIONAL FUNCTIONS

Interval	Test Number	$R(x)$	Graph of R
$x < -3$	$x = -6$	$\dfrac{36}{36 - 6 - 6} = \dfrac{3}{2}$	Above the x-axis
$-3 < x < 0$	$x = -2$	$\dfrac{4}{4 - 2 - 6} = -1$	Below the x-axis
$0 < x < 2$	$x = 1$	$\dfrac{1}{1 + 1 - 6} = -\dfrac{1}{4}$	Below the x-axis
$x > 2$	$x = 3$	$\dfrac{9}{9 + 3 - 6} = \dfrac{3}{2}$	Above the x-axis

Step 6:

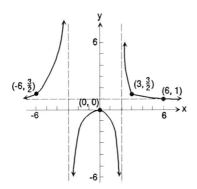

49. $G(x) = \dfrac{x}{x^2 - 4}$

Step 0: $p(x) = x;\ n = 1$
$q(x) = x^2 - 4 = (x + 2)(x - 2);\ m = 2$

Step 1: (a) The x-intercept is $x = 0$
(b) The y-intercept is $y = G(0) = 0$

The intercept is $(0, 0)$.

Step 2: $G(-x) = \dfrac{-x}{x^2 - 4} = -G(x)$
$G(x)$ is symmetric about the origin.

Step 3: The vertical asymptotes are $x = -2$, $x = 2$.

Step 4: The horizontal asymptote is $y = 0$; intersected at $(0, 0)$.

3.3 RATIONAL FUNCTIONS   163

<u>Step 5</u>: Zeros, -2, 0, 2; because of symmetry, we only check $x > 0$:

Interval	Test Number	$G(x)$	Graph of G
$0 < x < 2$	$x = 1$	$\dfrac{1}{1 - 4} = -\dfrac{1}{3}$	Below the x-axis
$x > 2$	$x = 3$	$\dfrac{3}{9 - 4} = \dfrac{3}{5}$	Above the x-axis

<u>Step 6</u>:

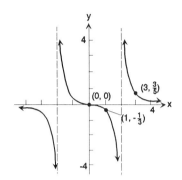

51. $R(x) = \dfrac{3}{(x - 1)(x^2 - 4)}$

<u>Step 0</u>: $p(x) = 3;\ n = 0$
$q(x) = (x - 1)(x^2 - 4) = (x - 1)(x + 2)(x - 2);\ m = 3$

<u>Step 1</u>: (a) No x-intercept.

(b) y-intercept is $y = R(0) = \dfrac{3}{4}$

<u>Step 2</u>: No symmetry.

<u>Step 3</u>: Vertical asymptotes: $x = -2,\ x = 1,\ x = 2$

<u>Step 4</u>: Horizontal asymptote: $y = 0$; not intersected.

<u>Step 5</u>:

Interval	Test Number	$R(x)$	Graph of R
$x < -2$	$x = -3$	$\dfrac{3}{(-4)(5)} = \dfrac{-3}{20}$	Below the x-axis
$-2 < x < 1$	$x = 0$	$\dfrac{3}{4}$	Above the x-axis
$1 < x < 2$	$x = \dfrac{3}{2}$	$\dfrac{3}{\left(\dfrac{1}{2}\right)\left(\dfrac{-7}{4}\right)} = \dfrac{-24}{7}$	Below the x-axis
$x > 2$	$x = 3$	$\dfrac{3}{(2)(5)} = \dfrac{3}{10}$	Above the x-axis

3 POLYNOMIAL AND RATIONAL FUNCTIONS

Step 6:

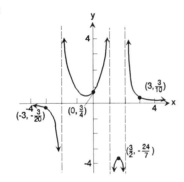

$(3, \frac{3}{10})$

$(-3, -\frac{3}{20})$

$(0, \frac{3}{4})$

$(\frac{3}{2}, -\frac{24}{7})$

53. $H(x) = \dfrac{4(x^2 - 1)}{x^4 - 16}$

Step 0: $p(x) = 4(x^2 - 1) = 4(x + 1)(x - 1); \ n = 2$
$q(x) = x^4 - 16 = (x^2 + 4)(x^2 - 4)$
$\qquad = (x^2 + 4)(x + 2)(x - 2); \ m = 4$

Step 1: (a) The x-intercepts are -1 and 1.
(b) The y-intercept is $y = H(0) = \dfrac{1}{4}$

Step 2: $H(-x) = \dfrac{4(x^2 - 1}{x^4 - 16} = H(x)$, so the graph is symmetric with
respect to the y-axis.

Step 3: The vertical asymptote are: $x = -2$, $x = 2$.

Step 4: The horizontal asymptote is $y = 0$; intersected at $(-1, 0)$
and $(1, 0)$.

Step 5: The zeros are -2, -1, 1, 2; we check the portion $x \geq 0$:

Interval	Test Number	$H(x)$	Graph of H
$-1 < x < 1$	$x = 0$	$\dfrac{1}{4}$	Above the x-axis
$1 < x < 2$	$x = \dfrac{3}{2}$	$\dfrac{4\left(\dfrac{9}{4} - 1\right)}{\dfrac{81}{16} - 16} = \dfrac{-16}{35}$	Below the x-axis
$x > 2$	$x = 3$	$\dfrac{4(9 - 1)}{81 - 16} \approx 0.49$	Above the x-axis

<u>Step 6</u>:

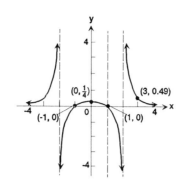

55. $F(x) = \dfrac{x^2 - 3x - 4}{x + 2}$

<u>Step 0</u>: (a) Since $n = m + 1$, perform division:

$$
\begin{array}{r}
x - 5 \\
x + 2 {\overline{\smash{\big)}\,x^2 - 3x - 4}} \\
\underline{x^2 + 2x} \\
-5x - 4 \\
\underline{-5x - 10} \\
6
\end{array}
$$

Therefore, $F(x) = x - 5 + \dfrac{6}{x + 2}$, so we have an oblique asymptote, $y = x - 5$.

(b) $p(x) = x^2 - 3x - 4 = (x + 1)(x - 4)$
$q(x) = x + 2$

<u>Step 1</u>: (a) The x-intercepts are -1, 4.
(b) The y-intercept is -2

<u>Step 2</u>: No symmetry.

<u>Step 3</u>: The vertical asymptote is $x = -2$.

<u>Step 4</u>: The oblique asymptote is $y = x - 5$; not intersected.

<u>Step 5</u>: The zeros are -2, -1, and 4.

3 POLYNOMIAL AND RATIONAL FUNCTIONS

Interval	Test Number	$F(x)$	Graph of F
$x < -2$	$x = -3$	$\dfrac{9 + 9 - 4}{-3 + 2} = -14$	Below the x-axis
$-2 < x < -1$	$x = -\dfrac{3}{2}$	$\dfrac{\dfrac{9}{4} + \dfrac{9}{2} - 4}{-\dfrac{3}{2} + 2} = \dfrac{11}{2}$	Above the x-axis
$-1 < x < 4$	$x = 0$	-2	Below the x-axis
$x > 4$	$x = 5$	$\dfrac{25 - 15 - 4}{5 + 2} = \dfrac{6}{7}$	Above the x-axis

Step 6:

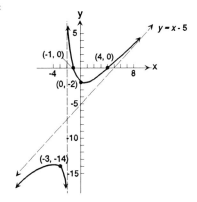

57. $R(x) = \dfrac{x^2 + x - 12}{x - 4}$

Step 0: (a) $n = m + 1$, so we do long division:

$$
\begin{array}{r}
x + 5 \\
x + 2 \overline{)\,x^2 + x - 12} \\
\underline{x^2 - 4x} \\
5x - 12 \\
\underline{5x - 20} \\
8
\end{array}
$$

So $R(x) = x + 5 + \dfrac{8}{x - 4}$, and we have an oblique asymptote, $y = x + 5$

(b) $p(x) = x^2 + x - 12 = (x + 4)(x - 3)$
$q(x) = x - 4$

Step 1: (a) The x-intercepts are -4 and 3
(b) The y-intercept is $y = R(0) = 3$

3.3 RATIONAL FUNCTIONS

167

Step 2: No symmetry.

Step 3: The vertical asymptote is $x = 4$.

Step 4: Since $n = m + 1$, we have only the oblique asymptote, $y = x + 5$, found above; not intersected.

Step 5: The zeros are -4, 3, and 4.

Interval	Test Number	$R(x)$	Graph of R
$x < -4$	-5	$\dfrac{-8}{9}$	Below x-axis
$-4 < x < 3$	1	$\dfrac{10}{3}$	Above x-axis
$3 < x < 4$	$\dfrac{7}{2}$	$\dfrac{-15}{2}$	Below x-axis
$x > 4$	5	18	Above x-axis

Step 6:

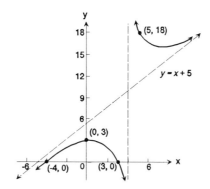

59. $F(x) = \dfrac{x^2 + x - 12}{x + 2}$

Step 0: (a) $n = m + 1$, so we do long division:

$$
\begin{array}{r}
x - 1 \\
x + 2 \overline{\smash{)}\, x^2 + x - 12} \\
\underline{x^2 + 2x} \\
-x - 12 \\
\underline{-x - 2} \\
-10
\end{array}
$$

So $F(x) = x - 1 + \dfrac{-10}{x + 2}$, and we have an oblique asymptote, $y = x - 1$

(b) $p(x) = x^2 + x - 12 = (x + 4)(x - 3)$
$q(x) = x + 2$

Step 1: (a) The x-intercepts are -4 and 3
(b) The y-intercept is $y = F(0) = -6$

Step 2: No symmetry

Step 3: The vertical asymptote is $x = -2$

Step 4: Since $n = m + 1$, we have only the oblique asymptote, $y = x - 1$, found above; not intersected.

Step 5: The zeros are -4, -2, and 3.

Interval	Test Number	$F(x)$	Graph of F
$x < -4$	-5	$\dfrac{-8}{3}$	Below x-axis
$-4 < x < -2$	-3	6	Above x-axis
$-2 < x < 3$	1	$\dfrac{-10}{3}$	Below x-axis
$x > 3$	4	$\dfrac{4}{3}$	Above x-axis

Step 6:

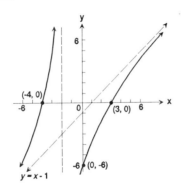

61. $R(x) = \dfrac{x(x - 1)^2}{(x + 3)^3}$

Step 0: $p(x) = x(x - 1)^2$; $n = 3$
$q(x) = (x + 3)^3$; $m = 3$

Step 1: (a) The x-intercepts are 0 and 1
(b) The y-intercept is $y = R(0) = 0$

Step 2: No symmetry

Step 3: The vertical asymptote is $x = -3$

Step 4: Since $n = m$, the horizontal asymptote is the line $y = 1$; not intersected.

3.3 RATIONAL FUNCTIONS

<u>Step 5</u>: The zeros are -3, 0, and 1.

Interval	Test Number	$R(x)$	Graph of R
$x < -3$	-4	100	Above x-axis
$-3 < x < 0$	-1	$\dfrac{-1}{2}$	Below x-axis
$0 < x < 1$	$\dfrac{1}{2}$	$\dfrac{1}{343}$	Above x-axis
$x > 1$	2	$\dfrac{2}{125}$	Above x-axis

<u>Step 6</u>:

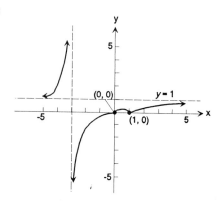

63. A vertical asymptote can only occur at a point where the denominator $q(x)$ equals 0. In other words, if $x = 4$ is a vertical asymptote for $R(x) = \dfrac{p(x)}{q(x)}$, then $q(4) = 0$. But then, by the Factor Theorem, if $q(4) = 0$, then $x - 4$ is a factor of $q(x)$.

65. No, $x = 1$ is not a vertical asymptote because each of the functions is not written in lowest terms. Each of the functions is a quotient of polynomials and is undefined for $x = 1$.

67. $f(1) = 2$

69. $f(0.79) = 1.89$

71. $f(1.32) = 1.75$

73. c, d

3 POLYNOMIAL AND RATIONAL FUNCTIONS

1. $f(x) = x^3 - x^2 + 2x + 4;\quad g(x) = x - 1$
   $f(1) = 1^3 - 1^2 + 2(1) + 4 = 6$

3. $f(x) = 3x^3 + 2x^2 - x + 3;\quad g(x) = x - 3$
   $f(3) = (3)^3 - 2(3)^2 - 3 + 3 = 81 + 18 - 3 + 3 = 99$

5. $f(x) = x^5 - 4x^3 + x;\quad g(x) = x + 3$
   $f(-3) = (-3)^5 - 4(-3)^3 + (-3) = -729 + 108 - 3 = -624$

7. $f(x) = 4x^6 - 3x^4 + x^2 + 5;\quad g(x) = x - 1$
   $f(1) = 4(1)^6 - 3(1)^4 + (1)^2 + 5 = 7$

9. $f(x) = 0.1x^3 + 0.2x;\quad g(x) = x + 1.1$
   $f(-1.1) = 0.1(-1.1)^3 + 0.2(-1.1) = -0.3531$

11. $f(x) = x^3 - x^2 + 2x + 4;\quad g(x) = x - 1$

$$
\begin{array}{r|rrrr}
1 & 1 & -1 & 2 & 4 \\
  &   & 1 & 0 & 2 \\
\hline
  & 1 & 0 & 2 & 6
\end{array}
$$

$q(x) = x^2 + 2$
$R = 6$

13.
$$
\begin{array}{r|rrrr}
3 & 3 & 2 & -1 & 3 \\
  &   & 3 & 9 & 33 & 96 \\
\hline
  & 3 & 11 & 32 & 99
\end{array}
$$
← $f(x) = 3x^3 + 2x^2 - x + 3$

← Remainder = 99

Quotient = $3x^2 + 11x + 32$

15.
$$
\begin{array}{r|rrrrrr}
-3 & 1 & 0 & -4 & 0 & 1 & 0 \\
   &   & -3 & 9 & -15 & 45 & -138 \\
\hline
   & 1 & -3 & 5 & -15 & 46 & -138
\end{array}
$$
← $f(x) = x^5 - 4x^3 + x$

← Remainder = -138

Quotient = $x^4 - 3x^3 + 5x^2 - 15x + 46$

17.
$$
\begin{array}{r|rrrrrrr}
1 & 4 & 0 & -3 & 0 & 1 & 0 & 5 \\
  &   & 4 & 4 & 1 & 1 & 2 & 2 \\
\hline
  & 4 & 4 & 1 & 1 & 2 & 2 & 7
\end{array}
$$
← Remainder = 7

Quotient = $4x^5 + 4x^4 + x^3 + x^2 + 2x + 2$

19.

$$-1.1\overline{)\begin{array}{cccc} 0.1 & 0 & 0.2 & 0 \\ & -0.11 & 0.121 & -0.3531 \\ \hline 0.1 & -0.11 & 0.321 & -0.3531 \end{array}}$$ ← Remainder = -0.3531

Quotient = $0.1x^2 - 0.11x + 0.321$

21.

$$1\overline{)\begin{array}{cccccc} 1 & 0 & 0 & 0 & 0 & -1 \\ & 1 & 1 & 1 & 1 & 1 \\ \hline 1 & 1 & 1 & 1 & 1 & 0 \end{array}}$$ ← $f(x) = x^5 - 1$

← Remainder = 0

Quotient = $x^4 + x^3 + x^2 + x + 1$

In Problems 23-32, we use the following facts: the remainder in synthetic division when $f(x)$ is divided by $x - c$, is $f(c)$; and $x - c$ is a factor of $f(x)$ only if $f(c) = 0$.

23.   We divide by $x - c = x - 2$:

$$2\overline{)\begin{array}{cccc} 3 & 2 & -3 & 3 \\ & 6 & 16 & 26 \\ \hline 3 & 8 & 13 & \boxed{29} \end{array}}$$

Remainder = 29 ≠ 0; therefore, $x - 2$ is <u>not</u> a factor of $f(x)$.

25. $x - c = x - 2$

$$2\overline{)\begin{array}{ccccc} 3 & -6 & 0 & -5 & 10 \\ & 6 & 0 & 0 & -10 \\ \hline 3 & 0 & 0 & -5 & \boxed{0} \end{array}}$$

The remainder = 0; therefore, $x - 2$ <u>is</u> a factor of $f(x)$.

27. $x - c = x - (-3) = x + 3$

$$-3\overline{)\begin{array}{ccccccc} 3 & 0 & 0 & 82 & 0 & 0 & 27 \\ & -9 & 27 & -81 & -3 & 9 & -27 \\ \hline 3 & -9 & 27 & 1 & -3 & 9 & \boxed{0} \end{array}}$$

Remainder = 0; therefore, $x + 3$ is a factor.

29. $x - c = x - (-4) = x + 4$

$$-4\overline{)\begin{array}{ccccccc} 4 & 0 & -64 & 0 & 1 & 0 & -15 \\ & -16 & 64 & 0 & 0 & -4 & 16 \\ \hline 4 & -16 & 0 & 0 & 1 & -4 & \boxed{1} \end{array}}$$

$x + 4$ is not a factor, since the remainder = 1 ≠ 0.

3 POLYNOMIAL AND RATIONAL FUNCTIONS

31. $x - c = x - \dfrac{1}{2}$

$$\dfrac{1}{2}\overline{)\,2 \ \ -1 \ \ 0 \ \ 2 \ \ -1}$$
$$\phantom{\dfrac{1}{2})} \ \ \ \ \ \ 1 \ \ 0 \ \ 0 \ \ \ \ 1$$
$$\overline{\phantom{\dfrac{1}{2})} \ \ 2 \ \ \ 0 \ \ 0 \ \ 2 \ \ \boxed{0}}$$

Since the remainder = 0; therefore, $x - \dfrac{1}{2}$ __is__ a factor of $f(x)$.

For Problems 33-38, recall that the remainder that results when $f(x)$ is divided by $(x - c)$ is precisely $f(c)$.

33. Divide $f(x) = 3x^4 = 2x^2 + 1$ by $x - c = x - 2$

$$2\overline{)\,3 \ \ \ 0 \ \ -2 \ \ \ 0 \ \ \ \ 1}$$
$$ \ \ \ \ \ 6 \ \ \ 12 \ \ 20 \ \ 40$$
$$\overline{ \ \ 3 \ \ \ 6 \ \ 10 \ \ 20 \ \ \boxed{41}} \ \ \leftarrow f(c) = f(2) = 41$$

35. Divide $f(x) = 4x^5 = 3x^3 + 2x - 1$ by $x - c = x + 1$

$$-1\overline{)\,4 \ \ \ 0 \ \ -3 \ \ \ 0 \ \ \ 2 \ \ -1}$$
$$ \ \ \ \ -4 \ \ \ 4 \ \ -1 \ \ \ 1 \ \ -3$$
$$\overline{ \ \ 4 \ \ -4 \ \ \ 1 \ \ -1 \ \ \ 3 \ \ \boxed{-4}} \ \ \leftarrow f(c) = f(-1) = -4$$

37. Divide $f(x) = 9x^{17} - 8x^{10} + 9x^8 + 5$ by $x - c = x - 1$

$$1\overline{)\,9 \ \ 0 \ \ 0 \ \ 0 \ \ 0 \ \ 0 \ \ 0 \ \ -8 \ \ 0 \ \ 9 \ \ \ 0 \ \ \ 0 \ \ \ 0 \ \ \ 0 \ \ \ 0 \ \ \ 0 \ \ \ 0 \ \ \ 5}$$
$$ \ \ \ \ \ 9 \ \ 9 \ \ 9 \ \ 9 \ \ 9 \ \ 9 \ \ \ 9 \ \ \ 1 \ \ 1 \ \ 10 \ \ 10 \ \ 10 \ \ 10 \ \ 10 \ \ 10 \ \ 10 \ \ 10$$
$$\overline{ \ \ 9 \ \ 9 \ \ 9 \ \ 9 \ \ 9 \ \ 9 \ \ 9 \ \ \ 1 \ \ 1 \ \ 10 \ \ 10 \ \ 10 \ \ 10 \ \ 10 \ \ 10 \ \ 10 \ \ \boxed{15}} \ \ \leftarrow f(1)$$

Note: In this case, it would be much easier to find $f(1)$ by substitution:

$$f(1) = 9(1)^{17} - 8(1)^{10} + 9(1)^8 + 5 = 9 - 8 + 9 + 5 = 15$$

39. $f(x) = 3x^3 + 2x^2 - 3x + 3$
$$= (3x^2 + 2x - 3)x + 3$$
$$= [(3x + 2)x - 3]x + 3, \text{ Nested form}$$

41. $f(x) = 3x^4 - 6x^3 - 5x + 10$
$$= (3x^3 - 6x^2 - 5)x + 10$$
$$= [(3x^2 - 6x)x - 5]x + 10$$
$$= \left\{[(3x - 6)x]x - 5\right\}x + 10$$

43.　$f(x) = 3x^6 - 82x^3 + 27$
　　　　$= (3x^5 - 82x^2)x + 27$
　　　　$= [(3x^4 - 82x)x]x + 27$
　　　　$= (3x^3 - 82)x \cdot x \cdot x + 27$
　　　　$= (3 \cdot x \cdot x \cdot x - 82)x \cdot x \cdot x + 27$

45.　$f(x) = 4x^6 - 64x^4 + x^2 - 15$
　　　　$= (4x^4 - 64x^2 + 1)x^2 - 15$
　　　　$= [(4x^2 - 64)x^2 + 1]x^2 - 15$
　　　　$= [(4x \cdot x - 64)x \cdot x + 1]x \cdot x - 15$

47.　$f(x) = 2x^4 - x^3 + 2x - 1$
　　　　$= (2x^3 - x^2 + 2)x - 1$
　　　　$= [(2x - 1)x^2 + 2]x - 1$
　　　　$= [(2x - 1)x \cdot x + 2]x - 1$

In Problems 49-58, first notice that the polynomials are the same as the ones in Problems 39-48, so we will start with the nested form which we obtained already. Then we let x = 1.2 and perform the calculations from left-to-right.

49.　$f(x) = 3x^3 + 2x^2 - 3x + 3 = ([3x + 2]x - 3)x + 3.$　Let $x = 1.2.$

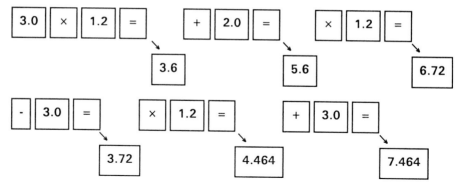

3 POLYNOMIAL AND RATIONAL FUNCTIONS

51. $f(x) = 3x^4 - 6x^3 - 5x + 10 = \{[(3x - 6)x]x - 5\}x + 10$
Let $x = 1.2$

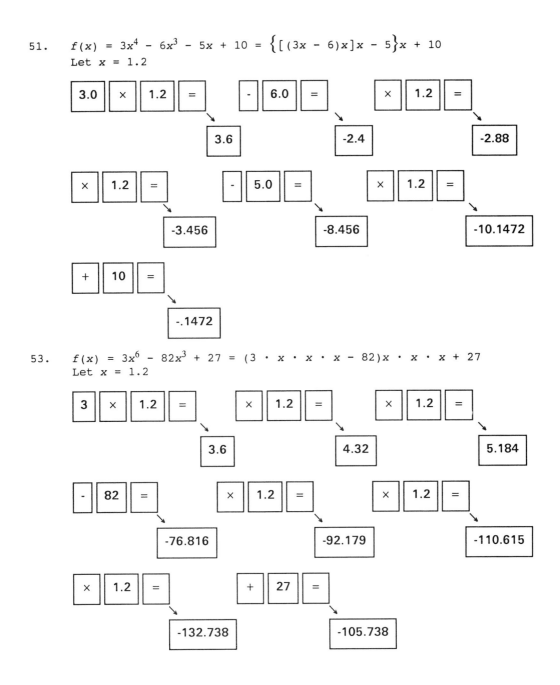

53. $f(x) = 3x^6 - 82x^3 + 27 = (3 \cdot x \cdot x \cdot x - 82)x \cdot x \cdot x + 27$
Let $x = 1.2$

55. $f(x) = 4x^6 = 64x^4 + x^2 - 15$
$= ([4 \cdot x \cdot x - 64]x \cdot x + 1)x \cdot x - 15$

Let $x = 1.2$

| 4 | × | 1.2 | = | | × | 1.2 | = | | - | 64 | = |

↘ 4.8 ↘ 5.76 ↘ -58.24

| × | 1.2 | = | | × | 1.2 | = | | + | 1 | = |

↘ -69.888 ↘ -83.866 ↘ -82.866

| × | 1.2 | = | | × | 1.2 | = | | - | 15 | = |

↘ -99.439 ↘ -119.326 ↘ -134.326

57. $f(x) = 2x^4 - x^3 + 2x - 1 = ([2x - 1]x \cdot x + 2)x - 1$
Let $x = 1.2$

| 2 | × | 1.2 | = | | - | 1 | = | | × | 1.2 | = |

↘ 2.4 ↘ 1.4 ↘ 1.68

| × | 1.2 | = | | + | 2 | = | | × | 1.2 | = |

↘ 2.016 ↘ 4.016 ↘ 4.8192

| - | 1 | = |

↘ 3.8192

59. $(x - 2)$ will be a factor of $f(x) = x^3 - kx^2 + kx + 2$ only if the remainder that results when $f(x)$ is divided by $x - 2$ is 0:

$$
\begin{array}{r|rrrr}
2) & 1 & -k & k & 2 \\
  &   & 2 & -2k + 4 & -2k + 8 \\
\hline
  & 1 & -k + 2 & -k + 4 & \boxed{-2k + 10} \leftarrow \text{Remainder}
\end{array}
$$

Therefore, we want $-2k + 10 = 0$ or $-2k = -10$, or $k = 5$

61. Either long division <u>or</u> synthetic division would be a tedious way to find the remainder in this problem, but we know by the Remainder Theorem that if $f(x)$ is divided by $x - c$, then the remainder must be $f(c)$. Here, $x - c = x - 1$, so that $c = 1$, so we want $f(1)$. This can be found by substitution:

$$f(1) = 2 - 8 + 1 - 2 = -7$$

Thus, if we divided $f(x)$ by $x - 1$, the remainder would be -7.

63. We want to prove that $x - c$ is a factor of $x^n - c^n$, for any positive integer n. By the Factor Theorem, $x - c$ will be a factor of $f(x)$ provided $f(c) = 0$. Here, $f(x) = x^n - c^n$ so that $f(c) = c^n - c^n = 0$, and therefore, $x - c$ is indeed a factor of $x^n - c^n$.

65. We are given $f(x) = 2x^3 - 6x^2 + 4x - 10$, and wish to find the time required to evaluate $f(x)$ at $x = 2.013$. To do this, we must count the number of multiplications, additions and subtractions.

(a) If we use the form $f(x) = 2x^3 - 6x^2 + 4x - 10$, then we have powers to compute, which the computer handles by repeated multiplication.

$$f(x) = 2 \cdot x \cdot x \cdot x - 6 \cdot x \cdot x + 4 \cdot x - 10$$

Here we count 6 multiplications (at 33,333 nanoseconds each) and 3 additions and subtractions (at 500 nanoseconds each). Thus, the total time required would be $T_1 = 6(33,333) + 3(500) = 201,498$ nanoseconds.

(b) For this part, we first write $f(x)$ in Nested Form:

$$
\begin{aligned}
f(x) &= 2x^3 - 6x^2 + 4x - 10 \\
&= (2x^2 - 6x + 4) \cdot x - 10 \\
&= ([2 \cdot x - 6] \cdot x + 4) \cdot x - 10
\end{aligned}
$$

Now we count 3 multiplications and 3 additions and subtractions, so the total time required is:

$$T_2 = 3(33,333) + 3(500) = 101,499 \text{ nanoseconds.}$$

3.4 REMAINDER AND FACTOR THEOREMS; SYNTHETIC DIVISION   177

☰ EXERCISE 3.5 THE ZEROS OF A POLYNOMIAL FUNCTION

In Problems 1-12, we must count the number of sign changes in both $f(x)$ and $f(-x)$.

1. $f(x) = -4x^7 - x^3 - 1$

    The maximum number of zeros is 7.

    For $f(x) = -4x^7 + x^3 - 1$, there are two variations in sign of

    $- \text{ to } + \quad + \text{ to } -$

    the coefficients, so there will be either two positive zeros, or none.

    To find $f(-x)$, replace x in $f(x)$ by $-x$:

    $$f(-x) = 4x^7 - x^3 - 1$$

    $+ \text{ to } -$

    There is only one variation of sign in $f(-x)$, so there will be exactly one negative zero.

3. $f(x) = 2x^6 - 3x^2 - x + 1$

    The maximum number of zeros is 6.

    Here, $f(x) = 2x^6 - 3x^2 - x + 1$, and

    $+ \text{ to } - \quad - \text{ to } +$

    $$f(-x) = 2x^6 - 3x^2 + x + 1, \text{ and}$$

    $+ \text{ to } - \quad - \text{ to } +$

    Thus, there are either two positive zeros, or none; and either two negative zeros or none.

5. $f(x) = 3x^3 - 2x^2 + x + 2$

    The maximum number of zeros is 3.

    For $f(x) = 3x^3 - 2x^2 + x + 2$,

    $+ \text{ to } - \quad - \text{ to } +$

    there are two variations in sign of the coefficients, so there will be either two positive zeros, or none.

    To find $f(-x)$, replace x in $f(x)$ by $-x$:

    $$f(-x) = -3x^3 - 2x^2 - x + 2$$

    $- \text{ to } +$

    There is only one variation of sign in $f(-x)$, so there will be exactly one negative zero.

3 POLYNOMIAL AND RATIONAL FUNCTIONS

7. $f(x) = -x^4 + x^2 - 1$

   The maximum number of zeros is 4.

   Here, $f(x) = -x^4 + x^2 - 1$, and

   $$- \text{ to } + \quad + \text{ to } -$$

   $$f(-x) = -x^4 + x^2 - 1$$

   $$- \text{ to } + \quad + \text{ to } -$$

   Thus, there are either two positive zeros, or none; and either two negative zeros, or none.

9. $f(x) = x^5 + x^4 + x^2 + x + 1$

   The maximum number of zeros is 5.

   We have $f(x) = x^5 + x^4 + x^2 + x + 1$, which has <u>no</u> changes of signs; hence, there will be <u>no</u> positive zeros.

   Meanwhile, $f(-x) = -x^5 + x^4 + x^2 - x + 1$ has three changes in sign; so $f(x)$ will have either three negative zeros, or one negative zero.

11. $f(x) = x^6 - 1$

   The maximum number of zeros is 6.

   $f(x) = x^6 - 1$; $f(-x) = x^6 - 1$

   We will have exactly one positive zero and one negative zero.

For Problems 13-24, use the Rational Zeros Theorem. The possible rational (whole number or fractional) zeros must be of the form $\frac{p}{q}$, where p is a factor of the constant term, and q is a factor of the coefficient of the highest power of x.

13.   For $f(x) = x^4 - 3x^3 + x^2 - x + 1$,

   p must be a factor of +1: $p = +1$ or -1
   q must be a factor of +1: $q = +1$ or -1

   So the possible zeros $\frac{p}{q}$ are ± 1.

15.   For $f(x) = x^5 - 6x^2 + 9x - 3$,

   p must be a factor of -3: $p = \pm 1$ or ± 3
   q must be a factor of +1: $q = \pm 1$

   So the possible rational zeros are ± 1, ± 3

3.5 THE ZEROS OF A POLYNOMIAL FUNCTION   179

17. For $f(x) = -2x^3 - x^2 + x + 1$,

> p must be a factor of $+1$: $p = \pm 1$
> q must be a factor of -2: $q = \pm 1, \pm 2$

Hence, the possible rational zeros, $\dfrac{p}{q}$, are ± 1, $\pm \dfrac{1}{2}$

19. For $f(x) = 3x^4 - x^2 + 2$,

> p must be a factor of $+2$: $p = \pm 1, \pm 2$
> q must be a factor of $+3$: $q = \pm 1, \pm 3$

Possible rational zeros: ± 1, $\pm \dfrac{1}{3}$, ± 2, $\pm \dfrac{2}{3}$

21. For $f(x) = 2x^5 - x^3 + 2x^2 + 4$, we have the following possibilities:

> $p = \pm 1, \pm 2, \pm 4$
> $q = \pm 1, \pm 2$

Rational zeros, $\dfrac{p}{q}$: ± 1, $\pm \dfrac{1}{2}$, ± 2, ± 4

23. For $f(x) = 6x^4 + 2x^3 - x^2 + 2$, we have the following possibilities:

> p: $\pm 1, \pm 2$
> q: $\pm 1, \pm 2, \pm 3, \pm 6$
> $\dfrac{p}{q}$: ± 1, $\pm \dfrac{1}{2}$, $\pm \dfrac{1}{3}$, $\pm \dfrac{1}{6}$, ± 2, $\pm \dfrac{2}{3}$

25. $f(x) = x^3 + 2x^2 - 5x - 6$; $f(-x) = -x^3 + 2x^2 + 5x - 6$

(a) $f(x)$ has at most three zeros.

(b) $f(x)$ has one positive zero, and either two negative zeros, or none.

(c) Possible rational zeros: p: $\pm 1, \pm 2, \pm 3, \pm 6$; q: ± 1
$\dfrac{p}{q}$: $\pm 1, \pm 2, \pm 3, \pm 6$

(d) Synthetic Division:

```
-1)1   2  -5  -6
      -1  -1   6
    1   1  -6  | 0 |
```

Since the remainder is 0, $x - (-1) = x + 1$ is a factor. The other factor is the quotient: $x^2 + x - 6$.

Thus, $f(x) = (x + 1)(x^2 + x - 6)$
$= (x + 1)(x + 3)(x - 2)$

and the zeros are -1, -3, and 2.

27. $f(x) = 2x^3 - x^2 + 2x - 1$; $f(-x) = -2x^3 - x^2 - 2x - 1$

   (a) $f(x)$ has at most three zeros.

   (b) $f(x)$ has either three positive zeros, or just one, and no negative zeros.

   (c) Possible rational zeros: p: ±1; q: ±1, ±2;
       $\frac{p}{q}$: ±1, ±$\frac{1}{2}$

   (d) Synthetic Division:

$$-1)\overline{\begin{array}{cccc} 2 & -1 & 2 & -1 \\ & -2 & 3 & -5 \end{array}}$$
$$\begin{array}{cccc} 2 & -3 & 5 & \boxed{-6} \end{array} \leftarrow (x + 1) \text{ is } \underline{not} \text{ a factor.}$$

   So we try $x - 1$

$$1)\overline{\begin{array}{cccc} 2 & -1 & 2 & -1 \\ & 2 & 1 & 3 \end{array}}$$
$$\begin{array}{cccc} 2 & 1 & 3 & \boxed{2} \end{array} \leftarrow (x - 1) \text{ is } \underline{not} \text{ a factor.}$$

   Let's turn to fractions, say $x - \frac{1}{2}$:

$$\tfrac{1}{2})\overline{\begin{array}{cccc} 2 & -1 & 2 & -1 \\ & 1 & 0 & 1 \end{array}}$$
$$\phantom{\tfrac{1}{2})}\begin{array}{cccc} 2 & 0 & 2 & \boxed{0} \end{array} \leftarrow \left(x - \frac{1}{2}\right) \text{ is a factor.}$$

   Quotient = $2x^2 + 2$

   Thus, $f(x) = \left(x - \frac{1}{2}\right)(2x^2 + 2)$

   $ = 2\left(x - \frac{1}{2}\right)(x^2 + 1)$

   The only real zero is $\frac{1}{2}$, since $x^2 + 1$ has no real zeros.

29. $f(x) = x^4 + x^2 - 2$; $f(-x) = x^4 + x^2 - 2$

   (a) $f(x)$ has at most four zeros.

   (b) $f(x)$ has one positive zero and one negative zero.
       (The other two zeros must be complex numbers.)

   (c) Possible rational zeros: p: ±1, ±2; q: ±1;
       $\frac{p}{q}$: ±1, ±2.

3.5 THE ZEROS OF A POLYNOMIAL FUNCTION   181

(d) Synthetic Division:

Try $x + 1$:

$$-1)\overline{1 \quad 0 \quad 1 \quad 0 \quad -2}$$
$$\underline{-1 \quad 1 \quad -2 \quad 2}$$
$$1 \quad -1 \quad 2 \quad -2 \quad \boxed{0} \quad (x + 1) \text{ is a factor!}$$

Quotient: $x^3 - x^2 + 2x - 2$

We have $f(x) = (x + 1)(x^3 - x^2 + 2x - 2)$

We can factor $x^3 - x^2 + 2x - 2$ by grouping terms:

$$x^3 - x^2 + 2x - 2 = x^2(x - 1) + 2(x - 1) = (x^2 + 2)(x - 1)$$

Thus, $f(x) = (x + 1)(x - 1)\underbrace{(x^2 + 2)}_{\text{no real zeros}}$

We have two real zeros: -1, 1, which agrees with what we discovered in part (b).

Note: $f(x) = x^4 + x^2 - 2$ could have been factored at the beginning: $x^4 + x^2 - 2 = (x^2 + 2)(x^2 - 1)$
$$= (x^2 + 2)(x + 1)(x - 1)$$

31. $f(x) = x^4 + 7x^2 - 2;\ f(-x) = 4x^4 + 7x^2 - 2$

(a) There are at most four zeros.

(b) There is one positive zero and one negative zero.

(c) Possible rational zeros: p: ± 1, ± 2; q: ± 1, ± 2, ± 4;
$\frac{p}{q}$: ± 1, $\pm\frac{1}{2}$, $\pm\frac{1}{4}$, ± 2.

We can start by factoring:

$$f(x) = 4x^4 + 7x^2 - 2 = (4x^2 - 1)(x^2 + 2)$$
$$= 4\left(x^2 - \frac{1}{4}\right)(x^2 + 2)$$
$$= 4\left(x + \frac{1}{2}\right)\left(x - \frac{1}{2}\right)\underbrace{(x^2 + 2)}_{\text{no real zeros}}$$

Thus, $f(x)$ has the two real zeros $-\frac{1}{2}$, $\frac{1}{2}$.

33. $f(x) = x^4 + x^3 - 3x^2 - x + 2;\ f(-x) = x^4 - x^3 - 3x^2 + x + 2$

(a) There are at most four zeros.

(b) We will have either two positive zeros, or none, and either two negative zeros, or none.

(c) Possible rational zeros: p: ± 1, ± 2; q: ± 1;
$\frac{p}{q}$: ± 1, ± 2.

(d) We cannot factor $f(x)$ by normal methods, so we start with synthetic division:

$$-1{\overline{)\begin{array}{rrrr} 1 & 1 & -3 & -1 & 2 \\ & -1 & 0 & 3 & -2 \end{array}}}$$

$$\begin{array}{rrrr} 1 & 0 & -3 & 2 \end{array} \boxed{0} \quad \text{so } x + 1 \text{ is a factor!}$$

$$x^3 - 3x + 2$$

We now work on the depressed equation, $x^3 - 3x + 2 = 0$. This is cubic, and not easily factored. We try synthetic division again, using $x + 1$ once more (since -1 could be a root of multiplicity two).

$$-1{\overline{)\begin{array}{rrrr} 1 & 0 & -3 & 2 \\ & -1 & 1 & 2 \end{array}}}$$

$$\begin{array}{rrr} 1 & -1 & -2 \end{array} \boxed{4} \qquad \begin{array}{l}\text{Therefore, } x + 1 \text{ is not} \\ \text{a factor of } x^3 - 3x + 2.\end{array}$$

But, we do know that there is another negative zero, and if it is rational, it must be -2. So we divide by $x - (-2) = x + 2$:

$$-2{\overline{)\begin{array}{rrrr} 1 & 0 & -3 & 2 \\ & -2 & 4 & -2 \end{array}}}$$

$$\begin{array}{rrr} 1 & -2 & 1 \end{array} \boxed{0} \quad \text{Thus, } (x + 2) \text{ is a factor!}$$

$$x^2 - 2x + 1$$

We now have:

$$\begin{aligned} f(x) &= (x + 1)(x^3 - 3x + 2) \\ &= (x + 1)(x + 2)(x^2 - 2x + 1) \\ &= (x + 1)(x + 2)(x - 1)(x - 1) \end{aligned}$$

and the zeros are -1, -2, 1, 1.

35. $f(x) = 4x^5 - 8x^4 - x + 2;\ f(-x) = -4x^5 - 8x^4 + x + 2$

(a) There are at most five zeros.

(b) There are either two or no positive zeros, and there is one negative zero.

(c) Possible rational zeros: p: ± 1, ± 2; q: ± 1, ± 2, ± 4;
$\dfrac{p}{q}$: ± 1, $\pm \dfrac{1}{2}$, $\pm \dfrac{1}{4}$, ± 2.

(d) Here we can factor by grouping:

$$f(x) = 4x^5 - 8x^4 - x + 2 = 4x^4(x - 2) - 1(x - 2)$$
$$= (4x^4 - 1)(x - 2)$$
$$= 4\left(x^4 - \frac{1}{4}\right)(x - 2)$$
$$= 4\left(x^2 - \frac{1}{2}\right)\left(x^2 + \frac{1}{2}\right)(x - 2)$$
$$= 4\left(x - \sqrt{\frac{1}{2}}\right)\left(x + \sqrt{\frac{1}{2}}\right)(x - 2)\left(x^2 + \frac{1}{2}\right)$$

We have $f(x) = 4\left(x - \dfrac{1}{\sqrt{2}}\right)\left(x + \dfrac{1}{\sqrt{2}}\right)(x - 2)\underbrace{\left(x^2 + \dfrac{1}{2}\right)}_{\text{no real zeros}}$

and the real zeros of $f(x)$ are:

$$\frac{1}{\sqrt{2}} = \frac{\sqrt{2}}{2}, \ \frac{-\sqrt{2}}{2}, \text{ and } 2$$

37. $x^4 - x^3 + 2x^2 - 4x - 8 = 0$

The solutions of this equation are the zeros of the polynomial function $f(x)$.

$$f(x) = x^4 - x^3 + 2x^2 - 4x - 8$$
$$f(-x) = x^4 + x^3 + 2x^2 + 4x - 8$$

(a) There are at most four zeros.

(b) There are either three or one positive zero, and there is one negative zero.

(c) p: $\pm 1, \ \pm 2, \ \pm 4, \ \pm 8$
$\ \ \ q$: ± 1
$\ \ \ \dfrac{p}{q}$: $\pm 1, \ \pm 2, \ \pm 4, \ \pm 8$

(d) We start with synthetic division:

$$-1\overline{)1 \quad -1 \quad 2 \quad -4 \quad -8}$$

```
-1)1  -1   2  -4  -8
     -1   2  -4   8
    _____
    1  -2   4  -8  |0|   Thus, x + 1 is a factor.
```

$$x^3 - 2x^2 + 4x - 8 = 0$$

We now have:

$$(x + 1)(x^3 - 2x^2 + 4x - 8) = 0$$
$$(x + 1)[x^2(x - 2) + 4(x - 2)] = 0$$
$$(x + 1)\underbrace{(x^2 + 4)}_{\text{no real zeros}}(x - 2) = 0$$

The real zeros are $-1, \ 2$.

3 POLYNOMIAL AND RATIONAL FUNCTIONS

39. $3x^3 + 4x^2 - 7x + 2 = 0$

The solutions of this equation are the zeros of the polynomial function $f(x)$.

$f(x) = 3x^3 + 4x^2 - 7x + 2$
$f(-x) = -3x^3 + 4x^2 + 7x + 2$

(a) There are at most three zeros.

(b) There are two or no positive zeros, and there is one negative zero.

(c) p: ± 1, ± 2
q: ± 1, ± 3
$\dfrac{p}{q}$: ± 1, $\pm\dfrac{1}{3}$, ± 2, $\pm\dfrac{2}{3}$

(d)
$$-1 \overline{)\begin{array}{cccc} 3 & 4 & -7 & 2 \\ & -3 & -1 & 8 \end{array}}$$
$$\,\begin{array}{cccc} 3 & 1 & -8 & \boxed{10} \end{array}$$
so $x + 1$ is not a factor.

$$1 \overline{)\begin{array}{cccc} 3 & 4 & -7 & 2 \\ & 3 & 7 & 0 \end{array}}$$
$$\,\begin{array}{cccc} 3 & 7 & 0 & \boxed{2} \end{array}$$
$x - 1$ is not a factor.

$$-2 \overline{)\begin{array}{cccc} 3 & 4 & -7 & 2 \\ & -6 & 4 & 6 \end{array}}$$
$$\,\begin{array}{cccc} 3 & -2 & -3 & \boxed{8} \end{array}$$
$x + 2$ is not a factor.

$$2 \overline{)\begin{array}{cccc} 3 & 4 & -7 & 2 \\ & 6 & 20 & 26 \end{array}}$$
$$\,\begin{array}{cccc} 3 & 10 & 13 & \boxed{28} \end{array}$$
$x - 2$ is not a factor.

Now to the fractions:

$$-\tfrac{1}{3} \overline{)\begin{array}{cccc} 3 & 4 & -7 & 2 \\ & -1 & -1 & \tfrac{8}{3} \end{array}}$$
$$\phantom{-\tfrac{1}{3})}\,\begin{array}{cccc} 3 & 3 & -8 & \boxed{\tfrac{14}{3}} \end{array}$$
$x + \dfrac{1}{3}$ is not a factor.

$$\tfrac{1}{3} \overline{)\begin{array}{cccc} 3 & 4 & -7 & 2 \\ & 1 & \tfrac{5}{3} & -\tfrac{16}{9} \end{array}}$$
$$\phantom{\tfrac{1}{3})}\,\begin{array}{cccc} 3 & 5 & \tfrac{-16}{3} & \boxed{\tfrac{2}{9}} \end{array}$$
$x - \dfrac{1}{3}$ is not a factor.

$$-\frac{2}{3} \overline{)\, 3 \quad 4 \quad -7 \quad 2}$$

$$\qquad \quad -2 \quad -\frac{4}{3} \quad \frac{50}{9}$$

$$\overline{\quad 3 \quad 2 \quad \frac{-25}{3} \quad \boxed{\frac{68}{9}}} \qquad x + \frac{2}{3} \text{ is not a factor.}$$

Sometimes you have days like this! We are down to our last possible rational root, $\frac{2}{3}$, so we try dividing by $x - \frac{2}{3}$:

$$\frac{2}{3} \overline{)\, 3 \quad 4 \quad -7 \quad 2}$$

$$\qquad \quad 2 \quad 4 \quad -2$$

$$\overline{\quad 3 \quad 6 \quad -3 \quad \boxed{0}} \quad \text{At last!} \left(x - \frac{2}{3}\right) \text{ is a factor.}$$

$$3x^2 + 6x - 3 = 0$$

$$f(x) = \left(x - \frac{2}{3}\right)(3x^2 + 6x - 3)$$

$$= 3\left(x - \frac{2}{3}\right)(x^2 + 2x - 1)$$

Now, $x^2 + 2x - 1$ cannot be factored over the integers, so we use the quadratic formula:

$$x = \frac{-2 \pm \sqrt{4 - 4(-1)}}{2} = \frac{-2 \pm \sqrt{8}}{2} = \frac{-2 \pm 2\sqrt{2}}{2}$$

The three roots are $\frac{2}{3}$, $-1 + \sqrt{2}$, $-1 - \sqrt{2}$.

Recall that once a zero, c, is known, then $x - c$ is a factor; so in factored form:

$$3\left(x - \frac{2}{3}\right)\left(x - (-1 + \sqrt{2})\right)\left(x - (-1 - \sqrt{2})\right) = 0$$

$$3\left(x - \frac{2}{3}\right)\left(x + 1 - \sqrt{2}\right)\left(x + 1 + \sqrt{2}\right) = 0$$

41. $\quad 3x^3 - x^2 - 15x + 5 = 0$

The solutions of this equation are the zeros of the polynomial function $f(x)$.

$$f(x) = 3x^3 - x^2 - 15x + 5$$
$$f(-x) = -3x^3 - x^2 + 15x + 5$$

Let's just start factoring:

$$3x^3 - x^2 - 15x + 5 = 0$$
$$x^2(3x - 1) = 5(3x - 1) = 0$$
$$(x^2 - 5)(3x - 1) = 0$$

$$\left(x + \sqrt{5}\right)\left(x - \sqrt{5}\right)(3x - 1) = 0$$

$$3\left(x - \frac{1}{3}\right)\left(x + \sqrt{5}\right)\left(x - \sqrt{5}\right) = 0$$

The zeros of $f(x)$ are $\frac{1}{3}$, $-\sqrt{5}$, $\sqrt{5}$.

43. $x^4 + 4x^3 + 2x^2 - x + 6 = 0$

The solutions of this equation are the zeros of the polynomial function $f(x)$.

$f(x) = x^4 + 4x^3 + 2x^2 - x + 6$
$f(-x) = x^4 - 4x^3 + 2x^2 + x + 6$

(a) There are at most four zeros.

(b) There are either two positive zeros or none, and either two negative zeros or none.

(c) p: ± 1, ± 2, ± 3, ± 6
q: ± 1

$\dfrac{p}{q}$: ± 1, ± 2, ± 3, ± 6

(d) Synthetic Division:

```
-1)1   4   2  -1   6
      -1  -3   1   0
    1   3  -1   0  | 6 |    x + 1 is not a factor.
```

```
 1)1   4   2  -1   6
       1   5   7   6
    1   5   7   6  | 12 |   x - 1 is not a factor.
```

```
-2)1   4   2  -1   6
      -2  -4   4  -6
    1   2  -2   3 · | 0 |   x + 2 is a factor!
```

$x^3 + 2x^2 - 2x + 3 = 0$ ← Depressed equation

We continue with synthetic division. We do not need to try 1 or −1, since they did not work in the original polynomial, and the depressed equation is a factor of the original.

Use the Rational Zeros Theorem on the cubic polynomial $x^3 + 2x^2 - 2x + 3$:

p: ± 1, ± 3
q: ± 1
$\dfrac{p}{q}$: ± 1, ± 3

3.5 THE ZEROS OF A POLYNOMIAL FUNCTION   187

But we have ruled out ± 1, since they did not work in the original equation. Notice that $f(x)$ has at least one negative root ($x = -2$), since $x + 2$ is a factor. Therefore, by part (b), there must be two negative zeros. So let's try -3. We divide by $x - (-3) = x + 3$:

$$
\begin{array}{r|rrrr}
-3 & 1 & 2 & -2 & 3 \\
   &   & -3 & 3 & -3 \\
\hline
   & 1 & -1 & 1 & \boxed{0}
\end{array}
\qquad \text{So, } x + 3 \text{ is a factor.}
$$

We now have:

$$(x + 2)(x^3 + 2x^2 - 2x + 3) = 0$$
$$(x + 2)(x + 3)(x^2 - x + 1) = 0$$

Now $x^2 - x + 1$ has no real zeros, since its discriminant is negative: $b^2 - 4ac = 1 - 4 = -3$. The real roots are -2 and -3.

45. $x^3 - \dfrac{2}{3}x^2 + \dfrac{8}{3}x + 1 = 0$

The solutions of this equation are the zeros of the polynomial function $f(x)$.

$$f(x) = x^3 - \frac{2}{3}x^2 + \frac{8}{3}x + 1$$

$$f(-x) = -x^3 - \frac{2}{3}x^2 - \frac{8}{3}x + 1$$

(a) There are at most three zeros.

(b) There are either two positive zeros, or none, and there is one negative zero.

(c) There is a trick to finding the possible rational zeros: All coefficients must be integers in order to use the Rational Zeros Theorem. But

$$x^3 - \frac{2}{3}x^2 + \frac{8}{3}x + 1 = 0$$

is equivalent to

$$3x^3 - 2x^2 + 8x + 3 = 0$$

and we can determine the possible rational zeros:

$$
\begin{array}{ll}
p: & \pm 1, \ \pm 3 \\
q: & \pm 1, \ \pm 3 \\
\dfrac{p}{q}: & \pm 1, \ \pm \dfrac{1}{3}, \ \pm 3
\end{array}
$$

3 POLYNOMIAL AND RATIONAL FUNCTIONS

It is easy to check that 1 and −1 are not zeros of $f(x)$, by substitution.

$$f(1) = 1 - \frac{2}{3} + \frac{8}{3} + 1 \neq 0$$

$$f(-1) = -1 - \frac{2}{3} - \frac{8}{3} + 1 \neq 0$$

So let's try the fractions:

$$-\frac{1}{3}\overline{)1 \quad -\frac{2}{3} \quad \frac{8}{3} \quad 1}$$
$$\phantom{-\frac{1}{3})1} -\frac{1}{3} \quad \frac{1}{3} \quad -1$$
$$\overline{\phantom{-\frac{1}{3})}1 \quad -1 \quad 3 \quad \boxed{0}} \quad \text{So } \left(x + \frac{1}{3}\right) \text{ is a factor.}$$

$$x^2 - x + 3$$

$$\left(x + \frac{1}{3}\right)(x^2 - x + 3) = 0$$

For $x^2 - x + 3$, we have:

$$x = \frac{1 \pm \sqrt{1 - 12}}{2} = \frac{1 \pm \sqrt{-11}}{2},$$

so we only have one real zero, $x = -\frac{1}{3}$

47. $f(x) = 2x^3 - x^2 + 2x - 1$

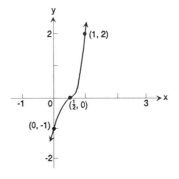

(a) To find the y-intercept, set $x = 0$:

$y = f(0) = -1$ is the y-intercept.

(b) To find x-intercepts, use the factored form (Problem 27):

$$y = f(x) = 2\left(x - \frac{1}{2}\right)(x^2 + 1)$$

Set $y = 0$ and solve for x:

$$2\left(x - \frac{1}{2}\right)(x^2 + 1) = 0$$

So $x = \frac{1}{2}$ is the only x-intercept.

The intercepts are $\left(\frac{1}{2}, 0\right)$ and $(0, -1)$.

3.5 THE ZEROS OF A POLYNOMIAL FUNCTION   189

This divides the x-axis into 2 intervals:

Interval	Test Number	$f(x)$	Graph of $f(x)$
$-\infty < x < \dfrac{1}{2}$	$x = -1$	-6	Below the x-axis
$\dfrac{1}{2} < x < \infty$	$x = 1$	2	Above the x-axis

49. $f(x) = x^4 + x^2 - 2$
$\quad\quad = (x + 1)(x - 1)(x^2 + 2)$

(See Problem 29.)

(a) Set $x = 0$: $y = f(0) = -2$ is the
$\quad$ y-intercept.

(b) The x-intercepts are the zeros of
$\quad$ $f(x)$: -1, 1

$\quad$ The intercepts are $(-1, 0)$, $(1, 0)$,
$\quad$ $(0, -2)$.

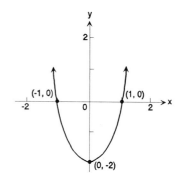

Interval	Test Number	$f(x)$	Graph of $f(x)$
$-\infty < x < -1$	$x = -2$	18	Above the x-axis
$-1 < x < 1$	$x = \dfrac{1}{2}$	$-\dfrac{27}{16}$	Below the x-axis
$1 < x < \infty$	$x = 2$	18	Above the x-axis

51. $f(x) = 4x^4 + 7x^2 - 2$
$\quad\quad = 4\left(x + \dfrac{1}{2}\right)\left(x - \dfrac{1}{2}\right)(x + 2)$

(See Problem 31.)

(a) Set $x = 0$: then $y = f(0) = -2$ is
$\quad$ the y-intercept.

(b) The x-intercepts are the zeros:
$\quad$ $x = -\dfrac{1}{2}, \dfrac{1}{2}$

$\quad$ The intercepts are $\left(-\dfrac{1}{2}, 0\right)$, $\left(-\dfrac{1}{2}, 0\right)$,
$\quad$ $(0, -2)$.

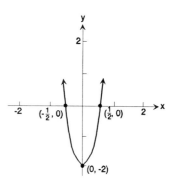

3 POLYNOMIAL AND RATIONAL FUNCTIONS

Interval	Test Number	$f(x)$	Graph of $f(x)$
$-\infty < x < -\dfrac{1}{2}$	$x = -1$	9	Above the x-axis
$-\dfrac{1}{2} < x < \dfrac{1}{2}$	$x = 0$	-2	Below the x-axis
$\dfrac{1}{2} < x < \infty$	$x = 1$	9	Above the x-axis

53. $f(x) = x^4 + x^3 - 3x^2 - x + 2$
    $ = (x + 1)(x + 2)(x - 1)(x - 1)$

(See Problem 33.)

(a) Set $x = 0$: Then $y = f(0) = 2$ is the y-intercept.

(b) The x-intercepts are the zeros:
    -2, -1, 1

    The intercepts are $(-1, 0)$, $(-2, 0)$, $(1, 0)$, $(0, 2)$.

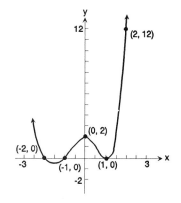

Interval	Test Number	$f(x)$	Graph of $f(x)$
$-\infty < x < -2$	$x = -3$	32	Above the x-axis
$-2 < x < -1$	$x = -\dfrac{3}{2}$	$-\dfrac{25}{16}$	Below the x-axis
$-1 < x < 1$	$x = 0$	2	Above the x-axis
$1 < x < \infty$	$x = 2$	12	Above the x-axis

55. $f(x) = 4x^5 - 8x^4 - x + 2$
    $ = 4\left(x - \dfrac{\sqrt{2}}{2}\right)\left(x + \dfrac{\sqrt{2}}{2}\right)(x - 2)\left(x^2 + \dfrac{1}{2}\right)$

(See Problem 35.)

(a) Set $x = 0$: Then $y = f(0) = 2$ is the y-intercept.

(b) The x-intercepts are the zeros:
    $-\dfrac{\sqrt{2}}{2} \approx -.707$, $\dfrac{\sqrt{2}}{2} \approx .707$, 2

    The intercepts are $(2, 0)$,
    $\left(-\dfrac{\sqrt{2}}{2}, 0\right), \left(\dfrac{\sqrt{2}}{2}, 0\right)$, $(0, 2)$.

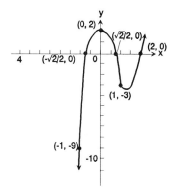

3.5 THE ZEROS OF A POLYNOMIAL FUNCTION

Interval	Test Number	$f(x)$	Graph of $f(x)$
$-\infty < x < \dfrac{-\sqrt{2}}{2}$	$x = -1$	-9	Below the x-axis
$\dfrac{-\sqrt{2}}{2} < x < \dfrac{\sqrt{2}}{2}$	$x = 0$	2	Above the x-axis
$\dfrac{\sqrt{2}}{2} < x < 2$	$x = 1$	-3	Below the x-axis
$2 < x < \infty$	$x = 3$	323	Above the x-axis

57. $x^4 + 3x^2 - 4 = 0$

The solutions of this equation are the zeros of the polynomial function $f(x)$.

$f(x) = x^4 + 3x^2 - 4$
$f(-x) = x^4 + 3x^2 - 4$

(a) There are at most 4 zeros.

(b) There is one positive zero and one negative zero.

(c) p: ± 1, ± 2, ± 4
q: ± 1
$\dfrac{p}{q}$: ± 1, ± 2, ± 4

(d) We start with synthetic division:

```
-1)1   0   3   0  -4
     -1   1  -4   4
    ──────────────────
    1  -1   1  -4  |0|   x + 1 is a factor.
```

$x^3 - x^2 + 4x - 4 = 0$ Depressed equation

We continue with synthetic division. We have one negative zero, -1. Let's try a positive zero, 1.

```
1)1  -1   4  -4
     1   0   4
    ──────────────
    1   0   4  |0|   Thus, x - 1 is a factor.
```

$x^2 + 4 = 0$

We now have:

$(x + 1)(x^3 - x^2 + 4x - 4) = 0$
$(x + 1)(x - 1)(x^2 + 4) = 0$

In the complex number system, the zeros are -1, 1, $2i$, and $-2i$.

3 POLYNOMIAL AND RATIONAL FUNCTIONS

59. $x^4 + x^3 - x - 1 = 0$

The solutions of this equation are the zeros of the polynomial function $f(x)$.

$f(x) = x^4 + x^3 - x - 1$
$f(-x) = x^4 + x^3 + x - 1$

(a) There are at most 4 zeros.

(b) There is one positive zero and three or one negative zeros.

(c) p: ± 1
     q: ± 1
     $\dfrac{p}{q}$: ± 1

(d) We start with synthetic division:

```
1)1   1   0  -1  -1
      1   2   2   1
   1   2   2   1  |0|   Therefore, x - 1 is a factor.
```

The depressed equation is $x^3 + 2x^2 + 2x + 1 = 0$

Let's try -1.

```
-1)1   2   2   1
     -1  -1  -1
    1   1   1  |0|   Therefore, x + 1 is a factor.
```

The depressed equation is $x^2 + x + 1 = 0$

We now have:

$(x - 1)(x^3 + 2x^2 + 2x + 1) = 0$
$(x - 1)(x + 1)(x^2 + x + 1) = 0$

For $x^2 + x + 1 = 0$, we have

$$x = \frac{-1 \pm \sqrt{1 - 4}}{2} = \frac{-1}{2} \pm \frac{\sqrt{3}\,i}{2}$$

In the complex number system, the zeros are -1, 1,

$-\dfrac{1}{2} + \dfrac{\sqrt{3}\,i}{2}$, and $-\dfrac{1}{2} - \dfrac{\sqrt{3}\,i}{2}$

61. $x^4 + 3x^3 - x^2 - 12x - 12 = 0$

The solutions of this equation are the zeros of the polynomial function $f(x)$.

$f(x) = x^4 + 3x^3 - x^2 - 12x - 12$
$f(-x) = x^4 + 3x^3 - x^2 + 12x - 12$

3.5 THE ZEROS OF A POLYNOMIAL FUNCTION   193

(a) There are at most 4 zeros.

(b) There is one positive zero and three or one negative zero(s).

(c) p: ±1, ±2, ±3, ±4, ±6, ±12
q: ±1
$\dfrac{p}{q}$: ±1, ±2, ±3, ±4, ±6, ±12

(d) We start with synthetic division:

$$-1\overline{)\begin{array}{rrrr} 1 & 3 & -1 & -12 & -12 \\ & -1 & -2 & 3 & 9 \\ \hline 1 & 2 & -3 & -9 & \boxed{-3} \end{array}}$$ Therefore, $x + 1$ is not a factor.

$$1\overline{)\begin{array}{rrrr} 1 & 3 & -1 & -12 & -12 \\ & 1 & 4 & 3 & -9 \\ \hline 1 & 4 & 3 & -9 & \boxed{-21} \end{array}}$$ Therefore, $x - 1$ is not a factor.

$$-2\overline{)\begin{array}{rrrr} 1 & 3 & -1 & -12 & -12 \\ & -2 & -2 & 6 & 12 \\ \hline 1 & 1 & -3 & -6 & \boxed{0} \end{array}}$$ Therefore, $x + 2$ <u>is</u> a factor.

The depressed equation is $x^3 + x^2 - 3x - 6 = 0$

We will try -2 again.

$$-2\overline{)\begin{array}{rrr} 1 & 1 & -3 & -6 \\ & -2 & 2 & 2 \\ \hline 1 & -1 & -1 & \boxed{-4} \end{array}}$$ Therefore, $x + 4$ is not a factor.

$$2\overline{)\begin{array}{rrr} 1 & 1 & -3 & -6 \\ & 2 & 6 & 6 \\ \hline 1 & 3 & 3 & \boxed{0} \end{array}}$$ Therefore, $x - 2$ is a factor.

The depressed equation is $x^2 + 3x + 3 = 0$

We now have:

$(x + 2)(x^3 + x^2 - 3x - 6) = 0$
$(x + 2)(x - 2)(x^2 + 3x + 3) = 0$

For $x^2 + 3x + 3 = 0$, we have

$$x = \frac{-3 \pm \sqrt{9 - 12}}{2} = -\frac{3}{2} \pm \frac{\sqrt{3}\,i}{2}$$

In the complex number system, the zeros are -2, 2,
$-\dfrac{3}{2} + \dfrac{\sqrt{3}\,i}{2}$, and $-\dfrac{3}{2} - \dfrac{\sqrt{3}\,i}{2}$

3 POLYNOMIAL AND RATIONAL FUNCTIONS

63. $x^5 - x^4 + 2x^3 - 2x^2 + x - 1 = 0$

The solutions of this equation are the zeros of the polynomial function $f(x)$.

$f(x) = x^5 - x^4 + 2x^3 - 2x^2 + x - 1$
$f(-x) = x^5 - x^4 - 2x^3 - 2x^2 - x - 1$

(a) There are at most 5 zeros.

(b) There is 5, 3, or 1 positive zeros and no negative zeros.

(c) p: ± 1
    q: ± 1
    $\dfrac{p}{q}$: ± 1

(d) We start with synthetic division:

```
-1)1  -1   2   -2   1   -1
      -1   2   -4   6   -7
   ─────────────────────────
    1  -2   4   -6   7  |-8|   Therefore, x + 1 is not a factor.
```

```
 1)1  -1   2   -2   1   -1
       1   0   2    0   1
   ─────────────────────────
    1   0   2   0    1  | 0|   Therefore, x - 1 is a factor.
```

$x^4 + 2x^2 + 1 = 0$

For $x^4 + 2x^2 + 1 = (x^2 + 1)(x^2 + 1)$

$x^2 = -1$ or $x = \pm\sqrt{-1} = i$ or $-i$

$(x - 1)(x^4 + 2x^2 + 1) = 0$
$(x - 1)(x^2 + 1)(x^2 + 1) = 0$

In the complex number system, the zeros are 1, i, and $-i$.

65. If 3 is a solution of $x^3 - 8x^2 + 16x - 3 = 0$, then $x - 3$ is a factor of $f(x) = x^3 - 8x^2 + 16x - 3$. Thus, using synthetic division we find

```
3)1  -8   16   -3
     3  -15    3
  ───────────────
   1  -5    1    0
```

Hence, $x^3 - 8x^2 + 16x - 3 = (x - 3)(x^2 - 5x + 1)$

Now, we solve $x^2 - 5x + 1 = 0$

The solutions are

$\dfrac{5 \pm \sqrt{25 - 4}}{2} = \dfrac{5 \pm \sqrt{21}}{2}$

Their sum is $\dfrac{5 + \sqrt{21}}{2} + \dfrac{5 - \sqrt{21}}{2} = 5$

3.5 THE ZEROS OF A POLYNOMIAL FUNCTION

67. $f(x) = 2x^3 + 3x^2 - 6x + 7$

By the Rational Zero Theorem, the only <u>possible</u> rational zeros of $f(x)$ are:

$$\frac{p}{q}: \quad \pm 1, \ \pm \frac{1}{2}, \ \pm 7, \ \pm \frac{7}{2}$$

Therefore, $\frac{1}{3}$ is <u>not</u> a zero of $f(x)$.

69. By the Rational Zero Theorem, the <u>possible</u> rational zeros of $f(x) = 2x^6 - 5x^4 + x^3 - x + 1$ are:

$$\frac{p}{q}: \quad \pm 1, \ \pm \frac{1}{2},$$

so $\frac{3}{5}$ is <u>not</u> a zero.

71. To start with, a cube has all three dimensions of equal length.

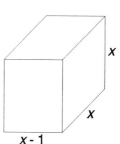

Let x = length of each side of the original cube. After removing the slice, we will have this situation:

Now the volume is width times length times height, so we have

$$
\begin{aligned}
(x - 1)x \cdot x &= 294 \\
(x - 1)x^2 &= 294 \\
x^3 - x^2 &= 294 \\
x^3 - x^2 - 294 &= 0
\end{aligned}
$$

We see that there is exactly one positive zero and we can list the possibilities if we take the time to factor 294:

$$294 = 2 \cdot 147 = 2 \cdot 3 \cdot 7 \cdot 7$$

Therefore, the possibilities for p are ± 1, ± 2, ± 3, ± 6, ± 7, ± 14, ± 21, ± 42, ± 49, ± 98, ± 147, ± 294. These are also the possible rational zeros, since $q = \pm 1$. Let's try 6 (divide by $x - 6$).

```
6)1  -1   0   -294
       6  30    180
   1   5  30   -114
```

How about 7 (divide by $x - 7$)?

```
7)1  -1   0   -294
       7  42    294
   1   6  42      0    Therefore, x - 7 is a factor.
```

Since $x - 7$ is a factor, 7 is a zero, and we know there is only one positive solution, by Descartes' Rule of signs.

Thus, the original length of each edge was $x = 7$ inches.

73. Since all we know about $f(x)$ is that its leading coefficient is 1, we can write

$$f(x) = x^n = a_{n-1}x^{n-1} + \dots + a_1x + a_0$$

where each coefficient is an integer. (See the Rational Zeros Theorem.)

Now let r be a real zero of $f(x)$. We need to show that r is either an integer or an irrational number.

Since r is a real number, it is either rational or irrational. But if r is rational (i.e., of the form $\frac{p}{q}$, where p and q are integers), then q must be a factor of the leading coefficient, which is 1. Therefore, $q = \pm1$, and $r = \frac{p}{q} = \pm p$, where p is a factor of a_0. Thus, if r is rational, then r is an <u>integer</u>!

Therefore, r is either an integer, or r is irrational.

75. $y^3 + by^2\ cy + d = 0$

Using the substitution, $y = x - \frac{b}{3}$, we have

$$\left(x - \frac{b}{3}\right)^3 + b\left(x - \frac{b}{3}\right)^2 + c\left(x - \frac{b}{3}\right) + d = 0$$

$$x^3 - 3\frac{b}{3}x^2 + 3\left(\frac{b^2}{9}\right)x - \frac{b^3}{27} + b\left(x^2 - 2\frac{b}{3}x + \frac{b^2}{9}\right) + cx - \frac{bc}{3} + d = 0$$

$$x^3 - \frac{b}{3}x + cx - \frac{b^3}{27} + \frac{b^3}{9} - \frac{bc}{3} + d = 0$$

$$x^3 + \left(c - \frac{b}{3}\right)x + \left(\frac{2b^3}{27} - \frac{bc}{3} + d\right) = 0$$

77. Based on Problem 74, we have two equations
$$3HK = -p \text{ and } H^3 + K^3 = -q$$

Thus, $K = \frac{-p}{3H}$

so that

$$H^3 + K^3 = -q$$

$$H^3 + \left(\frac{-p}{3H}\right)^3 = -q$$

$$H^3 - \frac{p^3}{qH^3} = -q$$

$$H^6 + qH^3 - \frac{p^3}{27} = 0$$

$$H^3 = \frac{-q \pm \sqrt{q^2 + \frac{4p^3}{27}}}{2} ; \quad \text{(choose + sign)}$$

$$H = \sqrt[3]{\frac{-q}{2} + \sqrt{\frac{q^2}{4} + \frac{p^3}{27}}}$$

3.5 THE ZEROS OF A POLYNOMIAL FUNCTION 197

79. $x = H + K$

Now use the results of Problems 77 and 78 to obtain

$$x = \sqrt[3]{\frac{-q}{2} + \sqrt{\frac{q^2}{4} + \frac{p^3}{27}}} + \sqrt[3]{\frac{-q}{2} - \sqrt{\frac{q^2}{4} + \frac{p^3}{27}}}$$

81. $p = 3,\ q = -14$

$$x = \sqrt[3]{\frac{14}{2} + \sqrt{\frac{196}{4} + \frac{27}{27}}} + \sqrt[3]{\frac{14}{2} - \sqrt{\frac{196}{4} + \frac{27}{27}}}$$

$$= \sqrt[3]{7 + \sqrt{50}} + \sqrt[3]{7 - \sqrt{50}}$$

$$= \sqrt[3]{7 + 5\sqrt{2}} + \sqrt[3]{7 - 5\sqrt{2}}$$

83. $p = -6,\ q = 4$

$$x = \sqrt[3]{\frac{-4}{2} + \sqrt{\frac{16}{4} + \frac{-216}{27}}} + \sqrt[3]{\frac{-4}{2} - \sqrt{\frac{16}{4} + \frac{-216}{27}}}$$

$$= \sqrt[3]{-2 + \sqrt{4 - 8}} + \sqrt[3]{-2 - \sqrt{4 - 8}}$$

$$= \sqrt[3]{-2 + 2i} + \sqrt[3]{-2 - 2i}$$

$$x^3 - 6x + 4 = (x - 2)(x^2 + 2x - 2) = 0$$

$$x = 2,\ x = \frac{-2 \pm \sqrt{4 + 8}}{2} = -1 \pm \sqrt{3}$$

≡ EXERCISE 3.6 APPROXIMATING THE REAL ZEROS OF A POLYNOMIAL FUNCTION

1. $f(x) = 8x^4 - 2x^2 + 5x - 1,\ [0, 1]$

We need to evaluate $f(x)$ at the endpoints, $x = 0$ and $x = 1$. In this case, substitution is probably easiest:

$f(0) = -1$
$f(1) = 8 - 2 + 5 - 1 = 10$

Since $f(0) = -1 < 0$ and $f(1) = 10 > 0$, $f(x)$ must have a zero between 0 and 1.

3. $f(x) = 2x^3 + 6x^2 - 8x + 2$; $[-5, -4]$

In this case, synthetic division is probably the easiest way to compute $f(-5)$ and $f(-4)$.

(a) To find $f(-5)$, divide by $x - (-5) = x + 5$:

```
-5)2    6   -8     2
      -10   20   -60
    ─────────────────
     2   -4   12  |-58|   ← f(-5) = -58
```

(See the Remainder Theorem.)

(b) To find $f(-4)$, divide by $x - (-4) = x + 4$:

```
-4)2   6   -8    2
      -8    8    0
    ───────────────
    2  -2    0  |2|   ← f(-4) = 2
```

Since $f(-5) < 0$ and $f(-4) > 0$, there must be a zero of $f(x)$ somewhere between -5 and -4.

5. $f(x) = x^5 - x^4 + 7x^3 - 7x^2 - 18x + 18$; $[1.4, 1.5]$

To find $f(1.4)$ and $f(1.5)$, it is probably best to write $f(x)$ in nested form and to use a calculator:

$$\begin{aligned}
f(x) &= x^5 - x^4 + 7x^3 - 7x^2 - 18x + 18 \\
&= (x^4 - x^3 + 7x^2 - 7x - 18)x + 18 \\
&= ([x^3 - x^2 + 7x - 7]x - 18)x + 18 \\
&= \left([\{x^2 - x + 7\}x - 7]x - 18\right)x + 18
\end{aligned}$$

or $f(x) = \left([\{(x - 1)x + 7\}x - 7]x - 18\right)x + 18$

Then $f(1.4) = -0.17536$, and
$f(1.5) = 1.40625$

Since $f(1.4) < 0$ and $f(1.5) > 0$, $f(x)$ must have a zero somewhere between 1.4 and 1.5.

For Problems 7-12, we must find upper and lower bounds to the zeros of each polynomial.

(a) *For upper bounds, we check 1, 2, 3, … , until we find the bound. Thus, we divide $f(x)$ by $x - 1$, $x - 2$, $x - 3$, … , until the third row in synthetic division contains only positive numbers or zero.*

(b) *For lower bounds, we check -1, -2, -3, … , so we must perform synthetic division using $x - (-1)$, $x - (-2)$, $x - (-3)$, … ; in other words, $x + 1$, $x + 2$, $x + 3$, … , until the third row alternates between positive (or zero) and negative (or zero).*

3.6 APPROXIMATING THE REAL ZEROS OF A
POLYNOMIAL FUNCTION

7. $f(x) = 2x^3 + x^2 - 1$

   (a) Upper bound:

   ```
   1)2  1  0  -1
        2  3   3
      ─────────────
      2  3  3  │2│  ← All positive.
   ```

   Therefore, 1 is an upper bound.

   (b) Lower bound:

   ```
   -1)2   1   0  -1
       -2   1  -1
      ──────────────
      2  -1   1  │-2│  ← Signs alternate.
   ```

   Therefore, -1 is a lower bound. Any real zeros of $f(x)$ must lie between -1 and 1.

9. $f(x) = x^3 - 5x^2 - 11x + 11$

   UPPER BOUND:

   ```
   1)1  -5  -11   11
        1   -4  -15
     ─────────────────
     1  -4  -15   -4
   ```

   ```
   2)1  -5  -11   11
        2   -6  -34
     ─────────────────
     1  -3  -17  -23
   ```

   ```
   3)1  -5  -11   11
        3   -6  -51
     ─────────────────
     1  -2  -17  -40
   ```

   ```
   5)1  -5  -11   11
        5    0  -55
     ─────────────────
     1   0  -11  -44
   ```

   ```
   6)1  -5  -11   11
        6    6  -30
     ─────────────────
     1   1   -5  -19
   ```

   ```
   7)1  -5  -11   11
        7   14   21
     ─────────────────
     1   2    3   32
   ```

   LOWER BOUND:

   ```
   -1)1  -5  -11   11
        -1    6    5
      ─────────────────
      1  -6   -5   16
   ```

   ```
   -2)1  -5  -11   11
        -2   14   -6
      ─────────────────
      1  -7    3    5
   ```

   ```
   -3)1  -5  -11   11
        -3   24  -39
      ─────────────────
      1  -8   13  -28
   ```

   UPPER BOUND: 7
   LOWER BOUND: -3

11. $f(x) = x^4 + 3x^3 - 5x^2 + 9$

    (a) Upper bound:

    ```
    1)1  3  -5   0   9
         1   4
      ────────────────
      1  4  -1         ← Stop!
    ```

    ```
    2)1  3  -5   0    9
         2  10  10   20
      ──────────────────
      1  5   5  10   29
    ```

    Therefore, 2 is an upper bound.

(b) Lower bound:

```
-1)1  3  -5   0   9
      -1
    ─────────────
    1  2                    ← Stop.  Try 3 next:

-3)1  3  -5   0   9
      -3  -0
    ─────────────
    1  0  -5

    +   -   -  Signs do not alternate!

-4)1  3  -5   0   9
      -4   4
    ─────────────
    1 -1  -1 ← Stop!

-5)1  3  -5   0     9
      -5  10  -25  125
    ──────────────────
    1 -2   5  -25  134
```

Therefore, -5 is a lower bound, and all zeros lie between -5 and 2.

13. $f(x) = x^3 + x^2 + x - 4$ has exactly one positive root (because there is one change in sign of the coefficients).

We start by finding two integers which the zero must lie between:

$f(0) = -4 < 0$
$f(1) = 1 + 1 + 1 - 4 = -1 < 0$
$f(2) = 8 + 4 + 2 - 4 = 10 > 0$

Therefore, the zero lies between 1 and 2. We next divide the interval [1, 2] into ten subintervals, and evaluate $f(x)$ at each endpoint: 1.0, 1.1, 1.2, 1.3, ... , 1.9, 2.0. To do this, the nested form of $f(x)$ will be best to work with:

$$f(x) = x^3 + x^2 + x - 4$$
$$= (x^2 + x + 1)x - 4$$

or $f(x) = ([x + 1]x + 1)x - 4$

Then $f(1.0) = -1 < 0$
$f(1.1) = -0.359 < 0$
$f(1.2) = 0.368 > 0$ Stop!

The zero must lie between 1.1 and 1.2. We now evaluate $f(x)$ at 1.10, 1.11, 1.12, ... , 1.19, 1.20.

$f(1.10) = -0.359 < 0$
$f(1.11) \approx -0.290 < 0$
$f(1.12) \approx -0.221 < 0$
$f(1.13) \approx -0.150 < 0$
$f(1.14) \approx -0.079 < 0$
$f(1.15) \approx -0.007 < 0$
$f(1.16) \approx 0.066 > 0$

Therefore, the zero must lie between 1.15 and 1.16.

3.6 APPROXIMATING THE REAL ZEROS OF A
POLYNOMIAL FUNCTION

Note: If we consider the possible <u>rational</u> zeros, we have:

$$p: \quad \pm 1, \ \pm 2, \ \pm 4$$
$$q: \quad \pm 1$$
$$\frac{p}{q}: \quad \pm 1, \ \pm 2, \ \pm 4$$

The only <u>possible</u> <u>rational</u> zeros are whole numbers! Therefore, the zero that lies between 1.15 and 1.16 <u>must</u> be irrational!

15.   Here $f(x) = 2x^4 - 3x^3 - 4x^2 - 8$

To get started:

$$f(0) = -8 < 0$$
$$f(1) = 2 - 3 - 4 - 8 = -13 < 0$$
$$f(2) = 32 - 24 - 16 - 8 = -16 < 0$$
$$f(3) = 162 - 81 - 36 - 8 = 37 > 0$$

Thus, the zero lies between 2 and 3. Let's find the nested form of $f(x)$:

$$f(x) = 2x^4 - 3x^3 - 4x^2 - 8$$
$$= (2x^2 - 3x - 4)x \cdot x - 8$$

or $f(x) = ([2x - 3]x - 4)x \cdot x - 8$

Then
$$f(2.0) = -16 < 0$$
$$f(2.1) = -14.5268 < 0$$
$$f(2.2) = -12.4528 < 0$$

Let's skip ahead to $x = 2.6$:

$$f(2.6) = 3.6272 > 0$$

Backup: $f(2.5) = -1.75 < 0$

Thus, the zero lies between 2.5 and 2.6.

We know:
$$f(2.5) = -1.75 < 0$$
$$f(2.6) = 3.6272 > 0$$

Try: $f(2.55) \approx 0.8109 > 0$ That narrows our search to the interval 2.5 to 2.55.

Then, say $f(2.53) \approx -0.2434 < 0$
$$f(2.54) \approx 0.2787 > 0$$

Thus, the zero lies between 2.53 and 2.54.

Can you see why this zero <u>must</u> be irrational?

17. $f(x) = 8x^4 - 2x^2 + 5x - 1 = (8x^3 - 2x + 5)x - 1$
   $= [(8x^2 - 2)x + 5]x - 1$

   $f(0) = -1; \ f(1) = 10$
   $f(0.1) = -0.5192$
   $f(0.2) = -0.0672$
   $f(0.3) = 0.3848$
   $f(0.21) = -0.0226$
   $f(0.22) = 0.0219$

   Thus, the zero lies between 0.21 and 0.22.

19. $f(x) = 2x^3 + 6x^2 - 8x + 2 = [(2x + 6)x - 8]x + 2$

   $f-5) = -58; \ f(-4) = 2$
   $f(-4.1) = -2.182$
   $f(-4.01) = 1.598198$
   $f(-4.02) = 1.192784$
   $f(-4.03) = 0.783746$
   $f(-4.04) = 0.371072$
   $f(-4.05) = -0.04525$

   Thus, the zero lies between -4.05 and -4.04.

21. $f(x) = x^3 + x^2 + x - 4$

   $f(1.15) \approx -0.007 < 0$
   $f(1.16) \approx 0.066 > 0$

   The zero lies between 1.15 and 1.16.

23. $f(x) = 2x^4 - 3x^3 - 4x^2 - 8$

   $f(2.53) \approx -0.2434 < 0$
   $f(2.54) \approx 0.2787 > 0$

   The zero lies between 2.53 and 2.54.

25. $f(x) = 8x^4 - 2x^2 + 5x - 1$

   $f(0.21) = -0.226 < 0$
   $f(0.22) = 0.0219 > 0$

   The zero lies between 0.21 and 0.22.

27. $f(x) = 2x^3 + 6x^2 - 8x + 2 = 0$

   $f(-4.04) = 0.371072 > 0$
   $f(-4.05) = -0.04525 < 0$

   The zero lies between -4.05 and -4.04.

3.6 APPROXIMATING THE REAL ZEROS OF A
POLYNOMIAL FUNCTION

≡ EXERCISE 3.7 COMPLEX POLYNOMIALS; FUNDAMENTAL THEOREM OF ALGEBRA

1. Since complex zeros appear as conjugate pairs, it follows that $1 + i$, the conjugate of $1 - i$, is the remaining zero of f.

3. Since complex zeros appear as conjugate pairs, it follows that $-i$, the conjugate of i, and $1 - i$, the conjugate of $1 + i$, are the remaining zeros of f.

5. Since complex zeros appear as conjugate pairs, it follows that $-i$, the conjugate of i, and $-2i$, the conjugate of $2i$, are the remaining zeros of f.

7. Since complex zeros appear as conjugate pairs, it follows that $-i$, the conjugate of i, is the remaining zero of f.

9. Since complex zeros appear as conjugate pairs, it follows that $2 - i$, the conjugate of $2 + i$, and $-3 + i$, the conjugate of $-3 - i$, are the remaining zeros of f.

11. If the coefficients are real and $2 + i$ is a zero, then $2 - i$ would also be a zero.

13. If the coefficients are real, complex zeros must come in <u>pairs</u>; i.e., there will always be an <u>even</u> number of complex zeros. If the remaining zero was complex, $f(z)$ would have three complex zeros, which is impossible. Thus, the remaining zero must be real.

15. $f(z) = z^3 - 1$ is a difference of cubes:
    $f(z) = z^3 - 1 = (z - 1)(z^2 + z + 1)$

    The zeros of $z^2 + z + 1$ are

    $$z = \frac{-1 \pm \sqrt{1 - 4}}{2} = \frac{-1 \pm \sqrt{-3}}{2}$$

    $$= -\frac{1}{2} + \frac{\sqrt{3}}{2}i \ \text{ or } \ -\frac{1}{2} - \frac{\sqrt{3}}{2}i$$

    Thus, the zeros of $z^3 - 1$ are 1, $-\frac{1}{2} + \frac{\sqrt{3}}{2}i$, and $-\frac{1}{2} - \frac{\sqrt{3}}{2}i$.

17. $f(z) = iz - 2$. Let $z = 1 + i$:

    $f(1 + i) = i(1 + i) - 2$
    $= i + i^2 - 2$
    $= i - 1 - 2, \text{ since } i^2 = -1$
    $= i - 3$
    $= -3 + i$

19. $f(z) = 3z^2 - z$. Let $z = 1 + i$:

$$\begin{aligned}
f(1 + i) &= 3(1 + i^2) - (1 + i) \\
&= 3(1 + 2i + i^2) - 1 - i \\
&= 3(1 + 2i - 1) - 1 - i \\
&= 6i - 1 - i \\
&= 5i - 1 \\
&= -1 + 5i
\end{aligned}$$

21. $f(z) = z^3 + iz - 1 + i$

We can evaluate $f(z)$ at $z = 1 + i$ by using synthetic division to divide $f(z)$ by $z - (1 + i) = z - 1 - i$:

```
1 + i )1      0        i      -1 + i
            1 + i     2i      -3 + 3i
       ─────────────────────────────
        1   1 + i     3i      -4 + 4i
```

Note: $(-1 - i)(1 + i) = -1 - i - i - i^2 = -2i$

and $(-1 - i)(3i) = -3i - 3i^2 = 3 - 3i$

Since the Remainder is $-4 + 4i$, we have $f(1 + i) = -4 + 4i$.

23. Here, $f(z) = 5z^5 - iz^4 + 2$. To find $f(r) = f(1 + i)$, divide $f(z)$ by $z - r = z - (1 + i) = z - 1 - i$:

```
1 + i )5     -i         0          0          0          2
           5 + 5i    1 + 9i     -8 + 10i   -18 + 2i   -20 - 16i
      ──────────────────────────────────────────────────────────
       5   5 + 4i    1 + 9i     -8 + 10i   -18 + 2i   -18 - 16i
```

We have computed the following products:

$$\begin{aligned}
(1 + i)(5 + 4i) &= 5 + 4i + 5i + 4i^2 = 1 + 9i \\
(1 + i)(1 + 9i) &= 1 + 9i + i + 9i^2 = -8 + 10i \\
(1 + i)(-8 + 10i) &= -8 + 10i - 8i + 10i^2 = -18 + 2i \\
(1 + i)(-18 + 2i) &= -18 + 2i - 18i + 2i^2 = -20 - 16i
\end{aligned}$$

Then $f(1 + i)$ is the Remainder, $-18 - 16i$.

25. Divide $f(z) = (1 + i)z^4 - z^3 + iz$ by $z - r = z - (2 - i) = z - 2 + i$:

```
2 - i )1 + i    -1       0         i          0
              3 + i      5      10 - 5i    16 - 18i
      ──────────────────────────────────────────────
       1 + i   2 + i     5      10 - 4i    16 - 18i
```

We have computed the following products:

$$\begin{aligned}
(2 - i)(1 + i) &= 2 + 2i - i - i^2 = 3 + i \\
(2 - i)(2 + i) &= 4 + 2i - 2i - i^2 = 5 \\
(2 - i)(10 - 4i) &= 20 - 8i - 10i + 4i^2 = 16 - 18i
\end{aligned}$$

27. Divide $f(z) = iz^5 + iz^3 + iz$ by $z - r = z - (1 + 2i) = z - 1 - 2i$:

$$
\begin{array}{r|rrrrrr}
1 + 2i) & i & 0 & i & 0 & i & 0 \\
& & -2 + i & -4 - 3i & -10i & 20 - 10i & 38 + 31i \\
\hline
& i & -2 + i & -4 - 2i & -10i & 20 - 9i & 38 + 31i
\end{array}
$$

Therefore, $f(r) = f(1 + 2i) = 38 + 31i$.

29. We know all three zeros: $1 + 2i$, 3, and 3.

Therefore,

$$
\begin{aligned}
f(z) &= (z - (1 + 2i))(z - 3)(z - 3) \\
&= (z - 1 - 2i)(z^2 - 6z + 9) \\
&= z^3 - 6z^2 + 9z - z^2 + 6z - 9 - 2iz^2 + 12iz - 18i \\
&= z^3 + (-7 - 2i)z^2 + (15 + 12i)z - 9 - 18i
\end{aligned}
$$

31. We have:

$$
\begin{aligned}
f(z) &= (z - 2)(z - (-i)(z - (1 + i)) \\
&= (z - 2)(z + i)(z - 1 - i) \\
&= (z^2 + iz - 2z - 2i)(z - 1 - i) \\
&= z^3 - z^2 - iz^2 + iz^2 - iz + z - 2z^2 + 2z + 2iz - 2iz \\
&\quad + 2i - 2 \\
&= z^3 - 3z^2 + (3 - i)z - 2 + 2i
\end{aligned}
$$

33. We have:

$$
\begin{aligned}
f(z) &= (z - 3)(z - 3)(z - (-i)(z - (-i)) \\
&= (z^2 - 6z + 9)(z^2 + 2iz - 1) \\
&= z^4 + 2iz^3 - z^2 - 6z^3 - 12iz^2 + 6z + 9z^2 + 18iz - 9 \\
&= z^4 + (2i - 6)z^3 + (8 - 12i)z^2 + (6 + 18i)z - 9
\end{aligned}
$$

1. $f(x) = (x - 2)^2 + 2$

Since the equation is in the form $f(x) = a(x - h)^2 + k$, we can read off a lot of information easily. Since $a = 1 > 0$, the parabola opens upward. Also, the vertex is at $(h, k) = (2, 2)$, and the axis of symmetry is $x = h$, or $x = 2$.

To find x-intercepts, set $y = 0$ and solve for x:

$$y = (x - 2)^2 + 2$$
$$0 = (x - 2)^2 + 2$$
$$-2 = (x- 2)^2$$

A number squared can never be negative, so this last equation has no solution; i.e., there are no x-intercepts.

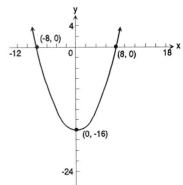

To find the y-intercept, set $x = 0$ and find $y = f(0) = (0 - 2)^2 + 2 = 6$.
Therefore, the y-intercept is 6.
We can now sketch the graph.

3. The function $f(x) = \frac{1}{4}x^2 - 16$ is in the form $f(x) = a(x - h)^2 + k$,

with $a = \frac{1}{4}$, $h = 0$, and $k = -16$. Therefore, the parabola opens

upward, since $a = \frac{1}{4} > 0$, and the vertex is at $(h, k) = (0, -16)$.

The axis of symmetry is $x = h$, or $x = 0$ (i.e., the y-axis). To find the x-intercepts, set $y = 0$ and solve for x:

$$y = \frac{1}{4}x^2 - 16$$

$$0 = \frac{1}{4}x^2 - 16$$

$$16 = \frac{1}{4}x^2$$

$$64 = x^2$$

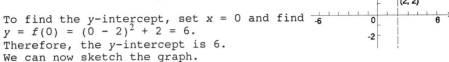

Therefore, $x = -8$ and $x = 8$ are the two x-intercepts. For the y-intercept, set $x = 0$: Then $y = f(0) = -16$. We are now ready to sketch the graph.

5. For $f(x) = -4x^2 + 4x$, since $a = -4 < 0$, the parabola opens downward. We can use the formulas in Display (2) of the text to find the vertex.

Here, $a = -4$, $b = 4$, and $c = 0$. Then the vertex is at (h, k), where

$$h = \frac{-b}{2a} = \frac{-4}{2(-4)} = \frac{1}{2} \text{ and}$$

$$k = f(h) = f\left(\frac{1}{2}\right) = -4\left(\frac{1}{2}\right)^2 + 4\left(\frac{1}{2}\right) = 1$$

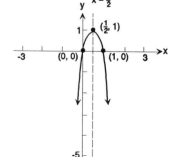

Thus, the vertex is $\left(\frac{1}{2}, 1\right)$, and the axis of symmetry is $x = \frac{1}{2}$.

The y-intercept is $y = f(0) = 0$.

To find the x-intercepts, set $y = 0$ and solve for x:

$$y = -4x^2 + 4x$$
$$0 = -4x^2 + 4x$$
$$0 = -4x(x - 1)$$

Therefore, $x = 0$ and $x = 1$ are the x-intercepts.

7. For $f(x) = \frac{9}{2}x^2 + 3x + 1$, $a = \frac{9}{2}$, $b = 3$, and $c = 1$. The parabola opens upward, since $a = \frac{9}{2} > 0$. We have $h = \frac{-b}{2a} = \frac{-3}{2\left(\frac{9}{2}\right)} = \frac{-3}{9} = \frac{-1}{3}$,

and $k = f(h) = f\left(-\frac{1}{3}\right) = \frac{9}{2}\left(\frac{1}{9}\right) + 3\left(-\frac{1}{3}\right) + 1 = \frac{1}{2}$. So the vertex is at $(h, k) = \left(-\frac{1}{3}, \frac{1}{2}\right)$, and the axis of symmetry is the line $x = -\frac{1}{3}$.

The y-intercept is $y = f(0) = 1$. Since the vertex is above the x-axis (at $\left(-\frac{1}{3}, \frac{1}{2}\right)$), and the parabola opens upward, there are no x-intercepts.

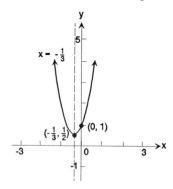

9. For $f(x) = 3x^2 + 4x - 1$, we have $a = 3$, $b = 4$, and $c = -1$. The parabola open upward, since $a = 3 > 0$. Also, $h = \frac{-b}{2a} = \frac{-4}{6} = \frac{-2}{3}$ and $k = f(h) = f\left(-\frac{2}{3}\right) = 3\left(\frac{4}{9}\right) + 4\left(-\frac{2}{3}\right) - 1 = -\frac{7}{3}$. Hence, the vertex is at $\left(-\frac{2}{3}, -\frac{7}{3}\right)$, and the axis of symmetry is $x = -\frac{2}{3}$.

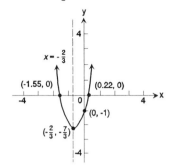

The y-intercept is $y = f(0) = -1$.

We now find the x-intercepts:

$$y = 3x^2 + 4x - 1$$
$$0 = 3x^2 + 4x - 1$$
$$(a = 3, \ b = 4, \ c = -1)$$

We first check the discriminant:

$$b^2 - 4ac = 16 - 4(3)(-1)$$
$$= 28$$

This is not a perfect square, so the x-intercepts are <u>irrational</u>:

$$x = \frac{-b \pm \sqrt{b^2 - 4ac}}{2a} = \frac{-4 \pm \sqrt{28}}{6} = \frac{-4 \pm 2\sqrt{7}}{6} = \frac{-2 \pm \sqrt{7}}{3}$$

or $x \approx 0.22$ and $x \approx -1.55$ are the x-intercepts.

11. For $f(x) = (x + 2)^3$, start with the graph of $y = x^3$, and then perform a <u>horizontal</u> shift, two units to the <u>left</u>.

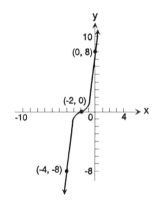

13. For $f(x) = -(x - 1)^4$, start with the graph of $y = x^4$, reflect about the x-axis (to obtain $y = -x^4$), and then shift one unit to the right.

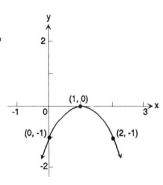

15. For $f(x) = (x - 1)^4 + 2$, start with the graph of $y = x^4$ (see Problem 13). Then shift to the <u>right</u> one unit, and <u>up</u> two units.

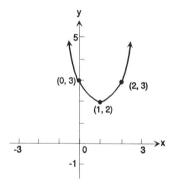

17. $f(x) = 2x^2 - 4x + 3$

Since $a = 2 > 0$, the parabola opens upward and the quadratic function has a minimum value which occurs at $x = \dfrac{-b}{2a} = \dfrac{4}{4} = 1$. The minimum value is $f(1) = 2(1)^2 - 4(1) + 3(1) = 1$.

19. $f(x) = -x^2 + 8x - 4$

Since $a = -1 < 0$, the parabola opens downward and the quadratic function has a maximum value which occurs at $x = \dfrac{-b}{2a} = \dfrac{-8}{-2} = 4$. The maximum value is $f(4) = -(4)^2 + 8(4) - 4 = -16 + 32 - 4 = 12$.

21. $f(x) = -3x^2 + 12x + 4$

Since $a = -3 < 0$, the parabola opens downward and the quadratic function has a maximum value which occurs at $x = \dfrac{-b}{2a} = \dfrac{-12}{-6} = 2$. The maximum value is $f(2) = -3(2)^2 + 12(2) + 4 = -12 + 24 + 4 = 16$.

23. $f(x) = x(x + 2)(x + 4)$

(a) x-intercepts: -4, -2, 0; y-intercept: 0
(b) crosses the x-axis at -4, -2, and 0
(c) $y = x^3$
(d) 2
(e)

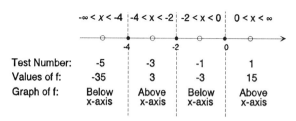

	$-\infty < x < -4$	$-4 < x < -2$	$-2 < x < 0$	$0 < x < \infty$
Test Number:	-5	-3	-1	1
Values of f:	-35	3	-3	15
Graph of f:	Below x-axis	Above x-axis	Below x-axis	Above x-axis

3 POLYNOMIAL AND RATIONAL FUNCTIONS

(f)

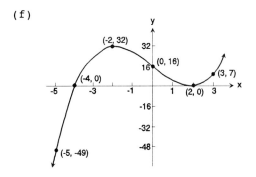

25. $f(x) = (x - 2)^2(x + 4)$

(a) x-intercepts: -4, 2; y-intercept: 16
(b) crosses at -4; touches at 2
(c) $y = x^3$
(d) 2
(e)

	$-\infty < x < -4$	$-4 < x < 2$	$2 < x < \infty$
Test Number:	-5	-2	3
Values of f:	-49	32	7
Graph of f:	Below x-axis	Above x-axis	Above x-axis

(f)

27. $f(x) = x^3 - 4x^2 = x^2(x - 4)$

(a) x-intercepts: 0, 4; y-intercept: 0
(b) touches at 0; crosses at 4
(c) $y = x^3$
(d) 2

(e)

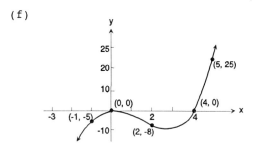

	$-\infty < x < 0$	$0 < x < 4$	$4 < x < \infty$
Test Number:	-1	2	5
Values of f:	-5	-8	25
Graph of f:	Below x-axis	Below x-axis	Above x-axis

(f)

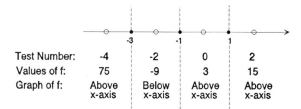

29. $f(x) = (x - 1)^2(x + 3)(x + 1)$

(a) x-intercepts: $-3, -1, 1$; y-intercept: 3
(b) crosses at -3 and -1; touches at 1
(c) $y = x^4$
(d) 3
(e)

	-4	-2	0	2
Test Number:	-4	-2	0	2
Values of f:	75	-9	3	15
Graph of f:	Above x-axis	Below x-axis	Above x-axis	Above x-axis

(f)

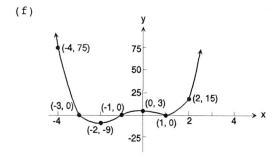

31. $R(x) = \dfrac{2x - 6}{x}$

Here, $p(x) = 2x - 6$, $n = 1$
      $q(x) = x$, $m = 1$

Step 1: (a) The x-intercept is the zero of $p(x)$, $x = 3$.

        (b) To find the y-intercept, we would let $x = 0$, but $R(0)$
            is undefined. Thus, there is no y-intercept.

Step 2: $R(-x) = \dfrac{-2x - 6}{-x} = \dfrac{2x + 6}{x}$

        This is neither $R(x)$ nor $-R(x)$, so there is no symmetry
        present.

Step 3: The vertical asymptotes are the zeros of $q(x)$: $x = 0$.

Step 4: Since $n = m$, the horizontal asymptote is the line
        $y = \dfrac{2}{1} = 2$; not intersected.

Step 5: The zeros of $p(x)$ and $q(x)$, $(x = 0, 3)$ divide the x-axis
        into three intervals:

Interval	Test Number	$R(x)$	Graph of $R(x)$
$x < 0$	$x = -2$	$\dfrac{-4 - 6}{-2} = 5$	Above the x-axis
$0 < x < 3$	$x = 1$	$\dfrac{2 - 6}{1} = -4$	Below the x-axis
$x > 3$	$x = 4$	$\dfrac{8 - 6}{4} = \dfrac{1}{2}$	Above the x-axis

Step 6: Sketch the graph:

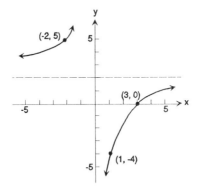

33. $H(x) = \dfrac{x + 2}{x(x - 2)}$

Here, $p(x) = x + 2$, $(n = 1)$
   $q(x) = x(x - 2)$ $(m = 2)$

Step 1: (a) The x-intercept is the zero of $p(x)$, $x = -2$.

   (b) $H(0)$ is undefined, so there is no y-intercept.

Step 2: $H(-x) = \dfrac{-x + 2}{-x(-x - 2)} = \dfrac{-x + 2}{x(x + 2)}$

   There is no symmetry.

Step 3: The vertical asymptotes are the zeros of $q(x)$: $x = 0$ and $x = 2$.

Step 4: Since $m > n$, the horizontal asymptote is the line $y = 0$ (the x-axis); intersected at $(-2, 0)$.

Step 5: $p(x)$ and $q(x)$ have a total of three zeros: $x = -2$, 0, and 2.

Interval	Test Number	$H(x)$	Graph of $H(x)$
$x < -2$	$x = -3$	$\dfrac{-3 + 2}{-3(-5)} = -\dfrac{1}{15}$	Below the x-axis
$-2 < x < 0$	$x = -1$	$\dfrac{-1 + 2}{-1(-3)} = \dfrac{1}{3}$	Above the x-axis
$0 < x < 2$	$x = 1$	$\dfrac{1 + 2}{1(1 - 2)} = -3$	Below the x-axis
$x > 2$	$x = 3$	$\dfrac{3 + 2}{3(1)} = \dfrac{5}{3}$	Above the x-axis

Step 6:

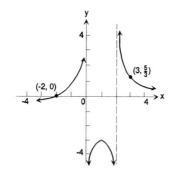

3 POLYNOMIAL AND RATIONAL FUNCTIONS

35.　$R(x) = \dfrac{x^2}{(x-1)^2}$

$p(x) = x^2 \quad (n = 2)$
$q(x) = (x-1)^2 \quad (m = 2)$

Step 1: (a) The x-intercept is the zero of $p(x)$, $x = 0$.

　　　　(b) The y-intercept is $y = R(0) = 0$.

Step 2: $R(-x) = \dfrac{x^2}{(-x-1)^2} = \dfrac{x^2}{(x+1)^2}$

　　　　No symmetry present.

Step 3: The vertical asymptote is the zero of $q(x)$: $\ x = 1$.

Step 4: Since $m = n$, the line $y = \dfrac{1}{1} = 1$ is the horizontal

　　　　asymptote; intersected at $\left(\dfrac{1}{2},\ 1\right)$.

Step 5: We have a total of two zeros: $\ x = 0,\ 1$.

Interval	Test Number	$R(x)$	Graph of $R(x)$
$x < 0$	$x = -1$	$\dfrac{1}{4}$	Above the x-axis
$0 < x < 1$	$x = \dfrac{1}{2}$	$\dfrac{\frac{1}{4}}{\frac{1}{4}} = 1$	Above the x-axis
$x > 1$	$x = 3$	$\dfrac{9}{4}$	Above the x-axis

Step 6:

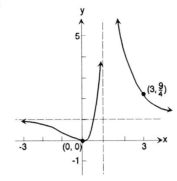

37.　$F(x) = \dfrac{x^3}{x^2 - 4}$

$p(x) = x^3$　$(n = 3)$
$q(x) = x^2 - 4 = (x + 2)(x - 2)$　$(m = 2)$

Step 1: (a) The x-intercept is $x = 0$.
　　　　(b) The y-intercept is $y = F(0) = 0$.

Step 2: $F(-x) = \dfrac{-x^3}{x^2 - 4} = -F(x)$, so the graph is symmetric about
　　　　the origin.

Step 3: The vertical asymptotes are $x = -2$ and $x = 2$.

Step 4: Since $n = m + 1$, perform long division:

$$
\begin{array}{r}
x \\
x^2 - 4 \overline{)x^3 \quad 0x^2 \quad 0x \quad 0} \\
\underline{x^3 -4x } \\
4x
\end{array}
$$

$$F(x) = \frac{x^3}{x^2 - 4} = x + \frac{4x}{x^2 - 4}$$

Thus, we have an oblique asymptote, $y = x$; intersected at
$(0, 0)$.

Step 5: We have a total of three zeros: $x = -2$, 0, and 2;
　　　　intersected at $(0, 0)$.

Interval	Test Number	$F(x)$	Graph of $F(x)$
$x < -2$	$x = -3$	$\dfrac{-27}{5}$	Below the x-axis
$-2 < x < 0$	$x = -1$	$\dfrac{-1}{-3} = \dfrac{1}{3}$	Above the x-axis
$0 < x < 2$	$x = 1$	$-\dfrac{1}{3}$	Below the x-axis
$x > 2$	$x = 3$	$\dfrac{27}{5}$	Above the x-axis

Step 6:

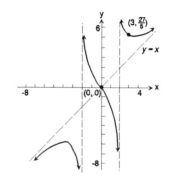

3 POLYNOMIAL AND RATIONAL FUNCTIONS

39. $R(x) = \dfrac{2x^4}{(x-1)^2}$

$p(x) = 2x^4 \quad (n = 4)$
$q(x) = (x-1)^2 \quad (m = 2)$

Step 1: (a) The x-intercept is $x = 0$.
        (b) The y-intercept is $y = 0$.

Step 2: $R(-x) = \dfrac{2x^4}{(-x-1)^2} = \dfrac{2x^4}{(x+1)^2}$

No symmetry.

Step 3: The vertical asymptote is the zero of $q(x)$: $x = 1$.

Step 4: Since $n > m + 1$, there is no horizontal nor oblique asymptote.

Step 5: We have a total of two zeros: $x = 0, 1$.

Interval	Test Number	$R(x)$	Graph of $R(x)$
$x < 0$	$x = -2$	$\dfrac{32}{9}$	Above the x-axis
$0 < x < 1$	$x = \dfrac{1}{2}$	$\dfrac{\frac{1}{8}}{\frac{1}{4}} = \dfrac{1}{2}$	Above the x-axis
$x > 1$	$x = 2$	32	Above the x-axis

Step 6:

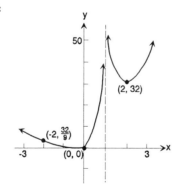

41.   Divide $f(x) = 8x^3 - 2x^2 + x - 4$ by $g(x) = x - 1$:

```
1)8  -2   1  -4
      8   6   7
   ─────────────
   8   6   7  |3|   Remainder, R = 3
```

Quotient, $q(x) = 8x^2 + 6x + 7$

43. Divide $f(x) = x^4 - 2x^3 + x - 1$ by $g(x) = x + 2$:

$$
\begin{array}{r}
-2)\overline{\begin{array}{rrrrr} 1 & -2 & 0 & 1 & -1 \\ & -2 & 8 & -16 & 30 \end{array}} \\
\overline{\begin{array}{rrrrr} 1 & -4 & 8 & -15 & \boxed{29} \end{array}}
\end{array}
$$

Remainder, $R = 29$

Quotient, $q(x) = x^3 - 4x^2 + 8x - 15$

45. To find $f(4)$, we can divide $x - 4$ and find the remainder, using synthetic division:

$$
\begin{array}{r}
4)\overline{\begin{array}{rrrrrrr} 12 & 0 & -8 & 0 & 0 & 0 & 1 \\ & 48 & 192 & 736 & 2944 & 11776 & 47104 \end{array}} \\
\overline{\begin{array}{rrrrrrr} 12 & 48 & 184 & 736 & 2944 & 11776 & \boxed{47105} \end{array}}
\end{array}
$$

$f(4)$ is the remainder: $f(4) = 47{,}105$

47. $f(x) = 12x^8 - x^7 + 6x^4 - x^3 + x - 3$

   + to - - to + + to - - to + + to -

   $f(-x) = 12x^8 + x^7 + 6x^4 + x^3 - x - 3$

   + to -

   We have exactly one negative zero, and either five, three, or one positive zero(s).

49. $f(x) = 12x^8 - x^7 + 6x^4 - x^3 + x - 3$

   p: ±1, ±3
   q: ±1, ±2, ±3, ±4, ±6, ±12

   Possible rational zeros:

   $$\frac{p}{q}: \ \pm 1, \ \pm\frac{1}{2}, \ \pm\frac{1}{3}, \ \pm\frac{1}{4}, \ \pm\frac{1}{6}, \ \pm\frac{1}{12}, \ \pm 3, \ \pm\frac{3}{2}, \ \pm\frac{3}{4}$$

51. First apply Descartes' Rule of Signs:

   $f(x) = x^3 - 3x^2 - 6x + 8$

   + to -   - to +

   $f(-x) = -x^3 - 3x^2 + 6x + 8$

   - to +

   We see that there are either two positive zeros, or none; and exactly one negative zero.

   To find the possible <u>rational</u> zeros, p must be a factor of 8:

   p: ±1, ±2, ±4, ±8

   and q must be a factor of 1:

   q: ±1

3 POLYNOMIAL AND RATIONAL FUNCTIONS

Therefore, the <u>possible</u> <u>rational</u> zeros are:

$\dfrac{p}{q}$: ± 1, ± 2, ± 4, ± 8

We now proceed to synthetic division:

$$-1)\overline{\begin{array}{rrrr} 1 & -3 & -6 & 8 \\ & -1 & 4 & 2 \end{array}}$$
$$\overline{\begin{array}{rrr|r} 1 & -4 & -2 & \boxed{10} \end{array}} \leftarrow (x + 1) \text{ is not a factor.}$$

$$1)\overline{\begin{array}{rrrr} 1 & -3 & -6 & 8 \\ & 1 & -2 & -8 \end{array}}$$
$$\overline{\begin{array}{rrr|r} 1 & -2 & -8 & \boxed{0} \end{array}} \leftarrow (x - 1) \text{ is a factor.}$$

We now have:

$$\begin{aligned} f(x) &= x^3 - 3x^2 - 6x + 8 \\ &= (x - 1)(x^2 - 2x - 8) \\ &= (x - 1)(x - 4)(x + 2) \end{aligned}$$

Therefore, the real zeros of $f(x)$ are 1, 4, and -2.

53. We have:

$$\begin{aligned} f(x) &= 4x^3 + 4x^2 - 7x + 2 \quad (2 \text{ sign changes}) \\ \text{and } f(-x) &= -4x^3 + 4x^2 + 7x + 2 \quad (1 \text{ sign change}) \end{aligned}$$

By Descartes' Rule of Signs, $f(x)$ must have one negative zero, and either two positive zeros or none.

Check for possible <u>rational</u> zeros:

$$\begin{aligned} p&: \quad \pm 1, \ \pm 2 \\ q&: \quad \pm 1, \ \pm 2, \ \pm 4 \\ \dfrac{p}{q}&: \quad \pm 1, \ \pm \dfrac{1}{2}, \ \pm \dfrac{1}{4}, \ \pm 2 \end{aligned}$$

We begin synthetic division:

$$-1)\overline{\begin{array}{rrrr} 4 & 4 & -7 & 2 \\ & -4 & 0 & 7 \end{array}}$$
$$\overline{\begin{array}{rrr|r} 4 & 0 & -7 & \boxed{9} \end{array}} \leftarrow (x + 1) \text{ is not a factor.}$$

$$1)\overline{\begin{array}{rrrr} 4 & 4 & -7 & 2 \\ & 4 & 8 & 1 \end{array}}$$
$$\overline{\begin{array}{rrr|r} 4 & 8 & 1 & \boxed{3} \end{array}} \leftarrow (x - 1) \text{ is not a factor.}$$

$$-2)\overline{\begin{array}{rrrr} 4 & 4 & -7 & 2 \\ & -8 & 8 & -2 \end{array}}$$
$$\overline{\begin{array}{rrr|r} 4 & -4 & 1 & \boxed{0} \end{array}} \leftarrow (x + 2) \ \underline{is} \text{ a factor.}$$

Quotient: $q(x) = 4x^2 - 4x + 1$

We now have:

$$f(x) = 4x^3 + 4x^2 - 7x + 2$$
$$= (x + 2)(4x^2 - 4x + 1)$$

Let's find the zeros of the depressed equation, $4x^2 - 4x + 1 = 0$, by the quadratic formula:

$$x = \frac{-b \pm \sqrt{b^2 - 4ac}}{2a} = \frac{4 \pm \sqrt{16 - 4(4)(1)}}{8}$$

or $x = \frac{4 \pm 0}{8} = \frac{1}{2}$ is a double root.

Thus, $\left(x - \frac{1}{2}\right)$ is a repeated factor. Since the leading coefficient is 4, we have,

$$4x^2 - 4x + 1 = 4\left(x - \frac{1}{2}\right)^2$$

and

$$f(x) = (x + 2)(4x^2 - 4x + 1)$$
$$= 4(x + 2)\left(x - \frac{1}{2}\right)^2$$

The real zeros of $f(x)$ are -2 and $\frac{1}{2}$ (with multiplicity 2).

55. $f(x) = x^4 - 4x^3 + 9x^2 - 20x + 20$
$f(-x) = x^4 + 4x^3 + 9x^2 + 20x + 20$

We see that we either have four positive zeros, or two positive zeros, or none, and we have no negative zeros (since there are no changes of sign in $f(-x)$).

Possible rational zeros:

p: $\pm 1, \pm 2, \pm 4, \pm 5, \pm 10, \pm 20$
q: ± 1
$\frac{p}{q}$: $\pm 1, \pm 2, \pm 4, \pm 5, \pm 10, \pm 20$

(But we can exclude the negative possibilities, since $f(x)$ has no negative zeros).

To see if $x = 1$ is a zero, we check to see if $(x - 1)$ is a factor:

```
1)1  -4   9  -20   20
      4   0    9  -11
   ─────────────────────
   1   0   9  -11  | 9 |  ←  (x - 1) is not a factor.
```

Now try $(x - 2)$:

$$
\begin{array}{r|rrrrr}
2) & 1 & -4 & 9 & -20 & 20 \\
   &   & 2 & -4 & 10 & -20 \\
\hline
   & 1 & -2 & 5 & -10 & \boxed{0}
\end{array}
$$
$\quad\leftarrow$ $(x - 2)$ <u>is</u> a factor.

Quotient: $q(x) = x^3 - 2x^2 + 5x - 10$

We now have:

$$
\begin{aligned}
f(x) &= x^4 - 4x^3 + 9x^2 - 20x + 20 \\
&= (x - 2)(x^3 - 2x^2 + 5x - 10)
\end{aligned}
$$

factor by grouping

$$
\begin{aligned}
&= (x - 2)[x^2(x - 2) + 5(x - 2)] \\
&= (x - 2)(x - 2)(x^2 + 5)
\end{aligned}
$$

no real zeros

The real zeros of $f(x)$ are 2 (multiplicity two).

57. $2x^4 + 2x^3 - 11x^2 + x - 6 = 0$

The solutions of this equation are the zeros of the polynomial function $f(x)$.

$$
\begin{aligned}
f(x) &= 2x^4 + 2x^3 - 11x^2 + x - 6 \\
f(-x) &= 2x^4 - 2x^3 - 11x^2 - x - 6
\end{aligned}
$$

We have either three positive zeros, or one; and exactly one negative zero.

Possible <u>rational</u> zeros:

$$
\begin{aligned}
p&: \quad \pm 1, \ \pm 2, \ \pm 3, \ \pm 6 \\
q&: \quad \pm 1, \ \pm 2 \\
\frac{p}{q}&: \quad \pm 1, \ \pm\frac{1}{2}, \ \pm 2, \ \pm 3, \ \pm\frac{3}{2}, \ \pm 6
\end{aligned}
$$

Start synthetic division:

$$
\begin{array}{r|rrrrr}
-1) & 2 & 2 & -11 & 1 & -6 \\
    &   & -2 & 0 & 11 & -12 \\
\hline
    & 2 & 0 & -11 & 12 & \boxed{-18}
\end{array}
$$
$\quad\leftarrow$ $(x + 1)$ is <u>not</u> a factor.

$$
\begin{array}{r|rrrrr}
1) & 2 & 2 & -11 & 1 & -6 \\
   &   & 2 & 4 & -7 & -6 \\
\hline
   & 2 & 4 & -7 & -6 & \boxed{-12}
\end{array}
$$
$\quad\leftarrow$ $(x - 1)$ is not a factor.

$$
\begin{array}{r|rrrrr}
-2) & 2 & 2 & -11 & 1 & -6 \\
    &   & -4 & 4 & 14 & -30 \\
\hline
    & 2 & -2 & -7 & 15 & \boxed{-36}
\end{array}
$$
$\quad\leftarrow$ $(x + 2)$ is not a factor.

$$
\begin{array}{r|rrrrr}
2) & 2 & 2 & -11 & 1 & -6 \\
   &   & 4 & 12 & 2 & 6 \\
\hline
   & 2 & 6 & 1 & 3 & \boxed{0}
\end{array}
$$
$\quad\leftarrow$ $(x - 2)$ <u>is</u> a factor.

We now have:

$$2x^4 + 2x^3 - 11x^2 + x - 6 = 0$$
$$(x - 2)(2x^3 + 6x^2 + x + 3) = 0$$

factor by grouping

$$(x - 2)[2x^2(x + 3) + 1(x = 3)] = 0$$
$$(x - 2)(x + 3)(2x^2 + 1) = 0$$

The real zeros are -3 and 2.

59. $2x^4 + 7x^3 + x^2 - 7x - 3 = 0$

The solutions of this equation are the zeros of the polynomial function $f(x)$.

$$f(x) = 2x^4 + 7x^3 + x^2 - 7x - 3$$
$$f(-x) = 2x^4 - 7x^3 + x^2 + 7x - 3$$

We have exactly one positive zero, and either three or one negative zero(s).

For possible <u>rational</u> zeros:

p: ±1, ±3
q: ±1, ±2
$\dfrac{p}{q}$: ±1, ±$\dfrac{1}{2}$, ±3, ±$\dfrac{3}{2}$

Synthetic division:

```
1)2   7    1   -7   -3
      2    9   10    3
   ─────────────────────
   2   9   10    3   |0|
```

Therefore, $(x - 1)$ is a factor, so $x = 1$ is a zero. All other real zeros <u>must</u> be negative.

We have:

$$f(x) = (x - 1)(2x^3 + 9x^2 + 10x + 3)$$

We now concentrate on the depressed equation, $q(x) = (2x^3 + 9x^2 + 10x + 3)$. The <u>possible</u> rational zeros are the same as before, but since we have found the only positive zero, we only need to consider the negative ones: -1, -3, $-\dfrac{1}{2}$, $-\dfrac{3}{2}$

To see if $x = -1$ is a zero, divide by $x - (-1) = x + 1$:

```
-1)2   9   10    3
      -2   -7   -3
   ─────────────────
    2   7    3   |0|
```

So $(x + 1)$ <u>is</u> a factor, and we have:

$$f(x) = (x - 1)(2x^3 + 9x^2 + 10x + 3)$$
$$= (x - 1)(x + 1)(2x^2 + 7x + 3)$$

Now $2x^2 + 7x + 3$ can be factored by trial and error:

$$2x^2 + 7x + 3 = (2x + 1)(x + 3)$$

We have:

$$2\left(x + \frac{1}{2}\right)$$

$$(x - 1)(x + 1)(2x + 1)(x + 3) = 0$$

$$2(x - 1)(x + 1)\left(x + \frac{1}{2}\right)(x + 3) = 0$$

The real zeros are: -3, -1, $-\frac{1}{2}$, and 1.

61.   From Problem 51:

$$f(x) = x^3 - 3x^2 - 6x + 8$$
$$= (x - 1)(x - 4)(x + 2)$$

Thus, $f(x)$ has three real zeros, -2, 1, and 4, which divide the x-axis into 4 intervals. Recall that the zeros <u>are</u> the x-intercepts, and the y-intercept is $y = f(0) = 8$.

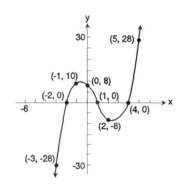

Interval	Test Number	$f(x)$	Graph of $f(x)$
$-\infty < x < -2$	$x = -3$	-28	Below the x-axis
$-2 < x < 1$	$x = -1$	10	Above the x-axis
$1 < x < 4$	$x = 2$	-8	Below the x-axis
$4 < x < \infty$	$x = 5$	28	Above the x-axis

63.   From Problem 53, we have:

$$f(x) = 4x^3 + 4x^2 - 7x + 2$$
$$= 4(x + 2)\left(x - \frac{1}{2}\right)^2$$

The y-intercept is $y = f(0) = 2$.

The x-intercepts are the real zeros: -2, $\frac{1}{2}$.

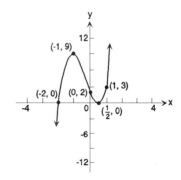

3 - CHAPTER REVIEW

Interval	Test Number	$f(x)$	Graph of $f(x)$
$-\infty < x < -2$	$x = -3$	-49	Below the x-axis
$-2 < x < \dfrac{1}{2}$	$x = -1$	9	Above the x-axis
$x > \dfrac{1}{2}$	$x = 1$	3	Above the x-axis

65. From Problem 55:

$$f(x) = x^4 - 4x^3 + 9x^2 - 20x + 20$$
$$= (x - 2)(x - 2)(x^2 + 5)$$

The y-intercept is $y = f(0) = 20$.

The only x-intercept is the one zero of $f(x)$, $x = 2$.

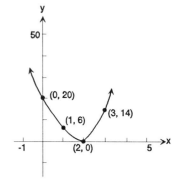

Interval	Test Number	$f(x)$	Graph of $f(x)$
$x < 2$	$x = 1$	6	Above the x-axis
$x > 2$	$x = 3$	14	Above the x-axis

67. From Problem 57,

$$f(x) = 2x^4 + 2x^3 - 11x^2 + x - 6$$
$$= (x - 2)(x + 3)(2x^2 + 1)$$

The y-intercept is $y = f(0) = -6$.

The x-intercepts are the zeros, -3 and 2.

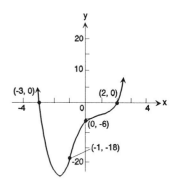

Interval	Test Number	$f(x)$	Graph of $f(x)$
$-\infty < x < -3$	$x = -4$	198	Above the x-axis
$-3 < x < 2$	$x = -1$	-18	Below the x-axis
$2 < x < \infty$	$x = 3$	171	Above the x-axis

69. From Problem 59,

$$f(x) = 2x^4 + 7x^3 + x^2 - 7x - 3$$
$$= 2(x - 1)(x + 1)\left(x + \frac{1}{2}\right)(x + 3)$$

The y-intercept is $y = f(0) = -3$.

The x-intercepts are the one zeros, -3, -1, $-\frac{1}{2}$ and 1.

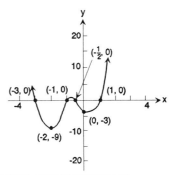

Interval	Test Number	$f(x)$	Graph of $f(x)$
$-\infty < x < -3$	$x = -4$	105	Above the x-axis
$-3 < x < -1$	$x = -2$	-9	Below the x-axis
$-1 < x < -\frac{1}{2}$	$x = -\frac{3}{4}$	$\frac{63}{128}$	Above the x-axis
$-\frac{1}{2} < x < 1$	$x = 0$	-3	Below the x-axis
$1 < x < \infty$	$x = 2$	75	Above the x-axis

71. We consider $f(x) = 3x^3 - x - 1$ on the interval $[0, 1]$:

$$f(0) = -1$$
$$f(1) = 3 - 1 - 1 = 1$$

Since $f(0)$ is below the x-axis, and $f(1)$ is above the x-axis, $f(x)$ must have a zero in the interval $[0, 1]$.

73. For $f(x) = 8x^4 - 4x^3 - 2x - 1$ on $[0, 1]$, we have:

$$f(0) = -1$$
$$f(1) = 8 - 4 - 2 - 1 = 1$$

Therefore, $f(x)$ must have a zero between 0 and 1.

75. $f(x) = 2x^3 - x^2 - 4x + 2$

To find an upper bound to the zeros of $f(x)$, we divide $f(x)$ by $x - 1$, $x - 2$, … , until the third row in the synthetic division process contains no negative numbers:

```
1)2  -1  -4   2
      2   1
   2   1  -3    ← Stop!
```

```
2)2  -1  -4   2
      4   6   4
   2   3   2   6
```

The entries in the last row are all positive, so there is no zero of $f(x)$ greater than 2.

To find a lower bound, divide $f(x)$ by $x + 1$, $x + 2$, ... , until the third row contains numbers that are alternately positive (or zero) and negative (or zero).

```
-1)2  -1  -4   2
       -2   3
    2  -3  -1        ← Stop!
```

```
-2)2  -1  -4    2
       -4  10  -12
    2  -5   6  -10
```

There are no zeros less than −2. Therefore, all real zeros must lie between −2 and 2.

77. $f(x) = 2x^3 - 7x^2 - 10x + 35$

(a) Upper bound:

```
1)2  -7  -10  35
      2
   2  -5             ← Stop
```

Now 2 and 3 will not work either (do you see why?), so we try 4:

```
4)2  -7  -10  35
      8    4
   2   1   -6        ← Stop
```

```
5)2  -7  -10  35
     10   15  25
   2   3    5  60
```

So, $x = 5$ is an upper bound.

(b) Lower bound:

```
-1)2  -7  -10  35
      -2    9
   2  -9   -1        ← Stop
```

```
-2)2   -7  -10   35
       -4   22  -24
    2  -11   12   11      (Not good)
```

```
-3)2   -7  -10   35
       -6   39  -87
    2  -13   29  -52
```

So $x = -3$ is a lower bound. All real zeros must lie between −3 and 5.

79. We are trying to locate the positive zero of $f(x) = x^3 - x - 2$. We start by finding the two consecutive whole numbers on either side of the zero:

$$f(0) = -2$$
$$f(1) = 1 - 1 - 2 = -2 < 0$$
$$f(2) = 8 - 2 - 2 = 4 > 0$$

Therefore, the zero lies between 1 and 2. We now check $x = 1.1$, 1.2, 1.3, … , 1.9.

First, write $f(x)$ in nested form:

$$f(x) = x^3 - x - 2$$
$$= (x^2 - 1)x - 2$$
$$= (x \cdot x - 1)x - 2$$

Then, $f(1.1) = -1.769 < 0$
$\qquad f(1.2) = -1.472 < 0$

Let's skip to: $f(1.5) = -.125 < 0$

(see how that saved some time?)

Now, $f(1.6) = .496 > 0$.

The zero must lie between 1.5 and 1.6. We continue once more, to isolate the zero to one of the intervals:

$$[1.5, 1.51], [1.51, 1.52], … , [1.59, 1.6].$$

We have $f(1.5) = -.125 < 0$

Then $\qquad f(1.51) = -.06705 < 0$
$\qquad\qquad f(1.52) = -.00819 < 0$
$\qquad\qquad f(1.53) = .05158 > 0$

The zero lies between 1.52 and 1.53, so we have approximated it to within .01.

81. We write $f(x)$ in nested form:

$$f(x) = 8x^4 - 4x^3 - 2x - 1$$
$$= (8x^3 - 4x^2 - 2)x - 1$$
$$= ([8x - 4]x \cdot x - 2)x - 1$$

We start with whole numbers:

$$f(0) = -1 < 0$$
$$f(1) = 1 > 0$$

The zero lies in the interval [0, 1].

Now proceed by tenths:

$$f(0) = -1 < 0$$
$$f(0.1) = -1.2032 < 0$$

Skip to:
$$f(0.5) = -2 < 0$$

How about:
$$f(0.7) = -1.8512 < 0$$
$$f(0.8) = -1.3712 < 0$$
$$f(0.9) = -0.4672 < 0$$

The zero lies between 0.9 and 1:

We go again:
$$f(0.9) = -.4672 < 0$$
$$f(0.91) = -.34829 < 0$$
$$\vdots$$
$$f(0.95) = 0.18655 > 0$$

This time I went too far, so I back up:

$$f(0.94) = .04366 > 0$$
$$f(0.93) = -.09301 < 0$$

So the zero lies between 0.93 and 0.94.

83. Since complex zeros appear as conjugate pairs, it follows that $1 - i$, the conjugate of $1 + i$, is the remaining zero of f.

85. $-i$, the conjugate of i, and $1 - i$, the conjugate of $1 + i$, are the remaining zeros of f.

87.
$$\begin{aligned}
f(z) &= (z - 1)(z - 1)(z - i)(z - 2) \\
&= (z^2 - 2z + 1)(z^2 - 2z - iz + 2i) \\
&= (z^4 - 2z^3 - iz^3 + 2iz^2 - 2z^3 + 4z^2 + 2iz^2 - 4iz \\
&\quad + z^2 - 2z - iz + 2i \\
&= z^4 + (-4 - i)z^3 + (5 + 4i)z^2 - (2 + 5i)z + 2i
\end{aligned}$$

89.
$$\begin{aligned}
f(z) &= (z - 2)(z - 3)(z - (1 + i)) \\
&= (z^2 - 5z + 6)(z - 1 - i) \\
&= (z^3 - z^2 - iz^2 - 5z^2 + 5z + 5iz + 6z - 6 - 6i \\
&= (z^3 - (6 + i)z^2 + (11 + 5i)z - 6 - 6i
\end{aligned}$$

91. We can first divide by $(x - 1)$, and then divide by $(x - 2)$:

```
1)1   2   -7   -8   12
      1    3   -4  -12
   ─────────────────────
   1   3   -4  -12   |0|
```

Thus, $\dfrac{x^4 + 2x^3 - 7x^2 - 8x + 12}{x - 1} = x^3 + 3x^2 - 4x - 12$

Now divide by $x - 2$:

```
2)1   3   -4  -12
      2   10   12
   ──────────────────
   1   5    6   |0|
```

Thus, $\dfrac{x^4 + 2x^3 - 7x^2 - 8x + 12}{(x - 2)(x - 1)} = \dfrac{x^3 + 3x^2 - 4x - 12}{x - 2} = x^2 + 5x + 6$

93. $x^3 - x^2 - 8x + 12 = 0$

The solutions of this equation are the zeros of the polynomial function $f(x)$.

$f(x) = x^3 - x^2 - 8x + 12$

Determine the possible <u>rational</u> zeros:

p: $\pm 1, \pm 2, \pm 3, \pm 4, \pm 6, \pm 12$
q: ± 1
$\dfrac{p}{q}$: $\pm 1, \pm 2, \pm 3, \pm 4, \pm 6, \pm 12$

Begin with synthetic division:

```
-1)1  -1  -8   12
      -1   2    6
    ───────────────
    1  -2  -6   18    ← (x + 1) is not a factor.
```

```
 1)1  -1  -8   12
       1   0   -8
    ───────────────
    1   0  -8    4
```

```
-2)1  -1  -8   12
      -2   6    4
    ───────────────
    1  -3  -2   16
```

```
 2)1  -1  -8   12
       2   2  -12
    ───────────────
    1   1  -6    0    ← (x - 2) is a factor.
```

So $x^3 - x^2 - 8x + 12 = 0$
       $(x - 2)(x^2 + x - 6) = 0$
    $(x - 2)(x + 3)(x - 2) = 0$

The zeros are $x = 2$ (multiplicity two), and $x = -3$.

95. $3x^4 - 4x^3 + 4x^2 + 1 = 0$

The solutions of this equation are the zeros of the polynomial function $f(x)$.

$f(x) = 3x^4 - 4x^3 + 4x^2 - 4x + 1$

For possible rational zeros:

p: ± 1
q: $\pm 1, \pm 3$
$\dfrac{p}{q}$: $\pm 1, \pm \dfrac{1}{3}$

```
-1)3  -4   4   -4    1
      -3   7  -11  -15
    ──────────────────────
    3  -7  11  -15   16
```

```
-1)3  -4   4   -4    1
       3  -1    3   -1
    ──────────────────────
    3  -1   3   -1    0    ← (x - 1) is a factor.
```

We have:

$$(x - 1)(3x^3 - x^2 + 3x - 1) = 0$$
$$(x - 1)[x^2(3x - 1) + 1(3x - 1)] = 0$$
$$(x - 1)(3x - 1)(x^2 + 1) = 0$$
$$3(x - 1)\left(x - \frac{1}{3}\right)(x + i)(x - i) = 0$$

The zeros are 1, $\frac{1}{3}$, $-i$, i

97. $f(x) = 8x^3 - 2x^2 + x - 4$
$$= (8x^2 - 2x + 1)x - 4$$
$$= (([8x - 2]x + 1)x - 4$$

Now let $x = 1.5$, and perform the following calculator operations:

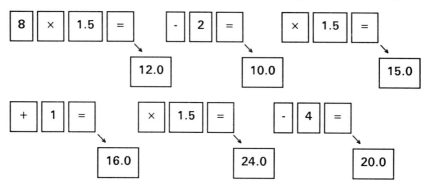

Therefore, $f(1.5) = 20$

99. $f(x) = x^4 - 2x^3 + x - 1$
$$= (x^3 - 2x^2 + 1)x - 1$$
$$= (([x - 2]x^2 + 1)x - 1$$
$$= (([x - 2]x \cdot x + 1)x - 1$$

Now let $x = 1.5$:

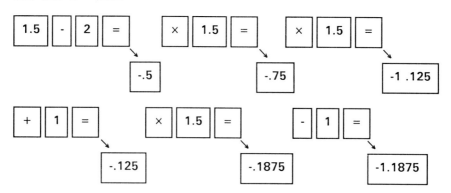

101. (a) To find an upper bound on the zeros of $f(x)$, divide $f(x)$ by
$x - 1, x - 2, \ldots$:

$$
\begin{array}{r|rrrrrr}
1) & 4 & -3 & 0 & 8 & 1 & 2 \\
   &   & 4 & 1 & 1 & 9 & 10 \\
\hline
   & 4 & 1 & 1 & 9 & 10 & 12
\end{array}
$$

Since the last row contains only positive entries, there is no
zero greater than 1.

(b) Lower bound:

$$
\begin{array}{r|rrrrrr}
-1) & 4 & -3 & 0 & 8 & 1 & 2 \\
    &   & -4 & 7 & -7 & \\
\hline
    & 4 & -7 & 7 & 1 & \leftarrow & \text{Stop}
\end{array}
$$

$$
\begin{array}{r|rrrrrr}
-2) & 4 & -3 & 0 & 8 & 1 & 2 \\
    &   & -8 & 22 & -44 & 72 & -146 \\
\hline
    & 4 & -11 & 22 & -36 & 73 & -144
\end{array}
$$

There is no zero less than -2.

Therefore, all zeros lie between -2 and 1.

103. Let (x, y) be any point on the line $y = x$. Then the distance from
(x, y) to $(3, 1)$ is given by the distance formula.

$$
d = \sqrt{(x - 3)^2 + (y - 1)^2} = \sqrt{x^2 - 6x + 9 + y^2 - 2y + 1}
$$

But since (x, y) in on the line $y = x$, we can replace all y's in
the above formula by x:

$$
d = \sqrt{x^2 - 6x + 9 + x^2 - 2x + 1}
$$

or

$$
d = \sqrt{2x^2 - 8x + 10}
$$

Now d is a fairly complicated function, but notice:

$$
d = 2x^2 - 8x + 10
$$

so d^2 is a parabola that opens upward, so we can find the minimum
value of d^2, and then it will be easy to find the minimum value of
d by taking a square root.

Consider the parabola $y = 2x^2 - 8x + 10$. We have $a = 2$, $b = -8$,
$c = 10$, so that $h = \dfrac{-b}{2a} = \dfrac{8}{4} = 2$, and $k = f(2) = 2$.

Thus, the vertex is at $(2, 2)$. That means that the minimum value
of d^2 is 2, and that value occurs when $x = 2$. (So the minimum
value of d is $\sqrt{2}$.) Since $y = x$, and $x = 2$, the point on the line
is the point $(2, 2)$.

105. Let the origin be the highest point on the parabolic arch (i.e., the vertex). Then the equation of the parabola is $y = ax^2$, where $a < 0$ (since the parabola opens downward). From the illustration, we see that when $x = 10$, $y = -10$. Therefore:

$$-10 = a(10)^2$$
$$-10 = 100a$$
$$a = -\frac{1}{10}$$

Now find y when $x = -8$: $y = -\frac{1}{10}(-8)^2 = -\frac{64}{10} = -6.4$

Since the water is 10 feet below the x-axis, $h = 10 - 6.4 = 3.6$ ft.

3 POLYNOMIAL AND RATIONAL FUNCTIONS

CHAPTER 4

EXPONENTIAL AND LOGARITHMIC FUNCTIONS

≡ EXERCISE 4.1 EXPONENTIAL FUNCTIONS

1. (a) $3^{2.2} \approx 11.211578$ (b) $3^{2.23} \approx 11.587251$

    (c) $3^{2.236} \approx 11.663882$ (d) $3^{\sqrt{5}} \approx 11.664753$

3. (a) $2^{3.14} \approx 8.8152409$ (b) $2^{3.141} \approx 8.8213533$

    (c) $2^{3.1415} \approx 8.8244111$ (d) $2^{\pi} \approx 8.8249778$

5. (a) $3.1^{2.7} \approx 21.216638$ (b) $3.14^{2.71} \approx 22.216690$

    (c) $3.141^{2.718} \approx 22.440403$ (d) $\pi^{e} \approx 22.459158$

7. B   9. D   11. A   13. E

15. $y = e^{-x}$

Using the graph of $y = e^{x}$, reflect about the y-axis.

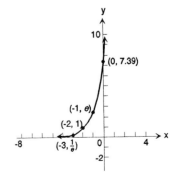

17. $y = e^{x+2}$

Using the graph of $y = e^{x}$, shift 2 units to the left.

19.

$$2^{2x+1} = 4$$
$$2^{2x+1} = 2^2$$

Then, $2x + 1 = 2$
$$2x = 1$$
$$x = \frac{1}{2}$$

21.

$$3^{x^3} = 9^x$$
$$3^{x^3} = (3^2)^x$$
$$3^{x^3} = 3^{2x}$$

Then, $x^3 = 2x$
$$x^3 - 2x = 0$$
$$x(x^2 - 2) = 0$$
$$x(x + \sqrt{2})(x - \sqrt{2}) = 0$$
$$x = 0 \quad \text{or} \quad x = -\sqrt{2} \quad \text{or} \quad x = \sqrt{2}$$
$$\left\{ 0, -\sqrt{2}, \sqrt{2} \right\}$$

23.

$$8^{x^2-2x} = \frac{1}{2}$$
$$2^{3(x^2-2x)} = 2^{-1}$$
$$3x^2 - 6x = -1$$
$$3x^2 - 6x + 1 = 0$$

Here, $a = 3$, $b = -6$, $c = 1$, and $b^2 - 4ac = 24$.

Then, $x = \dfrac{-(-6) \pm \sqrt{24}}{2(3)}$

$$= \frac{6 \pm 2\sqrt{6}}{6} = 1 \pm \frac{\sqrt{6}}{3}$$

$$\left\{ 1 + \frac{\sqrt{6}}{3},\ 1 - \frac{\sqrt{6}}{3} \right\}$$

25.

$$2^x \cdot 8^{-x} = 4^x$$
$$2^x \cdot (2^3)^{-x} = 4^x$$
$$2^x \cdot 2^{-3x} = 4^x$$
$$2^{x-3x} = 4^x$$
$$2^{-2x} = 4^x$$
$$(2^2)^{-x} = 4^x$$
$$4^{-x} = 4^x$$

Then, $-x = x$
$$0 = 2x$$
$$0 = x$$

27.

$$2^{2x} - 2^x - 12 = 0$$
$$(2^x)^2 - 2^x - 12 = 0$$

Let $u = 2^x$; then

$$u^2 - u - 12 = 0$$
$$(u - 4)(u + 3) = 0$$
$$u = 4 \quad \text{or} \quad u = -3$$
$$2^x = 4 \quad \text{or} \quad 2^x = -3$$
$$\qquad\qquad\qquad \text{Not possible.}$$
$$2^x = 2^2$$
$$x = 2$$

29.

$$3^{2x} + 3^{x+1} - 4 = 0$$
$$(3^x)^2 + 3^x \cdot 3^1 - 4 = 0$$
$$(3^x)^2 + 3(3^x) - 4 = 0$$

Let $u = 3^x$; then

$$u^2 + 3u - 4 = 0$$
$$(u + 4)(u - 1) = 0$$
$$u = -4 \quad \text{or} \quad u = 1$$
$$3^x = -4 \quad \text{or} \quad 3^x = 1$$
$$\text{(Not possible.)}$$
$$3^x = 3^0$$
$$x = 0$$

31.
$$4^{2x} = 25$$
$$(4^x)^2 = 5^2$$
$$4^x = 5$$
$$(4^x)^{-1} = (5)^{-1} = \frac{1}{5}$$
$$4^{-x} = \frac{1}{5}$$

33. $f(x) = \left(\sqrt{2}\right)^x$

x	-4	-2	-1	0	1	2	4
$f(x) = \left(\sqrt{2}\right)^x$	$\frac{1}{4}$	$\frac{1}{2}$	$\frac{\sqrt{2}}{2}$	1	$\sqrt{2}$	2	4

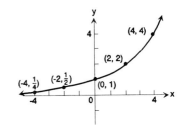

35. $f(x) = \left(\sqrt{2}\right)^x + 2$

Using the graph of $f(x) = \left(\sqrt{2}\right)^x$, vertically shift upward 2 units.

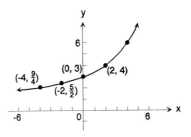

37. $f(x) = \left(\sqrt{2}\right)^{x+2}$

Using the graph of $f(x) = \left(\sqrt{2}\right)^x$, horizontally shift to the left 2 units.

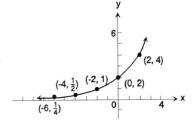

39. $f(x) = \left(\sqrt{2}\right)^{-x}$

Reflect the graph of $f(x) = \left(\sqrt{2}\right)^x$ about the y-axis.

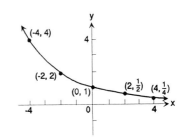

4.1 EXPONENTIAL FUNCTIONS

235

41. $f(x) = \left(2\sqrt{2}\right)^x$

Using the graph of $f(x) = \left(\sqrt{2}\right)^x$, vertically stretch the graph so (2, 2) becomes (2, 8).

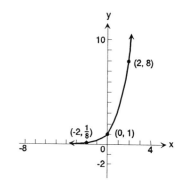

43.

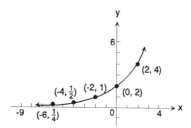

45.

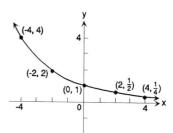

47.

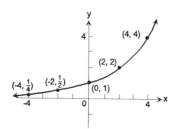

49.

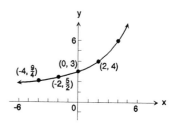

51.

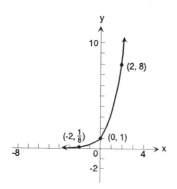

4 EXPONENTIAL AND LOGARITHMIC FUNCTIONS

53. (a) $R = 0.7 - e^{-0.2t}$

At $t = 10$,

$R = 0.7 - e^{-0.2(10)}$
$= 0.7 - e^{-2}$
$= 0.5647$
$= 56.47\%$

At $t = 20$,

$R = 0.7 - e^{-0.2(20)}$
$= 0.7 - e^{-4}$
$= 0.6817$
$= 68.17\%$

(b) 70%

55. (a) $R = 0.9 - e^{-0.2t}$

At $t = 10$,

$R = 0.9 - e^{-0.2(10)}$
$= 0.9 - e^{-2}$
$= 0.7647$
$= 76.47\%$

At $t = 20$,

$R = 0.9 - e^{-0.2(20)}$
$= 0.9 - e^{-4}$
$= 0.8817$
$= 88.17\%$

(b) 90%

57. $I = \dfrac{E}{R}\left(1 - e^{-(R/L)t}\right)$

(a) If $E = 120$, $R = 10$, and $L = 5$,

$$I = \frac{120}{10}\left(1 - e^{-\left(\frac{10}{5}\right)t}\right)$$

$$= 12\left(1 - e^{-2t}\right)$$

At $t = 0.01$

$I = 12(1 - e^{-2(0.01)}) = 0.2376$
$= 23.76\%$

At $t = 0.05$

$I = 12(1 - e^{-2(0.5)}) = 7.5854$

(b) 12 amps

(c)

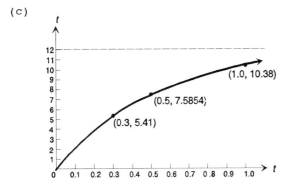

59. (a) If $t = 0$, $I = \dfrac{E}{R}e^0 = \dfrac{10}{2000} = \dfrac{1}{200} = 5 \times 10^{-3}$ milliamperes

If $t = 1000$, $I = \dfrac{1}{200}e^{-\frac{1000}{2000}} = 3.033 \times 10^{-3}$ milliamperes

If $t = 3000$, $I = \dfrac{1}{200}e^{-\frac{3000}{2000}} = 1.116 \times 10^{-3}$ milliamperes

(b)

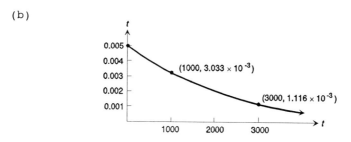

61. (a) $P = 100{,}000 - 60{,}000\left(\dfrac{1}{2}\right)^x$

At $x = 5$, $P = 100{,}000 - 60{,}000\left(\dfrac{1}{2}\right)^5 = \$98{,}125$

At $x = 10$, $P = 100{,}000 - 60{,}000\left(\dfrac{1}{2}\right)^{10} = \$99{,}941.41$

(b) \$100,000 from graph.

(c)

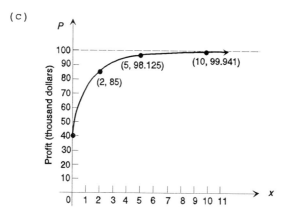

63. If $f(x) = e^x$, then $f(3) - f(2) = e^3 - e^2 = 12.696481$.

65. The x-intercepts occur between 0.4362 and 0.4468.

67. The point of intersection occurs between 0.4362 and 0.4468.

69.   (a) $\sinh (-x) = \frac{1}{2}\left(e^{-x} - e^{-(-x)}\right) = \frac{1}{2}\left(e^{-x} - e^{x}\right) = -\frac{1}{2}\left(e^{x} - e^{-x}\right) = -\sinh x$
       so that $\sinh x$ is an odd function.

       (b)

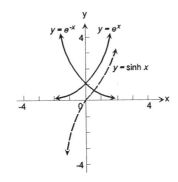

71.   For $f(x) = a^{x}$,

$$\frac{f(x + h) - f(x)}{h} = \frac{a^{x+h} - a^{x}}{h} = \frac{a^{x}a^{h} - a^{x}}{h} = a^{x}\left(\frac{a^{h} - 1}{h}\right)$$

73.   For $f(x) = a^{x}$, $f(-x) = a^{-x} = \dfrac{1}{a^{x}} = \dfrac{1}{f(x)}$

75. $f(x) = 2^{(2x)} + 1$

       $f(1) = 2^{(2^{1})} + 1 = 2^{2} + 1 = 4 + 1 = 5$

       $f(2) = 2^{(2^{2})} + 1 = 2^{4} + 1 = 16 + 1 = 17$

       $f(3) = 2^{(2^{3})} + 1 = 2^{8} + 1 = 256 + 1 = 257$

       $f(4) = 2^{(2^{4})} + 1 = 2^{16} + 1 = 65,536 + 1 = 65,537$

       $f(5) = 2^{(2^{5})} + 1 = 2^{32} + 1 = 4,294,967,296 + 1$
       $= 4,294,967,297$
       $= 641 \times 6,700,417$

≡ EXERCISE 4.2 COMPOUND INTEREST

1.   Here, $P = \$100$, $r = 0.08$, $n = 12$ and $t = 2$ in the formula

$$A = P\left(1 + \frac{r}{n}\right)^{nt} = 100\left(1 + \frac{0.08}{12}\right)^{(12)(2)} = \$117.29$$

3.   Here, $P = \$500$, $r = 0.10$, $n = 4$ and $t = 2.5$ in

$$A = P\left(1 + \frac{r}{n}\right)^{nt} = 500\left(1 + \frac{0.10}{4}\right)^{4(2.5)} = \$640.04$$

5. Here, $P = \$600$, $r = 0.05$, $n = 365$ and $t = 3$ in

$$A = P\left(1 + \frac{r}{n}\right)^{nt} = 600\left(1 + \frac{0.05}{365}\right)^{(365)(3)} = \$697.09$$

7. Here $P = \$10$, $r = 0.11$, and $t = 2$ in

$$A = Pe^{rt} = 10e^{(0.11)(2)} = \$12.46$$

9. Here $P = \$100$, $r = 0.10$, and $t = 2.25$ in

$$A = Pe^{rt} = 100e^{(0.10)(2.25)} = \$125.23$$

11. Here, $A = \$100$, $t = 2$, $r = 0.08$, and $n = 12$ in

$$V = A\left(1 + \frac{r}{n}\right)^{-nt} = 100\left(1 + \frac{0.08}{12}\right)^{-12(2)} = \$85.26$$

13. Here, $A = \$1000$, $t = 2.5$, $r = 0.06$, and $n = 365$ in

$$V = A\left(1 + \frac{r}{n}\right)^{-nt} = 1000\left(1 + \frac{0.06}{365}\right)^{-365(2.5)} = \$860.72$$

15. Here, $A = \$600$, $t = 2$, $r = 0.12$, and $n = 4$ in

$$V = A\left(1 + \frac{r}{n}\right)^{-nt} = 600\left(1 + \frac{0.12}{4}\right)^{-4(2)} = \$473.65$$

17. Here $A = \$80$, $t = 3.25$, and $r = 0.09$ in

$$V = Ae^{-rt} = 80e^{-0.09(3.25)} = \$59.71$$

19. Here $A = \$400$, $t = 1$, and $r = 0.10$ in

$$V = Ae^{-rt} = 400e^{-0.10(1)} = \$361.93$$

21. (a) We are asked for <u>present</u> <u>value</u> with <u>monthly</u> compounding. For $t = 20$, $A = \$10,000$, $r = 0.10$, and $n = 12$, we use

$$V = A\left(1 + \frac{r}{n}\right)^{-nt} = 10,000\left(1 + \frac{0.10}{12}\right)^{-12(20)} = \$1364.62$$

(b) For $r = 0.12$ and continuous compounding, we use

$$V = Ae^{-rt} = 10,000e^{-0.12(20)} = \$907.18$$

23. For you: $P = \$2000$, $r = 0.09$, $n = 2$, and $t = 20$ in

$$A = P\left(1 + \frac{r}{n}\right)^{nt} = 2000\left(1 + \frac{0.09}{2}\right)^{2(20)} = \$11,632.73$$

For your friend: $P = \$2000$, $r = 0.085$, and $t = 20$ in

$$A = Pe^{rt} = 2000e^{0.085(20)} = \$10,947.89$$

You have more money after 20 years.

25. With $1000 now, you can invest it for three years with $r = 0.10$, $P = 1000$, and $t = 3$ to get

$$A = Pe^{rt} = 1000e^{0.10(3)} = \$1349.86$$

You do better to get the $1000 now and invest it.

27. Here $P = \$50,000$ and $t = 5$ in each of the following:

(a) Now $r = 0.12$ and $n = \frac{1}{5}$ (no compounding) in

$$A = P\left(1 + \frac{r}{n}\right)^{nt} = 50,000\left(1 + \frac{0.12}{\frac{1}{5}}\right)^{1/5(5)} = \$80,000.00$$

(b) Now $r = 0.115$ and $n = 12$ in

$$A = P\left(1 + \frac{r}{n}\right)^{nt} = 50,000\left(1 + \frac{0.115}{12}\right)^{12(5)} = \$88,613.59$$

(c) Now, $r = 0.1125$ in

$$A = 50,000e^{0.1125(5)} = \$87,752.73$$

Subtracting the original 50,000 from each, we get the interest of
- (a) $30,000.00
- (b) $38,613.59
- (c) $37,752.73

Option (a) results in the least interest.

29. We let $P = \$100$, $r = 0.0950$, and $t = 1$ in all cases:

(a) For <u>continuous</u> compounding, $A = Pe^{rt} = 100e^{0.0950(1)} = \$109.97$ or an <u>effective</u> <u>rate</u> of 9.97%.

(b) For <u>daily</u> compounding, $n = 365$ in

$$A = P\left(1 + \frac{r}{n}\right)^{nt} = 100\left(1 + \frac{0.0950}{365}\right)^{365(1)} = \$109.96$$

or an <u>effective</u> <u>rate</u> of 9.96%.

(c) For <u>monthly</u> compounding, $n = 12$ in

$$A = P\left(1 + \frac{r}{n}\right)^{nt} = 100\left(1 + \frac{0.0950}{12}\right)^{12(1)} = \$109.92$$

or an <u>effective</u> <u>rate</u> of 9.92%.

(d) For <u>quarterly</u> compounding, $n = 4$ in

$$A = P\left(1 + \frac{r}{n}\right)^{nt} = 100\left(1 + \frac{0.0950}{4}\right)^{4(1)} = \$109.84$$

or an effective rate of 9.84%.

The yield is for quarterly compounding.

4.2 COMPOUND INTEREST   241

31. After one year, at 5.6% compounded continuously, you'll have $A = 1000e^{.056} = \$1057.60$. You will not have enough money to buy a computer system that costs $1,060. After one year, at 5.9% compounded monthly, you'll have $A = 1000\left(1 + \dfrac{.059}{12}\right)^{12} = \$1060.62$. This is a better deal. You will have enough money to buy the computer system.

33. $A = Pe^{rt}$

$$\dfrac{A}{e^{rt}} = \dfrac{Pe^{rt}}{e^{rt}}$$

$Ae^{-rt} = P$

$P = Ae^{-rt}$

≡ EXERCISE 4.3 LOGARITHMIC FUNCTIONS

1. For $f(x) = \ln(3 - x)$, the domain is all x such that

$$3 - x > 0$$
$$-x > -3$$
$$\{x \mid x < 3\}$$

3. For $F(x) = \log_2 x^2$, the domain is all x such that $x^2 > 0$, or all real numbers except zero.

5. For $h(x) = \log_{1/2}(x^2 - x - 6)$, the domain is all x such that

$$x^2 - x - 6 > 0$$
$$(x - 3)(x + 2) > 0$$
$$\{x \mid x < -2 \text{ or } x > 3\}$$

7. For $f(x) = \dfrac{1}{\ln x}$ the domain is all x such that $x > 0$ except that $\ln x$ cannot be zero. Recall that $\ln 1 = 0$ so that $x \neq 1$. The domain is $\{x \mid x > 0, x \neq 1\}$.

9. For $g(x) = \log_5\left(\dfrac{x + 1}{x}\right)$, the domain is all x such that $\dfrac{x + 1}{x} > 0$.

	$x + 1$	x	$\dfrac{x + 1}{x}$
$x < -1$	$-$	$-$	$+$
$-1 < x < 0$	$+$	$-$	$-$
$x > 0$	$+$	$+$	$+$

$\dfrac{x + 1}{x} > 0$ when $x < -1$ or $x > 0$.

The domain is $\{x \mid x < -1 \text{ or } x > 0\}$

11. $\log_2 8 = 3$ is equivalent to $2^3 = 8$

13. $\log_a 3 = 6$ is equivalent to $a^6 = 3$

15. $\log_3 2 = x$ is equivalent to $3^x = 2$

17. $\log_2 M = 1.3$ is equivalent to $2^{1.3} = M$

19. $\log_{\sqrt{2}} \pi = x$ is equivalent to $\left(\sqrt{2}\right)^x = \pi$

21. $\ln 4 = x$ is equivalent to $e^x = 4$

In Problems 23-33, we use the equivalence of $a^x = M$ and $x = \log_a M$.

23. $9 = 3^2$ is equivalent to $2 = \log_3 9$

25. $a^2 = 1.6$ is equivalent to $2 = \log_a 1.6$

27. $1.1^2 = M$ is equivalent to $2 = \log_{1.1} M$

29. $2^x = 7.2$ is equivalent to $x = \log_2 7.2$

31. $x^{\sqrt{2}} = \pi$ is equivalent to $\sqrt{2} = \log_x \pi$

33. $e^x = 8$ is equivalent to $x = \ln 8$

In Problems 35-45, we use (3) - (6):

35. $\log_2 1 = 0$ by (3)

37. $\log_5 25 = \log_5 5^2 = 2$ by (6)

39. $\log_{1/2} 16 = \log_{1/2} 2^4 = \log_{1/2}\left(\frac{1}{2}\right)^{-4} = -4$ by (6)

41. $\log_{10} \sqrt{10} = \log_{10} 10^{1/2} = \frac{1}{2}$ by (6)

43. $\log_{\sqrt{2}} 4 = \log_{\sqrt{2}}\left(\sqrt{2}\right)^4$ since $4 = \left(\sqrt{2}\right)^4$

$= 4$ by (6)

45. $\ln\sqrt{e} = \ln e^{1/2} = \frac{1}{2}$ by (6)

47. For $f(x) = \log_a x$, we want to find a so that $f(2) = \log_a 2 = 2$ or $a^2 = 2$ or $a = \sqrt{2}$. Recall that $a > 0$ by definition.

49. B   51. D   53. A   55. E

4.3 LOGARITHMIC FUNCTIONS   243

57.

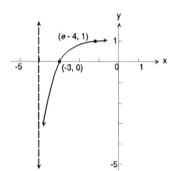

59.

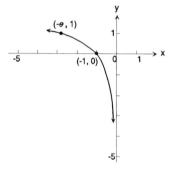

61.

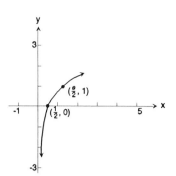

63.

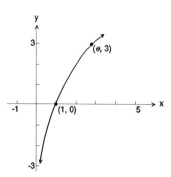

65.

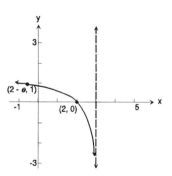

67 - 75. See answers for Problems 57 - 65.

4 EXPONENTIAL AND LOGARITHMIC FUNCTIONS

≡ EXERCISE 4.4 PROPERTIES OF LOGARITHMS

For Problems 1-11, we use $\ln 2 = a$ *and* $\ln 3 = b$.

1. $\ln 6 = \ln(3 \cdot 2) = \ln 3 + \ln 2 = b + a$

3. $\ln 1.5 = \ln \dfrac{3}{2} = \ln 3 - \ln 2 = b - a$

5. $\ln 2e = \ln 2 + \ln e = a + 1$

7. $\ln 12 = \ln(3 \cdot 4) = \ln 3 + \ln 4 = \ln 3 + \ln 2^2 = \ln 3 + 2 \ln 2$
$$= b + 2a$$

9. $\ln \sqrt[5]{18} = \ln(18)^{1/5} = \dfrac{1}{5}\ln(2 \cdot 3^2) = \dfrac{1}{5}(\ln 2 + \ln 3^2)$

$$= \dfrac{1}{5}(\ln 2 + 2 \ln 3) = \dfrac{1}{5}[a + 2b]$$

11. $\log_2 3 = \dfrac{\log_e 3}{\log_e 2} = \dfrac{\ln 3}{\ln 2} = \dfrac{b}{a}$

13. $\ln\left[x^2\sqrt{1 - x}\right] = \ln x^2 + \ln\sqrt{1 - x}$ by (7)

$$= \ln x^2 + \ln(1 - x)^{1/2}$$

$$= 2 \ln x + \dfrac{1}{2}\ln(1 - x)\ \text{by}(10)$$

15. $\log_2\left(\dfrac{x^3}{x - 3}\right) = \log_2 x^3 - \log_2(x - 3)\quad$ by (8)
$$= 3 \log_2 x - \log_2(x - 3)\quad \text{by (10)}$$

17. $\log\left[\dfrac{x(x + 2)}{(x + 3)^2}\right] = \log x(x + 2) - \log(x + 3)^2\quad$ by (8)
$$= \log x + \log(x + 2) - 2 \log(x + 3)\quad \text{by (7) and (10)}$$

19. $\ln\left[\dfrac{x^2 - x - 2}{(x + 4)^2}\right]^{1/3} = \dfrac{1}{3}\ln\left(\dfrac{(x + 1)(x - 2)}{(x + 4)^2}\right)\quad$ by (10)

$$= \dfrac{1}{3}\left(\ln(x + 1)(x - 2) - \ln(x + 4)^2\right)\ \text{by (8)}$$

$$= \dfrac{1}{3}\left(\ln(x + 1) + \ln(x - 2) - 2 \ln(x + 4)\right)\ \text{by (7) and (10)}$$

$$= \dfrac{1}{3}\ln(x + 1) + \dfrac{1}{3}\ln(x - 2) - \dfrac{2}{3}\ln(x + 4)$$

21. $\ln \dfrac{5x\sqrt{1-3x}}{(x-4)^3} = \ln\left(5x\sqrt{1-3x}\right) - \ln(x-4)^3$ by (8)

$$= \ln 5 + \ln x + \ln\sqrt{1-3x} - \ln(x-4)^3 \quad \text{by (7)}$$

$$= \ln 5 + \ln x + \ln(1-3x)^{1/2} - \ln(x-4)^3$$

$$= \ln 5 + \ln x + \frac{1}{2}\ln(1-3x) - 3\ln(x-4) \quad \text{by (10)}$$

23. $3\log_5 u + 4\log_5 v = \log_5 u^3 + \log_5 v^4$ by (10)

$$= \log_5 u^3 v^4 \quad \text{by (7)}$$

25. $\log_{1/2}\sqrt{x} - \log_{1/2} x^3 = \log_{1/2}\dfrac{\sqrt{x}}{x^3}$ by (8)

$$= \log_{1/2} x^{(1/2)-3}$$

$$= \log_{1/2} x^{-5/2}$$

$$= -\frac{5}{2}\log_{1/2} x \quad \text{by (10)}$$

27. $\ln\left(\dfrac{x}{x-1}\right) + \ln\left(\dfrac{x+1}{x}\right) - \ln(x^2-1) = \ln\dfrac{x}{x-1}\cdot\dfrac{x+1}{x} - \ln(x^2-1)$ by (7)

$$= \ln\frac{x+1}{x-1} \div (x^2-1) \quad \text{by (8)}$$

$$= \ln\frac{1}{(x-1)^2}$$

$$= \ln(x-1)^2$$

$$= -2\ln(x-1) \quad \text{by (10)}$$

29. $8\log_2\sqrt{3x-2} - \log_2\left(\dfrac{4}{x}\right) + \log_2 4 = \log_2\left(\sqrt{3x-2}\right)^8 - \log_2\dfrac{4}{x} + \log_2 4$ by (10)

$$= \log_2\frac{\left(\sqrt{3x-2}\right)^8 \cdot 4}{\dfrac{4}{x}} \quad \text{by (7) and (8)}$$

$$= \log_2\left[x\left(\sqrt{3x-2}\right)^8\right]$$

$$= \log_2\left[x\left((3x-2)^{1/2}\right)^8\right]$$

$$= \log_2\left[x(3x-2)^4\right]$$

31. $2 \log_a 5x^3 - \frac{1}{2} \log_a (2x + 3) = \log_a (5x^3)^2 - \log_a (2x + 3)^{1/2}$ by (10)

$$= \log_a 25x^6 - \log_a \sqrt{2x + 3}$$

$$= \log_a \frac{25x^6}{\sqrt{2x + 3}} \quad \text{by (8)}$$

$$= \log_a \frac{25x^6}{(2x + 3)^{1/2}}$$

33. $\frac{1}{2} \log_3 x = 2$

$\log_3 x^{1/2} = 2$

$x^{1/2} = 3^2$

$x = 3^4 = 81$

35. $3 \log_2 (x - 1) + \log_2 4 = 5$

$\log_2 (x - 1)^3 + \log_2 4 = 5$

$\log_2 (x - 1)^3 \cdot 4 = 5$

$4(x - 1)^3 = 2^5$

$(x - 1)^3 = \frac{2^5}{4}$

$(x - 1)^3 = 8$

$x - 1 = 2$

$x = 3$

37. $\log_{10} x + \log_{10} x(x + 15) = 2$

$\log_{10} x(x + 15) = 2$

$x(x + 15) = 10^2$

$x^2 + 15x - 100 = 0$

$(x + 20)(x - 5) = 0$

$x = -20 \quad \text{or} \quad x = 5$

Since $\log_{10} (-20)$ is undefined, we choose only $x = 5$.

39. $\log_x 4 = 2$

$4 = x^2$

$\pm 2 = x$

Since $\log_x (-2)$ is undefined, we have only $x = 2$.

41. $\log_3 (x - 1)^2 = 2$

$(x - 1)^2 = 3^2$

$x - 1 = 3 \quad \text{or} \quad x = 1 - 3$

$x = 3 \quad \text{or} \quad x = -2$

$\{-2, 4\}$

43. $\log_{1/2} (3x + 1)^{1/3} = -2$

$$(3x + 1)^{1/3} = \left(\frac{1}{2}\right)^{-2}$$

$$3x + 1 = \left(\frac{1}{2}\right)^{-6}$$

$$3x = 2^6 - 1$$

$$\frac{3x}{6} = \frac{63}{3}$$

$$x = 21$$

4.4 PROPERTIES OF LOGARITHMS

247

45. $\log_a(x - 1) - \log_a(x + 6) = \log_a(x - 2) - \log_a(x + 3)$

$$\log_a\frac{x - 1}{x + 6} = \log_a\frac{x - 2}{x + 3}$$

$$\frac{x - 1}{x + 6} = \frac{x - 2}{x + 3}$$

$$(x - 1)(x + 3) = (x - 2)(x + 6)$$

$$x^2 + 2x - 3 = x^2 + 4x - 12$$

$$9 = 2x$$

$$\frac{9}{2} = x$$

47. $\log_{1/3}(x^2 + x) - \log_{1/3}(x^2 - x) = -1$

$$\log_{1/3}\frac{x^2 + x}{x^2 - x} = -1$$

$$\frac{x^2 + x}{x^2 - x} = \left(\frac{1}{3}\right)^{-1}$$

$$\frac{x(x + 1)}{x(x - 1)} = 3$$

$$x + 1 = 3(x - 1)$$

$$x + 1 = 3x - 3$$

$$4 = 2x$$

$$2 = x$$

49. $\log_2 8^x = -3$

$x \log_2 8 = -3$

$$x = \frac{-3}{\log_2 8}$$

$$= \frac{-3}{\log_2 2^3}$$

$$= \frac{-3}{3}$$

$$= -1$$

51. $\log_2(x^2 + 1) - \log_4 x^2 = 1$

$$\log_2(x^2 + 1) - \frac{\log_2 x^2}{\log_2 4} = 1$$

$$\log_2(x^2 + 1) - \frac{\log_2 x^2}{2} = 1$$

$$\log_2(x^2 + 1) - \frac{1}{2}\left(\log_2 x^2\right) = 1$$

$$\log_2(x^2 + 1) - \log_2 x = 1$$

$$\log_2\frac{x^2 + 1}{x} = 1$$

$$\frac{x^2 + 1}{x} = 2$$

$$x^2 - 2x + 1 = 0$$

$$(x - 1)^2 = 0$$

$$x = 1$$

53. $\log_{16}x + \log_4x + \log_2x = 7$

$$\frac{\log_2x}{\log_216} + \frac{\log_2x}{\log_24} + \log_2x = 7$$

$$\frac{\log_2x}{4} + \frac{\log_2x}{2} + \log_2x = 7$$

$$\left(\frac{1}{4} + \frac{1}{2} + 1\right)\log_2x = 7$$

$$\frac{7}{4}\log_2x = 7$$

$$\log_2x^{\frac{7}{4}} = 7$$

$$x^{\frac{7}{4}} = 2^7$$

$$x^{\frac{1}{4}} = 2$$

$$x = 2^4 = 16$$

55. $\log_a\left(x + \sqrt{x^2 - 1}\right) + \log_a\left(x - \sqrt{x^2 - 1}\right) = \log_a\left(x + \sqrt{x^2 - 1}\right)\left(x - \sqrt{x^2 - 1}\right)$

$$= \log_a\left(x^2 - \left(\sqrt{x^2 - 1}\right)^2\right)$$

$$= \log_a\left(x^2 - \left(x^2 - 1\right)\right)$$

$$= \log_a\left(x^2 - x^2 + 1\right)$$

$$= \log_a1$$

$$= 0$$

57. $\log_321 = \dfrac{\log 21}{\log 3} = \dfrac{1.32222}{0.47712} = 2.7712437$

59. $\log_{1/3}71 = \dfrac{\log 71}{\log_{1/3}} = \dfrac{\log 7}{-\log 3} = \dfrac{1.85126}{-0.47712} = -3.880058$

61. $\log_{\sqrt{2}}7 = \dfrac{\log 7}{\log \sqrt{2}} = \dfrac{\log 7}{\log 2^{1/2}} = \dfrac{\log 7}{\frac{1}{2}\log 2} = \dfrac{0.84510}{0.5(0.30103)} = 5.6147098$

63. $\log_\pi e = \dfrac{\ln e}{\ln \pi} = \dfrac{1}{1.14473} = 0.8735685$

65.

$$2^x = 10$$

$$\log 2^x = \log 10$$

$$x \log 2 = 1$$

$$x = \frac{1}{\log 2}$$

$$= 3.3219281$$

67.

$$8^{-x} = 1.2$$

$$\log 8^{-x} = \log 1.2$$

$$-x \log 8 = \log 1.2$$

$$-x = \frac{\log 1.2}{\log 8}$$

$$-x = 0.0877$$

$$x = -0.0876781$$

69.

$$3^{1-2x} = 4^x$$

$$\log 3^{1-2x} = \log 4^x$$

$$(1 - 2x)\log 3 = x \log 4$$

$$\log 3 - 2x \log 3 = x \log 4$$

$$\log 3 = x \log 4 + 2x \log 3$$

$$\log 3 = x \log (4 + 2 \log 3)$$

$$\frac{\log 3}{\log 4 + 2 \log 3} = x$$

$$x = \frac{0.47712}{0.60206 + 2(0.47712)}$$

$$= 0.3065736$$

71.

$$\left(\frac{3}{5}\right)^x = 7^{1-x}$$

$$\log\left(\frac{3}{5}\right)^x = \log 7^{1-x}$$

$$x \log \frac{3}{5} = (1 - x)\log 7$$

$$x \log \frac{3}{5} = \log 7 - x \log 7$$

$$x \log \frac{3}{5} + x \log 7 = \log 7$$

$$x\left(\log \frac{3}{5} + \log 7\right) = \log 7$$

$$x = \frac{\log 7}{\log \frac{3}{5} + \log 7} = \frac{\log 7}{\log 3 - \log 5 + \log 7}$$

$$= 1.3559551$$

73.
$$1.2^x = (0.5)^{-x}$$
$$\log 1.2^x = \log (0.5)^{-x}$$
$$x \log 1.2 = -x \log 0.5$$
$$x \log 1.2 + x \log 0.5 = 0$$
$$x(\log 1.2 + \log 0.5) = 0$$
$$x = 0$$

75.
$$\pi^{1-x} = e^x$$
$$\ln \pi^{1-x} = \ln e^x$$
$$(1 - x)\ln \pi = x \ln e$$
$$\ln \pi - x \ln \pi = x(1)$$
$$\ln \pi = x + x \ln \pi$$
$$\ln \pi = x(1 + \ln \pi)$$
$$\frac{\ln \pi}{1 + \ln \pi} = x$$
$$0.5337408 = x$$

77.
$$5\left(2^{3x}\right) = 8$$
$$2^{3x} = \frac{8}{5}$$
$$\ln 2^{3x} = \ln \frac{8}{5}$$
$$3x \ln 2 = \ln \frac{8}{5}$$
$$x = \frac{\ln \frac{8}{5}}{3 \ln 2}$$
$$x = 0.2260239$$

79. $400e^{0.2x} = 600$
$$e^{0.2x} = \frac{600}{400}$$
$$e^{0.2x} = \frac{3}{2}$$
$$0.2x = \ln \frac{3}{2}$$
$$x = \frac{\ln \frac{3}{2}}{0.2}$$
$$= 2.0273255$$

81. Using the graphing calculator, the solution to $\ln x + x = 2$ is between 1.5532 and 1.5638.

83. Using the graphing calculator, the solution to $\ln x = x^2 - 1$ is between 0.4468 and 0.4681.

85. The domain of $f(x) = \log_a x^2$ is all real numbers except zero. The domain of $g(x) = 2 \log_a x$ is all positive real numbers. $f(x) = g(x)$ only for $x > 0$.

87. If $y = f(x) = \log_a x$,
$$a^y = x$$
$$\left(\frac{1}{a}\right)^{-y} = x$$
$$-y = \log_{1/a} x$$

Thus, $-f(x) = \log_{1/a} x$

89. If $f(x) = \log_a x$,
$$f(AB) = \log_a AB$$
$$= \log_a A + \log_a B$$
$$= f(A) + f(B)$$

91.
$$\ln y = \ln x = \ln C$$
$$\ln y = \ln xC$$
$$y = xC = Cx$$

93.
$$\ln y = \ln x + \ln(x + 1) + \ln C$$
$$\ln y = \ln x(x + 1)C$$
$$y = x(x + 1)C = Cx(x + 1)$$

4.4 PROPERTIES OF LOGARITHMS

95.
$$\ln y = 3x + \ln C$$
$$\ln y = \ln e^{3x} = \ln C$$
$$\ln y = \ln Ce^{3x}$$
$$y = Ce^{3x}$$

97.
$$\ln(y - 3) = -4x + \ln C$$
$$\ln(y - 3) = \ln e^{-4x} + \ln C$$
$$\ln(y - 3) = \ln Ce^{-4x}$$
$$y - 3 = Ce^{-4x}$$
$$y = Ce^{-4x} + 3$$

99.
$$3 \ln y = \frac{1}{2} \ln(2x + 1) - \frac{1}{3} \ln(x + 4) + \ln C$$

$$\ln y^3 = \ln(2x + 1)^{1/2} - \ln(x + 4)^{1/3} + \ln C$$

$$\ln y^3 = \ln\left[\frac{C(2x + 1)^{1/2}}{(x + 4)^{1/3}}\right]$$

$$y^3 = \left[\frac{C(2x + 1)^{1/2}}{(x + 4)^{1/3}}\right]$$

$$y = \left[\frac{C(2x + 1)^{1/2}}{(x + 4)^{1/3}}\right]^{1/3}$$

$$= \frac{\sqrt[3]{C}(2x + 1)^{1/6}}{(x + 4)^{1/9}}$$

101.
$$\log_2 3 \cdot \log_3 4 \cdot \log_4 5 \cdot \log_5 6 \cdot \log_6 7 \cdot \log_7 8$$

$$= \frac{\log 3}{\log 2} \cdot \frac{\log 4}{\log 3} \cdot \frac{\log 5}{\log 4} \cdot \frac{\log 6}{\log 5} \cdot \frac{\log 7}{\log 6} \cdot \frac{\log 8}{\log 7}$$

$$= \frac{\log 8}{\log 2} = \frac{\log 2^3}{\log 2} = \frac{3 \log 2}{\log 2} = 3$$

103.
$$\log_2 3 \cdot \log_3 4 \cdot \ldots \cdot \log_n(n + 1) \cdot \log_{n+1} 2$$

$$= \frac{\log 3}{\log 2} \cdot \frac{\log 4}{\log 3} \cdot \ldots \cdot \frac{\log n + 1}{\log n} \cdot \frac{\log 2}{\log n + 1}$$

$$= \frac{\log 2}{\log 2} = 1$$

105. We want to find a time t in years. Let the initial amount be $100 so that the new value will be $200. Now $r = 0.08$ and

(a) for <u>monthly</u> compounding $n = 12$ in

$$A = P\left(1 + \frac{r}{n}\right)^{nt}$$

$$200 = 100\left(1 + \frac{0.08}{12}\right)^{12t}$$

$$2 = (1.00667)^{12t}$$

$$\log 2 = \log(1.00667)^{12t}$$

$$\log 2 = 12t \log(1.00667)$$

$$t = \frac{\log 2}{12 \log(1.00667)}$$

$$= 8.6932 \text{ years}$$

$$= 104.32 \text{ months}$$

4 EXPONENTIAL AND LOGARITHMIC FUNCTIONS

(b) for <u>continuous</u> compounding

$$A = Pe^{rt}$$
$$200 = 100e^{0.08t}$$
$$2 = e^{0.08t}$$
$$\ln 2 = \ln e^{0.08t}$$
$$\ln 2 = 0.08t$$
$$\frac{\ln 2}{0.08} = t$$
$$8.6643 = t$$
$$t = 8.6643 \text{ years}$$
$$= 103.97 \text{ months}$$

107. We want to find a time t in years. Here $A = \$150$, $P = \$100$, $r = 0.08$, and

(a) for <u>monthly</u> compounding $n = 12$ in

$$A = P\left(1 + \frac{r}{n}\right)^{nt}$$
$$150 = 100\left(1 + \frac{0.08}{12}\right)^{12t}$$
$$1.5 = (1.00667)^{12t}$$
$$\log 1.5 = \log(1.00667)^{12t}$$
$$\log 1.5 = 12t \log (1.00667)$$
$$t = \frac{\log 1.5}{12 \log(1.00667)}$$
$$= 5.0852 \text{ years}$$
$$= 61.02 \text{ months}$$

(b) for <u>continuous</u> compounding

$$A = Pe^{rt}$$
$$150 = 100e^{0.08t}$$
$$1.5 = e^{0.08t}$$
$$\ln 1.5 = \ln e^{0.08t}$$
$$\ln 1.5 = 0.08t$$
$$t = \frac{\ln 1.5}{0.08} = 5.0683 \text{ years} = 60.82 \text{ months}$$

109. If $A = \log_a M$ and $B = \log_a N$, then $a^A = M$ and $a^B = N$.

Then, $\log_a\left(\dfrac{M}{N}\right) = \log_a\left(\dfrac{a^A}{a^B}\right) = \log_a a^{A-B} = A - B = \log_a M - \log_a N$

1. $L(10^{-5}) = 10 \log \dfrac{10^{-5}}{10^{-12}} = 10 \log 10^7 = 10(7) = 70$ decibels

3. $L(0.15) = 10 \log \dfrac{0.15}{10^{-12}} = 10 \log (0.15(10^{12}))$

    $$= 10(\log 0.15 + \log 10^{12})$$
    $$= 10(-0.8239 + 12)$$
    $$= 111.76 \text{ decibels}$$

5. $L(x) = 10 \log \dfrac{x}{10^{-12}} = 130$

    $$\log (x(10^{12})) = 13$$
    $$\log x + \log 10^{12} = 13$$
    $$\log x + 12 = 13$$
    $$\log x = 1$$
    $$x = 10^1$$
    $$x = 10 \text{ Watts per square meter}$$

7. For $M(x) = \log \dfrac{x}{x_0}$ we have $x = 10.0$ and $x_0 = 10^{-3}$:

    $$M(10.0) = \log \dfrac{10.0}{10^{-3}} = \log 10^4 = 4.0 \text{ on the Richter scale.}$$

9. Using $M(x) = \log \dfrac{x}{x_0}$ we have $M(x) = 7.85$ and $x_0 = 10^{-3}$:

    $$7.85 = \log \dfrac{x}{10^{-3}}$$
    $$\dfrac{x}{10^{-3}} = 10^{7.85}$$
    $$x = 10^{4.85} = 70,794.58 \text{ mm}$$

    For $M(x) = 8.9 = \log \dfrac{x}{10^{-3}}$
    $$\dfrac{x}{10^{-3}} = 10^{8.9}$$
    $$x = 10^{5.9}$$

    The San Francisco earthquake was $\dfrac{10^{5.9}}{10^{4.85}} = 10^{1.05} = 11.22$ times as intense as the one in Mexico City.

11. Using $h(x) = (30T + 8000) \log \dfrac{P_0}{x}$ where $T = 0^0$, $P_0 = 760$, and $x = 300$, we want to find $h(300)$:

    $$h(300) = (30(0) + 8000) \log \dfrac{760}{300} = 8000(0.404)$$
    $$= 3229.54 \text{ m above sea level}$$

≡ EXERCISE 4.6 GROWTH AND DECAY

1. For $P = 5000e^{0.02t}$ we want to find t (in days)

    when $P = 1000$: $1000 = 500e^{0.02t}$
    $$2 = e^{0.02t}$$
    $$0.02t = \ln 2$$
    $$t = \frac{\ln 2}{0.02} = 34.7 \text{ days}$$

    when $P = 2000$: $2000 = 500e^{0.02t}$
    $$4 = e^{0.02t}$$
    $$0.02t = \ln 4$$
    $$t = \frac{\ln 4}{0.02} = 69.3 \text{ days}$$

3. For $A = A_0 = e^{-0.0244t}$, the half-life is the time until $\frac{1}{2}A_0$ remains,

    so that $A = \frac{1}{2}A_0 = A_0e^{-0.0244t}$
    $$\frac{1}{2} = e^{-0.0244t}$$
    $$-0.0244t = \ln \frac{1}{2}$$
    $$t = \frac{-\ln 2}{-0.0244} = 28.4 \text{ years}$$

5. We have $A = A_0e^{-0.0244t}$, with $A_0 = 100$ and $A = 10$:
    $$10 = 100e^{-0.0244t}$$
    $$\frac{1}{10} = e^{-0.0244t}$$
    $$-0.0244t = \ln\frac{1}{10}$$
    $$t = \frac{-\ln 10}{-0.0244} = 94.4 \text{ years}$$

7. Using $N(t) = N_0e^{kt}$ where $N_0 = 1000$, $N(t) = 1800$, and $t = 1$, we get
    $$1800 = 1000e^{k1}$$
    $$1.8 = e^k$$
    $$k = \ln 1.8$$
    $$= 0.5878$$

    Thus, $N(t) = N_0e^{0.5878t}$ and when $t = 3$
    $$N(3) = 1000e^{0.5878(3)} = 1000e^{1.7634} = 5832 \text{ mosquitoes}$$

    When $N(t) = 10,000$, we have $10,000 = 1000e^{0.5878t}$
    $$10 = e^{0.5878t}$$
    $$0.5878t = \ln 10$$
    $$t = \frac{\ln 10}{0.5878} = 3.9 \text{ days}$$

9. Using $A = A_0 e^{kt}$ if after 18 months (= 1.5 years) we have $A = 2A_0$, then

$$\begin{aligned} 2A_0 &= A_0 e^{k(1.5)} \\ 2 &= e^{1.5k} \\ 1.5k &= \ln 2 \\ k &= \frac{\ln 2}{1.5} = 0.4621 \end{aligned}$$

Here $A_0 = 10,000$ and when $t = 2$,

$$A = 10,000e^{0.4621(2)} = 10,000e^{0.9242} = 25,198$$

is the population 2 years from now.

11. Using $A = A_0 e^{kt}$ where the half-life, when $\frac{1}{2}A_0$ is present, is $t = 1690$ years, gives

$$\begin{aligned} \frac{1}{2}A_0 &= A_0 e^{k(1690)} \\ \frac{1}{2} &= e^{1690k} \\ 1690k &= \ln \frac{1}{2} \\ k &= \frac{-\ln 2}{1690} = -0.00041 \end{aligned}$$

If $A_0 = 10$, the amount present after 50 years is

$$A = 10e^{-0.00041(50)} = 10e^{-0.0205} = 9.797 \text{ grams}$$

13. Using $A_0 = A_0 e^{kt}$ the half-life $t = 5600$ years when $\frac{1}{2}A_0$ will be present:

$$\begin{aligned} \frac{1}{2}A_0 &= A_0 e^{k(5600)} \\ \frac{1}{2} &= e^{5600k} \\ 5600k &= \ln \frac{1}{2} \\ k &= \frac{-\ln 2}{5600} = -0.000124 \end{aligned}$$

Thus, $A = A_0 e^{-0.000124t}$ and let $A_0 = 100$ so that $A = 30$:

$$\begin{aligned} 30 &= 100e^{-0.000124t} \\ 0.3 &= e^{-0.000124t} \\ -0.000124t &= \ln 0.3 \\ t &= \frac{\ln 0.3}{-0.000124} = 9727 \text{ years ago} \end{aligned}$$

15. Using $u = T + (u_0 - T)e^{kt}$ where $t = 5$, $T = 70$, $u_0 = 450$, and $u = 300$:

$$300 = 70 + (450 - 70)e^{k(5)}$$
$$230 = 380e^{5k}$$
$$0.6053 = e^{5k}$$
$$5k = \ln 0.6053$$
$$k = -0.1004$$

Thus, $u = T + (u_0 - T)e^{-0.1004t} = 70 + (450 - 70)e^{-0.1004t}$
$$= 70 + 380e^{-0.1004t}$$

And, when $u = 135$ (with $T = 70$, $u_0 = 450$ still)

$$135 = 70 + (450 - 70)e^{-0.1004t}$$
$$65 = 380e^{-0.1004t}$$
$$0.17105 = e^{-0.1004t}$$
$$-0.1004t = \ln 0.17105$$
$$t = 17.59 \text{ minutes past 5:00 p.m.}, \approx 5{:}18 \text{ p.m.}$$

17. Using $u = T + (u_0 - T)e^{kt}$ where $T = 35$, $u_0 = 8$, and $u = 15$, when $t = 3$:

$$15 = 35 + (8 - 35)e^{k(3)}$$
$$-20 = -27e^{3k}$$
$$0.7407 = e^{3k}$$
$$3k = \ln 0.7407$$
$$k = -0.100035$$

Thus, $u = T + (u_0 - T)e^{kt} = 35 + (8 - 35)e^{-0.100035t}$

When $t = 5$, $u = 35 - 27e^{-0.100035(5)} = 18.63°C$

When $t = 10$, $u = 35 - 27e^{-0.100035(10)} = 25.1°C$

19. Using $A = A_0 e^{kt}$ where $A_0 = 25$ and $A = 15$ when $t = 10$:

$$15 = 25e^{k(10)}$$
$$0.6 = e^{10k}$$
$$10k = \ln 0.6$$
$$k = \ = -0.0511$$

Thus, $A = A_0 e^{kt} = 25e^{-0.0511t}$

When $t = 24$, $A = 25e^{-0.0511(24)} = 7.34$ kg. remain.

When $A = \frac{1}{2}$ kg, $\frac{1}{2} = 25e^{-0.0511t}$

$$.02 = e^{-0.0511t}$$
$$-0.0511t = \ln .02$$
$$t = 76.6 \text{ hours have passed.}$$

21. Using $R = 1.5e^{ka}$ where $R = 10\%$ and $a = 0.11$, we find k:

$$10 = 1.5e^{k(0.11)}$$
$$6.67 = e^{0.11k}$$
$$0.11k = \ln 6.67$$
$$k = 17.25$$

Thus, $R = 1.5e^{17.25a}$

If $a = 0.17$, then $R = 1.5e^{17.25(0.17)} = 28.15\%$

If $R = 100 = 1.5e^{17.25a}$

$$66.67 = e^{17.25a}$$
$$17.25a = \ln 66.67$$
$$a = 0.24$$

If $R = 20 = 1.5e^{17.25a}$

$$13.33 = e^{17.25a}$$
$$17.25a = \ln 13.33$$
$$a = 0.15$$

23. For $I = \dfrac{E}{R}\left(1 - e^{-(R/L)t}\right)$ where $E = 12$, $R = 10$, $L = 5$, and $I = 0.5$:

$$0.5 = \frac{12}{10}\left(1 - e^{-(10/5)t}\right)$$
$$e^{-2t} = 0.5833$$
$$-2t = \ln 0.5833$$
$$t = 0.2695 \text{ seconds}$$

For $I = 1.0 = \dfrac{12}{10}\left(1 - e^{-(10/15)t}\right)$

$$e^{-2t} = 0.1667$$
$$-2t = \ln 0.1667$$
$$t = 0.8959 \text{ seconds}$$

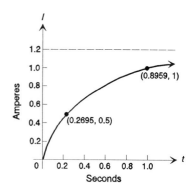

25. Using $L(t) = A\left(1 - e^{-kt}\right)$ where $A = 200$ and $L = 20$ when $t = 5$:

$$20 = 200\left(1 - e^{-k(5)}\right)$$
$$e^{-5k} = 0.9$$
$$-5k = \ln 0.9$$
$$k \approx 0.02107 \text{ is the rate of learning.}$$

When $t = 10$, $L(10) = 200\left(1 - e^{-0.02107(10)}\right) = 38$ words

When $t = 15$, $L(15) = 200\left(1 - e^{-0.02107(15)}\right) = 54$ words

When $L(t) = 180 = 200\left(1 - e^{-0.02107t}\right)$

$$e^{-0.02107t} = 0.1$$
$$-0.02107t = \ln 0.1$$
$$t \approx 109 \text{ minutes}$$

4 EXPONENTIAL AND LOGARITHMIC FUNCTIONS

27.   Using
$$A = A_0 e^{-0.087t}$$
$$.10A_0 = A_0 e^{-0.087t}$$
$$.10 = e^{-0.087t}$$
$$\ln .10 = -0.087t$$
$$t = \frac{\ln(.10)}{-0.087}$$
$$t \approx 26.5 \text{ days}$$

Farmers need to wait 26.5 days to use the hay.

≡ 4 - CHAPTER REVIEW

1. $\log_2 \frac{1}{8} = \log_2 (2)^{-3} = -3$ 3. $\ln e^{\sqrt{2}} = \sqrt{2}$

5. $2^{\log_2 0.4} = 0.4$

7. $3 \log_4 x^2 + \frac{1}{2} \log_4 \sqrt{x} = \log_4 (x^2)^3 + \log_4 \left(\sqrt{x}\right)^{1/2}$

$$= \log_4 x^6 + \log_4 x^{1/4}$$

$$= \log_4 x^6 \cdot x^{1/4}$$

$$= \log_4 x^{25/4}$$

$$= \frac{25}{4} \log_4 x$$

9. $\ln\left(\frac{x-1}{x}\right) + \ln\left(\frac{x}{x+1}\right) - \ln(x^2-1) = \ln \dfrac{\dfrac{x-1}{x} \cdot \dfrac{x}{x+1}}{x^2-1}$

$$= \ln \frac{x-1}{(x^2-1)(x+1)}$$

$$= \ln \frac{1}{(x+1)^2}$$

$$= \ln(x+1)^{-2}$$

$$= -2 \ln(x+1)$$

11. $2 \log 2 + 3 \log x - \frac{1}{2}[\log(x+3) + \log(x-2)]$

$$= \log 2^2 + \log x^3 - \frac{1}{2}[\log(x+3)(x-2)]$$

$$= \log 2^2 x^3 - \log\left((x+3)(x-2)\right)^{1/2}$$

$$= \log \frac{4x^3}{((x+3)(x-2))^{1/2}}$$

13. $\ln y = 2x^2 + \ln C$

$\ln y = \ln e^{2x^2} + \ln C$

$\ln y = \ln Ce^{2x^2} \quad y = Ce^{2x^2}$

15. $\frac{1}{2} \ln y = 3x^2 + \ln C$

$\ln y^{1/2} = \ln e^{3x^2} + \ln C$

$\ln y^{1/2} = \ln Ce^{3x^2}$

$y^{1/2} = Ce^{3x^2}$

$y = \left(Ce^{3x^2}\right)^2$

17. $\ln(y - 3) + \ln(y + 3) = x + C$

$\ln(y - 3)(y + 3) = x + C$

$(y - 3)(y + 3) = e^{x+C}$

$y^2 - 9 = e^{x+C}$

$y^2 = e^{x+C} + 9$

$y = \sqrt{e^{x+C} + 9}$

19. $e^{y+C} = x^2 + 4$

$\ln e^{y+C} = \ln(x^2 + 4)$

$y + C = \ln(x^2 + 4)$

$y = \ln(x^2 + 4) - C$

21. The graph of $f(x) = e^{-x}$ is that of the reciprocal values of $y = e^x$.

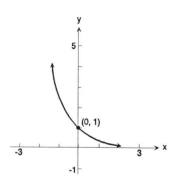

23. The graph of $f(x) = 1 - e^x$ is that of $y = -e^x$ raised one unit. Note that the graph of $y = -e^x$ is the reflection of that of $y = e^x$ in the x-axis.

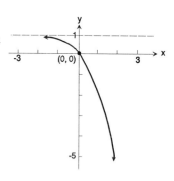

25. The graph of $f(x) = 3e^x$ is that of $y = e^x$ with each y-value multiplied by 3.

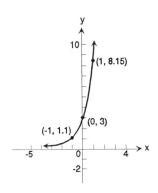

27. The graph of $f(x) = e^{|x|} = \begin{cases} e^x & x \geq 0 \\ e^{-x} & x < 0 \end{cases}$ is that of $y = e^x$ in either direction from $x = 0$.

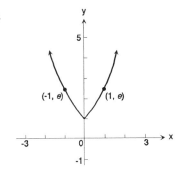

29. See discussion of Problem 23.

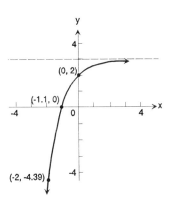

31.
$$2^{2(1-2x)} = 2^1$$
$$2(1 - 2x) = 1$$
$$2 - 4x = 1$$
$$-4x = -1$$
$$x = \frac{1}{4}$$

33.

$$3^{x^2+x} = \sqrt{3}$$

$$\log_3 3^{x^2+x} = \log_3 \sqrt{3}$$

$$(x^2 + x)\log_3 3 = \log_3 3^{1/2}$$

$$x^2 + x = \frac{1}{2}$$

$$x^2 + x - \frac{1}{2} = 0$$

$$2x^2 + 2x - 1 = 0$$

Here $a = 2$, $b = 2$, $c = -1$ and $b^3 - 4ac = 12$, so that

$$x = \frac{-2 \pm \sqrt{12}}{2(2)}$$

$$= \frac{-2 \pm 2\sqrt{3}}{4} = \frac{-1 \pm \sqrt{3}}{2}$$

$$x = \frac{-1 - \sqrt{3}}{2} \quad \text{or} \quad x = \frac{-1 + \sqrt{3}}{2}$$

$$\left\{ \frac{-1 - \sqrt{3}}{2}, \ \frac{-1 + \sqrt{3}}{2} \right\}$$

35.

$$\log_x 64 = -3$$
$$64 = x^{-3}$$
$$64^{-1/3} = (x^{-3})^{-1/3}$$
$$\frac{1}{\sqrt[3]{64}} = x$$
$$\frac{1}{4} = x$$

37.

$$\log_{\sqrt{3}} \left(9\sqrt{3}\right) = x$$
$$\log_{\sqrt{3}} \left(\sqrt{3}\right)^5 = x$$
$$5 = x$$

39.

$$5x = 3^{x+2}$$
$$\log 5^x = \log 3^{x+2}$$
$$x \log 5 = (x + 2)\log 3$$
$$x \log 5 = x \log 3 + 2 \log 3$$
$$x \log 5 - x \log 3 = \log 3^2$$
$$x(\log 5 - \log 3) = \log 9$$
$$x = \frac{\log 9}{\log 5 - \log 3} = 4.301$$

41.

$$3^{2(2x)} = 3^{3(3x-4)}$$
$$2(2x) = 3(3x - 4)$$
$$4x = 9x - 12$$
$$-5x = -12$$
$$x = \frac{12}{5}$$

43.
$$8 = 4^{x^2} \cdot 2^{5x}$$
$$8 = 2^{2x^2} \cdot 2^{5x}$$
$$2^3 = 2^{2x^2+5x}$$
$$3 = 2x^2 + 5x$$
$$0 = 2x^2 + 5x - 3$$
$$0 = (2x - 1)(x + 3)$$
$$2x - 1 = 0 \quad \text{or} \quad x + 3 = 0$$
$$2x = 1$$
$$x = \frac{1}{2} \quad \text{or} \quad x = -3$$
$$\left\{-3, \frac{1}{2}\right\}$$

45. $\log_6(x + 3) + \log_6(x + 4) = 1$
$$\log_6(x + 3)(x + 4) = 1$$
$$(x + 3)(x + 4) = 6^1$$
$$x^2 + 7x + 12 = 6$$
$$x^2 + 7x + 6 = 0$$
$$(x + 1)(x + 6) = 0$$
$$x = -1 \quad \text{or} \quad x = -6$$

But, -6 does not check, so $x = -1$ is the only solution.

47.
$$e^{1-x} = 5$$
$$\ln e^{1-x} = \ln 5$$
$$(1 - x)\ln e = \ln 5$$
$$1 - x = \ln 5$$
$$1 - \ln 5 = x$$
$$-0.6094 = x$$

49.
$$2^{3x} = 3^{2x+1}$$
$$\log 2^{3x} = \log 3^{2x+1}$$
$$3x \log 2 = (2x + 1)\log 3$$
$$3x \log 2 = 2x \log 3 + \log 3$$
$$3x \log 2 - 2x \log 3 = \log 3$$
$$x(3 \log 2) - 2 \log 3 = \log 3$$
$$x = \frac{\log 3}{3 \log 2 - 2 \log 3}$$
$$x = -9.3274$$

51. Using $A = Pe^{rt}$ where $A = \$5,000$, $P = \$620.17$ and $t = 20$:
$$5000 = 620.17e^{r(20)}$$
$$8.06 = e^{20r}$$
$$\ln 8.06 = \ln e^{20r}$$
$$\ln 8.06 = 20r$$
$$r = 0.10436 = 10.436\%$$

53. Using $L(x) = 10 \log \dfrac{x}{I^0}$, where $I_0 = 10^{-12}$

$$L(10^{-4}) = 10 \log \dfrac{10^{-4}}{10^{-12}}$$
$$= 10 \log 10^8$$
$$= 10(8)$$
$$= 80 \text{ decibels}$$

55. Using $A = A_0 e^{kt}$ where $A = \dfrac{1}{2}A_0$ when $t = 5600$:

$$\dfrac{1}{2}A_0 = A_0 e^{k(5600)}$$

$$\dfrac{1}{2} = e^{5600k}$$

$$\ln \dfrac{1}{2} = \ln e^{5600k}$$

$$-\ln 2 = 5600k$$

$$k = \dfrac{-\ln 2}{5600} = -0.000124$$

Thus, $A = A_0 e^{-0.000124t}$

In this case, $A = 0.05A_0$:

$$0.05A_0 = A_0 e^{-0.000124t}$$
$$\ln 0.05 = \ln e^{-0.000124t}$$
$$\ln 0.05 = -0.000124t$$
$$t = \dfrac{\ln 0.05}{-0.000124t} = 24{,}203 \text{ years ago}$$

4 EXPONENTIAL AND LOGARITHMIC FUNCTIONS

C H A P T E R
5

TRIGONOMETRIC
FUNCTIONS

≡ EXERCISE 5.1 ANGLES AND THEIR MEASURE

1.

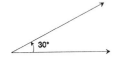

3.

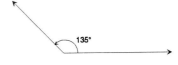

5.

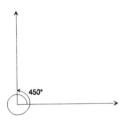

7.

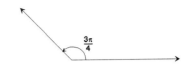

9.

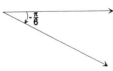

11.

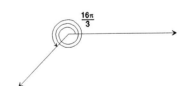

13. $30° = 30 \cdot 1 \text{ degree} = 30 \cdot \frac{\pi}{180} \text{ radian} = \frac{\pi}{6} \text{ radians}$

15. $240° = 240 \cdot 1 \text{ degree} = 240 \cdot \frac{\pi}{180} \text{ radian} = \frac{4\pi}{3} \text{ radians}$

17. $-60° = -60 \cdot 1 \text{ degree} = -60 \cdot \frac{\pi}{180} \text{ radian} = -\frac{\pi}{3} \text{ radians}$

19. $50° = 50 \cdot 1 \text{ degree} = 50 \cdot \frac{\pi}{180} \text{ radian} = \frac{5\pi}{18} \text{ radians}$

21. $225° = 225 \cdot 1 \text{ degree} = 225 \cdot \frac{\pi}{180} \text{ radian} = \frac{5\pi}{4} \text{ radians}$

23. $\frac{\pi}{3} \text{ radian} = \frac{\pi}{3} \cdot 1 \text{ radian} = \frac{\pi}{3} \cdot \frac{180}{\pi} \text{ degrees} = 60°$

25. $\frac{-5\pi}{4}$ radian = $\frac{-5\pi}{4} \cdot 1$ radian = $\frac{-5\pi}{4} \cdot \frac{180}{\pi}$ degrees = $-225°$

27. $\frac{7\pi}{2}$ radians = $\frac{7\pi}{2} \cdot 1$ radian = $\frac{7\pi}{2} \cdot \frac{180}{\pi}$ degrees = $630°$

29. $\frac{\pi}{12}$ radian = $\frac{\pi}{12} \cdot 1$ radian = $\frac{\pi}{12} \cdot \frac{180}{\pi}$ degrees = $15°$

31. $\frac{2\pi}{3}$ radian = $\frac{2\pi}{3} \cdot 1$ radian = $\frac{2\pi}{3} \cdot \frac{180}{\pi}$ degrees = $120°$

33. $r = 10$ meters
$\theta = \frac{1}{2}$ radian
$s = ?$

Use $s = r\theta$, $s = 10\left(\frac{1}{2}\right)$
$= 5$ meters

35. $\theta = \frac{1}{3}$ radian
$s = 2$ feet
$\theta = ?$

Use $s = r\theta$, $2 = r\left(\frac{1}{3}\right)$
$6 = r$
$r = 6$ feet

37. $r = 5$ miles
$s = 3$ miles
$\theta = ?$

Use $s = r\theta$ or $\theta = \frac{s}{r}$
$\theta = \frac{3}{5} = 0.6$ radian

39. $r = 2$ inches
$\theta = 30°$
$s = ?$

Convert $30°$ to $\frac{\pi}{6}$ radian.
Use $s = r\theta$
$s = 2\left(\frac{\pi}{6}\right) = \frac{\pi}{3}$ inches
≈ 1.047 inches

41. $17° = 17 \cdot 1$ degree = $17 \cdot \frac{\pi}{180}$ radians = $\frac{17\pi}{180}$ radians = 0.2967

43. $-40° = -40 \cdot 1$ degree = $-40 \cdot \frac{\pi}{180}$ radians = $\frac{-2\pi}{9}$ radians = -0.6981

45. $125° = 125 \cdot 1$ degree = $125 \cdot \frac{\pi}{180}$ radians = $\frac{25\pi}{36}$ radians = 2.1817

47. $340° = 340 \cdot 1$ degree = $340 \cdot \frac{\pi}{180}$ radians = $\frac{17\pi}{9}$ radians = 5.9341

49. 3.14 radians = $3.14 \cdot 1$ radian = $3.14 \cdot \frac{180}{\pi}$ degrees = $179.9°$

51. 10.25 radians = $10.25 \cdot 1$ radian = $10.25 \cdot \frac{180}{\pi}$ degrees = $587.3°$

53. 2 radians = $2 \cdot 1$ radian = $2 \cdot \frac{180}{\pi}$ degrees = $114.6°$

55. 6.32 radians = $6.32 \cdot 1$ radian = $6.32 \cdot \frac{180}{\pi}$ degrees = $362.1°$

57. $40°10'25" = \left(40 + 10 \cdot \frac{1}{60} + 25 \cdot \frac{1}{60} \cdot \frac{1}{60}\right)°$
$= (40 + 0.16667 + 0.00694)° = 40.1736$
$= 40.1736°$

5 TRIGONOMETRIC FUNCTIONS

59. $1°2'3" = \left(1 + 2 \cdot \dfrac{1}{60} + 3 \cdot \dfrac{1}{60} \cdot \dfrac{1}{60}\right)°$

$\qquad\qquad = (1 + 0.03333 + 0.00083)°$

$\qquad\qquad = 1.03416$

$\qquad\qquad = 1.0342°$

61. $9°9'9" = \left(9 + 9 \cdot \dfrac{1}{60} + 9 \cdot \dfrac{1}{60} \cdot \dfrac{1}{60}\right)$

$\qquad\qquad = (9 + 0.15 + 0.0025)$

$\qquad\qquad = 9.1525°$

63. $40.32° = ?$

$\qquad 0.32° = (0.32)(1°) = (0.32)(60') = 19.2'$

$\qquad\ \ 0.2' = (0.2)(1') = (0.2)(60") = 12"$

$\qquad 40.32° = 40° + 0.32° = 40° + 19.2' = 40° + 19' + 0.2'$

$\qquad\qquad = 40° + 19' + 12" = 40°19'12"$

65. $18.255° = ?$

$\qquad 0.255° = (0.255)(1\ \text{degree}) = (0.255)(60') = 15.3'$

$\qquad\ \ 0.3' = (0.3)(1') = (0.3)(60") = 18"$

$\qquad 18.255° = 18° + 0.255° = 18° + 15.3' = 18° + 15' + 0.3'$

$\qquad\qquad = 18° + 15' + 18" = 18°15'18"$

67. $19.99° = ?$

$\qquad 0.99° = (0.99)(1\ \text{degree}) = (0.99)(60') = 59.4'$

$\qquad\ \ 0.4' = (0.4)(1') = (0.4)(60') = 24"$

$\qquad 19.99° = 19° + 0.99° = 19° + 59.4' = 19° + 59' + 0.4'$

$\qquad\qquad = 19° + 59' + 24" = 19°59'24"$

69. $s = r\theta \qquad \theta = 90° = \dfrac{\pi}{2}$ in 15 minutes

$\quad s = (6)\left(\dfrac{\pi}{2}\right)$

$\quad s = 3\pi$ inches ≈ 9.4248 in.

$\qquad\qquad \theta = \dfrac{25}{60} = \dfrac{5}{12} \cdot \dfrac{360}{1} = 150° = \dfrac{5}{6}\pi$ in 25 minutes

$\quad s = 6\left(\dfrac{5\pi}{6}\right)$

$\quad s = 5\pi$ inches ≈ 15.7080 in.

71. $r = 5$ cm

$\quad t = 20$ sec

$\quad \theta = \dfrac{1}{3}$ rad

$\quad \omega = \dfrac{\theta}{t}$

$\quad \omega = \dfrac{\frac{1}{3}}{20} = \dfrac{1}{3} \cdot \dfrac{1}{20} = \dfrac{1}{60}$ rad/sec

$\quad v = \dfrac{s}{t}$ where $s = r\theta$

$\qquad\qquad s = 5\left(\dfrac{1}{3}\right)$

$\quad v = \dfrac{\frac{5}{3}}{20} = \dfrac{5}{3} \cdot \dfrac{1}{20} = \dfrac{1}{12}$ cm/sec

73. d = 26 inches
v = 35 mi/hr

$$\frac{35 \text{ mi}}{\text{hr}} \cdot \frac{5280 \text{ ft}}{\text{mi}} \cdot \frac{12 \text{ in.}}{\text{ft}} \cdot \frac{\text{rev}}{\pi(26 \text{ in.})} \cdot \frac{1 \text{ hr}}{60 \text{ min}} = 452.5 \text{ rpm}$$

75. $s = r\theta$
r = 18 inches
$\theta = \frac{1}{3} \cdot 2 = \frac{2\pi}{3}$
$s = 18 \cdot \frac{2\pi}{3} = 12\pi \approx 37.7$ inches

77. $v = r\omega$
$= 2.39 \times 10^5 \cdot \frac{1 \text{ rev}}{27.3 \text{ days}} \cdot \frac{1 \text{ day}}{24 \text{ hrs}} \cdot \frac{2\pi \text{ rad}}{\text{rev}}$
$= 2292$ mph

79. Find distance (d) traveled by 2" pulley in 3 revolutions
$d = \pi D \cdot N = \pi \cdot 2 \cdot 3 = 6\pi$"

This distance is the same traveled by 8" pulley. Solve for N.
$6\pi = \pi \cdot 8 \cdot N$
$N = \frac{6}{8} = \frac{3}{4}$ rpm

81. Find linear speed v using $v = r\omega$
$v = r\omega = 4 \cdot \left(\frac{2\pi \text{ rad}}{\text{rev}}\right)\left(\frac{10 \text{ rev}}{\text{min}}\right)\left(\frac{1 \text{ mile}}{5280 \text{ ft}}\right)\left(\frac{60 \text{ min}}{\text{hr}}\right) = 2.86$ mph

83. Find distance (s) on circle using $s = r\theta$
$s = r\theta \qquad \theta = \frac{1 \text{ deg}}{60 \text{ min}} \times 1 \text{ min} \times \frac{\pi}{180} \frac{\text{rad}}{\text{deg}} = 0.00029$ rad
$s = (3960)(0.00029) = 1.152$ miles

≡ EXERCISE 5.2 TRIGONOMETRIC FUNCTIONS: UNIT CIRCLE APPROACH

1. $(-3, 4)$

For $(a, b) = (-3, 4)$, we find $a = -3$, $b = 4$, and
$r = \sqrt{a^2 + b^2} = \sqrt{9 + 16} = \sqrt{25} = 5.$ Thus,

$\sin \theta = \frac{b}{r} = \frac{4}{5}$ $\qquad$ $\csc \theta = \frac{r}{b} = \frac{5}{4}$

$\cos \theta = \frac{a}{r} = -\frac{3}{5}$ $\qquad$ $\sec \theta = \frac{r}{a} = -\frac{5}{3}$

$\tan \theta = \frac{b}{a} = -\frac{4}{3}$ $\qquad$ $\cot \theta = \frac{a}{b} = -\frac{3}{4}$

3. (2, −3)

For $(a, b) = (2, -3)$, we find $a = 2$, $b = -3$, and
$$r = \sqrt{a^2 + b^2} = \sqrt{4 + 9} = \sqrt{13}. \quad \text{Thus,}$$

$\sin \theta = \dfrac{b}{r} = \dfrac{-3}{\sqrt{13}} \cdot \dfrac{\sqrt{13}}{\sqrt{13}} = -\dfrac{3\sqrt{13}}{13}$ $\qquad$ $\csc \theta = \dfrac{r}{b} = -\dfrac{\sqrt{13}}{3}$

$\cos \theta = \dfrac{a}{r} = \dfrac{2}{\sqrt{13}} \cdot \dfrac{\sqrt{13}}{\sqrt{13}} = \dfrac{2\sqrt{13}}{13}$ $\qquad$ $\sec \theta = \dfrac{r}{a} = \dfrac{\sqrt{13}}{2}$

$\tan \theta = \dfrac{b}{a} = -\dfrac{3}{2}$ $\qquad$ $\cot \theta = \dfrac{a}{b} = -\dfrac{2}{3}$

5. (−2, −2)

For $(a, b) = (2, -2)$, we find $a = -2$, $b = -2$, and
$$r = \sqrt{a^2 + b^2} = \sqrt{4 + 4} = \sqrt{8} = 2\sqrt{2}. \quad \text{Thus,}$$

$\sin \theta = \dfrac{b}{r} = \dfrac{-2}{2\sqrt{2}} = -\dfrac{1}{\sqrt{2}} \cdot \dfrac{\sqrt{2}}{\sqrt{2}} = -\dfrac{\sqrt{2}}{2}$ $\qquad$ $\csc \theta = \dfrac{r}{b} = \dfrac{2\sqrt{2}}{-2} = -\sqrt{2}$

$\cos \theta = \dfrac{a}{r} = \dfrac{-2}{2\sqrt{2}} = -\dfrac{1}{\sqrt{2}} \cdot \dfrac{\sqrt{2}}{\sqrt{2}} = -\dfrac{\sqrt{2}}{2}$ $\qquad$ $\sec \theta = \dfrac{r}{a} = \dfrac{2\sqrt{2}}{-2} = -\sqrt{2}$

$\tan \theta = \dfrac{b}{a} = \dfrac{-2}{-2} = 1$ $\qquad$ $\cot \theta = \dfrac{a}{b} = \dfrac{-2}{-2} = 1$

7. (−3, −2)

For $(a, b) = (-3, -2)$, we find $a = -3$, $b = -2$, and
$$r = \sqrt{a^2 + b^2} = \sqrt{9 + 4} = \sqrt{13}. \quad \text{Thus,}$$

$\sin \theta = \dfrac{b}{r} = \dfrac{-2}{\sqrt{13}} = -\dfrac{2}{\sqrt{13}} \cdot \dfrac{\sqrt{13}}{\sqrt{13}} = -\dfrac{2\sqrt{13}}{13}$ $\qquad$ $\csc \theta = \dfrac{r}{b} = \dfrac{\sqrt{13}}{-2} = -\dfrac{\sqrt{13}}{2}$

$\cos \theta = \dfrac{a}{r} = \dfrac{-3}{\sqrt{13}} = -\dfrac{3}{\sqrt{13}} \cdot \dfrac{\sqrt{13}}{\sqrt{13}} = -\dfrac{3\sqrt{13}}{13}$ $\qquad$ $\sec \theta = \dfrac{r}{a} = \dfrac{\sqrt{13}}{-3} = -\dfrac{\sqrt{13}}{3}$

$\tan \theta = \dfrac{b}{a} = \dfrac{-2}{-3} = \dfrac{2}{3}$ $\qquad$ $\cot \theta = \dfrac{a}{b} = \dfrac{-3}{-2} = \dfrac{3}{2}$

9. $\left(\dfrac{1}{3}, \dfrac{-1}{4}\right)$

For $(a, b) = \left(\dfrac{1}{3}, \dfrac{-1}{4}\right)$, we find $a = \dfrac{1}{3}$, $b = \dfrac{-1}{4}$, and
$$r = \sqrt{\dfrac{1}{9} + \dfrac{1}{16}} = \sqrt{\dfrac{25}{144}} = \dfrac{5}{12}.$$

$$\sin\theta = \frac{\frac{-1}{4}}{\frac{5}{12}} = \frac{-1}{4} \cdot \frac{12}{5} = \frac{-3}{5} \qquad \csc\theta = \frac{\frac{5}{12}}{\frac{-1}{4}} = \frac{5}{12} \cdot \frac{-4}{1} = \frac{-5}{3}$$

$$\cos\theta = \frac{\frac{1}{3}}{\frac{5}{12}} = \frac{1}{3} \cdot \frac{12}{5} = \frac{4}{5} \qquad \sec\theta = \frac{\frac{5}{12}}{\frac{1}{3}} = \frac{5}{12} \cdot \frac{3}{1} = \frac{5}{4}$$

$$\tan\theta = \frac{\frac{-1}{4}}{\frac{1}{3}} = \frac{-1}{4} \cdot \frac{3}{1} = \frac{-3}{4} \qquad \cot\theta = \frac{\frac{1}{3}}{\frac{-1}{4}} = \frac{1}{3} \cdot \frac{-4}{1} = \frac{-4}{3}$$

11. $\sin 45° + \cos 60° = \dfrac{\sqrt{2}}{2} + \dfrac{1}{2} = \dfrac{1}{2}\left(\sqrt{2} + 1\right)$

13. $\sin 90° + \tan 45° = 1 + 1 = 2$

15. $\sin 45° \cos 45° = \dfrac{\sqrt{2}}{2} \cdot \dfrac{\sqrt{2}}{2} = \dfrac{2}{4} = \dfrac{1}{2}$

17. $\csc 45° \tan 60° = \sqrt{2} \cdot \sqrt{3} = \sqrt{6}$

19. $4 \sin 90° - 3 \tan 180° = 4(1) = 3(0) = 4$

21. $2 \sin \dfrac{\pi}{3} - 3 \tan \dfrac{\pi}{6} = 2\left(\dfrac{\sqrt{3}}{2}\right) - 3\left(\dfrac{\sqrt{3}}{3}\right) = 0$

23. $\sin \dfrac{\pi}{4} - \cos \dfrac{\pi}{4} = \dfrac{\sqrt{2}}{2} - \dfrac{\sqrt{2}}{2} = 0$

25. $2 \sec \dfrac{\pi}{4} + 4 \cot \dfrac{\pi}{3} = 2\sqrt{2} + 4\left(\dfrac{\sqrt{3}}{3}\right)$

27. $\tan \pi - \cos 0 = 0 - 1 = -1$ $\qquad$ 29. $\quad \csc \dfrac{\pi}{2} + \cot \dfrac{\pi}{2} = 1 + 0 = 1$

31. Using Figure 27, we see that the point $P = \left(\dfrac{-1}{2}, \dfrac{\sqrt{3}}{2}\right)$ lies on the terminal side of $\theta = \dfrac{2\pi}{3}$. Thus,

$$\sin \frac{2\pi}{3} = \frac{\sqrt{3}}{2} \qquad\qquad \csc \frac{2\pi}{3} = \frac{1}{\frac{\sqrt{3}}{2}} = \frac{2\sqrt{3}}{3}$$

$$\cos \frac{2\pi}{3} = \frac{-1}{2} \qquad\qquad \sec \frac{2\pi}{3} = \frac{1}{\frac{-1}{2}} = -2$$

$$\tan \frac{2\pi}{3} = \frac{\frac{\sqrt{3}}{2}}{\frac{-1}{2}} = -\sqrt{3} \qquad\qquad \cot \frac{2\pi}{3} = \frac{\frac{-1}{2}}{\frac{\sqrt{3}}{2}} = \frac{-\sqrt{3}}{3}$$

33. Using Figure 27, see that the point $P = \left(\frac{-\sqrt{3}}{2}, \frac{1}{2}\right)$ lies on the terminal side of $\theta = 150° = \frac{5\pi}{6}$. Thus,

$$\sin 150° = \frac{1}{2}$$

$$\csc 150° = \frac{1}{\frac{1}{2}} = 2$$

$$\cos 150° = \frac{-\sqrt{3}}{2}$$

$$\sec 150° = \frac{1}{\frac{-\sqrt{3}}{2}} = \frac{-2\sqrt{3}}{3}$$

$$\tan 150° = \frac{\frac{1}{2}}{\frac{-\sqrt{3}}{2}} = \frac{-\sqrt{3}}{3}$$

$$\cot 150° = \frac{\frac{-\sqrt{3}}{2}}{\frac{1}{2}} = -\sqrt{3}$$

35. Using Figure 27, we see that the point $P = \left(\frac{\sqrt{3}}{2}, \frac{-1}{2}\right)$, lies on the terminal side of $\theta = \frac{-\pi}{6}$. Thus,

$$\sin \frac{-\pi}{6} = \frac{-1}{2}$$

$$\csc \frac{-\pi}{6} = \frac{1}{\frac{-1}{2}} = -2$$

$$\cos \frac{-\pi}{6} = \frac{\sqrt{3}}{2}$$

$$\sec \frac{-\pi}{6} = \frac{1}{\frac{\sqrt{3}}{2}} = \frac{2\sqrt{3}}{3}$$

$$\tan \frac{-\pi}{6} = \frac{\frac{-1}{2}}{\frac{\sqrt{3}}{2}} = \frac{-\sqrt{3}}{3}$$

$$\cot \frac{-\pi}{6} = \frac{\frac{\sqrt{3}}{2}}{\frac{-1}{2}} = -\sqrt{3}$$

37. We see that the point $P = \left(\frac{-\sqrt{2}}{2}, \frac{-\sqrt{2}}{2}\right)$ lies on the terminal side of $\theta = 225° = \frac{5\pi}{4}$. Thus,

$$\sin 225° = \frac{-\sqrt{2}}{2}$$

$$\csc 225° = \frac{1}{\frac{-\sqrt{2}}{2}} = -\sqrt{2}$$

$$\cos 225° = \frac{-\sqrt{2}}{2}$$

$$\sec 225° = \frac{1}{\frac{-\sqrt{2}}{2}} = -\sqrt{2}$$

$$\tan 225° = \frac{\frac{-\sqrt{2}}{2}}{\frac{-\sqrt{2}}{2}} = 1$$

$$\cot 225° = \frac{\frac{-\sqrt{2}}{2}}{\frac{-\sqrt{2}}{2}} = 1$$

39. Since $\frac{5\pi}{2} = 2\pi + \frac{\pi}{2}$ the point $P = (0, 1)$ lies on the terminal side of $\theta = \frac{5\pi}{2}$. Thus,

$$\sin \frac{5\pi}{2} = 1 \qquad\qquad \csc \frac{5\pi}{2} = 1$$

$$\cos \frac{5\pi}{2} = 0 \qquad\qquad \sec \frac{5\pi}{2} \text{ is not defined}$$

$$\tan \frac{5\pi}{2} \text{ is not defined} \qquad\qquad \cot \frac{5\pi}{2} = 0$$

41. The point $(-1, 0)$ lies on the terminal side of $\theta = -180°$. Thus,

$$\sin (-180°) = 0 \qquad\qquad \csc (-180°) \text{ is not defined}$$
$$\cos (-180°) = -1 \qquad\qquad \sec (-180°) = -1$$
$$\tan (-180°) = 0 \qquad\qquad \cot (-180°) \text{ is not defined}$$

43.

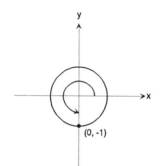

(0, -1)

$$\sin \frac{3\pi}{2} = \frac{-1}{1} = -1 \qquad\qquad \csc \frac{3\pi}{2} = -1$$

$$\cos \frac{3\pi}{2} = \frac{0}{1} = 0 \qquad\qquad \sec \frac{3\pi}{2} = \text{undefined}$$

$$\tan \frac{3\pi}{2} = \frac{1}{0} = \text{undefined} \qquad\qquad \cot \frac{3\pi}{2} = \frac{0}{1} = 0$$

45.

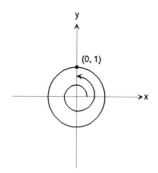

(0, 1)

$$\sin 450° = \frac{1}{1} = 1 \qquad\qquad \csc 450° = \frac{1}{1} = 1$$

$$\cos 450° = \frac{0}{1} = 0 \qquad\qquad \sec 450° = \frac{1}{0} = \text{undefined}$$

$$\tan 450° = \frac{1}{0} = \text{undefined} \qquad \cot 450° = \frac{0}{1} = 0$$

5 TRIGONOMETRIC FUNCTIONS

47. Using your calculator

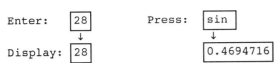

Enter: 28   Press: sin
         ↓   ↓
Display: 28   0.4694716

Thus, sin 28° ≈ 0.4694716 ≈ 0.4695.

49. Using your calculator

Enter: 21   Press: tan
         ↓   ↓
Display: 21   0.383864

Thus, tan 21° ≈ 0.383864 ≈ 0.3839.

51. Using your calculator

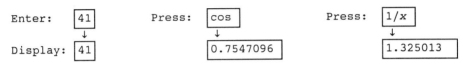

Enter: 41   Press: cos   Press: 1/x
         ↓   ↓   ↓
Display: 41   0.7547096   1.325013

Thus, sec 41° ≈ 1.3250.

53. Using your calculator

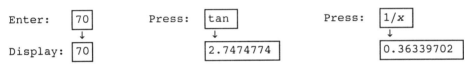

Enter: 70   Press: tan   Press: 1/x
         ↓   ↓   ↓
Display: 70   2.7474774   0.36339702

Thus, cot 70° ≈ 0.3639702 ≈ 0.3640.

55. Set the mode to receive radians. Then, to find $\sin \frac{\pi}{10}$,

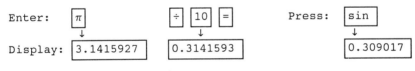

Enter: π   ÷ 10 =   Press: sin
         ↓   ↓   ↓
Display: 3.1415927   0.3141593   0.309017

Thus, $\sin \frac{\pi}{10} \approx 0.309017 \approx 0.3090$.

57. Set the mode to receive radians. Then, to find $\tan \frac{5\pi}{12}$,

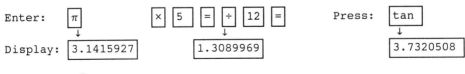

Enter: π   × 5 = ÷ 12 =   Press: tan
         ↓   ↓   ↓
Display: 3.1415927   1.3089969   3.7320508

Thus, $\tan \frac{5\pi}{12} \approx 3.7320508 \approx 3.7321$.

59. Set the mode to receive radians. Then, to find sec $\frac{\pi}{12}$,

Enter: $\boxed{\pi}$ $\boxed{\div}$ $\boxed{12}$ $\boxed{=}$ Press: $\boxed{\cos}$ Press: $\boxed{1/x}$

Display: $\boxed{3.1415927}$ $\boxed{0.2617994}$ $\boxed{0.9659258}$ $\boxed{1.0352762}$

Thus, sec $\frac{\pi}{12}$ ≈ 1.0352762 ≈ 1.0353.

61. Set the mode to receive radians. Then, to find cot $\frac{\pi}{18}$,

Enter: $\boxed{\pi}$ $\boxed{\div}$ $\boxed{18}$ $\boxed{=}$ Press: $\boxed{\tan}$ Press: $\boxed{1/x}$

Display: $\boxed{3.1415927}$ $\boxed{0.1745329}$ $\boxed{0.176327}$ $\boxed{5.6712819}$

Thus, cot $\frac{\pi}{18}$ ≈ 5.6712819 ≈ 5.6713.

63. Set the mode to receive radians. Then, to find sin 1,

Enter: $\boxed{1}$ Press: $\boxed{\sin}$

Display: $\boxed{1}$ $\boxed{0.841471}$

Thus, sin 1 ≈ 0.841471 ≈ 0.8415.

65. Use the (regular) degree mode. Then, to find sin 1°,

Enter: $\boxed{1}$ Press: $\boxed{\sin}$

Display: $\boxed{1}$ $\boxed{0.0174524}$

Thus, sin 1° ≈ 0.0174524 ≈ 0.0175.

67. Use the degree mode. Then, to find cos 21.5°,

Enter: $\boxed{21.5}$ Press: $\boxed{\cos}$

Display: $\boxed{21.5}$ $\boxed{0.9304176}$

Thus, cos 21.5° ≈ 0.9304176 ≈ 0.9304.

69. Set the mode to receive radians. Then, to find tan 0.3,

Enter: $\boxed{0.3}$ Press: $\boxed{\tan}$

Display: $\boxed{0.3}$ $\boxed{0.3093362}$

Thus, tan 0.3, ≈ 0.3093362 ≈ 0.3903.

71. $\sin 60° = \dfrac{\sqrt{3}}{2}$

73. $\sin \dfrac{60°}{2} = \sin 30° = \dfrac{1}{2}$

75. $(\sin 60°)^2 = \left(\dfrac{\sqrt{3}}{2}\right)^2 = \dfrac{3}{4}$

77. $\sin(2 \cdot 60°) = \sin 120°$
$$= \dfrac{\sqrt{3}}{2}$$

79. $2\sin 60° = 2\left(\dfrac{\sqrt{3}}{2}\right) = \sqrt{3}$

81. $\dfrac{\sin 60°}{2} = \dfrac{\dfrac{\sqrt{3}}{2}}{2} = \dfrac{\sqrt{3}}{4}$

83. $\sin 45° + \sin 135° + \sin 225° + \sin 315°$
$$= \dfrac{\sqrt{2}}{2} + \dfrac{\sqrt{2}}{2} - \dfrac{\sqrt{2}}{2} - \dfrac{\sqrt{2}}{2} = 0$$

85. If $\sin \theta = 0.1$, then $\sin(\theta + \pi) = -0.1$

87. If $\tan \theta = 3$, then $\tan(\theta + \pi) = 3$

89. If $\sin \theta = \dfrac{1}{5}$, then $\csc \theta = \dfrac{1}{\dfrac{1}{5}} = 5$

91. Using the formula $R = \dfrac{v_0^2 \sin 2\theta}{g}$ and $g \approx 32.2$ ft/sec^2, and given $\theta = 45°$ and $v_0 = 100$ ft/sec, we get

$R = \dfrac{(100)^2 \sin 2(45°)}{32.2}$

$R = \dfrac{(10,000)(\sin 90°)}{32.2}$, using calculator

$R = \dfrac{(10,000)(1)}{32.2} \approx 310.559$

$R \approx 310.56$ feet

Using the formula $H = \dfrac{v_0^2 \sin^2 \theta}{2g}$ and $g \approx 32.2$ ft/sec^2, and given $\theta = 45°$ and $v_0 = 100$ ft/sec, we get

$H = \dfrac{(100)^2(\sin 45°)^2}{2(32.2)}$, using calculator or table sin 45°

$\quad = 0.7071$

$H = \dfrac{(10,000)(0,7071)^2}{(64.4)}$, using calculator

$H \approx 77.638262$

$H \approx 77.64$ feet

93. Using the formula $R = \dfrac{v_0^2 \sin 2\theta}{g}$ and $g \approx 9.8$ m/sec^2, and given $\theta = 25°$ and $v_0 = 500$ m/sec, we get

$$R = \frac{(500)^2 \sin 2(25°)}{9.8}$$

$$R = \frac{(250{,}000)(\sin 50°)}{9.8}$$

$$R = \frac{(250{,}000)(0.7660444)}{9.8}$$

$$R \approx 19{,}541.95$$

$$R \approx 19{,}542 \text{ meters}$$

Using the formula $H = \dfrac{v_0^2 \sin^2 \theta}{2g}$ and $g \approx 9.8$ m/sec^2, and given $\theta = 25°$ and $v_0 = 500$ m/sec, we get

$$H = \frac{(500)^2 \sin^2(25°)}{2(9.8)}$$

$$H \approx 2278 \text{ m}$$

95. We use the formula, $t = \sqrt{\dfrac{2a}{g \sin \theta \cos \theta}}$,

where $g \approx 32$ ft/sec/sec and $a = 10$. Then

(a) $t = \sqrt{\dfrac{20}{32 \sin 30° \cos 30°}} = \sqrt{\dfrac{20}{32\left(\dfrac{1}{2}\right)\left(\dfrac{\sqrt{3}}{2}\right)}} = \sqrt{\dfrac{20}{8\sqrt{3}}} = \sqrt{\dfrac{5}{2\sqrt{3}}}$

≈ 1.2 sec

(b) $t = \sqrt{\dfrac{20}{32 \sin 45° \cos 45°}} = \sqrt{\dfrac{20}{32\left(\dfrac{\sqrt{2}}{2}\right)\left(\dfrac{\sqrt{2}}{2}\right)}} = \sqrt{\dfrac{20}{16}} = \sqrt{\dfrac{5}{4}}$

≈ 1.12 sec

(c) $t = \sqrt{\dfrac{20}{32 \sin 60° \cos 60°}} = \sqrt{\dfrac{20}{32\left(\dfrac{\sqrt{3}}{2}\right)\left(\dfrac{1}{2}\right)}} = \sqrt{\dfrac{5}{2\sqrt{3}}} \approx 1.2$ sec

97. Slope of $L^* = \dfrac{\sin \theta - 0}{\cos \theta - 0} = \dfrac{\sin \theta}{\cos \theta} = \dfrac{b}{a} = \tan \theta$

Since L is parallel to L^*, then slope of $L = \tan \theta$.

5 TRIGONOMETRIC FUNCTIONS

1. $\sin 405° = \sin(360° + 45°) = \sin 45° = \dfrac{\sqrt{2}}{2}$

3. $\tan 405° = \tan(180° + 180° + 45°) = \tan 45° = 1$

5. $\csc 450° = \csc(360° + 90°) = \csc 90° = 1$

7. $\cot 390° = \cot(180° + 180° + 30°) = \cot 30° = \sqrt{3}$

9. $\cos \dfrac{33\pi}{4} = \cos\left(\dfrac{\pi}{4} + \dfrac{32\pi}{4}\right) = \cos\left(\dfrac{\pi}{4} + 8\pi\right)$

$\qquad = \cos\left(\dfrac{\pi}{4} + 2\pi \cdot 4\right) = \cos \dfrac{\pi}{4} = \dfrac{\sqrt{2}}{2}$

11. $\tan 21\pi = \tan(\pi + 20\pi) = \tan(\pi + \pi \cdot 20)$
$\qquad = \tan \pi = 0$

13. $\sec \dfrac{17\pi}{4} = \sec\left(\dfrac{\pi}{4} + \dfrac{16\pi}{4}\right) = \sec\left(\dfrac{\pi}{4} + 4\pi\right)$

$\qquad = \sec\left(\dfrac{\pi}{4} + 2\pi \cdot 2\right) = \sec \dfrac{\pi}{4} = \sqrt{2}$

15. $\tan \dfrac{19\pi}{6} = \tan\left(\dfrac{\pi}{6} + \dfrac{18\pi}{6}\right) = \tan\left(\dfrac{\pi}{6} + 3\pi\right)$

$\qquad = \tan\left(\dfrac{\pi}{6} + \pi \cdot 3\right) = \tan \dfrac{\pi}{6} = \dfrac{\sqrt{3}}{3}$

17. Using Table 5, we find that $\sin \theta > 0$ for points P in quadrant I and II, and $\cos \theta < 0$ for points P in quadrants II and III. Both conditions are satisfied only if P lies in quadrant II.

19. Since $\sin \theta < 0$ for points P in quadrants III and IV, and $\tan \theta < 0$ for points P in quadrants II and IV, P lies in quadrant IV.

21. Since $\cos \theta > 0$ for points P in quadrants I and IV, and $\tan \theta < 0$ for points P in quadrants II and IV, P lies in quadrant IV.

23. Since $\sec \theta < 0$ for points P quadrants II and III, and $\sin \theta > 0$ for points P in quadrants I and II, P lies in quadrant II.

25. $\sin \theta = \dfrac{2}{\sqrt{5}}$, $\cos \theta = \dfrac{1}{\sqrt{5}}$

Based on the fundamental identities,

$$\tan \theta = \frac{\sin \theta}{\cos \theta} = \frac{\frac{2}{\sqrt{5}}}{\frac{1}{\sqrt{5}}} = 2$$

$$\csc \theta = \frac{1}{\sin \theta} = \frac{1}{\frac{2}{\sqrt{5}}} = \frac{\sqrt{5}}{2}$$

$$\sec \theta = \frac{1}{\cos \theta} = \frac{1}{\frac{1}{\sqrt{5}}} = \sqrt{5}$$

$$\cot \theta = \frac{1}{\tan \theta} = \frac{1}{2}$$

27. $\sin \theta = \dfrac{1}{2}$, $\cos \theta = \dfrac{\sqrt{3}}{2}$

Based on the fundamental identities,

$$\tan \theta = \frac{\sin \theta}{\cos \theta} = \frac{\frac{1}{2}}{\frac{\sqrt{3}}{2}} = \frac{1}{\sqrt{3}} = \frac{\sqrt{3}}{3}$$

$$\csc \theta = \frac{1}{\sin \theta} = \frac{1}{\frac{1}{2}} = 2$$

$$\sec \theta = \frac{1}{\cos \theta} = \frac{1}{\frac{\sqrt{3}}{2}} = \frac{2}{\sqrt{3}} = \frac{2\sqrt{3}}{3}$$

$$\cot \theta = \frac{1}{\tan \theta} = \frac{1}{\frac{1}{\sqrt{3}}} = \sqrt{3}$$

29. $\sin \theta = -\dfrac{1}{3}$, $\cos \theta = \dfrac{2\sqrt{2}}{3}$

Based on the fundamental identities,

$$\tan \theta = \frac{\sin \theta}{\cos \theta} = \frac{\frac{-1}{3}}{\frac{2\sqrt{2}}{3}} = \frac{-1}{2\sqrt{2}} = \frac{-\sqrt{2}}{4}$$

$$\csc \theta = \frac{1}{\sin \theta} = \frac{1}{\frac{-1}{3}} = -3$$

$$\sec \theta = \frac{1}{\cos \theta} = \frac{1}{\frac{2\sqrt{2}}{3}} = \frac{3}{2\sqrt{2}} = \frac{3\sqrt{2}}{4}$$

$$\cot \theta = \frac{1}{\tan \theta} = \frac{1}{\frac{-1}{2\sqrt{2}}} = -2\sqrt{2}$$

5 TRIGONOMETRIC FUNCTIONS

31. $\sin \theta = 0.2588$, $\cos \theta = 0.9659$

Based on the fundamental identities,

$$\tan \theta = \frac{\sin \theta}{\cos \theta} = \frac{0.2588}{0.9659} \approx 0.2679$$

$$\csc \theta = \frac{1}{\sin \theta} = \frac{1}{0.2588} \approx 3.8640$$

$$\sec \theta = \frac{1}{\cos \theta} = \frac{1}{0.9659} \approx 1.0353$$

$$\cot \theta = \frac{\cos \theta}{\sin \theta} = \frac{0.9659}{0.2588} \approx 3.7322$$

33. $\sin \theta = \frac{12}{13}$, $90° < \theta < 180°$

First, we solve for $\cos \theta$:

$$\sin^2 \theta + \cos^2 \theta = 1$$
$$\cos^2 = 1 - \sin^2 \theta$$
$$\cos \theta = \pm\sqrt{1 - \sin^2 \theta}$$

Because $90° < \theta < 180°$, $\cos \theta < 0$.

$$\cos \theta = -\sqrt{1 - \sin^2 \theta} = -\sqrt{1 - \frac{144}{169}} = -\sqrt{\frac{25}{169}} = \frac{-5}{13}$$

$$\tan \theta = \frac{\sin \theta}{\cos \theta} = \frac{\frac{13}{-5}}{13} = \frac{-12}{5}$$

$$\csc \theta = \frac{1}{\sin \theta} = \frac{1}{\frac{12}{13}} = \frac{13}{12}$$

$$\sec \theta = \frac{1}{\cos \theta} = \frac{1}{\frac{-5}{13}} = \frac{-13}{5}$$

$$\cot \theta = \frac{1}{\tan \theta} = \frac{1}{\frac{-12}{5}} = \frac{-5}{12}$$

35. $\cos \theta = \frac{-4}{5}$, $\pi < \theta < \frac{3\pi}{2}$

$$\sin^2 \theta = 1 - \cos^2 \theta$$
$$\sin \theta = \pm\sqrt{1 - \cos^2 \theta}$$

Because $\pi < \theta < \frac{3\pi}{2}$, $\sin \theta < 0$

$$\sin \theta = -\sqrt{1 - \cos^2 \theta} = -\sqrt{1 - \frac{16}{25}} = -\sqrt{\frac{9}{25}} = \frac{-3}{5}$$

$$\tan \theta = \frac{\sin \theta}{\cos \theta} = \frac{\frac{-3}{5}}{\frac{-4}{5}} = \frac{3}{4}$$

$$\cot \theta = \frac{1}{\tan \theta} = \frac{1}{\frac{3}{4}} = \frac{4}{3}$$

$$\csc \theta = \frac{1}{\sin \theta} = \frac{1}{\frac{-3}{5}} = \frac{-5}{3}$$

5.3 PROPERTIES OF THE TRIGONOMETRIC FUNCTIONS   279

$$\sec \theta = \frac{1}{\cos \theta} = \frac{1}{\frac{-4}{5}} = \frac{-5}{4}$$

37. $\sin \theta = \frac{5}{13}$, $\cos \theta < 0$

First, we solve for cos θ:
$$\sin^2 t + \cos^2 \theta = 1$$
$$\cos^2 \theta = 1 - \sin^2 \theta$$
$$\cos \theta = \pm\sqrt{1 - \sin^2 \theta}$$

Because cos θ < 0, we use the minus sign:
$$\cos \theta = -\sqrt{1 - \sin^2 \theta} = -\sqrt{1 - \left(\frac{5}{13}\right)^2}$$
$$= -\sqrt{1 - \frac{25}{169}} = -\sqrt{\frac{144}{169}} = \frac{-12}{13}$$

$$\tan \theta = \frac{\sin \theta}{\cos \theta} = \frac{\frac{5}{13}}{\frac{-12}{13}} = \frac{-5}{12}$$

$$\csc \theta = \frac{1}{\sin \theta} = \frac{1}{\frac{5}{13}} = \frac{13}{5}$$

$$\sec \theta = \frac{1}{\cos \theta} = \frac{1}{\frac{-12}{13}} = \frac{-13}{12}$$

$$\cot \theta = \frac{1}{\tan \theta} = \frac{1}{\frac{-5}{12}} = \frac{-12}{5}$$

39. $\cos \theta = \frac{-1}{3}$, $\csc \theta > 0$
$$\sin^2 \theta = 1 - \cos^2 \theta$$
$$\sin \theta = \pm\sqrt{1 - \cos^2 \theta}$$

Because csc θ > 0, and $\sin \theta = \frac{1}{\csc \theta}$, it follows that sin θ > 0.
$$\sin \theta = \sqrt{1 - \cos^2 \theta} = \sqrt{1 - \left(\frac{-1}{3}\right)^2}$$
$$= \sqrt{1 - \frac{1}{9}} = \frac{8}{9} = \frac{2\sqrt{2}}{3}$$

$$\tan \theta = \frac{\sin \theta}{\cos \theta} = \frac{\frac{2\sqrt{2}}{3}}{\frac{-1}{3}} = -2\sqrt{2}$$

$$\cot \theta = \frac{1}{\tan \theta} = \frac{1}{-2\sqrt{2}} = \frac{-\sqrt{2}}{4}$$

$$\csc \theta = \frac{1}{\sin \theta} = \frac{1}{\frac{2\sqrt{2}}{3}} = \frac{3\sqrt{2}}{4}$$

$$\sec\theta = \frac{1}{\cos\theta} = \frac{1}{\frac{-1}{3}} = -3$$

41. $\sin\theta = \frac{2}{3}$, $\tan\theta < 0$

$$\cos^2\theta = 1 - \sin^2\theta$$

Because $\tan\theta = \frac{\sin\theta}{\cos\theta} < 0$ and $\sin\theta > 0$, it follows that $\cos\theta < 0$.

Therefore, we use the minus sign:

$$\cos\theta = -\sqrt{1 - \sin^2\theta} = -\sqrt{1 - \left(\frac{2}{3}\right)^2}$$

$$= -\sqrt{1 - \frac{4}{9}} = -\sqrt{\frac{5}{9}} = \frac{-\sqrt{5}}{3}$$

$$\csc\theta = \frac{1}{\sin\theta} = \frac{1}{\frac{2}{3}} = \frac{3}{2}$$

$$\sec\theta = \frac{1}{\cos\theta} = \frac{1}{\frac{-\sqrt{5}}{3}} = \frac{-3\sqrt{5}}{5}$$

$$\tan\theta = \frac{\sin\theta}{\cos\theta} = \frac{\frac{2}{3}}{\frac{-\sqrt{5}}{3}} = \frac{-2\sqrt{5}}{5}$$

$$\cot\theta = \frac{1}{\tan\theta} = \frac{1}{\frac{-2}{\sqrt{5}}} = \frac{-\sqrt{5}}{2}$$

43. $\sec\theta = 2$, $\sin\theta < 0$

Because $\sec\theta = \frac{1}{\cos\theta}$ and $\sec\theta = 2$, than $\cos\theta = \frac{1}{2}$.

$$\cos\theta = \frac{1}{2}$$
$$\sin^2\theta = 1 - \cos^2\theta$$

$$\sin\theta = \pm\sqrt{1 - \cos^2\theta}$$

Because $\sin\theta < 0$, we use the minus sign:

$$\sin\theta = -\sqrt{1 - \cos^2\theta} = -\sqrt{1 - \left(\frac{1}{2}\right)^2}$$

$$= -\sqrt{1 - \frac{1}{4}} = -\sqrt{\frac{3}{4}} = \frac{-\sqrt{3}}{2}$$

$$\csc\theta = \frac{1}{\sin\theta} = \frac{1}{\frac{-\sqrt{3}}{2}} = \frac{-2\sqrt{3}}{3}$$

5.3 PROPERTIES OF THE TRIGONOMETRIC FUNCTIONS 281

$$\tan\theta = \frac{\sin\theta}{\cos\theta} = \frac{\frac{-\sqrt{3}}{2}}{\frac{1}{2}} = -\sqrt{3}$$

$$\cot\theta = \frac{1}{\tan\theta} = \frac{1}{-\sqrt{3}} = \frac{-\sqrt{3}}{3}$$

45. $\tan\theta = \dfrac{3}{4}$, $\sin\theta < 0$

$$\cot\theta = \frac{1}{\tan\theta} = \frac{1}{\frac{3}{4}} = \frac{4}{3}$$

Because $\tan\theta = \dfrac{\sin\theta}{\cos\theta} = \dfrac{3}{4} > 0$ and $\sin\theta < 0$, it follows that $\cos\theta < 0$. We know that $\tan^2\theta + 1 = \sec^2\theta$.

$$\sec\theta = \pm\sqrt{\tan^2\theta + 1}$$

Because $\cos\theta = \dfrac{1}{\sec\theta} < 0$, it follows that $\sec\theta < 0$. Therefore, we use the minus sign:

$$\sec\theta = -\sqrt{\tan^2\theta + 1} = -\sqrt{\left(\frac{3}{4}\right)^2 + 1}$$

$$= \sqrt{\frac{9}{16} + 1} = -\sqrt{\frac{25}{16}} = \frac{-5}{4}$$

$$\cos\theta = \frac{1}{\sec\theta} = \frac{1}{\frac{-5}{4}} = \frac{-4}{5}$$

$$\sin\theta = -\sqrt{1 - \left(\frac{-4}{5}\right)^2} = -\sqrt{\frac{9}{25}} = \frac{-3}{5}$$

$$\csc\theta = \frac{1}{\sin\theta} = \frac{1}{\frac{-3}{5}} = \frac{-5}{3}$$

47. $\tan\theta = \dfrac{-1}{3}$, $\sin\theta > 0$

Because $\tan\theta = \dfrac{\sin\theta}{\cos\theta} < 0$ and $\sin\theta > 0$, it follows that $\cos\theta < 0$.

$$\sec^2\theta = \tan^2\theta + 1$$

$$\sec\theta = \pm\sqrt{\tan^2\theta + 1}$$

Because $\cos\theta = \dfrac{1}{\sec\theta}$, it follows that $\sec\theta < 0$. Therefore, we use the minus sign.

$$\sec\theta = \pm\sqrt{-\tan^2\theta + 1} = -\sqrt{\left(\frac{-1}{3}\right)^2 + 1}$$

$$= -\sqrt{\frac{1}{9} + 1} = \frac{-\sqrt{10}}{3}$$

$$\cos \theta = \frac{1}{\sec \theta} = \frac{1}{\frac{-\sqrt{10}}{3}} = \frac{-3\sqrt{10}}{10}$$

$$\sin \theta = \sqrt{1 - \left(\frac{-3}{\sqrt{10}}\right)^2} = \sqrt{1 - \frac{9}{10}} = \frac{1}{\sqrt{10}} = \frac{\sqrt{10}}{10}$$

$$\csc \theta = \frac{1}{\sin \theta} = \frac{1}{\frac{\sqrt{10}}{10}} = \sqrt{10}$$

$$\cot \theta = \frac{1}{\tan \theta} = \frac{1}{\frac{-1}{3}} = -3$$

49. $\sin(-60°) = -\sin 60° = \frac{-\sqrt{3}}{2}$ 51. $\tan(-30°) = -\tan 30° = \frac{-\sqrt{3}}{3}$

53. $\sec(-60°) = \sec 60° = 2$ 55. $\sin(-90°) = -\sin 90° = -1$

57. $\tan\left(\frac{-\pi}{4}\right) = -\tan \frac{\pi}{4} = -1$ 59. $\cos\left(\frac{-\pi}{4}\right) = \cos \frac{\pi}{4} = \frac{\sqrt{2}}{2}$

61. $\tan(-\pi) = -\tan \pi = 0$ 63. $\csc\left(\frac{-\pi}{4}\right) = -\csc \frac{\pi}{4} = -\sqrt{2}$

65. $\sec\left(\frac{-\pi}{6}\right) = \sec \frac{\pi}{6} = \frac{2\sqrt{3}}{3}$

67. $\sin(-\pi) + \cos 5\pi = -\sin \pi + \cos(\pi + 4\pi)$
$$= 0 + \cos \pi = -1$$

69. $\sec(-\pi) + \csc\left(-\frac{\pi}{2}\right) = \sec \pi - \csc \frac{\pi}{2}$
$$= -1 - 1 = -2$$

71. $\sin\left(\frac{-9\pi}{4}\right) - \tan\left(\frac{-9\pi}{4}\right) = -\sin \frac{9\pi}{4} + \tan \frac{9\pi}{4}$
$$= -\sin\left(\frac{\pi}{4} + \frac{8\pi}{4}\right) + \tan\left(\frac{\pi}{4} + \frac{8\pi}{4}\right)$$
$$= -\sin \frac{\pi}{4} + \tan \frac{\pi}{4}$$
$$= \frac{-\sqrt{2}}{2} + 1 = 1 - \frac{\sqrt{2}}{2}$$

73. $\sin^2 40° + \cos^2 40° = 1$

75. $\sin 80° \csc 80° = \sin 80° \cdot \frac{1}{\sin 80°} = \frac{\sin 80°}{\sin 80°} = 1$

77. $\tan 40° - \frac{\sin 40°}{\cos 40°} = \tan 40° - \tan 40° = 0$

79. If $\sin \theta = 0.3$, then
$\sin \theta + \sin(\theta + 2\pi) + \sin(\theta + 4\pi) = \sin \theta + \sin \theta + \sin \theta$
$$= 0.3 + 0.3 + 0.3 = 0.9$$

5.3 PROPERTIES OF THE TRIGONOMETRIC FUNCTIONS

81. If $\tan \theta = 3$, then $\tan \theta + \tan(\theta + \pi) + \tan (\theta + 2\pi)$
$= \tan \theta + \tan \theta + \tan \theta = 3 + 3 + 3 = 9$

83. $f(\theta) = \tan \theta$ is defined for all real numbers except odd multiples of $\frac{\pi}{2}$.

85. $f(\theta) = \sec \theta$ is defined for all real numbers except odd multiples of $\frac{\pi}{2}$.

87. The value of $\sin k\pi$ where k is any integer is 0.

89. Let $P = (x, y)$ be the point on the unit circle that corresponds to an angle θ.

Consider the equation $\tan \theta = \frac{y}{x} = a$. Then $y = ax$. But $x^2 + y^2 = 1$ so that $x^2 + a^2x^2 = 1$. Thus, $x = \pm\dfrac{1}{\sqrt{1 + a^2}}$ and $y = \pm\dfrac{a}{\sqrt{1 + a^2}}$; that is, for any real number a, there is a point $P = (x, y)$ on the unit circle for which $\tan \theta = a$. In other words, $-\infty < \tan \theta < +\infty$, and the range of the tangent function is the set of all real numbers.

91. Suppose there is a number p, $0 < p < 2\pi$, for which $\sin (\theta + p) = \sin \theta$ for all θ. If $\theta = 0$, then $\sin (0 + p) = \sin p = \sin 0 = 0$; so that $p = \pi$. If $\theta = \frac{\pi}{2}$, then $\sin\left(\frac{\pi}{2} + p\right) = \sin\left(\frac{\pi}{2}\right)$. But $p = \pi$.
Thus, $\sin\left(\frac{3\pi}{2}\right) = -1 = \sin\left(\frac{\pi}{2}\right) = 1$. This is impossible. The smallest positive number p for which $\sin(\theta + p) = \sin \theta$ for all θ is therefore $p = 2\pi$.

93. $\sec \theta = \dfrac{1}{\cos \theta}$; since $\cos \theta$ has period 2π, so does $\sec \theta$.

95. If $P = (a, b)$ is the point on the unit circle corresponding to θ, then $Q = (-a, -b)$ is the point on the unit circle corresponding to $\theta + \pi$. Thus, $\tan(\theta + \pi) = \dfrac{-b}{-a} = \dfrac{b}{a} = \tan \theta$; that is, the period of the tangent function is π.

97. Let $P = (a, b)$ be the point on the unit circle corresponding to θ.

Then $\csc \theta = \dfrac{1}{b} = \dfrac{1}{\sin \theta}$; $\sec \theta = \dfrac{1}{a} = \dfrac{1}{\cos \theta}$;
$\cot \theta = \dfrac{a}{b} = \dfrac{1}{\left(\dfrac{b}{a}\right)} = \dfrac{1}{\tan \theta}$

99. $(\sin \theta \cos \phi)^2 + (\sin \theta \sin \phi)^2 + \cos^2 \theta$
$= \sin^2 \theta \cos^2 \phi + \sin^2 \theta \sin^2 \phi + \cos^2 \theta$
$= \sin^2 \theta(\cos^2 \phi + \sin^2 \phi) + \cos^2 \theta$
$= \sin^2 \theta + \cos^2 \theta = 1$

　　　　　　　　　　　5 TRIGONOMETRIC FUNCTIONS

☰ EXERCISE 5.4 RIGHT TRIANGLE TRIGONOMETRY

1.   opposite = 5
       adjacent = 12

   By the Pythagorean Theorem:

   $$5^2 + 12^2 = (\text{hypotenuse})^2$$
   $$25 + 144 = (\text{hypotenuse})^2$$
   $$169 = (\text{hypotenuse})^2$$
   $$13 = \text{hypotenuse}$$

   $\sin\theta = \dfrac{\text{opp}}{\text{hyp}} = \dfrac{5}{13}$ $\cos\theta = \dfrac{\text{adj}}{\text{hyp}} = \dfrac{12}{13}$ $\tan\theta = \dfrac{\text{opp}}{\text{adj}} = \dfrac{5}{12}$

   $\csc\theta = \dfrac{\text{hyp}}{\text{opp}} = \dfrac{13}{5}$ $\sec\theta = \dfrac{\text{hyp}}{\text{adj}} = \dfrac{13}{12}$ $\cot\theta = \dfrac{\text{adj}}{\text{opp}} = \dfrac{12}{5}$

3.   opposite = 2
       adjacent = 3

   By the Pythagorean Theorem:

   $$2^2 + 3^2 = (\text{hypotenuse})^2$$
   $$4 + 9 = (\text{hypotenuse})^2$$
   $$13 = \text{hypotenuse}$$
   $$\sqrt{13} = \text{hypotenuse}$$

   $\sin\theta = \dfrac{\text{opp}}{\text{hyp}} = \dfrac{2}{\sqrt{13}} \cdot \dfrac{\sqrt{13}}{\sqrt{13}} = \dfrac{2\sqrt{13}}{13}$ $\csc\theta = \dfrac{\text{hyp}}{\text{opp}} = \dfrac{\sqrt{13}}{2}$

   $\cos\theta = \dfrac{\text{adj}}{\text{hyp}} = \dfrac{3}{\sqrt{13}} \cdot \dfrac{\sqrt{13}}{\sqrt{13}} = \dfrac{3\sqrt{13}}{13}$ $\sec\theta = \dfrac{\text{hyp}}{\text{adj}} = \dfrac{\sqrt{13}}{3}$

   $\tan\theta = \dfrac{\text{opp}}{\text{adj}} = \dfrac{2}{3}$ $\cot\theta = \dfrac{\text{adj}}{\text{opp}} = \dfrac{3}{2}$

5.   adjacent = 2
       hypotenuse = 4

   By the Pythagorean Theorem:

   $$2^2 + (\text{opp})^2 = 4^2$$
   $$4 + (\text{opp})^2 = 16$$
   $$(\text{opp})^2 = 12$$
   $$\text{opp} = \sqrt{12} = 2\sqrt{3}$$

   $\sin\theta = \dfrac{\text{opp}}{\text{hyp}} \; \dfrac{2\sqrt{3}}{4} = \dfrac{\sqrt{3}}{2}$ $\csc\theta = \dfrac{\text{hyp}}{\text{opp}} = \dfrac{4}{2\sqrt{3}} \cdot \dfrac{\sqrt{3}}{\sqrt{3}} = \dfrac{4\sqrt{3}}{6} = \dfrac{2\sqrt{3}}{3}$

   $\cos\theta = \dfrac{\text{adj}}{\text{hyp}} = \dfrac{2}{4} = \dfrac{1}{2}$ $\sec\theta = \dfrac{\text{hyp}}{\text{adj}} = \dfrac{4}{2} = 2$

   $\tan\theta = \dfrac{\text{opp}}{\text{adj}} = \dfrac{2\sqrt{3}}{2} = \sqrt{3}$ $\cot\theta = \dfrac{\text{adj}}{\text{opp}} = \dfrac{2}{2\sqrt{3}} \cdot \dfrac{\sqrt{3}}{\sqrt{3}} = \dfrac{\sqrt{3}}{3}$

7. opposite = $\sqrt{2}$
    adjacent = 1

    By the Pythagorean Theorem:

    $$\left(\sqrt{2}\right)^2 + 1^2 = (\text{hypotenuse})^2$$
    $$2 + 1 = (\text{hypotenuse})^2$$
    $$3 = (\text{hypotenuse})^2$$
    $$\sqrt{3} = (\text{hypotenuse})$$

    $\sin\theta = \dfrac{\text{opp}}{\text{hyp}} = \dfrac{\sqrt{2}}{\sqrt{3}} \cdot \dfrac{\sqrt{3}}{\sqrt{3}} = \dfrac{\sqrt{6}}{3}$ $\qquad$ $\csc\theta = \dfrac{\text{hyp}}{\text{opp}} = \dfrac{\sqrt{3}}{\sqrt{2}} \cdot \dfrac{\sqrt{2}}{\sqrt{2}} = \dfrac{\sqrt{6}}{2}$

    $\cos\theta = \dfrac{\text{adj}}{\text{hyp}} = \dfrac{1}{\sqrt{3}} \cdot \dfrac{\sqrt{3}}{\sqrt{3}} = \dfrac{\sqrt{3}}{3}$ $\qquad$ $\sec\theta = \dfrac{\text{hyp}}{\text{adj}} = \dfrac{\sqrt{3}}{1} = \sqrt{3}$

    $\tan\theta = \dfrac{\text{opp}}{\text{adj}} = \dfrac{\sqrt{2}}{1} = \sqrt{2}$ $\qquad$ $\cot\theta = \dfrac{\text{adj}}{\text{opp}} = \dfrac{1}{\sqrt{2}} \cdot \dfrac{\sqrt{2}}{\sqrt{2}} = \dfrac{\sqrt{2}}{2}$

9. opposite = 1

    hypotenuse = $\sqrt{5}$

    By the Pythagorean Theorem:

    $$1^2 + (\text{adjacent})^2 = \left(\sqrt{5}\right)^2$$
    $$1 + (\text{adjacent})^2 = 5$$
    $$(\text{adjacent})^2 = 4$$
    $$\text{adjacent} = 2$$

    $\sin\theta = \dfrac{\text{opp}}{\text{hyp}} = \dfrac{1}{\sqrt{5}} \cdot \dfrac{\sqrt{5}}{\sqrt{5}} = \dfrac{\sqrt{5}}{5}$ $\qquad$ $\csc\theta = \dfrac{\text{hyp}}{\text{opp}} = \dfrac{\sqrt{5}}{1} = \sqrt{5}$

    $\cos\theta = \dfrac{\text{adj}}{\text{hyp}} = \dfrac{2}{\sqrt{5}} \cdot \dfrac{\sqrt{5}}{\sqrt{5}} = \dfrac{2\sqrt{5}}{5}$ $\qquad$ $\sec\theta = \dfrac{\text{hyp}}{\text{adj}} = \dfrac{\sqrt{5}}{2}$

    $\tan\theta = \dfrac{\text{opp}}{\text{adj}} = \dfrac{1}{2}$ $\qquad$ $\cot\theta = \dfrac{\text{adj}}{\text{opp}} = \dfrac{2}{1} = 2$

11. The reference angle for
    $-30°$ is $0° - (-30°) = 30°$.

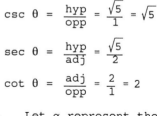

13. Let α represent the
    reference angle.

    $$120° + \alpha = 180°$$
    $$\alpha = 60°$$

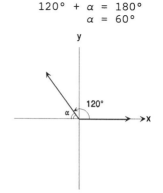

15. $180° + \alpha = 210°$
$\alpha = 30°$

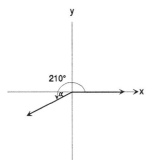

17. Remember that the reference angle is the acute angle formed by the terminal side of θ and the positive or negative x-axis.

Hence, $\pi + \alpha = \dfrac{5\pi}{4}$

$\alpha = \dfrac{5\pi}{4} - \pi = \dfrac{\pi}{4}$

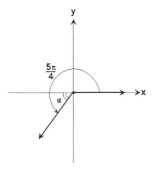

19. $\dfrac{8\pi}{3} + \alpha = 3\pi$

$\alpha = 3\pi - \dfrac{8\pi}{3}$

$= \dfrac{9\pi}{3} - \dfrac{8\pi}{3}$

$\alpha = \dfrac{\pi}{3}$

21. $180° - 135° = \alpha$
$45° = \alpha$

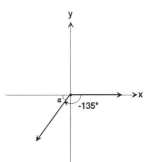

23. $\pi - \dfrac{2\pi}{3} = \alpha$

$\dfrac{\pi}{3} = \alpha$

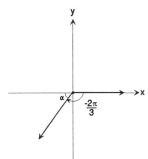

25. $420° - 360° = \alpha$
$60° = \alpha$

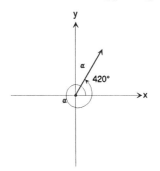

27. sin 150°

The angle 150° is in quadrant II, where the sin θ is positive. The reference angle for 150° is 30°. Thus,

$$\sin 150° = \sin 30° = \frac{1}{2}$$

29. cos 315°

The angle 315° is in quadrant IV where the cos θ is positive. The reference angle for 315° is 45°. Thus,

$$\cos 315° = \cos 45° = \frac{\sqrt{2}}{2}$$

31. sec 240°

The angle 240° is in quadrant III where sec θ is negative. The reference angle for 240° is 60°. Thus,

$$\sec 240° = -\sec 60° = -2$$

33. cot 330°

The angle 330° is in quadrant IV where cot θ is negative. The reference angle for 330° is 30°. Thus,

$$\cot 330° = -\cot 30° = -\frac{\sqrt{3}}{1} = -\sqrt{3}$$

35. $\sin \frac{3\pi}{4}$

The angle $\frac{3\pi}{4}$ is in quadrant II where sin is positive. The reference angle for $\frac{3\pi}{4}$ is $\frac{\pi}{4}$. Thus,

$$\sin \frac{3\pi}{4} = \sin \frac{\pi}{4} = \frac{\sqrt{2}}{2}$$

37. $\cot \frac{7\pi}{6}$

The angle $\frac{7\pi}{6}$ is in quadrant III where cot θ is positive. The reference angle for $\frac{7\pi}{6}$ is $\frac{\pi}{6}$. Thus,

$$\cot \frac{7\pi}{6} = \cot \frac{\pi}{6} = \sqrt{3}$$

39. cos (−60°)

The angle (−60°) is in quadrant IV where cos θ is positive. The reference angle for −60° is 60°. Thus,

$$\cos (-60°) = \cos 60° = \frac{1}{2}$$

41. $\sin\left(-\dfrac{2\pi}{3}\right)$

The angle $-\dfrac{2\pi}{3}$ is in quadrant III where $\sin\theta$ is negative. The reference angle for $\left(-\dfrac{2\pi}{3}\right)$ is $\dfrac{\pi}{3}$. Thus,

$$\sin\left(-\dfrac{2\pi}{3}\right) = -\sin\dfrac{\pi}{3} = -\dfrac{\sqrt{3}}{2}$$

43. $\tan\dfrac{14\pi}{3}$

$$\tan\dfrac{14\pi}{3} = \tan\left(\dfrac{12\pi}{3} + \dfrac{2\pi}{3}\right) = \tan\left(4\pi + \dfrac{2\pi}{3}\right)$$

The angle $\dfrac{14\pi}{3}$ is in quadrant II where $\tan\theta$ is negative. The reference angle for $\dfrac{14\pi}{3}$ is $\dfrac{\pi}{3}$. Thus,

$$\tan\dfrac{14\pi}{3} = -\tan\dfrac{\pi}{3} = -\sqrt{3}$$

45. $\csc(-315°)$

The angle $-315°$ is in quadrant I where $\csc\theta$ is positive. The reference angle for $-315°$ is $45°$. Thus,

$$\csc(-315°) = \csc 45° = \sqrt{2}$$

47. $\sin 38° - \cos 52° = 0$

Since $\sin(90° - 52°) = \cos 52°$ or $\sin 38°$ by the cofunction formula.

49. $\dfrac{\cos 10°}{\sin 80°} = 1$

Since $\cos(90° - 80°) = \sin 80°$ or $\cos 10° = \sin 80°$ by the cofunction formula.

51. $1 - \cos^2 20° - \cos^2 70°$

Since by the fundamental identity,

$$\sin^2 20° + \cos^2 20° = 1, \text{ then}$$
$$\sin^2 20° = 1 - \cos^2 20°$$

Also, $\cos^2(90° - 20°) = \sin^2 20°$ by the cofunction formula so
$$\cos^2 70° = \sin^2 20°$$

By substitution, $1 - \cos^2 20° - \cos^2 70°$
$$= (1 - \cos^2 20°) - \cos^2 70°$$
$$= \sin^2 20° - \sin^2 20°$$
$$= 0$$

53. $\tan 20° - \dfrac{\cos 70°}{\cos 20°} = \tan 20° - \dfrac{\sin 20°}{\cos 20°}$
$$= \tan 20° - \tan 20° = 0$$

5.4 RIGHT TRIANGLE TRIGONOMETRY   289

55. cos 35° sin 55° + sin 35° cos 55°
$$= \cos 35° \cos 35° + \sin 35° \sin 35°$$
$$= \cos^2 35° + \sin^2 35° = 1$$

57. (a) Since $\sin \theta = \cos(90° - \theta)$ (they are cofunctions),
$$\cos(90° - \theta) = \frac{1}{3}$$

(b) Find $\cos^2 \theta$. By the fundamental identity, $\sin^2 \theta + \cos^2 \theta$
$$= 1, \text{ so } \cos^2 \theta = 1 - \sin^2 \theta = 1 - \left(\frac{1}{3}\right)^2 = 1 - \frac{1}{9} = \frac{8}{9}$$

(c) Since $\csc \theta = \dfrac{1}{\sin \theta} = \dfrac{1}{\frac{1}{3}} = 3$

(d) $\sec\left(\dfrac{\pi}{2} - \theta\right) = \csc \theta = 3$ (from part (c))
$$= \csc \theta = 3$$

59. (a) $1 + \tan^2 \theta = \sec^2 \theta$ (fundamental identity)
given $\tan \theta = 4$ so
$$1 + (4)^2 = \sec^2 \theta$$
$$1 + 16 = \sec^2 \theta$$
$$17 = \sec^2 \theta$$

(b) $\cot \theta = \dfrac{1}{\tan \theta}$ (fundamental identity)
given $\tan \theta = 4$ so
$$\cot \theta = \frac{1}{4}$$

(c) $\cot\left(\dfrac{\pi}{2} - \theta\right) = \tan \theta$ (cofunctions)

$\cot\left(\dfrac{\pi}{2} - \theta\right) = 4$ since $\tan \theta = 4$ is given

(d) Using part (b), $\cot \theta = \dfrac{1}{4}$ and
the fundamental identity $1 + \cot^2 \theta = \csc^2$, we have
$$1 + \left(\frac{1}{4}\right)^2 = \csc^2 \theta$$
$$1 + \frac{1}{16} = \csc^2 \theta$$
$$\frac{17}{16} = \csc^2 \theta$$

61. (a) Using $\csc \theta = 4$ (given),
and the fundamental identity $\sin \theta = \dfrac{1}{\csc \theta}$ we have
$$\sin \theta = \frac{1}{4}$$

(b) Using csc θ = 4 (given) and the fundamental identity
$$1 + \cot^2 \theta \csc^2, \text{ we have}$$
$$1 + \cot^2 \theta = (4)^2$$
$$1 + \cot^2 \theta = 16$$
$$\cot^2 \theta = 15$$

(c) sec($90°$ - θ) = csc θ (cofunctions) and csc θ = 4 (given) so
sec($90°$ - θ) = 4

(d) Using part (b), $\cot^2 \theta$ = 15 and the fundamental identities
$$\tan \theta = \frac{1}{\cot \theta} \text{ and } 1 + \tan^2 = \sec^2 \theta,$$

we have $\tan^2 \theta = \dfrac{1}{\cot^2 \theta} = \dfrac{1}{15}$

so $1 + \dfrac{1}{15} = \sec^2 \theta$

$\dfrac{16}{15} = \sec^2 \theta$

63. $\cos\left(\dfrac{\pi}{2} - \theta\right)$ = sin θ since they are cofunctions. Thus,

sin θ + $\cos\left(\dfrac{\pi}{2} - \theta\right)$ = sin θ + sin θ = 0.3 + 0.3 = 0.6

65. sin 1° + sin 2° + sin 3° + … + sin 358° + sin 359°
$$= (\sin 1° + \sin 359°) + (\sin 2° + \sin 358°) + …$$
$$= [\sin 1° + \sin (-1°)] + [\sin 2° + (\sin -2°)] + …$$
$$\quad + [\sin 179° + \sin (-179°)] + \sin 90° + \sin 270° + \sin 180°$$
$$= (\sin 1° - \sin 1°) + (\sin 2° - \sin 2°) + …$$
$$\quad + (\sin 179° - \sin 179°) + 1 + (-1) + 0$$
$$= 0 + 0 + … + 0 + 0 = 0$$

67. sin θ = cos(2θ + 30°)

Since sin θ = cos(90° - θ), then

$$2\theta + 30° = 90° - \theta$$
$$3\theta = 60°$$
$$\theta = 20°$$

69.

θ	0.5	0.4	0.2	0.1	0.01	0.001	0.0001	0.00001
sin θ	0.4794	0.3894	0.1987	0.0998	0.0100	0.0010	0.0001	0.00001
$\dfrac{\sin \theta}{\theta}$	0.9589	0.9735	0.9933	0.9983	1.0000	1.0000	1.0000	1.0000

$\dfrac{\sin \theta}{\theta}$ approaches 1 as θ approaches 0

71. (a) $|OA| = |OC| = 1$; Angle OAC = Angle OCA. Thus,
Angle OAC + Angle OAC + 180° - θ = 180°

$$2(\text{Angle } OAC) = \theta$$
$$\text{Angle } OAC = \frac{\theta}{2}$$

5.4 RIGHT TRIANGLE TRIGONOMETRY

(b) $\sin \theta = \dfrac{|CD|}{|OC|} = |CD|$, since $|OC| = 1$

$\cos \theta = \dfrac{|OD|}{|OC|} = |OD|$

(c) $\tan \dfrac{\theta}{2} = \dfrac{|CD|}{|AD|}$, since angle $OAC = \dfrac{\theta}{2}$ by part (a)

$\quad = \dfrac{\sin \theta}{1 + |OD|}$, since $|CD| = \sin \theta$ by part (b) and

$\quad\quad |AD| = |AO| + |OD| = 1 + |OD|$, since $|AO| = 1$

$\quad = \dfrac{\sin \theta}{1 + \cos \theta}$, since $|OD| = \cos \theta$ by part (b)

73. $\quad h = x \cdot \dfrac{h}{x} = x \tan \theta$ and $h = (1 - x)\dfrac{h}{1 - x} = (1 - x) \tan n\theta$

Thus, $\qquad\qquad x \tan \theta = (1 - x) \tan n\theta$
$$x \tan \theta = \tan n\theta - x \tan n\theta$$
$$x(\tan \theta + \tan n\theta) = \tan n\theta$$
$$x = \dfrac{\tan n\theta}{\tan \theta + \tan n\theta}$$

75. (a) Area $\triangle OAC = \dfrac{1}{2}|OC|\,|AC| = \dfrac{1}{2}\dfrac{|OC|}{1} \cdot \dfrac{|AC|}{1} = \dfrac{1}{2}\sin \alpha \cos \alpha$

(b) Area $\triangle OCB = \dfrac{1}{2}|OC|\,|BC| = \dfrac{1}{2}|OB|^2\dfrac{|BC|}{|OB|} \cdot \dfrac{|OC|}{|OB|}$

$\qquad\qquad = \dfrac{1}{2}|OB|^2 \sin \beta \cos \beta$

(c) Area $\triangle OAB = \dfrac{1}{2}|BD|\,|OA| = \dfrac{1}{2}|BD| \cdot 1 = \dfrac{1}{2}|OB|\dfrac{|BD|}{|OB|}$

$\qquad\qquad = \dfrac{1}{2}|OB|\sin(\alpha + \beta)$

(d) $\dfrac{\cos \alpha}{\cos \beta} = \dfrac{\dfrac{|OC|}{|OA|}}{\dfrac{|OC|}{|OB|}} = \dfrac{|OC|}{1} \cdot \dfrac{|OB|}{|OC|} = |OB|$

(e) Area $\triangle OAB =$ Area $\triangle OAC +$ Area $\triangle OCB$

$\dfrac{1}{2}|OB|\sin(\alpha + \beta) = \dfrac{1}{2}\sin \alpha(\cos \alpha + \dfrac{1}{2}|OB|^2 \sin \beta \cos \beta$

$\dfrac{\cos \alpha}{\cos \beta}\sin(\alpha + \beta) = \sin \alpha \cos \alpha + \dfrac{\cos^2 \alpha}{\cos^2 \beta}\sin \beta \cos \beta$

$\sin(\alpha + \beta) = \dfrac{\cos \beta}{\cos \alpha}(\sin \alpha \cos \alpha) + \dfrac{\cos \alpha}{\cos \beta}(\sin \beta \cos \beta)$

$\sin(\alpha + \beta) = \cos \beta \sin \alpha + \cos \alpha \sin \beta$

5 TRIGONOMETRIC FUNCTIONS

77. $\sin \alpha = \dfrac{\sin \alpha}{\cos \alpha} \cos \alpha = \tan \alpha \cos \alpha = \cos \beta \cos \alpha$

$\qquad = \cos \beta \tan \beta = \cos \beta \cdot \dfrac{\sin \beta}{\cos \beta} = \sin \beta$

$\sin^2 \alpha + \cos^2 \alpha = 1$, thus

$\sin^2 \alpha + \tan^2 \beta = 1$

$\sin^2 \alpha + \dfrac{\sin^2 \beta}{\cos^2 \beta} = 1$

$\sin^2 \alpha + \dfrac{\sin^2 \alpha}{1 - \sin^2 \alpha} = 1$

$\sin^2 \alpha - \sin^4 \alpha + \sin^2 \alpha = 1 - \sin^2 \alpha$

$\sin^4 \alpha - 3 \sin^2 \alpha + 1 = 0$

$\sin^2 \alpha = \dfrac{3 \pm \sqrt{5}}{2}$

$\sin^2 \alpha = \dfrac{3 - \sqrt{5}}{2}$

$\sin \alpha = \sqrt{\dfrac{3 - \sqrt{5}}{2}}$

≡ EXERCISE 5.5 APPLICATIONS

1. $b = 5$
    $\beta = 10°$

    $\cot \beta = \dfrac{a}{b}$ $\qquad\qquad\qquad$ $\csc \beta = \dfrac{c}{b}$

    $\cot 10° = \dfrac{a}{5}$ $\qquad\qquad\qquad$ $\csc 10° = \dfrac{c}{5}$

    $5(\cot 10°) = a$ $\qquad\qquad\quad$ $5(\csc 10°) = c$
    $\quad 5(5.6713) = a$ $\qquad\qquad\quad$ $28.793852 = c$
    $\quad\; 28.3565 = a$ $\qquad\qquad\qquad\;\; 28.79 \approx c$
    $\qquad 28.36 \approx a$

    $\alpha = 90° - \beta = 90° - 10° = 80°$

3. $a = 6$
    $\beta = 40°$

    $\tan \beta = \dfrac{b}{a}$ $\qquad\qquad\qquad$ $\sec \beta = \dfrac{c}{a}$

    $\tan 40° = \dfrac{b}{6}$ $\qquad\qquad\qquad$ $\sec 40° = \dfrac{c}{6}$

    $6(\tan 40°) = b$ $\qquad\qquad\quad$ $6(\sec 40°) = c$
    $5.0345978 \approx b$ $\qquad\qquad\;\; 7.8324437 \approx c$
    $\qquad 5.03 \approx b$ $\qquad\qquad\qquad\;\; 7.83 \approx c$

    $\alpha = 90° - \beta = 90° - 40° = 50°$

5.　$b = 4$
　$\alpha = 10°$

$$\tan \alpha = \frac{a}{b} \qquad\qquad \sec \alpha = \frac{c}{b}$$

$$\tan 10° = \frac{a}{4} \qquad\qquad \sec 10° = \frac{c}{4}$$

$4(\tan 10°) = a$ $\qquad\qquad$ $4(\sec 10°) = c$
$0.7053079 \approx a$ $\qquad\qquad$ $4.0617064 \approx c$
$0.705 \approx a$ $\qquad\qquad$ $4.06 \approx c$

$\beta = 90° - \alpha = 90° - 10° = 80°$

7.　$a = 5$
　$\alpha = 25°$

$$\cot \alpha = \frac{b}{a} \qquad\qquad \csc \alpha = \frac{c}{a}$$

$$\cot 25° = \frac{b}{5} \qquad\qquad \csc 25° = \frac{c}{5}$$

$5(\cot 25°) = b$ $\qquad\qquad$ $5(\csc 25°) = c$
$10.722535 \approx b$ $\qquad\qquad$ $11.831008 \approx c$
$10.72 \approx b$ $\qquad\qquad$ $11.83 \approx c$

$\beta = 90° - 25° = 65°$

9.　$c = 9$
　$\beta = 20°$

$$\sin \beta = \frac{b}{c} \qquad\qquad \cos \beta = \frac{a}{c}$$

$$\sin 20° = \frac{b}{9} \qquad\qquad \cos 20° = \frac{a}{9}$$

$9(\sin 20°) = b$ $\qquad\qquad$ $9(\cos 20°) = a$
$3.0781813 \approx b$ $\qquad\qquad$ $8.4572336 \approx a$
$3.08 \approx b$ $\qquad\qquad$ $8.46 \approx a$

$\alpha = 90° - 20° = 70°$

11.　$c^2 = a^2 + b^2 = 5^2 + 3^2 = 25 + 9 = 34$

$c = \sqrt{34} \approx 5.83$

$\sin \alpha = \frac{a}{c} = \frac{5}{5.83}$

$\quad \alpha \approx 59.0°$
$\quad \beta = 90° - \alpha \approx 90 - 59° \approx 31.0°$

13.　$b^2 = c^2 - a^2 = 5^2 - 2^2 = 21$

$b = \sqrt{21} \approx 4.58$

$\sin \alpha = \frac{a}{c} = \frac{2}{5}$

$\quad \alpha = 23.6°$
$\quad \beta = 90° - \alpha \approx 66.4°$

15.

$\sin 35° = \frac{a}{3}$ $\qquad$ $\cos 35° = \frac{b}{3}$
$3(\sin 35°) = a$ $\qquad$ $3(\cos 35°) = b$
$1.7207293 \approx a$ $\qquad$ $2.4574561 \approx b$
1.72 in. $\approx a$ $\qquad$ 2.46 in. $\approx b$

294 $\qquad\qquad\qquad\qquad\qquad\qquad$ 5 TRIGONOMETRIC FUNCTIONS

17.

$$\text{csc } 25° = \frac{c}{5} \qquad \text{or} \qquad \text{sec } 25° = \frac{c}{5}$$

$$5(\text{csc } 25°) = c \qquad \text{or} \qquad 5(\text{sec } 25°) = c$$

$$11.831008 \approx c \qquad \text{or} \qquad 5.5168896 \approx c$$

$$11.83 \text{ in.} \approx c \qquad \text{or} \qquad 5.52 \text{ in.} \approx c$$

19. $c = 5$. Suppose $a = 2$. Then,

$$\sin \alpha = \frac{a}{c} = \frac{2}{5} \text{ so } \alpha = 23.6° \text{ and } \beta = 90° - \alpha \approx 66.4°$$

21. $$\tan 35° = \frac{a}{100}$$

$$100(\tan 35°) = a$$
$$70.020753 \approx a$$
$$70 \text{ ft.} \approx a$$

23. $$\tan 85.361° = \frac{x}{80}$$

$$80(\tan 85.361°) = x$$
$$985.91117 \approx x$$
$$985.9 \text{ feet} \approx x$$

25.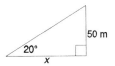

$$\frac{t}{10} = \frac{1}{\frac{1}{2}}$$

$$\frac{1}{2}t = 10$$

$$t = 20 \text{ meters}$$

27. $$\cot 15° = \frac{x}{30}$$

$$30(\cot 15°) = x$$
$$111.96153 \approx x$$

Hence, he is traveling at 111.96 ft/sec. Convert to mi/hr by:

$$\frac{1 \text{ mile}}{5280 \text{ ft}} \cdot \frac{111.96 \text{ ft}}{1 \text{ sec}} \cdot \frac{3600 \text{ sec}}{\text{hr}} \approx 76.336 \text{ mi/hr}$$

$$\approx 76.34 \text{ mph}$$

29.

$$\cot 20° = \frac{x}{50}$$

$$50(\cot 20°) = x$$
$$137.37387 \approx x$$
$$137 \text{ m} \approx x$$

31.

$$\sin 70° = \frac{x}{22}$$

$$22(\sin 70°) = x$$
$$20.673238 \approx x$$
$$20.67 \text{ ft.} \approx x$$

5.5 APPLICATIONS

33. We want to solve for x:

$$\cot 46.27° = \frac{100 + y}{x} \qquad \text{so} \qquad x \cot 46.27 = 100 + y$$
$$x \cot 46.27 - 100 = y$$

$$\cot 40.3° = \frac{200 + y}{x} \qquad \text{so} \qquad x \cot 40.3 = 200 + y$$
$$x \cot 40.3 - 200 = y$$

Since $y = y$, we have

$$x \cot 46.27 - 100 = x \cot 40.3 - 200$$
$$100 = x \cot 40.3 - x \cot 46.27$$
$$100 = x(\cot 40.3 - \cot 46.27)$$

$$\frac{100}{\cot 40.3 - \cot 46.27} = x$$

$$\frac{100}{1.1791595 - 0.956623} = x$$

$$\frac{100}{0.2225365} = x$$

$$449.36449 \approx x$$

$$449.36 \text{ ft.} \approx x$$

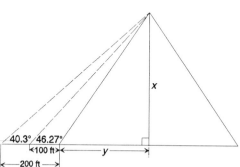

35. If α is the angle of elevation, then $\tan \alpha = \dfrac{300}{50} = 6$, so $\alpha \approx 80.5°$.

37.

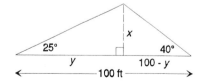

We want to solve for x. Therefore, we relabel the base of the triangle in two parts, y and $100 - y$.

$$\cot 25° = \frac{y}{x} \qquad \cot 40° = \frac{100 - y}{x}$$
$$x \cot 25° = y \qquad x \cot 40° = 100 - y$$
$$\qquad\qquad\qquad y = 100 - x \cot 40°$$

Since $y = y$, we have

$$x \cot 25° = 100 - x \cot 40°$$
$$x \cot 25° + x \cot 40° = 100$$
$$x(\cot 25° + \cot 40°) = 100$$
$$x = \frac{100}{\cot 25° + \cot 40°}$$
$$x = \frac{100}{2.1445069 + 1.1917536}$$
$$x = \frac{100}{3.3362605}$$
$$x \approx 29.973679$$
$$x \approx 30 \text{ ft.}$$

5 TRIGONOMETRIC FUNCTIONS

39.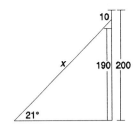

$\sin 21° = \dfrac{190}{x}$

$x \sin 21° = 190$

$x = \dfrac{190}{\sin 21°}$

$x \approx 530$ feet

41.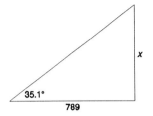

$\tan 35.1° = \dfrac{x}{789}$

$x = 789 \tan 35.1°$

$x \approx 555$ *feet*

43. If α is the angle of elevation, then $\tan \alpha = \dfrac{10 - 6}{15} = \dfrac{4}{15}$, so
$\alpha \approx 14.9°$.

☰ 5 - CHAPTER REVIEW

1. $135° = 135 \cdot 1$ degree $= 135 \cdot \dfrac{\pi}{180}$ radian $= \dfrac{3\pi}{4}$ radians.

3. $18° = 18 \cdot 1$ degree $= 18 \cdot \dfrac{\pi}{180}$ radian $= \dfrac{\pi}{10}$ radians.

5. $\dfrac{3\pi}{4}$ radian $= \dfrac{3\pi}{4} \cdot 1$ radian $= \dfrac{3\pi}{4} \cdot \dfrac{180}{\pi}$ degrees $= 135°$

7. $-\dfrac{5\pi}{2}$ radian $= -\dfrac{5\pi}{2} \cdot 1$ radian $= -\dfrac{5\pi}{2} \cdot \dfrac{180}{\pi}$ degrees $= -450°$

9. $\tan \dfrac{\pi}{4} - \sin \dfrac{\pi}{6} = 1 - \dfrac{1}{2} = \dfrac{1}{2}$

11. $3 \sin 45° - 4 \tan \dfrac{\pi}{6} = 3\left(\dfrac{\sqrt{2}}{2}\right) - 4\left(\dfrac{\sqrt{3}}{3}\right) = \dfrac{3\sqrt{2}}{2} - \dfrac{4\sqrt{3}}{3}$

13. $6 \cos \dfrac{3\pi}{4} + 2 \tan\left(-\dfrac{\pi}{3}\right)$

Using reference angles, $\cos \dfrac{3\pi}{4} = -\cos \dfrac{\pi}{4} = -\dfrac{\sqrt{2}}{2}$

$\tan\left(-\dfrac{\pi}{3}\right) = -\tan \dfrac{\pi}{3} = -\sqrt{3}$

Hence, $6\left(\dfrac{-\sqrt{2}}{2}\right) + 2\left(-\sqrt{3}\right) = -3\sqrt{2} - 2\sqrt{3}$

15. $\sec\left(-\dfrac{\pi}{3}\right) - \cot\left(-\dfrac{5\pi}{4}\right)$

Using reference angles:

$$\sec\left(-\dfrac{\pi}{3}\right) = \sec\dfrac{\pi}{3} = 2$$

$$\cot\left(-\dfrac{5\pi}{4}\right) = -\cot\left(\dfrac{\pi}{4}\right) = -1$$

Hence, $\sec\left(-\dfrac{\pi}{3}\right) - \cot\left(-\dfrac{5\pi}{4}\right) = 2 - (-1) = 3$

17. $\tan \pi + \sin \pi = 0 + 0 = 0$

19. $\cos 180° - \tan(-45°) = -1 - (-1) = -1 + 1 = 0$

21. $\sin^2 20° + \dfrac{1}{\sec^2 20°} = \sin^2 20° + \cos^2 20° = 1$

23. $\sec 50° \cdot \cos 50° = \dfrac{1}{\cos 50°} \cdot \cos 50° = 1$

25. $\dfrac{\sin 50°}{\cos 40°} = \dfrac{\sin 50°}{\sin(90° - 40°)} = \dfrac{\sin 50°}{\sin 50°} = 1$

27. $\dfrac{\sin(-40°)}{\cos 50°} = \dfrac{-\sin 40°}{\sin(90° - 50°)} = \dfrac{-\sin 40°}{\sin 40°} = -1$

29. $\sin 400° \sec(-50°) = \sin(400° - 360°)\sec 50°$
$$= \sin 40° \csc(90° - 50°)$$
$$= \sin 40° \csc 40°$$
$$= \sin 40° \cdot \dfrac{1}{\sin 40°} = 1$$

31. $\sin \theta = \dfrac{-4}{5}$, $\cos \theta > 0$

First we solve for $\cos \theta$:

$$\cos^2 \theta = 1 - \sin^2 \theta$$

$$\cos \theta = \sqrt{1 - \sin^2 \theta} = \sqrt{1 - \left(\dfrac{-4}{5}\right)^2}$$

$$= \sqrt{1 - \dfrac{16}{25}} = \dfrac{3}{5}$$

$$\tan \theta = \dfrac{\sin \theta}{\cos \theta} = \dfrac{\dfrac{-4}{5}}{\dfrac{3}{5}} = \dfrac{-4}{3}$$

$$\csc \theta = \dfrac{1}{\sin \theta} = \dfrac{1}{\dfrac{-4}{5}} = \dfrac{-5}{4}$$

$$\sec \theta = \dfrac{1}{\cos \theta} = \dfrac{1}{\dfrac{3}{5}} = \dfrac{5}{3}$$

$$\cot \theta = \dfrac{1}{\tan \theta} = \dfrac{1}{\dfrac{-4}{3}} = \dfrac{-3}{4}$$

5 TRIGONOMETRIC FUNCTIONS

33. $\tan \theta = \dfrac{12}{5}$, $\sin \theta < 0$

Because $\tan \theta = \dfrac{\sin \theta}{\cos \theta} > 0$ and $\sin \theta < 0$, $\cos \theta < 0$. Since

$\cos \theta = \dfrac{1}{\sec \theta} < 0$, $\sec \theta < 0$.

$$\sec^2 \theta = \tan^2 \theta + 1$$

$$\sec \theta = -\sqrt{\tan^2\theta + 1} = -\sqrt{\left(\dfrac{12}{5}\right)^2 + 1}$$

$$= -\sqrt{\dfrac{144}{25} + 1} = -\sqrt{\dfrac{169}{25}} = \dfrac{-13}{5}$$

$$\cos \theta = \dfrac{1}{\sec \theta} = \dfrac{1}{\dfrac{-13}{5}} = \dfrac{-5}{13}$$

$$\sin \theta = -\sqrt{1 - \cos^2 \theta} = -\sqrt{1 - \left(\dfrac{-5}{13}\right)^2}$$

$$= -\sqrt{1 - \dfrac{25}{169}} = -\sqrt{\dfrac{144}{169}} = \dfrac{-12}{13}$$

$$\csc \theta = \dfrac{1}{\sin \theta} = \dfrac{1}{\dfrac{-12}{13}} = \dfrac{-13}{12}$$

$$\cot \theta = \dfrac{1}{\tan \theta} = \dfrac{1}{\dfrac{12}{5}} = \dfrac{5}{12}$$

35. $\sec \theta = \dfrac{-5}{4}$, $\tan \theta < 0$

$$\tan^2 \theta + 1 = \sec^2 \theta$$
$$\tan^2 \theta = \sec^2 \theta - 1$$

$$\tan \theta = \pm\sqrt{\sec^2 \theta - 1}$$

Because $\tan t < 0$, we use the minus sign:

$$\tan \theta = -\sqrt{\left(\dfrac{-5}{4}\right)^2 - 1} = -\sqrt{\dfrac{25}{16} - 1}$$

$$= -\sqrt{\dfrac{9}{16}} = \dfrac{-3}{4}$$

$$\cot \theta = \dfrac{1}{\tan \theta} = \dfrac{1}{\dfrac{-3}{4}} = \dfrac{-4}{3}$$

$$\cos \theta = \dfrac{1}{\sec \theta} = \dfrac{1}{\dfrac{-5}{4}} = \dfrac{-4}{5}$$

$$\sin \theta = \pm\sqrt{1 - \cos^2 \theta}$$

Since $\tan \theta = \dfrac{\sin \theta}{\cos \theta} < 0$ and $\cos \theta < 0$, $\sin \theta > 0$. Therefore, we use the plus sign:

$$\sin \theta = \sqrt{1 - \left(\frac{-4}{5}\right)^2} = \sqrt{1 - \frac{16}{25}}$$

$$= \sqrt{\frac{9}{25}} = \frac{3}{5}$$

$$\csc \theta = \frac{1}{\sin \theta} = \frac{1}{\frac{3}{5}} = \frac{5}{3}$$

37. $\sin \theta = \dfrac{12}{13}$, θ in quadrant II

$\sin \theta = \dfrac{b}{r} = \dfrac{12}{13}$ so $b = 12$ and $r = 13$

Since $a < 0$ in quadrant II and using $r = \sqrt{a^2 + b^2}$, we get

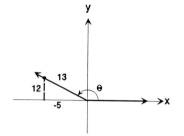

$$13 = \sqrt{a^2 + 12^2}$$
$$169 = a^2 + 144$$
$$25 = a^2$$
$$\pm = a$$
$$-5 = a$$

$$\cos \theta = \frac{a}{5} = \frac{-5}{13} = -\frac{5}{13} \qquad \csc \theta = \frac{r}{b} = \frac{13}{12}$$

$$\tan \theta = \frac{b}{a} = \frac{12}{-5} = -\frac{12}{5} \qquad \sec \theta = \frac{r}{a} = \frac{13}{-5} = -\frac{13}{5}$$

$$\cot \theta = \frac{a}{b} = \frac{-5}{12} = -\frac{5}{12}$$

39. $\sin \theta = -\dfrac{5}{13}$, $\dfrac{3\pi}{2} < \theta < 2\pi$

$\sin \theta = \dfrac{b}{r} = \dfrac{-5}{13}$ so $b = -5$ and $r = 13$

Since $a > 0$ in quadrant IV and using $r = \sqrt{a^2 + b^2}$, we get

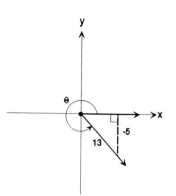

$$\sqrt{a^2 + (-5)^2} = 13$$
$$a^2 + 25 = 169$$
$$a^2 = 144$$
$$a = \pm 12$$
$$a = 12$$

$$\cos \theta = \frac{a}{r} = \frac{12}{13} \qquad \csc \theta = \frac{r}{b} = \frac{13}{-5} = -\frac{13}{5}$$

$$\tan \theta = \frac{b}{a} = \frac{-5}{12} \qquad \sec \theta = \frac{r}{a} = \frac{13}{12}$$

$$\cot \theta = \frac{a}{b} = \frac{12}{-5} = -\frac{12}{5}$$

5 TRIGONOMETRIC FUNCTIONS

41. $\tan \theta = \dfrac{1}{3}$, $180° < \theta < 270°$

$\tan \theta = \dfrac{b}{a}$ but $a < 0$ and $b < 0$ in

quadrant III so

$\tan \theta = \dfrac{b}{a} = \dfrac{1}{3}$ so $a = -3$ and $b = -1$

Using $r = \sqrt{a^2 + b^2}$, we find

$$\sqrt{(-3)^2 + (-1)^2} = r$$
$$\sqrt{9 + 1} = r$$
$$\sqrt{10} = r$$

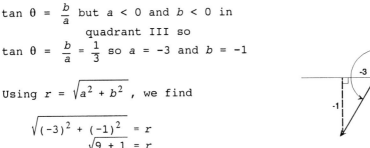

$\sin \theta = \dfrac{b}{r} = -\dfrac{1}{\sqrt{10}} = -\dfrac{1}{\sqrt{10}} \cdot \dfrac{\sqrt{10}}{\sqrt{10}} = -\dfrac{\sqrt{10}}{10}$ $\qquad$ $\csc \theta = \dfrac{r}{b} = \dfrac{\sqrt{10}}{-1} = -\sqrt{10}$

$\cos \theta = \dfrac{a}{r} = \dfrac{-3}{\sqrt{10}} = -\dfrac{3}{\sqrt{10}} \cdot \dfrac{\sqrt{10}}{\sqrt{10}} = -\dfrac{3\sqrt{10}}{10}$ $\qquad$ $\sec \theta = \dfrac{r}{a} = \dfrac{\sqrt{10}}{-3} = -\dfrac{\sqrt{10}}{3}$

$\cot \theta = \dfrac{a}{b} = \dfrac{-3}{-1} = 3$

43. $\sec \theta = 3$, $\dfrac{3\pi}{2} < \theta < 2\pi$

$\sec \theta = \dfrac{r}{a} = \dfrac{3}{1}$ so $a = 1$ and $r = 3$

Sin $b < 0$ in quadrant IV and using

$\sqrt{a^2 + b^2} = r$, we get

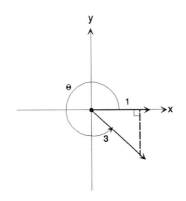

$$\sqrt{1^2 + b^2} = 3$$
$$1 + b^2 = 9$$
$$b^2 = 8$$
$$b = \pm\sqrt{8}$$
$$b = -2\sqrt{2}$$

$\sin \theta = \dfrac{b}{r} = \dfrac{-2\sqrt{2}}{3}$ $\qquad$ $\csc \theta = \dfrac{r}{b} = \dfrac{3}{-2\sqrt{2}} = -\dfrac{3}{2\sqrt{2}} \cdot \dfrac{\sqrt{2}}{\sqrt{2}} = -\dfrac{3\sqrt{2}}{4}$

$\cos \theta = \dfrac{a}{r} = \dfrac{1}{3}$

$\tan \theta = \dfrac{b}{a} = \dfrac{-2\sqrt{2}}{1} = -2\sqrt{2}$ $\qquad$ $\cot \theta = \dfrac{a}{b} = \dfrac{1}{-2\sqrt{2}} = -\dfrac{1}{2\sqrt{2}} \cdot \dfrac{\sqrt{2}}{\sqrt{2}} = -\dfrac{\sqrt{2}}{4}$

45. $\cot \theta = -2, \quad \frac{\pi}{2} < \theta < \pi$

$\cot \theta = \frac{a}{b}$ so $a < 0$ and $b > 0$

in quadrant II

$\cot \theta = \frac{a}{b} = \frac{-2}{1}$ so $a = -2$ and $b = 1$

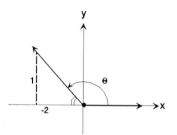

Using $\sqrt{a^2 + b^2} = r$, we get

$$\sqrt{(-2)^2 + 1^2} = r$$
$$\sqrt{5} = r$$

$\sin \theta = \frac{b}{r} = \frac{1}{\sqrt{5}} = \frac{\sqrt{5}}{5} \qquad \csc \theta = \frac{r}{b} = \sqrt{5}$

$\cos \theta = \frac{a}{r} = \frac{-2}{\sqrt{5}} = \frac{-2\sqrt{5}}{5} \qquad \sec \theta = \frac{r}{a} = \frac{\sqrt{5}}{-2} = \frac{-\sqrt{5}}{2}$

$\tan \theta = \frac{b}{a} = \frac{1}{-2} = \frac{-1}{2}$

47. $\alpha = 90° - 20° = 70°$

$\sin 20° = \frac{b}{10}$, so $b = 10 \sin 20° \approx 3.42$

$a^2 = c^2 - b^2 = 10^2 - (3.42)^2 \approx 88.3$

$a = \sqrt{88.3} \approx 9.4$

49. $a^2 = c^2 - b^2 = 5^2 - 2^2 = 21$

$a = 4.58$

$\cos \alpha = \frac{2}{5}$, so $\alpha = 66.4°$

$\sin \beta = \frac{2}{5}$, so $\beta = 23.6°$

51. $\theta = 30°$ or $\theta = \frac{\pi}{6}$

radius = 2 feet

Using $s = r\theta$, we get

$$s = 2\left(\frac{\pi}{6}\right)$$
$$s = \frac{\pi}{3} \text{ feet}$$

53. v = 180 mi/hr.

diameter = $\frac{1}{2}$ mi. so r = $\frac{1}{4}$ mi.

Find angular speed: ω

Using v = rω, we get

$$180 = \frac{1}{4}\omega$$

720 rad/hr. = ω (Remember that ω is expressed in **radians** per unit time.)

$$\frac{720 \text{ rad}}{\text{hr}} = \frac{1 \text{ rev}}{2\pi \text{ rad}} = \omega$$

$$\frac{720}{2\pi} \text{ rev/hr} = \omega$$

114.59 rev/hr. = ω

55.
$$\cot 65° = \frac{b}{500}$$
$$500(0.4663) = b$$
$$233.15 = b$$
$$\cot 25° = \frac{a + b}{500}$$
$$500(\cot 25°) = a + b$$
$$500(2.1445) = a + 233.15$$
$$1072.25 - 233.15 = a$$
$$839.1 = a$$
839 feet

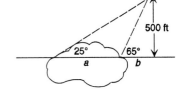

57.
$$\tan 25° = \frac{b}{50}$$
$$50(0.4663) = b$$
$$23.315 = b$$
23.32 feet

59.
$$1454 = 0.2754 \text{ mi}$$
$$\cot 5° = \frac{a + 1}{0.2754}$$
$$0.2754(\cot 5°) = a + 1$$
$$0.2754(11.43) = a + 1$$
$$3.147822 = a + 1$$
$$3.15 = a + 1$$
$$3.15 - 1 = a$$
2.15 mi

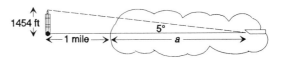

CHAPTER 6

GRAPHS OF TRIGONOMETRIC FUNCTIONS

≡ EXERCISE 6.1   GRAPHS OF THE SIX TRIGONOMETRIC
             FUNCTIONS

1.   0   3.   0

5.   The graph of $y = \sin x$ is increasing for $\frac{-\pi}{2} \le x \le \frac{\pi}{2}$

7. $y = \cos x$, $y = \sec x$

9.   The largest value of $y = \sin x$ is 1.

11. $\sin x = 0$ when $x = 0$, π, 2π

13. $\sin x = 1$ for $x = -\frac{3\pi}{2}$, $\frac{\pi}{2}$ if $-2\pi \le x \le 2\pi$

     $\sin x = -1$ for $x = -\frac{\pi}{2}$, $\frac{3\pi}{2}$ if $-2\pi \le x \le 2\pi$

15. $\sec x = 1$ for $x = -2\pi$, 0, 2π if $-2\pi \le x \le 2\pi$
     $\sec x = -1$ for $x = -\pi$, π if $-2\pi \le x \le 2\pi$

17. $y = \sec x$ has vertical asymptotes for $x = -\frac{3\pi}{2}$, $-\frac{\pi}{2}$, $\frac{\pi}{2}$, $\frac{3\pi}{2}$ if
     $-2\pi \le x \le 2\pi$.

19. $y = \tan x$ has vertical asymptotes for $x = -\frac{3\pi}{2}$, $-\frac{\pi}{2}$, $\frac{\pi}{2}$, $\frac{3\pi}{2}$ if
     $-2\pi \le x \le 2\pi$.

21.   23.

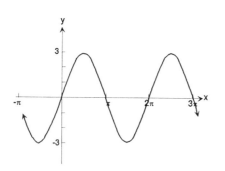   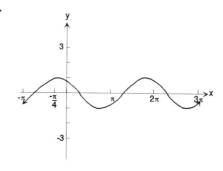

25.

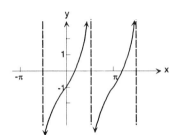

27.

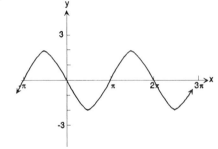

29.

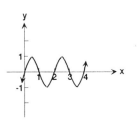

31.

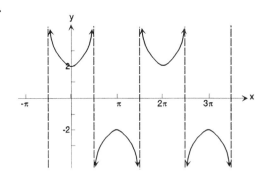

33.

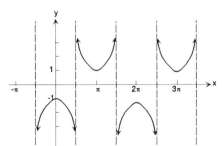

35.

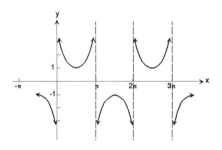

37.

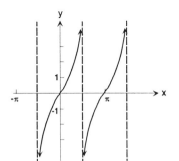

39.

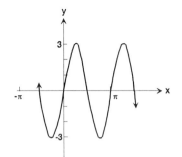

41.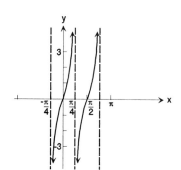

43.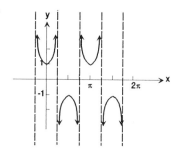

45.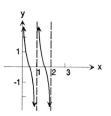

47.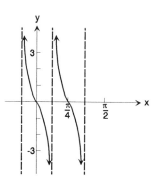

49.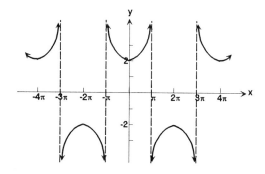

51. It would appear that $\tan x = -\cot\left(x + \frac{\pi}{2}\right)$ because if the cotangent function was reflected through the origin it would have the same shape as the tangent function. Also, if the cotangent function was shifted to the left $\frac{\pi}{2}$ units, it asymptotes would match up with the asymptotes of the tangent function. The graphs are identical.

6 GRAPHS OF TRIGONOMETRIC FUNCTIONS

53. The graph of $y = A \sin x$, $A > 0$, lies between $-A$ and A.

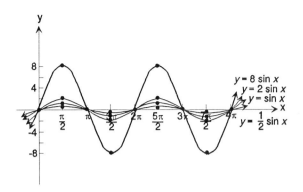

55. The graph of $y = \sin(x - \phi)$, $\phi > 0$, is the graph of $y = \sin x$ shifted horizontally to the right ϕ units.

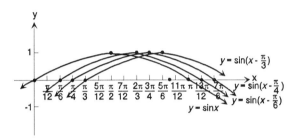

≡ EXERCISE 6.2 SINUSOIDAL GRAPHS

1. $y = 2 \sin x$

   Comparing $y = 2 \sin x$ to
   $\quad$ $y = A \sin \omega x$, we find $A = 2$ and $\omega = 1$.

   Thus, the amplitude is $|A| = |2| = 2$.

   The period is $T = \dfrac{2\pi}{\omega} = \dfrac{2\pi}{1} = 2\pi$.

3. $y = -4 \cos 2x$

   Comparing $y = -4 \cos 2x$ to
   $\quad$ $y = A \cos \omega x$, we find $A = -4$ and $\omega = 2$.

   Thus, the amplitude is $|A| = |-4| = 4$.

   The period is $T = \dfrac{2\pi}{\omega} = \dfrac{2\pi}{2} = \pi$.

5. $y = 6 \sin \pi x$

Comparing $y = 6 \sin x\pi$ to
$y = A \sin \omega x$, we find $A = 6$ and $\omega = \pi$.

Thus, the amplitude is $|A| = |6| = 6$.

The period is $T = \dfrac{2\pi}{\omega} = \dfrac{2\pi}{\pi} = 2$.

7. $y = -\dfrac{1}{2} \cos \dfrac{3}{2}x$

Comparing $y = -\dfrac{1}{2} \cos \dfrac{3}{2}x$ to

$\qquad y = A \cos \omega x$, we find $A = -\dfrac{1}{2}$ and $\omega = \dfrac{3}{2}$.

Thus, the amplitude is $|A| = \left|-\dfrac{1}{2}\right| = \dfrac{1}{2}$.

The period is $T = \dfrac{2\pi}{\omega} = \dfrac{2\pi}{\dfrac{3}{2}} = \dfrac{2\pi}{1} \cdot \dfrac{2}{3} = \dfrac{4\pi}{3}$.

9. $y = \dfrac{5}{3} \sin\left(\dfrac{-2\pi}{3}x\right)$

Comparing $y = \dfrac{5}{3}\sin\left(\dfrac{-2\pi}{3}x\right) = \dfrac{-5}{3}\sin\left(\dfrac{2\pi}{3}x\right)$

$\qquad y = A \sin \omega x$, we find $A = \dfrac{-5}{3}$ and $\omega = \dfrac{2\pi}{3}$.

Thus, the amplitude is $|A| = \left|\dfrac{-5}{3}\right| = \dfrac{5}{3}$.

The period is $T = \dfrac{2\pi}{\omega} = \dfrac{2\pi}{\dfrac{2\pi}{3}} = 3$.

11. $y = 2 \sin \dfrac{\pi}{2}x$

Amplitude: $|A| = |2| = 2$

Period: $T = \dfrac{2\pi}{\dfrac{\pi}{2}} = \dfrac{2\pi}{1} \cdot \dfrac{2}{\pi} = 4$

Hence, graph F is a sin graph with amplitude = 2 and period = 4.

6 GRAPHS OF TRIGONOMETRIC FUNCTIONS

13. $y = 2 \cos \frac{1}{2}x$

Amplitude: $|A| = |2| = 2$

Period: $T = \dfrac{2\pi}{\frac{1}{2}} = \dfrac{2\pi}{1} \cdot \dfrac{2}{1} = 4\pi$

Hence, graph A is a cos graph with amplitude 2 and period 4π.

15. $y = -3 \sin 2x$

Amplitude: $|A| = |-3| = 3$

Period: $T = \dfrac{2\pi}{2} = \pi$

Hence, graph H is a reflected sin graph with amplitude 3 and period π.

17. $y = -2 \cos \frac{1}{2}x$

Amplitude: $|A| = |-2| = 2$

Period: $T = \dfrac{2\pi}{\frac{1}{2}} = 4\pi$

Hence, graph C is a reflected cos graph with amplitude 2 and period 4π.

19. $y = 3 \sin 2x$

Amplitude: $|A| = |3| = 3$

Period: $T = \dfrac{2\pi}{2} = \pi$

Hence, graph J is a sin graph with amplitude 3 and period π.

21. $y = 5 \sin 4x$

Amplitude: $|A| = |5| = 5$

Period: $T = \dfrac{2\pi}{4} = \dfrac{\pi}{2}$

Use the amplitude to scale the y-axis and the period to scale the x-axis. Then, fill in the graph of the sine function.

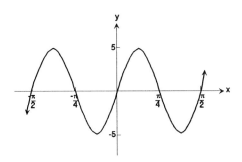

6.2 SINUSOIDAL GRAPHS

23. $y = 5 \cos \pi x$

Amplitude: $|A| = |5| = 5$

Period: $T = \dfrac{2\pi}{\pi} = 2$

Use the amplitude to scale the y-axis and the period to scale the x-axis. Then, fill in the graph of the cosine function.

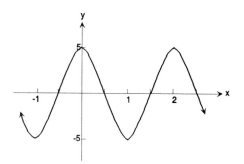

25. $y = -2 \cos 2\pi x$

Amplitude: $|A| = |-2| = 2$

Period: $T = \dfrac{2\pi}{2\pi} = 1$

Use the amplitude to scale the y-axis and the period to scale the x-axis. Then, fill in the graph of the reflected cosine function. $(A = -2 < 0)$

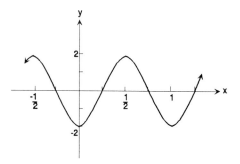

27. $y = -4 \sin \dfrac{1}{2}x$

Amplitude: $|A| = |-4| = 4$

Period: $T = \dfrac{2\pi}{\dfrac{1}{2}} = 4\pi$

Use the amplitude to scale the y-axis and the period to scale the x-axis. Then, fill in the graph of the reflected sine function. $(A = -4 < 0)$

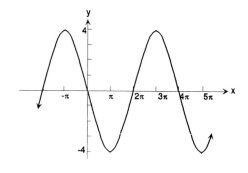

29. $y = \dfrac{3}{2}\sin\left(\dfrac{-2}{3}x\right) = \dfrac{-3}{2}\sin\left(\dfrac{2}{3}x\right)$

Amplitude: $|A| = \left|\dfrac{-3}{2}\right| = \dfrac{3}{2}$

Period: $T = \dfrac{2\pi}{\dfrac{2}{3}} = \dfrac{2\pi}{1} \cdot \dfrac{3}{2} = 3\pi$

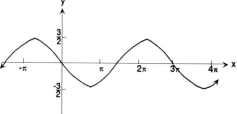

Use the amplitude to scale the y-axis and the period to scale the x-axis. Then, fill in the graph of the reflected sine function. $(A = \dfrac{-3}{2} < 0)$

6 GRAPHS OF TRIGONOMETRIC FUNCTIONS

31. Since the graph starts at $(0, 5)$, we have a cos graph with $A = 5$. Since the period is 8, we have

$$8 = \frac{2\pi}{\omega} \text{ so}$$
$$8\omega = 2\pi$$
$$\omega = \frac{2\pi}{8} = \frac{\pi}{4}$$

Hence, the function is $y = 5 \cos \frac{\pi}{4}x$.

33. Since the graph starts at $(0, -3)$, we have a reflected cosine function with $A = -3$. Since the period is 4π, we have

$$4\pi = \frac{2\pi}{\omega}, \text{ so}$$
$$4\pi\omega = 2\pi$$
$$\omega = \frac{2\pi}{4\pi} = \frac{1}{2}$$

Hence, the function is $y = -3 \cos \frac{1}{2}x$.

35. Since the graph starts at $(0, 0)$, we have the sine function. Since it increases first to $\frac{3}{4}$, we have $A = \frac{3}{4}$. Since the period is 1, we have

$$1 = \frac{2\pi}{\omega}, \text{ so}$$
$$\omega = 2\pi$$

Hence, the function is $y = \frac{3}{4} \sin 2\pi x$.

37. Since the graph starts at $(0, 0)$, we have the sine function. Since it decreases first to -1, we have a reflected sine function with $A = -1$. The period is $\frac{4\pi}{3}$ so

$$\frac{4\pi}{3} = \frac{2\pi}{\omega}$$
$$4\pi\omega = 6\pi$$
$$\omega = \frac{6\pi}{4\pi} = \frac{6}{4} = \frac{3}{2}$$

Hence, the function is $y = -1 \sin \frac{3}{2}x$

$$y = -\sin \frac{3}{2}x$$

39. Since the graph starts at $(0, -2)$, we have the reflected cosine function with $A = -2$. The period is $\frac{4}{3}$, so

$$\frac{4}{3} = \frac{2\pi}{\omega}$$
$$4\omega = 6\pi$$
$$\omega = \frac{6\pi}{4} = \frac{3\pi}{2}$$

Hence, the function is $y = -2 \cos \frac{3\pi}{2}x$.

41. $y = 4 \sin(2x - \pi)$ compared to
    $y = A \sin(\omega x - \phi)$.

Amplitude: $|A| = |4| = 4$

Period: $T = \dfrac{2\pi}{\omega} = \dfrac{2\pi}{2} = \pi$

Phase Shift: $\dfrac{\phi}{\omega} = \dfrac{\pi}{2}$

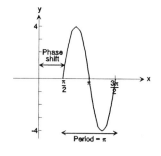

43. $y = 2 \cos\left(3x + \dfrac{\pi}{2}\right)$ compared to
    $y = A \cos(\omega x - \phi)$.

$y = 2 \cos\left[3x - \left(-\dfrac{\pi}{2}\right)\right]$

Amplitude: $|A| = |2| = 2$

Period: $T = \dfrac{2\pi}{\omega} = \dfrac{2\pi}{3}$

Phase Shift: $\dfrac{\phi}{\omega} = \dfrac{-\dfrac{\pi}{2}}{3} = -\dfrac{\pi}{6}$

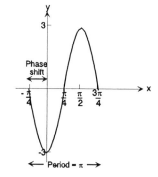

45. $y = -3 \sin\left(2x + \dfrac{\pi}{2}\right)$ compared to
    $y = A \sin(\omega x - \phi)$.

$y = -3 \sin\left[2x - \left(-\dfrac{\pi}{2}\right)\right]$

Amplitude: $|A| = |-3| = 3$

Period: $T = \dfrac{2\pi}{\omega} = \dfrac{2\pi}{3} = \pi$

Phase Shift: $\dfrac{\phi}{\omega} = \dfrac{-\dfrac{\pi}{2}}{2} = -\dfrac{\pi}{4}$

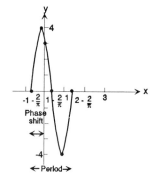

47. $y = 4 \sin(\pi x + 2)$
    $y = A \sin(\omega x - \phi)$.
    $A = 4,\ \omega = \pi,\ \phi = -2$

Amplitude: 4

Period: $\dfrac{2\pi}{\omega} = \dfrac{2\pi}{\pi} = 2$

Phase Shift: $\dfrac{\phi}{\omega} = \dfrac{-2}{\pi}$

49. $y = 3 \cos (\pi x - 2)$
$y = A \cos (\omega x - \phi).$
$A = 3, \omega = \pi, \phi = 2$

Amplitude: 3

Period: $\dfrac{2\pi}{\omega} = \dfrac{2\pi}{\pi} = 2$

Phase Shift: $\dfrac{\phi}{\omega} = \dfrac{2}{\pi}$

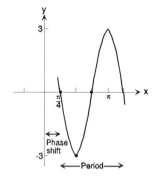

51. $y = 3 \sin \left(-2x + \dfrac{\pi}{2}\right)$

$y = -3 \sin \left(2x - \dfrac{\pi}{2}\right)$

$y = A \cos (\omega x - \phi).$

Amplitude: 3

Period: $\dfrac{2\pi}{\omega} = \dfrac{2\pi}{2} = \pi$

Phase Shift: $\dfrac{\phi}{\omega} = \dfrac{\dfrac{\pi}{2}}{2} = \dfrac{\pi}{4}$

≡ EXERCISE 6.3 ADDITIONAL GRAPHING TECHNIQUES

1.

x	$\dfrac{-\pi}{2}$	0	$\dfrac{\pi}{2}$	π	$\dfrac{3\pi}{2}$	2π
$f_1(x) = x$	$\dfrac{-\pi}{2}$	0	$\dfrac{\pi}{2}$	π	$\dfrac{3\pi}{2}$	2π
$f_2(x) = \cos x$	0	1	0	-1	0	1
$f(x) = x + \cos x$	$\dfrac{-\pi}{2}$	1	$\dfrac{\pi}{2}$	$\pi - 1$	$\dfrac{3\pi}{2}$	$2\pi + 1$
Point on graph of f	$\left(\dfrac{-\pi}{2}, \dfrac{-\pi}{2}\right)$	$(0, 1)$	$\left(\dfrac{\pi}{2}, \dfrac{\pi}{2}\right)$	$(\pi, \pi - 1)$	$\left(\dfrac{3\pi}{2}, \dfrac{3\pi}{2}\right)$	$(2\pi, 2\pi + 1)$

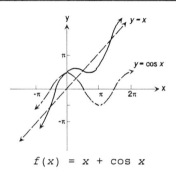

$f(x) = x + \cos x$

3.

x	$-\pi$	0	$\dfrac{\pi}{2}$	π	$\dfrac{3\pi}{2}$	2π
$f_1(x) = x$	$-\pi$	0	$\dfrac{\pi}{2}$	π	$\dfrac{3\pi}{2}$	2π
$f_2(x) = -\sin x$	0	0	-1	0	1	0
$f(x) = x - \sin x$	$-\pi$	0	$\dfrac{\pi}{2} - 1$	π	$\dfrac{3\pi}{2} + 1$	2π
Point on graph of f	$(-\pi, -\pi)$	$(0, 0)$	$\left[\dfrac{\pi}{2}, \dfrac{\pi}{2} - 1\right]$	(π, π)	$\left[\dfrac{3\pi}{2}, \dfrac{3\pi}{2} + 1\right]$	$(2\pi, 2\pi)$

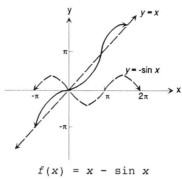

$$f(x) = x - \sin x$$

5.

x	$\dfrac{-\pi}{2}$	0	$\dfrac{\pi}{2}$	π	$\dfrac{3\pi}{2}$	2π
$f_1(x) = \sin x$	-1	0	1	0	-1	0
$f_2(x) = \cos x$	0	1	0	-1	0	1
$f(x) = \sin x + \cos x$	-1	1	1	-1	-1	1
Point on graph of f	$\left[\dfrac{-\pi}{2}, -1\right]$	$(0, 1)$	$\left[\dfrac{\pi}{2}, 1\right]$	$(\pi, -1)$	$\left[\dfrac{3\pi}{2}, -1\right]$	$(2\pi, 1)$

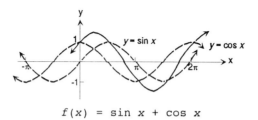

$$f(x) = \sin x + \cos x$$

7.

x	0	$\dfrac{\pi}{2}$	π	$\dfrac{3\pi}{2}$	2π
$g_1(x) = \sin x$	0	1	0	-1	0
$g_2(x) = \sin 2x$	0	0	0	0	0
$g(x) = \sin x + \sin 2x$	0	1	0	-1	0
Point on graph of g	$(0, 0)$	$\left(\dfrac{\pi}{2}, 1\right)$	$(\pi, 0)$	$\left(\dfrac{3\pi}{2}, -1\right)$	$(2\pi, 0)$

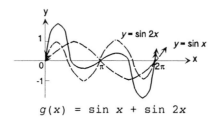

$$g(x) = \sin x + \sin 2x$$

9.

x	0	$\dfrac{\pi}{2}$	π	$\dfrac{3\pi}{2}$	2π
$h_1(x) = \sqrt{x}$	0	$\sqrt{\dfrac{\pi}{2}}$	$\sqrt{\pi}$	$\sqrt{\dfrac{3\pi}{2}}$	$\sqrt{2\pi}$
$h_2(x) = \sin x$	0	1	0	-1	0
$h(x) = \sqrt{x} + \sin x$	0	$\sqrt{\dfrac{\pi}{2}} + 1$	$\sqrt{\pi}$	$\sqrt{\dfrac{3\pi}{2}} - 1$	$\sqrt{2\pi}$
Point on graph of h	$(0, 0)$	$\left(\dfrac{\pi}{2}, 2.25\right)$	$(\pi, 1.17)$	$\left(\dfrac{3\pi}{2}, 1.17\right)$	$(2\pi, 2.51)$

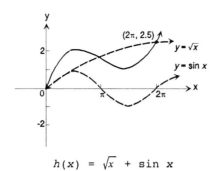

$$h(x) = \sqrt{x} + \sin x$$

11.

x	0	$\dfrac{\pi}{2}$	π	$\dfrac{3\pi}{2}$	2π
$F_1(x) = 2 \sin x$	0	2	0	-2	0
$F_2(x) = -\cos 2x$	-1	1	-1	1	-1
$F(x) = 2 \sin x - \cos 2x$	-1	3	-1	-1	-1
Point on graph of F	(0, -1)	$\left[\dfrac{\pi}{2}, 3\right]$	$(\pi, -1)$	$\left[\dfrac{3\pi}{2}, -1\right]$	$(2\pi, -1)$

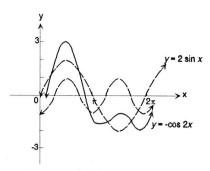

$$F(x) = 2 \sin x - \cos 2x$$

13.

x	0	$\dfrac{1}{2}$	1	$\dfrac{3}{2}$	2
$f_1(x) = 2 \sin \pi x$	0	2	0	-2	0
$f_2(x) = \cos \pi x$	1	0	-1	0	1
$f(x) = 2 \sin \pi x + \cos \pi x$	1	2	-1	-2	1
Point on graph of f	(0, 1)	$\left[\dfrac{1}{2}, 2\right]$	(1, -1)	$\left[\dfrac{3}{2}, -2\right]$	(2, 1)

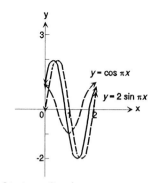

$$f(x) = 2 \sin \pi x + \cos \pi x$$

15.

x	$\dfrac{-\pi}{2}$	0	$\dfrac{\pi}{2}$	π
$f_1(x) = \dfrac{x^2}{\pi^2}$	$\dfrac{1}{4}$	0	$\dfrac{1}{4}$	1
$f_2(x) = \sin 2x$	0	0	0	0
$f(x) = \dfrac{x^2}{\pi^2} + \sin 2x$	$\dfrac{1}{4}$	0	$\dfrac{1}{4}$	1
Point on graph of f	$\left(\dfrac{-\pi}{2}, \dfrac{1}{4}\right)$	$(0, 0)$	$\left(\dfrac{\pi}{2}, \dfrac{1}{4}\right)$	$(\pi, 1)$

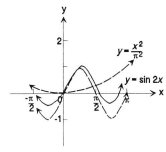

$$f(x) = \frac{x^2}{\pi^2} + \sin 2x$$

17.

x	-1	0	1	2	3	4
$f_1(x) = |x|$	1	0	1	2	3	4
$f_2(x) = \sin \dfrac{\pi}{2}x$	-1	0	1	0	-1	0
$f(x) = |x| + \sin \dfrac{\pi}{2}x$	0	0	2	2	2	4
Point on graph of f	(-1, 0)	(0, 0)	(1, 2)	(2, 2)	(3, 2)	(4, 4)

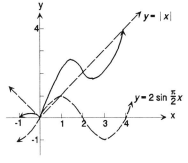

$$f(x) = |x| + \sin \frac{\pi}{2} x$$

19.

x	0	$\dfrac{\pi}{6}$	$\dfrac{\pi}{3}$	$\dfrac{\pi}{2}$	$\dfrac{2\pi}{3}$	$\dfrac{5\pi}{6}$	π
$f_1(x) = 3\sin 2x$	0	$\dfrac{3\sqrt{3}}{2}$	$\dfrac{3\sqrt{3}}{2}$	0	$\dfrac{-3\sqrt{3}}{2}$	$\dfrac{-3\sqrt{3}}{2}$	0
$f_2(x) = 2\cos 3x$	2	0	-2	0	2	0	-2
$f(x) = 3\sin 2x + 2\cos 3x$	2	$\dfrac{3\sqrt{3}}{2}$	$\dfrac{3\sqrt{3}}{2} - 2$	0	$2 - \dfrac{3\sqrt{3}}{2}$	$\dfrac{-3\sqrt{3}}{2}$	-2
Point on graph of f	(0, 2)	$\left[\dfrac{\pi}{6},\ 2.6\right]$	$\left[\dfrac{\pi}{3},\ .6\right]$	$\left[\dfrac{\pi}{2},\ 0\right]$	$\left[\dfrac{2\pi}{3},\ -.6\right]$	$\left[\dfrac{5\pi}{6},\ -2.6\right]$	$(\pi, -2)$

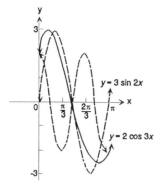

$$f(x) = 3 \sin 2x + 2 \cos 3x$$

21.

x	$\dfrac{-\pi}{2}$	0	$\dfrac{\pi}{2}$	π	$\dfrac{3\pi}{2}$	2π
$f_1(x) = x$	$\dfrac{-\pi}{2}$	0	$\dfrac{\pi}{2}$	π	$\dfrac{3\pi}{2}$	2π
$f_2(x) = \cos x$	0	1	0	-1	0	1
$f(x) = x\cos x$	0	0	0	$-\pi$	0	2π
Point on graph of f	$\left[\dfrac{-\pi}{2},\ 0\right]$	(0, 0)	$\left[\dfrac{\pi}{2},\ 0\right]$	$(\pi, -\pi)$	$\left[\dfrac{3\pi}{2},\ 0\right]$	$(2\pi, 2\pi)$

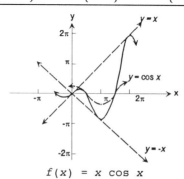

$$f(x) = x \cos x$$

6 GRAPHS OF TRIGONOMETRIC FUNCTIONS

23.

x	$-\pi$	$\dfrac{-\pi}{2}$	0	$\dfrac{\pi}{2}$	π	$\dfrac{3\pi}{2}$	2π
$f_1(x) = x^2$	π^2	$\dfrac{\pi^2}{4}$	0	$\dfrac{\pi^2}{4}$	π^2	$\dfrac{9\pi^2}{4}$	$4\pi^2$
$f_2(x) = \sin x$	0	-1	0	1	0	-1	0
$f(x) = x^2 \sin x$	0	$\dfrac{-\pi^2}{4}$	0	$\dfrac{\pi^2}{4}$	0	$\dfrac{-9\pi^2}{4}$	0
Point on graph of f	$(-\pi,\,0)$	$\left[\dfrac{-\pi}{2},\,\dfrac{-\pi^2}{4}\right]$	$(0,\,0)$	$\left[\dfrac{\pi}{2},\,\dfrac{\pi^2}{4}\right]$	$(\pi,\,0)$	$\left[\dfrac{3\pi}{2},\,\dfrac{-9\pi^2}{4}\right]$	$(2\pi,\,0)$

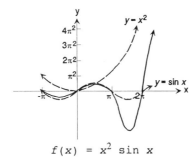

$$f(x) \;=\; x^2 \, \sin x$$

25.

x	0	$\dfrac{\pi}{2}$	π	$\dfrac{3\pi}{2}$	2π
$f_1(x) = \lvert x \rvert$	0	$\dfrac{\pi}{2}$	π	$\dfrac{3\pi}{2}$	2π
$f_2(x) = \cos x$	1	0	-1	0	1
$f(x) = \lvert x \rvert \cos x$	0	0	$-\pi$	0	2π
Point on graph of f	$(0,\,0)$	$\left[\dfrac{\pi}{2},\,0\right]$	$(\pi,\,-\pi)$	$\left[\dfrac{3\pi}{2},\,0\right]$	$(2\pi,\,2\pi)$

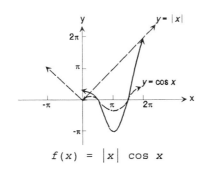

$$f(x) \;=\; \lvert x \rvert \, \cos x$$

27.

x	$\dfrac{-\pi}{2}$	$\dfrac{-\pi}{4}$	0	$\dfrac{\pi}{4}$	$\dfrac{\pi}{2}$	π
$f_1(x) = e^{-x}$	4.8	2.2	1	.46	.21	.04
$f_2(x) = \cos 2x$	-1	0	1	0	-1	1
$f(x) = e^{-x} \cos 2x$	-4.8	0	1	0	-.21	.04
Point on graph of f	$\left[\dfrac{-\pi}{2},\ -4.8\right]$	$\left[\dfrac{-\pi}{4},\ 0\right]$	(0, 1)	$\left[\dfrac{\pi}{4},\ 0\right]$	$\left[\dfrac{\pi}{2},\ -.21\right]$	$(\pi, .04)$

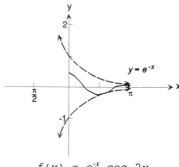

$$f(x) = e^{-x} \cos 2x$$

29.

x	0	$\dfrac{1}{4}$	$\dfrac{1}{2}$	$\dfrac{3}{4}$	1	$\dfrac{5}{4}$	$\dfrac{3}{2}$	$\dfrac{7}{4}$	2
$f_1(x) = \dfrac{1}{2} \sin 2\pi x$	0	$\dfrac{1}{2}$	0	$\dfrac{-1}{2}$	0	$\dfrac{1}{2}$	0	$\dfrac{-1}{2}$	0
$f_2(x) = \dfrac{1}{4} \sin 4\pi x$	0	0	0	0	0	0	0	0	0
$f(x) = \dfrac{1}{2} \sin 2\pi x + \dfrac{1}{4} \sin 4\pi x$	0	$\dfrac{1}{2}$	0	$\dfrac{-1}{2}$	0	$\dfrac{1}{2}$	0	$\dfrac{-1}{2}$	0
Point on graph of f	(0, 0)	$\left[\dfrac{1}{4},\ \dfrac{1}{2}\right]$	$\left[\dfrac{1}{2},\ 0\right]$	$\left[\dfrac{3}{4},\ \dfrac{-1}{2}\right]$	(1, 0)	$\left[\dfrac{5}{4},\ \dfrac{1}{2}\right]$	$\left[\dfrac{3}{2},\ 0\right]$	$\left[\dfrac{7}{4},\ \dfrac{-1}{2}\right]$	(2, 0)

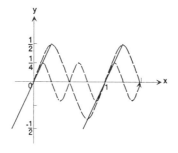

$$f(x) = \dfrac{1}{2} \sin 2\pi x + \dfrac{1}{4} \sin 4\pi x$$

6 GRAPHS OF TRIGONOMETRIC FUNCTIONS

31.

t	0	$\frac{1}{2}$	1	$\frac{3}{2}$	2
$V_1(t) = e^{-10t}$	1	$\approx .007$	$\approx .00005$	≈ 0	≈ 0
$V_2(t) = \cos \pi t$	1	0	-1	0	1
$V(t) = e^{-10t} \cos \pi t$	1	0	$\approx -.00005$	0	0
Point on graph of V	(0, 1)	$\left(\frac{1}{2}, 0\right)$	(1, -.00005)	$\left(\frac{3}{2}, 0\right)$	(2, 0)

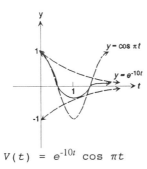

$$V(t) = e^{-10t} \cos \pi t$$

33. The graph will lie between the bounding curves $y = \pm x$, $y = \pm x^2$, and $y = \pm x^3$, respectively, touching them at odd multiples of $\frac{\pi}{2}$. The x-intercepts of each graph are the multiples of π.

x	-2π	$\frac{-3\pi}{2}$	$-\pi$	$\frac{-\pi}{2}$	0	$\frac{\pi}{2}$	π	$\frac{3\pi}{2}$	2π
x^2	$4\pi^2$	$\frac{9\pi^2}{4}$	π^2	$\frac{\pi^2}{4}$	0	$\frac{\pi^2}{4}$	π^2	$\frac{9\pi^2}{4}$	$4\pi^2$
x^3	$-8\pi^3$	$\frac{-27\pi^3}{8}$	$-\pi^3$	$\frac{-\pi^3}{8}$	0	$\frac{\pi^3}{8}$	π^3	$\frac{27\pi^3}{8}$	$8\pi^3$
$\sin x$	0	1	0	-1	0	1	0	-1	0
$x \sin x$	0	$\frac{-3\pi}{2} \approx -4.7$	0	$\frac{\pi}{2} \approx 1.57$	0	$\frac{\pi}{2} \approx 1.57$	0	$\frac{-3\pi}{2} \approx -4.7$	0
$x^2 \sin x$	0	$\frac{9\pi^2}{4} \approx 22.2$	0	$\frac{-\pi^2}{4} \approx -2.5$	0	$\frac{\pi^2}{4} \approx 2.5$	0	$\frac{-9\pi^2}{4} \approx -22.2$	0
$x^3 \sin x$	0	$\frac{-27\pi^3}{8} \approx -104.6$	0	$\frac{\pi^3}{8} \approx 3.9$	0	$\frac{\pi^3}{8} \approx 3.9$	0	$\frac{-27\pi^3}{8} \approx -104.6$	0

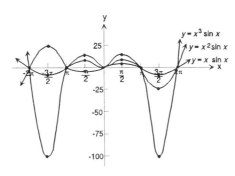

☰ EXERCISE 6.4 THE INVERSE TRIGONOMETRIC FUNCTIONS

1. $\sin^{-1} 0$

We seek the angle θ, $-\dfrac{\pi}{2} \leq \theta \leq \dfrac{\pi}{2}$, whose sine equals 0.

$\sin \theta = 0 \qquad\qquad -\dfrac{\pi}{2} \leq \theta \leq \dfrac{\pi}{2}$

$\theta = 0$

$\sin^{-1} 0 = 0$

3. $\sin^{-1} (-1)$

We seek the angle θ, $-\dfrac{\pi}{2} \leq \theta \leq \dfrac{\pi}{2}$, whose sine equals (-1).

$\sin \theta = -1 \qquad -\dfrac{\pi}{2} \leq \theta \leq \dfrac{\pi}{2}$

$\theta = -\dfrac{\pi}{2}$

$\sin^{-1} (-1) = -\dfrac{\pi}{2}$

5. $\tan^{-1} 0$

We seek the angle θ, $-\dfrac{\pi}{2} < \theta < \dfrac{\pi}{2}$, whose tangent equals 0.

$\tan \theta = 0 \qquad\qquad -\dfrac{\pi}{2} < \theta < \dfrac{\pi}{2}$

$\theta = 0$

$\tan^{-1} 0 = 0$

7. $\sin^{-1} \dfrac{\sqrt{2}}{2}$

We seek the angle θ, $-\dfrac{\pi}{2} \leq \theta \leq \dfrac{\pi}{2}$, whose sine equals $\dfrac{\sqrt{2}}{2}$.

$\sin \theta = \dfrac{\sqrt{2}}{2} \qquad -\dfrac{\pi}{2} \leq \theta \leq \dfrac{\pi}{2}$

$\theta = \dfrac{\pi}{4}$

$\sin^{-1} \dfrac{\sqrt{2}}{2} = \dfrac{\pi}{4}$

9. $\tan^{-1} \sqrt{3}$

We seek the angle θ, $-\frac{\pi}{2} < \theta < \frac{\pi}{2}$, whose tangent equals $\sqrt{3}$.

$$\tan \theta = \sqrt{3} \qquad\qquad -\frac{\pi}{2} < \theta < \frac{\pi}{2}$$
$$\theta = \frac{\pi}{3}$$
$$\tan^{-1} \sqrt{3} = \frac{\pi}{3}$$

11. $\cos^{-1}\left(-\frac{\sqrt{3}}{2}\right)$

We seek the angle θ, $0 \le \theta \le \pi$, whose tangent equals $-\frac{\sqrt{3}}{2}$.

$$\cos \theta = \left(\frac{-\sqrt{3}}{2}\right) \qquad 0 \le \theta \le \pi$$
$$\theta = \frac{5\pi}{6}$$
$$\cos^{-1}\left(-\frac{\sqrt{3}}{2}\right) = \frac{5\pi}{6}$$

13. $\sin^{-1} 0.1$ (Calculator used was Slimline TI35.)

Set the mode of the calculator to radians.

Enter: $\boxed{0.1}$

    Display: $\boxed{0.1}$

Press: $\boxed{\text{INV}}$ Press: $\boxed{\sin}$

                Display: $\boxed{0.1001674}$

Hence, $\sin^{-1} 0.1 = 0.1002$

15. $\tan^{-1} 5$

Set the mode of the calculator to radians.

Enter: $\boxed{5}$

    Display: $\boxed{5}$

Press: $\boxed{\text{INV}}$ Press: $\boxed{\tan}$

                Display: $\boxed{1.3734008}$

Hence, $\tan^{-1} 5 = 1.3734$

6.4 THE INVERSE TRIGONOMETRIC FUNCTIONS

17. $\cos^{-1} \dfrac{7}{8}$

Set the mode of the calculator to radians.

Enter: $\boxed{7}\ \boxed{\div}\ \boxed{8}\ \boxed{=}$

   Display: $\boxed{0.875}$

Press: $\boxed{\text{INV}}$ Press: $\boxed{\cos}$

         Display: $\boxed{0.5053605}$

Hence, $\cos^{-1} \dfrac{7}{8} = 0.5054$

19. $\tan^{-1} (-0.4)$

Set the mode of the calculator to radians.

Enter: $\boxed{0.4}$ Press: $\boxed{+/-}$

   Display: $\boxed{0.4}$ Display: $\boxed{-0.4}$

Press: $\boxed{\text{INV}}$ Press: $\boxed{\tan}$

         Display: $\boxed{-0.3805064}$

Hence, $\tan^{-1} (-0.4) = -0.3805$

21. $\sin^{-1} (-0.12)$

Set the mode of the calculator to radians.

Enter: $\boxed{0.12}$ Press: $\boxed{+/-}$

   Display: $\boxed{0.12}$ Display: $\boxed{-0.12}$

Press: $\boxed{\text{INV}}$ Press: $\boxed{\sin}$

         Display: $\boxed{-0.1202899}$

Hence, $\sin^{-1} (-0.12) = -0.1203$

23. $\cos^{-1} \dfrac{\sqrt{2}}{3}$

Set the mode of the calculator to radians.

Enter: [2] [INV] [$\sqrt{x}$] [÷] [3] [=]

Display: [0.4714045]

Press: [INV] Press: [cos]

Display: [1.0799136]

Hence, $\cos^{-1} \dfrac{\sqrt{2}}{3} = 1.0799$

25. $\cos\left(\sin^{-1} \dfrac{\sqrt{2}}{2}\right)$

First find the angle θ, $-\dfrac{\pi}{2} \leq \theta \leq \dfrac{\pi}{2}$, whose sine equals $\dfrac{\sqrt{2}}{2}$.

$$\sin \theta = \dfrac{\sqrt{2}}{2} \qquad -\dfrac{\pi}{2} \leq \theta \leq \dfrac{\pi}{2}$$
$$\theta = \dfrac{\pi}{4}$$

Now, $\cos\left(\sin^{-1} \dfrac{\sqrt{2}}{2}\right) = \cos \theta = \cos \dfrac{\pi}{4} = \dfrac{\sqrt{2}}{2}$.

27. $\tan\left[\cos^{-1}\left(-\dfrac{\sqrt{3}}{2}\right)\right]$

First find the angle θ, $0 \leq \theta \leq \pi$, whose cosine equals $-\dfrac{\sqrt{3}}{2}$.

$$\cos \theta = -\dfrac{\sqrt{3}}{2} \qquad 0 \leq \theta \leq \pi$$
$$\theta = \dfrac{5\pi}{4}$$

Now, $\tan\left[\cos^{-1}\left(-\dfrac{\sqrt{3}}{2}\right)\right] = \tan \theta = \tan \dfrac{5\pi}{6} = -\dfrac{\sqrt{3}}{3}$.

29. $\sec\left(\cos^{-1} \dfrac{1}{2}\right)$

First find the angle θ, $0 \leq \theta \leq \pi$, whose cosine equals $\dfrac{1}{2}$.

$$\cos \theta = \dfrac{1}{2} \qquad 0 \leq \theta \leq \pi$$
$$\theta = \dfrac{\pi}{3}$$

Now, $\sec\left(\cos^{-1} \dfrac{1}{2}\right) = \sec \theta = \sec \dfrac{\pi}{3} = 2$.

6.4 THE INVERSE TRIGONOMETRIC FUNCTIONS

31. $\csc(\tan^{-1} 1)$

First find the angle θ, $-\frac{\pi}{2} < \theta < \frac{\pi}{2}$, whose tangent equals 1.

$\tan \theta = 1$ $\qquad\qquad -\frac{\pi}{2} < \theta < \frac{\pi}{2}$

$\theta = \frac{\pi}{4}$

Now, $\csc(\tan^{-1} 1) = \csc \theta = \cos \frac{\pi}{4} = \sqrt{2}$.

33. $\sin(\tan^{-1} (-1))$

First find the angle θ, $-\frac{\pi}{2} < \theta < \frac{\pi}{2}$, whose tangent equals -1.

$\tan \theta = -1$ $\qquad\qquad -\frac{\pi}{2} < \theta < \frac{\pi}{2}$

$\theta = -\frac{\pi}{4}$

Now, $\sin[\tan^{-1} (-1)] = \sin \theta = \sin \left(-\frac{\pi}{4}\right) = -\frac{\sqrt{2}}{2}$.

35. $\sec\left[\sin^{-1}\left(-\frac{1}{2}\right)\right]$

First find the angle θ, $-\frac{\pi}{2} \le \theta \le \frac{\pi}{2}$, whose sine equals $-\frac{1}{2}$.

$\sin \theta = -\frac{1}{2}$

$\theta = -\frac{\pi}{6}$

Now, $\sec\left[\sin^{-1}\left(-\frac{1}{2}\right)\right] = \sec \theta = \sec\left(-\frac{\pi}{6}\right) = \frac{2\sqrt{3}}{3}$.

37. $\tan\left(\sin^{-1} \frac{1}{3}\right)$

First we know that

$\sin \theta = \frac{1}{3}$, $-\frac{\pi}{2} \le \theta \le \frac{\pi}{2}$, so we have

By the Pythagorean Theorem, the missing side of the triangle is

$x^2 + 1 = 9$

$x^2 = 8$

$x = \pm\sqrt{8}$

but x is positive in quadrant I, so

$x = \sqrt{8} = 2\sqrt{2}$

Now, $\tan\left(\sin^{-1} \frac{1}{3}\right) = \tan \theta = \frac{1}{2\sqrt{2}}$ (using $\frac{\text{opp}}{\text{adj}}$ in triangle)

$\tan \theta = \frac{1}{2\sqrt{2}} \cdot \frac{\sqrt{2}}{\sqrt{2}} = \frac{\sqrt{2}}{4}$

39. $\sec\left(\tan^{-1}\dfrac{1}{2}\right)$

First we know that

$\tan\theta = \dfrac{1}{2}$, $-\dfrac{\pi}{2} < \theta < \dfrac{\pi}{2}$, so we have

By the Pythagorean Theorem, the hypotenuse is

$$1^2 + 2^2 = r^2$$
$$1 + 4 = r^2$$
$$5 = r^2$$
$$\sqrt{5} = r$$

Now, $\sec\left(\tan^{-1}\dfrac{1}{2}\right) = \sec\theta = \dfrac{\text{hyp}}{\text{adj}}$

$$= \sec\theta = \dfrac{\sqrt{5}}{2}$$

41. $\cot\left[\sin^{-1}\left(-\dfrac{\sqrt{2}}{3}\right)\right]$

First draw the angle θ, $-\dfrac{\pi}{2} \le \theta \le \dfrac{\pi}{2}$, whose sine equals $-\dfrac{\sqrt{2}}{3}$

$\sin\theta = -\dfrac{\sqrt{2}}{3}$

By the Pythagorean
Theorem, the missing side is:

$$x^2 + \left(-\sqrt{2}\right)^2 = 3^2$$
$$x^2 + 2 = 9$$
$$x^2 = 7$$
$$x = \pm\sqrt{7}, \text{ but } x \text{ is positive in quadrant IV}$$
$$x = \sqrt{7}$$

Now, $\cot\left[\sin^{-1}\left(-\dfrac{\sqrt{2}}{3}\right)\right] = \cot\theta = \dfrac{\text{adj}}{\text{opp}}$

$$\cot\theta = \dfrac{\sqrt{7}}{-\sqrt{2}}$$

$$= -\dfrac{\sqrt{7}}{\sqrt{2}} \cdot \dfrac{\sqrt{2}}{\sqrt{2}}$$

$$= -\dfrac{\sqrt{14}}{2}$$

43. $\sin\left[\tan^{-1}(-3)\right]$

First draw the angle θ, $-\dfrac{\pi}{2} < \theta < \dfrac{\pi}{2}$, whose tangent is -3.

$\tan \theta = -3$

By the Pythagorean Theorem, the hypotenuse is

$r^2 = 1^2 + (-3)^2$
$r^2 = 1 + 9$
$r^2 = 10$
$r = \sqrt{10}$

Now, $\sin\left[\tan^{-1}(-3)\right] = \sin \theta = \dfrac{\text{opp}}{\text{hyp}}$

$$\sin \theta = \dfrac{-3}{\sqrt{10}}$$

$$\sin \theta = \dfrac{-3}{\sqrt{10}} \cdot \dfrac{\sqrt{10}}{\sqrt{10}}$$

$$\sin \theta = \dfrac{-3\sqrt{10}}{10}$$

45. $\sec\left(\sin^{-1} \dfrac{2\sqrt{5}}{5}\right)$

First draw the angle θ, $-\dfrac{\pi}{2} \le \theta \le \dfrac{\pi}{2}$, whose sine is $\dfrac{2\sqrt{5}}{5}$

$\sin \theta = \dfrac{2\sqrt{5}}{5}$

By the Pythagorean Theorem, we find the missing side,

$\left(2\sqrt{5}\right)^2 + x^2 = (5)^2$
$20 + x^2 = 25$
$x^2 = 5$

$x = \pm\sqrt{5}$ but $x > 0$ in quadrant I
$x = \sqrt{5}$

Now, $\sec\left(\sin^{-1} \dfrac{2\sqrt{5}}{5}\right) = \sec \theta = \dfrac{\text{hyp}}{\text{adj}}$

$$\sec \theta = \dfrac{5}{\sqrt{5}}$$

$$\sec \theta = \dfrac{5}{\sqrt{5}} \cdot \dfrac{\sqrt{5}}{\sqrt{5}}$$

$$\sec \theta = \dfrac{5\sqrt{5}}{5}$$

$$\sec \theta = \sqrt{5}$$

47. $\sin^{-1}(\tan 0.5)$

Use radian mode on calculator.

Enter: $\boxed{0.5}$ Press: $\boxed{\tan}$

    Display: $\boxed{0.5}$ Display: $\boxed{0.5463025}$

Press: $\boxed{\text{INV}}$ Press: $\boxed{\sin}$

       Display: $\boxed{0.5779434}$

Hence, $\sin^{-1}(\tan 0.5) = 0.5779$

49. $\tan^{-1}(\sin 0.1)$

Use radian mode on calculator.

Enter: $\boxed{0.1}$ Press: $\boxed{\sin}$

    Display: $\boxed{0.1}$ Display: $\boxed{0.0998334}$

Press: $\boxed{\text{INV}}$ Press: $\boxed{\tan}$

       Display: $\boxed{0.0995037}$

Hence, $\tan^{-1}(\sin 0.1) = 0.0995$

51. $\cos^{-1}(\sin 1)$

Use radian mode on calculator.

Enter: $\boxed{1}$ Press: $\boxed{\sin}$

    Display: $\boxed{1}$ Display: $\boxed{0.841471}$

Press: $\boxed{\text{INV}}$ Press: $\boxed{\cos}$

       Display: $\boxed{0.5707963}$

Hence, $\cos^{-1}(\sin 1) = 0.5708$

53. $\sin^{-1}\left(\tan \dfrac{\pi}{8}\right)$

Use radian mode on calculator.

Enter: $\boxed{\pi}\,\boxed{\div}\,\boxed{8}\,\boxed{=}$ Press: $\boxed{\tan}$

    Display: $\boxed{0.3926991}$ Display: $\boxed{0.4142136}$

Press: [INV]   Press: [sin]

Display: [0.4270786]

Hence, $\sin^{-1}\left(\tan\frac{\pi}{8}\right) = 0.4271$

55. $\tan^{-1}\left(\sin\frac{\pi}{8}\right)$

Use radian mode on calculator.

Enter: [π] [÷] [8] [=]   Press: [sin]

Display: [0.3926991]   Display: [0.3826834]

Press: [INV]   Press: [tan]

Display: [0.3654898]

Hence, $\tan^{-1}\left(\sin\frac{\pi}{8}\right) = 0.3655$.

57. $\sec(\tan^{-1}v) = \sqrt{1+v^2}$

Let $\theta = \tan^{-1}v$

Then, $\tan\theta = v$, $-\frac{\pi}{2} < \theta < \frac{\pi}{2}$.

Hence, $\sec\theta > 0$ and $\tan^2\theta + 1 = \sec^2\theta$
$$v^2 + 1 = \sec^2\theta$$
$$\sqrt{v^2 + 1} = \sec^2\theta$$

Thus, $\sec(\tan^{-1}v) = \sec\theta = \sqrt{v^2+1} = \sqrt{1+v^2}$

59.   Let $\theta = \cos^{-1}v$.   Then $\cos\theta = v$, $0 \le \theta \le \pi$

$$\tan(\cos^{-1}v) = \tan\theta = \frac{\sin\theta}{\cos\theta} = \frac{\sqrt{1-\cos^2\theta}}{\cos\theta}$$
$$= \frac{\sqrt{1-v^2}}{v}$$

61.   Let $\theta = \sin^{-1}v$.   Then $\sin\theta = v$, $\frac{-\pi}{2} \le \theta \le \frac{\pi}{2}$

$$\cos(\sin^{-1}v) = \cos\theta = \sqrt{1-\sin^2\theta} = \sqrt{1-v^2}$$

63. Let $\alpha = \sin^{-1} v$ and $\beta = \cos^{-1} v$

Then $\sin \alpha = v = \cos \beta$, so α, β are complementary. Thus,
$\alpha + \beta = \frac{\pi}{2}$.

65. $\sec^{-1} 4$

Let $v = \sec^{-1} 4$. Then $\sec \theta = 3$, $0 \le v \le \pi$, $\theta \ne \frac{\pi}{2}$.

Thus, $\cos \theta = \frac{1}{4}$ and
$$\sec^{-1} 4 = \theta = \cos^{-1} \frac{1}{4} \approx 1.32.$$

67. $\cot^{-1} 2$

Let $\theta = \cot^{-1} 2$. Then $\cot \theta = 2$, $0 < v < \pi$.

Thus, $\tan \theta = \frac{1}{2}$ and

$$\cot^{-1} 2 = \theta = \tan^{-1} \frac{1}{2} \approx 0.46.$$

69. $\csc^{-1} (-3)$

Let $\theta = \csc^{-1} (-3)$. Then $\csc \theta = -3$, $\frac{-\pi}{2} \le \theta \le \frac{\pi}{2}$, $\theta \ne 0$.

Thus, $\sin \theta = \frac{-1}{3}$ and

$$\csc^{-1} (-3) = \theta = \sin^{-1}\left(\frac{-1}{3}\right) \approx -0.34.$$

71. $\cot^{-1}\left(-\sqrt{5}\right)$

Let $\theta = \cot^{-1}\left(-\sqrt{5}\right)$. Then $\cot \theta = -\sqrt{5}$, $0 < \theta < \pi$.

Thus, $\tan \theta = \frac{-1}{\sqrt{5}}$ and $\cot^{-1}\left(-\sqrt{5}\right) = \theta = \tan^{-1}\left(\frac{-1}{\sqrt{5}}\right) \approx 2.72$

73. Let $\theta = \csc^{-1}\left(-\frac{3}{2}\right)$. Then $\csc \theta = -\frac{3}{2}$, so $\sin \theta = -\frac{2}{3}$
and $\theta = \sin^{-1}\left(-\frac{2}{3}\right) \approx -0.73$.

75. Let $\theta = \cot^{-1}\left(-\frac{3}{2}\right)$. Then $\cot \theta = -\frac{3}{2}$, $0 < \theta < \pi$. Thus,
$\cos \theta = \frac{-3}{\sqrt{13}}$, $\frac{\pi}{2} < \theta < \pi$, and $\theta = \cos^{-1}\left(\frac{-3}{\sqrt{13}}\right) \approx 2.55$.

6.4 THE INVERSE TRIGONOMETRIC FUNCTIONS

77.

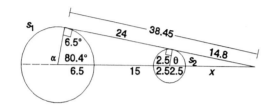

$$\frac{6.5}{2.5} = \frac{26.5 + x}{2.5 + x}$$

$$(2.5)(26.5) + 2.5x = 6.5x + (6.5)(2.5)$$
$$4x = 2.5(20)$$
$$x = 12.5$$

$$\cos \theta = \frac{2.5}{15}$$

$$\theta = 1.4 \text{ radians} = 80.4°, \ \alpha = 180 - 80.4° = 99.6° = 1.73$$

$$s_1 = r_1\alpha = 6.5(1.73) = 11.3$$
$$s_2 = r_2\alpha = 2.5(1.4) = 3.5$$

Length of belt = 2(11.3 + 24 + 3.5)
= 77.6 inches

79. $\sin(\sin^{-1}x) = x$

Let $\theta = \sin^{-1} x$

$\sin \theta = x$ where $-\frac{\pi}{2} \le \theta \le \frac{\pi}{2}$ and $-1 \le x \le 1$

Hence, $-1 \le x \le 1$.

81. $\sin^{-1}(\sin x) = x$

Then x is the angle whose sine equals the sin x, i.e., $\sin x = x$, $\frac{-\pi}{2} \le x \le \frac{\pi}{2}$

This is true only at $x = 0$.

83. Draw the graph of $y = \cot^{-1} x$.

By definition, $\cot y = x$, $0 < y < \pi$ and $-\infty < x < \infty$.

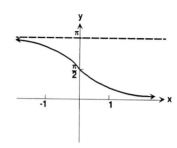

85. Draw the graph of $y = \csc^{-1} x$.

By definition $\csc y = x$, $-\dfrac{\pi}{2} \le y \le \dfrac{\pi}{2}$, $y \ne 0$, and $|x| \ge 1$.

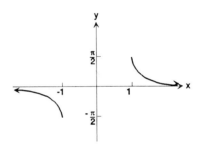

≡ EXERCISE 6.5 SIMPLE HARMONIC MOTION

1. $d = 5 \cos \pi t$

3. $d = 6 \cos 2t$

5. $d = 5 \sin \pi t$

7. $d = 6 \sin 2t$

9. (a) Simple harmonic
    (b) 5 m
    (c) $\dfrac{2\pi}{3}$ sec
    (d) $\dfrac{3}{2\pi}$ oscillations/sec

11. (a) Simple harmonic
    (b) 6 m
    (c) 2 sec
    (d) $\dfrac{1}{2}$ oscillation/sec

13. (a) Simple harmonic
    (b) 3 m
    (c) 4π sec
    (d) $\dfrac{1}{4\pi}$ oscillation/sec

15. (a) Simple harmonic
    (b) 2 m
    (c) 1 sec
    (d) 1 oscillation/sec

17. (a) 120 volts (the amplitude)
    (b) $f = \dfrac{\omega}{2\pi} = \dfrac{120\pi}{2\pi} = 60$ oscillations/sec
    (c) Period $= \dfrac{2\pi}{\omega} = \dfrac{2\pi}{120\pi} = \dfrac{1}{60}$ sec

1. $y = 4 \cos x$

    Amplitude = 4
    Period = 2π

3. $y = -8 \sin \dfrac{\pi}{2}x$

    Amplitude = 8
    Period = 4

5. $y = 4 \sin 3x$ compared to
    $y = A \sin(\omega x - \phi)$ so $A = 4$, $\omega = 3$, and $\phi = 0$

    Amplitude: $\left|A\right| = \left|4\right| = 4$

    Period: $T = \dfrac{2\pi}{\omega} = \dfrac{2\pi}{3}$

    Phase Shift: $\dfrac{\phi}{\omega} = \dfrac{0}{3} = 0$

    Use amplitude to scale y-axis and the period to scale x-axis.

    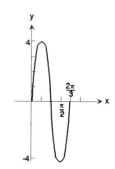

7. $y = -2 \sin\left(\dfrac{\pi}{2}x + \dfrac{1}{2}\right)$ compare to $y = A$

    $\sin(\omega x - \phi)$ so $A = -2$, $\omega = \dfrac{\pi}{2}$, and $\phi = -\dfrac{1}{2}$

    Amplitude: $\left|A\right| = \left|-2\right| = 2$

    Period: $T = \dfrac{2\pi}{\omega} = \dfrac{2\pi}{\dfrac{\pi}{2}} = 4$

    Phase Shift: $\dfrac{\phi}{\omega} = \dfrac{-\dfrac{1}{2}}{\dfrac{\pi}{2}} = -\dfrac{1}{2} \cdot \dfrac{2}{\pi} = -\dfrac{1}{\pi}$

    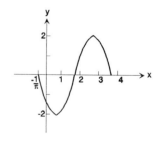

    Since $A < 0$, graph is reflected about x-axis.

9. $y = \dfrac{1}{2} \sin\left(\dfrac{3}{2}x - \pi\right)$ compare to

    $y = A \sin(\omega x - \phi)$ so $A = \dfrac{1}{2}$, $\omega = \dfrac{3}{2}$,
    and $\theta = \pi$

    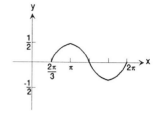

    Amplitude: $\left|A\right| = \left|\dfrac{1}{2}\right| = \dfrac{1}{2}$

    Period: $T = \dfrac{2\pi}{\omega} = \dfrac{2\pi}{\dfrac{3}{2}} = 2\pi \cdot \dfrac{2}{3} = \dfrac{4\pi}{3}$

    Phase Shift: $\dfrac{\phi}{\omega} = \dfrac{\pi}{\dfrac{3}{2}} = \pi \cdot \dfrac{2}{3} = \dfrac{2\pi}{3}$

11. $y = -\dfrac{2}{3}\cos(\pi x - 6)$ compare to $y = A\sin(\omega x - \phi)$ so $A = -\dfrac{2}{3}$, $\omega = \pi$, and $\theta = 6$

Amplitude: $|A| = \left|-\dfrac{2}{3}\right| = \dfrac{2}{3}$

Period: $T = \dfrac{2\pi}{\omega} = \dfrac{2\pi}{\pi} = 2$

Phase Shift: $\dfrac{\phi}{\omega} = \dfrac{6}{\pi}$

Since $A < 0$, graph is reflected about x-axis.

(Problems 13 and 15 could have other answers besides those given.)

13. Since the graph starts one cycle at (0, 5), it is a cosine function with the amplitude is 5 so $A = 5$. The period is 8π.

$$\dfrac{2\pi}{\omega} = 8\pi$$
$$8\pi\omega = 2\pi$$
$$\omega = \dfrac{2\pi}{8\pi} = \dfrac{1}{4}$$

There is no phase shift.

Hence, we get $y = 5\cos\dfrac{1}{4}x$ or

$$y = 5\cos\dfrac{x}{4}$$

15. Since the graph goes through (0, -6), we see that we have a reflected cosine function with $A = -6$ and period = 8, so

$$\dfrac{2\pi}{\omega} = 8$$
$$8\omega = 2\pi$$
$$\omega = \dfrac{2\pi}{8} = \dfrac{\pi}{4}$$

Hence, $y = -6\cos\dfrac{\pi}{4}x$

17. $y = \tan(x + \pi)$

We have tangent function shifted to the left π units which looks exactly like $y = \tan x$.

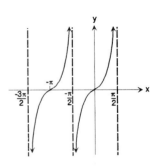

19. $y = -2 \tan 3x$

We have a reflected tangent function since
$A = -2 < 0$. The period is $\frac{\pi}{3}$ so the asymptotes
are now at ... $-\frac{\pi}{6}$, $\frac{\pi}{6}$, $\frac{\pi}{2}$, ...

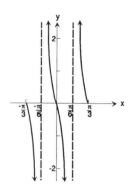

21.

x	$-\pi$	$\dfrac{-\pi}{2}$	0	$\dfrac{\pi}{2}$	π	$\dfrac{3\pi}{2}$	2π
$f_1(x) = 2x$	-2π	$-\pi$	0	π	2π	3π	4π
$f_2(x) = \sin 2x$	0	0	0	0	0	0	0
$f(x) = 2x + \sin 2x$	-2π	$-\pi$	0	$\dfrac{\pi}{2}$	2π	3π	4π
Point on graph of f	$(-\pi, -2\pi)$	$\left[\dfrac{-\pi}{2}, -\pi\right]$	$(0, 0)$	$\left[\dfrac{\pi}{2}, \dfrac{\pi}{2}\right]$	$(\pi, 2\pi)$	$\left[\dfrac{3\pi}{2}, 3\pi\right]$	$(2\pi, 4\pi)$

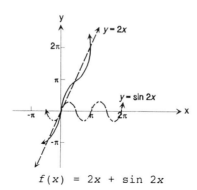

$f(x) = 2x + \sin 2x$

23.

x	0	$\dfrac{1}{2}$	1	$\dfrac{3}{2}$	2	$\dfrac{5}{2}$	3
$f_1(x) = \sin \pi x$	0	1	0	-1	0	1	0
$f_2(x) = \cos \dfrac{\pi}{2}x$	1	$\dfrac{\sqrt{2}}{2}$	0	$\dfrac{-\sqrt{2}}{2}$	-1	$\dfrac{-\sqrt{2}}{2}$	0
$f(x) = \sin \pi x + \cos \dfrac{\pi}{2}x$	1	$1 + \dfrac{\sqrt{2}}{2}$	0	$-1 - \dfrac{\sqrt{2}}{2}$	-1	$1 - \dfrac{\sqrt{2}}{2}$	0
Point on graph of f	$(0, 1)$	$\left[\dfrac{1}{2},\ 1 + \dfrac{\sqrt{2}}{2}\right]$	$(1, 0)$	$\left[\dfrac{3}{2},\ -1 - \dfrac{\sqrt{2}}{2}\right]$	$(2, -1)$	$\left[\dfrac{5}{2},\ 1 - \dfrac{\sqrt{2}}{2}\right]$	$(3, 0)$

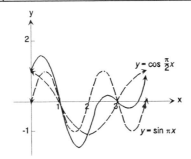

$$f(x) \;=\; \sin \pi x \;+\; \cos \frac{\pi}{2}x$$

25.

x	0	$\dfrac{1}{2}$	1	$\dfrac{3}{2}$	2
$f_1(x) = 3 \sin \pi x$	0	3	0	-3	0
$f_2(x) = 2 \cos \pi x$	2	0	-2	0	2
$f(x) = 3 \sin \pi x + 2 \cos \pi x$	2	3	-2	-3	2
Point on graph of f	$(0, 2)$	$\left[\dfrac{1}{2},\ 3\right]$	$(1, -2)$	$\left[\dfrac{3}{2},\ -3\right]$	$(2, 2)$

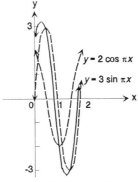

$$f(x) \;=\; 3 \sin \pi x \;+\; 2 \cos \pi x$$

27.

x	0	$\dfrac{\pi}{4}$	$\dfrac{\pi}{2}$	$\dfrac{3\pi}{4}$	π
$f_1(x) = x$	0	$\dfrac{\pi}{4}$	$\dfrac{\pi}{2}$	$\dfrac{3\pi}{4}$	π
$f_2(x) = \cos 2x$	1	0	-1	0	1
$f(x) = x\cos 2x$	0	0	$\dfrac{-\pi}{2}$	0	π
Point on graph of f	$(0,0)$	$\left[\dfrac{\pi}{4}, 0\right]$	$\left[\dfrac{\pi}{2}, \dfrac{-\pi}{2}\right]$	$\left[\dfrac{3\pi}{4}, 0\right]$	(π, π)

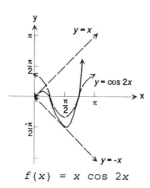

$$f(x) = x \cos 2x$$

29.

x	0	$\dfrac{1}{4}$	$\dfrac{1}{2}$	$\dfrac{3}{4}$	1	$\dfrac{5}{4}$	$\dfrac{3}{2}$	$\dfrac{7}{4}$	2
$f_1(x) = e^{-x}$	1	.78	.61	.47	.37	.29	.22	.17	.14
$f_2(x) = \sin \pi x$	0	$\dfrac{\sqrt{2}}{2}$	1	$\dfrac{\sqrt{2}}{2}$	0	$\dfrac{-\sqrt{2}}{2}$	-1	$\dfrac{-\sqrt{2}}{2}$	0
$f(x) = e^{-x}\sin \pi x$	0	.55	.61	.33	0	-.21	-.22	-.12	0
Point on graph of f	$(0,0)$	$\left[\dfrac{1}{4}, .55\right]$	$\left[\dfrac{1}{2}, .61\right]$	$\left[\dfrac{3}{4}, .33\right]$	$(1,0)$	$\left[\dfrac{5}{4}, -.21\right]$	$\left[\dfrac{3}{2}, -.22\right]$	$\left[\dfrac{7}{4}, -.12\right]$	$(2,0)$

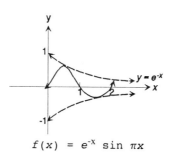

$$f(x) = e^{-x} \sin \pi x$$

6 GRAPHS OF TRIGONOMETRIC FUNCTIONS

31.

x	0	$\frac{1}{2}$	1	$\frac{3}{2}$	2
$f_1(x) = e^x$	1	1.6	2.7	4.5	7.4
$f_2(x) = \sin \pi x$	0	1	01	-1	0
$f(x) = e^x \sin \pi x$	0	1.6	0	-4.5	0
Point on graph of f	(0, 0)	$\left[\frac{1}{2}, 1.6\right]$	(1, 0)	$\left[\frac{3}{2}, -4.5\right]$	(2, 0)

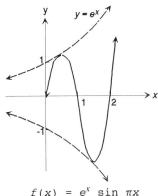

$$f(x) = e^x \sin \pi x$$

33. $\sin^{-1} 1$

We are looking for the angle θ, $-\frac{\pi}{2} \le \theta \le \frac{\pi}{2}$, whose sine is 1.

$$\sin \theta = 1 \qquad\qquad -\frac{\pi}{2} \le \theta \le \frac{\pi}{2}$$

$$\theta = \frac{\pi}{2}$$

Hence, $\sin^{-1} 1 = \frac{\pi}{2}$

35. $\tan^{-1} 1$

We are looking for the angle θ, $-\frac{\pi}{2} < \theta < \frac{\pi}{2}$, whose tangent is 1.

$$\tan \theta = 1 \qquad\qquad -\frac{\pi}{2} < \theta < \frac{\pi}{2}$$

$$\theta = \frac{\pi}{4}$$

Hence, $\tan^{-1} 1 = \frac{\pi}{4}$

37. $\cos^{-1}\left(-\dfrac{\sqrt{3}}{2}\right)$

We are looking for the angle θ, $0 \le \theta \le \pi$, whose cosine is $-\dfrac{\sqrt{3}}{2}$.

$$\cos\theta = -\dfrac{\sqrt{3}}{2} \qquad 0 \le \theta \le \pi$$
$$\theta = \dfrac{5\pi}{6}$$

Hence, $\cos^{-1}\left(-\dfrac{\sqrt{3}}{2}\right) = \dfrac{5\pi}{6}$

39. $\sin\left(\cos^{-1}\dfrac{\sqrt{2}}{2}\right)$

First find the angle θ, $0 \le \theta \le \pi$, whose cosine equals $\dfrac{\sqrt{2}}{2}$.

$$\cos\theta = \dfrac{\sqrt{2}}{2}, \ 0 \le \theta \le \pi$$
$$\theta = \dfrac{\pi}{4}$$

Now, $\sin\left(\cos^{-1}\dfrac{\sqrt{2}}{2}\right) = \sin\theta = \sin\dfrac{\pi}{4} = \dfrac{\sqrt{2}}{2}$

41. $\tan\left[\sin^{-1}\left(-\dfrac{\sqrt{3}}{2}\right)\right]$

First find the angle θ, $-\dfrac{\pi}{2} \le \theta \le \dfrac{\pi}{2}$, whose sine equals $-\dfrac{\sqrt{3}}{2}$.

$$\sin\theta = -\dfrac{\sqrt{3}}{2}, \ -\dfrac{\pi}{2} \le \theta \le \dfrac{\pi}{2}$$
$$\theta = -\dfrac{\pi}{3}$$

Now, $\tan\left[\sin^{-1}\left(-\dfrac{\sqrt{3}}{2}\right)\right] = \tan\theta = \tan\left(-\dfrac{\pi}{3}\right) = -\sqrt{3}$

43. $\sec\left(\tan^{-1}\dfrac{\sqrt{3}}{3}\right)$

First find the angle θ, $-\dfrac{\pi}{2} < \theta < \dfrac{\pi}{2}$, whose tangent equals $\dfrac{\sqrt{3}}{3}$.

$$\tan\theta = \dfrac{\sqrt{3}}{3} \quad -\dfrac{\pi}{2} < \theta < \dfrac{\pi}{2}$$
$$\theta = \dfrac{\pi}{6}$$

Now, $\sec\left(\tan^{-1}\dfrac{\sqrt{3}}{3}\right) = \sec\theta = \sec\dfrac{\pi}{6} = \dfrac{2\sqrt{3}}{3}$

45. $\sin\left(\tan^{-1}\dfrac{3}{4}\right)$

Let $\theta = \tan^{-1}\dfrac{3}{4}$

Then, $\tan\theta = \dfrac{3}{4} = \dfrac{\text{opp}}{\text{adj}}$, $-\dfrac{\pi}{2} < \theta < \dfrac{\pi}{2}$

The hypotenuse is $r = \sqrt{x^2 + y^2}$

$$r = \sqrt{16 + 9}$$
$$r = \sqrt{25}$$
$$r = 5$$

Thus, $\sin\left(\tan^{-1}\dfrac{3}{4}\right) = \sin\theta = \dfrac{3}{5}$

47. $\tan\left[\sin^{-1}\left(-\dfrac{4}{5}\right)\right]$

Let $\theta = \sin^{-1}\left(-\dfrac{4}{5}\right)$

Then, $\sin\theta = -\dfrac{4}{5} = \dfrac{\text{opp}}{\text{hyp}}$, $-\dfrac{\pi}{2} \le \theta \le \dfrac{\pi}{2}$.

Using the Pythagorean Theorem, we find the missing side:

$$x^2 + (-4)^2 = 5^2$$
$$x^2 + 16 = 25$$
$$x^2 = 9$$
$$x = \pm 3$$
$$x = 3, \ x > 0 \text{ in quadrant IV}$$

Hence, $\tan\left[\sin^{-1}\left(-\dfrac{4}{5}\right)\right] = \tan\theta = -\dfrac{4}{3}$

49. (a) Simple harmonic
     (b) 6 ft
     (c) π sec
     (d) $\dfrac{1}{\pi}$ oscillation/sec

51. (a) Simple harmonic
     (b) 2 ft
     (c) 2 sec
     (d) $\dfrac{1}{2}$ oscillation/sec

ANALYTIC TRIGONOMETRY

≡ EXERCISE 7.1 BASIC TRIGONOMETRIC IDENTITIES

1. To prove that

$$\csc \theta \cos \theta = \cot \theta,$$

we start with the left side and apply a reciprocal identity:

$$\csc \theta \cdot \cos \theta = \frac{1}{\sin \theta} \cdot \cos \theta$$

$$= \frac{\cos \theta}{\sin \theta}$$

$$= \cot \theta$$

3. To prove that

$$1 + \tan^2(-\theta) = \sec^2 \theta,$$

we begin with the left side and apply an even-odd identity:

$$1 + \tan^2(-\theta) = 1 + (-\tan \theta)^2$$
$$= 1 + \tan^2 \theta$$
$$= \sec^2 \theta$$

5. To prove that

$$\cos \theta(\tan \theta + \cot \theta) = \csc \theta,$$

we begin with the left side because it contains the more complicated expression, and write the expression so it contains only sines and cosines:

$$\cos \theta(\tan \theta + \cot \theta) = \csc \theta\left(\frac{\sin \theta}{\cos \theta} + \frac{\cos \theta}{\sin \theta}\right)$$

$$= \cos \theta\left(\frac{\sin^2 \theta + \cos^2 \theta}{\cos \theta \sin \theta}\right)$$

$$= \frac{1}{\sin \theta}$$
$$= \csc \theta$$

7. To prove that

$$\tan \theta \cot \theta - \cos^2 \theta = \sin^2 \theta,$$

we write the left side expression in terms of sines and cosines:

$$\begin{aligned}
\tan \theta \cot \theta - \cos^2 \theta &= \frac{\sin \theta}{\cos \theta} \cdot \frac{\cos \theta}{\sin \theta} - \cos^2 \theta \\
&= 1 - \cos^2 \theta \\
&= \sin^2 \theta
\end{aligned}$$

9. To prove that

$$(\sec \theta - 1)(\sec \theta + 1) = \tan^2 \theta,$$

we bein with the left side, multiply, and then apply a form of a Pythagorean identity:

$$\begin{aligned}
(\sec \theta - 1)(\sec \theta + 1) &= \sec^2 \theta - 1 \\
&= \tan^2 \theta
\end{aligned}$$

11. To prove that

$$(\sec \theta + \tan \theta)(\sec \theta - \tan \theta) = 1,$$

we begin with the left side, multiply, and then apply a form of a Pythagorean identity:

$$\begin{aligned}
(\sec \theta + \tan \theta)(\sec \theta - \tan \theta) &= \sec^2 \theta - \tan^2 \theta \\
&= 1
\end{aligned}$$

13. To prove that

$$\sin^2 \theta (1 + \cot^2 \theta) = 1,$$

we apply a form of a Pythagorean identity:

$$\begin{aligned}
\sin^2 \theta (1 + \cot^2 \theta) &= \sin^2 \theta (\cos^2 \theta) \\
&= \sin^2 \theta \left(\frac{1}{\sin^2 \theta} \right) \\
&= 1
\end{aligned}$$

15. To prove that

$$(\sin \theta + \cos \theta)^2 + (\sin \theta - \cos \theta)^2 = 2,$$

we begin with the complicated left side and proceed by carrying out the exponents and simplifying:

$$\begin{aligned}
&(\sin \theta + \cos \theta)^2 + (\sin \theta - \cos \theta)^2 \\
&= \sin^2 \theta + 2 \sin \theta \cos \theta + \cos^2 \theta + \sin^2 \theta - 2 \sin \theta \cos \theta + \cos^2 \theta \\
&= \sin^2 \theta + \cos^2 \theta + \sin^2 \theta + \cos^2 \theta \\
&= 1 + 1 \\
&= 2
\end{aligned}$$

7.1 BASIC TRIGONOMETRIC IDENTITIES

17. To prove that

$$\sec^4 \theta - \sec^2 \theta = \tan^4 \theta + \tan^2 \theta$$

we begin with the left side and factor:

$$\begin{aligned}
\sec^4 \theta - \sec^2 \theta &= \sec^2 \theta(\sec^2 \theta - 1) \\
&= (1 + \tan^2 \theta) \tan^2 \theta \\
&= \tan^2 \theta + \tan^4 \theta \\
&= \tan^4 \theta + \tan^2 \theta
\end{aligned}$$

19. To prove that

$$\sec \theta - \tan \theta = \frac{\cos \theta}{1 + \sin \theta},$$

we put the left side's expression in terms of sines and cosines:

$$\begin{aligned}
\sec \theta - \tan \theta &= \frac{1}{\cos \theta} - \frac{\sin \theta}{\cos \theta} \\
&= \frac{1 - \sin \theta}{\cos \theta} \cdot \frac{1 + \sin \theta}{1 + \sin \theta} \\
&= \frac{1 - \sin^2 \theta}{\cos \theta(1 + \sin \theta)} \\
&= \frac{\cos^2 \theta}{\cos \theta(1 + \sin \theta)} \\
&= \frac{\cos \theta}{1 + \sin \theta}
\end{aligned}$$

21. To prove that

$$3 \sin^2 \theta + 4 \cos^2 \theta = 3 + \cos^2 \theta,$$

we begin with the left side and let $4 \cos^2 \theta$
$= 3 + \cos^2 \theta + 1 \cos^2 \theta$:

$$\begin{aligned}
3 \sin^2 \theta + 4 \cos^2 \theta &= 3 \sin^2 \theta + 3 \cos^2 \theta + \cos^2 \theta \\
&= 3(\sin^2 \theta + \cos^2 \theta) + \cos^2 \theta \\
&= 3 + \cos^2 \theta
\end{aligned}$$

23. To prove that

$$1 - \frac{\cos^2 \theta}{1 + \sin \theta} = \sin \theta,$$

we begin with the complicated left side and use a form of a
Phythagorean identity:

$$\begin{aligned}
1 - \frac{\cos^2 \theta}{1 + \sin \theta} &= 1 - \frac{1 - \sin^2 \theta}{1 + \sin \theta} \\
&= 1 - \frac{(1 - \sin \theta)(1 + \sin \theta)}{1 + \sin \theta} \\
&= 1 - 1 + \sin \theta \\
&= \sin \theta
\end{aligned}$$

7 ANALYTIC TRIGONOMETRY

25. To prove that

$$\frac{1 + \tan \theta}{1 - \tan \theta} = \frac{\cot \theta + 1}{\cot \theta - 1},$$

$$\frac{1 + \tan \theta}{1 - \tan \theta} = \frac{1 + \dfrac{1}{\cot \theta}}{1 - \dfrac{1}{\cot \theta}}$$

$$= \frac{\dfrac{\cot \theta + 1}{\cot \theta}}{\dfrac{\cot \theta - 1}{\cot \theta}}$$

$$= \frac{\cot \theta + 1}{\cot \theta - 1}$$

27. To prove that

$$\frac{\sec \theta}{\csc \theta} + \frac{\sin \theta}{\cos \theta} = 2 \tan \theta,$$

$$\frac{\sec \theta}{\csc \theta} + \frac{\sin \theta}{\cos \theta} = \frac{\dfrac{1}{\cos \theta}}{\dfrac{1}{\sin \theta}} + \tan \theta$$

$$= \frac{\sin \theta}{\cos \theta} + \tan \theta$$

$$= \tan \theta + \tan \theta$$

$$= 2 \tan \theta$$

29. To prove that

$$\frac{1 + \sin \theta}{1 - \sin \theta} = \frac{\csc \theta + 1}{\csc \theta - 1},$$

$$\frac{1 + \sin \theta}{1 - \sin \theta} = \frac{1 + \dfrac{1}{\csc \theta}}{1 - \dfrac{1}{\csc \theta}}$$

$$= \frac{\dfrac{\csc \theta + 1}{\csc \theta}}{\dfrac{\csc \theta - 1}{\csc \theta}}$$

$$= \frac{\csc \theta + 1}{\csc \theta - 1}$$

31. To prove that

$$\frac{1 - \sin \theta}{\cos \theta} + \frac{\cos \theta}{1 - \sin \theta} = 2 \sec \theta,$$

$$\frac{1 - \sin \theta}{\cos \theta} + \frac{\cos \theta}{1 - \sin \theta} = \frac{(1 - \sin \theta)^2 + \cos^2 \theta}{\cos \theta (1 - \sin \theta)}$$

$$= \frac{1 - 2 \sin \theta + \sin^2 \theta + \cos^2 \theta}{\cos \theta (1 - \sin \theta)}$$

$$= \frac{2 - 2 \sin \theta}{\cos \theta (1 - \sin \theta)}$$

$$= \frac{2(1 - \sin \theta}{\cos \theta (1 - \sin \theta)}$$

$$= \frac{2}{\cos \theta}$$

$$= 2 \sec \theta$$

7.1 BASIC TRIGONOMETRIC IDENTITIES

33. To prove that

$$\frac{\sin\theta}{\sin\theta - \cos\theta} = \frac{1}{1 - \cot\theta},$$

$$\frac{\sin\theta}{\sin\theta - \cos\theta} = \frac{1}{\dfrac{\sin\theta - \cos\theta}{\sin\theta}}$$

$$= \frac{1}{1 - \dfrac{\cos\theta}{\sin\theta}}$$

$$= \frac{1}{1 - \cot\theta}$$

35. To prove that

$$\frac{1 - \sin\theta}{1 + \sin\theta} = (\sec\theta - \tan\theta)^2,$$

we begin with the right side and square the expression:

$$(\sec\theta - \tan\theta)^2 = \sec^2\theta - 2\sec\theta\tan\theta + \tan^2\theta$$

$$= \frac{1}{\cos^2\theta} - \frac{2\sin\theta}{\cos^2\theta} + \frac{\sin^2\theta}{\cos^2\theta}$$

$$= \frac{1 - 2\sin\theta + \sin^2\theta}{\cos^2\theta}$$

$$= \frac{(1 - \sin\theta)^2}{1 - \sin^2\theta}$$

$$= \frac{(1 - \sin\theta)^2}{(1 - \sin\theta)(1 + \sin\theta)}$$

$$= \frac{1 - \sin\theta}{1 + \sin\theta}$$

37. To prove that

$$\frac{\cos\theta}{1 - \tan\theta} + \frac{\sin\theta}{1 - \cot\theta} = \sin\theta + \cos\theta,$$

$$\frac{\cos\theta}{1 - \tan\theta} + \frac{\sin\theta}{1 - \cot\theta} = \frac{\cos\theta}{1 - \dfrac{\sin\theta}{\cos\theta}} + \frac{\sin\theta}{1 - \dfrac{\cos\theta}{\sin\theta}}$$

$$= \frac{\cos\theta}{\dfrac{\cos\theta - \sin\theta}{\cos\theta}} + \frac{\sin\theta}{\dfrac{\sin\theta - \cos\theta}{\sin\theta}}$$

$$= \frac{\cos^2\theta}{\cos\theta - \sin\theta} + \frac{\sin^2\theta}{\sin\theta - \cos\theta}$$

$$= \frac{\cos^2\theta - \sin^2\theta}{\cos\theta - \sin\theta}$$

$$= \frac{(\cos\theta - \sin\theta)(\cos\theta + \sin\theta)}{\cos - \sin\theta}$$

$$= \cos\theta + \sin\theta$$

$$= \sin\theta + \cos\theta$$

39. To prove that

$$\tan \theta + \frac{\cos \theta}{1 + \sin \theta} = \sec \theta,$$

$$\tan \theta + \frac{\cos \theta}{1 + \sin \theta} = \frac{\sin \theta}{\cos \theta} + \frac{\cos \theta}{(1 + \sin \theta)}$$

$$= \frac{\sin \theta(1 + \sin^2 \theta) + \cos^2 \theta}{\cos \theta(1 + \sin \theta)}$$

$$= \frac{\sin \theta + \sin^2 \theta + \cos^2 \theta}{\cos \theta(1 + \sin \theta)}$$

$$= \frac{\sin \theta + 1}{\cos \theta(1 + \sin \theta)}$$

$$= \frac{1}{\cos \theta}$$

$$= \sec \theta$$

41. To prove that

$$\frac{\tan \theta + \sec \theta - 1}{\tan \theta - \sec \theta + 1} = \tan \theta + \sec \theta,$$

$$\frac{\tan \theta + \sec \theta - 1}{\tan \theta - \sec \theta + 1} = \frac{\tan \theta + (\sec \theta - 1)}{\tan \theta - (\sec \theta - 1)} \cdot \frac{\tan \theta + (\sec \theta - 1)}{\tan \theta + (\sec \theta - 1)}$$

$$= \frac{\tan^2 \theta + 2 \tan \theta(\sec \theta - 1) + \sec^2 \theta - 2 \sec \theta + 1}{\tan^2 \theta - (\sec^2 \theta - 2 \sec \theta + 1)}$$

$$= \frac{\sec^2 \theta - 1 + 2 \tan \theta(\sec \theta - 1) + \sec^2 \theta - 2 \sec \theta + 1}{\sec^2 \theta - 1 - \sec^2 \theta + 2 \sec \theta - 1}$$

$$= \frac{2 \sec^2\theta - 2 \sec \theta + 2 \tan \theta(\sec \theta - 1)}{-2 + 2 \sec \theta}$$

$$= \frac{2 \sec \theta(\sec \theta - 1) + 2 \tan \theta(\sec \theta - 1)}{2(\sec \theta - 1)}$$

$$= \frac{2 (\sec \theta - 1)(\sec \theta + \tan \theta)}{2(\sec \theta - 1)}$$

$$= \sec \theta + \tan \theta$$

$$= \tan \theta + \sec \theta$$

43. To prove that

$$\frac{\tan \theta - \cot \theta}{\tan \theta + \cot \theta} = \sin^2 \theta - \cos^2 \theta,$$

$$\frac{\tan \theta - \cot \theta}{\tan \theta + \cot \theta} = \frac{\dfrac{\sin \theta}{\cos \theta} - \dfrac{\cos \theta}{\sin \theta}}{\dfrac{\sin \theta}{\cos \theta} + \dfrac{\cos \theta}{\sin \theta}}$$

$$= \frac{\dfrac{\sin^2 \theta - \cos^2 \theta}{\cos \theta \sin \theta}}{\dfrac{\sin^2 \theta + \cos^2 \theta}{\cos \theta \sin \theta}}$$

$$= \frac{\sin^2 \theta - \cos^2 \theta}{1}$$

$$= \sin^2 \theta - \cos^2 \theta$$

45. To prove that

$$\frac{\tan \theta - \cot \theta}{\tan \theta + \cot \theta} = 2 \sin^2 \theta - 1,$$

$$\frac{\tan \theta - \cot \theta}{\tan \theta + \cot \theta} = \frac{\dfrac{\sin \theta}{\cos \theta} - \dfrac{\cos \theta}{\sin \theta}}{\dfrac{\sin \theta}{\cos \theta} + \dfrac{\cos \theta}{\sin \theta}}$$

$$= \frac{\dfrac{\sin^2 \theta - \cos^2 \theta}{\cos \theta \sin \theta}}{\dfrac{\sin^2 \theta + \cos^2 \theta}{\cos \theta \sin \theta}}$$

$$= \sin^2 \theta - \cos^2 \theta$$

$$= \sin^2 \theta - (1 - \sin^2 \theta)$$

$$= 2 \sin^2 \theta - 1$$

47. To prove that

$$\frac{\sec \theta + \tan \theta}{\cot \theta + \cos \theta} = \tan \theta \sec \theta,$$

$$\frac{\sec \theta + \tan \theta}{\cot \theta + \cos \theta} = \frac{\dfrac{1}{\cos \theta} + \dfrac{\sin \theta}{\cos \theta}}{\dfrac{\cos \theta}{\sin \theta} + \cos \theta}$$

$$= \frac{\dfrac{1 + \sin \theta}{\cos \theta}}{\dfrac{\cos \theta + \cos \theta \sin \theta}{\sin \theta}}$$

$$= \frac{1 + \sin \theta}{\cos \theta} \cdot \frac{\sin \theta}{\cos \theta (1 + \sin \theta)}$$

$$= \frac{\sin \theta}{\cos \theta} \cdot \frac{1}{\cos \theta}$$

$$= \tan \theta \sec \theta$$

49. To prove that

$$\frac{1 - \tan^2 \theta}{1 + \tan^2 \theta} = 2 \cos^2 \theta - 1,$$

$$\frac{1 - \tan^2 \theta}{1 + \tan^2 \theta} = \frac{1 - \tan^2 \theta}{\sec^2 \theta}$$

$$= \frac{1}{\sec^2 \theta} - \frac{\tan^2 \theta}{\sec^2 \theta}$$

$$= \cos^2 \theta - \frac{\dfrac{\sin^2 \theta}{\cos^2 \theta}}{\dfrac{1}{\cos^2 \theta}}$$

$$= \cos^2 \theta - \sin^2 \theta$$

$$= \cos^2 \theta - (1 - \cos^2 \theta)$$

$$= 2 \cos^2 \theta - 1$$

51. To prove that

$$\frac{\sec\theta - \csc\theta}{\sec\theta \csc\theta} = \sin\theta - \cos\theta,$$

$$\frac{\sec\theta - \csc\theta}{\sec\theta \csc\theta} = \frac{\dfrac{1}{\cos\theta} - \dfrac{1}{\sin\theta}}{\dfrac{1}{\cos\theta} \cdot \dfrac{1}{\sin\theta}}$$

$$= \frac{\dfrac{\sin\theta - \cos\theta}{\cos\theta \cdot \sin\theta}}{\dfrac{1}{\cos\theta \cdot \sin\theta}}$$

$$= \sin\theta - \cos\theta$$

53. To prove that

$$\sec\theta - \cos\theta = \sin\theta \tan\theta,$$

$$\sec\theta - \cos\theta = \frac{1}{\cos\theta} - \cos\theta$$

$$= \frac{1 - \cos^2\theta}{\cos\theta}$$

$$= \frac{\sin^2\theta}{\cos\theta}$$

$$= \sin\theta \cdot \frac{\sin\theta}{\cos\theta}$$

$$= \sin\theta \tan\theta$$

55. To prove that

$$\frac{1}{1 - \sin\theta} + \frac{1}{1 + \sin\theta} = 2\sec^2\theta,$$

$$\frac{1}{1 - \sin\theta} + \frac{1}{1 + \sin\theta} = \frac{1 + \sin\theta + 1 - \sin\theta}{(1 + \sin\theta)(1 - \sin\theta)}$$

$$= \frac{2}{1 - \sin^2\theta}$$

$$= \frac{2}{\cos^2\theta}$$

$$= 2\sec^2\theta$$

57. To prove that

$$\frac{\sec\theta}{1 - \sin\theta} = \frac{1 + \sin\theta}{\cos^3\theta},$$

$$\frac{\sec\theta}{1 - \sin\theta} = \frac{\sec\theta}{1 - \sin\theta} \cdot \frac{1 + \sin\theta}{1 + \sin\theta}$$

$$= \frac{\sec\theta(1 + \sin\theta)}{1 - \sin^2\theta}$$

$$= \frac{\sec\theta(1 + \sin\theta)}{\cos^2\theta}$$

$$= \frac{1 + \sin\theta}{\cos^3\theta}$$

7.1 BASIC TRIGONOMETRIC IDENTITIES   349

59. To prove that

$$\frac{(\sec\theta - \tan\theta)^2 + 1}{\csc\theta(\sec\theta - \tan\theta)} = 2\tan\theta,$$

$$\frac{(\sec\theta - \tan\theta)^2 + 1}{\csc\theta(\sec\theta - \tan\theta)} = \frac{\sec^2\theta - 2\sec\theta\tan\theta + \tan^2\theta + 1}{\csc\theta\sec\theta - \csc\theta\tan\theta}$$

$$= \frac{2\sec^2\theta - 2\sec\theta\tan\theta}{\csc\theta\sec\theta - \csc\theta\tan\theta}$$

$$= \frac{\dfrac{2}{\cos^2\theta} - \dfrac{2\sin\theta}{\cos^2\theta}}{\dfrac{1}{\sin\theta\cos\theta} - \dfrac{\sin\theta}{\sin\theta\cos\theta}}$$

$$= \frac{\dfrac{2 - 2\sin\theta}{\cos^2\theta}}{\dfrac{1 - \sin\theta}{\sin\theta\cos\theta}}$$

$$= \frac{2(1 - \sin\theta)}{\cos^2\theta} \cdot \frac{\sin\theta\cos\theta}{1 - \sin\theta}$$

$$= \frac{2\sin\theta}{\cos\theta}$$

$$= 2\tan\theta$$

61. To prove that

$$\frac{\sin\theta + \cos\theta}{\cos\theta} - \frac{\sin\theta - \cos\theta}{\sin\theta} = \sec\theta\csc\theta,$$

$$\frac{\sin\theta + \cos\theta}{\cos\theta} - \frac{\sin\theta - \cos\theta}{\sin\theta}$$

$$= \frac{\sin\theta(\sin\theta + \cos\theta) - \cos\theta(\sin\theta - \cos\theta)}{\cos\theta\sin\theta}$$

$$= \frac{\sin^2\theta + \sin\theta\cos\theta - \sin\theta\cos\theta + \cos^2\theta}{\cos\theta\sin\theta}$$

$$= \frac{1}{\cos\theta\sin\theta}$$

$$= \sec\theta\csc\theta$$

63. To prove that

$$\frac{\sin^3\theta + \cos^3\theta}{\sin\theta + \cos\theta} = 1 - \sin\theta\cos\theta,$$

$$\frac{\sin^3\theta + \cos^3\theta}{\sin\theta + \cos\theta} = \frac{(\sin\theta + \cos\theta)(\sin^2\theta - \sin\theta\cos\theta + \cos^2\theta)}{\sin\theta + \cos\theta}$$

$$= \sin^2\theta + \cos^2\theta - \sin\theta\cos\theta$$

$$= 1 - \sin\theta\cos\theta$$

7 ANALYTIC TRIGONOMETRY

65. To prove that

$$\frac{\cos^2 \theta - \sin^2 \theta}{1 - \tan^2 \theta} = \cos^2 \theta,$$

$$\frac{\cos^2 \theta - \sin^2 \theta}{1 - \tan^2 \theta} = \frac{\cos^2 \theta - \sin^2 \theta}{1 - \dfrac{\sin^2 \theta}{\cos^2 \theta}}$$

$$= \frac{\cos^2 \theta - \sin^2 \theta}{\dfrac{\cos^2 \theta - \sin^2 \theta}{\cos^2 \theta}}$$

$$= \cos^2 \theta$$

67. To prove that

$$\frac{(2 \cos^2 \theta - 1)^2}{\cos^4 \theta - \sin^4 \theta} = 1 - 2 \sin^2 \theta,$$

$$\frac{(2 \cos^2 \theta - 1)^2}{\cos^4 \theta - \sin^4 \theta} = \frac{\left[2 \cos^2 \theta (\sin^2 \theta + \cos^2 \theta)\right]^2}{(\cos^2 \theta - \sin^2 \theta)(\cos^2 \theta + \sin^2 \theta)}$$

$$= \frac{(\cos^2 \theta - \sin^2 \theta)^2}{(\cos^2 \theta - \sin^2 \theta)(\cos^2 \theta + \sin^2 \theta)}$$

$$= \frac{\cos^2 \theta - \sin^2 \theta}{\cos^2 \theta + \sin^2 \theta}$$

$$= \cos^2 \theta - \sin^2 \theta$$

$$= (1 - \sin^2 \theta) - \sin^2 \theta$$

$$= 1 - 2 \sin^2 \theta$$

69. To prove that

$$\frac{1 + \sin \theta + \cos \theta}{1 + \sin \theta - \cos \theta} = \frac{1 + \cos \theta}{\sin \theta},$$

$$\frac{1 + \sin \theta + \cos \theta}{1 + \sin \theta - \cos \theta} = \frac{(1 + \sin \theta) + \cos \theta}{(1 + \sin \theta) - \cos \theta} \cdot \frac{(1 + \sin \theta) + \cos \theta}{(1 + \sin \theta) + \cos \theta}$$

$$= \frac{1 + 2 \sin \theta + \sin^2 \theta + 2(1 + \sin \theta)(\cos \theta) + \cos^2 \theta}{1 + 2 \sin \theta + \sin^2 \theta - \cos^2 \theta}$$

$$= \frac{1 + 2 \sin \theta + \sin^2 \theta + 2(1 + \sin \theta)(\cos \theta) + (1 - \sin^2 \theta)}{1 + 2 \sin \theta + \sin^2 \theta - (1 - \sin^2 \theta)}$$

$$= \frac{2 + 2 \sin \theta + 2(1 + \sin \theta)(\cos \theta)}{2 \sin \theta + 2 \sin^2 \theta}$$

$$= \frac{2(1 + \sin \theta) + 21 + \sin \theta)(\cos \theta)}{2 \sin \theta(1 + \sin \theta)}$$

$$= \frac{2(1 + \sin \theta)(1 + \cos \theta)}{2 \sin \theta(1 + \sin \theta)}$$

$$= \frac{1 + \cos \theta}{\sin \theta}$$

71. To prove that
$$(a \sin \theta + b \cos \theta)^2 + (a \cos \theta - b \sin \theta)^2 = a^2 + b^2,$$

$(a \sin \theta + b \cos \theta)^2 + (a \cos \theta - b \sin \theta)^2$
$\quad = a^2 \sin^2 \theta + 2ab \sin \theta \cos \theta + b^2 \cos^2 \theta + a^2 \cos^2 \theta$
$\qquad - 2ab \sin \theta \cos \theta + b^2 \sin^2 \theta$
$\quad = a^2(\sin^2 \theta + \cos^2 \theta) + b^2(\cos^2 \theta + \sin^2 \theta)$
$\quad = a^2 + b^2$

73. To prove that
$$\frac{\tan \alpha + \tan \beta}{\cot \alpha + \cot \beta} = \tan \alpha \tan \beta,$$

$\dfrac{\tan \alpha + \tan \beta}{\cot \alpha + \cot \beta} = \dfrac{\tan \alpha + \tan \beta}{\dfrac{1}{\tan \alpha} + \dfrac{1}{\tan \beta}}$

$\qquad = \dfrac{\tan \alpha + \tan \beta}{\dfrac{\tan \beta + \tan \alpha}{\tan \alpha \tan \beta}}$

$\qquad = (\tan \alpha + \tan \beta) \cdot \dfrac{\tan \alpha \tan \beta}{\tan \alpha + \tan \beta}$

$\qquad = \tan \alpha \tan \beta$

75. To prove that
$$(\sin \alpha + \cos \beta)^2 + (\cos \beta + \sin \alpha)(\cos \beta - \sin \alpha) = 2 \cos \beta(\sin \alpha + \cos \beta),$$

$(\sin \alpha + \cos \beta)^2 + (\cos \beta + \sin \alpha)(\cos \beta - \sin \alpha)$
$\quad = (\sin^2 \alpha + 2 \sin \alpha \cos \beta + \cos^2 \beta) + (\cos^2 \beta - \sin^2 \alpha)$
$\quad = 2 \cos^2 \beta + 2 \sin \alpha \cos \beta$
$\quad = 2 \cos \beta(\cos \beta + \sin \alpha)$

77. To prove that
$$\ln|\sec \theta| = -\ln|\cos \theta|,$$
$$\ln|\sec \theta| = \ln|\cos \theta|^{-1} = -\ln|\cos \theta|$$

79. To prove that
$$\ln|1 + \cos \theta| + \ln|1 - \cos \theta| = 2 \ln|\sin \theta|,$$

$\ln|1 + \cos \theta| + \ln|1 - \cos \theta| = \ln(|1 + \cos \theta||1 - \cos \theta|)$
$\qquad = \ln|1 - \cos^2 \theta|$
$\qquad = \ln|\sin^2 \theta|$
$\qquad = 2 \ln|\sin \theta|$

1. $\dfrac{5\pi}{12} = \sin\left(\dfrac{3\pi}{12} + \dfrac{2\pi}{12}\right)$

   $= \sin \dfrac{\pi}{4} \cos \dfrac{\pi}{6} + \cos \dfrac{\pi}{4} \sin \dfrac{\pi}{6}$

   $= \dfrac{\sqrt{2}}{2} \cdot \dfrac{\sqrt{3}}{2} + \dfrac{\sqrt{2}}{2} \cdot \dfrac{1}{2}$

   $= \dfrac{1}{4}\left(\sqrt{6} + \sqrt{2}\right)$

3. $\cos \dfrac{7\pi}{12} = \cos\left(\dfrac{4\pi}{12} + \dfrac{3\pi}{12}\right)$

   $= \cos \dfrac{\pi}{3} \cos \dfrac{\pi}{4} - \sin \dfrac{\pi}{3} \sin \dfrac{\pi}{4}$

   $= \dfrac{1}{2} \cdot \dfrac{\sqrt{2}}{2} - \dfrac{\sqrt{3}}{2} \cdot \dfrac{\sqrt{2}}{2}$

   $= \dfrac{1}{4}\left(\sqrt{2} - \sqrt{6}\right)$

5. $\cos 165° = \cos(120° + 45°)$

   $= \cos 120° \cos 45° - \sin 120° \sin 45°$

   $= -\dfrac{1}{2} \cdot \dfrac{\sqrt{2}}{2} - \dfrac{\sqrt{3}}{2} \cdot \dfrac{\sqrt{2}}{2}$

   $= \dfrac{-1}{4}\left(\sqrt{2} + \sqrt{6}\right)$

7. $\tan 15° = \tan (45° - 30°)$

   $= \dfrac{\tan 45° - \tan 30°}{1 + \tan 45° \tan 30°}$

   $= \dfrac{1 - \dfrac{1}{\sqrt{3}}}{1 + 1 \cdot \dfrac{1}{\sqrt{3}}}$

   $= \dfrac{\dfrac{\sqrt{3} - 1}{\sqrt{3}}}{\dfrac{\sqrt{3} + 1}{\sqrt{3}}}$

   $= \dfrac{\sqrt{3} - 1}{1 + \sqrt{3}} \cdot \dfrac{1 - \sqrt{3}}{1 - \sqrt{3}}$

   $= \dfrac{2\sqrt{3} - 4}{-2} = \dfrac{-2\left(2 - \sqrt{3}\right)}{-2} = 2 - \sqrt{3}$

9. $\sin \dfrac{17\pi}{12} = \sin\left(\dfrac{15\pi}{12} + \dfrac{2\pi}{12}\right)$

$= \sin \dfrac{5\pi}{4} \cos \dfrac{\pi}{6} + \cos \dfrac{5\pi}{4} \sin \dfrac{\pi}{6}$

$= \dfrac{-\sqrt{2}}{2} \cdot \dfrac{\sqrt{3}}{2} + \dfrac{-\sqrt{2}}{2} \cdot \dfrac{1}{2}$

$= \dfrac{-1}{4}\left(\sqrt{6} + \sqrt{2}\right)$

11. $\sec\left(-\dfrac{\pi}{12}\right) = \dfrac{1}{\cos\left(\dfrac{-\pi}{12}\right)} = \dfrac{1}{\cos\left(\dfrac{3\pi}{12} - \dfrac{4\pi}{12}\right)}$

$= \dfrac{1}{\cos \dfrac{\pi}{4} \cos \dfrac{\pi}{3} + \sin \dfrac{\pi}{4} \sin \dfrac{\pi}{3}}$

$= \dfrac{1}{\dfrac{\sqrt{2}}{2} \cdot \dfrac{1}{2} + \dfrac{\sqrt{2}}{2} \cdot \dfrac{\sqrt{3}}{2}}$

$= \dfrac{1}{\dfrac{\sqrt{2} + \sqrt{2}\sqrt{3}}{4}} = \dfrac{4}{\sqrt{2} + \sqrt{6}}$

$= \dfrac{4}{\sqrt{2} + \sqrt{6}} \cdot \dfrac{\sqrt{2} - \sqrt{6}}{\sqrt{2} - \sqrt{6}} = \dfrac{4\left(\sqrt{2} - \sqrt{6}\right)}{-4}$

$= \sqrt{6} - \sqrt{2}$

13. $\sin 20° \cos 10° + \cos 20° \sin 10° = \sin(20° + 10°)$

$= \sin 30° = \dfrac{1}{2}$

15. $\cos 70° \cos 20° - \sin 70° \sin 20° = \cos(70° + 20°)$

$= \cos 90° = 0$

17. $\dfrac{\tan 20° + \tan 25°}{1 - \tan 20° \tan 25°} = \tan(20° + 25°) = \tan 45° = 1$

19. $\sin \dfrac{\pi}{12} \cos \dfrac{7\pi}{12} - \cos \dfrac{\pi}{12} \sin \dfrac{7\pi}{12} = \sin\left(\dfrac{\pi}{12} - \dfrac{7\pi}{12}\right)$

$= \sin\left(\dfrac{-\pi}{2}\right) = -1$

21. $\sin\dfrac{\pi}{12} \cos\dfrac{5\pi}{12} - \sin\dfrac{5\pi}{12} \cos \dfrac{\pi}{12} = \sin\left(\dfrac{\pi}{12} - \dfrac{5\pi}{12}\right)$

$= \sin\left(\dfrac{-\pi}{3}\right) = -\dfrac{\sqrt{3}}{2}$

23. $\sin \alpha = \dfrac{3}{5}$, $0 < \alpha < \dfrac{\pi}{2}$; $\cos \beta = \dfrac{2}{\sqrt{5}}$, $\dfrac{-\pi}{2} < \beta < 0$.

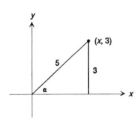

Notice that $y = 3$ and $r = 5$ so that $x^2 + 3^2 = 5^2$, $x > 0$

$$x^2 = 25 - 9 = 16, \; x > 0$$
$$x = 4$$

Thus, $\cos \alpha = \dfrac{4}{5}$

$$\tan \alpha = \dfrac{3}{4}$$

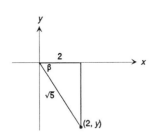

Notice that $x = 2$ and $r = \sqrt{5}$ so that $2^2 + y^2 = \left(\sqrt{5}\right)^2$, $y < 0$

$$y^2 = 5 - 4 = 1, \; y < 0$$
$$y = -1$$

Thus, $\sin \beta = \dfrac{-1}{\sqrt{5}}$

$$\tan \beta = \dfrac{-1}{2}$$

(a) $\sin(\alpha + \beta) = \sin \alpha \cos \beta + \cos \alpha \sin \beta$

$$= \dfrac{3}{5} \cdot \dfrac{2}{\sqrt{5}} + \dfrac{4}{5} \cdot \dfrac{-1}{\sqrt{5}}$$

$$= \dfrac{6 - 4}{5\sqrt{5}} = \dfrac{2}{5\sqrt{5}} = \dfrac{2\sqrt{5}}{25}$$

(b) $\cos(\alpha + \beta) = \cos \alpha \cos \beta - \sin \alpha \sin \beta$

$$= \dfrac{4}{5} \cdot \dfrac{2}{\sqrt{5}} - \dfrac{3}{5} \cdot \dfrac{-1}{\sqrt{5}}$$

$$= \dfrac{8 + 3}{5\sqrt{5}} = \dfrac{11}{5\sqrt{5}} = \dfrac{11\sqrt{5}}{25}$$

(c) $\sin(\alpha - \beta) = \sin \alpha \cos \beta - \cos \alpha \sin \beta$

$$= \dfrac{3}{5} \cdot \dfrac{2}{\sqrt{5}} - \dfrac{4}{5} \cdot \dfrac{-1}{\sqrt{5}}$$

$$= \dfrac{6 + 4}{5\sqrt{5}} = \dfrac{10}{5\sqrt{5}} = \dfrac{2\sqrt{5}}{5}$$

(d) $\tan(\alpha - \beta) = \dfrac{\tan \alpha - \tan \beta}{1 + \tan \alpha \tan \beta}$

$$= \dfrac{\dfrac{3}{4} - \dfrac{-1}{2}}{1 + \dfrac{3}{4} \cdot \dfrac{-1}{2}}$$

$$= \dfrac{\dfrac{5}{4}}{\dfrac{5}{8}} = 2$$

25. $\tan \alpha = \dfrac{-4}{3}, \dfrac{\pi}{2} < \alpha < \pi; \cos \beta = \dfrac{1}{2}, 0 < \beta < \dfrac{\pi}{2}$

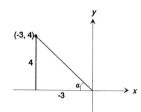

$\tan \alpha = \dfrac{-4}{3}$

$(-3)^2 + (4)^2 = 5^2$

$\sin \alpha = \dfrac{4}{5}, \quad \cos \alpha = \dfrac{-3}{5}$

$\cos \beta = \dfrac{1}{2}$

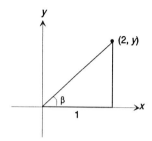

$1^2 + y^2 = 2^2, \ y > 0$
$y^2 = 4 - 1 = 3, \ y > 0$
$y = \sqrt{3}$

$\sin \beta = \dfrac{\sqrt{3}}{2}, \quad \tan \beta = \sqrt{3}$

(a) $\sin(\alpha + \beta) = \sin \alpha \cos \beta + \cos \alpha \sin \beta$

$= \dfrac{4}{5} \cdot \dfrac{1}{2} + \dfrac{-3}{5} \cdot \dfrac{\sqrt{3}}{2}$

$= \dfrac{4 - 3\sqrt{3}}{10}$

(b) $\cos(\alpha + \beta) = \cos \alpha \cos \beta - \sin \alpha \sin \beta$

$= \dfrac{-3}{5} \cdot \dfrac{1}{2} - \dfrac{4}{5} \cdot \dfrac{\sqrt{3}}{2}$

$= \dfrac{-3 - 4\sqrt{3}}{10}$

(c) $\sin(\alpha - \beta) = \sin \alpha \cos \beta - \cos \alpha \sin \beta$

$= \dfrac{4}{5} \cdot \dfrac{1}{2} - \dfrac{-3}{5} \cdot \dfrac{\sqrt{3}}{2}$

$= \dfrac{4 + 3\sqrt{3}}{10}$

(d) $\tan(\alpha - \beta) = \dfrac{\tan \alpha - \tan \beta}{1 + \tan \alpha \tan \beta}$

$= \dfrac{\dfrac{-4}{3} - \sqrt{3}}{1 + \dfrac{-4}{3}\sqrt{3}} = \dfrac{\dfrac{-4 - 3\sqrt{3}}{3}}{\dfrac{3 - 4\sqrt{3}}{3}}$

$= \dfrac{-4 - 3\sqrt{3}}{3 - 4\sqrt{3}} = \dfrac{4 + 3\sqrt{3}}{4\sqrt{3} - 3} \cdot \dfrac{4\sqrt{3} + 3}{4\sqrt{3} + 3} = \dfrac{25\sqrt{3} + 48}{39}$

27. $\sin \alpha = \dfrac{5}{13}$, $\dfrac{-3\pi}{2} < \alpha < -\pi$; $\tan \beta = -\sqrt{3}$, $\dfrac{\pi}{2} < \beta < \pi$

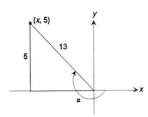

$\sin \alpha = \dfrac{5}{15}$

$x^2 + 5^2 = 13^2$, $x < 0$
$\qquad x^2 = 169 - 25 = 144$, $x < 0$
$\qquad x = -12$

$\cos \alpha = \dfrac{-12}{13}$, $\tan \alpha = \dfrac{-5}{12}$

$\left(\sqrt{3}\right)^2 + (-1)^2 = 2^2$

$\sin \beta = \dfrac{\sqrt{3}}{2}$, $\cos \beta = \dfrac{-1}{2}$

(a) $\sin(\alpha + \beta) = \sin \alpha \cos \beta + \cos \alpha \sin \beta$

$\qquad = \dfrac{5}{13} \cdot \dfrac{-1}{2} + \dfrac{-12}{13} \cdot \dfrac{\sqrt{3}}{2}$

$\qquad = \dfrac{-5 - 12\sqrt{3}}{26} = \dfrac{-1}{26}\left(5 + 12\sqrt{3}\right)$

(b) $\cos(\alpha + \beta) = \cos \alpha \cos \beta - \sin \alpha \sin \beta$

$\qquad = \dfrac{-12}{13} \cdot \dfrac{-1}{2} - \dfrac{5}{13} \cdot \dfrac{\sqrt{3}}{2}$

$\qquad = \dfrac{12 - 5\sqrt{3}}{26} = \dfrac{1}{26}\left(12 - 5\sqrt{3}\right)$

(c) $\sin(\alpha - \beta) = \sin \alpha \cos \beta - \cos \alpha \sin \beta$

$\qquad = \dfrac{5}{13} \cdot \dfrac{-1}{2} - \dfrac{-12}{13} \cdot \dfrac{\sqrt{3}}{2}$

$\qquad = \dfrac{-5 + 12\sqrt{3}}{26} = \dfrac{-1}{26}\left(5 - 12\sqrt{3}\right)$

(d) $\tan(\alpha - \beta) = \dfrac{\tan \alpha - \tan \beta}{1 + \tan \alpha \tan \beta}$

$\qquad = \dfrac{\dfrac{-5}{12} - -\sqrt{3}}{1 + \dfrac{-5}{12} \cdot -\sqrt{3}} = \dfrac{\dfrac{-5 + 12\sqrt{3}}{12}}{\dfrac{12 + 5\sqrt{3}}{12}}$

$\qquad = \dfrac{-5 + 12\sqrt{3}}{12 + 5\sqrt{3}} \cdot \dfrac{12 - 5\sqrt{3}}{12 - 5\sqrt{3}} = \dfrac{-240 + 169\sqrt{3}}{69}$

29. $\sin\left(\dfrac{\pi}{2} + \theta\right) = \sin\ \dfrac{\pi}{2}\ \cos\theta + \cos\ \dfrac{\pi}{2}\ \sin\theta$

$\qquad\qquad\quad = 1 \cdot \cos\theta + 0 \cdot \sin\theta$

$\qquad\qquad\quad = \cos\theta$

31. $\sin(\pi - \theta) = \sin\pi\ \cos\theta - \cos\pi\ \sin\theta$

$\qquad\qquad\quad = 0 \cdot \cos\theta - (-1)\ \sin\theta$

$\qquad\qquad\quad = \sin\theta$

33. $\sin(\pi + \theta) = \sin\pi\ \cos\theta + \cos\pi\ \sin\theta$

$\qquad\qquad\quad = 0 \cdot \cos\theta + (-1)\ \sin\theta$

$\qquad\qquad\quad = -\sin\theta$

35. $\tan(\pi - \theta) = \dfrac{\tan\pi - \tan\theta}{1 + \tan\pi\ \tan\theta}$

$\qquad\qquad\quad = \dfrac{0 - \tan\theta}{1 + 0}$

$\qquad\qquad\quad = -\tan\theta$

37. $\sin\left(\dfrac{3\pi}{2} + \theta\right) = \sin\ \dfrac{3\pi}{2}\ \cos\theta + \cos\ \dfrac{3\pi}{2}\ \sin\theta$

$\qquad\qquad\quad = -1 \cdot \cos\theta + 0 \cdot \sin\theta$

$\qquad\qquad\quad = -\cos\theta$

39. $\sin(\alpha + \beta) + \sin(\alpha - \beta)$

$\quad = \sin\alpha\ \cos\beta + \cos\alpha\ \sin\beta + \sin\alpha\ \cos\beta - \cos\alpha\ \sin\beta$

$\quad = 2\ \sin\alpha\ \cos\beta$

41. $\dfrac{\sin(\alpha + \beta)}{\sin\alpha\ \cos\beta} = \dfrac{\sin\alpha\ \cos\beta + \cos\alpha\ \sin\beta}{\sin\alpha\ \cos\beta}$

$\qquad\qquad = \dfrac{\sin\alpha\ \cos\beta}{\sin\alpha\ \cos\beta} + \dfrac{\cos\alpha\ \sin\beta}{\sin\alpha\ \cos\beta}$

$\qquad\qquad = 1 + \cot\alpha\ \tan\beta$

43. $\dfrac{\cos(\alpha + \beta)}{\cos\alpha\ \cos\beta} = \dfrac{\cos\alpha\ \cos\beta - \sin\alpha\ \sin\beta}{\cos\alpha\ \cos\beta}$

$\qquad\qquad = \dfrac{\cos\alpha\ \cos\beta}{\cos\alpha\ \cos\beta} - \dfrac{\sin\alpha\ \sin\beta}{\cos\alpha\ \cos\beta}$

$\qquad\qquad = 1 - \tan\alpha\ \tan\beta$

45. $\dfrac{\sin(\alpha + \beta)}{\sin(\alpha - \beta)} = \dfrac{\sin\alpha\ \cos\beta + \cos\alpha\ \sin\beta}{\sin\alpha\ \cos\beta - \cos\alpha\ \sin\beta}$

$\qquad\qquad = \dfrac{\dfrac{\sin\alpha\ \cos\beta + \cos\alpha\ \sin\beta}{\cos\alpha\ \cos\beta}}{\dfrac{\sin\alpha\ \cos\beta - \cos\alpha\ \sin\beta}{\cos\alpha\ \cos\beta}}$

$\qquad\qquad = \dfrac{\dfrac{\sin\alpha\ \cos\beta}{\cos\alpha\ \cos\beta} + \dfrac{\cos\alpha\ \sin\beta}{\cos\alpha\ \cos\beta}}{\dfrac{\sin\alpha\ \cos\beta}{\cos\alpha\ \cos\beta} - \dfrac{\cos\alpha\ \sin\beta}{\cos\alpha\ \cos\beta}}$

$\qquad\qquad = \dfrac{\tan\alpha + \tan\beta}{\tan\alpha - \tan\beta}$

7 ANALYTIC TRIGONOMETRY

47. $\cot(\alpha + \beta) = \dfrac{\cos(\alpha + \beta)}{\sin(\alpha + \beta)}$

$$= \dfrac{\cos \alpha \cos \beta - \sin \alpha \sin \beta}{\sin \alpha \cos \beta + \cos \alpha \sin \beta}$$

$$= \dfrac{\dfrac{\cos \alpha \cos \beta - \sin \alpha \sin \beta}{\sin \alpha \sin \beta}}{\dfrac{\sin \alpha \cos \beta + \cos \alpha \sin \beta}{\sin \alpha \sin \beta}}$$

$$= \dfrac{\dfrac{\cos \alpha \cos \beta}{\sin \alpha \sin \beta} - \dfrac{\sin \alpha \sin \beta}{\sin \alpha \sin \beta}}{\dfrac{\sin \alpha \cos \beta}{\sin \alpha \sin \beta} + \dfrac{\cos \alpha \sin \beta}{\sin \alpha \sin \beta}}$$

$$= \dfrac{\cot \alpha \cot \beta - 1}{\cot \beta + \cot \alpha}$$

49. $\sec(\alpha + \beta) = \dfrac{1}{\cos(\alpha + \beta)}$

$$= \dfrac{1}{\cos \alpha \cos \beta - \sin \alpha \sin \beta}$$

$$= \dfrac{\dfrac{1}{\sin \alpha \sin \beta}}{\dfrac{\cos \alpha \cos \beta - \sin \alpha \sin \beta}{\sin \alpha \sin \beta}}$$

$$= \dfrac{\dfrac{1}{\sin \alpha} \cdot \dfrac{1}{\sin \beta}}{\dfrac{\cos \alpha \cos \beta}{\sin \alpha \sin \beta} - \dfrac{\sin \alpha \sin \beta}{\sin \alpha \sin \beta}}$$

$$= \dfrac{\csc \alpha \csc \beta}{\cot \alpha \cot \beta - 1}$$

51. $\sin(\alpha - \beta)\sin(\alpha + \beta)$

$= (\sin \alpha \cos \beta - \cos \alpha \sin \beta)(\sin \alpha \cos \beta + \cos \alpha \sin \beta)$

$= \sin^2 \alpha \cos^2 \beta - \cos^2 \alpha \sin^2 \beta$

$= (\sin^2 \alpha)(1 - \sin^2 \beta) - (1 - \sin^2 \alpha)(\sin^2 \beta)$

$= \sin^2 \alpha - \sin^2 \alpha \sin^2 \beta - \sin^2 \beta + \sin^2 \alpha \sin^2 \beta$

$= \sin^2 \alpha - \sin^2 \beta$

53. $\sin(\theta + k\pi) = \sin \theta \cos k\pi + \cos \theta \sin k\pi$

$= (\sin \theta)(-1)^k + (\cos \theta)(0)$

$= (-1)^k \cdot \sin \theta, \ k$ any integer

55. $\sin(\sin^{-1} \frac{1}{2} + \cos^{-1} 0) = \sin\left(\dfrac{\pi}{6} + \dfrac{\pi}{2}\right)$

$= \sin \dfrac{\pi}{6} \cos \dfrac{\pi}{2} + \cos \dfrac{\pi}{6} \sin \dfrac{\pi}{2}$

$= \dfrac{1}{2} \cdot 0 + \dfrac{\sqrt{3}}{2} \cdot 1$

$= \dfrac{\sqrt{3}}{2}$

57.

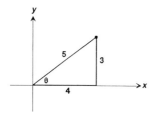

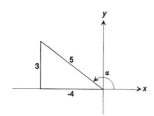

Let $\theta = \sin^{-1} \frac{3}{5}$ and $\alpha = \cos^{-1}\left(-\frac{4}{5}\right)$

Then,

$\sin\left[\sin^{-1} \frac{3}{5} - \cos^{-1}\left(-\frac{4}{5}\right)\right]$

$\quad = \sin(\theta - \alpha)$

$\quad = \sin \theta \cos \alpha - \cos \theta \sin \alpha$

$\quad = \left(\frac{3}{5}\right)\left(\frac{-4}{5}\right) - \left(\frac{4}{5}\right)\left(\frac{3}{5}\right)$

$\quad = \frac{-12}{25} - \frac{12}{25} = \frac{-24}{25}$

59.

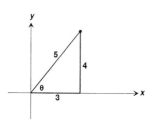

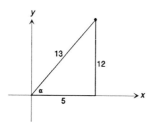

Let $\theta = \tan^{-1} \frac{4}{3}$ and $\alpha = \cos^{-1} \frac{5}{13}$

Then,

$\cos\left(\tan^{-1} \frac{4}{3} + \cos^{-1} \frac{5}{13}\right)$

$\quad = \cos(\theta + \alpha)$

$\quad = \cos \theta \cos \alpha - \sin \theta \sin \alpha$

$\quad = \left(\frac{3}{5}\right)\left(\frac{5}{13}\right) - \left(\frac{4}{5}\right)\left(\frac{12}{13}\right)$

$\quad = \frac{15}{65} - \frac{48}{65} = \frac{-33}{65}$

61.

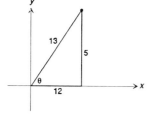

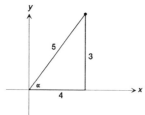

Let $\theta = \sin^{-1} \frac{5}{13}$ and $\alpha = \cos^{-1} \frac{3}{4}$

Then,

$\sec\left(\sin^{-1} \frac{5}{13} - \tan^{-1} \frac{3}{4}\right)$

$\quad = \sec(\theta - \alpha)$

$\quad = \frac{1}{\cos(\theta - \alpha)}$

$\quad = \frac{1}{\cos \theta \cos \alpha + \sin \theta \sin \alpha}$

$\quad = \frac{1}{\left(\frac{12}{13}\right)\left(\frac{4}{5}\right) + \left(\frac{5}{13}\right)\left(\frac{3}{5}\right)}$

$\quad = \frac{1}{\frac{48}{65} + \frac{15}{65}} = \frac{1}{\frac{63}{65}} = \frac{65}{63}$

7 ANALYTIC TRIGONOMETRY

63.

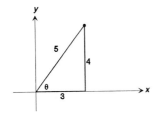

Let $\theta = \sec^{-1} \dfrac{5}{3}$

Then,

$$\cot\left(\sec^{-1} \frac{5}{3} + \frac{\pi}{6}\right)$$

$$= \cot\left(\theta + \frac{\pi}{6}\right)$$

$$= \frac{1}{\tan\left(\theta + \frac{\pi}{6}\right)}$$

$$= \frac{1}{\dfrac{\tan\theta + \tan\frac{\pi}{6}}{1 - \tan\theta\tan\frac{\pi}{6}}}$$

$$= \frac{1 - \tan\theta\tan\frac{\pi}{6}}{\tan\theta + \tan\frac{\pi}{6}} = \frac{1 - \left(\frac{4}{3}\right)\left(\frac{\sqrt{3}}{3}\right)}{\frac{4}{3} + \frac{\sqrt{3}}{3}} = \frac{1 - \frac{4\sqrt{3}}{9}}{\frac{4 + \sqrt{3}}{3}}$$

$$= \frac{9 - 4\sqrt{3}}{9} \cdot \frac{3}{4 + \sqrt{3}} = \frac{9 - 4\sqrt{3}}{3\left(4 + \sqrt{3}\right)} = \frac{1}{3}\left[\frac{9 - 4\sqrt{3}}{4 + \sqrt{3}} \cdot \frac{4 - \sqrt{3}}{4 - \sqrt{3}}\right]$$

$$= \frac{1}{3}\left(\frac{48 - 25\sqrt{3}}{13}\right) = \frac{1}{39}\left[48 - 25\sqrt{3}\right]$$

65. Let $\alpha = \cos^{-1} u$ and $\beta = \sin^{-1} v$. Then, $\cos\alpha = u$, $0 \le \alpha \le \pi$, and $\sin\beta = v$, $\frac{-\pi}{2} \le \beta \le \frac{\pi}{2}$. Since $\sin\alpha \ge 0$ and $\cos\beta \ge 0$, we have

$$\sin\alpha = \sqrt{1 - \cos^2\alpha} = \sqrt{1 - u^2} \quad \text{and} \quad \cos\beta = \sqrt{1 - \sin^2\beta} = \sqrt{1 - v^2}$$

Thus, $\cos(\cos^{-1} u + \sin^{-1} v) = \cos(\alpha + \beta)$

$$= \cos\alpha\cos\beta - \sin\alpha\sin\beta$$

$$= u\sqrt{1 - v^2} - v\sqrt{1 - u^2}$$

67. Let $\alpha = \tan^{-1} u$ and $\beta = \sin^{-1} v$. Then, $\tan\alpha = u$, $\frac{-\pi}{2} < \alpha < \frac{\pi}{2}$ and $\sin\beta = v$, $\frac{-\pi}{2} \le \beta \le \frac{\pi}{2}$. Now, $\sin\alpha = \dfrac{u}{\sqrt{1 + u^2}}$, $\cos\alpha = \dfrac{1}{\sqrt{1 + u^2}}$,

and $\cos\beta = \sqrt{1 - v^2}$

Thus, $\sin(\tan^{-1} u - \sin^{-1} v) = \sin(\alpha - \beta)$

$$= \sin\alpha\cos\beta - \cos\alpha\sin\beta$$

$$= \frac{u\sqrt{1 - v^2}}{\sqrt{1 + u^2}} - \frac{v}{\sqrt{1 + u^2}}$$

$$= \frac{u\sqrt{1 - v^2} - v}{\sqrt{1 + u^2}}$$

7.2 SUM AND DIFFERENCE FORMULAS

69. Let $\alpha = \sin^{-1} u$ and $\beta = \cos^{-1} v$. Then $\sin \alpha = u$, $\frac{-\pi}{2} \le \alpha \le \frac{\pi}{2}$ and

$\cos \beta = v$, $0 \le \beta \le \pi$. Then, $\cos \alpha = \sqrt{1 - u^2}$ and $\sin \beta = \sqrt{1 - v^2}$

Thus, $\tan(\sin^{-1} u - \cos^{-1} v) = \tan(\alpha - \beta)$

$$= \frac{\tan \alpha - \tan \beta}{1 + \tan \alpha \tan \beta}$$

$$= \frac{\dfrac{u}{\sqrt{1 - u^2}} - \dfrac{\sqrt{1 - v^2}}{v}}{1 + \dfrac{u\sqrt{1 - v^2}}{v\sqrt{1 - u^2}}}$$

$$= \frac{uv - \sqrt{1 - u^2}\sqrt{1 - v^2}}{v\sqrt{1 - u^2} + u\sqrt{1 - v^2}}$$

71. $\dfrac{\sin(x + h) - \sin x}{h} = \dfrac{\sin x \cos h + \cos x \sin h - \sin x}{h}$

$$= \frac{\cos x \sin h - (\sin x)(1 - \cos h)}{h}$$

$$= \cos x \cdot \frac{\sin h}{h} - \sin x \cdot \frac{1 - \cos h}{h}$$

73. $\sin(\sin^{-1} u + \cos^{-1} u)$

$= \sin(\sin^{-1} u) \cos(\cos^{-1} u) + \cos(\sin^{-1} u) \sin(\cos^{-1} u)$

$= (u)(u) + \sqrt{1 - u^2}\sqrt{1 - u^2}$
$= u^2 + 1 - u^2$
$= 1$

75. $\tan\left(\dfrac{\pi}{2} - \theta\right) = \dfrac{\tan \dfrac{\pi}{2} - \tan \theta}{1 + \tan \dfrac{\pi}{2} \tan \theta}$

Impossible because $\tan \dfrac{\pi}{2}$ is not defined.

Therefore,

$$\tan\left(\frac{\pi}{2} - \theta\right) = \frac{\sin\left(\dfrac{\pi}{2} - \theta\right)}{\cos\left(\dfrac{\pi}{2} - \theta\right)} = \frac{\cos \theta}{\sin \theta} = \cot \theta$$

77. $\tan \theta = \tan(\theta_2 - \theta_1) = \dfrac{\tan \theta_2 - \tan \theta_1}{1 + \tan \theta_2 \tan \theta_1}$

$$= \frac{m_2 - m_1}{1 + m_1 m_2}$$

362   7 ANALYTIC TRIGONOMETRY

EXERCISE 7.3 DOUBLE-ANGLE AND HALF-ANGLE FORMULAS

1. $\sin \theta = \frac{3}{5}$, $0 < \theta < \frac{\pi}{2}$. Thus, $0 < \frac{\theta}{2} < \frac{\pi}{4}$, or $\frac{\theta}{2}$ lies in quadrant I.

(a) Because $\sin 2\theta = 2 \sin \theta \cos \theta$ and because we know $\sin \theta = \frac{3}{5}$, we only need to find $\cos \theta$. It is given that $0 < \theta < \frac{\pi}{2}$, which means $\cos \theta > 0$. Using a form of Pythagorean identity, we find that

$$\cos \theta = \sqrt{1 - \sin^2 \theta} = \sqrt{1 - \frac{9}{25}} = \frac{4}{5}$$

$$\sin 2\theta = 2 \sin \theta \cos \theta = 2\left(\frac{3}{5}\right)\left(\frac{4}{5}\right) = \frac{24}{25}$$

(b) $\cos 2\theta = 1 - 2 \sin^2 \theta = 1 - 2\left(\frac{9}{25}\right) = \frac{7}{25}$

(c) $\sin \frac{1}{2} \theta = \sqrt{\frac{1 - \cos \theta}{2}} = \sqrt{\frac{1}{10}} = \frac{\sqrt{10}}{10}$

(d) $\cos \frac{1}{2} \theta = \sqrt{\frac{1 + \cos \theta}{2}} = \sqrt{\frac{9}{10}} = \frac{3\sqrt{10}}{10}$

3. $\tan \theta = \frac{4}{3}$, $\pi < \theta < \frac{3\pi}{2}$. Thus, $\frac{\pi}{2} < \frac{\theta}{2} < \frac{3\pi}{4}$, or $\frac{\theta}{2}$ lies in quadrant II.

$$-(3)^2 + (-4)^2 = 5^2$$

$$\sin 2\theta = \frac{-4}{5} \quad \cos \theta = \frac{-3}{5}$$

(a) $\sin 2\theta = 2 \sin \theta \cos \theta = 2\left(\frac{-4}{5}\right)\left(\frac{-3}{5}\right) = \frac{24}{25}$

(b) $\cos 2\theta = 1 - 2 \sin^2 \theta = 1 - 2\left(\frac{16}{25}\right) = \frac{-7}{25}$

(c) $\sin \frac{1}{2} \theta = +\sqrt{\frac{1 - \cos \theta}{2}} = \sqrt{\frac{\frac{8}{5}}{2}} = \sqrt{\frac{4}{5}} = \frac{2\sqrt{5}}{5}$

(d) $\cos \frac{1}{2} \theta = -\sqrt{\frac{1 + \cos \theta}{2}} = -\sqrt{\frac{\frac{2}{5}}{2}} = -\sqrt{\frac{1}{5}} = -\frac{\sqrt{5}}{5}$

5. $\cos \theta = \dfrac{-\sqrt{2}}{\sqrt{3}}$, $\dfrac{\pi}{2} < \theta < \pi$. Thus, $\dfrac{\pi}{4} < \dfrac{\theta}{2} < \dfrac{\pi}{2}$, or $\dfrac{\theta}{2}$ lies in quadrant I.

(a) $\sin \theta = \sqrt{1 - \cos^2 \theta} = \sqrt{1 - \dfrac{2}{3}} - \dfrac{1}{\sqrt{3}} = \dfrac{\sqrt{3}}{3}$

$\sin 2\theta = 2 \sin \theta \cos \theta = 2\left(\dfrac{\sqrt{3}}{3}\right)\left(\dfrac{-\sqrt{2}}{\sqrt{3}}\right) = \dfrac{-2\sqrt{2}}{3}$

(b) $\cos 2\theta = 2 \cos^2 \theta - 1 = 2\left(\dfrac{2}{3}\right) - 1 = \dfrac{1}{3}$

(c) $\sin \dfrac{1}{2} \theta = \sqrt{\dfrac{1 - \cos \theta}{2}} = \sqrt{\dfrac{1 - \dfrac{-\sqrt{2}}{\sqrt{3}}}{2}} = \sqrt{\dfrac{\dfrac{3 + \sqrt{6}}{3}}{2}} = \sqrt{\dfrac{3 + \sqrt{6}}{6}}$

(d) $\cos \dfrac{1}{2} \theta = \sqrt{\dfrac{1 + \cos \theta}{2}} = \sqrt{\dfrac{1 + \dfrac{-\sqrt{6}}{3}}{2}} = \sqrt{\dfrac{3 - \sqrt{6}}{6}}$

7. $\sec \theta = 3$, $\sin \theta > 0$. Thus, θ lies in quadrant I or $0 < \theta < \dfrac{\pi}{2}$, and $0 < \dfrac{\theta}{2} < \dfrac{\pi}{4}$, or $\dfrac{\theta}{2}$ lies in quadrant I.

Therefore, $\cos \theta = \dfrac{1}{3}$, $\sin \theta = \sqrt{1 - \cos^2 \theta} = \sqrt{\dfrac{9 - 1}{9}} = \dfrac{\sqrt{8}}{3} = \dfrac{2\sqrt{2}}{3}$

(a) $\sin 2\theta = 2 \sin \theta \cos \theta = 2\left(\dfrac{\sqrt{8}}{3}\right)\left(\dfrac{1}{3}\right) = \dfrac{4\sqrt{2}}{9}$

(b) $\cos 2\theta = 2 \cos^2 - 1 = 2\left(\dfrac{1}{9}\right) - 1 = -\dfrac{7}{9}$

(c) $\sin \dfrac{1}{2}\theta = \sqrt{\dfrac{1 - \cos \theta}{2}} = \sqrt{\dfrac{1 - \dfrac{1}{3}}{2}} = \dfrac{\sqrt{3}}{3}$

(d) $\cos \dfrac{1}{2}\theta = \sqrt{\dfrac{1 + \cos \theta}{2}} = \sqrt{\dfrac{1 + \dfrac{1}{3}}{2}} = \sqrt{\dfrac{3 + 1}{6}} = \sqrt{\dfrac{4}{6}} = \dfrac{2}{\sqrt{6}} \cdot \dfrac{\sqrt{6}}{\sqrt{6}}$

$= \dfrac{2\sqrt{6}}{6} = \dfrac{\sqrt{6}}{3}$

7 ANALYTIC TRIGONOMETRY

9. $\cot \theta = -2$, $\sec \theta < 0$. Thus, $\cos \theta < 0$ and $\sin \theta < 0$ and θ lies in quadrant II, or $\frac{\pi}{2} < \theta < \pi$. Hence, $\frac{\pi}{4} < \frac{\theta}{2} < \frac{\pi}{2}$, or $\frac{\theta}{2}$ lies in quadrant I.

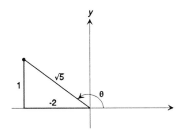

$$1^2 + (-2)^2 = \left(\sqrt{5}\right)^2$$

$$\cos \theta = \frac{-2}{\sqrt{5}} = \frac{-2\sqrt{5}}{5}$$

$$\sin \theta = \frac{1}{\sqrt{5}} = \frac{\sqrt{5}}{5}$$

(a) $\sin 2\theta = 2 \sin \theta \cos \theta = 2\left(\frac{\sqrt{5}}{5}\right)\left(\frac{-2\sqrt{5}}{5}\right) = \frac{-20}{25} = \frac{-4}{5}$

(b) $\cos 2\theta = \cos^2 \theta - \sin^2 \theta = \frac{4}{5} - \frac{1}{5} = \frac{3}{5}$

(c) $\sin \frac{1}{2}\theta = \sqrt{\dfrac{1 - \cos \theta}{2}} = \sqrt{\dfrac{1 + \dfrac{2\sqrt{5}}{5}}{2}} = \sqrt{\dfrac{5 + 2\sqrt{5}}{10}}$

(d) $\cos \frac{1}{2}\theta = \sqrt{\dfrac{1 + \cos \theta}{2}} = \sqrt{\dfrac{1 - \dfrac{2\sqrt{5}}{5}}{2}} = \sqrt{\dfrac{5 - 2\sqrt{5}}{10}}$

11. Because $22.5° = \dfrac{45°}{2}$, we can use the half-angle formula for sin $\left(\dfrac{\alpha}{2}\right)$ with $\alpha = 45°$. Also, because $22.5°$ is in quadrant I, sin $22.5° > 0$, and we choose the $+$ sign in using the half-angle formula:

$$\sin 22.5° = \sin \frac{45°}{2} = \sqrt{\dfrac{1 - \cos 45°}{2}} = \sqrt{\dfrac{1 - \dfrac{\sqrt{2}}{2}}{2}} = \sqrt{\dfrac{2 - \sqrt{2}}{4}}$$

$$= \frac{\sqrt{2 - \sqrt{2}}}{2}$$

13. Because $\frac{7\pi}{8}$ is in quadrant II, $\tan \frac{7\pi}{8} < 0$, and we choose the $-$ sign in using the half-angle formula:

$$\tan \frac{7\pi}{8} = \tan\frac{\frac{7\pi}{4}}{2} = -\sqrt{\frac{1 - \cos \alpha}{1 + \cos \alpha}} = -\sqrt{\frac{1 - \frac{1}{\sqrt{2}}}{1 + \frac{1}{\sqrt{2}}}} = -\sqrt{\frac{\sqrt{2} - 1}{1 + \sqrt{2}}\left(\frac{\sqrt{2} - 1}{\sqrt{2} - 1}\right)}$$

$$= -\left(\frac{\sqrt{2} - 1}{1}\right) = 1 - \sqrt{2}$$

15. Because $165°$ is in quadrant II, $\cos 165° < 0$, and we choose the $-$ sign in using the half-angle formula:

$$\cos 165° = \cos \frac{330°}{2} = -\sqrt{\frac{1 + \cos 330°}{2}} = -\sqrt{\frac{1 + \frac{\sqrt{3}}{2}}{2}} = -\frac{\sqrt{2 + \sqrt{3}}}{2}$$

17. Because $\frac{15\pi}{8}$ is in quadrant IV, $\sec \frac{15\pi}{8} > 0$, and we choose the $+$ sign in using the formula

$$\sec \frac{15\pi}{8} = \frac{1}{\cos \frac{15\pi}{8}} = \frac{1}{\cos \frac{\frac{15\pi}{4}}{2}} = \frac{1}{\sqrt{1 + \frac{\cos \frac{15\pi}{4}}{2}}} = \frac{1}{\sqrt{1 + \frac{\frac{\sqrt{2}}{2}}{2}}}$$

$$= \frac{1}{\sqrt{\frac{2 + \sqrt{2}}{4}}} = \frac{2}{\sqrt{2 + \sqrt{2}}} \cdot \frac{\sqrt{2 + \sqrt{2}}}{\sqrt{2 + \sqrt{2}}} = \frac{2\left(\sqrt{2 + \sqrt{2}}\right)}{2 + \sqrt{2}} \cdot \frac{2 - \sqrt{2}}{2 - \sqrt{2}}$$

$$= \frac{2\left(2 - \sqrt{2}\right)\left(\sqrt{2 + \sqrt{2}}\right)}{4 - 2} = \left(2 - \sqrt{2}\right)\left(\sqrt{2 + \sqrt{2}}\right)$$

19. Because $\frac{-\pi}{8}$ is in quadrant IV, $\sin\left(\frac{-\pi}{8}\right) < 0$, and we choose the $-$ sign in using the half-angle formula:

$$\sin\left(\frac{-\pi}{8}\right) = \sin\frac{\frac{-\pi}{4}}{2} = -\sqrt{\frac{1 - \cos\left(\frac{-\pi}{4}\right)}{2}}$$

$$= -\sqrt{\frac{1 - \frac{\sqrt{2}}{2}}{2}} = -\sqrt{\frac{2 - \sqrt{2}}{4}} = -\frac{\sqrt{2 - \sqrt{2}}}{2}$$

7 ANALYTIC TRIGONOMETRY

21. To show that $\sin^4 \theta = \dfrac{3}{8} - \dfrac{1}{2} \cos 2\theta + \dfrac{1}{8} \cos 4\theta,$

$$
\begin{aligned}
\sin^4 \theta = (\sin^2 \theta)^2 &= \left(\dfrac{1 - \cos 2\theta}{2}\right)^2 \\
&= \dfrac{1}{4}\left(1 - 2 \cos 2\theta + \cos^2 2\theta\right) \\
&= \dfrac{1}{4} - \dfrac{1}{2} \cos 2\theta + \dfrac{1}{4} \cos^2 2\theta \\
&= \dfrac{1}{4} - \dfrac{1}{2} \cos 2\theta + \dfrac{1}{4}\left(\dfrac{1 + \cos 4\theta}{2}\right) \\
&= \dfrac{1}{4} - \dfrac{1}{2} \cos 2\theta + \dfrac{1}{8} + \dfrac{1}{8} \cos 4\theta \\
&= \dfrac{3}{8} - \dfrac{1}{2} \cos 2\theta + \dfrac{1}{8} \cos 4\theta
\end{aligned}
$$

23. $\begin{aligned}[t] \sin 4\theta &= \sin 2(2\theta) \\ &= 2 \sin 2\theta \cos 2\theta \\ &= (4 \sin \theta \cos \theta)(1 - 2 \sin^2 \theta) \\ &= 4 \sin \theta \cos \theta - 8 \sin^3 \theta \cos \theta \\ &= (\cos \theta)(4 \sin \theta - 8 \sin^3 \theta) \end{aligned}$

25. $\begin{aligned}[t] \sin(5\theta) &= \sin(2\theta + 3\theta) = \sin 2\theta \cos 3\theta + \cos 2\theta \sin 3\theta \\ &= (2 \sin \theta \cos \theta)\cos(2\theta + \theta) + (1 - 2 \sin^2 \theta) \sin(2\theta + \theta) \\ &= (2 \sin \theta \cos \theta)(\cos 2\theta \cos \theta - \sin 2\theta \sin \theta) \\ &\quad + (1 - 2 \sin^2 \theta)(\sin 2\theta \cos \theta + \cos 2\theta \sin \theta) \\ &= 2 \cos 2\theta \sin \theta \cos^2 \theta - 2 \sin 2\theta \sin^2 \theta \cos \theta \\ &\quad + \sin 2\theta \cos \theta + \cos 2\theta \sin \theta - 2 \sin 2\theta \sin^2 \theta \cos \theta \\ &\quad - 2 \cos 2\theta \sin^3 \theta \\ &= 2(1 - 2 \sin^2 \theta) \sin \theta(1 - \sin^2 \theta) \\ &\quad - 4(2 \sin \theta \cos \theta) \sin^2 \theta \cos \theta + (2 \sin \theta \cos \theta) \cos \theta \\ &\quad + (1 - 2 \sin^2 \theta) \sin \theta - 2(1 - 2 \sin^2 \theta) \sin^3 \theta \\ &= (2 - 4 \sin^2 \theta)(\sin \theta - \sin^3 \theta) - 8 \sin^3 \theta \cos^2 \theta \\ &\quad + 2 \sin \theta \cos^2 \theta + \sin \theta - 2 \sin^3 \theta - 2 \sin^3 \theta \\ &\quad + 4 \sin^5 \theta \\ &= 2 \sin \theta = 6 \sin^3 \theta + 4 \sin^5 \theta - 8 \sin^3 \theta (1 - \sin^2 \theta) \\ &\quad + 2 \sin \theta(1 - \sin^2 \theta) + \sin \theta - 4 \sin^3 \theta + 4 \sin^5 \theta \\ &= 3 \sin \theta - 10 \sin^3 \theta + 8 \sin^5 \theta - 8 \sin^3 \theta + 8 \sin^5 \theta \\ &\quad + 2 \sin \theta - 2 \sin^3 \theta \\ &= 16 \sin^5 \theta - 20 \sin^3 \theta + 5 \sin \theta \end{aligned}$

27. To show that $\tan \dfrac{\theta}{2} = \dfrac{1 - \cos \theta}{\sin \theta}$,

$$\text{let } \frac{1 - \cos \theta}{\sin \theta} = \frac{1 - \cos 2\left(\frac{\theta}{2}\right)}{\sin 2\left(\frac{\theta}{2}\right)}$$

$$= \frac{1 - \left[1 - 2 \sin^2 \frac{\theta}{2}\right]}{2 \sin \frac{\theta}{2} \cos \frac{\theta}{2}}$$

$$= \frac{2 \sin^2 \frac{\theta}{2}}{2 \sin \frac{\theta}{2} \cos \frac{\theta}{2}}$$

$$= \frac{\sin \frac{\theta}{2}}{\cos \frac{\theta}{2}}$$

$$= \tan \frac{\theta}{2}$$

29. Using the difference of two squares,

$$\cos^4 \theta - \sin^4 \theta = (\cos^2 \theta + \sin^2 \theta)(\cos^2 \theta - \sin^2 \theta)$$
$$= \cos^2 \theta - \sin^2 \theta$$
$$= \cos 2\theta$$

31. $\cot 2\theta = \dfrac{1}{\tan 2\theta}$

$$= \frac{1}{\dfrac{2 \tan \theta}{1 - \tan^2 \theta}}$$

$$= \frac{1 - \tan^2 \theta}{2 \tan \theta}$$

$$= \frac{1 - \dfrac{1}{\cot^2 \theta}}{\dfrac{2}{\cot \theta}}$$

$$= \frac{\dfrac{\cot^2 \theta - 1}{\cot^2 \theta}}{\dfrac{2}{\cot \theta}}$$

$$= \frac{\cot^2 \theta - 1}{\cot^2 \theta} \cdot \frac{\cot \theta}{2}$$

$$= \frac{\cot^2 \theta - 1}{2 \cot \theta}$$

7 ANALYTIC TRIGONOMETRY

33. $\sec 2\theta = \dfrac{1}{\cos 2\theta}$

$$= \dfrac{1}{2 \cos^2 \theta - 1}$$

$$= \dfrac{1}{\dfrac{2}{\sec^2 \theta} - 1}$$

$$= \dfrac{1}{\dfrac{2 - \sec^2 \theta}{\sec^2 \theta}}$$

$$= \dfrac{\sec^2 \theta}{2 - \sec^2 \theta}$$

35. $\cos^2 2\theta - \sin^2 2\theta = \cos 2(2\theta)$
$$= \cos 4\theta$$

37. $\dfrac{\cos 2\theta}{1 + \sin 2\theta} = \dfrac{\cos^2 \theta - \sin^2 \theta}{1 + 2 \sin \theta \cos \theta}$

$$= \dfrac{(\cos \theta - \sin \theta)(\cos \theta + \sin \theta)}{\sin^2 \theta + \cos^2 \theta + 2 \sin \theta \cos \theta}$$

$$= \dfrac{(\cos \theta - \sin \theta)(\cos \theta + \sin \theta)}{(\sin \theta + \cos \theta)^2}$$

$$= \dfrac{\cos \theta - \sin \theta}{\cos \theta + \sin \theta}$$

$$= \dfrac{\dfrac{\cos \theta - \sin \theta}{\sin \theta}}{\dfrac{\cos \theta + \sin \theta}{\sin \theta}}$$

$$= \dfrac{\dfrac{\cos \theta}{\sin \theta} - \dfrac{\sin \theta}{\sin \theta}}{\dfrac{\cos \theta}{\sin \theta} + \dfrac{\sin \theta}{\sin \theta}}$$

$$= \dfrac{\cot \theta - 1}{\cot \theta + 1}$$

39. $\sec^2 \dfrac{\theta}{2} = \dfrac{1}{\cos^2\left(\dfrac{\theta}{2}\right)}$

$$= \dfrac{1}{\dfrac{1 + \cos \theta}{2}}$$

$$= \dfrac{2}{1 + \cos \theta}$$

41. $\cot^2 \dfrac{\theta}{2} = \dfrac{1}{\tan^2\left(\dfrac{\theta}{2}\right)}$

$$= \dfrac{1}{\dfrac{1 - \cos\theta}{1 + \cos\theta}}$$

$$= \dfrac{1 + \cos\theta}{1 - \cos\theta}$$

$$= \dfrac{1 + \dfrac{1}{\sec\theta}}{1 - \dfrac{1}{\sec\theta}}$$

$$= \dfrac{\dfrac{\sec\theta + 1}{\sec\theta}}{\dfrac{\sec\theta - 1}{\sec\theta}}$$

$$= \dfrac{\sec\theta + 1}{\sec\theta - 1}$$

43. Let $\dfrac{1 - \tan^2\left(\dfrac{\theta}{2}\right)}{1 + \tan^2\left(\dfrac{\theta}{2}\right)} = \dfrac{1 - \dfrac{1 - \cos\theta}{1 + \cos\theta}}{1 + \dfrac{1 - \cos\theta}{1 + \cos\theta}}$

$$= \dfrac{\dfrac{1 + \cos\theta - (1 - \cos\theta)}{1 + \cos\theta}}{\dfrac{1 + \cos\theta + 1 - \cos\theta}{1 + \cos\theta}}$$

$$= \dfrac{2\cos\theta}{1 + \cos\theta} \cdot \dfrac{1 + \cos\theta}{2}$$

$$= \cos\theta$$

Therefore, $\cos\theta = \dfrac{1 - \tan^2\left(\dfrac{\theta}{2}\right)}{1 + \tan^2\left(\dfrac{\theta}{2}\right)}$

45. $\dfrac{\sin 3\theta}{\sin\theta} - \dfrac{\cos 3\theta}{\cos\theta} = \dfrac{\sin 3\theta \cos\theta - \cos 3\theta \sin\theta}{\sin\theta \cos\theta}$

$$= \dfrac{\sin(3\theta - \theta)}{\dfrac{1}{2}(2\sin\theta\cos\theta)}$$

$$= \dfrac{2\sin 2\theta}{\sin 2\theta}$$

$$= 2$$

47. $\tan 3\theta = \tan(\theta + 2\theta)$

$$= \frac{\tan 2\theta + \tan \theta}{1 - \tan 2\theta \tan \theta}$$

$$= \frac{\dfrac{2\tan \theta}{1 - \tan^2 \theta} + \tan \theta}{1 - \dfrac{(2 \tan \theta) \tan \theta}{1 - \tan^2 \theta}}$$

$$= \frac{2 \tan \theta + \tan \theta(1 - \tan^2 \theta)}{1 - \tan^2 \theta - 2 \tan^2 \theta}$$

$$= \frac{3 \tan \theta - \tan^3 \theta}{1 - 3 \tan^2 \theta}$$

49. To graph $f(x) = \sin^2(x)$

$= \dfrac{(1 - \cos 2x)}{2}$ for $0 \le x \le 2\pi$,

begin with the graph of $y = \cos x$.
Apply a horizontal compression to
obtain the graph of $y = \cos 2x$.
Reflect about the x-axis to obtain
the graph of $y = -\cos 2x$. Shift up
1 unit to obtain the graph of
$y = 1 - \cos 2x$. Apply a vertical
compression to obtain the final
graph:

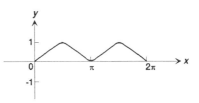

$$y = \frac{1}{2}(1 - \cos 2x).$$

51. $\sin\left(\dfrac{\pi}{24}\right) = \sin\left(\dfrac{\dfrac{\pi}{12}}{2}\right) = \sqrt{\dfrac{1 - \cos\dfrac{\pi}{12}}{2}}$

$$= \sqrt{\frac{1 - \dfrac{\sqrt{2}}{4}\left(\sqrt{3} + 1\right)}{2}}$$

$$= \sqrt{\frac{4 - \sqrt{2}\left(\sqrt{3} + 1\right)}{8}}$$

$$= \sqrt{\frac{4 - \sqrt{6} - \sqrt{2}}{8}}$$

$$= \frac{1}{2\sqrt{2}}\sqrt{4 - \sqrt{6} - \sqrt{2}}$$

$$= \frac{\sqrt{2}}{4}\sqrt{4 - \sqrt{6} - \sqrt{2}}$$

7.3 DOUBLE-ANGLE AND HALF-ANGLE FORMULAS

$$\cos\left(\frac{\pi}{24}\right) = \cos\left(\frac{\frac{\pi}{12}}{2}\right) = \sqrt{\frac{1 + \cos\frac{\pi}{12}}{2}}$$

$$= \sqrt{\frac{1 + \frac{\sqrt{2}}{4}\left(\sqrt{3} + 1\right)}{2}}$$

$$= \sqrt{\frac{4 + \sqrt{6} + \sqrt{2}}{4} \cdot \frac{1}{2}}$$

$$= \frac{1}{2\sqrt{2}}\sqrt{4 + \sqrt{6} + \sqrt{2}}$$

$$= \frac{\sqrt{2}}{4}\sqrt{4 + \sqrt{6} + \sqrt{2}}$$

53. $$\sin\left(2 \sin^{-1}\frac{1}{2}\right) = \sin\left(2 \cdot \frac{\pi}{6}\right)$$

$$= \sin\left(\frac{\pi}{3}\right)$$

$$= \frac{\sqrt{3}}{2}$$

55. $$\cos\left(2 \sin^{-1}\frac{3}{5}\right) = 1 - 2\sin^2\left(\sin^{-1}\frac{3}{5}\right)$$

$$= 1 - 2\left(\frac{3}{5}\right)^2$$

$$= 1 - 2\left(\frac{9}{25}\right)$$

$$= \frac{25 - 18}{25} = \frac{7}{25}$$

57. $$\tan\left[2 \cos^{-1}\left(-\frac{3}{5}\right)\right] = \frac{2\tan\left[\cos^{-1}\left(-\frac{3}{5}\right)\right]}{1 - \tan^2\left[\cos^{-1}\left(-\frac{3}{5}\right)\right]}$$

$$= \frac{2\left(\frac{-4}{3}\right)}{1 - \left(\frac{-4}{3}\right)^2}$$

$$= \frac{\frac{-8}{3}}{\frac{9 - 16}{9}}$$

$$= \frac{24}{7}$$

59. $$\sin\left(2 \cos^{-1}\frac{4}{5}\right) = 2\sin\left(\cos^{-1}\frac{4}{5}\right)\cos\left(\cos^{-1}\frac{4}{5}\right)$$

$$= 2\left(\frac{3}{5}\right)\left(\frac{4}{5}\right)$$

$$= \frac{24}{25}$$

7 ANALYTIC TRIGONOMETRY

61. $\sin^2\left(\dfrac{1}{2}\,\cos^{-1}\dfrac{3}{5}\right) = \dfrac{1 - \cos\left(\cos^{-1}\dfrac{3}{5}\right)}{2}$

$\qquad\qquad = \dfrac{1 - \dfrac{3}{5}}{2}$

$\qquad\qquad = \dfrac{2}{10} = \dfrac{1}{5}$

63. $\sec\left[2\,\tan^{-1}\dfrac{3}{4}\right] = \dfrac{1}{\cos\left[2\,\tan^{-1}\dfrac{3}{4}\right]}$

$\qquad\qquad = \dfrac{1}{1 - 2\,\sin^2\left(\tan^{-1}\dfrac{3}{4}\right)}$

$\qquad\qquad = \dfrac{1}{1 - 2\left(\dfrac{3}{5}\right)^2}$

$\qquad\qquad = \dfrac{1}{1 - 2\left(\dfrac{9}{25}\right)}$

$\qquad\qquad = \dfrac{1}{\dfrac{25 - 18}{25}} = \dfrac{1}{\dfrac{7}{25}} = \dfrac{25}{7}$

65. $\cot^2\left[\dfrac{1}{2}\,\tan^{-1}\dfrac{4}{3}\right] = \dfrac{1}{\tan^2\left[\dfrac{1}{2}\,\tan^{-1}\dfrac{4}{3}\right]}$

$\qquad\qquad = \dfrac{1}{\dfrac{1 - \cos 2\left(\dfrac{1}{2}\,\tan^{-1}\dfrac{4}{3}\right)}{1 + \cos 2\left(\dfrac{1}{2}\,\tan^{-1}\dfrac{4}{3}\right)}}$

$\qquad\qquad = \dfrac{1 + \cos\left(\tan^{-1}\dfrac{4}{3}\right)}{1 - \cos\left(\tan^{-1}\dfrac{4}{3}\right)}$

$\qquad\qquad = \dfrac{1 + \dfrac{3}{5}}{1 - \dfrac{3}{5}} = \dfrac{5}{\dfrac{2}{5}} = 4$

67. $\sin^3\theta + \sin^3(\theta + 120°) + \sin^3(\theta + 240°)$

$\quad = \sin^3\theta + [\sin\theta\cos 120° + \cos\theta\sin 120°]^3$
$\qquad + [\sin\theta\cos 240° + \cos\theta\sin 240°]^3$

$\quad = \sin^3\theta + \left[-\dfrac{1}{2}\sin\theta + \dfrac{\sqrt{3}}{2}\cos\theta\right]^3 + \left[-\dfrac{1}{2}\sin\theta - \dfrac{\sqrt{3}}{2}\cos\theta\right]^3$

$\quad = \sin^3\theta + \dfrac{1}{8}\left[3\sqrt{3}\cos^3\theta - 9\cos^2\theta\sin\theta + 3\sqrt{3}\cos\theta\sin^2\theta\right.$

$\qquad\qquad \left. - \sin^3\theta\right]$

$\qquad\qquad -\dfrac{1}{8}\left[\sin^3\theta + 3\sqrt{3}\sin^2\theta\cos\theta + 9\sin\theta\cos^2\theta + 3\sqrt{3}\cos^3\theta\right]$

$\quad = \dfrac{3}{4}\sin^3\theta - \dfrac{9}{4}\cos^2\theta\sin\theta = \dfrac{3}{4}\left[\sin^3\theta - 3\sin\theta(1 - \sin^2\theta)\right]$

$\quad = \dfrac{3}{4}\left[4\sin^3\theta - 3\sin\theta\right] = \dfrac{-3}{4}\sin 3\theta$

7.3 DOUBLE-ANGLE AND HALF-ANGLE FORMULAS

69. Let $\frac{1}{2}(\ln|1 - \cos 2\theta| - \ln 2) = \ln\left(\frac{|1 - \cos 2\theta|}{2}\right)^{1/2}$

$$= \ln|\sin^2 \theta|^{1/2}$$

$$= \ln|\sin \theta|$$

≡ EXERCISE 7.4 PRODUCT-TO-SUM AND SUM-TO-PRODUCT FORMULAS

For Problems 1-10, we are using the formulas:

$$\sin \alpha \sin \beta = \frac{1}{2}[\cos(\alpha - \beta) - \cos(\alpha + \beta)]$$

$$\cos \alpha \cos \beta = \frac{1}{2}[\cos(\alpha - \beta) + \cos(\alpha + \beta)]$$

$$\sin \alpha \cos \beta = \frac{1}{2}[\sin(\alpha + \beta) + \sin(\alpha - \beta)]$$

1. $\sin 4\theta \sin 2\theta = \frac{1}{2}[\cos(4\theta - 2\theta) - \cos(4\theta + 2\theta)]$

$$= \frac{1}{2}(\cos 2\theta - \cos 6\theta)$$

3. $\sin 4\theta \sin 2\theta = \frac{1}{2}[\sin(4\theta + 2\theta) + \sin(4\theta - 2\theta)]$

$$= \frac{1}{2}(\sin 6\theta + \sin 2\theta)$$

5. $\cos 3\theta \cos 5\theta = \frac{1}{2}[\cos(3\theta - 5\theta) + \cos(3\theta + 5\theta)]$

$$= \frac{1}{2}[\cos(-2\theta) + \cos(8\theta)]$$

$$= \frac{1}{2}(\cos 2\theta + \cos 8\theta)$$

7. $\sin \theta \sin 2\theta = \frac{1}{2}[\cos(\theta - 2\theta) - \cos(\theta + 2\theta)]$

$$= \frac{1}{2}[\cos(-\theta) - \cos(3\theta)]$$

$$= \frac{1}{2}(\cos \theta - \cos 3\theta)$$

9. $\sin\frac{3\theta}{2} \cos\frac{\theta}{2} = \frac{1}{2}\left[\sin\left(\frac{3\theta}{2} + \frac{\theta}{2}\right) + \sin\left(\frac{3\theta}{2} - \frac{\theta}{2}\right)\right]$

$$= \frac{1}{2}(\sin 2\theta + \sin \theta)$$

In Problems 11-18, we are using the formulas:

$$\sin \alpha + \sin \beta = 2 \sin \frac{\alpha + \beta}{2} \cos \frac{\alpha - \beta}{2}$$

$$\sin \alpha - \sin \beta = 2 \sin \frac{\alpha - \beta}{2} \cos \frac{\alpha + \beta}{2}$$

$$\cos \alpha + \cos \beta = 2 \cos \frac{\alpha + \beta}{2} \cos \frac{\alpha - \beta}{2}$$

$$\cos \alpha - \cos \beta = -2 \sin \frac{\alpha + \beta}{2} \cos \frac{\alpha - \beta}{2}$$

7 ANALYTIC TRIGONOMETRY

11. $\sin 4\theta - \sin 2\theta = 2 \sin \dfrac{4\theta - 2\theta}{2} \cos \dfrac{4\theta + 2\theta}{2}$

$\qquad\qquad\qquad = 2 \sin \theta \cos 3\theta$

13. $\cos 2\theta + \cos 4\theta = 2 \cos \dfrac{2\theta + 4\theta}{2} \cos \dfrac{2\theta - 4\theta}{2}$

$\qquad\qquad\qquad = 2 \cos 3\theta \cos \theta$

15. $\sin \theta + \sin 3\theta = 2 \sin \dfrac{\theta + 3\theta}{2} \cos \dfrac{\theta - 3\theta}{2}$

$\qquad\qquad\qquad = 2 \sin 2\theta \cos \theta$

17. $\cos \dfrac{\theta}{2} - \cos \dfrac{3\theta}{2} = -2 \sin \dfrac{\frac{\theta}{2} + \frac{3\theta}{2}}{2} \sin \dfrac{\frac{\theta}{2} - \frac{3\theta}{2}}{2}$

$\qquad\qquad\qquad = -2 \sin \theta \left(-\sin \dfrac{\theta}{2}\right)$

$\qquad\qquad\qquad = 2 \sin \theta \sin \dfrac{\theta}{2}$

19. $\dfrac{\sin \theta + \sin 3\theta}{2 \sin 2\theta} = \dfrac{2 \sin 2\theta \cos(-\theta)}{2 \sin 2\theta}$

$\qquad\qquad\qquad = \cos (-\theta)$

$\qquad\qquad\qquad = \cos \theta$

21. $\dfrac{\sin 4\theta + \sin 2\theta}{\cos 4\theta + \cos 2\theta} = \dfrac{2 \sin 3\theta \cos \theta}{2 \cos 3\theta \cos \theta}$

$\qquad\qquad\qquad = \dfrac{\sin 3\theta}{\cos 3\theta}$

$\qquad\qquad\qquad = \tan 3\theta$

23. $\dfrac{\cos \theta - \cos 3\theta}{\sin \theta + \sin 3\theta} = \dfrac{-2 \sin 2\theta \sin(-\theta)}{2 \sin 2\theta \cos(-\theta)}$

$\qquad\qquad\qquad = \dfrac{2 \sin 2\theta \sin \theta}{2 \sin 2\theta \cos \theta}$

$\qquad\qquad\qquad = \dfrac{\sin \theta}{\cos \theta}$

$\qquad\qquad\qquad = \tan \theta$

25. $\sin \theta(\sin \theta + \sin 3\theta) = \sin \theta[2 \sin 2\theta \cos(-\theta)]$

$\qquad\qquad\qquad = 2 \sin 2\theta \sin \theta \cos \theta$

$\qquad\qquad\qquad = \cos \theta(2 \sin 2\theta \sin \theta)$

$\qquad\qquad\qquad = \cos \theta\left[2 \cdot \dfrac{1}{2}(\cos \theta - \cos 3\theta)\right]$

$\qquad\qquad\qquad = \cos \theta(\cos \theta - \cos 3\theta)$

27. $\dfrac{\sin 4\theta + \sin 8\theta}{\cos 4\theta + \cos 8\theta} = \dfrac{2 \sin 6\theta \cos(-2\theta)}{2 \cos 6\theta \cos(-2\theta)}$

$\qquad\qquad\qquad = \dfrac{\sin 6\theta}{\cos 6\theta}$

$\qquad\qquad\qquad = \tan 6\theta$

29.
$$\frac{\sin 4\theta + \sin 8\theta}{\sin 4\theta - \sin 8\theta} = \frac{2 \sin 6\theta \cos(-2\theta)}{2 \sin(-2\theta) \cos 6\theta}$$

$$= \frac{\sin 6\theta}{\cos 6\theta} \cdot \frac{\cos 2\theta}{-\sin \theta}$$

$$= (\tan 6\theta)(-\cot 2\theta)$$

$$= -\frac{\tan 6\theta}{\tan 2\theta}$$

31.
$$\frac{\sin \alpha + \sin \beta}{\sin \alpha - \sin \beta} = \frac{2 \sin \dfrac{\alpha + \beta}{2} \cos \dfrac{\alpha - \beta}{2}}{2 \sin \dfrac{\alpha - \beta}{2} \cos \dfrac{\alpha + \beta}{2}}$$

$$= \frac{\sin \dfrac{\alpha + \beta}{2}}{\cos \dfrac{\alpha + \beta}{2}} \cdot \frac{\cos \dfrac{\alpha - \beta}{2}}{\sin \dfrac{\alpha - \beta}{2}}$$

$$= \tan \frac{\alpha + \beta}{2} \cot \frac{\alpha - \beta}{2}$$

33.
$$\frac{\sin \alpha + \sin \beta}{\cos \alpha + \cos \beta} = \frac{2 \sin \dfrac{\alpha + \beta}{2} \cos \dfrac{\alpha - \beta}{2}}{2 \cos \dfrac{\alpha + \beta}{2} \cos \dfrac{\alpha - \beta}{2}}$$

$$= \frac{\sin \dfrac{\alpha + \beta}{2}}{\cos \dfrac{\alpha + \beta}{2}}$$

$$= \tan \frac{\alpha + \beta}{2}$$

35. $1 + \cos 2\theta + \cos 4\theta + \cos 6\theta$

$$= (1 \cos 0\theta + \cos 6\theta) + (\cos 2\theta + \cos 4\theta)$$
$$= 2 \cos 3\theta \cos(-3\theta) + 2 \cos 3\theta \cos(-\theta)$$
$$= 2 \cos^2 3\theta + 2 \cos 3\theta \cos \theta$$
$$= 2 \cos 3\theta(\cos 3\theta + \cos \theta)$$
$$= 2 \cos 3\theta(2 \cos 2\theta \cos \theta)$$
$$= 4 \cos \theta \cos 2\theta \cos 3\theta$$

37. $\sin 2\alpha + \sin 2\beta + \sin 2\gamma$

$$= 2 \sin 2 \cdot \left(\frac{\alpha + \beta}{2}\right) \cos 2 \cdot \left(\frac{\alpha - \beta}{2}\right) + \sin 2\gamma$$
$$= 2 \sin(\alpha + \beta) \cos(\alpha - \beta) + 2 \sin \gamma \cos \gamma$$

(Because $\alpha + \beta + \gamma = \pi$)

$$= 2 \sin(\pi - \gamma)\cos(\alpha - \beta) + 2 \sin \gamma \cos \gamma$$
$$= 2 \sin \gamma \cos(\alpha - \beta) + 2 \sin \gamma \cos \gamma$$
$$= 2 \sin \gamma[\cos(\alpha - \beta) + \cos \gamma]$$
$$= 2 \sin \gamma\left(2 \cos \frac{\alpha - \beta + \gamma}{2} \cos \frac{\alpha - \beta - \gamma}{2}\right)$$
$$= 4 \sin \gamma \cos \frac{\pi - 2\beta}{2} \cos \frac{2\alpha - \pi}{2}$$
$$= 4 \sin \gamma \cos\left(\frac{\pi}{2} - \beta\right) \cos\left(\alpha - \frac{\pi}{2}\right)$$
$$= 4 \sin \gamma \sin \beta \sin \alpha$$
$$= 4 \sin \alpha \sin \beta \sin \gamma$$

7 ANALYTIC TRIGONOMETRY

39.
$$\sin(\alpha - \beta) = \sin \alpha \cos \beta - \cos \alpha \sin \beta$$
$$\sin(\alpha - \beta) = \sin \alpha \cos \beta + \cos \alpha \sin \beta$$
$$\sin(\alpha - \beta) + \sin(\alpha + \beta) = 2 \sin \alpha \cos \beta$$
$$2 \sin \alpha \cos \beta = \frac{1}{2}[\sin(\alpha + \beta) + \sin(\alpha - \beta)]$$

41.
$$2 \cos \frac{\alpha + \beta}{2} \cos \frac{\alpha - \beta}{2} = 2 \cdot \frac{1}{2}\left[\cos\left(\frac{\alpha + \beta}{2} - \frac{\alpha - \beta}{2}\right) + \cos\left(\frac{\alpha + \beta}{2} + \frac{\alpha - \beta}{2}\right)\right]$$
$$= \cos \frac{2\beta}{2} + \cos \frac{2\alpha}{2}$$
$$= \cos \beta + \cos \alpha$$

Therefore, $\cos \alpha + \cos \beta = 2 \cos \frac{\alpha + \beta}{2} \cos \frac{\alpha - \beta}{2}$

≡ EXERCISE 7.5 TRIGONOMETRIC EQUATIONS

1. The period of the sine function is 2π; and, on the interval
    $[0, 2\pi)$, the sine function has the value $\frac{1}{2}$ at $\frac{\pi}{6}$ and $\frac{5\pi}{6}$.
    Because the sine function has period 2π, all the solutions of
    $\sin \theta = \frac{1}{2}$ may be given by

    $$\theta = \frac{\pi}{6} + 2k\pi \text{ or } \theta = \frac{5\pi}{6} + 2k\pi, \ k \text{ any integer}$$

    On the interval $[0, 2\pi)$, the solutions of $\sin \theta = \frac{1}{2}$ are
    $\theta = \frac{\pi}{6}, \frac{5\pi}{6}$

3. The period of the tangent function is π; and, in the interval
    $[0, \pi]$, the tangent function has the value $-\frac{1}{\sqrt{3}}$ at $\frac{5\pi}{6}$.

    $$\theta = \frac{5\pi}{6} + k\pi, \ k \text{ any integer}$$

    On the interval $[0, 2\pi)$, the solutions of $\tan \theta = \frac{-1}{\sqrt{3}}$ are
    $\theta = \frac{5\pi}{6}, \frac{11\pi}{6}$

5. $\cos \theta = 0$
    The solutions on the interval $[0, 2\pi)$ are $\theta = \frac{\pi}{2}$ or $\theta = \frac{3\pi}{2}$.

7. $\sin 3\theta = -1$

    $$3\theta = \frac{3\pi}{2} + 2k\pi, \ k \text{ any integer}$$
    $$\theta = \frac{\pi}{2} + \frac{2}{3}k\pi, \ k \text{ any integer}$$

9. $\cos\left(2\theta - \dfrac{\pi}{2}\right) = -1$

$\quad 2\theta - \dfrac{\pi}{2} = \pi + 2k\pi$, k any integer

$\quad\quad 2\theta = \dfrac{3\pi}{2} + 2k\pi$, k any integer

$\quad\quad\quad \theta = \dfrac{3\pi}{4} + k\pi$, k any integer

The solutions on the interval $[0, 2\pi)$ are $\theta = \dfrac{3\pi}{4}, \dfrac{7\pi}{4}$

11. $\sec \dfrac{3\theta}{2} = -2$, since $\sec \theta = \dfrac{1}{\cos \theta}$, $\cos \dfrac{3\theta}{2} = -\dfrac{1}{2}$

$\quad \dfrac{3\theta}{2} = \dfrac{2\pi}{3} + 2k\pi$ or $\dfrac{3\theta}{2} = \dfrac{4\pi}{3} + 2k\pi$, k any integer

$\quad\quad \theta = \dfrac{4\pi}{9} + \dfrac{4}{3}k\pi$ or $\theta = \dfrac{8\pi}{9} + \dfrac{4}{3}k\pi$, k any integer

The solutions on the interval $[0, 2\pi)$ are $\theta = \dfrac{4\pi}{9}, \dfrac{16\pi}{9}$ or $\theta = \dfrac{8\pi}{9}$

13. $\sin \theta = 0.4$
$\quad \theta = 0.4115168$ or $\pi - 0.4115168$

15. $\tan \theta = 5$
$\quad \theta = 1.3734008$ or $\pi + 1.3734008$

17. $\cos \theta = -0.9$
$\quad \theta = 2.6905658$ or $2\pi - 2.6905658$

19. $\sec \theta = -4$. Therefore, $\cos \theta = -\dfrac{1}{4}$
$\quad \theta = 1.8234766$ or $2\pi - 1.8234766$

21. $\quad 2 \cos^2 \theta + \cos \theta = 0$
$\quad\quad \cos \theta(2 \cos \theta + 1) = 0$
$\quad \cos \theta = 0 \quad\quad$ or $\quad 2 \cos \theta + 1 = 0$

$\quad\quad\quad\quad\quad\quad\quad\quad\quad\quad \cos \theta = -\dfrac{1}{2}$

$\quad\quad \theta = \dfrac{\pi}{2}, \dfrac{3\pi}{2}$ or $\quad\quad\quad \theta = \dfrac{2\pi}{3}, \dfrac{4\pi}{3}$

23. $\quad 2 \sin^2 \theta - \sin \theta - 1 = 0$
$\quad (2 \sin \theta + 1)(\sin \theta - 1) = 0$
$\quad 2 \sin \theta + 1 = 0 \quad\quad$ or $\sin \theta - 1 = 0$

$\quad\quad \sin \theta = -\dfrac{1}{2} \quad\quad$ or $\quad\quad \sin \theta = 1$

$\quad\quad\quad \theta = \dfrac{7\pi}{6}, \dfrac{11\pi}{6}$ or $\quad\quad\quad \theta = \dfrac{\pi}{2}$

25. $\quad (\tan \theta - 1)(\sec \theta - 1) = 0$
$\quad\quad \tan \theta - 1 = 0 \quad\quad$ or $\quad \sec \theta - 1 = 0$
$\quad\quad\quad \tan \theta = 1 \quad\quad$ or $\quad\quad \sec \theta = 1$
$\quad\quad \theta = \dfrac{\pi}{4}, \dfrac{5\pi}{4} \quad\quad$ or $\quad\quad\quad \theta = 0$

27. $\cos \theta = \sin \theta$
$\quad \tan \theta = 1$
$\quad\quad \theta = \dfrac{\pi}{4}$ or $\theta = \dfrac{5\pi}{4}$

7 ANALYTIC TRIGONOMETRY

29. $\tan \theta = 2 \sin \theta$

$\dfrac{\sin \theta}{\cos \theta} = 2 \sin \theta$
$\sin \theta = 2 \sin \theta \cos \theta$
$\quad 0 = 2 \sin \theta \cos \theta - \sin \theta$
$\quad 0 = \sin \theta (2 \cos \theta - 1)$
$\qquad \sin \theta = 0 \qquad$ or $\qquad 2 \cos \theta - 1 = 0$
$\qquad\qquad\qquad\qquad\qquad\qquad \cos \theta = \dfrac{1}{2}$

$\theta = 0, \ \pi, \text{ or } \theta = \dfrac{\pi}{3}, \ \dfrac{5\pi}{3}$

31. $\sin \theta = \csc \theta$
$\sin \theta - \csc \theta = 0$

$\sin \theta - \dfrac{1}{\sin \theta} = 0$

$\dfrac{\sin^2 \theta - 1}{\sin \theta} = 0$
$\sin^2 \theta - 1 = 0$
$\cos^2 \theta = 0$
$\cos \theta = 0$
$\theta = \dfrac{\pi}{2} \quad$ or $\quad \theta = \dfrac{3\pi}{2}$

33. $\cos 2\theta = \cos \theta$
$2 \cos^2 \theta - 1 = \cos \theta$
$2 \cos^2 \theta - \cos \theta - 1 = 0$
$(\cos \theta - 1)(2 \cos \theta + 1) = 0$
$\cos \theta - 1 = 0 \qquad$ or $\qquad 2 \cos \theta + 1 = 0$
$\cos \theta = 1 \qquad$ or $\qquad \cos \theta = \dfrac{-1}{2}$

$\theta = 0, \ \dfrac{2\pi}{3}, \ \dfrac{4\pi}{3}$

35. $\sin 2\theta + \sin 4\theta = 0$
$2 \sin \theta \cos \theta + 2 \sin 2\theta \cos 2\theta = 0$
$\sin 2\theta (1 + 2 \cos 2\theta) = 0$
$\sin 2\theta = 0 \qquad$ or $\qquad 1 + 2 \cos 2\theta = 0$

$2\theta = 0 + 2k\pi, \ 2\theta = \pi + 2k\pi, \text{ or } \cos 2\theta = -\dfrac{1}{2}$

$2\theta = 0 + 2k\pi, \ 2\theta = \pi + 2k\pi, \text{ or } 2\theta = \dfrac{2\pi}{3} + 2k\pi, \ 2\theta = \dfrac{4\pi}{3} + 2k\pi$

$\theta = k\pi, \ \theta = \dfrac{\pi}{2} + k\pi \text{ or } \theta = \dfrac{\pi}{3} + k\pi, \ \theta = \dfrac{2\pi}{3} + k\pi, \ k \text{ any integer}$

The solutions on the interval $[0, \ 2\pi)$ are

$\theta = 0, \ \dfrac{\pi}{2}, \ \dfrac{\pi}{3}, \ \dfrac{2\pi}{3}, \ \dfrac{4\pi}{3}, \ \pi, \ \dfrac{5\pi}{3}, \ \dfrac{3\pi}{2}$

7.5 TRIGONOMETRIC EQUATIONS   379

37. $\cos 4\theta - \cos 6\theta = 0$

$-2 \sin \dfrac{10\theta}{2} \sin \dfrac{-2\theta}{2} = 0$

$\sin 5\theta = 0 \quad$ or $\sin \theta = 0$

$\qquad 5\theta = k\pi \qquad\qquad \theta = k\pi$

$\theta = k\dfrac{\pi}{5},\ k$ any integer

The solutions on the interval $[0,\ 2\pi)$ are

$\theta = 0, \quad \dfrac{\pi}{5},\ \dfrac{2\pi}{5},\ \dfrac{3\pi}{5},\ \dfrac{4\pi}{5},\ \pi,\ \dfrac{6\pi}{5},\ \dfrac{7\pi}{5},\ \dfrac{8\pi}{5},\ \dfrac{9\pi}{5}$

39.
$$1 + \sin \theta = 2 \cos^2 \theta$$
$$1 + \sin \theta = 2(1 - \sin^2 \theta)$$
$$1 + \sin \theta = 2 - 2 \sin^2 \theta$$
$$2 \sin^2 \theta + \sin \theta - 1 = 0$$
$$(2 \sin \theta - 1)(\sin \theta + 1) = 0$$
$2 \sin \theta - 1 = 0 \quad$ or $\quad \sin \theta + 1 = 0$

$\qquad \sin \theta = \dfrac{1}{2} \quad$ or $\qquad \sin \theta = -1$

$\theta = \dfrac{\pi}{6},\ \dfrac{5\pi}{6},\ \dfrac{3\pi}{2}$

41.
$$\tan^2 \theta = \dfrac{3}{2} \sec \theta$$
$$\sec^2 \theta - 1 = \dfrac{3}{2} \sec \theta$$
$$\sec^2 \theta - \dfrac{3}{2} \sec \theta - 1 = 0$$
$$\left(\sec \theta + \dfrac{1}{2}\right)(\sec \theta - 2) = 0$$

$\sec \theta = \dfrac{-1}{2} \quad$ or $\quad \sec \theta = 2$

$\cos \theta = -2 \quad$ or $\quad \cos \theta = \dfrac{1}{2}$

For any angle θ, $-1 \le \cos \theta \le 1$; thus, $\cos \theta = -2$ has no solution.

The solutions of $\cos \theta = \dfrac{1}{2}$ are $\theta = \dfrac{\pi}{3},\ \dfrac{5\pi}{3}$

43.
$$3 - \sin \theta = \cos 2\theta$$
$$3 - \sin \theta = 1 - 2 \sin^2 \theta$$
$$2 \sin^2 \theta - \sin \theta + 2 = 0$$

This is a quadratic equation in $\sin \theta$. The discriminant is $b^2 - 4ac = 1 - 16 = -15 < 0$. The equation, therefore, has no real solutions.

45.
$$\sec^2 \theta + \tan \theta = 0$$
$$(\tan^2 \theta + 1) + \tan \theta = 0$$
$$\tan^2 + \tan \theta + 1 = 0$$

The discriminate is $b^2 - 4ac = 1 - 4 = -3 < 0$. The equation, therefore, has no real solutions.

7 ANALYTIC TRIGONOMETRY

47. $\sin \theta - \sqrt{3} \cos \theta = 1$

We divide each side of the equation by 2. Then,

$$\frac{1}{2} \sin \theta - \frac{\sqrt{3}}{2} \cos \theta = \frac{1}{2}$$

There is a unique angle ϕ, $0 \le \phi < 2\pi$, for which

$$\cos \phi = \frac{1}{2} \text{ and } \sin \phi = \frac{\sqrt{3}}{2}$$
$$\phi = \frac{\pi}{3}$$

Thus, the equation may be written as

$$\sin \theta \cos \phi - \cos \theta \sin \phi = \frac{1}{2}$$

or $\qquad \sin(\theta - \phi) = \frac{1}{2}$

$$\theta - \phi = \frac{\pi}{6} \quad \text{or} \quad \theta - \phi = \frac{5\pi}{6}$$
$$\theta - \frac{\pi}{3} = \frac{\pi}{6} \quad \text{or} \quad \theta - \frac{\pi}{3} = \frac{5\pi}{6}$$
$$\theta = \frac{\pi}{2} \quad \text{or} \quad \theta = \frac{7\pi}{6}$$

49.
$$\tan 2\theta + 2 \sin \theta = 0$$
$$\frac{\sin 2\theta}{\cos 2\theta} + 2 \sin \theta = 0$$
$$\frac{2 \sin \theta \cos \theta + 2 \sin \theta \cos 2\theta}{\cos 2\theta} = 0$$
$$2 \sin \theta (\cos \theta + \cos 2\theta) = 0$$
$$2 \sin \theta [\cos \theta + (2 \cos^2 \theta - 1)] = 0$$
$$2 \sin \theta (2 \cos^2 \theta + \cos \theta - 1) = 0$$
$$2 \sin \theta (2 \cos \theta - 1)(\cos \theta + 1) = 0$$
$$2 \sin \theta = 0 \quad \text{or} \quad 2 \cos \theta - 1 = 0 \quad \text{or} \quad \cos \theta + 1 = 0$$
$$\sin \theta = 0 \quad \text{or} \quad \cos \theta = \frac{1}{2} \text{ or} \quad \cos \theta = -1$$

$$\theta = 0, \frac{\pi}{3}, \pi, \frac{5\pi}{3}$$

51.

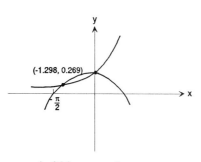

$x = -1.298, \ x = 0$

53.

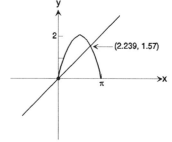

$x = 0, \ x = 2.239$

7.5 TRIGONOMETRIC EQUATIONS

55.

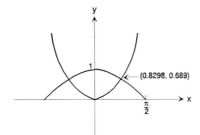

$x = -0.8298, \quad x = 0.8298$

57. The ratio $\dfrac{v_1}{v_2}$ is the index of refraction. The index of refraction of light in passing from a vacuum into water is 1.33.

Therefore, $\dfrac{v_1}{v_2} = 1.33$.

The angle of incidence, θ_1, is 40°.

Because $\dfrac{\sin \theta_1}{\sin \theta_2} = \dfrac{v_1}{v_2}$, $\dfrac{\sin 40°}{\sin \theta_2} = 1.33$.

The angle of refraction, θ_2, can be found as follows:

$$\frac{\sin 40°}{\sin \theta_2} = 1.33$$
$$\sin 40° = 1.33 \sin \theta_2$$
$$\frac{\sin 40°}{1.33} = \sin \theta_2$$
$$0.4833 = \sin \theta_2$$
$$\sin^{-1} 0.4833 = \sin_2$$
$$28.9° = \theta_2$$

59. To agree with Snell's Law, the measured values in the table for the angle of incidence θ and the angle of refraction θ_2 for a light beam passing from air into water must have the proportion of $\dfrac{\sin \theta_1}{\sin \theta_2}$ which is $\dfrac{v_1}{v_2}$, the index of refraction.

For $\theta = 10°$ and $\theta_2 = 7° \, 45'$, $\dfrac{\sin 10°}{\sin 7°45'} = \dfrac{\sin 10°}{\sin 7.75°}$.

Now, using a calculator, $\dfrac{\sin 10°}{\sin 7.75°} \approx 1.2877$.

For $\theta_1 = 20°$ and $\theta_2 = 15° \, 30'$, $\dfrac{\sin 20°}{\sin 15° \, 30'} = \dfrac{\sin 20°}{\sin 15.5°} \approx 1.2798$.

For $\theta_1 = 30°$ and $\theta_2 = 22° \, 30'$, $\dfrac{\sin 30°}{\sin 22° \, 30'} = \dfrac{\sin 30°}{\sin 22.5°} \approx 1.3066$.

For $\theta_1 = 40°$ and $\theta_2 = 29.0'$, $\dfrac{\sin 40°}{\sin 29°} \approx 1.3259$.

For $\theta_1 = 50°$ and $\theta_2 = 35° 0'$, $\dfrac{\sin 50°}{\sin 35°} \approx 1.3356$.

For $\theta_1 = 60°$ and $\theta_2 = 40° 30'$, $\dfrac{\sin 60°}{\sin 40° 30'} = \dfrac{\sin 60°}{\sin 40.5°} \approx 1.3335$.

For $\theta_1 = 70°$ and $\theta_2 = 45° 30'$, $\dfrac{\sin 70°}{\sin 45° 30'} = \dfrac{\sin 70°}{\sin 45.5°} \approx 1.3175$.

For $\theta_1 = 80°$ and $\theta_2 = 50° 0'$, $\dfrac{\sin 80°}{\sin 50°} \approx 1.2856$.

These values do agree with Snell's Law, and the index of refraction varies from 1.27 to 1.34.

61. $\dfrac{\sin \theta_1}{\sin \theta_2} = \dfrac{\sin 40°}{\sin 26°} = \dfrac{v_1}{v_2}$, the index of refraction

$\dfrac{\sin 40°}{\sin 26°} = \dfrac{v_1}{v_2} = 1.47$

The index of refraction of a beam of light of wave length 589 nanometers traveling in air is 1.47.

63. If θ is the original angle of incidence and ϕ is the angle of refraction, then $\dfrac{\sin \theta}{\sin \phi} = n_2$. The angle of incidence of the emerging beam is also ϕ, and the index of refraction is $\dfrac{1}{n_2}$. Thus, θ is the angle of refraction of the emerging beam.

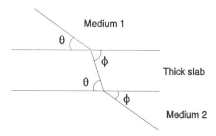

≡ 7 - CHAPTER REVIEW

1. $\tan \theta \cot \theta - \sin^2 \theta = 1 - \sin^2 \theta$
    $= \cos^2 \theta$

3. $\cos^2 \theta (1 + \tan^2 \theta) = (\cos^2 \theta)(\sec^2 \theta)$
    $= 1$

5. $4 \cos^2 \theta + 3 \sin^2 \theta = \cos^2 \theta + 3(\cos^2 \theta + \sin^2 \theta)$
    $= 3 + \cos^2 \theta$

7.
$$\frac{1 - \cos\theta}{\sin\theta} + \frac{\sin\theta}{1 - \cos\theta} = \frac{(1 - \cos\theta)^2 + \sin^2\theta}{\sin\theta(1 - \cos\theta)}$$

$$= \frac{1 - 2\cos\theta + \cos^2\theta + \sin^2\theta}{\sin\theta(1 - \cos\theta)}$$

$$= \frac{2 - 2\cos\theta}{\sin\theta(1 - \cos\theta)}$$

$$= \frac{2(1 - \cos\theta)}{\sin\theta(1 - \cos\theta)}$$

$$= \frac{2}{\sin\theta}$$

$$= 2\csc\theta$$

9.
$$\frac{\cos\theta}{\cos\theta - \sin\theta} = \frac{\dfrac{\cos\theta}{\cos\theta}}{\dfrac{\cos\theta - \sin\theta}{\cos\theta}}$$

$$= \frac{1}{1 - \dfrac{\sin\theta}{\cos\theta}}$$

$$= \frac{1}{1 - \tan\theta}$$

11.
$$\frac{\csc\theta}{1 + \csc\theta} = \frac{\dfrac{1}{\sin\theta}}{1 + \dfrac{1}{\sin\theta}}$$

$$= \frac{\dfrac{1}{\sin\theta}}{\dfrac{\sin\theta + 1}{\sin\theta}}$$

$$= \frac{1}{1 + \sin\theta}$$

$$= \frac{1}{1 + \sin\theta} \cdot \frac{1 - \sin\theta}{1 - \sin\theta}$$

$$= \frac{1 - \sin\theta}{1 - \sin^2\theta}$$

$$= \frac{1 - \sin\theta}{\cos^2\theta}$$

13.
$$\csc\theta - \sin\theta = \frac{1}{\sin\theta} - \sin\theta$$

$$= \frac{1 - \sin^2\theta}{\sin\theta}$$

$$= \frac{\cos^2\theta}{\sin\theta}$$

$$= \cos\theta \cdot \frac{\cos\theta}{\sin\theta}$$

$$= \cos\theta \cot\theta$$

15. $\dfrac{1 - \sin \theta}{\sec \theta} = \cos \theta (1 - \sin \theta$

$\qquad = (\cos \theta)(1 - \sin \theta) \cdot \dfrac{1 + \sin \theta}{1 + \sin \theta}$

$\qquad = \dfrac{\cos \theta (1 - \sin^2 \theta)}{1 + \sin \theta}$

$\qquad = \dfrac{(\cos \theta)(\cos^2 \theta)}{1 + \sin \theta}$

$\qquad = \dfrac{\cos^3 \theta}{1 + \sin \theta}$

17. $\cot \theta - \tan \theta = \dfrac{\cos \theta}{\sin \theta} - \dfrac{\sin \theta}{\cos \theta}$

$\qquad = \dfrac{\cos^2 \theta - \sin^2 \theta}{\sin \theta \cos \theta}$

$\qquad = \dfrac{1 - \sin^2 \theta - \sin^2 \theta}{\sin \theta \cos \theta}$

$\qquad = \dfrac{1 - 2 \sin^2 \theta}{\sin \theta \cos \theta}$

19. $\dfrac{\cos(\alpha + \beta)}{\cos \alpha \sin \beta} = \dfrac{\cos \alpha \cos \beta - \sin \alpha \sin \beta}{\cos \alpha \sin \beta}$

$\qquad = \dfrac{\cos \alpha \cos \beta}{\cos \alpha \sin \beta} - \dfrac{\sin \alpha \sin \beta}{\cos \alpha \sin \beta}$

$\qquad = \dfrac{\cos \beta}{\sin \beta} - \dfrac{\sin \alpha}{\cos \alpha}$

$\qquad = \cot \beta - \tan \alpha$

21. $\dfrac{\cos(\alpha - \beta)}{\cos \alpha \cos \beta} = \dfrac{\cos \alpha \cos \beta + \sin \alpha \sin \beta}{\cos \alpha \cos \beta}$

$\qquad = \dfrac{\cos \alpha \cos \beta}{\cos \alpha \cos \beta} + \dfrac{\sin \alpha \sin \beta}{\cos \alpha \cos \beta}$

$\qquad = 1 + \tan \alpha \tan \beta$

23. $(1 + \cos \theta)\left(\tan \dfrac{\theta}{2}\right) = \left(2 \cos^2 \dfrac{\theta}{2}\right) \dfrac{\sin \dfrac{\theta}{2}}{\cos \dfrac{\theta}{2}}$

$\qquad = 2 \sin \dfrac{\theta}{2} \cos \dfrac{\theta}{2}$

$\qquad = \sin \theta$

25. $2 \cot \theta \cot 2\theta = 2\left(\dfrac{\cos \theta}{\sin \theta}\right)\left(\dfrac{\cos 2\theta}{\sin 2\theta}\right)$

$\qquad = \dfrac{2 \cos \theta (\cos^2 \theta - \sin^2 \theta)}{\sin \theta (2 \sin \theta \cos \theta)}$

$\qquad = \dfrac{2 \cos \theta (\cos^2 \theta - \sin^2 \theta}{(2 \cos \theta) \sin^2 \theta}$

$\qquad = \dfrac{\cos^2 \theta - \sin^2 \theta}{\sin^2 \theta}$

$\qquad = \cot^2 \theta - 1$

27. $1 - 8 \sin^2 \theta \cos^2 \theta = 1 - 2(2 \sin \theta \cos \theta)^2$
$$= 1 - 2 \sin^2 2\theta$$
$$= \cos 2(2\theta)$$
$$= \cos 4\theta$$

29. $\dfrac{\sin 2\theta + \sin 4\theta}{\cos 2\theta + \cos 4\theta} = \dfrac{2 \sin 3\theta \cos(-\theta)}{2 \cos 3\theta \cos(-\theta)}$
$$= \dfrac{\sin 3\theta}{\cos 3\theta}$$
$$= \tan 3\theta$$

31. $\dfrac{\cos 2\theta - \cos 4\theta}{\cos 2\theta + \cos 4\theta} - \tan \theta \tan 3\theta = \dfrac{-2 \sin 3\theta \sin(-\theta)}{2 \cos 3\theta \cos(-\theta)} - \tan \theta \tan 3\theta$
$$= -\tan 3\theta \tan(-\theta) - \tan \theta \tan 3\theta$$
$$= \tan 3\theta \tan \theta - \tan \theta \tan 3\theta$$
$$= 0$$

33. $\sin 165° = \sin(30° + 135°)$
$$= \sin 30° \cos 135° + \cos 30° \sin 135°$$
$$= \frac{1}{2} \cdot \frac{-\sqrt{2}}{2} + \frac{\sqrt{3}}{2} \cdot \frac{\sqrt{2}}{2}$$
$$= \frac{-\sqrt{2}\left(1 - \sqrt{3}\right)}{4} = \frac{\sqrt{2}}{4}\left(\sqrt{3} - 1\right) = \frac{1}{4}\left(\sqrt{6} - \sqrt{2}\right)$$

35. $\cos \dfrac{5\pi}{12} = \cos\left(\dfrac{\pi}{6} + \dfrac{\pi}{4}\right)$
$$= \cos \frac{\pi}{6} \cos \frac{\pi}{4} - \sin \frac{\pi}{6} \sin \frac{\pi}{4}$$
$$= \frac{\sqrt{3}}{2} \cdot \frac{\sqrt{2}}{2} - \frac{1}{2} \cdot \frac{\sqrt{2}}{2}$$
$$= \frac{\sqrt{2}}{4}\left(\sqrt{3} - 1\right) = \frac{1}{4}\left(\sqrt{6} - \sqrt{2}\right)$$

37. $\cos 80° \cos 20° + \sin 80° \sin 20° = \cos(80° - 20°)$
$$= \cos 60°$$
$$= \frac{1}{2}$$

39. $\tan \dfrac{\pi}{8} = \tan \dfrac{\frac{\pi}{4}}{2} = \sqrt{\dfrac{1 - \cos \frac{\pi}{4}}{1 + \cos \frac{\pi}{4}}} = \sqrt{\dfrac{1 - \frac{\sqrt{2}}{2}}{1 + \frac{\sqrt{2}}{2}}} = \sqrt{\dfrac{\frac{2 - \sqrt{2}}{2}}{\frac{2 + \sqrt{2}}{2}}}$

$$= \sqrt{\dfrac{2 - \sqrt{2}}{2 + \sqrt{2}}} = \sqrt{\dfrac{2 - \sqrt{2}}{2 + \sqrt{2}} \cdot \left(\dfrac{2 - \sqrt{2}}{2 - \sqrt{2}}\right)}$$

$$= \sqrt{\dfrac{\left(2 - \sqrt{2}\right)^2}{2}} = \dfrac{2 - \sqrt{2}}{\sqrt{2}}\left(\dfrac{\sqrt{2}}{\sqrt{2}}\right)$$

$$= \dfrac{2\sqrt{2} - 2}{2} = \sqrt{2} - 1$$

7 ANALYTIC TRIGONOMETRY

41. $\sin \alpha = \frac{4}{5}, \ 0 < \alpha < \frac{\pi}{2}; \ \sin \beta = \frac{5}{13}, \ \frac{\pi}{2} < \beta < \pi$

Therefore, $\cos \alpha = \frac{3}{5}, \ \cos \beta = \frac{-12}{13}, \ 0 < \frac{\alpha}{2} < \frac{\pi}{4}, \ \frac{\pi}{4} < \frac{\beta}{2} < \frac{\pi}{2}$

(a) $\sin(\alpha + \beta) = \sin \alpha \cos \beta + \cos \alpha \sin \beta$

$= \frac{4}{5} \cdot \frac{-12}{13} + \frac{3}{5} \cdot \frac{5}{13}$

$= \frac{-48 + 15}{65} = \frac{-33}{65}$

(b) $\cos(\alpha + \beta) = \cos \alpha \cos \beta + \sin \alpha \sin \beta$

$= \frac{3}{5} \cdot \frac{-12}{13} - \frac{4}{5} \cdot \frac{5}{13}$

$= \frac{-36 - 20}{65} = \frac{-56}{65}$

(c) $\sin(\alpha - \beta) = \sin \alpha \cos \beta - \cos \alpha \sin \beta$

$= \frac{4}{5} \cdot \frac{-12}{13} - \frac{3}{5} \cdot \frac{5}{13}$

$= \frac{-48 - 15}{65} = \frac{-63}{65}$

(d) $\tan(\alpha + \beta) = \dfrac{\tan \alpha + \tan \beta}{1 - \tan \alpha \tan \beta}$

$= \dfrac{\frac{4}{3} + \frac{-5}{12}}{1 + \frac{5}{9}}$

$= \dfrac{\frac{11}{12}}{\frac{14}{9}} = \frac{33}{56}$

(e) $\sin 2\alpha = 2 \sin \alpha \cos \alpha$

$= 2 \cdot \frac{4}{5} \cdot \frac{3}{5}$

$= \frac{24}{25}$

(f) $\cos 2\beta = 1 - 2 \sin^2 \beta$

$= 1 - 2\left(\frac{5}{13}\right)^2$

$= 1 - \frac{50}{169}$

$= \frac{119}{169}$

(g) $\sin \dfrac{\beta}{2} = \sqrt{\dfrac{1 - \cos \beta}{2}}$

$= \sqrt{\dfrac{1 - \frac{-12}{13}}{2}}$

$= \sqrt{\dfrac{\frac{25}{13}}{2}}$

$= \sqrt{\dfrac{25}{26}} = \frac{5\sqrt{26}}{26}$

(h) $\cos \frac{\alpha}{2} = \sqrt{\frac{1 + \cos \alpha}{2}}$

$= \sqrt{\frac{1 + \frac{3}{5}}{2}}$

$= \sqrt{\frac{\frac{8}{5}}{2}}$

$= \sqrt{\frac{4}{5}} = \frac{2\sqrt{5}}{5}$

43. $\sin \alpha = -\frac{3}{5}$, $\pi < \alpha < \frac{3\pi}{2}$; $\cos \beta = \frac{12}{13}$, $\frac{3\pi}{2} < \beta < 2\pi$

Therefore, $\cos \alpha = \frac{-4}{5}$, $\tan \alpha = \frac{3}{4}$, $\frac{\pi}{2} < \frac{\alpha}{2} < \frac{3\pi}{4}$, $\frac{3\pi}{4} < \frac{\beta}{2} < \pi$

$\sin \beta = \frac{-5}{13}$, $\tan \beta = \frac{-5}{12}$

(a) $\sin(\alpha + \beta) = \sin \alpha \cos \beta + \cos \alpha \sin \beta$

$= \frac{-3}{5} \cdot \frac{12}{13} + \frac{-4}{5} \cdot \frac{-5}{13}$

$= \frac{-36 + 20}{65} = \frac{-16}{65}$

(b) $\cos(\alpha + \beta) = \cos \alpha \cos \beta - \sin \alpha \sin \beta$

$= \frac{-4}{5} \cdot \frac{12}{13} - \frac{3}{5} \cdot \frac{-5}{13}$

$= \frac{-48 - 15}{65} = \frac{-63}{65}$

(c) $\sin(\alpha - \beta) = \sin \alpha \cos \beta - \cos \alpha \sin \beta$

$= \frac{-3}{5} \cdot \frac{12}{13} - \frac{-4}{5} \cdot \frac{-5}{13}$

$= \frac{-36 - 20}{65} = \frac{-56}{65}$

(d) $\tan(\alpha + \beta) = \frac{\tan \alpha + \tan \beta}{1 - \tan \alpha \tan \beta}$

$= \frac{\frac{3}{4} - \frac{5}{12}}{1 + \frac{5}{16}}$

$= \frac{\frac{1}{3}}{\frac{21}{16}} = \frac{16}{63}$

(e) $\sin 2\alpha = 2 \sin \alpha \cos \alpha$

$= 2 \cdot \frac{-3}{5} \cdot \frac{-4}{5}$

$= \frac{24}{25}$

(f) $\cos 2\beta = 2 \cos^2 \beta - 1$

$$= 2\left(\frac{12}{13}\right)^2 - 1$$

$$= \frac{288}{169} - 1$$

$$= \frac{119}{169}$$

(g) $\sin \frac{\beta}{2} = \sqrt{\frac{1 - \cos \beta}{2}}$

$$= \sqrt{\frac{1 - \frac{12}{13}}{2}}$$

$$= \sqrt{\frac{\frac{1}{13}}{2}} = \sqrt{\frac{1}{26}} = \frac{\sqrt{26}}{26}$$

(h) $\cos \frac{\alpha}{2} = -\sqrt{\frac{1 + \cos \alpha}{2}}$

$$= -\sqrt{\frac{1 - \frac{4}{5}}{2}}$$

$$= -\sqrt{\frac{\frac{1}{5}}{2}}$$

$$= -\sqrt{\frac{1}{10}} = \frac{-\sqrt{10}}{10}$$

45. $\tan \alpha = \frac{3}{4}$, $\pi < \alpha < \frac{3\pi}{2}$; $\tan \beta = \frac{12}{5}$, $0 < \beta < \frac{\pi}{2}$

Therefore, $\sin \alpha = \frac{-3}{5}$, $\cos \alpha = \frac{-4}{5}$

$\sin \beta = \frac{12}{13}$, $\cos \beta = \frac{5}{13}$, $\frac{\pi}{2} < \frac{\alpha}{2} < \frac{3\pi}{4}$, $0 < \frac{\beta}{2} < \frac{\pi}{4}$

(a) $\sin(\alpha + \beta) = \sin \alpha \cos \beta + \cos \alpha \sin \beta$

$$= \frac{-3}{5} \cdot \frac{5}{13} + \frac{-4}{5} \cdot \frac{12}{13}$$

$$= \frac{-15 - 48}{65} = \frac{-63}{65}$$

(b) $\cos(\alpha + \beta) = \cos \alpha \cos \beta - \sin \alpha \sin \beta$

$$= \frac{-4}{5} \cdot \frac{5}{13} - \frac{-3}{5} \cdot \frac{12}{13}$$

$$= \frac{-20 + 36}{65} = \frac{16}{65}$$

(c) $\sin(\alpha - \beta) = \sin \alpha \cos \beta - \cos \alpha \sin \beta$

$$= \frac{-3}{5} \cdot \frac{5}{13} - \frac{-4}{5} \cdot \frac{12}{13}$$

$$= \frac{-15 + 48}{65} = \frac{33}{65}$$

(d) $\tan(\alpha + \beta) = \dfrac{\tan \alpha + \tan \beta}{1 - \tan \alpha \tan \beta}$

$$= \dfrac{\dfrac{3}{4} + \dfrac{12}{5}}{1 - \dfrac{9}{5}}$$

$$= \dfrac{\dfrac{15 + 48}{20}}{\dfrac{-4}{5}}$$

$$= \dfrac{63}{20}\left(-\dfrac{5}{4}\right) = \dfrac{-63}{16}$$

(e) $\sin 2\alpha = 2 \sin \alpha \cos \alpha$

$$= 2\left(\dfrac{-3}{5}\right)\left(\dfrac{-4}{5}\right)$$

$$= \dfrac{24}{25}$$

(f) $\cos 2\beta = \cos^2 \beta - \sin^2 \beta$

$$= \left(\dfrac{5}{13}\right)^2 - \left(\dfrac{12}{13}\right)^2$$

$$= \dfrac{25 - 144}{169} = \dfrac{-119}{169}$$

(g) $\sin \dfrac{\beta}{2} = \sqrt{\dfrac{1 - \cos \beta}{2}}$

$$= \sqrt{\dfrac{1 - \dfrac{5}{13}}{2}}$$

$$= \sqrt{\dfrac{8}{26}} = \dfrac{4\sqrt{13}}{26} = \dfrac{2\sqrt{13}}{13}$$

(h) $\cos \dfrac{\alpha}{2} = -\sqrt{\dfrac{1 + \cos \alpha}{2}}$

$$= -\sqrt{\dfrac{1 + \dfrac{-4}{5}}{2}}$$

$$= -\sqrt{\dfrac{1}{10}} = \dfrac{-\sqrt{10}}{10}$$

47. $\sec \alpha = 2$, $\dfrac{-\pi}{2} < \alpha < 0$; $\sec \beta = 3$, $\dfrac{3\pi}{2} < \beta < 2\pi$

Therefore, $\sin \alpha = \dfrac{-\sqrt{3}}{2}$, $\cos \alpha = \dfrac{1}{2}$, $\dfrac{-\pi}{4} < \dfrac{\alpha}{2} < 0$, $\dfrac{3\pi}{4} < \dfrac{\beta}{2} < \pi$

$\sin \beta = \dfrac{-2\sqrt{2}}{3}$, $\cos \beta = \dfrac{1}{3}$

(a) $\sin(\alpha + \beta) = \sin \alpha \cos \beta + \cos \alpha \sin \beta$

$$= \left(\frac{-\sqrt{3}}{2}\right)\left(\frac{1}{3}\right) + \left(\frac{1}{2}\right)\left(\frac{-2\sqrt{2}}{3}\right)$$

$$= \frac{-\sqrt{3}}{6} + \frac{-2\sqrt{2}}{6}$$

$$= \frac{-\left(\sqrt{3} + 2\sqrt{2}\right)}{6} = \frac{-\sqrt{3} - 2\sqrt{2}}{6}$$

(b) $\cos(\alpha + \beta) = \cos \alpha \cos \beta + \sin \alpha \sin \beta$

$$= \left(\frac{1}{2}\right)\left(\frac{1}{3}\right) - \left(\frac{-\sqrt{3}}{2}\right)\left(\frac{-2\sqrt{2}}{3}\right)$$

$$= \frac{1}{6} - \frac{2\sqrt{6}}{6}$$

$$= \frac{1 - 2\sqrt{6}}{6}$$

(c) $\sin(\alpha - \beta) = \sin \alpha \cos \beta - \cos \alpha \sin \beta$

$$= \left(\frac{-\sqrt{3}}{2}\right)\left(\frac{1}{3}\right) - \left(\frac{1}{2}\right)\left(\frac{-2\sqrt{2}}{3}\right)$$

$$= \frac{-\sqrt{3}}{6} + \frac{2\sqrt{2}}{6}$$

$$= \frac{2\sqrt{2} - \sqrt{3}}{6}$$

(d) $\tan(\alpha + \beta) = \dfrac{\tan \alpha + \tan \beta}{1 - \tan \alpha \tan \beta}$

$$= \frac{-\sqrt{3} + -2\sqrt{2}}{1 - \left(-\sqrt{3}\right)\left(-2\sqrt{2}\right)}$$

$$= \frac{-\sqrt{3} - 2\sqrt{2}}{1 - 2\sqrt{6}}$$

$$= \frac{-\sqrt{3} - 2\sqrt{2}}{1 - 2\sqrt{6}} \cdot \frac{1 + 2\sqrt{6}}{1 + 2\sqrt{6}}$$

$$= \frac{8\sqrt{2} + 9\sqrt{3}}{23}$$

(e) $\sin 2\alpha = 2 \sin \alpha \cos \alpha$

$$= 2\left(\frac{-\sqrt{3}}{2}\right)\left(\frac{1}{2}\right)$$

$$= \frac{-2\sqrt{3}}{4} = \frac{-\sqrt{3}}{2}$$

(f) $\cos 2\beta = \cos^2 \beta - \sin^2 \beta$

$$= \left(\frac{1}{3}\right)^2 - \left(\frac{-2\sqrt{2}}{3}\right)^2$$

$$= \frac{1}{9} - \frac{8}{9} = \frac{-7}{9}$$

(g) $\sin \frac{\beta}{2} = \sqrt{\frac{1 - \cos \beta}{2}}$

$= \sqrt{\frac{1 - \frac{1}{3}}{2}}$

$= \sqrt{\frac{2}{6}} = \frac{\sqrt{3}}{3}$

(h) $\cos \frac{\alpha}{2} = \sqrt{\frac{1 + \cos \alpha}{2}}$

$= \sqrt{\frac{1 + \frac{1}{2}}{2}}$

$= \sqrt{\frac{3}{4}} \frac{\sqrt{3}}{2}$

49. $\sin \alpha = -\frac{2}{3}, \ \pi < \alpha < \frac{3\pi}{2}; \ \cos \beta = \frac{-2}{3}, \ \pi < \beta < \frac{3\pi}{2}$

Therefore, $\cos \alpha = \frac{-\sqrt{5}}{3}, \ \sin \beta = \frac{-\sqrt{5}}{3}, \ \frac{\pi}{2} < \frac{\alpha}{2} < \frac{3\pi}{4}, \ \frac{\pi}{2} < \frac{\beta}{2} < \frac{3\pi}{4}$

(a) $\sin(\alpha + \beta) = \sin \alpha \cos \beta + \cos \alpha \sin \beta$

$= \left(\frac{-2}{3}\right)\left(\frac{-2}{3}\right) + \left(\frac{-\sqrt{5}}{3}\right)\left(\frac{-\sqrt{5}}{3}\right)$

$= \frac{4 + 5}{9} = 1$

(b) $\cos(\alpha + \beta) = \cos \alpha \cos \beta + \sin \alpha \sin \beta$

$= \left(\frac{-\sqrt{5}}{3}\right)\left(\frac{-2}{3}\right) + \left(\frac{-2}{3}\right)\left(\frac{-\sqrt{5}}{3}\right)$

$= \frac{2\sqrt{5} - 2\sqrt{5}}{9} = 0$

(c) $\sin(\alpha - \beta) = \sin \alpha \cos \beta - \cos \alpha \sin \beta$

$= \left(\frac{-2}{3}\right)\left(\frac{-2}{3}\right) - \left(\frac{-\sqrt{5}}{3}\right)\left(\frac{-\sqrt{5}}{3}\right)$

$= \frac{4 - 5}{9} = \frac{-1}{9}$

(d) $\tan(\alpha + \beta) = \frac{\tan \alpha + \tan \beta}{1 - \tan \alpha \tan \beta}$

$= \dfrac{\dfrac{2}{\sqrt{5}} + \dfrac{\sqrt{5}}{2}}{1 - \left(\dfrac{2}{\sqrt{5}}\right)\left(\dfrac{\sqrt{5}}{2}\right)}$

Not defined.

7 ANALYTIC TRIGONOMETRY

(e) $\sin 2\alpha = 2 \sin \alpha \cos \alpha$

$$= 2\left(\frac{-2}{3}\right)\left(\frac{-\sqrt{5}}{3}\right)$$

$$= \frac{4\sqrt{5}}{9}$$

(f) $\cos 2\beta = 2 \cos^2 \beta - 1$

$$= 2\left(\frac{-2}{3}\right)^2 - 1$$

$$= -\frac{1}{9}$$

(g) $\sin \frac{\beta}{2} = \sqrt{\frac{1 - \cos \beta}{2}}$

$$= \sqrt{\frac{1 + \frac{2}{3}}{2}}$$

$$= \sqrt{\frac{5}{6}} = \frac{\sqrt{30}}{6}$$

(h) $\cos \frac{\alpha}{2} = -\sqrt{\frac{1 + \cos \alpha}{2}}$

$$= -\sqrt{\frac{1 - \frac{\sqrt{5}}{3}}{2}}$$

$$= -\sqrt{\frac{3 - \sqrt{5}}{6}} = \frac{-\sqrt{6}\sqrt{3 - \sqrt{5}}}{6}$$

51. $\cos \theta = \frac{1}{2}$

On the interval $[0, 2\pi)$, the solutions are $\theta = \frac{\pi}{3}, \frac{5\pi}{3}$

53. $\cos \theta = \frac{-\sqrt{2}}{2}$

On the interval $[0, 2\pi)$, the solutions are $\theta = \frac{3\pi}{4}$, or $\theta = \frac{5\pi}{4}$

55. $\sin 2\theta = -1$

$$2\theta = \frac{3\pi}{2} + 2k\pi, \ k \text{ any integer}$$

$$\theta = \frac{3\pi}{4} + k\pi, \ k \text{ any integer}$$

On the interval $[0, 2\pi)$, the solutions are $\theta = \frac{3\pi}{4}, \frac{7\pi}{4}$

57. $\tan 2\theta = 0$

$2\theta = 0 + k\pi$, k any integer

$\theta = \dfrac{k\pi}{2}$, k any integer

On the interval $[0, 2\pi)$, the solutions are $\theta = 0$, $\dfrac{\pi}{2}, \pi$, and $\dfrac{3\pi}{2}$

59. $\sin \theta = 0.9$

$\theta = 1.1197695$ or $\theta = \pi - 1.1197695$

61.
$$\sin \theta = \tan \theta$$
$$\sin \theta = \frac{\sin \theta}{\cos \theta}$$
$$\sin \theta - \frac{\sin \theta}{\cos \theta} = 0$$
$$\sin \theta \left(1 - \frac{1}{\cos \theta}\right) = 0$$
$$\sin \theta \left(\frac{\cos \theta - 1}{\cos \theta}\right) = 0$$

$\sin \theta = 0$ or $\cos \theta - 1 = 0$

$\sin \theta = 0$ or $\cos \theta = 1$

$\theta = 0, \pi$

63.
$$\sin \theta + \sin 2\theta = 0$$
$$\sin \theta + 2 \sin \theta \cos \theta = 0$$
$$\sin \theta (1 + 2 \cos \theta) = 0$$

$\sin \theta = 0$ or $1 + 2 \cos \theta = 0$

$$\cos \theta = \frac{-1}{2}$$

$\theta = 0, \dfrac{2\pi}{3}, \dfrac{4\pi}{3}, \pi$

65.
$$\sin 2\theta - \cos \theta - 2 \sin \theta + 1 = 0$$
$$2 \sin \theta \cos \theta - \cos \theta - 2 \sin \theta + 1 = 0$$
$$\cos \theta (2 \sin \theta - 1) - 1(2 \sin \theta - 1) = 0$$
$$(\cos \theta - 1)(2 \sin \theta - 1) = 0$$

$\cos \theta - 1 = 0$ or $2 \sin \theta - 1 = 0$

$\cos \theta = 1$ or $\sin \theta = \dfrac{1}{2}$

$\theta = 0$ or $\theta = \dfrac{\pi}{6}, \dfrac{5\pi}{6}$

67.
$$2 \sin^2 - 3 \sin \theta + 1 = 0$$
$$(2 \sin \theta - 1)(\sin \theta - 1) = 0$$

$2 \sin \theta - 1 = 0$ or $\sin \theta - 1 = 0$

$\sin \theta = \dfrac{1}{2}$ or $\sin \theta = 1$

$\theta = \dfrac{\pi}{6}, \dfrac{5\pi}{6}, \dfrac{\pi}{2}$

7 ANALYTIC TRIGONOMETRY

69.
$$\sin \theta - \cos \theta = 1$$

$$\frac{1}{\sqrt{2}} \sin \theta - \frac{1}{\sqrt{2}} \cos \theta = \frac{1}{\sqrt{2}} \qquad (\text{Divide by } \sqrt{2})$$

Let $\cos \phi = \dfrac{1}{\sqrt{2}}$, $\sin \phi = \dfrac{1}{\sqrt{2}}$, $0 \le \phi \le 2\pi$

Then, $\phi = \dfrac{\pi}{4}$ and $\cos \phi \sin \theta - \sin \phi \cos \theta = \dfrac{1}{\sqrt{2}}$

$$\sin(\theta - \phi) = \sin\!\left(\theta - \frac{\pi}{4}\right) = \frac{1}{\sqrt{2}}$$

$$\theta - \frac{\pi}{4} = \frac{\pi}{4} \quad \text{or} \quad \theta - \frac{\pi}{4} = \frac{3\pi}{4}$$

$$\theta = \frac{\pi}{2} \quad \text{or} \qquad \theta = \pi$$

7 - CHAPTER REVIEW

C H A P T E R

ADDITIONAL APPLICATIONS OF TRIGONOMETRY

8

≡ EXERCISE 8.1 THE LAW OF SINES

1. The third angle α is easily found because the sum of the angles of a triangle equals 180°.

$$\alpha + \beta + \gamma = 180°$$
$$\alpha + 45° + 95° = 180°$$
$$\alpha = 40°$$

Now we use the Law of Sines (twice) to find the unknown sides a and b:

$$\frac{\sin \alpha}{a} = \frac{\sin \gamma}{c} \qquad\qquad \frac{\sin \alpha}{b} = \frac{\sin \gamma}{c}$$

$$\frac{\sin 40°}{a} = \frac{\sin 95°}{5} \qquad\qquad \frac{\sin 45°}{b} = \frac{\sin 95°}{5}$$

Thus,

$$a = \frac{5 \sin 40°}{\sin 95°} \approx 3.2262 \text{ (From table or calculator)}$$

and

$$b = \frac{5 \sin 45°}{\sin 95°} \approx 3.5490 \text{ (From table or calculator)}$$

3. Because we know two angles ($\alpha = 50°$ and $\gamma = 85°$), it is easy to find the third angle using the equation:

$$\alpha + \beta + \gamma = 180°$$
$$50° + \beta + 85° = 180°$$
$$\beta = 45°$$

Now we know the three angles and one side ($b = 3$) of the triangle. To find the remaining two sides a and c, we use the Law of Sines (twice):

$$\frac{\sin \alpha}{a} = \frac{\sin \gamma}{b} \qquad\qquad \frac{\sin \beta}{b} = \frac{\sin \gamma}{c}$$

$$\frac{\sin 50°}{a} = \frac{\sin 45°}{3} \qquad\qquad \frac{\sin 45°}{3} = \frac{\sin 85°}{c}$$

$$a = \frac{3 \sin 50°}{\sin 45°} \qquad\qquad c = \frac{3 \sin 85°}{\sin 45°}$$

$$a \approx 3.2501 \qquad\qquad c \approx 4.2265$$

5.

$$\alpha + \beta + \gamma = 180°$$
$$40° + 45° + \gamma = 180°$$
$$\gamma = 95°$$

$$\frac{\sin \alpha}{a} = \frac{\sin \gamma}{b} \qquad\qquad \frac{\sin \beta}{b} = \frac{\sin \gamma}{c}$$

$$\frac{\sin 40°}{a} = \frac{\sin 45°}{7} \qquad\qquad \frac{\sin 45°}{7} = \frac{\sin 95°}{c}$$

$$a = \frac{7 \sin 40°}{\sin 45°} \qquad\qquad c = \frac{7 \sin 95°}{\sin 45°}$$

$$a \approx 6.3633 \qquad\qquad c \approx 9.8618$$

7.

$$\alpha + \beta + \gamma = 180°$$
$$\alpha + 40° + 100° + \gamma = 180°$$
$$\alpha = 40°$$

$$\frac{\sin \alpha}{a} = \frac{\sin \gamma}{b} \qquad\qquad \frac{\sin \beta}{b} = \frac{\sin \gamma}{c}$$

$$\frac{\sin 40°}{a} = \frac{\sin 40°}{2} \qquad\qquad \frac{\sin 40°}{2} = \frac{\sin 100°}{c}$$

$$a = \frac{2 \sin 40°}{\sin 40°} \qquad\qquad c = \frac{2 \sin 100°}{\sin 40°}$$

$$a = 2.0000 \qquad\qquad c \approx 3.0642$$

9.

$$\alpha + \beta + \gamma = 180°$$
$$40° + 20° + \gamma = 180°$$
$$\gamma = 120°$$

$$\frac{\sin \alpha}{a} = \frac{\sin \beta}{b} \qquad\qquad \frac{\sin \alpha}{a} = \frac{\sin \gamma}{c}$$

$$\frac{\sin 40°}{2} = \frac{\sin 20°}{b} \qquad\qquad \frac{\sin 40°}{2} = \frac{\sin 120°}{c}$$

$$b = \frac{2 \sin 20°}{\sin 40°} \qquad\qquad c = \frac{2 \sin 120°}{\sin 40°}$$

$$b \approx 1.0642 \qquad\qquad c \approx 2.6946$$

11.

$$\alpha + \beta + \gamma = 180°$$
$$\alpha + 70° + 10° + \gamma = 180°$$
$$\alpha = 100°$$

$$\frac{\sin \alpha}{a} = \frac{\sin \gamma}{b} \qquad\qquad \frac{\sin \beta}{b} = \frac{\sin \gamma}{c}$$

$$\frac{\sin 100°}{a} = \frac{\sin 70°}{5} \qquad\qquad \frac{\sin 70°}{5} = \frac{\sin 10°}{c}$$

$$a \approx 5.2401 \qquad\qquad c \approx 0.9240$$

13.

$$\alpha + \beta + \gamma = 180°$$
$$110° + \beta + 30° + \gamma = 180°$$
$$\beta = 40°$$

$$\frac{\sin \alpha}{a} = \frac{\sin \gamma}{c} \qquad\qquad \frac{\sin \beta}{b} = \frac{\sin \gamma}{c}$$

$$\frac{\sin 110°}{a} = \frac{\sin 30°}{3} \qquad\qquad \frac{\sin 40°}{b} = \frac{\sin 30°}{3}$$

$$a = \frac{3 \sin 110°}{\sin 30°} \qquad\qquad b = \frac{3 \sin 40°}{\sin 30°}$$

$$a \approx 5.6382 \qquad\qquad b \approx 3.8567$$

8.1 THE LAW OF SINES

15.

$$\alpha + \beta + \gamma = 180°$$
$$40° + 40° + \gamma = 180°$$
$$\gamma = 100°$$

$$\frac{\sin \alpha}{a} = \frac{\sin \gamma}{c} \qquad\qquad \frac{\sin \beta}{b} = \frac{\sin \gamma}{c}$$

$$\frac{\sin 40°}{a} = \frac{\sin 100°}{2} \qquad\qquad \frac{\sin 40°}{b} = \frac{\sin 100°}{2}$$

$$a = \frac{2 \sin 40°}{\sin 100°} \qquad\qquad b = \frac{2 \sin 40°}{\sin 100°}$$

$$a \approx 1.3054 \qquad\qquad b \approx 1.3054$$

17. Because $a = 3$, $b = 2$, and $\alpha = 50°$ are known, we use the Law of Sines to find the angle β:

$$\frac{\sin \alpha}{a} = \frac{\sin \beta}{b}$$

Then, $$\frac{\sin 50°}{3} = \frac{\sin \beta}{2}$$

$$\sin \beta = \frac{2 \sin 50°}{3} \approx 0.5107$$

There are two angles β, $0° < \beta < 180°$, for which

$$\sin \beta \approx 0.5107, \text{ namely}$$
$$\beta \approx 30.7° \text{ or } \beta \approx 149.3$$

The second possibility is ruled out, because $\alpha = 50°$, making $\alpha + \beta \approx 199.3° > 180°$. Now, using $\beta \approx 30.7°$, we find $\gamma = 180° - \alpha - \beta \approx 99.3°$. The third side c is determined using the Law of Sines:

$$\frac{\sin \alpha}{a} = \frac{\sin \gamma}{c}$$

$$\frac{\sin 50°}{3} = \frac{\sin 99.3°}{c}$$

$$c = \frac{3 \sin 99.3°}{\sin 50°} \approx 3.8647$$

One triangle; $\beta \approx 30.7°$, $\gamma \approx 99.3°$, $c \approx 3.8647$

19. Because $b = 5$, $c = 3$, and $\beta = 100°$ are known, we use the Law of Sines to find the angle γ:

$$\frac{\sin \alpha}{b} = \frac{\sin \gamma}{c}$$

Then, $$\frac{\sin 100°}{5} = \frac{\sin \gamma}{3}$$

$$\sin \gamma = \frac{3 \sin 100°}{5} \approx 0.5909$$

$$\gamma \approx 36.2° \text{ or } \gamma \approx 143.8°$$

The second possibility is ruled out, because $\beta = 100°$, making $\beta + \gamma \approx 243.8° > 180°$. Now, using $\gamma \approx 36.2°$, we find

$$\alpha = 180° - \beta - \gamma \approx 43.8°$$

The third side is determined using the Law of Sines.

$$\frac{\sin \alpha}{a} = \frac{\sin \beta}{b}$$

$$\frac{\sin 43.8 \degree}{a} = \frac{\sin 100\degree}{5}$$

$$a = \frac{5 \sin 43.8\degree}{\sin 100\degree} \approx 3.5141$$

One triangle; $\gamma \approx 36.2\degree$, $\alpha \approx 43.8\degree$, $a \approx 3.5141$

21. Because $a = 4$, $b = 5$, and $\alpha = 60\degree$ are known, we use the Law of Sines to find the angle β:

$$\frac{\sin \alpha}{a} = \frac{\sin \beta}{b}$$

Then, $\quad \dfrac{\sin 60 \degree}{4} = \dfrac{\sin \beta}{5}$

$$\sin \beta = \frac{5 \sin 60\degree}{4} \approx 1.0825$$

There is no angle β for which $\sin \beta > 1$. Hence, there can be no triangle having the given measurements.

23. Because $b = 4$, $c = 6$, and $\beta = 20\degree$ are known, we use the Law of Sines to find the angle γ:

$$\frac{\sin \beta}{b} = \frac{\sin \gamma}{c}$$

$$\frac{\sin 20 \degree}{4} = \frac{\sin \gamma}{6}$$

$$\sin \gamma = \frac{6 \sin 20\degree}{4} \approx 0.5130$$

$$\gamma_1 \approx 30.9\degree \text{ or } \gamma_2 \approx 149.1\degree$$

For both possibilities we have $\beta + \gamma < 180\degree$. Hence, there are two triangles—one containing the angle $\gamma = \gamma_1 \approx 30.9\degree$, the other containing the angle $\gamma = \gamma_2 \approx 149.1\degree$. The third angle α is either:

$$\alpha_1 = 180\degree - \beta - \gamma_1 \approx 129.1\degree \text{ or } \alpha_2 = 180\degree - \beta - \gamma_2 \approx 10.9\degree$$

The third side a obeys the Law of Sines, so we have

$$\frac{\sin \alpha}{a} = \frac{\sin \beta}{b}$$

$$\frac{\sin 129.1\degree}{a_1} = \frac{\sin 20\degree}{4} \quad \text{or} \quad \frac{\sin 10.9\degree}{a_2} = \frac{\sin 20\degree}{4}$$

$$a_1 = \frac{4 \sin 129.1\degree}{\sin 20\degree} \quad \text{or} \quad a_2 = \frac{4 \sin 10.9\degree}{\sin 20\degree}$$

$$a_1 \approx 9.0760 \qquad\qquad a_2 \approx 2.2115$$

Two triangles: $\quad \gamma_1 \approx 30.9\degree$, $\alpha_1 \approx 129.1\degree$, $a_1 \approx 9.0760$
or $\quad \gamma_2 \approx 149.1\degree$, $\alpha_2 \approx 10.9\degree$, $a_2 \approx 2.2115$

8.1 THE LAW OF SINES

25. Because $a = 2$, $c = 1$, and $\gamma = 100°$, are known, we use the Law of Sines to find the angle α:

$$\frac{\sin \alpha}{a} = \frac{\sin \gamma}{c}$$

$$\frac{\sin \alpha}{2} = \frac{\sin 100°}{1}$$

There is no angle α for which $\sin \alpha > 1$. Hence, there can be no triangle having the given measurements.

27. Because $a = 2$, $c = 1$, and $\gamma = 25°$, we use the Law of Sines to find the angle α:

$$\frac{\sin \alpha}{a} = \frac{\sin \gamma}{c}$$

Then,

$$\frac{\sin \alpha}{2} = \frac{\sin 25°}{1}$$
$$\sin \alpha = 2 \sin 25° \approx 0.8452$$
$$\alpha_1 \approx 57.7° \text{ or } \alpha_2 = 122.3°$$

The third angle β is either

$$\beta_1 = 180° - 25° - 57.7° \approx 97.3° \text{ or}$$
$$\beta_2 = 180° - 25° - 122.3° \approx 32.7°$$

The third side b obeys the Law of Sines, so we have

$$\frac{\sin \beta}{b} = \frac{\sin \gamma}{c}$$

$$\frac{\sin 97.3°}{b} = \frac{\sin 25°}{1} \qquad \text{or} \qquad \frac{\sin 32.7°}{b_2} = \frac{\sin 25°}{1}$$

$$b_1 = \frac{\sin 97.3°}{\sin 25°} \qquad \text{or} \qquad b_2 = \frac{\sin 32.7°}{\sin 25°}$$

$$b_1 \approx 2.3470 \qquad\qquad\qquad b_2 \approx 1.2783$$

Two triangles: $\alpha_1 \approx 57.7°$, $\beta_1 \approx 97.3$, $b_1 \approx 2.3470$
or $\alpha_2 \approx 122.3°$, $\beta_2 \approx 32.7°$, $b_2 \approx 1.2783$

29.

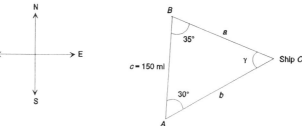

(a) The angle γ is found to be

$$\gamma = 180° - 30° - 35° = 115°$$

The Law of Sines can now be used to find the two distances a and b. We seek:

$$\frac{\sin 30°}{a} = \frac{\sin 115°}{150}$$

$$a = \frac{150 \sin 30°}{\sin 115°}$$

$$a \approx 82.8 \text{ miles}$$

$$\frac{\sin 35°}{b} = \frac{\sin 115°}{150}$$

$$b = \frac{150 \sin 35°}{\sin 115°}$$

$$b \approx 94.9 \text{ miles}$$

Thus, station Able is 94.9 miles and station B is 82.8 miles from the ship.

(b) The time t needed for the helicopter to reach the ship from Station Baker is found by using the formula:

(Velocity, v)(Time, t) = Distance

Then,

$$t = \frac{a}{v} = \frac{82.8}{200} \approx 0.41 \text{ hour} \approx 25 \text{ minutes}$$

31. Because angle DAB is supplementary to angle CAB, angle $CAB = 180° - 25° = 155°$. Angle $ABC = 180° - 155° - 15° = 10°$.

Let c denote the distance from A to B. Using the Law of Sines,

$$\frac{\sin 10°}{100} = \frac{\sin 15°}{c}$$

$$c = \frac{100 \sin 15°}{\sin 10°} \approx 149.0$$

The length of the span of a ski lift from A to B is 149 feet.

33.

$$\gamma_1 = 180° - 40° - 90° = 50°$$
$$\gamma_2 = 180° - 35° - 90° = 55°$$
$$\gamma = 180° - 40° - 35° = 50° + 55°$$
$$= 105°$$

Before we find the height of the airplane, we need to know either the distance from A to C or the distance from B to C. Let's find the distance from A to C, denote it by b, using the Law of Sines.

$$\frac{\sin 35°}{b} = \frac{\sin 105°}{1000}$$

$$b = \frac{1000 \sin 35°}{\sin 105°} \approx 593.8$$

Now we can find the height of the airplane. Let a denote the distance from C to E.

$$\frac{\sin 40°}{a} = \frac{\sin 90°}{593.8}$$

$$a = \frac{593.8 \sin 40°}{\sin 90°} \approx 381.7$$

The airplane is 381.7 feet high.

8.1 THE LAW OF SINES

35. (a) Angle $CBA = 180° - 40° = 140°$

Let the distance from city A to city B be denoted by c and let γ denote angle ACB. We can use the Law of Sines to find γ.

$$\frac{\sin 140°}{300} = \frac{\sin \gamma}{150}$$
$$\sin \gamma = \frac{150 \sin 140°}{300} \approx 0.3214$$
$$\gamma = 18.7°$$

Let α denote angle BAC.

$$\alpha = 180° - 140° - 18.7° = 21.3°$$

Using the Law of Sines to find the distance from city B to city C, denoted by a,

$$\frac{\sin 21.3°}{a} = \frac{\sin 140°}{300}$$
$$a = \frac{300 \sin 21.3°}{\sin 140°} \approx 169$$

It is 169.0 miles from city B to city C.

(b)

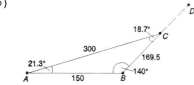

Angle $ACD = 180° - 18.7°$
$= 161.3°$

The pilot should turn through an angle 161.3° at city C to return to city A.

37.

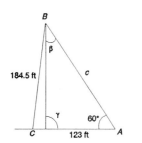

Let β = Angle ACB

$$\frac{\sin 60°}{184.5} = \frac{\sin \beta}{123}$$
$$\sin \beta = \frac{123 \sin 60°}{184.5} \approx 0.5774$$
$$\beta = 35.3°$$

Thus, Angle $CAB = 180° - 60° - 35.3° \approx 84.7°$

Let x be the perpendicular distance from C to AB. Then

$$\sin 84.7 = \frac{x}{184.5}$$
$$\sin 84.7(184.5) = x$$
$$183.7 \text{ ft} = x$$

39. We want to prove that

$$\frac{a+b}{c} = \frac{\cos \frac{1}{2}(\alpha - \beta)}{\sin \frac{1}{2}\gamma}$$

$$\frac{a+b}{c} = \frac{a}{c} + \frac{b}{c} = \frac{\sin \alpha}{\sin \gamma} + \frac{\sin \beta}{\sin \gamma} = \frac{\sin \alpha + \sin \beta}{\sin \gamma}$$

$$= \frac{2 \sin \frac{\alpha + \beta}{2} \cos \frac{\alpha - \beta}{2}}{\sin\left(\frac{\gamma}{2} + \frac{\gamma}{2}\right)} = \frac{2 \sin \frac{\alpha + \beta}{2} \cos \frac{\alpha - \beta}{2}}{\sin \frac{\gamma}{2} \cos \frac{\gamma}{2} + \cos \frac{\gamma}{2} \sin \frac{\gamma}{2}}$$

$$= \frac{2 \sin \frac{\alpha + \beta}{2} \cos \frac{\alpha - \beta}{2}}{2 \sin \frac{\gamma}{2} \cos \frac{\gamma}{2}} = \frac{\sin\left(\frac{\pi}{2} - \frac{\gamma}{2}\right) \cos \frac{\alpha - \beta}{2}}{\sin \frac{\gamma}{2} \cos \frac{\gamma}{2}}$$

$$= \frac{\cos \frac{\gamma}{2} \cos \frac{\alpha - \beta}{2}}{\sin \frac{\gamma}{2} \cos \frac{\gamma}{2}} = \frac{\cos \frac{\alpha - \beta}{2}}{\sin \frac{\gamma}{2}} = \frac{\cos \frac{1}{2}(\alpha - \beta)}{\sin \frac{1}{2}\gamma}$$

41. We want to derive the formula:

$$a = \frac{b \sin \alpha}{\sin \beta} = \frac{b \sin[180° - (\beta + \gamma)]}{\sin \beta}$$

$$= \frac{b}{\sin \beta}[\sin 180° \cos(\beta + \gamma) - \cos 180° \sin(\beta + \gamma)]$$

$$= \frac{b}{\sin \beta}[0 + \sin(\beta + \gamma)] = \frac{b}{\sin \beta}(\sin \beta \cos \gamma + \cos \beta \sin \gamma)$$

$$= b \cos \gamma + \frac{b \sin \gamma}{\sin \beta} \cos \beta = b \cos \gamma + c \cos \beta$$

43. $\sin \beta = \sin(\text{Angle } AB'C) = \frac{b}{2r}$;

Thus, $\frac{\sin \beta}{b} = \frac{1}{2r}$. By the Law of Sines,

$$\frac{\sin \alpha}{a} = \frac{\sin \beta}{b} = \frac{\sin \alpha}{c} = \frac{1}{2r}$$

≡ EXERCISE 8.2 THE LAW OF COSINES

1. $b^2 = a^2 + c^2 - 2ac \cos \beta$
   $b^2 = 4 + 16 - 2 \cdot 2 \cdot 4 \cos 45°$

   $b^2 = 20 - \left(16 \cdot \frac{\sqrt{2}}{2}\right)$

   $b^2 = 20 - 8\sqrt{2} \approx 8.6863$
   $b \approx 2.9473$

To find α:
$$a^2 = b^2 + c^2 = 2bc \cos \alpha$$
$$2bc \cos \alpha = b^2 + c^2 - a^2$$
$$\cos \alpha = \frac{b^2 + c^2 - a^2}{2bc}$$
$$\cos \alpha = \frac{8.6866 + 16 - 4}{2 \cdot 2.9473 \cdot 4}$$
$$= \frac{20.6866}{23.5784} \approx 0.8774$$
$$\alpha \approx 28.7°$$

To find γ:
$$c^2 = a^2 + b^2 - 2ab \cos \gamma$$
$$2ab \cos \gamma = a^2 + b^2 - c^2$$
$$\cos \gamma = \frac{a^2 + b^2 - c^2}{2ab}$$
$$\cos \gamma = \frac{4 + 8.6863 - 16}{2 \cdot 2 \cdot 2 \cdot 9473} = \frac{-3.3137}{11.7892} \approx -0.2811$$
$$\gamma \approx 106.3°$$

3. $$c^2 = a^2 + b^2 - 2ab \cos \gamma$$
$$c^2 = 4 + 9 - 12 \cos 95°$$
$$c^2 = 13 - 12(-0.0872) = 14.0464$$
$$c \approx 3.7478$$

$$\cos \alpha = \frac{b^2 + c^2 - a^2}{2bc}$$
$$\cos \alpha = \frac{9 + 14.0464 - 4}{2 \cdot 3 \cdot 3.7479} = \frac{19.0464}{22.4874} \approx 0.8470$$
$$\alpha = 32.1°$$
$$\cos \beta = \frac{a^2 + c^2 - b^2}{2ac}$$
$$\cos \beta = \frac{4 + 14.0464 - 9}{2 \cdot 2 \cdot 3.7479} = \frac{9.0464}{14.9916} \approx 0.6034$$
$$\beta \approx 52.9°$$

5. $$\cos \alpha = \frac{b^2 + c^2 - a^2}{2bc}$$
$$\cos \alpha = \frac{25 + 64 - 36}{2 \cdot 5 \cdot 8} = \frac{53}{80} = 0.6625$$
$$\alpha = 48.5°$$
$$\cos \beta = \frac{a^2 + c^2 - b^2}{2ac}$$
$$\cos \beta = \frac{36 + 64 - 25}{2 \cdot 6 \cdot 8} = \frac{75}{96} \approx 0.7813$$
$$\beta = 38.6°$$
$$\cos \gamma = \frac{a^2 + b^2 - c^2}{2ab}$$
$$\cos \gamma = \frac{36 + 25 - 64}{2 \cdot 6 \cdot 5} = \frac{-3}{60} = -0.05$$
$$\gamma = 92.9°$$

8 ADDITIONAL APPLICATIONS OF TRIGONOMETRY

7. $\cos \alpha = \dfrac{b^2 + c^2 - a^2}{2bc}$

$\cos \alpha = \dfrac{36 + 16 - 81}{2 \cdot 6 \cdot 4} = \dfrac{-29}{48} \approx -0.6042$

$\alpha = 127.2°$

$\cos \beta = \dfrac{a^2 + c^2 - b^2}{2ac}$

$\cos \beta = \dfrac{81 + 16 - 36}{2 \cdot 9 \cdot 4} = \dfrac{61}{72} \approx 0.8472$

$\beta = 32.1°$

$\cos \gamma = \dfrac{a^2 + b^2 - c^2}{2ab}$

$\cos \gamma = \dfrac{81 + 36 - 16}{2 \cdot 9 \cdot 6} = \dfrac{101}{108} \approx 0.9352$

$\gamma = 20.7°$

9. For c:

$c^2 = a^2 + b^2 - 2ab \cos \gamma$

$c^2 = 9 + 16 - 2 \cdot 3 \cdot 4 \cdot \cos 40°$

$c^2 = 25 - 24(0.76604) = 6.615$

$c \approx 2.5720$

For α:

$\cos \alpha = \dfrac{b^2 + c^2 - a^2}{2bc}$

$\cos \alpha = \dfrac{16 + 6.616 - 9}{2 \cdot 4 \cdot 2.5722} = \dfrac{13.616}{20.5776} \approx 0.6617$

$\alpha = 48.6°$

For β:

$\cos \beta = \dfrac{a^2 + c^2 - b^2}{2ac}$

$\cos \beta = \dfrac{9 + 6.616 - 16}{2 \cdot 3 \cdot 2.5722} = \dfrac{-0.384}{15.4332} \approx -0.0249$

$\beta \approx 91.4°$

11. $a^2 = b^2 + c^2 - 2bc \cos \alpha$

$a^2 = 1 + 9 - 2 \cdot 1 \cdot 3 \cdot \cos 80°$

$a^2 = 10 - 6(-0.17365) = 8.9581$

$a \approx 2.9930$

$\cos \beta = \dfrac{a^2 + c^2 - b^2}{2ac}$

$\cos \beta = \dfrac{8.9584 + 9 - 1}{2 \cdot 2.9931 \cdot 3} = \dfrac{16.9584}{17.9586} \approx 0.9443$

$\beta = 19.2°$

$\cos \gamma = \dfrac{a^2 + b^2 - c^2}{2ab}$

$\cos \gamma = \dfrac{8.9584 + 1 - 9}{2 \cdot 2.9931 \cdot 1} = \dfrac{0.9584}{5.9862} \approx 0.1601$

$\gamma = 80.8°$

8.2 THE LAW OF COSINES

13. $b^2 = a^2 + c^2 - 2ac \cos \beta$
$b^2 = 9 + 4 - 2 \cdot 3 \cdot 2 \cos 110°$
$b^2 = 13 - 12(-0.3420) = 17.104$
$b \approx 4.1357$

$$\cos \alpha = \frac{b^2 + c^2 - a^2}{2bc}$$

$$\cos \alpha = \frac{17.104 + 4 - 9}{2 \cdot 4.1357 \cdot 2} = \frac{12.104}{16.5428} \approx 0.7317$$

$$\alpha = 43.0°$$

$$\cos \gamma = \frac{a^2 + b^2 - c^2}{2ab}$$

$$\cos \gamma = \frac{9 + 17.104 - 4}{2 \cdot 3 \cdot 4.1357} = \frac{22.104}{24.8142} \approx 0.8908$$

$$\gamma \approx 27.0°$$

15. $c^2 = a^2 + b^2 - 2ab \cos \gamma$
$c^2 = 4 + 4 - 2 \cdot 2 \cdot 2 \cdot \cos 50°$
$c^2 = 8 - 8(0.64279) = 2.85768$
$c \approx 1.6905$

$$\cos \alpha = \frac{b^2 + c^2 - a^2}{2bc}$$

$$\cos \alpha = \frac{4 + 2.8576 - 4}{2 \cdot 2 \cdot 1.6904} = \frac{2.8576}{6.7616} \approx 0.4226$$

$$\alpha = 65.0°$$

$$\cos \beta = \frac{a^2 + c^2 - b^2}{2ac}$$

$$\cos \beta = \frac{4 + 2.8576 - 4}{2 \cdot 2 \cdot 1.6904} = \frac{2.8576}{6.7616} \approx 0.4226$$

$$\beta \approx 65.0°$$

17. For α:

$$\cos \alpha = \frac{b^2 + c^2 - a^2}{2bc}$$

$$\cos \alpha = \frac{169 + 25 - 144}{2 \cdot 13 \cdot 5} = \frac{50}{130} \approx 0.3846$$

$$\alpha \approx 67.4°$$

For β:

$$\cos \beta = \frac{a^2 + c^2 - b^2}{2ac}$$

$$\cos \beta = \frac{144 + 25 - 169}{2 \cdot 12 \cdot 5} = \frac{0}{120} \approx 0$$

$$\beta = 90°$$

For γ:

$$\cos \gamma = \frac{a^2 + b^2 - c^2}{2ab}$$

$$\cos \gamma = \frac{144 + 169 - 25}{2 \cdot 12 \cdot 13} = \frac{288}{312} \approx 0.9231$$

$$\gamma = 22.6°$$

19. $\cos \alpha = \dfrac{b^2 + c^2 - a^2}{2bc}$

$\cos \alpha = \dfrac{4 + 4 - 4}{2 \cdot 2 \cdot 2} = \dfrac{4}{8} \approx 0.5$

$\alpha \approx 60°$

$\cos \beta = \dfrac{a^2 + c^2 - b^2}{2ac}$

$\cos \beta = \dfrac{4 + 4 - 4}{2 \cdot 2 \cdot 3} = \dfrac{1}{2} = 0.5$

$\beta = 60°$

$\cos \gamma = \dfrac{a^2 + b^2 - c^2}{2ab}$

$\cos \gamma = \dfrac{4 + 4 - 4}{2 \cdot 2 \cdot 2} = \dfrac{1}{2} = 0.5$

$\gamma \approx 60°$

21. $\cos \alpha = \dfrac{b^2 + c^2 - a^2}{2bc}$

$\cos \alpha = \dfrac{64 + 81 - 25}{2 \cdot 8 \cdot 9} = \dfrac{120}{144} \approx 0.8333$

$\alpha = 33.6°$

$\cos \beta = \dfrac{a^2 + c^2 - b^2}{2ac}$

$\cos \beta = \dfrac{25 + 81 - 64}{2 \cdot 5 \cdot 9} = \dfrac{42}{90} \approx 0.4667$

$\beta \approx 62.2°$

$\cos \gamma = \dfrac{a^2 + b^2 - c^2}{2ab}$

$\cos \gamma = \dfrac{25 + 64 - 81}{2 \cdot 5 \cdot 8} = \dfrac{8}{80} \approx 0.1$

$\gamma \approx 84.3°$

23. $\cos \alpha = \dfrac{b^2 + c^2 - a^2}{2bc}$

$\cos \alpha = \dfrac{64 + 25 - 100}{2 \cdot 8 \cdot 5} = \dfrac{-11}{80} \approx 0.1375$

$\alpha = 97.9°$

$\cos \beta = \dfrac{a^2 + c^2 - b^2}{2ac}$

$\cos \beta = \dfrac{100 + 25 - 64}{2 \cdot 10 \cdot 5} = \dfrac{61}{100} \approx 0.61$

$\beta \approx 52.4°$

$\cos \gamma = \dfrac{a^2 + b^2 - c^2}{2ab}$

$\cos \gamma = \dfrac{100 + 64 - 25}{2 \cdot 10 \cdot 8} = \dfrac{139}{160} \approx 0.8688$

$\gamma \approx 29.7°$

25.   The distance c we seek is the third side of a triangle in which the other two sides and their included angle are known.

$$c^2 = a^2 + b^2 - 2ab \cos \gamma$$
$$c^2 = 4900 + 2500 - 2 \cdot 70 \cdot 50 \cdot \cos 70°$$
$$c^2 = 7400 - 7000(0.34202) = 5005.86$$
$$c \approx 70.7521 \text{ feet}$$

The houses are about 70.7521 feet apart.

27.   (a) Using the formula,

Velocity × Time = Distance
220 mi/hr × .25 hr. = 55 miles

We are looking for angle 180° − ?, the angle at which the pilot should turn to head toward city B. But first we need to find the measurement of the third side a. We know the other two sides and their included angle so we use the Law of Cosines:

$$a^2 = b^2 + c^2 - 2bc \cos \alpha$$
$$a^2 = 3,025 + 108,900 - 2 \cdot 55 \cdot 330 \cdot \cos 10°$$
$$a^2 = 111,925 - 36(0.9848) = 76,176.76$$
$$a \approx 276$$

Now we can find angle γ:

$$\cos \gamma = \frac{a^2 + b^2 - c^2}{2ab}$$
$$\cos \gamma = \frac{76,176.76 + 3,025 - 108,900}{2 \cdot 276 \cdot 55}$$
$$\cos \gamma = \frac{-29,698.24}{30,360} \approx -0.9782$$
$$\gamma = 168°$$

$$180° - \gamma = 12°$$

The pilot should turn through an angle of 12°.

(b) Velocity = $\dfrac{\text{Distance}}{\text{Time}}$

Velocity = $\dfrac{331 \text{ miles}}{1.5 \text{ m.}} \approx 220.8$ mph

The pilot should maintain an average speed of 220.7 miles per hour so that the total time of the trip is 90 minutes or 1.5 hours.

8 ADDITIONAL APPLICATIONS OF TRIGONOMETRY

29.

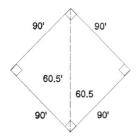

(a)

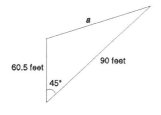

We know that $\alpha = 45°$ because the diagonal of the square diamond bisects the right angle. Because we know two sides and the included angle, we use the Law of Cosines to find the third side a.

$$a^2 = b^2 + c^2 - 2bc \cos \alpha$$
$$a^2 = 3660.25 + 8100 - 2 \cdot 60.5 \cdot 90 \cdot \cos 45°$$
$$a^2 = 11{,}760.5 - 10{,}890(0.7071) = 4060.181$$
$$a \approx 63.7 \text{ feet}$$

It is about 63.7 feet from the pitching rubber to first base.

(b)

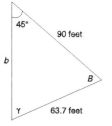

Using the Law of Sines, we find angle γ

$$\frac{\sin 45°}{63.7} = \frac{\sin \gamma}{90}$$
$$\sin \gamma = \frac{90 \sin 45°}{63.7} \approx 0.9991$$
$$\gamma = 87.5°$$
$$\beta = 180° - \alpha - \gamma = 47.5°$$

Now we can use the Law of Cosines to find side b:

$$b^2 = b^2 + c^2 - 2bc \cos \beta$$
$$b^2 = 4060.181 + 8100 - 2 \cdot 63.7 \cdot 90 \cdot \cos 47.5°$$
$$b^2 = 12{,}160.181 - 11{,}466(0.67559) = 4413.86606$$
$$b \approx 66.8 \text{ feet}$$

(c)

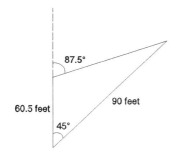

From (b), we know that $\gamma = 87.5°$. $180° - \gamma = 92.5°$. The pitcher needs to turn through an angle of 92.5° to face first base.

31. (a)

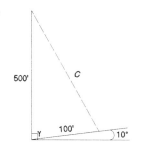

$\gamma = 90° - 10° = 80°$

Now we know two sides of the triangle and the included angle. Therefore, we use the Law of Cosines to find the third side:

$c^2 = a^2 + b^2 - 2ab \cos \gamma$
$c^2 = 10,000 + 250,000 - 2 \cdot 100 \cdot 500 \cdot \cos 80°$
$c^2 = 260,000 - 100,000(0.1736) = 242,640$
$c \approx 492.6$ feet

The guy wire should be 492.6 feet.

(b)

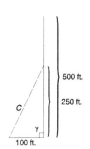

We know two sides of the triangle and the included angle. We use the Law of Cosines to find the third side.

$c^2 = a^2 + b^2 - 2ab \cos \gamma$
$c^2 = 10,000 + 62,500 - 2 \cdot 100 \cdot 250 \cdot \cos 90°$
$c^2 = 72,500 - 50,000(0) = 72,500$
$c \approx 269.3$ feet

33.

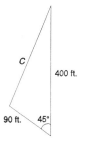

We know two sides and the included angle so we can use the Law of Cosines to find the third side.

$c^2 = a^2 + b^2 - 2ab \cos \gamma$
$c^2 = 8100 + 160,000 - 2 \cdot 90 \cdot 400 \cdot \cos 45°$
$c^2 = 168,100 - 72,000(0.7071)$
$c^2 \approx 117,188.8$
$c \approx 342.3$ feet

In Wrigley Field, it is 342.3 feet from dead center to third base.

35. $\cos \dfrac{\gamma}{2} = \sqrt{\dfrac{1 + \cos \gamma}{2}} = \sqrt{\dfrac{1 + \dfrac{a^2 + b^2 - c^2}{2ab}}{2}} = \sqrt{\dfrac{2ab + a^2 + b^2 - c^2}{4ab}}$

$= \sqrt{\dfrac{(a + b)^2 - c^2}{4ab}} = \sqrt{\dfrac{(a + b + c)(a + b - c)}{4ab}} = \sqrt{\dfrac{2s(2x - c - c)}{4ab}}$

$= \sqrt{\dfrac{4s(s - c)}{4ab}} = \sqrt{\dfrac{s(s - c)}{ab}}$

37. In order to prove that $\dfrac{\cos \alpha}{a} + \dfrac{\cos \beta}{b} + \dfrac{\cos \gamma}{c} = \dfrac{a^2 + b^2 + c^2}{2abc}$, we let

$\dfrac{\cos \alpha}{a} + \dfrac{\cos \beta}{b} + \dfrac{\cos \gamma}{c} = \dfrac{b^2 + c^2 - a^2}{2bca} + \dfrac{a^2 + c^2 - b^2}{2acb} + \dfrac{a^2 + b^2 - c^2}{2abc}$

$= \dfrac{b^2 + c^2 - a^2 + a^2 + c^2 - b^2 + a^2 + b^2 - c^2}{2abc}$

$= \dfrac{a^2 + b^2 + c^2}{2abc}$

≡ EXERCISE 8.3 THE AREA OF A TRIANGLE

1. $A = \dfrac{1}{2}ac \sin \beta$

$A = \dfrac{1}{2} \cdot 2 \cdot 4 \sin 45° = 2.8284$

3. $A = \dfrac{1}{2}ac \sin \gamma$

$A = \dfrac{1}{2} \cdot 2 \cdot 3 \sin 95° = 2.9886$

5. $a = 6$, $b = 5$, $c = 8$

$s = \dfrac{1}{2}(a + b + c) = \dfrac{1}{2}(6 + 5 + 8) = \dfrac{19}{2}$

Heron's Formula then gives the area A as:

$A = \sqrt{s(s - a)(s - b)(s - c)} = \sqrt{\dfrac{19}{2} \cdot \dfrac{7}{2} \cdot \dfrac{9}{2} \cdot \dfrac{3}{2}}$

$= \sqrt{\dfrac{3591}{16}} = \sqrt{224.4375} \approx 14.9812$

7. $s = \dfrac{1}{2}(a + b + c) = \dfrac{1}{2}(9 + 6 + 4) = \dfrac{19}{2}$

$A = \sqrt{s(s - a)(s - b)(s - c)} = \sqrt{\dfrac{19}{2} \cdot \dfrac{1}{2} \cdot \dfrac{7}{2} \cdot \dfrac{11}{2}}$

$= \sqrt{\dfrac{1463}{16}} = \sqrt{91.4375} \approx 9.5623$

9. $A = \frac{1}{2}ab \sin \gamma$

   $A = \frac{1}{2} \cdot 3 \cdot 4 \sin 40° = 3.8567$

11. $A = \frac{1}{2}ab \sin \alpha$

   $A = \frac{1}{2} \cdot 1 \cdot 3 \sin 80° = 1.4772$

13. $A = \frac{1}{2}ac \sin \beta$

   $A = \frac{1}{2} \cdot 3 \cdot 2 \sin 110° = 2.8191$

15. $A = \frac{1}{2}ab \sin \gamma$

   $A = \frac{1}{2} \cdot 2 \cdot 2 \sin 50° = 1.5321$

17. $s = \frac{1}{2}(a + b + c) = \frac{1}{2}(12 + 13 + 5) = 15$

   $A = \sqrt{s(s - a)(s - b)(s - c)} = \sqrt{15 \cdot 3 \cdot 2 \cdot 10} = \sqrt{900} = 30$

19. $s = \frac{1}{2}(a + b + c) = \frac{1}{2}(2 + 2 + 2) = 3$

   $A = \sqrt{s(s - a)(s - b)(s - c)} = \sqrt{3 \cdot 1 \cdot 1 \cdot 1} = \sqrt{3} \approx 1.7321$

21. $s = \frac{1}{2}(a + b + c) = \frac{1}{2}(5 + 8 + 9) = 11$

   $A = \sqrt{s(s - a)(s - b)(s - c)} = \sqrt{11 \cdot 6 \cdot 3 \cdot 2} = \sqrt{396} \approx 19.8997$

23. $s = \frac{1}{2}(a + b + c) = \frac{1}{2}(10 + 8 + 5) = \frac{23}{2}$

   $A = \sqrt{s(s - a)(s - b)(s - c)} = \sqrt{\frac{23}{2} \cdot \frac{3}{2} \cdot \frac{7}{2} \cdot \frac{13}{2}} = \sqrt{\frac{6279}{16}}$

   $= \sqrt{392.4375} \approx 19.8100$

25. We let $a = 100$, $b = 50$, and $c = 75$. The area of the lot can be found using Heron's Formula:

   $$s = \frac{1}{2}(a + b + c) = \frac{1}{2}(100 + 50 + 75) = \frac{225}{2}$$

   $$A = \sqrt{s(s - a)(s - b)(s - c)} = \sqrt{\frac{225}{2} \cdot \frac{25}{2} \cdot \frac{125}{2} \cdot \frac{75}{2}}$$

   $$= \sqrt{\frac{51,734,375}{16}} = \sqrt{3,295,898.438}$$

   $$\approx 1815.4609 \text{ square feet}$$

   If the price of the land is $3 per square foot, the triangular lot costs 1815.4609 × $3 = $5446.38.

27. We know that $A = \frac{1}{2}ab \sin \gamma$. We use this formula to prove that

$$A = \frac{a^2 \sin \beta \sin \gamma}{2 \sin \alpha}$$

$$A = \frac{1}{2}ab \sin \gamma = \frac{1}{2}a \sin \gamma\left(\frac{a \sin \beta}{\sin \alpha}\right) = \frac{a^2 \sin \beta \sin \gamma}{2 \sin \alpha}$$

29. $A = \frac{a^2 \sin \beta \sin \gamma}{2 \sin \alpha}$, $\gamma = 180° - \alpha - \beta = 120°$

$$A = \frac{2^2 \sin 20° \sin 120°}{2 \sin 40°} = \frac{4(0.3420)(0.8660)}{2(0.6428)} \approx 0.9216$$

31. $A = \frac{b^2 \sin \beta \sin \gamma}{2 \sin \beta}$, $\alpha = 180° - \beta - \gamma = 100°$

$$A = \frac{25 \sin 100° \sin 10°}{2 \sin 70°} = \frac{25(0.9848)(0.1736)}{2(0.9397)} \approx 2.2748$$

33. $A = \frac{c^2 \sin \alpha \sin \beta}{2 \sin \gamma}$, $\beta = 180° - \alpha - \gamma = 40°$

$$A = \frac{9 \sin 110° \sin 40°}{2 \sin 30°} = \frac{9(0.9397)(0.6428)}{2(0.5)} \approx 5.4362$$

35. $A = \frac{c^2 \sin \alpha \sin \beta}{2 \sin \gamma}$, $\gamma = 180° - \alpha - \beta = 100°$

$$A = \frac{4 \sin 40° \sin 40°}{2 \sin 100°} = \frac{4(0.6428)(0.6428)}{2(0.9848)} \approx 0.8391$$

37. The area A is the sum of the area of a triangle and a sector. Thus,

$$A = \frac{1}{2}r \cdot r \sin(\pi - \theta) + \frac{1}{2}r^2\theta$$
$$= \frac{1}{2}r^2[\sin(\pi - \theta) + \theta]$$
$$= \frac{1}{2}r^2[\sin \pi \cos \theta + \cos \pi \sin \theta + \theta]$$
$$= \frac{1}{2}r^2[0 - \sin \theta + \theta]$$
$$= \frac{1}{2}r^2(\theta - \sin \theta)$$

8.3 THE AREA OF A TRIANGLE

≡ EXERCISE 8.4 POLAR COORDINATES

1.

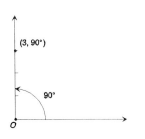

3.

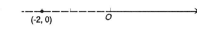

5.

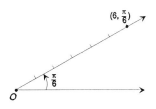

7.

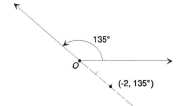

9.

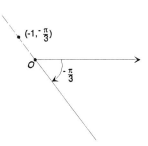

11.

13.

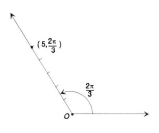

(a) $r > 0,\ -2\pi < \theta < 0$

$$\left(5,\ \frac{2\pi}{3} - 2\pi\right) = \left(5,\ \frac{-4\pi}{3}\right)$$

(b) $r < 0,\ 0 \le \theta < 2\pi$

$$\left(-5,\ \frac{2\pi}{3} + \pi\right) = \left(-5,\ \frac{5\pi}{3}\right)$$

(c) $r > 0,\ 2\pi \le \theta < 4\pi$

$$\left(5,\ \frac{2\pi}{3} + 2\pi\right) = \left(5,\ \frac{8\pi}{3}\right)$$

15.

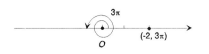

(a) $(2,\ 3\pi - \pi - 4\pi) = (2,\ -2\pi)$

(b) $(-2,\ 3\pi - 2\pi) = (-2,\ \pi)$

(c) $(2,\ 3\pi - \pi) = (2,\ 2\pi)$

17.

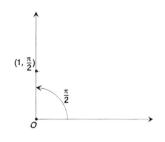

(a) $\left(1, \dfrac{\pi}{2} - 2\pi\right) = \left(1, \dfrac{-3\pi}{2}\right)$

(b) $\left(-1, \dfrac{\pi}{2} + \pi\right) = \left(-1, \dfrac{3\pi}{2}\right)$

(c) $\left(1, \dfrac{\pi}{2} + 2\pi\right) = \left(1, \dfrac{5\pi}{2}\right)$

19.

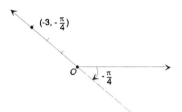

(a) $\left(3, \dfrac{-\pi}{4} - \pi\right) = \left(3, \dfrac{-5\pi}{4}\right)$

(b) $\left(-3, \dfrac{-\pi}{4} + 2\pi\right) = \left(-3, \dfrac{7\pi}{4}\right)$

(c) $\left(3, \dfrac{-\pi}{4} + \pi + 2\pi\right) = \left(3, \dfrac{11\pi}{4}\right)$

21. $x = r \cos \theta = 3 \cos \dfrac{\pi}{2} = 3 \cdot 0 = 0$

$y = r \sin \theta = 3 \sin \dfrac{\pi}{2} = 3 \cdot 1 = 3$

The rectangular coordinates of the point $\left(3, \dfrac{\pi}{2}\right)$ are (0, 3).

23. $x = r \cos \theta = -2 \cos 0 = -2 \cdot 0 = -2$
$y = r \sin \theta = -2 \sin 0 = -2 \cdot 0 = 0$

The rectangular coordinates of the point (-2, 0) are (-2, 0).

25. $x = r \cos \theta = 6 \cos 150° = 6 \cdot \dfrac{-\sqrt{3}}{2} = -3\sqrt{3}$

$y = r \sin \theta = 6 \sin 150° = 6 \cdot \dfrac{1}{2} = 3$

The rectangular coordinates of the point (6, 150°) are $\left(-3\sqrt{3}, 3\right)$.

27. $x = r \cos \theta = -2 \cos \dfrac{3\pi}{4} = -2 \cdot \dfrac{-\sqrt{2}}{2} = \sqrt{2}$

$y = r \sin \theta = -2 \sin \dfrac{3\pi}{4} = -2 \cdot \dfrac{\sqrt{2}}{2} = -\sqrt{2}$

The rectangular coordinates of the point $\left(-2, \dfrac{3\pi}{4}\right)$ are $\left(\sqrt{2}, -\sqrt{2}\right)$.

29. $x = r \cos \theta = -1 \cos \dfrac{-\pi}{3} = -1 \cdot \dfrac{1}{2} = \dfrac{-1}{2}$

$y = r \sin \theta = -1 \sin \dfrac{-\pi}{3} = -1 \cdot \dfrac{-\sqrt{3}}{2} = \dfrac{\sqrt{3}}{2}$

The rectangular coordinates of the point $\left(-1, \dfrac{-\pi}{3}\right)$ are $\left(\dfrac{-1}{2}, \dfrac{\sqrt{3}}{2}\right)$.

31. $x = r \cos \theta = -2 \cos -180° = -2 \cdot -1 = 2$
    $y = r \sin \theta = -2 \sin -180° = -2 \cdot 0 = 0$

    The rectangular coordinates of the point $(-2, -180°)$ are $(2, 0)$.

33. $x = r \cos \theta = 7.5 \cos 110° = -2.565$
    $y = r \sin \theta = 7.5 \sin 110° = 7.048$

    or

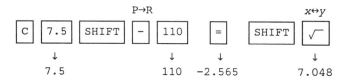

35. $x = 6.3 \cos 3.8 = -4.983$
    $y = 6.3 \sin 3.8 = -3.855$

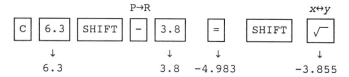

37. Since the point $(3, 0)$ lies on the positive x-axis, then

    $r = \sqrt{x^2 + y^2} = \sqrt{9 + 0} = 3$ and
    $\theta = \tan^{-1} \dfrac{y}{x} = \tan^{-1} \dfrac{0}{3} = 0.$

    Thus, polar coordinates are $(3, 0)$.

39. Since the point $(-1, 0)$ lies on the negative x-axis, then

    $r = -\sqrt{1 + 0} = -1$ and $\theta = \tan^{-1} \dfrac{0}{-1} = 0.$

    Thus, polar coordinates are $(-1, 0)$.

41. Since the point $(1, -1)$ lies in quadrant IV, then

    $r = \sqrt{1 + 1} = \sqrt{2}$ and $\theta = \tan^{-1}(-1) = \dfrac{-\pi}{4}.$

    Thus, polar coordinates are $\left(\sqrt{2}, \dfrac{-\pi}{4}\right)$.

43. Since the point $\left(\sqrt{3}, 1\right)$ lies in quadrant I, then
    $r = \sqrt{3 + 1} = 2$ and $\theta = \tan^{-1} \dfrac{1}{\sqrt{3}} = \dfrac{\pi}{6}.$

    Thus, polar coordinates are $\left(2, \dfrac{\pi}{6}\right)$.

45. $x = 1.3$, $y = -2.1$

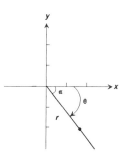

$$r = \sqrt{x^2 + y^2} = \sqrt{6.1} = 2.47$$
$$\alpha = \tan^{-1}\left|\frac{y}{x}\right| = \tan^{-1}(1.615) = 1.02$$
$$\theta = -\alpha = -1.02$$

Polar coordinates of (x, y) are $(2.47, -1.02)$.

C	1.3	SHIFT	+	2.1		+/-	=	SHIFT	√

↓   ↓   ↓   ↓

1.3   -2.1   2.47   -1.02

47. $x = 8.3$, $y = 4.2$

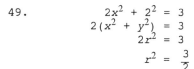

$$r = \sqrt{x^2 + y^2} = \sqrt{86.53} = 9.3$$
$$\alpha = \tan^{-1}\left|\frac{y}{x}\right| = \tan^{-1}(0.506) = 0.47$$
$$\theta = \alpha = 0.47$$

Polar coordinates of (x, y) are $(9.3, 0.47)$.

C	8.3	SHIFT	+	4.2	=	SHIFT	√

↓   ↓   ↓   ↓

8.3   4.2   8.3   4.2

49.
$$2x^2 + 2^2 = 3$$
$$2(x^2 + y^2) = 3$$
$$2r^2 = 3$$
$$r^2 = \frac{3}{2}$$

51.
$$x^2 = 4y$$
$$(r \cos \theta)^2 = 4r \sin \theta$$
$$r^2 \cos^2 \theta - 4r \sin \theta = 0$$

53.
$$2xy = 1$$
$$2(r \cos \theta)(r \sin \theta) = 1$$
$$2r^2 \cos \theta \sin \theta = 1$$
$$r^2 \sin 2\theta = 1$$

55.
$$x = 4$$
$$r \cos \theta = 4$$

8.4 POLAR COORDINATES

57.
$$r = \cos \theta$$
$$r^2 = r \cos \theta$$
$$x^2 + y^2 = x$$
$$x^2 + y^2 - x = 0$$
$$(x^2 - x) + y^2 = 0$$
$$\left(x^2 - x + \frac{1}{4}\right) + y^2 = 0 + \frac{1}{4}$$
$$\left(x - \frac{1}{2}\right)^2 + y^2 = \frac{1}{4}$$

This is the equation of a circle with center at $\left(\frac{1}{2},\ 0\right)$ and its radius $\frac{1}{2}$.

59.
$$r^2 = \cos \theta$$
$$r^3 = r \cos \theta$$
$$(x^2 + y^2)\left(\pm\sqrt{x^2 + y^2}\right) = x$$
$$(x^2 + y^2)^{3/2} - x = 0$$

61.
$$r = 2$$
$$\pm\sqrt{x^2 + y^2} = 2$$
$$x^2 + y^2 = 4$$

This is the equation of a circle with center $(0, 0)$ and radius 2.

63.
$$r = \frac{4}{1 - \cos \theta}$$
$$r(1 - \cos \theta) = 4$$
$$r - r \cos \theta = 4$$
$$\pm\sqrt{x^2 + y^2} - x = 4$$
$$\pm\sqrt{x^2 + y^2} = 4 + x$$
$$x^2 + y^2 = 16 + 8x + x^2$$
$$y^2 = 8(x + 2)$$

65. $P_1 = (r_1,\ \theta_1)$ and $P_2 = (r_2,\ \theta_2)$
or $P_1 = (r_1,\ \cos \theta_1,\ r_1 \sin \theta_1)$ and $P_2 = (r_2,\ \cos \theta_2,\ r_2,\ \sin \theta_2)$

$$d = \sqrt{(r_2 \cos \theta_2 - r_1 \cos \theta_1)^2 + (r_2 \sin \theta_2 - r_1 \sin \theta_1)^2}$$
$$= \sqrt{r_2^2 \cos^2 \theta_2 - 2r_1 r_2 \cos \theta_2 \cos \theta_1 + r_1^2 \cos^2 \theta_1 + r_2^2 \sin^2 \theta_2 - 2r_1 r_2 \sin \theta_2 \sin \theta_1 + r_1^2 \sin^2 \theta}$$
$$= \sqrt{r_2^2(\cos^2 \theta_2 + \sin^2 \theta_2) + r_1^2(\cos^2 \theta_1 + \sin^2 \theta_1) - 2r_1 r_2(\cos \theta_2 \cos \theta_1 + \sin \theta_2 \sin \theta_1)}$$
$$= \sqrt{r_1^2 + r_2^2 - 2r_1 r_2 \cos(\theta_2 - \theta_1)}$$

8 ADDITIONAL APPLICATIONS OF TRIGONOMETRY

≡ EXERCISE 8.5 POLAR EQUATIONS AND GRAPHS

1.　$r = 4$

This is of the form $r = a$, $a > 0$. Thus, by Table 7, the graph of $r = 4$ is a circle, center at the pole and radius 4.

If we convert the polar equation to a rectangular equation, then

$$r = 4$$
$$r^2 = 16$$
$$x^2 + y^2 = 16$$

Circle: Center at the pole and radius 4.

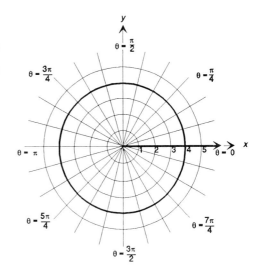

3.　$\theta = \dfrac{\pi}{3}$

This is of the form $\theta = \alpha$. Thus, by Table 7, the graph of $\theta = \dfrac{\pi}{3}$ is a line passing through the pole making an angle of $\dfrac{\pi}{3}$ with the polar axis.

If we convert the polar equation to a rectangular equation, then

$$\theta = \frac{\pi}{3}$$
$$\tan \theta = \tan \frac{\pi}{3} = \sqrt{3}$$
$$\frac{y}{x} = \sqrt{3}$$
$$y = \sqrt{3}\,x$$

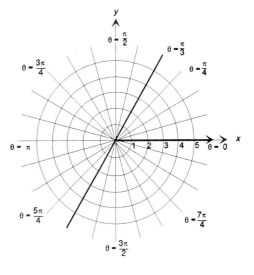

This is the graph of a line passing through the pole making an angle of $\dfrac{\pi}{3}$ with the polar axis.

5. $r \sin \theta = 4$

This is of the form
$r \sin \theta = b$. Thus, by Table
7, the graph of $r \sin \theta = 4$ is
a horizontal line. Since
$y = r \sin \theta$, we can write the
rectangular equation as

$$y = 4$$

We conclude that the graph of
$r \sin \theta = 4$ is a horizontal
line 4 units above the pole.

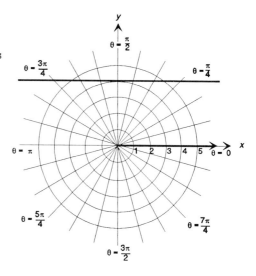

7. $r \cos \theta = -2$

This is of the form
$r \cos \theta = a$. Thus, by Table
7, the graph of $r \cos \theta = -2$
is a vertical line.

Since $x = r \cos \theta$, we can
write the rectangular equation
as

$$x = -2$$

which is the graph of a
vertical line 2 units to the
left of the pole.

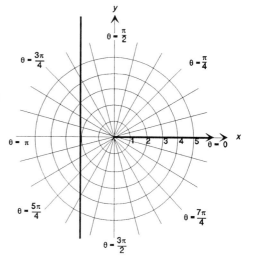

8 ADDITIONAL APPLICATIONS OF TRIGONOMETRY

9. $r = 2 \cos \theta$

This is of the form
$r = 2a \cos \theta$, $a > 0$.   Thus, by
Table 7, the graph of
$r = 2 \cos \theta$ is a circle,
passing through the pole,
tangent to the line $\theta = \frac{\pi}{2}$,
center on polar axis.

If we convert to a rectangular
equation,
$$r = 2 \cos \theta$$
$$r^2 = 2r \cos \theta$$
$$x^2 + y^2 = 2x$$
$$x^2 - 2x + y^2 = 0$$
$$(x - 1)^2 + y^2 = 1$$

This is the equation of a
circle, center at (1, 0) in
rectangular coordinates and
radius 1.

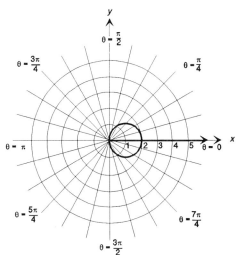

11. $r = -4 \sin \theta$ is the graph of a
circle which passes through
the pole, is tangent to the
polar axis with center on line
$\theta = \frac{\pi}{2}$.   Converting to
rectangular coordinates, we
have
$$r = -4 \sin \theta$$
$$r^2 = -4r \sin \theta$$
$$x^2 + y^2 = -4y$$
$$x^2 + y^2 + 4y = 0$$
$$x^2 + (y + 2)^2 = 4$$

This is the equation of a
circle, center at (0, -2) in
rectangular coordinates or
$\left(2, \frac{3\pi}{2}\right)$ in polar coordinates,
and radius 2.

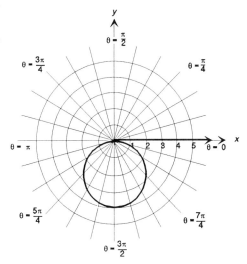

13. We convert the polar equation to a rectangular equation.

$$r \sec \theta = 4$$
$$r \frac{1}{\cos \theta} = 4$$
$$r = 4 \cos \theta$$
$$r^2 = 4r \cos \theta$$
$$x^2 + y^2 = 4x$$
$$x^2 - 4x + y^2 = 0$$
$$(x - 2)^2 + y^2 = 4$$

This is the equation of a circle, center (2, 0) in rectangular coordinates, (2, 0) in polar coordinates, and radius 2.

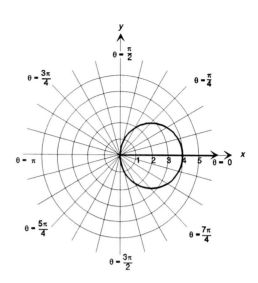

15. We convert to a rectangular equation.

$$r \csc \theta = -2$$
$$r \frac{1}{\sin \theta} = -2$$
$$r = -2 \sin \theta$$
$$r^2 = -2r \sin \theta$$
$$x^2 + y^2 = -2y$$
$$x^2 + y^2 + 2y = 0$$
$$x^2 + (y + 1)^2 = 1$$

This is the equation of a circle, center (0, -1) in rectangular coordinates, $\left(1, \frac{3\pi}{2}\right)$ in polar coordinates, and radius 1.

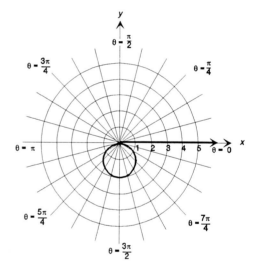

17. $r = 2$; E

19. $r = 2 \cos \theta$; F

21. $r = 1 + \cos \theta$; H

23. $\theta = \frac{3\pi}{4}$; D

25. $r = 2 + 2 \cos \theta$

We check for symmetry first.

Polar axis: Replace θ by $-\theta$. The result is
$$r = 2 + 2 \cos(-\theta) = 2 + 2 \cos \theta$$

Thus, the graph is symmetric with respect to the polar axis.

The line $\theta = \dfrac{\pi}{2}$:   Replace θ by $\pi - \theta$.

$$\begin{aligned}
r &= 2 + 2 \cos(\pi - \theta) \\
&= 2 + 2(\cos \pi \cos \theta + \sin \pi \sin \theta) \\
&= 2 + 2(-\cos \theta + 0) = 2 - 2 \cos \theta
\end{aligned}$$

The test fails.

The pole: Replace r by $-r$
$$-r = 2 + 2 \cos \theta$$

The test fails.

Next, we identify points on the graph by assigning values to the angle θ and calculating the corresponding values of r. Due to the symmetry with respect to the polar axis, we only need to assign values to θ from 0 to π.

θ	0	$\dfrac{\pi}{6}$	$\dfrac{\pi}{3}$	$\dfrac{\pi}{2}$	$\dfrac{2\pi}{3}$	$\dfrac{5\pi}{6}$	π
$r = 2 + 2 \cos \theta$	4	$2 + \sqrt{3} \approx 3.7$	3	2	1	$2 - \sqrt{3} \approx 0.3$	0
(r, θ)	$(4, 0)$	$\left(3.7, \dfrac{\pi}{6}\right)$	$\left(3, \dfrac{\pi}{3}\right)$	$\left(2, \dfrac{\pi}{2}\right)$	$\left(1, \dfrac{2\pi}{3}\right)$	$\left(0.3, \dfrac{5\pi}{6}\right)$	$(0, \pi)$

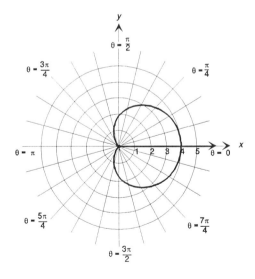

27. $r = 3 - 3 \sin \theta$

We check for symmetry first.

Polar axis: $r = 3 - 3 \sin(-\theta) = 3 + 3 \sin \theta$
        The test fails.

The line $\theta = \frac{\pi}{2}$: $r = 3 - 3 \sin(\pi - \theta)$

$$= 3 - 3(\sin \pi \cos \theta - \cos \pi \sin \theta)$$
$$= 3 - 3[0 - (-\sin \theta]$$
$$= 3 - 3 \sin \theta$$

Thus, the graph is symmetric with respect to the line $\theta = \frac{\pi}{2}$.

The pole: $-r = 3 - 3 \sin \theta$

The test fails.

Due to the symmetry with respect to the line $\theta = \frac{\pi}{2}$, we only need to assign values to θ from $\frac{-\pi}{2}$ to $\frac{\pi}{2}$.

θ	$\frac{-\pi}{2}$	$\frac{-\pi}{3}$	$\frac{-\pi}{6}$	0	$\frac{\pi}{6}$	$\frac{\pi}{3}$	$\frac{\pi}{2}$
$r = 3 - 3 \sin \theta$	6	$3 + \frac{3\sqrt{3}}{2} \approx 5.6$	$\frac{9}{2}$	3	$\frac{3}{2}$	$3 - \frac{3\sqrt{3}}{2} \approx 0.4$	0
(r, θ)	$\left(6, \frac{-\pi}{2}\right)$	$\left(5.6, \frac{-\pi}{3}\right)$	$\left(\frac{9}{2}, \frac{-\pi}{6}\right)$	$(3, 0)$	$\left(\frac{3}{2}, \frac{\pi}{6}\right)$	$\left(0.4, \frac{\pi}{3}\right)$	$\left(0, \frac{\pi}{2}\right)$

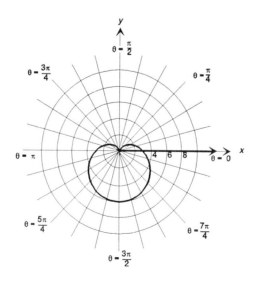

29. $r = 2 + \sin \theta$

We check for symmetry.

Polar axis: $r = 2 + \sin(-\theta) = 2 - \sin \theta$
The test fails.

The line $\theta = \dfrac{\pi}{2}$: $r = 2 + \sin(\pi - \theta)$
$$= 2 + (\sin \pi \cos \theta - \cos \pi \sin \theta)$$
$$= 2 + \sin \theta$$

Thus, the graph is symmetric with respect to the line $\theta = \dfrac{\pi}{2}$.

The pole: $-r = 2 + \sin \theta$

The test fails.

Due to the symmetry with respect to the line $\theta = \dfrac{\pi}{2}$, we only need to assign values to θ from $\dfrac{-\pi}{2}$ to $\dfrac{\pi}{2}$.

θ	$\dfrac{-\pi}{2}$	$\dfrac{-\pi}{3}$	$\dfrac{-\pi}{6}$	0	$\dfrac{\pi}{6}$	$\dfrac{\pi}{3}$	$\dfrac{\pi}{2}$
$r = 2 + \sin \theta$	1	$2 - \dfrac{\sqrt{3}}{2} \approx 1.1$	$\dfrac{3}{2}$	2	$\dfrac{5}{2}$	$2 + \dfrac{\sqrt{3}}{2} \approx 2.9$	3
(r, θ)	$\left(1, \dfrac{-\pi}{2}\right)$	$\left(1.1, \dfrac{-\pi}{3}\right)$	$\left(\dfrac{3}{2}, \dfrac{-\pi}{6}\right)$	$(2, 0)$	$\left(\dfrac{5}{2}, \dfrac{\pi}{6}\right)$	$\left(2.9, \dfrac{\pi}{3}\right)$	$\left(3, \dfrac{\pi}{2}\right)$

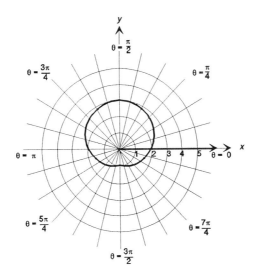

31. $r = 4 - 2 \cos \theta$

We check for symmetry.

Polar axis: $r = 4 - 2 \cos(-\theta) = 4 - 2 \cos \theta$
  Thus, the graph is symmetric with respect to the polar axis.

The line $\theta = \dfrac{\pi}{2}$: $r = 4 = -2 \cos(\pi - \theta)$

$$= 4 - 2(\cos \pi \cos \theta + \sin \pi \sin \theta)$$
$$= 4 - 2(-\cos \theta)$$
$$= 4 + 2 \cos \theta$$
The test fails.

The pole: $-r = 4 - 2 \cos \theta$
  The test fails.

Due to the symmetry with respect to the polar axis, we only need to assign values to θ from 0 to π.

θ	0	$\dfrac{\pi}{6}$	$\dfrac{\pi}{4}$	$\dfrac{\pi}{3}$	$\dfrac{\pi}{2}$	$\dfrac{2\pi}{3}$
$r = 4 - 2 \cos \theta$	2	$4 - \sqrt{3} \approx 2.3$	$4 - \sqrt{2} \approx 2.6$	3	4	5
(r, θ)	$(2, 0)$	$\left(2.3, \dfrac{\pi}{6}\right)$	$\left(2.6, \dfrac{\pi}{4}\right)$	$\left(3, \dfrac{\pi}{3}\right)$	$\left(4, \dfrac{\pi}{2}\right)$	$\left(5, \dfrac{2\pi}{3}\right)$

θ	$\dfrac{5\pi}{6}$	π
$r = 4 - 2 \cos \theta$	$4 + \sqrt{3} \approx 5.7$	6
(r, θ)	$\left(5.7, \dfrac{5\pi}{6}\right)$	$(6, \pi)$

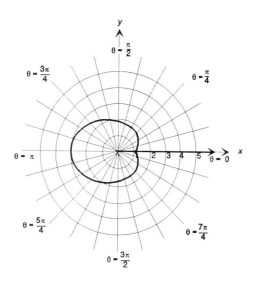

8 ADDITIONAL APPLICATIONS OF TRIGONOMETRY

33. $r = 1 + 2 \sin \theta$

We check for symmetry.

Polar axis: $r = 1 + 2 \sin(-\theta) = 1 - 2 \sin \theta$
The test fails.

The line $\theta = \dfrac{\pi}{2}$: $r = 1 + 2 \sin(\pi - \theta)$
$= 1 + 2 \sin \theta$

Thus, the graph is symmetric with respect to the line $\theta = \dfrac{\pi}{2}$.

The pole: $-r = 1 + 2 \sin \theta$
The test fails.

Due to the symmetry with respect to the line $\theta = \dfrac{\pi}{2}$, we only need to assign values to θ from $\dfrac{-\pi}{2}$ to $\dfrac{\pi}{2}$.

θ	$\dfrac{-\pi}{2}$	$\dfrac{-\pi}{3}$	$\dfrac{-\pi}{6}$	0	$\dfrac{\pi}{6}$	$\dfrac{\pi}{3}$	$\dfrac{\pi}{2}$
$r = 1 + 2 \sin \theta$	-1	$1 - \sqrt{3} \approx -0.7$	0	1	2	$1 + \sqrt{3} \approx 2.7$	3
(r, θ)	$\left(-1, \dfrac{-\pi}{2}\right)$	$\left(-0.7, \dfrac{-\pi}{3}\right)$	$\left(0, \dfrac{-\pi}{6}\right)$	$(1, 0)$	$\left(2, \dfrac{\pi}{6}\right)$	$\left(2.7, \dfrac{\pi}{3}\right)$	$\left(3, \dfrac{\pi}{2}\right)$

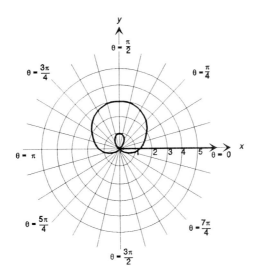

35. $r = 2 - 3 \cos \theta$

We check for symmetry.

Polar axis: $r = 2 - 3 \cos(-\theta) = 2 - 3 \cos \theta$
Thus, the graph is symmetric with respect to the polar axis.

The line $\theta = \dfrac{\pi}{2}$: $r = 2 - 3 \cos(\pi - \theta) = 2 + 3 \cos \theta$
The test fails.

The pole: $-r = 2 - 3 \cos \theta$
The test fails.

Due to the symmetry with respect to the polar axis, we only need to assign values to θ from 0 to π.

θ	0	$\dfrac{\pi}{6}$	$\dfrac{\pi}{3}$	$\dfrac{\pi}{2}$	$\dfrac{2\pi}{3}$	$\dfrac{5\pi}{6}$	π
$r = 2 - 3 \cos \theta$	-1	$2 - \dfrac{3\sqrt{3}}{2} \approx -0.6$	$\dfrac{1}{2}$	2	$\dfrac{7}{2}$	$2 + \dfrac{3\sqrt{3}}{2} \approx 4.6$	5
(r, θ)	$(-1, 0)$	$\left(-0.6, \dfrac{\pi}{6}\right)$	$\left(\dfrac{1}{2}, \dfrac{\pi}{3}\right)$	$\left(2, \dfrac{\pi}{3}\right)$	$\left(\dfrac{7}{2}, \dfrac{2\pi}{3}\right)$	$\left(4.6, \dfrac{5\pi}{6}\right)$	$(5, \pi)$

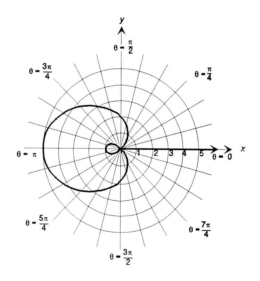

37.　$r = 3 \cos 2\theta$

We check for symmetry.

Polar axis: $r = 3 \cos 2(-\theta) = 3 \cos (-2\theta) = 3 \cos 2\theta$
　　　　Thus, the graph is symmetric with respect to the polar
　　　　axis.

The line $\theta = \frac{\pi}{2}$:　　$r = 3 \cos 2(\pi - \theta) = 2 \cos 2\theta$

Thus, the graph is symmetric with respect to
the line $\theta = \frac{\pi}{2}$. Consequently, since the
graph is symmetric with respect to both the
polar axis and the line $\theta = \frac{\pi}{2}$, it must be
symmetric with respect to the pole. Due to
the fact that $r = 3 \cos 2\theta$ has period π and is
symmetric with respect to the polar axis, the
line $\theta = \frac{\pi}{2}$, and the pole, we only consider
values of θ from 0 to $\frac{\pi}{2}$.

θ	0	$\frac{\pi}{6}$	$\frac{\pi}{4}$	$\frac{\pi}{3}$	$\frac{\pi}{2}$
$r = 3 \cos 2\theta$	3	$\frac{3}{2}$	0	$\frac{-3}{2}$	-3
(r, θ)	$(3, 0)$	$\left(\frac{3}{2}, \frac{\pi}{6}\right)$	$\left(0, \frac{\pi}{4}\right)$	$\left(\frac{-3}{2}, \frac{\pi}{3}\right)$	$\left(-3, \frac{\pi}{2}\right)$

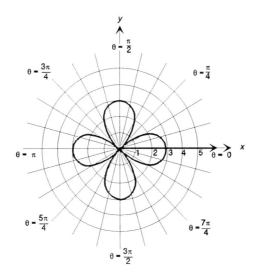

39. $r = 4 \sin 3\theta$

We check for symmetry.

Polar axis: $r = 4 \sin 3(-\theta) = -4 \sin 3\theta$
The test fails.

The line $\theta = \dfrac{\pi}{2}$: $r = -4 \sin 3(\pi - \theta) = 4 \sin 3\theta$

Thus, the graph is symmetric with respect to $\theta = \dfrac{\pi}{2}$.

The pole: $-r = 4 \sin 3\theta$
The test fails.

Due to the symmetry with respect to the line $\theta = \dfrac{\pi}{2}$, we only need to assign values to θ from $\dfrac{-\pi}{2}$ to $\dfrac{\pi}{2}$.

θ	$\dfrac{-\pi}{2}$	$\dfrac{-\pi}{3}$	$\dfrac{-\pi}{4}$	$\dfrac{-\pi}{6}$	0	$\dfrac{\pi}{6}$
$r = 4 \sin 3\theta$	4	0	$-2\sqrt{2} \approx -2.8$	-4	0	4
(r, θ)	$\left(4, \dfrac{-\pi}{2}\right)$	$\left(0, \dfrac{-\pi}{3}\right)$	$\left(-2.8, \dfrac{-\pi}{4}\right)$	$\left(-4, \dfrac{-\pi}{6}\right)$	$(0, 0)$	$\left(4, \dfrac{\pi}{6}\right)$

θ	$\dfrac{\pi}{4}$	$\dfrac{\pi}{3}$	$\dfrac{\pi}{2}$
$r = 4 \sin 3\theta$	$2\sqrt{2} \approx 2.8$	0	-4
(r, θ)	$\left(2.8, \dfrac{\pi}{4}\right)$	$\left(0, \dfrac{\pi}{3}\right)$	$\left(-4, \dfrac{\pi}{2}\right)$

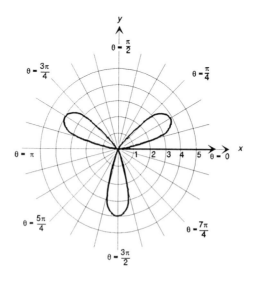

8 ADDITIONAL APPLICATIONS OF TRIGONOMETRY

41. $r^2 = 9 \cos 2\theta$

We check for symmetry.

Polar axis: $r^2 = 9 \cos 2(-\theta) = 9 \cos 2\theta$
           Thus, the graph is symmetric with respect to the polar axis.

The line $\theta = \dfrac{\pi}{2}$**:** $r^2 = 9 \cos 2(\pi - \theta) = 9 \cos 2\theta$
                 Thus, the graph is symmetric with respect to the line $\theta = \dfrac{\pi}{2}$.

The pole: $(-r)^2 = 9 \cos 2\theta$
               $r^2 = 9 \cos 2\theta$
      Thus, the graph is symmetric with respect to the pole.

Due to the fact that $r^2 = 9 \cos 2\theta$ has period π and is symmetric with respect to the polar axis, the line $\theta = \dfrac{\pi}{2}$, and the pole, we only consider values of θ from 0 to $\dfrac{\pi}{2}$.

θ	0	$\dfrac{\pi}{6}$	$\dfrac{\pi}{4}$
$r^2 = 9 \cos 2\theta$	9	$\dfrac{9}{2}$	0
$r = \pm\sqrt{9 \cos 2\theta}$	± 3	$\dfrac{\pm 3\sqrt{2}}{2} \approx \pm 2.1$	0
(r, θ)	(3, 9) (-3, 9)	$\left[2.1, \dfrac{\pi}{6}\right]$ $\left[-2.1, \dfrac{\pi}{6}\right]$	(0, 0)

Note there are no points on the graph for $\dfrac{\pi}{4} < \theta < \dfrac{3\pi}{4}$ since $\cos 2\pi < 0$ for such values.

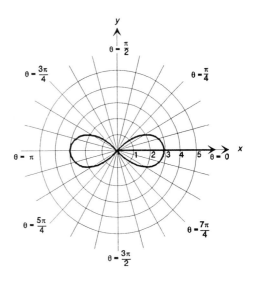

43. $r = 2^\theta$

We check for symmetry.

Polar axis: $r = 2^{-\theta}$
          The test fails.

The line $\theta = \dfrac{\pi}{2}$: $r = 2^{\pi-\theta}$

                    The test fails.

The pole: $-r = 2^\theta$
          The test fails.

There is no number θ for which $r = 0$. Hence, the graph does not pass through the pole. We observe that r is positive for all θ, that r increases as θ increases, that $r \to 0$ as $\theta \to -\infty$, and that $r \to \infty$ as $\theta \to \infty$.

θ	0	$\dfrac{\pi}{4}$	$\dfrac{\pi}{2}$	π	$\dfrac{3\pi}{2}$	2π
$r = 2^\theta$	1	1.7	3.0	8.8	26.2	77.9
(r, θ)	$(1, 0)$	$\left(1.7, \dfrac{\pi}{4}\right)$	$\left(3.0, \dfrac{\pi}{2}\right)$	$(8.8, \pi)$	$\left(26.2, \dfrac{3\pi}{2}\right)$	$(77.9, 2\pi)$

θ	$\dfrac{-\pi}{4}$	$\dfrac{-\pi}{2}$	$-\pi$	$\dfrac{-3\pi}{2}$
$r = 2^\theta$	0.6	0.3	0.1	0.0
(r, θ)	$\left(0.6, \dfrac{-\pi}{4}\right)$	$\left(0.3, \dfrac{-\pi}{2}\right)$	$(0.1, -\pi)$	$\left(0.0, \dfrac{-3\pi}{2}\right)$

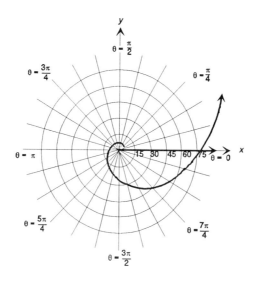

45. $r = 1 - \cos\theta$

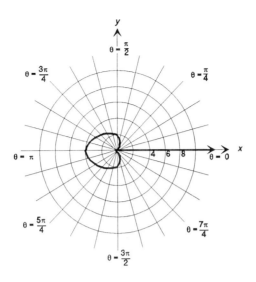

47. $r = 1 - 3\cos\theta$

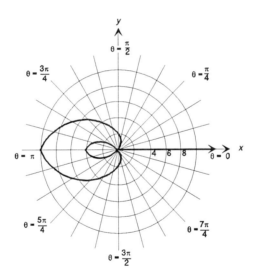

49. $r = \dfrac{2}{1 - \cos\theta}$

We check for symmetry.

Polar axis: $r = \dfrac{2}{1 - \cos(-\theta)} = \dfrac{2}{1 - \cos\theta}$

Thus, the graph is symmetric with respect to the polar axis.

8.5 POLAR EQUATIONS AND GRAPHS   433

The line $\theta = \frac{\pi}{2}$: $\qquad r = \dfrac{2}{1 - \cos(\pi - \theta)} = \dfrac{2}{1 + \cos\theta}$

$\qquad\qquad\qquad$ The test fails.

The pole: $\qquad -r = \dfrac{2}{1 - \cos\theta}$

$\qquad\qquad$ The test fails.

Due to symmetry with respect to the polar axis, we only need to assign values to θ from 0 to π. Since $1 - \cos 0 = 0$, $\theta \neq 0 \pm 2k\pi$, k any integer.

θ	$\frac{\pi}{6}$	$\frac{\pi}{4}$	$\frac{\pi}{3}$	$\frac{\pi}{2}$	$\frac{2\pi}{3}$
$r = \dfrac{2}{1 - \cos\theta}$	$\dfrac{2}{1 - \frac{\sqrt{3}}{2}} \approx 14.9$	$\dfrac{2}{1 - \frac{\sqrt{2}}{2}} \approx 6.8$	4	2	$\dfrac{4}{3}$
(r, θ)	$\left(14.9, \frac{\pi}{6}\right)$	$\left(6.8, \frac{\pi}{4}\right)$	$\left(4, \frac{\pi}{3}\right)$	$\left(2, \frac{\pi}{2}\right)$	$\left(\frac{4}{3}, \frac{2\pi}{3}\right)$

θ	$\frac{3\pi}{4}$	$\frac{5\pi}{6}$	π
$r = \dfrac{2}{1 - \cos\theta}$	$\dfrac{2}{1 + \frac{\sqrt{2}}{2}} \approx 1.2$	$\dfrac{2}{1 + \frac{\sqrt{3}}{2}} \approx 1.1$	1
(r, θ)	$\left(1.2, \frac{3\pi}{4}\right)$	$\left(1.1, \frac{5\pi}{6}\right)$	$(1, \pi)$

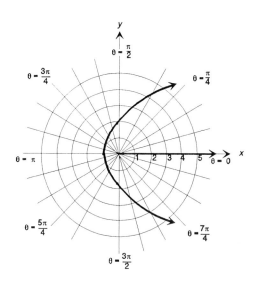

8 ADDITIONAL APPLICATIONS OF TRIGONOMETRY

51. $r = \dfrac{1}{3 - 2 \cos \theta}$

We check for symmetry.

Polar axis: $r = \dfrac{1}{3 - 2 \cos(-\theta)} = \dfrac{1}{3 - 2 \cos \theta}$

Thus, the graph is symmetric with respect to the polar axis.

The line $\theta = \dfrac{\pi}{2}$: $r = \dfrac{1}{3 - 2 \cos(\pi - \theta)} = \dfrac{1}{3 + 2 \cos \theta}$

The test fails.

The pole: $-r = \dfrac{1}{3 - 2 \cos \theta}$

The test fails.

Due to symmetry with respect to the polar axis, we only need to assign values to θ from 0 to π.

θ	0	$\dfrac{\pi}{6}$	$\dfrac{\pi}{4}$	$\dfrac{\pi}{3}$	$\dfrac{\pi}{2}$
$r = \dfrac{1}{3 - 2 \cos \theta}$	1	$\dfrac{1}{3 - \sqrt{3}} \approx 0.8$	$\dfrac{1}{3 - \sqrt{2}} \approx 0.6$	$\dfrac{1}{2}$	$\dfrac{1}{3}$
(r, θ)	$(1, 0)$	$\left[0.8, \dfrac{\pi}{6}\right]$	$\left[0.6, \dfrac{\pi}{4}\right]$	$\left[\dfrac{1}{2}, \dfrac{\pi}{3}\right]$	$\left[\dfrac{1}{3}, \dfrac{\pi}{2}\right]$

θ	$\dfrac{2\pi}{3}$	$\dfrac{3\pi}{4}$	$\dfrac{5\pi}{6}$	π
$r = \dfrac{1}{3 - 2 \cos \theta}$	$\dfrac{1}{4}$	$\dfrac{1}{3 + \sqrt{2}} \approx 0.2$	$\dfrac{1}{3 + \sqrt{3}} \approx 0.2$	$\dfrac{1}{5}$
(r, θ)	$\left[\dfrac{1}{4}, \dfrac{2\pi}{3}\right]$	$\left[0.2, \dfrac{3\pi}{4}\right]$	$\left[1.1, \dfrac{5\pi}{6}\right]$	$\left[\dfrac{1}{5}, \pi\right]$

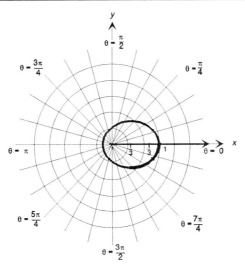

53. $r = \theta, \ \theta \geq 0$

The tests for symmetry with respect to the pole, polar axis, and the line $\theta = \dfrac{\pi}{2}$ fail. We observe that r increases as θ increases.

θ	0	$\dfrac{\pi}{4}$	$\dfrac{\pi}{2}$	π	$\dfrac{3\pi}{2}$	2π
$r = \theta$	0	$\dfrac{\pi}{4} \approx 0.8$	$\dfrac{\pi}{2} \approx 1.6$	$\pi \approx 3.1$	$\dfrac{3\pi}{2} \approx 4.7$	$2\pi \approx 6.3$
(r, θ)	$(0, 0)$	$\left(0.8, \dfrac{\pi}{4}\right)$	$\left(1.6, \dfrac{\pi}{2}\right)$	$(3.1, \pi)$	$\left(4.7, \dfrac{3\pi}{2}\right)$	$(6.3, 2\pi)$

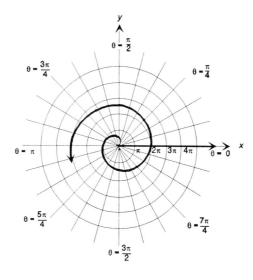

    8 ADDITIONAL APPLICATIONS OF TRIGONOMETRY

55. $r = \csc \theta - 2$, $0 < \theta < \pi$

   $r = \dfrac{1}{\sin \theta} - 2$

   The graph is symmetric with respect to the line $\theta = \dfrac{\pi}{2}$. We only

   need to assign values to θ from $\dfrac{-\pi}{2}$ to $\dfrac{\pi}{2}$. We observe that the

   points (r, θ) where $\theta = k\pi$, k any integer, are undefined.

θ	$\dfrac{-\pi}{2}$	$\dfrac{-\pi}{3}$	$\dfrac{-\pi}{4}$	$\dfrac{-\pi}{6}$
$r = \dfrac{1}{\sin \theta} - 2$	-3	$\dfrac{-2\sqrt{3}}{3} - 2 \approx -3.2$	$-\sqrt{2} - 2 \approx -3.4$	-4
(r, θ)	$\left(-3, \dfrac{-\pi}{2}\right)$	$\left(-3.2, \dfrac{-\pi}{3}\right)$	$\left(-3.4, \dfrac{-\pi}{4}\right)$	$\left(-4, \dfrac{-\pi}{6}\right)$
θ	$\dfrac{\pi}{6}$	$\dfrac{\pi}{4}$	$\dfrac{\pi}{3}$	$\dfrac{\pi}{2}$
$r = \dfrac{1}{\sin \theta} - 2$	0	$\sqrt{2} - 2 \approx -0.6$	$\dfrac{2\sqrt{3}}{3} - 2 \approx -0.8$	-1
(r, θ)	$\left(0, \dfrac{\pi}{6}\right)$	$\left(-0.6, \dfrac{\pi}{4}\right)$	$\left(-0.8, \dfrac{\pi}{3}\right)$	$\left(-1, \dfrac{\pi}{2}\right)$

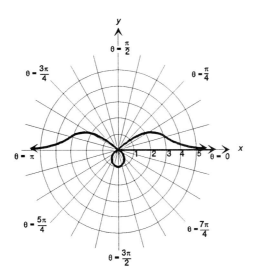

57. $r = \tan \theta$

Check for symmetry.

Polar axis: $r = \tan(-\theta) = -\tan \theta$
   The test fails.

The line $\theta = \dfrac{\pi}{2}$: $\quad r = \tan(\pi - \theta) = \dfrac{\tan \pi - \tan \theta}{1 + \tan \pi \tan \theta}$
$$= -\tan \theta$$
   The test fails.

The pole: $\quad -r = \tan \theta$
   The test fails.

θ	0	$\dfrac{\pi}{6}$	$\dfrac{\pi}{4}$	$\dfrac{\pi}{3}$	$\dfrac{2\pi}{3}$
$r = \tan \theta$	0	$\dfrac{1}{\sqrt{3}} \approx 0.6$	1	$\sqrt{3} \approx 1.7$	$-\sqrt{3} \approx -1.7$
(r, θ)	$(0, 0)$	$\left(0.6, \dfrac{\pi}{6}\right)$	$\left(1, \dfrac{\pi}{4}\right)$	$\left(1.7, \dfrac{\pi}{3}\right)$	$\left(-1.7, \dfrac{2\pi}{3}\right)$
θ	$\dfrac{3\pi}{4}$	$\dfrac{5\pi}{6}$	π	$\dfrac{7\pi}{6}$	$\dfrac{5\pi}{4}$
$r = \tan \theta$	-1	$\dfrac{-1}{\sqrt{3}} \approx -0.6$	0	$\dfrac{1}{\sqrt{3}} \approx 0.6$	1
(r, θ)	$\left(-1, \dfrac{3\pi}{4}\right)$	$\left(-0.6, \dfrac{5\pi}{6}\right)$	$(0, \pi)$	$\left(0.6, \dfrac{7\pi}{6}\right)$	$\left(1, \dfrac{5\pi}{4}\right)$
θ	$\dfrac{4\pi}{3}$	$\dfrac{5\pi}{3}$	$\dfrac{7\pi}{4}$	$\dfrac{11\pi}{6}$	2π
$r = \tan \theta$	$\sqrt{3} \approx 1.7$	$-\sqrt{3} \approx -1.7$	-1	$\dfrac{1}{-\sqrt{3}} \approx -0.6$	0
(r, θ)	$\left(1.7, \dfrac{4\pi}{3}\right)$	$\left(-1.7, \dfrac{5\pi}{3}\right)$	$\left(-1, \dfrac{7\pi}{4}\right)$	$\left(-0.6, \dfrac{11\pi}{6}\right)$	$(0, 2\pi)$

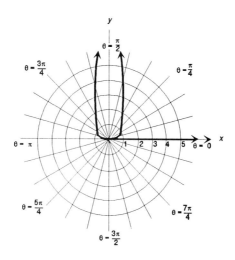

59.   We convert the polar equation to a rectangular equation.

$$r \sin \theta = a$$
$$y = a$$

Thus, the graph of $r \sin \theta = a$ is a horizontal line a units above the pole, if $a > 0$, and a units below the pole, if $a < 0$.

61.   We convert the polar equation to a rectangular equation.

$$r = 2a \sin \theta, \ a > 0$$
$$r^2 = 2ar \sin \theta$$
$$x^2 + y^2 = 2ay$$
$$x^2 + y^2 - 2ay = 0$$
$$x^2 + (y - a)^2 = a^2$$

Circle: radius a, center at rectangular coordinates $(0, a)$.

63.   We convert the polar equation to a rectangular equation.

$$r = 2a \cos \theta, \ a > 0$$
$$r^2 = 2ar \cos \theta$$
$$x^2 + y^2 = 2ax$$
$$x^2 - 2ax + y^2 = 0$$
$$(x - a)^2 + y^2 = a^2$$

Circle: radius a, center at rectangular coordinates $(a, 0)$.

65.   (a) $r^2 = \cos \theta$: $r^2 = \cos(\pi - \theta)$ $(-r^2) = \cos(-\theta)$
                          $r^2 = -\cos \theta$ $r^2 = \cos \theta$
                          Not equivalent; test fails   New test works

       (b) $r^2 = \sin \theta$: $r^2 = \sin(\pi - \theta)$ $(-r^2) = \sin(-\theta)$
                          $r^2 = \sin \theta$ $r^2 = -\sin \theta$
                          Test works   New test fails

≡ EXERCISE 8.6 THE COMPLEX PLANE; DEMOIVRE'S THEOREM

1. $r = \sqrt{x^2 + y^2} = \sqrt{(1)^2 + (1)^2} = \sqrt{2}$ and $\tan \theta = \dfrac{y}{x} = 1$

      Thus, $\theta = 45°$ and $r = \sqrt{2}$, so the polar form of $z = 1 + i$ is

          $z = r(\cos \theta + i \sin \theta) = \sqrt{2}(\cos 45° + i \sin 45°)$

3. $r = \sqrt{x^2 + y^2} = \sqrt{(\sqrt{3})^2 + (-1)^2} = \sqrt{4} = 2$ and $\tan \theta = \dfrac{y}{x} = \dfrac{-1}{\sqrt{3}}$

      Thus, $\theta = 330°$ and $r = 2$, so the polar form of $z = \sqrt{3} - i$ is

          $z = r(\cos \theta + i \sin \theta) = 2(\cos 330° + i \sin 330°)$

5. $r = \sqrt{x^2 + y^2} = \sqrt{(0)^2 + (-3)^2} = \sqrt{9} = 3$ and $\tan \theta = \dfrac{y}{x} = \dfrac{-3}{0}$ undefined.

Thus, $\theta = 270°$ and $r = 3$, so the polar form of $z = -3i$ is

$$z = r(\cos \theta + i \sin \theta) = 3(\cos 270° + i \sin 270°)$$

7. $r = \sqrt{x^2 + y^2} = \sqrt{(4)^2 + (-4)^2} = \sqrt{32} = 4\sqrt{2}$ and $\tan \theta = \dfrac{y}{x} = \dfrac{-4}{4} = -1$

Thus, $\theta = 315°$ and $r = 4\sqrt{2}$, so the polar form of $z = 4 - 4i$ is

$$z = r(\cos \theta + i \sin \theta) = 4\sqrt{2}(\cos 315° + i \sin 315°)$$

9. $r = \sqrt{x^2 + y^2} = \sqrt{(3)^2 + (-4)^2} = \sqrt{25} = 5$

$\tan \theta = \dfrac{y}{x} = \dfrac{-4}{3} \approx -1.3333$

$\theta = 306.9°$

The polar form of $z = 3 - 4i$ is

$$z = 5(\cos 306.9° + i \sin 306.9°)$$

11. $r = \sqrt{x^2 + y^2} = \sqrt{(-2)^2 + (3)^2} = \sqrt{13}$

$\tan \theta = \dfrac{y}{x} = \dfrac{-3}{2} \approx -1.5$

$\theta = 123.7°$

The polar form of $z = -2 + 3i$ is

$$z = \sqrt{13}(\cos 123.7° + i \sin 123.7°)$$

13. $2(\cos 120° + i \sin 120°) = 2\left(-\dfrac{1}{2} + \dfrac{\sqrt{3}}{2}i\right)$

$$= -1 + \sqrt{3}i$$

15. $4\left(\cos \dfrac{7\pi}{4} + i \sin \dfrac{7\pi}{4}\right) = 4\left(\dfrac{\sqrt{2}}{2} + -\dfrac{\sqrt{2}}{2}i\right)$

$$= 2\sqrt{2} - 2\sqrt{2}i$$

17. $3\left(\cos \dfrac{3\pi}{2} + i \sin \dfrac{3\pi}{2}\right) = 3(0 + -i)$

$$= -3i$$

19. $0.2(\cos 100° + i \sin 100°) = 0.2(-0.1736 + 0.9848i)$

$$= -0.035 + 0.197i$$

21. $2\left(\cos \dfrac{\pi}{18} + i \sin \dfrac{\pi}{18}\right) = 2(0.9848 + 0.1736i)$

$$= 1.97 + 0.347i$$

23. $\quad zw = 2 \cdot 4[\cos(40° + 20°) + i \sin(40° + 20°)$
$\quad\quad\quad = 8(\cos 60° + i \sin 60°)$

$\quad\quad \dfrac{z}{w} = \dfrac{2}{4}[\cos(40° - 20°) + i \sin(40° - 20°)]$

$\quad\quad \dfrac{z}{w} = \dfrac{1}{2}(\cos 20° + i \sin 20°)$

25. $\quad zw = 3 \cdot 4[\cos(130° + 270°) + i \sin(130° + 270°)]$
$\quad\quad zw = 12[\cos(400° - 360°) + i \sin(400° - 360°)]$
$\quad\quad zw = 12(\cos 40° + i \sin 40°)$

$\quad\quad \dfrac{z}{w} = \dfrac{3}{4}[\cos(130° - 270°) + i \sin(130° - 270°)]$

$\quad\quad \dfrac{z}{w} = \dfrac{3}{4}[\cos(-140°) + i \sin(-140°)]$

$\quad\quad \dfrac{z}{w} = \dfrac{3}{4}(\cos 220° + i \sin 220°)$

27. $\quad zw = 4\left[\cos\left(\dfrac{\pi}{8} + \dfrac{\pi}{10}\right) + i \sin\left(\dfrac{\pi}{8} + \dfrac{\pi}{10}\right)\right]$

$\quad\quad zw = 4\left(\cos \dfrac{9\pi}{40} + i \sin \dfrac{9\pi}{40}\right)$

$\quad\quad \dfrac{z}{w} = \dfrac{2}{2}\left[\cos\left(\dfrac{\pi}{8} - \dfrac{\pi}{10}\right) + i \sin\left(\dfrac{\pi}{8} - \dfrac{\pi}{10}\right)\right]$

$\quad\quad \dfrac{z}{3} = \cos \dfrac{\pi}{40} + i \sin \dfrac{\pi}{40}$

29. $\quad\quad\quad r = \sqrt{x^2 + y^2} = \sqrt{8} = 2\sqrt{2}$

$\quad\quad \tan \theta = \dfrac{y}{x} = 1, \ \theta = 45°$

$\quad\quad\quad z = 2\sqrt{2}(\cos 45° + i \sin 45°)$

$\quad\quad\quad r = \sqrt{x^2 + y^2} = \sqrt{4} = 2$

$\quad\quad \tan \theta = \dfrac{y}{x} = \dfrac{-1}{\sqrt{3}}, \ \theta = 330°$

$\quad\quad\quad w = 2(\cos 330° + i \sin 330°)$

$\quad\quad\quad zw = 2\sqrt{2} \cdot 2[\cos(45° + 330°) + i \sin(45° + 330°)]$

$\quad\quad\quad zw = 4\sqrt{2}[\cos(375° - 360°) + i \sin(375° - 360°)]$

$\quad\quad\quad zw = 4\sqrt{2}(\cos 15° + i \sin 15°)$

$\quad\quad\quad \dfrac{z}{w} = \dfrac{2\sqrt{2}}{2}[\cos(45° - 330°) + i \sin(45° - 330°)]$

$\quad\quad\quad \dfrac{z}{w} = \sqrt{2}(\cos 75° + i \sin 75°)$

31. $\quad [4(\cos 40° + i \sin 40°)]^3 = 4^3[\cos(3 \cdot 40°) + i \sin(3 \cdot 40°)]$
$\quad\quad\quad\quad\quad\quad\quad\quad\quad\quad\quad = 64(\cos 120° + i \sin 120°)$

$\quad\quad\quad\quad\quad\quad\quad\quad\quad\quad\quad = 64\left(\dfrac{-1}{2} + \dfrac{\sqrt{3}}{2}i\right)$

$\quad\quad\quad\quad\quad\quad\quad\quad\quad\quad\quad = -32 + 32\sqrt{3}i$

$\quad\quad\quad\quad\quad\quad\quad\quad\quad\quad\quad = 32\left(-1 + \sqrt{3}i\right)$

33. $\left[2\left(\cos\dfrac{\pi}{10}+i\sin\dfrac{\pi}{10}\right)\right]^5 = 2^5\left[\cos\left(5\cdot\dfrac{\pi}{10}\right)+i\sin\left(5\cdot\dfrac{\pi}{10}\right)\right]$

$$= 32\left(\cos\dfrac{\pi}{2}+i\sin\dfrac{\pi}{2}\right)$$

$$= 32i$$

35. $\left[\sqrt{3}\,(\cos 10° + i\sin 10°)\right]^6 = \sqrt{3}^6\,[\cos(6\cdot 10°)+i\sin(6\cdot 10°)]$

$$= 27(\cos 60° + i\sin 60°)$$

$$= 27\left(\dfrac{1}{2}+\dfrac{\sqrt{3}}{2}\,i\right)$$

$$= \dfrac{27}{2}\left(1+\sqrt{3}\,i\right)$$

37. $\left[\sqrt{5}\left(\cos\dfrac{3\pi}{16}+i\sin\dfrac{3\pi}{16}\right)\right]^4 = \sqrt{5}^4\left[\cos\left(4\cdot\dfrac{3\pi}{16}\right)+i\sin\left(4\cdot\dfrac{3\pi}{16}\right)\right]$

$$= 25\left(\cos\dfrac{3\pi}{4}+i\sin\dfrac{3\pi}{4}\right)$$

$$= 25\left(\dfrac{-\sqrt{2}}{2}+\dfrac{\sqrt{2}}{2}\,i\right)$$

$$= \dfrac{25\sqrt{2}}{2}\,(-1+i)$$

39. $(1-i)^5 = \left[\sqrt{2}\left(\cos\dfrac{7\pi}{4}+i\sin\dfrac{7\pi}{4}\right)\right]^5$

$$= 4\sqrt{2}\left(\cos\dfrac{35\pi}{4}+i\sin\dfrac{35\pi}{4}\right)$$

$$= 4\sqrt{2}\left(\dfrac{-\sqrt{2}}{2}+i\dfrac{\sqrt{2}}{2}\right)$$

$$= -4+4i$$

$$= 4(-1+i)$$

41. $$r = \sqrt{x^2+y^2} = \sqrt{3}$$
$$\tan\theta = \dfrac{y}{x} = \dfrac{-1}{\sqrt{2}},\ \theta = 324.7°$$

$$\sqrt{2}-i = \sqrt{3}\,(\cos 324.7 + i\sin 324.7)$$
$$\left(\sqrt{2}-i\right)^6 = \left[\sqrt{3}\right]^6[\cos(6\cdot 324.7)+i\sin(6\cdot 324.7)]$$

$$= 27(\cos 1948.2° + i\sin 1948.2°)$$

$$= -23+14.15i$$

43. $1+i = \sqrt{2}\,(\cos 45° + i\sin 45°)$

The three complex cube roots of $1 + i = \sqrt{2}\,(\cos 45° + i\sin 45°)$ are

$$z_k = \sqrt[3]{\sqrt{2}}\left[\cos\left(\dfrac{45°}{3}+\dfrac{360°k}{3}\right)+i\sin\left(\dfrac{45°}{3}+\dfrac{360°k}{3}\right)\right]$$

Thus,

$$z_0 = \sqrt[6]{2}(\cos 15° + i \sin 15°)$$

$$z_1 = \sqrt[6]{2}(\cos 135° + i \sin 135°)$$

$$z_2 = \sqrt[6]{2}(\cos 255° + i \sin 255°)$$

45. $4 - 4\sqrt{3}i = 8(\cos 300° + i \sin 300°)$

The four complex cube roots of

$4 - 4\sqrt{3}i = 8(\cos 300° + i \sin 300°)$ are

$$z_k = \sqrt[4]{8}\left[\cos\left(\frac{300°}{4} + \frac{360°k}{4}\right) + i \sin\left(\frac{300°}{4} + \frac{360°k}{4}\right)\right], \quad k = 0, 1, 2, 3$$

Thus,

$$z_0 = \sqrt[4]{8}(\cos 75° + i \sin 75°)$$

$$z_1 = \sqrt[4]{8}(\cos 165° + i \sin 165°)$$

$$z_2 = \sqrt[4]{8}(\cos 255° + i \sin 255°)$$

$$z_3 = \sqrt[4]{8}(\cos 345° + i \sin 345°)$$

47. $-16i = 16(\cos 270° = i \sin 270°)$

$$z_k = \sqrt[4]{16}\left[\cos\left(\frac{270°}{4} + \frac{360°k}{4}\right) + i \sin\left(\frac{270°}{4} + \frac{360°k}{4}\right)\right],$$
$$k = 0, 1, 2, 3$$

Thus,

$$z_0 = 2(\cos 67.5° + i \sin 67.5°)$$
$$z_1 = 2(\cos 157.5° + i \sin 157.5°)$$
$$z_2 = 2(\cos 247.5° + i \sin 247.5°)$$
$$z_3 = 2(\cos 337.5° + i \sin 377.5°)$$

49. $i = 1(\cos 90° = i \sin 90°)$, $k = 0, 1, 2, 3, 4$

$$z_k = \sqrt[5]{1}\left[\cos\left(\frac{90°}{5} + \frac{360°k}{5}\right) + i \sin\left(\frac{90°}{5} + \frac{360°k}{5}\right)\right]$$

Thus,

$$z_0 = \cos 18° + i \sin 18°$$
$$z_1 = \cos 90° + i \sin 90°$$
$$z_2 = \cos 162° + i \sin 162°$$
$$z_3 = \cos 234° + i \sin 234°$$
$$z_4 = \cos 306° + i \sin 306°$$

51.　$z = 1 + 0i = 1(\cos 0° + i \sin 0°)$　　$k = 0, 1, 2, 3$

$$z_k = \sqrt[4]{1}\left[\cos\left(\frac{0°}{4} + \frac{360°k}{4}\right) + i \sin\left(\frac{0°}{4} + \frac{360°k}{4}\right)\right]$$

Thus,

$z_0 = \cos 0° + i \sin 0° = 1$
$z_1 = \cos 90° + i \sin 90° = i$
$z_2 = \cos 180° + i \sin 180° = -1$
$z_3 = \cos 270° + i \sin 270° = -i$

　　　$1, i, -1, -i$

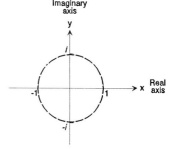

53.　Let $w = r(\cos θ + i \sin θ)$ be a complex number. If $w \neq 0$, there are n distinct complex nth roots of w, given by the formula:

$$z_k = \sqrt[n]{r}\left[\cos\left(\frac{θ}{n} + \frac{2kπ}{n}\right) + i \sin\left(\frac{θ}{n} + \frac{2kπ}{n}\right)\right]$$

where $k = 0, 1, 2, ..., n - 1$

$$\left|z_k\right| = \sqrt[n]{r}\ \text{for all } k.$$

55.　Looking at the formula for the number of distinct complex nth roots of the complex number $w = r(\cos θ + i \sin θ)$,

$$z_k = \sqrt[n]{r}\left[\cos\left(\frac{θ}{n} + \frac{2kπ}{n}\right) + i \sin\left(\frac{θ}{n} + \frac{2kπ}{n}\right)\right]$$

where $k = 0, 1, 2, ..., n - 1$, we see that the z_k are spaced apart by an angle of $\frac{2π}{n}$

≡ 8 - CHAPTER REVIEW

1.　$\dfrac{\sin α}{a} = \dfrac{\sin β}{b}$

$\dfrac{\sin 50°}{1} = \dfrac{\sin 30°}{b}$

$b = \dfrac{\sin 30°}{\sin 50°} \approx 0.6527$

$γ = 180° - α - β = 100°$
$c^2 = a^2 + b^2 - 2ab \cos γ$
$c^2 = 1 + 0.4260 - 2(1)(0.6527)\cos 100°$
　≈ 1.6527
　$c \approx 1.2856$

3. $\dfrac{\sin \alpha}{a} = \dfrac{\sin \gamma}{c}$

$\dfrac{\sin 100^\circ}{5} = \dfrac{\sin \gamma}{2}$

$\sin \gamma = \dfrac{2 \sin 100^\circ}{5} \approx 0.3939$

$\gamma = 23.2^\circ$
$\beta = 180^\circ - \alpha - \gamma = 56.8^\circ$
$b^2 = a^2 + c^2 - 2ac \cos \beta$
$b^2 = 25 + 4 - 2(5)(2)\cos 56.8^\circ$
$b^2 \approx 18.0487$
$b \approx 4.2484$

5. $\dfrac{\sin \alpha}{a} = \dfrac{\sin \gamma}{c}$

$\dfrac{\sin \alpha}{3} = \dfrac{\sin 100^\circ}{1}$
$\sin \alpha = 3 \sin 100^\circ = 2.8191$
Impossible! No Triangle.

7. $b^2 = a^2 + c^2 - 2ac \cos \beta$
$b^2 = 9 + 1 - 2(3)(1) \cos 100^\circ$
$b^2 = 10 - 6(-0.17365) \approx 11.0419$
$b \approx 3.3229$

$\cos \alpha = \dfrac{b^2 + c^2 - a^2}{2bc}$

$\cos \alpha = \dfrac{11.0419 + 1 - 9}{2(3.3229)(1)}$

$= \dfrac{3.0419}{6.6458} \approx 0.4577$

$\alpha \approx 62.8^\circ$
$\gamma \approx 180^\circ - \alpha - \beta \approx 17.2^\circ$

9. $\cos \alpha = \dfrac{b^2 + c^2 - a^2}{2bc}$

$\cos \alpha = \dfrac{9 + 1 - 4}{2(3)(1)} = 1$

$\alpha = 0^\circ$

No Triangle.

11. $c^2 = a^2 + b^2 - 2ab \cos \gamma$
$c^2 = 1 + 9 - 2(1)(3) \cos 40^\circ$
$= 10 - 6(0.76604) = 5.4037$
$c \approx 2.3246$

$\cos \alpha = \dfrac{b^2 + c^2 - a^2}{2bc}$

$\cos \alpha = \dfrac{9 + 5.4037 - 1}{2(3)(2.3246} = \dfrac{13.4037}{13.9476} \approx 0.9610$

$\alpha \approx 16.1^\circ$
$\beta \approx 180^\circ - \alpha - \gamma \approx 123.9^\circ$

13.
$$\frac{\sin \alpha}{a} = \frac{\sin \gamma}{b}$$

$$\frac{\sin 80°}{5} = \frac{\sin \beta}{3}$$

$$\sin \beta = \frac{3 \sin 80°}{5} \approx 0.5909$$

$$\beta = 36.2°$$
$$\gamma = 180° - \alpha - \beta \approx 63.8°$$
$$c^2 = a^2 + b^2 - 2ab \cos \gamma$$
$$c^2 = 25 + 9 - 2(5)(3)(0.4415) \approx 20.755$$
$$c \approx 4.5555$$

15.
$$\cos \alpha = \frac{b^2 + c^2 - a^2}{2bc}$$

$$\cos \alpha = \frac{\frac{1}{4} + \frac{16}{9} - 1}{2 \cdot \frac{1}{2} \cdot \frac{4}{3}} = \frac{\frac{9 + 64 - 36}{36}}{\frac{4}{3}} = \frac{\frac{37}{36}}{\frac{4}{3}} = \frac{37}{48} \approx 0.7708$$

$$\alpha = 39.6°$$
$$\frac{\sin \alpha}{a} = \frac{\sin \beta}{b}$$

$$\frac{\sin 39.6°}{1} = \frac{\sin \beta}{\frac{1}{2}}$$

$$\sin \beta = \frac{1}{2} \sin 39.6° \approx 0.3187$$

$$\beta = 18.5°$$
$$\gamma = 180° - \alpha - \beta = 121.9°$$

17.
$$\frac{\sin \alpha}{a} = \frac{\sin \beta}{b}$$

$$\frac{\sin 10°}{3} = \frac{\sin \beta}{4}$$

$$\sin \beta = \frac{4 \sin 10°}{3} \approx 0.2315 \quad \text{(Two Triangles)}$$

$$\beta_1 \approx 13.4° \qquad\qquad \beta_2 = 180° - 13.4° = 166.6°$$
$$\gamma_1 = 180° - \alpha - \beta_1 = 156.6° \qquad \gamma_2 = 180° - \alpha - \beta_2 \approx 3.4°$$
$$\frac{\sin 10°}{3} = \frac{\sin 156.6°}{c} \qquad\qquad \frac{\sin 10°}{3} = \frac{\sin 3.4°}{c}$$
$$c = \frac{3 \sin 156.6°}{\sin 10°} \qquad\qquad c = \frac{3 \sin 3.4°}{\sin 10°}$$
$$c_1 \approx 6.8613 \qquad\qquad c \approx 1.0246$$

19.
$$a^2 = b^2 + c^2 - 2bc \cos \alpha$$
$$a^2 = 16 + 25 - 2(4)(5)\cos 70°$$
$$a^2 = 41 - 40(5.3420) \approx 27.3192$$
$$a \approx 5.2268$$

$$\cos \beta = \frac{a^2 + c^2 - b^2}{2ac}$$

$$\cos \beta = \frac{27.3192 + 25 - 16}{2(5.2268)(5)} \approx \frac{36.3192}{52.268} \approx 0.6949$$

$$\beta = 46°$$
$$\gamma \approx 180° - \alpha - \gamma \approx 64°$$

21. $A = \frac{1}{2}ab \sin \gamma$

$A = \frac{1}{2} \cdot 2 \cdot 3 \cdot \sin 40°$

≈ 1.93

23. $A = bc \sin \frac{1}{2}\alpha$

$A = \frac{1}{2} \cdot 4 \cdot 10 \cdot \sin 70°$

≈ 18.79

25. $s = \frac{1}{2}(a + b + c)$

$s = \frac{1}{2}(4 + 3 + 5) = 6$

$A = \sqrt{s(s - a)(s - b)(s - c)}$

$= \sqrt{6(2)(3)(1)} = \sqrt{36} = 6$

27. $s = \frac{1}{2}(a + b + c)$

$s = \frac{1}{2}(4 + 2 + 5) = \frac{11}{2}$

$A = \sqrt{s(s - a)(s - b)(s - c)}$

$= \sqrt{\frac{11}{2}\left(\frac{3}{2}\right)\left(\frac{7}{2}\right)\left(\frac{1}{2}\right)}$

$= \sqrt{\frac{231}{16}} = \frac{\sqrt{231}}{4} \approx 3.80$

29. $A = \frac{a^2 \sin \beta \sin \gamma}{2 \sin \alpha}$

$\gamma = 180° - \alpha - \beta = 100°$

$A = \frac{1^2 \sin 30° \sin 100°}{2 \sin 50°}$

$= \frac{(.5)(0.9848)}{(1.5321)} \approx 0.32$

31. $x = r \cos \theta$ $\qquad$ $y = r \sin \theta$

$x = 3 \cos \frac{\pi}{6}$ $\qquad$ $y = 3 \sin \frac{\pi}{6}$

$x = \frac{3\sqrt{3}}{2}$ $\qquad$ $y = \frac{3}{2}$

$\left(\frac{3\sqrt{3}}{2}, \frac{3}{2}\right)$

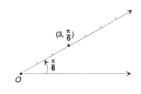

33. $x = r \cos \theta$ $\qquad$ $y = r \sin \theta$

$x = -2 \cos \frac{4\pi}{3}$ $\qquad$ $y = -2 \sin \frac{4\pi}{3}$

$x = 1$ $\qquad$ $y = \sqrt{3}$

$\left(1, \sqrt{3}\right)$

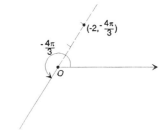

35. $x = r \cos \theta$ $\qquad$ $y = r \sin \theta$

$x = -3 \cos \frac{-\pi}{2}$ $\qquad$ $y = -3 \sin \frac{-\pi}{2}$

$x = 1$ $\qquad$ $y = 3$

$(0, 3)$

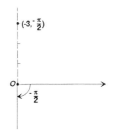

37. Since the point $(-3, 3)$ lies in quadrant II, then

$$r = -\sqrt{x^2 + y^2} = -\sqrt{9 + 9} = -\sqrt{18} = -3\sqrt{2}$$

and $\theta = \tan^{-1} \dfrac{y}{x} = \tan^{-1}(-1) = \dfrac{-\pi}{4}$. Hence, two pairs of polar coordinates are $\left(-3\sqrt{2}, \dfrac{-\pi}{4}\right)$ and $\left(3\sqrt{2}, \dfrac{3\pi}{4}\right)$

39. Since the point $(0, -2)$ lies on the negative y-axis, then

$$r = |y| = |-2| = 2 \text{ and } \theta = \dfrac{-\pi}{2}.$$ Hence, two pairs of polar coordinates are $\left(2, \dfrac{-\pi}{2}\right)$ and $\left(-2, \dfrac{\pi}{2}\right)$.

41. Since the point $(3, 4)$ lies in quadrant I, then

$$r = \sqrt{9 + 16} = \sqrt{25} = 5 \text{ and } \theta = \tan^{-1} \dfrac{y}{x} = \tan^{-1} \dfrac{4}{3} \approx 0.93.$$ Hence, two pairs of polar coordinates are $(5, 0.93)$ and $(-5, 4.07)$.

43.
$$3x^2 + 3y^2 = 6y$$
$$3(x^2 + y^2) = 6y$$
$$3r^2 = 6(r \sin \theta)$$
$$3r^2 - 6r \sin \theta = 0$$

45.
$$2x^2 - 3y^2 + 2y^2 = \dfrac{y}{x}$$
$$2(x^2 + y)^2 - 3y^2 = \tan \theta$$
$$2r^2 - 3(r \sin \theta)^2 = \tan \theta$$
$$2r^2 - 3r^2 \sin^2 \theta - \tan \theta = 0$$
$$r^2(2 - 3 \sin^2 \theta) - \tan \theta = 0$$

47.
$$x(x^2 + y^2) = 4$$
$$r \cos \theta(r^2) = 4$$
$$r^3 \cos \theta = 4$$

49.
$$r(r) = (2 \sin \theta)r$$
$$r^2 = 2r \sin \theta$$
$$x^2 + y^2 = 2y$$
$$x^2 + y^2 - 2y = 0$$

51.
$$r = 5$$
$$r^2 = 25$$
$$x^2 + y^2 = 25$$

53.
$$r \cos \theta + 3r \sin \theta = 6$$
$$x + 3y = 6$$

55. $r = 4 \cos \theta$

We test for symmetry:

With respect to the Pole: Replace r by $-r$. The test fails, so the graph may not be symmetric with respect to the pole.

With respect to the Polar Axis: Replace θ by $-\theta$. The result is $r = 4 \cos(-\theta) = 4 \cos \theta$. Thus, the graph is symmetric with respect to the polar axis.

With respect to the Line $\theta = \dfrac{\pi}{2}$: Replace θ by $\pi - \theta$. The result is $r = 4 \cos(\pi - \theta) = 4(\cos \pi \cos \theta + \sin \pi \sin \theta) = -4 \cos \theta$. The test fails.

Due to the periodicity of $\cos \theta$, we only consider values of θ between 0 and 2π.

θ	0	$\dfrac{\pi}{6}$	$\dfrac{\pi}{3}$	$\dfrac{\pi}{2}$
$r = 4 \cos \theta$	4	$4\left[\dfrac{\sqrt{3}}{2}\right] = 2\sqrt{3}$	$4\left[\dfrac{1}{2}\right] = 2$	$4(0) = 0$
(r, θ)	$(4, 0)$	$\left(2\sqrt{3}, \dfrac{\pi}{6}\right)$	$\left(2, \dfrac{\pi}{3}\right)$	$\left(0, \dfrac{\pi}{2}\right)$
θ	$\dfrac{2\pi}{3}$	$\dfrac{5\pi}{6}$	π	
$r = 4 \cos \theta$	$4\left[\dfrac{-1}{2}\right] = -2$	$4\left[\dfrac{-\sqrt{3}}{2}\right] = -2\sqrt{3}$	$4(-1) = -4$	
(r, θ)	$\left(-2, \dfrac{2\pi}{3}\right)$	$\left(-2\sqrt{3}, \dfrac{5\pi}{6}\right)$	$(-4, \pi)$	

The remaining points on the graph can be found by using symmetry with respect to the polar axis.

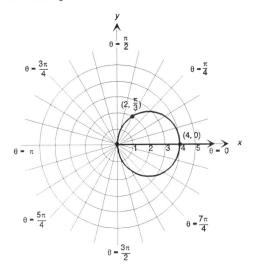

57. $r = 3 - 3 \sin \theta$

Test for symmetry:

With respect to the Pole: Test fails.

With respect to the Line $\theta = \dfrac{\pi}{2}$: $r = 3 - 3 \sin(\pi - \theta)$
$$= 3 - 3 \sin \theta$$

The graph is symmetric with respect to the line $\theta = \dfrac{\pi}{2}$

θ	$\dfrac{-\pi}{2}$	$\dfrac{-\pi}{3}$	$\dfrac{-\pi}{6}$	0	$\dfrac{\pi}{6}$	$\dfrac{\pi}{3}$	$\dfrac{\pi}{2}$
$r = 3 - 3\sin\theta$	6	$3 + \dfrac{3\sqrt{3}}{2} \approx 5.6$	$\dfrac{9}{2}$	3	$\dfrac{3}{2}$	$3 - \dfrac{3\sqrt{3}}{2} \approx 0.4$	0
(r, θ)	$\left(6, \dfrac{-\pi}{2}\right)$	$\left(5.6, \dfrac{-\pi}{3}\right)$	$\left(\dfrac{9}{2}, \dfrac{-\pi}{6}\right)$	$(3, 0)$	$\left(\dfrac{3}{2}, \dfrac{\pi}{6}\right)$	$\left(0.4, \dfrac{\pi}{3}\right)$	$\left(0, \dfrac{\pi}{2}\right)$

Due to the periodicity of sin θ, we only consider values of θ between 0 and 2π.

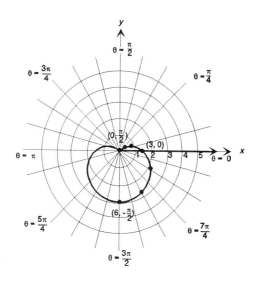

59. $r = 4 - \cos\theta$

Test for symmetry:

   With respect to the Pole: Test fails.

With respect to the Polar Axis: $r = 4 - \cos(-\theta) = 4 - \cos\theta$.   The graph is symmetric with respect to the polar axis.

With respect to the Line $\theta = \dfrac{\pi}{2}$:   The result is $r = 4 - \cos(\pi - \theta)$ $= 4 + \cos\theta$.   Test fails.

θ	0	$\dfrac{\pi}{6}$	$\dfrac{\pi}{3}$	$\dfrac{\pi}{2}$
$r = 4 - \cos\theta$	3	$4 - \dfrac{\sqrt{3}}{2} = \dfrac{8 - \sqrt{3}}{2}$	$4 - \dfrac{1}{2} = \dfrac{7}{2}$	$4 - 0 = 4$
(r, θ)	$(3, 0)$	$\left(\dfrac{8 - \sqrt{3}}{2}, \dfrac{\pi}{6}\right)$	$\left(\dfrac{7}{2}, \dfrac{\pi}{3}\right)$	$\left(4, \dfrac{\pi}{2}\right)$

θ	$\dfrac{2\pi}{3}$	$\dfrac{5\pi}{6}$	π
$r = 4 - \cos\theta$	$4 + \dfrac{1}{2} = \dfrac{9}{2}$	$4 + \dfrac{\sqrt{3}}{2} = \dfrac{8 + \sqrt{3}}{2}$	$4 - (-1) = -5$
(r, θ)	$\left(\dfrac{9}{2}, \dfrac{2\pi}{3}\right)$	$\left(\dfrac{8 + \sqrt{3}}{2}, \dfrac{5\pi}{6}\right)$	$(5, \pi)$

The remaining points on the graph can be found by using symmetry with respect to the polar axis.

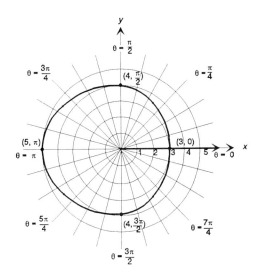

61. $r = \sqrt{x^2 + y^2} = \sqrt{2}$, $\tan\theta = \dfrac{-1}{-1} = 1$

$\theta = 225°$

Thus, $-1 - i = \sqrt{2}(\cos 225° + i \sin 225°)$

63. $r = \sqrt{x^2 + y^2} = \sqrt{(4)^2 + (-3)^2} = 5$
$\tan\theta = \dfrac{-3}{4}$, $\theta = 323.1°$

Thus, $4 - 3i = 5(\cos 323.1° + i \sin 323.1°)$

65. $2(\cos 150° + i \sin 150°) = 2\left(\frac{-\sqrt{3}}{2} + \frac{1}{2}i\right)$

$\qquad = -\sqrt{3} + i$

67. $3\left(\cos \frac{2\pi}{3} + i \sin \frac{2\pi}{3}\right) = 3\left(\frac{-1}{2} + \frac{\sqrt{3}}{2}i\right)$

$\qquad = \frac{-3}{2} + \frac{3\sqrt{3}}{2}i$

69. $0.1(\cos 350° + i \sin 350°) = 0.1(0.9848 + -0.1736i)$

$\qquad = 0.098 - 0.017i$

71. $zw = \cos(80° + 50°) + i \sin(80° + 50°)$
$zw = \cos 130° + i \sin 130°$

$\frac{z}{w} = \cos(80° - 50°) + i \sin(80° - 50°)$

$\frac{z}{w} = \cos 30° + i \sin 30°$

73. $zw = 6(\cos 0 + i \sin 0)$

$\frac{z}{w} = \frac{3}{2}\left(\cos \frac{8\pi}{5} + i \sin \frac{8\pi}{5}\right)$

75. $zw = \cos(10° + 355°) + i \sin(10° + 355°)$
$zw = 5(\cos 5° + i \sin 5°)$

$\frac{z}{w} = 5[\cos(10° - 355°) + i \sin(10° - 355°)]$

$\frac{z}{w} = 5(\cos(-345°) + i \sin(-345°))$

$\qquad = 5(\cos 15° + i \sin 15°)$

77. $[3(\cos 20° + i \sin 20°)]^3 = 3^3[3 \cdot 20°) + i \sin(3 \cdot 20°)]$

$\qquad = 27(\cos 60° + i \sin 60°)$

$\qquad = 27\left(\frac{1}{2} + i\frac{\sqrt{3}}{2}\right)$

$\qquad = \frac{27}{2}\left(1 + \sqrt{3}i\right)$

79. $\left[\sqrt{2}\left(\cos \frac{5\pi}{8} + i \sin \frac{5\pi}{8}\right)\right]^4 = \sqrt{2^4}\left[\cos\left(4 \cdot \frac{5\pi}{8}\right) + i \sin\left(4 \cdot \frac{5\pi}{8}\right)\right]$

$\qquad = 4\left(\cos \frac{5\pi}{2} + i \sin \frac{5\pi}{2}\right)$

$\qquad = 4\left(\cos \frac{\pi}{2} + i \sin \frac{\pi}{2}\right)$

$\qquad = 4(0 + i)$

$\qquad = 4i$

81. $\left(1 - \sqrt{3}i\right) = 2(\cos 300° + i \sin 300°)$

$\left(1 - \sqrt{3}i\right)^6 = 2^6[\cos(6 \cdot 300°) + i \sin(6 \cdot 300°)]$

$\qquad = 64(\cos 1800° + i \sin 1800°)$

$\qquad = 64(1 + 0i)$

$\qquad = 64$

83. $(3 + 4i) = 5(\cos 53.1° + i \sin 53.1°)$
$(3 + 4i)^4 = 5^4[\cos(4 \cdot 53.1°) + i \sin(4 \cdot 53.1°)]]$
$= 625(\cos 212.4° + i \sin 212.4°)$
$= -527.1 - 335.8i$

85. $z_k = \sqrt[3]{27} = 3\left[\cos\left(\dfrac{0°}{4} + \dfrac{360°k}{4}\right) + i \sin\left(\dfrac{0°}{4} + \dfrac{360°k}{4}\right)\right]$, $k - 0, 1, 2$

Thus, $z_0 = 3(\cos 0° + i \sin 0°) = 3$
$z_1 = 3(\cos 120° + i \sin 120°)$
$z_2 = 3(\cos 240° + i \sin 240°)$

87.

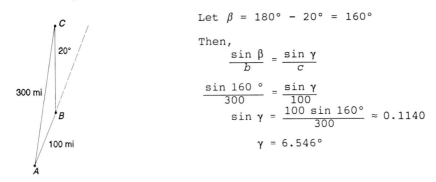

Let $\beta = 180° - 20° = 160°$

Then,
$$\frac{\sin \beta}{b} = \frac{\sin \gamma}{c}$$

$$\frac{\sin 160°}{300} = \frac{\sin \gamma}{100}$$

$$\sin \gamma = \frac{100 \sin 160°}{300} \approx 0.1140$$

$$\gamma = 6.546°$$

Therefore, $\alpha = 180° - \beta - \gamma \approx 13.45°$

Now, we can find a, the distance from city B to city C, using the Law of Cosines:

$a^2 = b^2 + c^2 - 2bc \cos \alpha$
$a^2 = 90,000 + 10,000 - 2(300)(100)\cos 13.45°$
$a^2 \approx 41,645.6$
$a \approx 204.1$

The distance from city B to city C is 204.1 miles.

89.   We divide the irregular parcel of land into two triangles. First, we find x using the Law of Cosines.

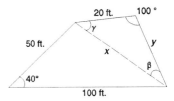

$x^2 = (100)^2 + (50)^2 - 2(50)(100) \cos 40°$
$x^2 = 4839.56$
$x = 69.567$

Now we can find the area of one triangle:

$$A = \frac{1}{2}(50)(100)\sin 40° = 1606.97$$

Using the Law of Sines, we can find β:

$$\frac{\sin 100°}{x} = \frac{\sin \beta}{20}$$

$$\frac{\sin 100°}{69.567} = \frac{\sin \beta}{20}$$

$$\sin \beta = \frac{20 \sin 100°}{69.567} \approx 0.2831$$

$$\beta = 16.447°$$

$$\gamma = 180° - 100° - \beta \approx 63.553°$$

Using the Law of sines, we find γ:

$$\frac{\sin 63.553°}{y} = \frac{\sin 100°}{69.567}$$

$$y = \frac{69.567 \sin 63.533°}{\sin 100°} = 63.247$$

Now we can find the area of the second triangle:

$$A = \frac{1}{2}(20(63.247)\sin 100° = 622.86.$$

Adding the areas of the two triangles, we find the area of the irregular parcel of land:

$$\text{Area} = 1606.97 + 622.86 \approx 2,229.8$$

If the land is being sold for $100 per square foot, the cost of the parcel is $2229.8 × $100 = $222,980.00.

CHAPTER
9

THE CONICS

≡ EXERCISE 9.2 THE PARABOLA

1. B   3. E   5. H   7. C

9. We want an equation of the parabola with focus at (2, 0) and vertex at (0, 0). The focus and vertex both lie on the horizontal line $y = 0$ (i.e., the x-axis). The distance, a, from (2, 0) to (0, 0) is $a = 2$. Also, the focus is to the right of the vertex, so the parabola opens to the right. Since the vertex is at (0, 0), the equation of the parabola is:

$$y^2 = 4ax \quad \text{(from Table 1)}$$
$$\text{or} \quad y^2 = 8x$$

Letting $x = 2$, we find $y^2 = 16$ or $y = \pm 4$. The points (2, 4) and (2, −4) define the latus rectum.

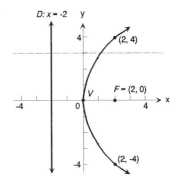

11. The focus, (0, −3), and the vertex, (0, 0), both lie on the vertical line $x = 0$ (the y-axis). We have $a = 3$, and since (0, −3) is <u>below</u> (0, 0), the parabola opens down. Therefore, from Table 1,

$$x^2 = -4ay$$
$$\text{or} \quad x^2 = -12y \quad (a = 3)$$

Letting $y = -3$, we find $x^2 = 36$ or $x = \pm 6$. The points (−6, −3) and (6, −3) define the latus rectum.

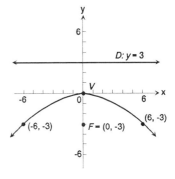

13. The vertex must be midway between the focus, $(-2, 0)$, and the directrix, the line $x = 2$. Therefore, the vertex is the point $(0, 0)$. The distance from the focus to the vertex is $a = 2$, and the parabola opens to the left, since $(-2, 0)$ is to the <u>left</u> of the vertex. Therefore, we have:

$$y^2 = -4ax$$
$$\text{or} \quad y^2 = -8x \quad (a = 2)$$

Letting $x = -2$, we find $y^2 = 16$ or $y = \pm 4$. The points $(-2, 4)$ and $(-2, -4)$ define the latus rectum.

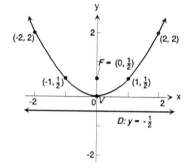

15. The directrix, $y = -\frac{1}{2}$, lies $\frac{1}{2}$ unit <u>below</u> the vertex, $(0, 0)$, so the focus must be $\frac{1}{2}$ unit <u>above</u> the vertex, at the point $\left(0, \frac{1}{2}\right)$. Therefore, $a = \frac{1}{2}$ and the parabola opens upward, so we have:

$$x^2 = 4ay$$
$$\text{or} \quad x^2 = 2y \quad \left(a = \frac{1}{2}\right)$$

Letting $y = \frac{1}{2}$, we find $x^2 = 1$ or $x = \pm 1$. The points $\left(1, \frac{1}{2}\right)$ and $\left(-1, \frac{1}{2}\right)$ define the latus rectum.

17. Here, the vertex, $(2, -3)$, and the focus, $(2, -5)$, both lie on the vertical line, $x = 2$. The distance between the vertex and the focus is $a = 2$, and the parabola opens down, since the focus is <u>below</u> the vertex. From Table 2, we have:

$$(x - h)^2 = -4a(y - k)$$
$$(x - 2)^2 = -4a(y - (-3)), \quad \text{since}$$
$$(h, k) = (2, -3)$$
$$(x - 2)^2 = -8(y + 3), \quad \text{since } a = 2$$

Letting $y = -5$, we find $(x - 2)^2 = 16$ or $x - 2 = \pm 4$ so that $x = 6$ or $x = -2$. The points $(-2, -5)$ and $(6, -5)$ define the latus rectum.

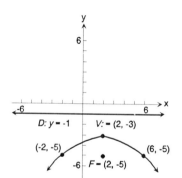

9 THE CONICS

19. Since the axis of symmetry, the y-axis, is vertical, the parabola opens up or down. The point (2, 3) is <u>above</u> the vertex, (0, 0), so the parabola opens <u>up</u>. Therefore, from Table 1,

$$x^2 = 4ay$$

But here we don't know what a is. But the point (2, 3) must satisfy the equation, since it lies on the graph.

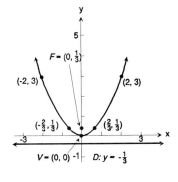

Therefore, $x^2 = 4ay$
$$4 = 4a(3) \text{ using } x = 2,$$
$$y = 3)$$
$$1 = 3a$$
$$a = \frac{1}{3}$$

The equation of the parabola is

$$x^2 = 4\left(\frac{1}{3}\right)y$$
$$\text{or } x^2 = \frac{4}{3}y$$

To help sketch the graph, note that the focus is at $(0, a) = \left(0, \frac{1}{3}\right)$. Letting $y = \frac{1}{3}$, we find $x^2 = \frac{4}{9}$ or $x = \pm\frac{2}{3}$. The points $\left(\frac{-2}{3}, \frac{1}{3}\right)$ and $\left(\frac{2}{3}, \frac{1}{3}\right)$ define the latus rectum.

21. The directrix, $y = 2$, is horizontal, so the parabola opens up or down. Also, the axis of symmetry must be vertical, and it contains the focus, (-3, 4), so it must be the line $x = -3$. The vertex lies on the axis of symmetry, midway between the focus and the directrix, so the vertex must be the point (-3, 3).

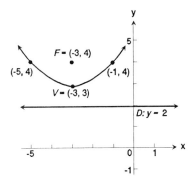

The focus, then, is one unit <u>above</u> the vertex, so we know that $a = 1$ and the parabola opens up. From Table 2, we have:

$$(x - h)^2 = 4a(y - k)$$
$$(x - (-3))^2 = 4a(y - 3), \text{ since}$$
$$(h, k) = (-3, 3)$$
$$(x + 3)^2 = 4(y - 3), \text{ since } a = 1$$

Letting $y = 4$, we find $(x + 3)^2 = 4$ or $x + 3 = \pm2$, so that $x = -1$ or $x = -5$. The points (-1, 4) and (-5, 4) define the latus rectum.

9.2 THE PARABOLA

457

23. Here, the directrix, $x = 1$, is vertical, so the axis of symmetry is horizontal, and the parabola opens to the left or right. Since the focus, $(-3, -2)$ is on the axis, the equation of the axis must be $y = -2$. Again, the vertex is on the axis, midway between focus and directrix, so the vertex is $(-1, -2)$. The parabola opens to the left, since the focus is to the left of the vertex. Finally, $a = 2$.

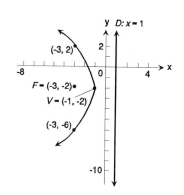

From 2,

$$(y - k)^2 = -4a(x - h)$$
or $(y + 2)^2 = -8(x + 1)$

Letting $x = -3$, $(y + 2)^2 = 16$ or $y + 2 = \pm 4$ so that $y = 2$ or $y = -6$. The points $(-3, -6)$ or $(-3, 2)$ define the latus rectum.

25. The equation $x^2 = 8y$ is in the form:

$$x^2 = 4ay$$
where $4a = 8$,
or $a = 2$

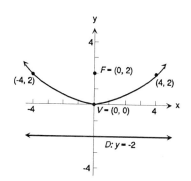

Thus, by Table 1, we have:

    Vertex: $(0, 0)$
    Focus: $(0, 2)$
    Directrix: $y = -2$

Letting $y = 2$, we find $x^2 = 16$ or $x = \pm 4$. The points $(-4, 2)$ and $(4, 2)$ define the latus rectum.

27. The equation $y^2 = -16x$ is in the form

$$y^2 = -4ax$$
where $-4a = -16$
or $a = 4$

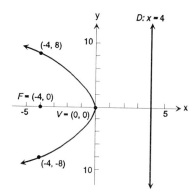

By Table 1, we have:

    Vertex: $(0, 0)$
    Focus: $(-4, 0)$
    Directrix: $x = 4$

Letting $x = -4$, we have $y^2 = 64$ or $y = \pm 8$. The points $(-4, 8)$ and $(-4, -8)$ define the latus rectum.

9 THE CONICS

29. The equation $(y - 2)^2 = 8(x + 1)$ is in the form:

$$(y - k)^2 = 4a(x - h),$$

where: (1) $4a = 8$, or $a = 2$
          (2) $y - k = y - 2$, or $k = 2$
          (3) $x - h = x + 1$, or $-h = 1$,
               or $h = -1$

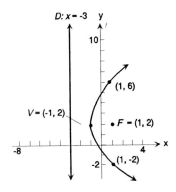

Thus, the vertex is at $(h, k) =$ $(-1, 2)$. From Table 2, the parabola opens to the right, so the focus is $a = 2$ units to the <u>right</u> of the vertex. The focus is $(1, 2)$. The directrix is a vertical line 2 units to the <u>left</u> of the vertex: $x = -3$.

Letting $x = 1$, we have $(y - 2)^2 = 16$ or $y - 2 = \pm 4$, so that $y = 6$ or $y = -2$. The points $(1, 6)$ and $(1, -2)$ define the latus rectum.

31. The equation $(x - 3)^2 = -(y + 1)$ is in the form $(x - h)^2 = -4a(y - k)$, where:

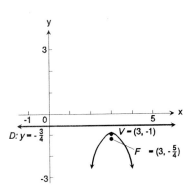

    (1) $-4a = -1$, or $a = \dfrac{1}{4}$

    (2) $x - h = x - 3$
           $-h = -3$
             $h = 3$
and (3) $y - k = y + 1$
           $-k = 1$
            $k = -1$

So, from Table 2, we have:

Vertex: $(h, k) = (3, -1)$

The parabola opens <u>down</u>, so the focus is <u>below</u> the vertex:

    Focus: $\left(3, -\dfrac{5}{4}\right)$

The directrix is a horizontal line $a = \dfrac{1}{4}$ unit <u>above</u> the vertex:

    Directrix: $y = -\dfrac{3}{4}$

Letting $y = \dfrac{-5}{4}$, we have $(x - 3)^2 = \dfrac{1}{4}$ or $(x - 3) = \pm\dfrac{1}{2}$ so that $x = \dfrac{7}{2}$ or $x = \dfrac{5}{2}$. The points $\left(\dfrac{7}{2}, \dfrac{-5}{2}\right)$ and $\left(\dfrac{5}{2}, \dfrac{-5}{2}\right)$ define the latus rectum.

33. The equation $(y + 3)^2 = 8(x - 2)$ is in the form:

$$(y - k) = 4a(x - h)$$

where:

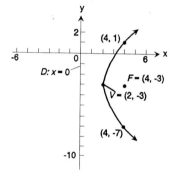

(1) $4a = 8$, or $a = 2$
(2) $y - k = y + 3$, or
   $k = -3$
(3) $x - h = x - 2$, or $h = 2$

Thus, from Table 2, we have

Vertex: $(h, k) = (2, -3)$

The parabola opens to the right, so the focus is $a = 2$ units to the right of the vertex:

Focus: $(4, -3)$

The directrix is the vertical line 2 units to the left of the vertex:

Directrix: $x = 0$

Letting $x = 4$, we have $(y + 3)^2 = 16$ or $y + 3 = \pm 4$ so that $y = 1$ or $y = -7$. The points $(4, 1)$ and $(4, -7)$ define the latus rectum.

35.
$$y^2 - 4y + 4x + 4 = 0$$
$$y^2 - 4y = -4x - 4$$
$$y^2 - 4y + 4 = -4x$$
$$(y - 2)^2 = -4(x + 0)$$

The equation is now in the form

$$(y - k)^2 = -4a(x - h), \text{ where:}$$

(1) $-4a = -4$, or $a = 1$
(2) $y - k = y - 2$, or $k = 2$
(3) $x - h = x + 0$, or $h = 0$

Thus, from Table 2, we have

Vertex: $(h, k) = (0, 2)$

The parabola opens to the left, so the focus is $a = 1$ unit to the left of the vertex:

Focus: $(-1, 2)$

The directrix is the vertical line 1 unit to the right of the vertex:

Directrix: $x = 1$

Letting $x = -1$, we have $(y - 2)^2 = 4$ or $y - 2 = \pm 2$ so that $y = 4$ or $y = 0$. The points $(-1, 4)$ and $(-1, 0)$ define the latus rectum.

37.

$$x^2 + 8x = 4y - 8$$
$$x^2 + 8x + 16 = 4y + 8$$
$$(x + 4)^2 = 4(y + 2)$$

The equation is now in the form
$(x - h)^2 = 4a(y - k)$, where:

(1) $4a = 4$, or $a = 1$
(2) $x - h = x + 4$, or $h = -4$
(3) $y - k = y + 2$, or $y = -2$

Thus, from Table 2, we have

Vertex: $(h, k) = (-4, -2)$

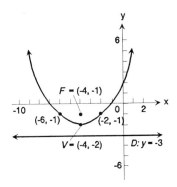

The parabola opens up, so the focus is
$a = 1$ unit above the vertex:

Focus: $(-4, -1)$

The directrix is the horizontal line 1 unit below the vertex:

Directrix: $y = -3$

Letting $y = -1$, we have $(x + 4)^2 = 4$ or $x + 4 = \pm 2$ so that $x = -2$
or $x = -6$. The points $(-2, -1)$ and $(-6, -1)$ define the latus
rectum.

39.   For $y^2 + 2y - x = 0$, complete the
square:

$$y^2 + 2y - x = 0$$
$$y^2 + 2y + \underline{\quad} = x + \underline{\quad}$$
$$y^2 + 2y + 1 = x + 1$$
$$(y + 1)^2 = (x + 1)$$

This is in the form:

$(y - k)^2 = 4a(x - h)$, where:

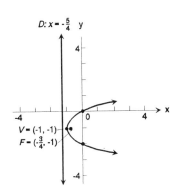

(1) $4a = 1$, or $a = \dfrac{1}{4}$
(2) $x - h = x + 1$
    $-h = -1$

and (3) $y - k = y + 1$
    $-k = 1$
    $k = -1$

From Table 2, we have:

Vertex: $(h, k) = (-1, -1)$

The parabola opens to the right, and has

Focus: $\left(-\dfrac{3}{4}, -1\right)$

The directrix is $x = -\dfrac{5}{4}$

Letting $x = \dfrac{-3}{4}$, we have $(y + 1)^2 = \dfrac{7}{4}$ or $y + 1 = \pm\dfrac{\sqrt{7}}{2}$ so that

$y = \dfrac{\sqrt{7}}{2} - 1$ or $y = \dfrac{-\sqrt{7}}{2} - 1$. The points $\left(\dfrac{-3}{4}, \dfrac{\sqrt{7}}{2} - 1\right)$ and

$\left(\dfrac{-3}{4}, \dfrac{-\sqrt{7}}{2} - 1\right)$ define the latus rectum.

9.2 THE PARABOLA

41. We complete the square:

$$x^2 - 4x + \underline{} = y + 4 + \underline{}$$
$$x^2 - 4x + 4 = y + 4 + 4$$
$$(x - 2)^2 = y + 8$$

which is in the form

$$(x - h)^2 = 4a(y - k)$$

where:

(1) $4a = 1$, $a = \dfrac{1}{4}$

(2) $x - h = x - 2$
    $h = 2$

and (3) $y - k = y + 8$
    $k = -8$

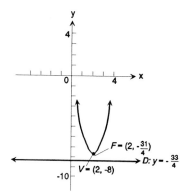

$F = (2, -\frac{31}{4})$
$D: y = -\frac{33}{4}$
$V = (2, -8)$

We have:

Vertex: $(2, -8)$

Focus: $\left(2, -7\dfrac{3}{4}\right) = \left(2, \dfrac{-31}{4}\right)$

Directrix: $y = -8\dfrac{1}{4}$

or $y = \dfrac{-33}{4}$

Letting $y = \dfrac{-31}{4}$, we have $(x - 2)^2 = \dfrac{1}{4}$ or $x - 2 = \pm\dfrac{1}{2}$ so that $x = \dfrac{5}{2}$ or $x = \dfrac{3}{2}$. The points $\left(\dfrac{5}{2}, \dfrac{-31}{4}\right)$ and $\left(\dfrac{3}{2}, \dfrac{-31}{4}\right)$ define the latus rectum.

43.
$$(y - 1)^2 = c(x - 0)$$
$$(y - 1)^2 = cx$$
$$(2 - 1)^2 = c(1)$$
$$1 = c$$
$$(y - 1)^2 = x$$

45.
$$(y - 1)^2 = c(x - 2)$$
$$(0 - 1)^2 = c(1 - 2)$$
$$1 = c(-1)$$
$$c = -1$$
$$(y - 1)^2 = -(x - 2)$$

47.
$$(x - 0)^2 = c(y - 1)$$
$$x^2 = c(y - 1)$$
$$2^2 = c(2 - 1)$$
$$4 = c$$
$$x^2 = 4(y - 1)$$

49.
$$(y - 0)^2 = c(x - -2)$$
$$y^2 = c(x + 2)$$
$$1^2 = c(0 + 2)$$
$$1 = 2c$$
$$\dfrac{1}{2} = c$$
$$y^2 = \dfrac{1}{2}(x + 2)$$

51. Situate the parabola so that its vertex is at $(0, 0)$, and it opens up. Then, we know

$$x^2 = 4ay$$

Since the parabola is 10 feet across and 3 feet deep, the points $(-5, 3)$ and $(5, 3)$ must satisfy the equation:

$$x^2 = 4ay$$
$$25 = 4a(3) \quad (x = 5, \ y = 3)$$
$$25 = 12a$$
$$a = \dfrac{25}{12} \approx 2.083$$

But a is, by definition, the distance from the vertex to the focus.

Therefore, the receiver (at the focus) is $\frac{25}{12} \approx 2.083$ feet, or 25 inches, from the vertex of the dish.

53.   Situate the parabola so that its vertex is at (0, 0), and it opens up.   Then, we know

$$x^2 = 4ay$$

Since the focus is 1 inch from the vertex, we have $a = 1$.   The depth is 2 inches, so $y = 2$.   Thus, we have

$$x^2 = 4(1)(2)$$
$$x^2 = 8$$

$$x = \pm\sqrt{8} = \pm 2\sqrt{2}$$

Thus, the diameter at the opening is $2\left(2\sqrt{2}\right)$ or $4\sqrt{2}$ inches.

55.

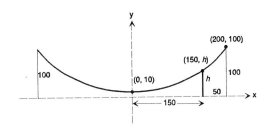

Vertex is at (0, 10); $(x - 10)^2 = cy$

Point on parabola is (200, 100):
$$(200 - 10)^2 = c(100)$$

$$\frac{190 \times 190}{100} = c$$

$$c = 19 \times 19$$

$$(x - 10)^2 = 19 \times 19y$$

When $x = 150$, $(150 - 10)^2 = 19 \times 19h$

$$h = \frac{140 \times 140}{19 \times 19}$$

$$= 54.29 \text{ feet}$$

57. $Ax^2 + Ey = 0$ $A \neq 0, E \neq 0$
$$Ax^2 = -Ey$$
$$x^2 = \frac{-E}{A}y$$

Parabola; vertex at (0, 0) and axis of symmetry the y-axis; focus at $\left(0, \frac{-E}{4A}\right)$; directrix the line $y = \frac{E}{4A}$.   The parabola opens up if $\frac{-E}{A} > 0$ and down if $\frac{-E}{A} < 0$.

9.2 THE PARABOLA   463

59. $Ax^2 + Dx + Ey + F = 0$ $A \neq 0$

(a) If $E \neq 0$, then

$$Ax^2 + Dx + Ey + F = 0$$

$$Ax^2 + Dx = -Ey - F$$

$$A\left(x^2 + \frac{D}{A}x + \frac{D^2}{4A^2}\right) = -Ey - F + \frac{D^2}{4A}$$

$$\left(x + \frac{D}{2A}\right)^2 = \frac{1}{A}\left(-Ey - F + \frac{D^2}{4A}\right)$$

$$\left(x + \frac{D}{2A^2}\right) = \frac{-E}{A}\left[y + \left(\frac{F}{E} - \frac{D^2}{4AE}\right)\right]$$

$$\left(x + \frac{D}{2A}\right)^2 = \frac{-E}{A}\left(y - \frac{D^2 - 4AF}{4AE}\right)$$

Parabola; vertex at $\left(\frac{-D}{2A}, \frac{D^2 - 4AF}{4AE}\right)$, axis of symmetry parallel to y-axis.

(b) If $E = 0$, then

$$Ax^2 + Dx + F = 0$$

$$x = \frac{-D \pm \sqrt{D^2 - 4AF}}{2A}$$

If $D^2 - 4AF = 0$, then

$$x = \frac{-D}{2A}$$

Vertical line

(c) If $E = 0$, then

$$Ax^2 + Dx + F = 0$$

$$x = \frac{-D \pm \sqrt{D^2 - 4AF}}{2A}$$

If $D^2 - 4AF > 0$, then

$$x = \frac{-D + \sqrt{D^2 - 4AF}}{2A} \quad \text{or} \quad x = \frac{-D - \sqrt{D^2 - 4AF}}{2A}$$

Two vertical lines

9 THE CONICS

(d) If $E = 0$, then

$$Ax^2 + Dx + F = 0$$

$$x = \frac{-D \pm \sqrt{D^2 - 4AF}}{2A}$$

If $D^2 - 4AF < 0$, there is no real solution. Hence, the graph contains no points.

≡ EXERCISE 9.3 THE ELLIPSE

1. C   3. B

In Problems 5-10, write the equation in the form shown in Table 3 in the text, so that you can identify (h, k), a, and b, and also tell whether the major axis is parallel to the x-axis or the y-axis.

5. $\dfrac{x^2}{9} + \dfrac{y^2}{4} = 1$

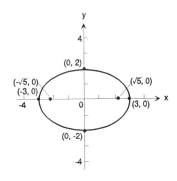

This is in the proper form, and we see that $h = 0$ and $k = 0$, so the center of the ellipse is at the origin. Since the larger denominator is associated with x^2, we see from Table 3 that the major axis is parallel to the x-axis, and:

$$a^2 = 9, \text{ so } a = 3$$
$$b^2 = 4, \text{ so } b = 2$$

We will be able to locate the vertices using $a = 3$, but to find the foci we will need c:

$$c^2 = a^2 - b^2 = 5$$

$$c = \sqrt{5}$$

From Table 3, the vertices are located at $(h \pm a, k) = (0 \pm 3, 0)$:

Vertices: $(-3, 0)$, $(3, 0)$

and the foci are at $(h \pm c, k)$:

Foci: $\left(-\sqrt{5}, 0\right), \left(\sqrt{5}, 0\right)$

(Notice that, since the major axis is parallel to the x-axis, we find the vertices by moving left and right of the center (h, k), a distance $a = 3$. Similarly, the foci are $c = \sqrt{5}$ units to the left and right of the center.)

Remember that we use b to locate the end-points of the <u>minor axis</u>, which passes through the center and is perpendicular to the major axis.

7. $\dfrac{x^2}{9} + \dfrac{y^2}{25} = 1$

This is in the proper form, and the fact that the larger denominator appears in the y^2-term means that the major axis is parallel to the y-axis. By Table 3,

(1) $h = 0$, $k = 0$, so the center is (0, 0)
(2) $a^2 = 25$, so a = 5
(3) $b^2 = 9$, so b = 3, and
(4) $c^2 = a^2 - b^2 = 16$, so c = 4

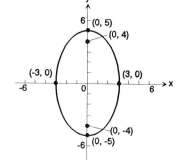

Therefore, we have:

Vertices: $(h, k \pm a)$:
            (0, -5) and (0, 5)
Foci: $(h, k \pm c)$:
            (0, -4) and (0, 4)

Use b to find the endpoints of the <u>minor</u> axis.

9. $4x^2 + y^2 = 16$ We must get a 1 on the right-hand side.

$\dfrac{4x^2}{16} + \dfrac{y^2}{16} = 1$ Divide both sides by 16. But we can't have the 4 in the numerator, so we simplify.

$\dfrac{x^2}{4} + \dfrac{y^2}{16} = 1$

Now we have the proper form, with $h = 0$, $k = 0$. By Table 3, the major axis is parallel to the y-axis, and:

$a^2 = 16$, a = 4
$b^2 = 4$, b = 2

$c^2 = a^2 - b^2 = 12$, $c = \sqrt{12} = 2\sqrt{3}$

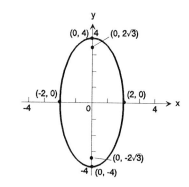

Also:

Foci: $(h, k \pm c)$:
            $\left(0, -2\sqrt{3}\right), \left(0, 2\sqrt{3}\right)$
Vertices: $(h, k \pm a)$:
            (0, -4), (0, 4)

11. $4y^2 + x^2 = 8$

$\dfrac{4y^2}{8} + \dfrac{x^2}{8} = 1$ Obtain a 1 on the right.

$\dfrac{y^2}{2} + \dfrac{x^2}{8} = 1$ Simplify.

We know:

(1) Major axis is parallel to the x-axis.

(2) $a^2 = 8$, so $a = \sqrt{8} = 2\sqrt{2}$

(3) $b^2 = 2$, so $b = \sqrt{2}$

(4) $c^2 = a^2 - b^2 = 6$, $c = \sqrt{6}$

(5) $h = 0$, $k = 0$, so the center is at $(0, 0)$

(6) Foci: $(h \pm c, k)$: $\left(-\sqrt{6},\ 0\right)$, $\left(\sqrt{6},\ 0\right)$

(7) Vertices: $(h \pm a, k)$: $\left(-2\sqrt{2},\ 0\right)$, $\left(2\sqrt{2},\ 0\right)$

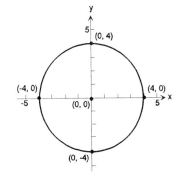

13. $x^2 + y^2 = 16$

This is the equation of a circle, with center at $(0, 0)$ and radius = 4.

Vertices: $(\pm 4, 0)$
  Foci: $(0, 0)$

For Problems 15-24, in order to find the <u>equation</u> of an ellipse, it is necessary to find a, b, h and k, and to know whether the major axis is parallel to the x-axis or the y-axis. Once we have these pieces of information, we can use Table 3 to write down the equation.

We will use the following facts:

(1) (h, k) are the coordinates of the <u>center</u> of the ellipse.
(2) a = the distance from the center to either vertex (half the length of the major axis)
(3) b = the distance from the center to either endpoint of the minor axis (half the length of the minor axis), <u>or</u>, if c is known, then
    $b^2 = a^2 - c^2$
(4) c = the distance from the center to either focus.
(5) Since $b^2 = a^2 - c^2$, if any two of the three numbers a, b, and c are known, we can find the third.
(6) The center, the foci, and the vertices all lie on the major axis.

Use these facts to find a, b, h, and k, and to determine if the major axis is vertical or horizontal.

15. We are given:

Center: $C(0, 0)$
Focus: $F(3, 0)$
Vertex: $V(6, 0)$

Please refer to the paragraph above.

By fact (1) above, $h = 0$, $k = 0$
By fact (2), $a = d(C, V) = 6$
By fact (4), $c = d(C, F) = 3$
By fact (3), $b^2 = a^2 - c^2$
$\qquad = 36 - 9 = 27$
By fact (6), since C, F and V all lie
on the horizontal line,
$y = 0$, the major axis is
parallel to the x-axis.

Then from Table 1, the equation is:

$$\frac{(x - h)^2}{a^2} + \frac{(y - k)^2}{b^2} = 1$$

or

$$\frac{x^2}{36} + \frac{y^2}{27} = 1$$

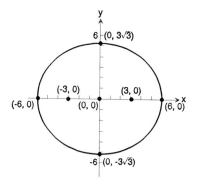

17. We are given:

Center: $C(0, 0)$
Focus: $F(0, -4)$
Vertex: $V(0, 5)$

The numbering below refers to the
paragraph preceding the solution to
Problem 11.

By (1), $h = 0$, $k = 0$
By (2), $a = d(C, V) = 5$
By (4), $c = d(C, F) = 4$
By (3), $b^2 = a^2 - c^2 = 9$
By (6), the major axis is the
vertical line $x = 0$

From Table 1, the equation is:

$$\frac{(x - h)^2}{b^2} + \frac{(y - k)^2}{a^2} = 1$$

or

$$\frac{x^2}{9} + \frac{y^2}{25} = 1$$

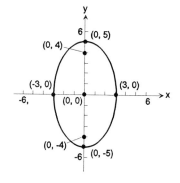

9 THE CONICS

19. We are given:

Foci: $F_1(-2, 0)$ and $F_2(2, 0)$

Length of Major Axis = 6.
First of all, the center is the
midpoint between F_1 and F_2: $C(0, 0)$.

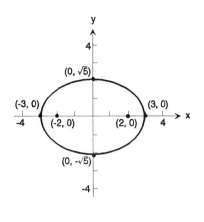

The numbers below refer to the
paragraph preceding the solution to
Problem 11.

By (1), $h = 0$, $k = 0$
By (2), a = half the length of the
        major axis,
    or $a = 3$
By (4), $c = d(F_1, C) = 2$
By (3), $b^2 = a^2 - c^2 = 5$
By (6), the major axis is the horizontal line, $y = 0$.

Then, by Table 3, we have

$$\frac{(x - h)^2}{a^2} + \frac{(y - k)^2}{b^2} = 1$$

or $$\frac{x^2}{9} + \frac{y^2}{5} = 1$$

21. We are given:

Foci: $F_1(0, -3)$ and $F_2(0, 3)$

x-intercepts: ± 2

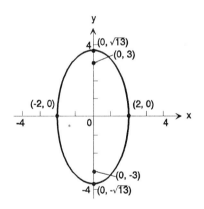

The center is $C(0, 0)$, and the major
axis is the vertical line $x = 0$, (the
y-axis). Therefore, the minor axis
lies on the x-axis, so the x-
intercepts are the endpoints of the
minor axis. Thus, we have $b = 2$. We
know $c = d(C, F) = 3$, and
$b^2 = a^2 - c^2$, or $a^2 = b^2 + c^2 = 13$.

From Table 3, we have

$$\frac{(x - h)^2}{b^2} + \frac{(y - k)^2}{a^2} = 1$$

or $$\frac{x^2}{4} + \frac{y^2}{13} = 1$$

9.3 THE ELLIPSE

23. Here we are given:

        Center: $C(0, 0)$
        Vertex: $V(0, 4)$

and $b = 1$

Therefore, $h = 0$, $k = 0$, and $a = d(C, V) = 4$, so we have all we need to write down the equation. Since the major axis is parallel to the y-axis ($x = 0$), we have

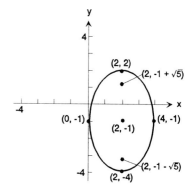

$$\frac{(x - h)^2}{b^2} + \frac{(y - k)^2}{a^2} = 1$$

or
$$\frac{x^2}{1} + \frac{y^2}{16} = 1$$

25. $\dfrac{(x + 1)^2}{4} + (y - 1)^2 = 1$

27. $(x - 1)^2 + \dfrac{y^2}{4} = 1$

29. $\dfrac{(x - 2)^2}{4} + \dfrac{(y + 1)^2}{9} = 1$

This is in the form

$$\frac{(x - h)^2}{b^2} + \frac{(y - k)^2}{a^2} = 1$$

where $h = 2$, $k = -1$, $a = 3$, $b = 2$.

Since the larger denominator belongs to the y^2-term, we see from Table 3 that the major axis is parallel to the y-axis, and we have:

      Center: (h, k), or $(2, -1)$
        Foci: $(h, k \pm c)$, where $c^2 = a^2 - b^2 = 5$,

           or $c = \sqrt{5}$, giving the points $(2, -1 - \sqrt{5})$ and

           $(2, -1 + \sqrt{5})$
    Vertices: $(h, k \pm a)$, or $(2, -4)$ and $(2, 2)$

31. $(x + 5)^2 + 4(y - 4)^2 = 16$

To put this into the form listed in Table 3, we need to obtain a 1 on the right-hand-side:

$$\frac{(x + 5)^2}{16} + \frac{4(y - 4)^2}{16} = 1$$

Now get rid of the 4 in the numerator of the y^2-term (multiply top and bottom by $\frac{1}{4}$).

$$\frac{(x + 5)^2}{16} + \frac{(y - 4)^2}{4} = 1$$

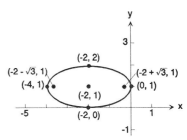

We now have the proper form:

$$\frac{(x - h)^2}{a^2} + \frac{(y - k)^2}{b^2} = 1$$

where $h = -5$, $k = 4$, $a = 4$, $b = 2$. Also, $c^2 = a^2 - b^2 = 12$, so $c = \sqrt{12}$.

Since the larger denominator is associated with the x^2 term, the major axis is parallel to the x-axis, and, from Table 3, we have:

Center: (h, k), or $(-5, 4)$
Foci: $(h \pm c, k)$, or $\left(-5 - 2\sqrt{3}, 4\right)$ and $\left(-5 + 2\sqrt{3}, 4\right)$
Vertices: $(h \pm a, k)$, or $(-9, 4)$ and $(-1, 4)$

33. Here we start by completing the square in both x and y:

$$x^2 + 4x + \underline{\quad} + 4(y^2 - 2y + \underline{\quad}) = -4$$
$$x^2 + 4x + 4 + 4(y^2 - 2y + 1) = -4 + 4 + 4$$
$$(4)(1) = 4$$

$$(x + 2)^2 + 4(y - 1)^2 = 4$$
$$\frac{(x + 2)^2}{4} + \frac{(y - 1)^2}{1} = 1$$

This is in the form:

$$\frac{(x - h)^2}{a^2} + \frac{(y - k)^2}{b^2} = 1$$

where $h = -2$, $k = 1$, $a = 2$, $b = 1$, and $c^2 = a^2 - b^2 = 3$, so $c = \sqrt{3}$.

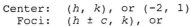

From Table 3, we have:

Center: (h, k), or $(-2, 1)$
Foci: $(h \pm c, k)$, or
$\left(-2 - \sqrt{3}, 1\right)$ and $\left(-2 + \sqrt{3}, 1\right)$
Vertices: $(h \pm a, k)$, or $(-4, 1)$ and $(0, 1)$

9.3 THE ELLIPSE

35.

$$2x^2 + 3y^2 - 8x + 6y + 5 = 0$$
$$2x^2 - 8x + 3y^2 + 6y = -5$$
$$2(x^2 - 4x + \underline{}) + 3(y^2 + 2y + \underline{}) = -5$$
$$2(x^2 - 4x + 4) + 3(y^2 + 2y + 1) = -5 + 8 + 3$$
$$2(x - 2)^2 + 3(y + 1)^2 = 6$$
$$\frac{(x - 2)^2}{3} + \frac{(y + 1)^2}{2} = 1$$

This is in the form

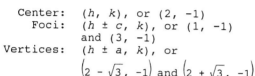

$$\frac{(x - h)^2}{a^2} + \frac{(y - k)^2}{b^2} = 1, \text{ where}$$

$h = 2$, $k = -1$, $a^2 = 3$, so $a = \sqrt{3}$,
$b^2 = 2$, so $b = \sqrt{2}$, and $c^2 = a^2 - b^2$
$= 1$, so that $c = 1$. Then we have:

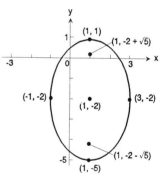

Center: (h, k), or $(2, -1)$
Foci: $(h \pm c, k)$, or $(1, -1)$
and $(3, -1)$
Vertices: $(h \pm a, k)$, or
$$\left(2 - \sqrt{3}, -1\right) \text{ and } \left(2 + \sqrt{3}, -1\right)$$

37.

$$9x^2 + 4y^2 - 18x + 16y - 11 = 0$$
$$9x^2 - 18x + 4y^2 + 16y = 11$$
$$9(x^2 - 2x + \underline{}) + 4(y^2 + 4y + \underline{}) = 11$$
$$9(x^2 - 2x + 1) + 4(y^2 + 4y + 4) = 11 + 9 + 16$$
$$9(x - 1)^2 + 4(y + 2)^2 = 36$$
$$\frac{(x - 1)^2}{4} + \frac{(y + 2)^2}{9} = 1$$

This is in the form

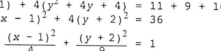

$$\frac{(x - h)^2}{b^2} + \frac{(y - k)^2}{a^2} = 1$$

where $h = 1$, $k = -2$, $a^2 = 9$, so $a = 3$,
$b^2 = 4$, so $b = 2$, and $c^2 = a^2 - b^2 = 5$,
so that $c = \sqrt{5}$.

Then we have:

Center: (h, k), or $(1, -2)$
Foci: $(h, k \pm c)$, or $(1, -2 + \sqrt{5})$ and $(1, -2 - \sqrt{5})$
Vertices: $(h, k \pm a)$, or $(1, 1)$ and $(1, -5)$

9 THE CONICS

39.
$$4x^2 + y^2 + 4y = 0$$
$$4x^2 + (y^2 + 4y + \underline{\quad}) = 0$$
$$4x^2 + (y^2 + 4y + 4) = 4$$
$$4x^2 + (y + 2)^2 = 4$$

$$x^2 + \frac{(y + 2)^2}{4} = 1$$

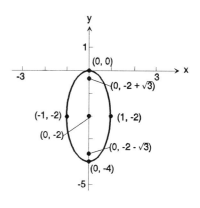

This is in the form

$$\frac{(x - h)^2}{b^2} + \frac{(y - k)^2}{a^2} = 1, \text{ where}$$

$h = 0$, $k = -2$, $a^2 = 4$, so $a = 2$,
$b^2 = 1$, so $b = 1$, and $c^2 = a^2 - b^2$

$= 3$, so that $c = \sqrt{3}$. Then we have:

Center: $(0, -2)$

Foci: $(h, k \pm c)$, or $(0, -2 + \sqrt{3})$ and $(0, -2 - \sqrt{3})$
Vertices: $(h, k \pm a)$, or $(0, 0)$ and $(0, -4)$

41.   We are given:

Center: $C(2, -2)$
Vertex: $V(5, -2)$
Focus: $F(4, -2)$

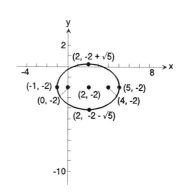

From the center, $h = 2$, $k = -2$

Also, $a = d(C, V) = 3$, and $c = d(C, F)$
$= 2$. Therefore, $b^2 = a^2 - c^2 = 5$, so

$b = \sqrt{5}$. Finally, C, V, and F all lie
on the horizontal line $y = -2$, so the
major axis is parallel to the x-axis.

From Table 3, the equation is of the
form:

$$\frac{(x - h)^2}{a^2} + \frac{(y - k)^2}{b^2} = 1$$

or $$\frac{(x - 2)^2}{9} + \frac{(y + 2)^2}{5} = 1$$

43.   We are given:

Vertices: $V_1(4, 3)$ and $V_2(4, 9)$
Focus: $F(4, 8)$

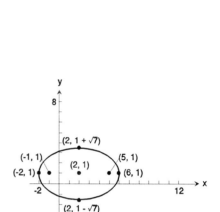

First of all, the center is the
midpoint between V_1 and V_2: $C(4, 6)$,
and $h = 4$, $k = 6$.

Then, $a = d(C, V) = 3$
    $c = d(C, F) = 2$
and $b^2 = a^2 - c^2 = 5$, so $b = \sqrt{5}$

Now V_1, V_2, F, and C all lie on the
vertical line $x = 4$, so the major
axis is parallel to the y-axis, and the equation is of the form:

$$\frac{(x - h)^2}{b^2} + \frac{(y - k)^2}{a^2} = 1$$

or $\frac{(x - 4)^2}{5} + \frac{(y - 6)^2}{9} = 1$

45.   We are given:

Foci: $F_1(5, 1)$ and $F_2(-1, 1)$

length of major axis: 8

Then the center is midway between
F_1 and F_2: $C(2, 1)$, so $h = 2$,
$k = 1$.   The length of the major
axis is $2a$, so $a = 4$, and
$c = d(C, F_1) = 3$.   Therefore,
$b^2 = a^2 - c^2 = 7$.   The major axis
(the line $y = 1$) is parallel to
the x-axis, so we have

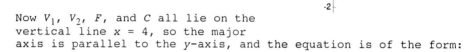

$$\frac{(x - h)^2}{a^2} + \frac{(y - k)^2}{b^2} = 1$$

or $\frac{(x - 2)^2}{16} + \frac{(y - 1)^2}{7} = 1$

47.   We are given:

> Center: $C(1, 2)$
> Focus: $F(4, 2)$
Contains the point: $(1, 3)$

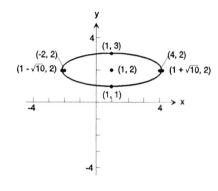

From the center $h = 1$, $k = 2$.
Also, $c = d(C, F) = 3$, so $c^2 = 9$.
The major axis is parallel to the
x-axis, so we have

$$\frac{(x - h)^2}{a^2} + \frac{(y - k)^2}{b^2} = 1$$

or $$\frac{(x - 1)^2}{a^2} + \frac{(y - 2)^2}{b^2} = 1$$

But the point (1, 3) must satisfy the equation since it lies on
the graph. Therefore:

$$\frac{0}{a^2} + \frac{1}{b^2} = 1 \text{ or } b^2 = 1, \text{ so } b = 1$$

and $a^2 = b^2 + c^2 = 10$.   Thus,

$$\frac{(x - 1)^2}{10} + \frac{(y - 2)^2}{1} = 1$$

49.   We are given:

> Center: $C(1, 2)$
> Focus: $F(4, 2)$
Contains the point: $(1, 3)$

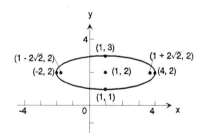

From the center $h = 1$, $k = 2$. Also,
$a = d(C, V) = 3$, so $a^2 = 9$. The
major axis is parallel to the x-axis,
so we have

$$\frac{(x - h)^2}{a^2} + \frac{(y - k)^2}{b^2} = 1$$

or $$\frac{(x - 1)^2}{a^2} + \frac{(y - 2)^2}{b^2} = 1$$

But the point (1, 3) must satisfy the equation since it lies on
the graph. Therefore:

$$\frac{0}{a^2} + \frac{1}{b^2} = 1 \text{ or } b^2 = 1, \text{ so } b = 1$$

and $c^2 = a^2 - b^2 = 8$.   Thus,

$$\frac{(x - 1)^2}{9} + \frac{(y - 2)^2}{1} = 1$$

9.3 THE ELLIPSE   475

51.
$$y = \sqrt{16 - 4x^2}$$
$$y^2 = 16 - 4x^2, \quad y \geq 0$$
$$4x^2 + y^2 = 16, \quad y \geq 0$$
$$\frac{x^2}{4} + \frac{y^2}{16} = 1, \quad y \geq 0$$

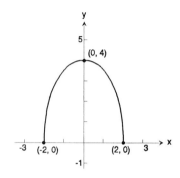

53.
$$y = -\sqrt{64 - 16x^2}$$
$$y^2 = 64 - 16x^2, \quad y \leq 0$$
$$16x^2 + y^2 = 64, \quad y \leq 0$$
$$\frac{x^2}{4} + \frac{y^2}{64} = 1, \quad y \leq 0$$

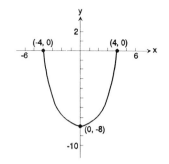

55.   The center of the ellipse will be at (0, 0) due to the nice positioning of the axes. The length of the major axis is 20, so we know $a = 10$. The length of <u>half</u> the minor axis is 6, i.e., $b = 6$. Finally, the major axis is horizontal, so we have:

$$\frac{(x - h)^2}{a^2} + \frac{(y - k)^2}{b^2} = 1$$

or
$$\frac{x^2}{100} + \frac{y^2}{36} = 1$$

57.   First we need to find an equation for the ellipse. Let the center of the ellipse be the origin, (0, 0), with the x-axis at ground-level. Then the major axis has length 40, so we know $a = 20$, and the length of <u>half</u> the minor axis is 15, i.e., $b = 15$. Since the major axis is horizontal, we have:

$$\frac{x^2}{a^2} + \frac{y^2}{b^2} = 1 \text{ or } \frac{x^2}{400} + \frac{y^2}{225} = 1$$

We wish to find y when $x = 0, \pm 10, \pm 20$, so we solve for y:

$$\frac{x^2}{400} + \frac{y^2}{225} = 1$$

$$\frac{y^2}{225} = 1 - \frac{x^2}{400}$$

$$\frac{y^2}{225} = \frac{400 - x^2}{400}$$

$$y^2 = 225\left(\frac{400 - x^2}{400}\right)$$

$$y^2 = \frac{225}{400}(400 - x^2)$$

$$y = \sqrt{\frac{225}{400}}\sqrt{400 - x^2}$$

$$\text{or} \quad y = \frac{15}{20}\sqrt{400 - x^2}$$

$$= \frac{3}{4}\sqrt{400 - x^2}$$

x	$y = \dfrac{3}{4}\sqrt{400 - x^2}$
0	$\dfrac{3}{4}\sqrt{400} = \dfrac{3}{4}(20) = 15$
± 10	$\dfrac{3}{4}\sqrt{400 - 100} \approx 12.99$
± 20	$\dfrac{3}{4}\sqrt{400 - 400} = 0$

59. (a) $e = \dfrac{c}{a}$ is close to zero when c is close to zero. Since $c^2 = a^2 - b^2$, c is close to zero when $a^2 \approx b^2$. Hence, the ellipse is close to a circle.

(b) $e = \dfrac{c}{a} = \dfrac{1}{2}$ when $c = 1$ and $a = 2$. Thus, $b^2 = a^2 - c^2 = 4 - 1 = 3$. Hence, the ellipse is oval.

(c) $e = \dfrac{c}{a}$ is close to 1 when $c \approx a$ or $c^2 \approx a^2$. Hence, $a^2 - c^2$ will be close to zero. Since $b^2 = a^2 - c^2$, then b^2 is close to zero. Thus, the ellipse is elongated with the length of the minor axis small in comparison to the major axis.

61. If the *x*-axis is placed along the 100 ft. portion and the *y*-axis along the 50 ft. portion, one equation for the ellipse is

$$\frac{x^2}{(50)^2} + \frac{y^2}{(25)^2} = 1$$

When $x = 40$, then

$$\frac{(40)^2}{(50)^2} + \frac{y^2}{(25^2)} = 1$$

$$\frac{y^2}{(25)^2} = 1 - \left(\frac{4}{5}\right)^2 = \frac{9}{25}$$

$$y^2 = (25)(9)$$

$$y = (5)(3) = 15$$

The width 10 feet from the side is 30 feet.

63. (a) $Ax^2 + Cy^2 + F = 0$
         $Ax^2 + Cy^2 = -F$

If *A* and *C* are the same sign and *F* is of the opposite sign, then the equation takes the form $x^2/(-F/A) + y^2/(-F/C) = 1$, where $-F/A$ and $-F/C$ are positive. This is the equation of an ellipse with center at (0, 0).

(b) If $A = C$, the equation may be written as $x^2 + y^2 = -F/A$. This is the equation of a circle with center at (0, 0) and radius equal to $\sqrt{-F/A}$.

≡ EXERCISE 9.4 THE HYPERBOLA

For Problems 1-10, refer to Table 4. There we see that in order to find the equation of a hyperbola, we need to determine, h, k, a, b, and to decide whether the transverse axis is horizontal or vertical.

We will use the following facts:

(1) (h, k) are the coordinates of the <u>center</u> of the hyperbola, which is midway between the vertices, and also midway between the foci.

(2) a = the distance from the center to either vertex.

(3) $b^2 = c^2 - a^2$, where c = the distance from the center to either focus.

(4) The center, the vertices and the foci all lie on the transverse axis.

1. B   3. A

5.   We are given:

Center: $C(0, 0)$
Focus: $F_2(4, 0)$
Vertex: $V_2(1, 0)$

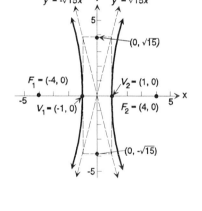

Please refer to the paragraph above.

By (1) we have $h = 0$, $k = 0$
By (2), $a = d(C, V_2) = 1$
By (3), $c = d(C, F_2) = 4$, and
$\quad\quad b^2 = c^2 - a^2 = 15$, so that

$\quad\quad b = \sqrt{15}$
By (4), since C, F_2 and V_2 lie on the horizontal line $y = 0$, the transverse axis is parallel to the x-axis.

Then by Table 4, the equation is

$$\frac{(x - h)^2}{a^2} - \frac{(y - k)^2}{b^2} = 1 \quad \text{or} \quad \frac{x^2}{1} - \frac{y^2}{15} = 1$$

As an aid in sketching the graph, we locate the asymptotes of the hyperbola. First, plot the points that lie on the conjugate axis (perpendicular to the transverse axis) a distance b from the

center: $\left(0, -\sqrt{15}\right)$ and $\left(0, \sqrt{15}\right)$. These two points, together with the vertices, determine a rectangle whose diagonals are the asymptotes of the hyperbola.

7.   Here we are given:

Center: $C(0, 0)$
Focus: $F_1(0, -6)$
Vertex: $V_2(0, 4)$

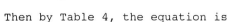

Please refer to the paragraph before the solution to Problem 1.

By (1), $h = 0$, $k = 0$
By (2), $a = d(C, V_2) = 4$
By (3), $c = d(C, F_1) = 6$, and
$\quad\quad b^2 = c^2 - a^2 = 20$,
$\quad\quad\quad$ so that

$\quad\quad b = \sqrt{20}$
By (4), the transverse axis is the vertical line $x = 0$.

Hence, from Table 4, we have

$$\frac{(y - k)^2}{a^2} - \frac{(x - h)^2}{b^2} = 1 \quad \text{or} \quad \frac{y^2}{16} - \frac{x^2}{20} = 1$$

9.4 THE HYPERBOLA   479

9.   We are given:

Foci: $F_1(-5, 0)$ and $F_2(5, 0)$
Vertex: $V_2(3, 0)$

Refer to the paragraph before the solution to Problem 1.

By (1), the center is the midpoint between F_1 and F_2: $C(0, 0)$, so $h = 0$, $k = 0$.

By (2), $a = d(C, V_2) = 3$
By (3), $c = d(C, F_1) = 5$, and
$b^2 = c^2 - a^2 = 16$
$b = 4$

By (4), the transverse axis is the horizontal line $y = 0$.

Therefore, the equation is

$$\frac{(x - h)^2}{a^2} - \frac{(y - k)^2}{b^2} = 1 \quad \text{or} \quad \frac{x^2}{9} - \frac{y^2}{16} = 1$$

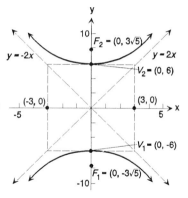

11.   We are given:

Vertices: $V_1(0, -6)$ and $V_2(0, 6)$
Asymptote: $y = 2x$

The center is the midpoint between V_1 and V_2: $C(0, 0)$.   Then $a = d(C, V_1) = 6$.   The transverse axis is the vertical line $x = 0$.   From Table 4, we see that the asymptotes of a hyperbola with a vertical transverse axis are:

$$y - k = \pm \frac{a}{b}(x - h)$$

Here, $h = 0$, $k = 0$, so one asymptote would be:

$$y = \frac{a}{b}x$$

Comparing this with the given asymptote, $y = 2x$, we find:

$$\frac{a}{b} = 2$$

$$a = 2b$$

$$b = \frac{1}{2}a$$

$$b = 3 \quad (\text{since } a = 6)$$

Then the equation of the hyperbola is:

$$\frac{(y - k)^2}{a^2} - \frac{(x - h)^2}{b^2} = 1 \quad \text{or} \quad \frac{y^2}{36} - \frac{x^2}{9} = 1$$

13. Here we are given:

Foci: $F_1(-4, 0)$
and $F_2(4, 0)$
Asymptote: $y = -x$

The Center is the midpoint between F_1 and F_2: $C(0, 0)$. The transverse axis is the <u>horizontal</u> line $y = 0$.

By Table 4, the asymptotes are

$$y - k = \pm \frac{b}{a}(x - h)$$

or $y = \pm \frac{b}{a}x$ (since $h = 0$, $k = 0$)

Comparing this with the given asymptote, $y = -x$, we see:

$$-\frac{b}{a} = -1$$

$$b = a$$

Now $c = d(C, F_1) = 4$, and

$$b^2 = c^2 - a^2$$
$$a^2 + b^2 = c^2$$
$$a^2 + b^2 = 16 \qquad (c = 4)$$
$$a^2 + a^2 = 16 \qquad (b = a)$$
$$2a^2 = 16$$
$$a^2 = 8$$

$$a = \sqrt{8}$$
$$b = \sqrt{8} \qquad (b = a)$$

The equation is: $\dfrac{(x - h)^2}{a^2} - \dfrac{(y - k)^2}{b^2} = 1$ or $\dfrac{x^2}{8} - \dfrac{y^2}{8} = 1$

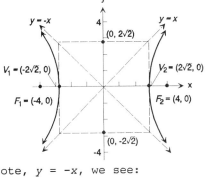

15. $\dfrac{x^2}{9} - \dfrac{y^2}{4} = 1$

This is in the form

$$\frac{(x - h)^2}{a^2} - \frac{(y - k)^2}{b^2} = 1$$
with $h = 0$, $k = 0$, $a = 3$, and $b = 2$.

From Table 4, we have:
$c^2 = a^2 + b^2 = 13$, so

$$c = \sqrt{13}, \text{ and}$$

Center: $(h, k) = (0, 0)$

Transverse axis: Parallel to x-axis, and contains the Center: $y = 0$

Foci: $(h \pm c, k)$: $\left(-\sqrt{13}, 0\right)$ and $\left(\sqrt{13}, 0\right)$
Vertices: $(h \pm a, k)$: $(-3, 0)$ and $(3, 0)$
Asymptotes: $y - k = \pm \dfrac{b}{a}(x - h)$, or $y = \pm \dfrac{2}{3}x$

(Lines through $(0, 0)$ with slopes $\dfrac{2}{3}$ and $-\dfrac{2}{3}$.)

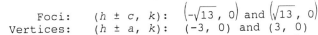

9.4 THE HYPERBOLA

17. $4x^2 - y^2 = 16$

$$\frac{4x^2}{16} - \frac{y^2}{16} = 1 \qquad \text{Obtain a 1 on the right-hand side.}$$

$$\frac{x^2}{4} - \frac{y^2}{16} = 1 \qquad \text{Simplify}$$

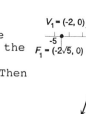

This is a hyperbola with transverse axis parallel to the x-axis (since the x^2-term is the positive one), with $h = 0$, $k = 0$, $a^2 = 4$ and $b^2 = 16$. Then $c^2 = a^2 + b^2 = 20$.

Therefore, $a = 2$
$\qquad\qquad b = 4$
$\qquad\qquad c = \sqrt{20}$

$\qquad\qquad$ Center: $(h, k) = (0, 0)$
Transverse axis: $y = 0$

$\qquad\qquad\qquad$ Foci: $(h \pm c, k)$: $(-2\sqrt{5}, 0)$ and $(2\sqrt{5}, 0)$
$\qquad\qquad$ Vertices: $(h \pm a, k)$: $(-2, 0)$ and $(2, 0)$
$\qquad$ Asymptotes: $y - k = \pm\dfrac{b}{a}(x - h)$, or $y = \pm 2x$

(Lines through $(0, 0)$ with slopes 2 and -2.)

19. $y^2 - 9x^2 = 9$

$$\frac{y^2}{9} - \frac{x^2}{1} = 1$$

This is a hyperbola with transverse axis parallel to the y-axis (since the y^2 term is positive), with $h = 0$, $k = 0$, $a^2 = 9$ and $b^2 = 1$.

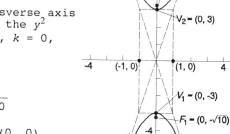

Therefore, $a = 3$
$\qquad\qquad b = 1$
$\qquad c^2 = a^2 + b^2 = 10$, so $c = \sqrt{10}$

$\qquad\qquad\qquad$ Center: $(h, k) = (0, 0)$
Transverse axis: $x = 0$

$\qquad\qquad\qquad$ Foci: $(h, k \pm c)$: $\left(0, -\sqrt{10}\right)$ and $\left(0, \sqrt{10}\right)$
$\qquad\qquad$ Vertices: $(h, k \pm a)$: $(0, -3)$ and $(0, 3)$
$\qquad$ Asymptotes: $y - k = \pm\dfrac{a}{b}(x - h)$, or $y = \pm 3x$

9 THE CONICS

21. $y^2 - x^2 = 25$

$$\frac{y^2}{25} - \frac{x^2}{25} = 1$$

This is a hyperbola with transverse axis parallel to the y-axis (since the y^2-term is positive), with $h = 0$, $k = 0$, $a^2 = 25$, $b^2 = 25$. Therefore, $c^2 = a^2 + b^2 = 50$, and we have

$a = 5$
$b = 5$
$c = \sqrt{50} = 5\sqrt{2}$

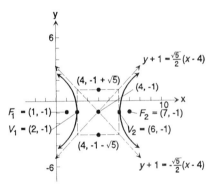

Center:	$(h, k) = (0, 0)$
Transverse axis:	$x = 0$
Foci:	$(h, k \pm c)$: $\left(0, -5\sqrt{2}\right)$ and $\left(0, 5\sqrt{2}\right)$
Vertices:	$(h, k \pm a)$: $(0, -5)$ and $(0, 5)$
Asymptotes:	$y - k = \pm\dfrac{a}{b}(x - h)$, or $y = \pm x$

23. $x^2 - y^2 = 1$

25. $\dfrac{y^2}{36} - \dfrac{x^2}{9} = 1$

27. We are given:

Center: $C(4, -1)$
Focus: $F_2(7, -1)$
Vertex: $V_2(6, -1)$

Please refer to the paragraph before the solution to Problem 1.

By (1), $h = 4$, $k = -1$
By (2), $a = d(C, V_2) = 2$
By (3), $c = d(C, F_2) = 3$, and $b^2 = c^2 - a^2 = 5$, so that
$b = \sqrt{5}$

By (4), the Transverse axis is the horizontal line $y = -1$

Then, by Table 4, we have:

$$\frac{(x - h)^2}{a^2} - \frac{(y - k)^2}{b^2} = 1 \quad \text{or} \quad \frac{(x - 4)^2}{4} - \frac{(y + 1)^2}{5} = 1$$

29.   We are given:

Center: $C(-3, -4)$
Focus: $F_1(-3, -8)$
Vertex: $V_2(-3, -2)$

Refer to the paragraph before the solution to Problem 1.

By (1), $h = -3$, $k = -4$
By (2), $a = d(C, V_2) = 2$
By (3), $c = d(C, F_1) = 4$
$b^2 = c^2 - a^2 = 12$
$b = \sqrt{12} = 2\sqrt{3}$

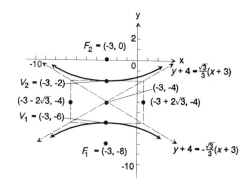

By (4), the transverse axis is the vertical line $x = -3$

Then, by Table 4, we have:

$$\frac{(y - k)^2}{a^2} - \frac{(x - h)^2}{b^2} = 1 \quad \text{or} \quad \frac{(y + 4)^2}{4} - \frac{(x + 3)^2}{12} = 1$$

31.   We are given:

Foci: $F_1(3, 7)$ and $F_2(7, 7)$
Vertex: $V_1(6, 7)$

Refer to the paragraph before the solution to Problem 1.

By (1) the Center is midway between F_1 and F_2: $C(5, 7)$, so $h = 5$, $k = 7$.

By (2), $a = d(C, V_1) = 1$
By (3), $c = d(C, F_1) = 2$
$b^2 = c^2 - a^2 = 3$
$b = \sqrt{3}$

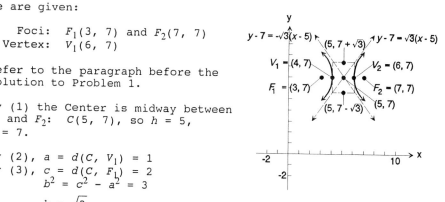

By (4), the transverse axis is the <u>horizontal</u> line $y = 7$

By Table 4,

$$\frac{(x - h)^2}{a^2} - \frac{(y - k)^2}{b^2} = 1 \quad \text{or} \quad \frac{(x - 5)^2}{1} - \frac{(y - 7)^2}{3} = 1$$

9 THE CONICS

33. We are given:

Vertices: $V_1(-1, -1)$ and
$\quad\quad\quad\quad V_2(3, -1)$

Asymptote: $\dfrac{(x - 1)}{2} = \dfrac{(y + 1)}{3}$

Refer to the paragraph before the solution to Problem 1.

By (1) the Center is $C(1, -1)$.
By (2), $a = d(C, V_1) = 2$
By (4), the transverse axis is the <u>horizontal</u> line $y = -1$.

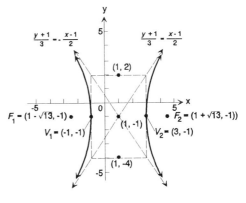

We still need b, and we don't know c. From Table 4, the asymptotes would be

$$y - k = \pm\frac{b}{a}(x - h), \text{ or } y + 1 = \pm\frac{b}{a}(x - 1)$$

Compare that formula with the <u>given</u> asymptote:

$$\frac{(x - 1)}{2} = \frac{(y + 1)}{3}, \text{ or } y + 1 = \frac{3}{2}(x - 1)$$

We see that $\dfrac{b}{a} = \dfrac{3}{2}$ or $b = \dfrac{3a}{2}$

$$b = \frac{6}{2} \quad \text{(since } a = 2\text{)}$$
$$b = 3$$

Therefore, the equation is:

$$\frac{(x - h)^2}{a^2} - \frac{(y - k)^2}{b^2} = 1 \text{ or } \frac{(x - 1)^2}{4} - \frac{(y + 1)^2}{9} = 1$$

35. $\dfrac{(x - 3)^2}{4} - \dfrac{(y + 2)^2}{9} = 1$

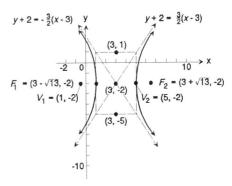

This is in the form found in Table 4. Since the x^2-term is positive, the transverse axis is parallel to the x-axis. We have $h = 3$, $k = -2$, $a^2 = 4$, and $b^2 = 9$. Therefore,

$a = 2$
$b = 3$
$c^2 = a^2 + b^2 = 13$, so that
$c = \sqrt{13}$

Center: $(h, k) = (3, -2)$

Foci: $(h \pm c, k)$: $\left(3 - \sqrt{13}, -2\right)$ and $\left(3 + \sqrt{13}, -2\right)$

Vertices: $(h \pm a, k)$: $(1, -2)$ and $(5, -2)$

Asymptotes: $y - k = \pm\dfrac{b}{a}(x - h)$, or $y + 2 = \pm\dfrac{3}{2}(x - 3)$

(Lines through $(3, -2)$ with slopes $\dfrac{3}{2}$ and $-\dfrac{3}{2}$.)

9.4 THE HYPERBOLA

37. $(y - 2)^2 - 4(x + 2)^2 = 4$

$$\frac{(y - 2)^2}{4} - \frac{(x + 2)^2}{1} = 1$$

This is a hyperbola with $h = -2$, $k = 2$; the transverse axis is parallel to the y-axis, and $a^2 = 4$, $b^2 = 1$. Then:

$a = 2$
$b = 1$
$c^2 = a^1 + b^2 = 5$, so
$c = \sqrt{5}$

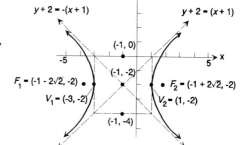

Center: (h, k): $(-2, 2)$
Foci: $(h, k \pm c)$: $(-2, 2 - \sqrt{5})$
        and $(-2, 2 + \sqrt{5})$
Vertices: $(h, k \pm a)$: $(-2, 0)$ and $(-2, 4)$
Asymptotes: $(y - k = \pm \frac{a}{b}(x - h)$, or $y - 2 = \pm 2(x + 2)$

(Lines through $(-2, 2)$ with slopes 2 and -2.)

39. $(x + 1)^2 - (y + 2)^2 = 4$

$$\frac{(x + 1)^2}{4} - \frac{(y + 2)^2}{4} = 1$$

This is a hyperbola with $h = -1$, $k = -2$; the transverse axis is parallel to the x-axis, and $a^2 = 4$, $b^2 = 4$. Then:

$a = 2$
$b = 2$
$c^2 = a^2 + b^2 = 8$, so
$c = 2\sqrt{2}$

Center: $(h, k) = (-1, -2)$
Foci: $(h \pm c, k)$: $\left(-1 - 2\sqrt{2}, -2\right)$ and $\left(-1 + 2\sqrt{2}, -2\right)$
Vertices: $(h \pm a, k)$: $(-3, -2)$ and $(1, -2)$
Asymptotes: $y - k = \pm \frac{b}{a}(x - h)$, or $y + 2 = \pm(x + 1)$

(Lines through $(-1, -2)$ with slopes 1 and -1.)

9 THE CONICS

41.

$$x^2 - y^2 - 2x - 2y - 1 = 0$$
$$(x^2 - 2x + 1) - (y^2 + 2y + 1) = 1 + 1 - 1$$
$$(x - 1)^2 - (y + 1)^2 = 1$$

This is a hyperbola with $h = -1$, $k = -1$; the transverse axis is parallel to the x-axis, and $a^2 = 1$, $b^2 = 1$. Then

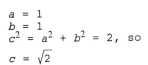
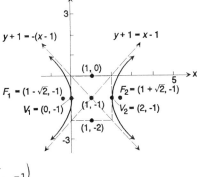

$$a = 1$$
$$b = 1$$
$$c^2 = a^2 + b^2 = 2, \text{ so}$$

$$c = \sqrt{2}$$

Center: $(h, k) = (-1, -1)$
Foci: $(h \pm c, k)$:

$\left(1 - \sqrt{2}, -1\right)$ and $\left(1 + \sqrt{2}, -1\right)$

Vertices: $(h \pm a, k)$: $(0, -1)$ and $(2, -1)$
Asymptotes: $y + 1 = \pm(x - 1)$

(Lines through $(1, -1)$ with slopes 1 and -1.)

43.

$$y^2 - 4x^2 - 4y - 8x - 4 = 0$$
$$(y^2 - 4y + 4) - 4(x^2 + 2x + 1) = 4 + 4 - 4$$
$$(y - 2)^2 - 4(x + 1)^2 = 4$$

$$\frac{(y - 2)^2}{4} - (x + 1)^2 = 1$$

This is a hyperbola with $h = 2$, $k = -1$; the transverse axis is parallel to the y-axis, and $a^2 = 4$, $b^2 = 1$. Then

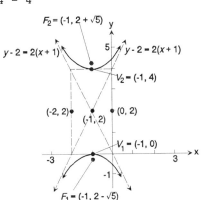

$$a = 2$$
$$b = 1$$
$$c^2 = a^2 + b^2 = 5, \text{ so}$$

$$c = \sqrt{5}$$

Center: $(h, k) = (-1, 2)$

Foci: $(h, k \pm c)$: $\left(-1, 2 - \sqrt{5}\right)$ and $\left(-1, 2 + \sqrt{5}\right)$
Vertices: $(h, k \pm a)$: $(-1, 0)$ and $(-1, 4)$
Asymptotes: $y - 2 = \pm 2(x + 1)$

(Lines through $(-1, 2)$ with slopes 2 and -2.)

45.

$$4x^2 - y^2 - 24x - 4y + 16 = 0$$
$$4x^2 - 24x - y^2 - 4y = -16$$
$$4(x^2 - 6x + \underline{\quad}) - 1(y^2 + 4y + \underline{\quad}) = -16$$
$$4(x^2 - 6x + 9) - 1(y^2 + 4y + 4) = -16 + 36 - 4$$
$$4(x - 3)^2 - (y + 2)^2 = 16$$
$$\frac{(x - 3)^2}{4} - \frac{(y + 2)^2}{16} = 1$$

This is now in a form we can recognize: A hyperbola with transverse axis parallel to the x-axis (since the x^2-term is positive), with center at $C(3, -2)$, and $a^2 = 4$, and $b^2 = 16$. Then $c^2 = a^2 + b^2 = 20$, and we have:

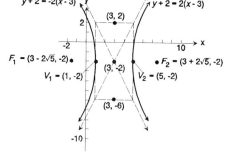

$a = 2$
$b = 4$
$c = \sqrt{20} = 2\sqrt{5}$

Center: $C(3, -2)$
Foci: $(h \pm c, k)$: $\left(3 - 2\sqrt{5}, -2\right)$ and $\left(3 + 2\sqrt{5}, -2\right)$
Vertices: $(h \pm a, k)$: $(1, -2)$ and $(5, -2)$
Asymptotes: $y - k = \pm\dfrac{b}{a}(x - h)$, or $y + 2 = \pm 2(x - 3)$

47.

$$y^2 - 4x^2 - 16x - 2y - 19 = 0$$
$$y^2 - 2y - 4x^2 - 16x = 19$$
$$(y^2 - 2y + \underline{\quad}) - 4(x^2 + 4x + \underline{\quad}) = 19$$
$$(y^2 - 2y + 1) - 4(x^2 + 4x + 4) = 19 + 1 - 16$$
$$(y - 1)^2 - 4(x + 2)^2 = 4$$
$$\frac{(y - 1)^2}{4} - \frac{(x + 2)^2}{1} = 1$$

This is the equation of a hyperbola with transverse axis parallel to the y-axis, with center at $C(-2, 1)$, with $a^2 = 4$ and $b^2 = 1$. Then $c^2 = a^2 + b^2 = 5$, and we have:

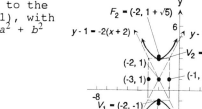

$a = 2$
$b = 1$
$c = \sqrt{5}$

Center: $C(-2, 1)$
Foci: $(h, k \pm c)$:
     $\left(-2, 1 - \sqrt{5}\right)$ and $\left(-2, 1 + \sqrt{5}\right)$
Vertices: $(h, k \pm a)$:
     $(-2, -1)$ and $(-2, 3)$
Asymptotes: $y - k = \pm\dfrac{a}{b}(x - h)$, or $y - 1 = \pm 2(x + 2)$

49.
$$y = \sqrt{16 + 4x^2}$$
$$y^2 = 16 + 4x^2, \qquad y \geq 0$$
$$y^2 - 4x^2 = 16, \qquad y \geq 0$$
$$\frac{y^2}{16} - \frac{x^2}{4} = 1, \qquad y \geq 0$$

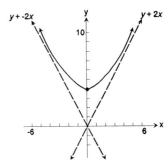

51.
$$y = -\sqrt{-25 + x^2}$$
$$y^2 = -25 + x^2, \qquad y \leq 0$$
$$y^2 - x^2 = -25, \qquad y \leq 0$$
$$\frac{y^2}{25} - \frac{x^2}{25} = -1, \qquad y \leq 0$$
$$\frac{x^2}{25} - \frac{y^2}{25} = 1, \qquad y \leq 0$$

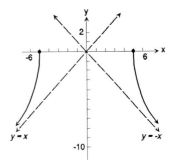

53. By definition of the eccentricity, e,

$$e = \frac{c}{a}, \text{ or } c = ae$$

Therefore, if $e \approx 1$, then $c \approx a$, and $b^2 = c^2 - a^2 \approx 0$, so that b is close to 0.

Assume, for the sake of simplicity, that we have a hyperbola, centered at $(0, 0)$, with transverse axis lying along the x-axis. The asymptotes are:

$$y = \pm\frac{b}{a}x,$$

i.e., lines through the origin with slope $\pm\frac{b}{a}$. Now, if $e \approx 1$, we have $b \approx 0$, so the slopes of the asymptotes are nearly 0. Hence, the asymptotes are nearly horizontal, so the hyperbola is very narrow.

On the other hand, if e is very large, we have:
$$c = ae \quad \text{and} \quad b^2 = c^2 - a^2$$
$$= e^2 a^2 - a^2$$
$$= (e^2 - 1)a^2, \text{ and}$$
$$b = \left(\sqrt{e^2 - 1}\right)a$$

If e is much larger than 1, then b will be much larger than a. In this case, a hyperbola with horizontal transverse axis will

have asymptotes with slopes $\pm\frac{b}{a} = \pm\frac{\left(\sqrt{e^2 - 1}\right)a}{a} = \pm\sqrt{e^2 - 1} > 1$.

9.4 THE HYPERBOLA

Thus, the asymptotes will be nearly vertical, producing a <u>wide</u> hyperbola. As an example, look at the graph of the hyperbola in Problem 1. There, $e = \dfrac{c}{a} = \dfrac{4}{1} = 4$, and the asymptotes have slopes $\pm\sqrt{e^2 - 1} = \pm\sqrt{15} \approx \pm 3.9$. As you can see, the hyperbola is very wide.

55.   (a) $\dfrac{x^2}{4} - y^2 = 1$ $(a^2 = 4,\ b^2 = 1)$

is a hyperbola with <u>horizontal</u> transverse axis, centered at $(0, 0)$ with asymptotes

$$y - k = \pm\dfrac{b}{a}(x - h), \text{ or } y = \pm\dfrac{1}{2}x$$

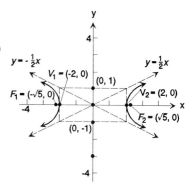

(b) $y^2 - \dfrac{x^2}{4} = 1$ $(a^2 = 1,\ b^2 = 4)$

is a hyperbola with <u>vertical</u> transverse axis, also is centered at $(0, 0)$ and has asymptotes

$$y - k = \pm\dfrac{a}{b}(x - h), \text{ or } y = \pm\dfrac{1}{2}x$$

Since the two hyperbolas have the same asymptotes, they are conjugate.

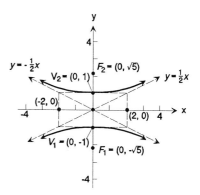

57. $Ax^2 + Cy^2 + F = 0$ $A \neq 0,\ C \neq 0,\ F \neq 0$
        $Ax^2 + Cy^2 = -F$

Since A and C are of opposite signs and $F \neq 0$, this equation may be written as $x^2/-F/A) + y^2/(-F/C) = 1$, where $-F/A$ and $-F/C$ are opposite in sign. This is the equation of a hyperbola with center at $(0, 0)$. The transverse axis is the x-axis if $-F/A > 0$. The transverse axis is the y-axis if $-F/A < 0$.

1. $x^2 + 4x + y + 3 = 0$

   Here $A = 1$, $B = 0$, and $C = 0$, so that $B^2 - 4AC = 0$.
   Since $B^2 - 4AC = 0$, the equation defines a parabola.

3. $6x^2 + 3y^2 - 12x + 6y = 0$

   Here $A = 6$, $B = 0$, and $C = 3$, so that $B^2 - 4AC = -72$.
   Since $B^2 - 4AC < 0$, the equation defines an ellipse.

5. $3x^2 - 2y^2 + 6x + 4 = 0$

   Here $A = 3$, $B = 0$, and $C = -2$, so that $B^2 - 4AC = 24$.
   Since $B^2 - 4AC > 0$, the equation defines a hyperbola.

7. $2y^2 - x^2 - y + x = 0$

   Here $A = -1$, $B = 0$, and $C = 2$, so that $B^2 - 4AC = 8 > 0$.
   The equation defines a hyperbola.

9. $x^2 + y^2 - 8x + 4y = 0$

   Here $A = 1$, $B = 0$, and $C = 1$, so that $B^2 - 4AC = -4 < 0$.
   The equation defines an ellipse, specifically a circle.

For Problems 11 - 12, we use the formulas $\cot 2\theta = \dfrac{A - C}{B}$,
$x = x' \cos \theta - y' \sin \theta$, *and* $y = x' \sin \theta + y' \cos \theta$.

11. $A = 1$, $B = 4$, $C = 1$, $\cot 2\theta = 0$ so that $\theta = \dfrac{\pi}{4}$

   $x = \dfrac{\sqrt{2}}{2}(x' - y')$; $y = \dfrac{\sqrt{2}}{2}(x' + y')$

13. $A = 5$, $B = 6$, $C = 5$, $\cot 2\theta = 0$ so that $\theta = \dfrac{\pi}{4}$

   $x = \dfrac{\sqrt{2}}{2}(x' - y')$; $y = \dfrac{\sqrt{2}}{2}(x' + y')$

15. $A = 13$, $B = -6\sqrt{3}$, $C = 7$, $\cot 2\theta = \dfrac{6}{-6\sqrt{3}} = \dfrac{-1}{\sqrt{3}}$; $\cos 2\theta = \dfrac{-1}{2}$

   $\sin \theta = \sqrt{\dfrac{1 + \frac{1}{2}}{2}} = \dfrac{\sqrt{3}}{2}$; $\cos \theta = \sqrt{\dfrac{1 - \frac{1}{2}}{2}} = \dfrac{1}{2}$

   $x = \dfrac{1}{2}x' - \dfrac{\sqrt{3}}{2}y' = \dfrac{1}{2}\left(x' - \sqrt{3}y'\right)$

   $y = \dfrac{\sqrt{3}}{2}x' + \dfrac{1}{2}y' = \dfrac{1}{2}\left(\sqrt{3}x' + y'\right)$

17. $A = 4$, $B = -4$, $C = 1$, $\cot 2\theta = \dfrac{-3}{4}$; $\cos 2\theta = \dfrac{-3}{5}$

$$\sin \theta = \sqrt{\dfrac{1 + \dfrac{3}{5}}{2}} = \dfrac{2}{\sqrt{5}} = \dfrac{2\sqrt{5}}{5}; \cos \theta = \sqrt{\dfrac{1 - \dfrac{3}{5}}{2}} = \dfrac{1}{\sqrt{5}} = \dfrac{\sqrt{5}}{5}$$

$$x = \dfrac{\sqrt{5}}{5}x' - \dfrac{2\sqrt{5}}{2}y' = \dfrac{\sqrt{5}}{5}(x' - 2y')$$

$$y = \dfrac{2\sqrt{5}}{5}x' + \dfrac{\sqrt{5}}{5}y' = \dfrac{\sqrt{5}}{5}(2x' + y')$$

19. $A = 25$, $B = -36$, $C = 40$, $\cot 2\theta = \dfrac{-15}{-36} = \dfrac{5}{12}$; $\cos 2\theta = \dfrac{5}{13}$

$$\sin \theta = \sqrt{\dfrac{1 - \dfrac{5}{13}}{2}} = \dfrac{2}{\sqrt{13}} = \dfrac{2\sqrt{13}}{13}; \cos \theta = \sqrt{\dfrac{1 + \dfrac{5}{13}}{2}} = \dfrac{3}{\sqrt{13}} = \dfrac{3\sqrt{13}}{13}$$

$$x = \dfrac{3\sqrt{13}}{13}x' - \dfrac{2\sqrt{13}}{13}y' = \dfrac{\sqrt{13}}{13}(3x' - 2y')$$

$$y = \dfrac{2\sqrt{13}}{13}x' + \dfrac{3\sqrt{13}}{13}y' = \dfrac{\sqrt{13}}{13}(2x' + 3y')$$

21. $\theta = 45°$ (see Problem 11)

$$x^2 + 4xy + y^2 - 3 = 0$$

$$\left[\dfrac{\sqrt{2}}{2}(x' - y')\right]^2 + 4\left[\dfrac{\sqrt{2}}{2}(x' - y')\right]\left[\dfrac{\sqrt{2}}{2}(x' + y')\right] + \left[\dfrac{\sqrt{2}}{2}(x' + y')\right]^2 - 3 = 0$$

$$\dfrac{1}{2}\left(x'^2 - 2x'y' + y'^2\right) + \dfrac{4}{2}\left(x'^2 - y'^2\right) + \dfrac{1}{2}\left(x'^2 + 2x'y' + y'^2\right) = 3$$

$$6x'^2 - 2y'^2 = 6$$

$$x'^2 - \dfrac{y'^2}{3} = 1$$

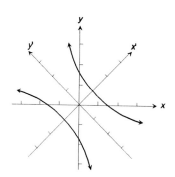

Hyperbola; center at origin; transverse axis the x'-axis; vertices at $(\pm 1, 0)$.

23. $\theta = 45°$ (see Problem 13)

$5x^2 + 6xy + 5y^2 - 8 = 0$

$$5\left[\frac{\sqrt{2}}{2}(x' - y')\right]^2 + 6\left[\frac{\sqrt{2}}{2}(x' - y')\right]\left[\frac{\sqrt{2}}{2}(x' + y')\right] + 5\left[\frac{\sqrt{2}}{2}(x' + y')\right]^2 - 8 = 0$$

$$\frac{5}{2}(x'^2 - 2x'y' + y'^2) + \frac{6}{2}(x'^2 - y'^2) + \frac{5}{2}(x'^2 + 2x'y' + y'^2) = 8$$

$16x'^2 + 4y'^2 = 16$

$x'^2 + \dfrac{y'^2}{4} = 1$

Ellipse; center at $(0, 0)$; major axis the y'-axis; vertices at $(0, \pm 2)$.

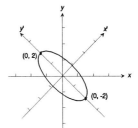

25. $\theta = 60°$ (see Problem 15)

$13x^2 - 6\sqrt{3}\,xy + 7y^2 - 16 = 0$

$$13\left[\frac{1}{2}(x' - \sqrt{3}y')\right]^2 - 6\sqrt{3}\left[\frac{1}{2}(x' - \sqrt{3}y')\right]\left[\frac{1}{2}(\sqrt{3}x' + y')\right]$$
$$+ 7\left[\frac{1}{2}(\sqrt{3}x' + y')\right]^2 = 16$$

$$13(x'^2 - 2\sqrt{3}x'y^2 + 3y'^2) - 6\sqrt{3}(\sqrt{3}x'^2 - 2x'y' - \sqrt{3}y'^2)$$
$$+ 7(3x'^2 + 2\sqrt{3}x'y' + y'^2) = 64$$

$16x'^2 + 64y'^2 = 64$

$\dfrac{x'^2}{4} + \dfrac{y'^2}{1} = 1$

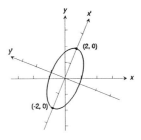

Ellipse; center at $(0, 0)$; major axis is the x'-axis; vertices at $(\pm 2, 0)$.

27. $\theta = 63°$ (see Problem 17)

$$4x^2 - 4xy + y^2 - 8\sqrt{5}x - 16\sqrt{5}y = 0$$

$$4\left[\frac{\sqrt{5}}{5}(x' - 2y')\right]^2 - 4\left[\frac{\sqrt{5}}{5}(x' - 2y')\right]\left[\frac{\sqrt{5}}{5}(2x' + y')\right] + \left[\frac{\sqrt{5}}{5}(2x' + y')\right]^2$$
$$- 8\sqrt{5}\left[\frac{\sqrt{5}}{5}(x' - 2y')\right] - 16\sqrt{5}\left[\frac{\sqrt{5}}{5}(2x' + y')\right] = 0$$

$$\frac{4}{5}\left(x'^2 - 4x'y' + 4y'^2\right) - \frac{4}{5}\left(2x'^2 - 3x'y' - 2y'^2\right) + \frac{1}{5}\left(4x'^2 + 4x'y' + y'^2\right)$$
$$- 8(x' - 2y') - 16(2x' + y') = 0$$

$$5y'^2 - 40x' = 0$$
$$y'^2 = 8x'$$

Parabola; vertex at (0, 0); focus at (2, 0).

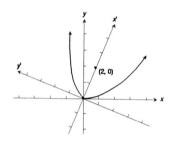

29. $\theta \approx 34°$ (see Problem 19)

$$25x^2 - 36xy + 40y^2 - 12\sqrt{13}x - 8\sqrt{13}y = 0$$

$$25\left[\frac{\sqrt{13}}{13}(3x' - 2y')\right]^2 - 36\left[\frac{\sqrt{13}}{13}(3x' - 2y')\right]\left[\frac{\sqrt{13}}{13}(2x' + 3y')\right]$$
$$+ 40\left[\frac{\sqrt{13}}{13}(2x' + 3y')\right]^2 - 12\sqrt{13}\left[\frac{\sqrt{13}}{13}(3x' - 2y')\right]$$
$$- 8\sqrt{13}\left[\frac{\sqrt{13}}{13}(2x' + 3y')\right] = 0$$

$$\frac{25}{13}\left(9x'^2 - 12x'y' + 4y'^2\right) - \frac{36}{13}\left(6x'^2 + 5x'y' - 6y'^2\right)$$
$$+ \frac{40}{13}\left(4x'^2 + 12x'y' + 9y'^2\right) - 12(3x' - 2y')$$
$$- 8(2x' + 3y') = 0$$

$$\frac{169}{13}x'^2 + \frac{676}{13}y'^2 - 52x' = 0$$

$$13x'^2 + 52y'^2 - 52x' = 0$$
$$x'^2 - 4x' + 4y'^2 = 0$$
$$(x' - 2)^2 + 4y'^2 = 4$$
$$\frac{(x' - 2)^2}{4} + y'^2 = 1$$

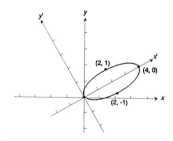

Ellipse; center at (2, 0); major axis the
x'-axis; vertices at (4, 0) and (0, 0).

9 ANALYTIC GEOMETRY

31. $A = 16$, $B = 24$, $C = 9$, $\cot 2\theta = \dfrac{7}{24}$; $\cos 2\theta = \dfrac{7}{25}$

$\sin \theta = \sqrt{\dfrac{1 - \dfrac{7}{25}}{2}} = \dfrac{3}{5}$; $\cos \theta = \sqrt{\dfrac{1 + \dfrac{7}{25}}{2}} = \dfrac{4}{5}$; $\theta \approx 37°$

$x = \dfrac{4}{5}x' - \dfrac{3}{5}y' = \dfrac{1}{5}(4x' - 3y')$

$y = \dfrac{3}{5}x' + \dfrac{4}{5}y' = \dfrac{1}{5}(3x' + 4y')$

$16x^2 + 24xy + 9y^2 - 130x + 90y = 0$

$16\left[\dfrac{1}{5}(4x' - 3y')\right]^2 + 24\left[\dfrac{1}{5}(4x' - 3y')\right]\left[\dfrac{1}{5}(3x' + 4y')\right]$
$\quad 9\left[\dfrac{1}{5}(3x' + 4y')\right]^2 - 130\left[\dfrac{1}{5}(4x' - 3y')\right] + 90\left(\dfrac{1}{5}(3x' + 4y')\right) = 0$

$\dfrac{16}{25}(16x'^2 - 24x'y' + 9y'^2) + \dfrac{24}{25}(12x'^2 + 7x'y' - 12y'^2)$
$\quad + \dfrac{9}{25}(9x'^2 + 24x'y' + 16y'^2) - 26(4x' - 3y')$
$\quad + 18(3x' + 4y') = 0$

$\dfrac{625}{25}x'^2 - 50x' + 150y' = 0$

$x'^2 - 2x' = -6y'$

$(x' - 1)^2 = -6\left(y' - \dfrac{1}{6}\right)$

Parabola; vertex at $\left(1, \dfrac{1}{6}\right)$; focus at

$\left(1, \dfrac{-4}{3}\right)$

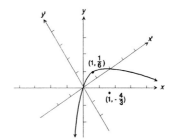

33. $B^2 - 4AC = 9 + 8 = 17 > 0$; hyperbola

35. $B^2 - 4AC = 49 - 12 = 37 > 0$; hyperbola

37. $B^2 - 4AC = 144 - 144 = 0$; parabola

39. $B^2 - 4AC = 144 - 160 = -16 < 0$; ellipse

41. $B^2 - 4AC = 4 - 12 = -8 < 0$; ellipse

43. Refer to Equation (6)

$A' = A \cos^2 \theta + B \sin \theta \cos \theta + C \sin^2 \theta$
$B' = B(\cos^2 \theta - \sin^2 \theta) + 2(C - A \sin \theta \cos \theta)$
$C' = A \sin^2 \theta - B \sin \theta \cos \theta + C \cos^2 \theta$
$D' = D \cos \theta + E \sin \theta$
$E' = -D \sin \theta + E \cos \theta$
$F' = F$

45.
$$B'^2 - 4A'C' = [B(\cos^2\theta - \sin^2\theta) + 2(C - A)\sin\theta\cos\theta]^2$$
$$- 4[A\cos^2\theta + B\sin\theta\cos\theta + C\sin^2\theta)]$$
$$[A\sin^2\theta - B\sin\theta\cos\theta + C\sin^2\theta]$$
$$= B^2(\cos^4\theta - 2\sin^2\theta\cos^2\theta + \sin^4\theta)$$
$$+ 4B(C - A)\sin\theta\cos\theta(\cos^2\theta - \sin^2\theta)$$
$$+ 4(C - A)^2\sin^2\theta\cos^2\theta$$
$$- 4[A^2\sin^2\theta\cos^2\theta - AB\sin\theta\cos^3\theta + AC\cos^4\theta$$
$$- B^2\sin^2\theta\cos^2\theta + BC\sin\theta\cos^3\theta$$
$$+ AB\sin^3\theta\cos\theta + AC\sin^4\theta - BC\sin^3\theta\cos\theta$$
$$+ C^2\sin^2\theta\cos^2\theta]$$
$$= B^2[\cos^4\theta - 2\sin^2\theta\cos^2\theta + \sin^4\theta + 4\sin^2\theta\cos^2\theta]$$
$$+ BC[4\sin\theta\cos\theta(\cos^2\theta - \sin^2\theta)$$
$$- 4\sin\theta\cos\theta(\cos^2\theta - \sin^2\theta)]$$
$$- AB[4\sin\theta\cos\theta(\cos^2\theta - \sin^2\theta) - 4\sin\theta\cos^3\theta$$
$$+ 4\sin^3\theta\cos\theta]$$
$$+ 4C^2[\sin^2\theta\cos^2\theta - \sin^2\theta\cos^2\theta]$$
$$- 4AC[2\sin^2\theta\cos^2\theta + \cos^4\theta\sin^4\theta]$$
$$+ 4A^2[\sin^2\theta\cos^2\theta - \sin^2\theta\cos^2\theta]$$
$$= B^2[\cos^4\theta + 2\sin^2\theta\cos^2\theta + \sin^4\theta]$$
$$- 4AC[\cos^4\theta + 2\sin^2\theta\cos^2\theta + \sin^4\theta]$$
$$= B^2[\sin^2\theta + \cos^2\theta]^2 - 4AC[\cos^2\theta + \sin^2\theta]^2$$
$$= B^2 - 4AC$$

47. Refer to Equation (5)
$$d^2 = (y_2 - y_1)^2 + (x_2 - x_1)^2$$
$$= [x_2'\sin\theta + y_2'\cos\theta - x_1'\sin\theta - y_1'\cos\theta]^2$$
$$+ [x_2'\cos\theta - y_2'\sin\theta - x_1'\cos\theta + y_1'\sin\theta]^2$$
$$= [(x_2' - x_1')\sin\theta + (y_2' - y_1')\cos\theta]^2$$
$$+ [(x_2' - x_1')\cos\theta - (y_2' - y_1')^2\sin\theta]^2$$
$$= (x_2' - x_1')^2\sin^2\theta + 2(x_2' - x_1')^2(y_2' - y_1')\sin\theta\cos\theta$$
$$+ (y_2' - y_1')^2\cos^2\theta + (x_2' - x_1')^2\cos^2\theta$$
$$- 2(x_2' - x_1')^2(y_2' - y_1')\sin\theta\cos\theta + (y_2' - y_1')^2\sin^2\theta$$
$$= (x_2' - x_1')^2(\sin^2\theta + \cos^2\theta) + (y_2' - y_1')^2(\cos^2\theta + \sin^2\theta)$$
$$= (x_2' - x_1')^2 + (y_2' - y_1')^2$$

≡ EXERCISE 9.6 POLAR EQUATIONS OF CONICS

For Problems 1-18, use Formulas 4 and 5.

1. $e = 1$; $p = 1$; parabola; directrix is perpendicular to the polar axis 1 unit to the right of the pole.

3. $r = \dfrac{4}{2\left(1 - \dfrac{3}{2}\sin\theta\right)} = \dfrac{2}{1 - \dfrac{3}{2}\sin\theta}$; $ep = 2$, $e = \dfrac{3}{2}$; $p = \dfrac{4}{3}$

Hyperbola; directrix is parallel to the polar axis $\dfrac{4}{3}$ units below the pole.

5. $r = \dfrac{3}{4\left(1 - \dfrac{1}{2}\cos\theta\right)} = \dfrac{\dfrac{3}{4}}{1 - \dfrac{1}{2}\cos\theta}$; $ep = \dfrac{3}{4}$, $e = \dfrac{1}{2}$; $p = \dfrac{3}{2}$

Ellipse; directrix is perpendicular to the polar axis $\dfrac{3}{2}$ units to the left of the pole.

7. $r = \dfrac{1}{1 + \cos\theta}$
   $ep = 1$, $e = 1$, $p = 1$

Parabola; directrix is perpendicular to the polar axis 1 unit to the right of the pole; vertex is $\left(\dfrac{1}{2},\ 0\right)$.

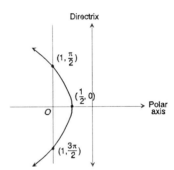

9. $r = \dfrac{8}{4\left(1 + \dfrac{3}{4}\sin\theta\right)} = \dfrac{2}{1 + \dfrac{3}{4}\sin\theta}$
   $ep = 2$, $e = \dfrac{3}{4}$, $p = \dfrac{8}{3}$

Ellipse; directrix parallel to the polar axis $\dfrac{8}{3}$ units above the pole. Vertices are at $\left(\dfrac{8}{7},\ \dfrac{\pi}{2}\right)$ and $\left(8,\ \dfrac{3\pi}{2}\right)$.

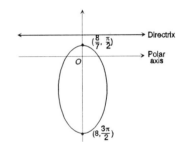

11. $r = \dfrac{9}{3(1 - 2\cos\theta)} = \dfrac{3}{1 - 2\cos\theta}$
    $ep = 3$, $e = 2$, $p = \dfrac{3}{2}$

Hyperbola, directrix is perpendicular to the polar axis $\dfrac{3}{2}$ units to the left of the pole. Vertices are at $(1,\ \pi)$ and $(-3,\ 0)$.

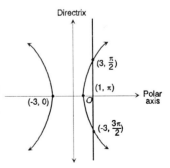

13. $r = \dfrac{8}{2 - \sin \theta}$

$r = \dfrac{8}{2\left(1 - \dfrac{1}{2} \sin \theta\right)}$

$= \dfrac{4}{1 - \dfrac{1}{2} \sin \theta}$

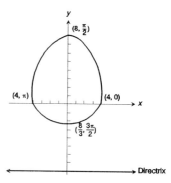

$e = \dfrac{1}{2}$; $ep = 4$, so $p = \dfrac{4}{\dfrac{1}{2}} = 8$

The conic is an ellipse; major axis is perpendicular to the directrix. The directrix is parallel to the polar axis at a distance 8 units below the pole.

15. $r = \dfrac{6}{(3 - 2 \sin \theta)} = \dfrac{2}{1 - \dfrac{2}{3} \sin \theta}$

$ep = 2$, $e = \dfrac{2}{3}$, $p = 3$

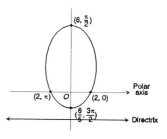

Ellipse; directrix parallel to the polar axis 3 units below the pole. Vertices are at $\left(6, \dfrac{\pi}{2}\right)$ and $\left(\dfrac{6}{5}, \dfrac{3\pi}{2}\right)$.

17. $r = \dfrac{6 \sec \theta}{2 \sec \theta - 1} = \dfrac{6}{2 - \cos \theta} = \dfrac{3}{1 - \dfrac{1}{2} \cos \theta}$

$ep = 3$, $e = \dfrac{1}{2}$, $p = 6$

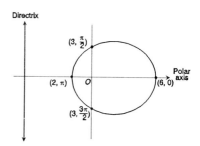

Ellipse; directrix is perpendicular to the polar axis 6 units to the left of the pole. Vertices are at (6, 0) and (2, π).

19. $r = \dfrac{1}{1 + \cos \theta}$
$r + r \cos \theta = 1$
$r = 1 - r \cos \theta$
$r^2 = (1 - r \cos \theta)^2$
$x^2 + y^2 = (1 - x)^2$
$x^2 + y^2 = 1 - 2x + x^2$
$y^2 + 2x - 1 = 0$

21. $r = \dfrac{8}{4 + 3 \sin \theta}$
$4r + 3r \sin \theta = 8$
$4r = 8 - 3r \sin \theta$
$16r^2 = (8 - 3r \sin \theta)^2$
$16(x^2 + y^2) = (8 - 3y)^2$
$16x^2 + 16y^2 = 64 - 48y + 9y^2$
$16x^2 + 7y^2 + 48y - 64 = 0$

9 ANALYTIC GEOMETRY

23. $r = \dfrac{9}{3 - 6 \cos \theta}$

$3r - 6r \cos \theta = 9$

$3r = 9 + 6r \cos \theta$

$r = 3 + 2r \cos \theta$

$r^2 = (3 + 2r \cos \theta)^2$

$x^2 + y^2 = (3 + 2x)^2$

$x^2 + y^2 = 9 + 12x + 4x^2$

$3x^2 - y^2 + 12x + 9 = 0$

25. $r = \dfrac{8}{2 - \sin \theta}$

$r(2 - \sin \theta) = 8$

$2r - r \sin \theta = 8$

$2r = 8 + r \sin \theta$

$4r^2 = (8 + r \sin \theta)^2$

$4(x^2 + y^2) = (8 + y)^2$

$4x^2 + 4y^2 = (8 + y)^2$

$4x^2 + 4y^2 = 64 + 16y + y^2$

$4x^2 + 3y^2 - 16y - 64 = 0$

27. $r(3 - 2 \sin \theta) = 6$

$3r - 2r \sin \theta = 6$

$3r = 6 + 2r \sin \theta$

$9r^2 = (6 + 2r \sin \theta)^2$

$9(x^2 + y^2) = (6 + 2y)^2$

$9x^2 + 9y^2 = 36 + 24y + 4y^2$

$9x^2 + 5y^2 - 24y - 36 = 0$

29. $r = \dfrac{6 \sec \theta}{2 - \sec \theta - 1}$

$r = \dfrac{6}{2 - \cos \theta}$

$2r - r \cos \theta = 6$

$2r = 6 + r \cos \theta$

$4r^2 = (6 + r \cos \theta)^2$

$4(x^2 + y^2) = (6 + x)^2$

$4x^2 + 4y^2 = 36 + 12x + x^2$

$3x^2 + 4y^2 - 12x - 36 = 0$

31. $r = \dfrac{ep}{1 + e \sin \theta}$

$e = 1, \ p = 1$

$r = \dfrac{1}{1 + \sin \theta}$

33. $r = \dfrac{ep}{1 - e \cos \theta}$

$e = \dfrac{4}{5}, \ p = 3$

$r = \dfrac{\dfrac{12}{5}}{1 - \dfrac{4}{5} \cos \theta} = \dfrac{12}{5 - 4 \cos \theta}$

35. $r = \dfrac{ep}{1 - e \sin \theta}$

$e = 6, \ p = 2$

$r = \dfrac{12}{1 - 6 \sin \theta}$

37. $d(F, P) = e \cdot d(D, P)$

$d(D, P) = p - r \cos \theta$

$\therefore r = e(p - r \cos \theta)$

$r = ep - er \cos \theta$

$r + er \cos \theta = ep$

$r = \dfrac{ep}{1 + e \cos \theta}$

39. $d(FP) = e \cdot d(D, P)$

$d(D, P) = p + r \sin \theta$

$\therefore r = e(p + r \sin \theta)$

$r = ep + er \sin \theta$

$r - er \sin \theta = ep$

$r = \dfrac{ep}{1 - e \sin \theta}$

1. $x = 3t + 2$, $y = t + 1$
$x = 3(y - 1) + 2$
$x = 3y - 1$
$x - 3y + 1 = 0$

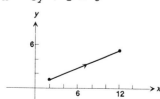

3. $x = t + 2$, $y = \sqrt{t}$
$y = \sqrt{x - 2}$

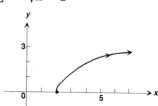

5. $x = t^2 + 4$, $y = t^2 - 4$
$x = (y + 4) + 4$
$x = y + 8$

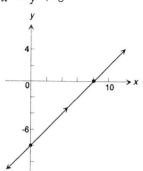

7. $x = 3t^2$, $y = t + 1$
$x = 3(y - 1)^2$

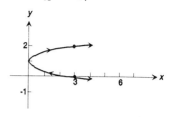

9. $x = 2e^t$, $y = 1 + e^t$
$y = 1 + \dfrac{x}{2}$
$2y = 2 + x$

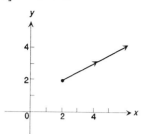

11. $x = \sqrt{t}$, $y = t^{3/2}$
$y = (x^2)^{3/2}$
$y = x^3$

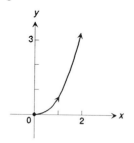

13. $x = 2 \cos t$, $y = 3 \sin t$;
$0 \le t \le 2\pi$

$$\left(\frac{x}{2}\right)^2 + \left(\frac{y}{3}\right)^2 = \cos^2 t + \sin^2 t$$
$$\frac{x^2}{4} + \frac{y^2}{9} = 1$$

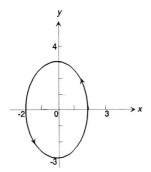

15. $x = 2 \cos t$, $y = 3 \sin t$;
$-\pi \le t \le 0$

$$\left(\frac{x}{2}\right)^2 + \left(\frac{y}{3}\right)^2 = \cos^2 t + \sin^2 t$$
$$\frac{x^2}{4} + \frac{y^2}{9} = 1$$

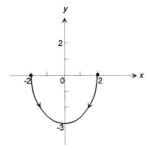

17. $x = \sec t$, $y = \tan t$
$1 + \tan^2 t = \sec^2 t$
$1 + y^2 = x^2$
$x^2 - y^2 = 1$

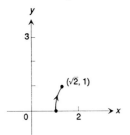

19. $x = \sin^3 t$, $y = \cos^3 t$
$\sin^2 t + \cos^2 t = 1$
$(x^{1/3})^2 + (y^{1/3})^2 = 1$
$x^{2/3} + y^{2/3} = 1$

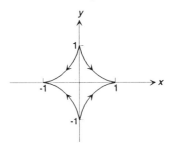

21. $x = t$, $y = t^3$

$x = \sqrt[3]{t}$, $y = t$

23. $x = t$, $y = t^{2/3}$
$x = t^{3/2}$, $y = t$

25. $x = 2 \cos \omega t$, $y = -3 \sin \omega t$
$\frac{2\pi}{\omega} = 2$, $\omega = \pi$
$x = 2 \cos \pi t$, $y = -3 \sin \pi t$,
$0 \le t \le 2$

27. $x = -2 \sin \omega t$, $y = 3 \cos \omega t$
$\frac{2\pi}{\omega} = 1$, $\omega = 2\pi$
$x = -2 \sin^2 \pi t$,
$y = 3 \cos 2\pi t$, $0 \le t \le 1$

29.

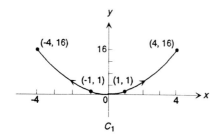

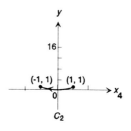

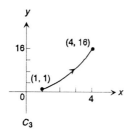

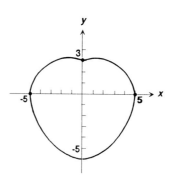

31. $x = (x_2 - x_1)t + x_1$
      $y = (y_2 - y_1)t + y_1$

      $\dfrac{x - x_1}{x_2 - x_1} = t$

      $y = (y_2 - y_1)\left(\dfrac{x - x_1}{x_2 - x_1}\right) + y_1$

      $y - y_1 = \left(\dfrac{y_2 - y_1}{x_2 - x_1}\right)(x - x_1)$

      This is the equation of a line. Its orientation is from $(x_1,\ y_1)$
      to $(x_2,\ y_2)$.

33.

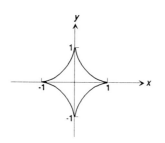

35.

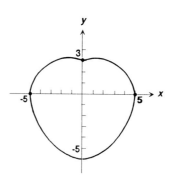

9 ANALYTIC GEOMETRY

For Problems 1-20, use the following rules:

  I.   *If only <u>one</u> variable is squared, the equation represents a parabola.*
  II. *If <u>both</u> variables are squared, the equation is either an ellipse or a hyperbola.*

   A. *If both the x^2-term and the y^2-term are positive, the graph is an ellipse.*
   B. *If one of the two squared terms is negative, the graph is a hyperbola.*

1. $y^2 = -16x$. This is a parabola, with equation of the form:

$$y^2 = -4ax$$

where $a = 4$. By Table 1, in Section 2, we have:

Vertex:	(0, 0)
Focus:	(-4, 0)
Directrix:	$x = 4$

3. $\dfrac{x^2}{4} - y^2 = 1$

This is a hyperbola in the form:

$$\frac{(x - h)^2}{a^2} - \frac{(y - k)^2}{b^2} = 1,$$

where $a = 2$, $b = 1$, $h = 0$, $k = 0$. By Table 4 in Section 4, $c^2 = a^2 + b^2 = 5$, so $c = \sqrt{5}$, and we have:

Center:	(0, 0)	
Foci:	$(h \pm c,\ k)$:	$\left(-\sqrt{5},\ 0\right)$ and $\left(\sqrt{5},\ 0\right)$
Vertices:	$(h \pm a,\ k)$:	(-2, 0) and (2, 0)
Asymptotes:	$y - k = \pm\dfrac{b}{a}(x - h)$ or $y = \pm\dfrac{1}{2}x$	

5. $\dfrac{y^2}{25} + \dfrac{x^2}{16} = 1$

This is an ellipse since both variables are squared, and both terms are positive. The equation is already in the form

$$\frac{(y - k)^2}{a^2} + \frac{(x - h)^2}{b^2} = 1$$

where $h = 0$, $k = 0$, $a = 5$, $b = 4$

By Table 3, we have $c^2 = a^2 - b^2 = 9$, so that $c = 3$, and:

Center:	$(h,\ k)$, or (0, 0)
Foci:	$(h,\ k \pm c)$, or (0, -3) and (0, 3)
Vertices:	$(h,\ k \pm a)$, or (0, -5) and (0, 5)

7. $x^2 + 4y = 4$ is a <u>parabola</u>.

$$x^2 + 4y = 4$$
$$x^2 = -4y + 4$$
$$x^2 = -4(y - 1)$$

This is in the form $(x - h)^2 = -4a(y - k)$, where

(1) $a = 1$
(2) $x - h = x$
$\qquad h = 0$
(3) $y - k = y - 1$
$\qquad k = 1$

From Table 2, Section 2, we have:

$$\begin{array}{ll} \text{Vertex:} & (0, 1) \\ \text{Focus:} & (0, 0) \\ \text{Directrix:} & y = 2 \end{array}$$

9. $4x^2 - y^2 = 8$

This is a hyperbola, since it consists of a <u>difference</u> of squared terms:

$$4x^2 - y^2 = 8$$
$$\frac{x^2}{2} - \frac{y^2}{8} = 1$$

From Table 4, $a^2 = 2$, so $a = \sqrt{2}$, and $b^2 = 8$, so $b = \sqrt{8} = 2\sqrt{2}$.

Also, $c^2 = a^2 + b^2 = 10$, so $c = \sqrt{10}$. Then we have:

$$\begin{array}{ll} \text{Transverse axis:} & \text{horizontal:} \quad y = 0 \\ \text{Center:} & (0, 0) \\ \text{Foci:} & \left(-\sqrt{10}, 0\right) \text{ and } \left(\sqrt{10}, 0\right) \\ \text{Vertices:} & \left(-\sqrt{2}, 0\right) \text{ and } \left(\sqrt{2}, 0\right) \\ \text{Asymptotes:} & y - k = \pm\frac{b}{a}(x - h), \text{ or} \\ & y = \pm 2x \end{array}$$

11. $x^2 - 4x = 2y$ is a parabola:

$$x^2 - 4x + \underline{\quad} = 2y + \underline{\quad}$$
$$x^2 - 4x + 4 = 2y + 4$$
$$(x - 2)^2 = 2(y + 2)$$

We have:

(1) $4a = 2$, or $a = \dfrac{1}{2}$
(2) $h = 2$
(3) $k = -2$

and:

$$\begin{array}{ll} \text{Vertex:} & (2, -2) \\ \text{Focus:} & \left(2, -\dfrac{3}{2}\right) \\ \text{Directrix:} & y = -\dfrac{5}{2} \end{array}$$

13. $y^2 - 4y - 4x^2 + 8x = 4$

Complete the square:

$$(y^2 - 4y + \underline{\hphantom{xx}}) - 4(x^2 - 2x + \underline{\hphantom{xx}}) = 4$$
$$(y^2 - 4y + 4) - 4(x^2 - 2x + 1) = 4 + 4 - 4$$
$$(y - 2)^2 - 4(x - 1)^2 = 4$$
$$\frac{(y - 2)^2}{4} - \frac{(x - 1)^2}{1} = 1$$

This is a hyperbola with vertical transverse axis, center at (1, 2), $a^2 = 4$, $b^2 = 1$, and $c^2 = a^2 + b^2 = 5$.

Then, $a = 2$
$b = 1$
$c = \sqrt{5}$

Center: $(h, k) = (1, 2)$
Foci: $(h, k \pm c) = \left(1, 2 - \sqrt{5}\right)$ and $\left(1, 2 + \sqrt{5}\right)$
Vertices: $(h, k \pm a) = (1, 0)$ and $(1, 4)$
Asymptotes: $y - 2 = \pm 2(x - 1)$

(Lines through (1, 2) with slopes 2 and -2.)

15. $4x^2 + 9y^2 - 16x - 18y = 11$

Since both variables are squared, this is either an ellipse or a hyperbola. Since both squared terms are positive, the graph <u>must</u> be an ellipse. We start by completing the square, to put the equation in a recognizable form:

$$4x^2 - 16x + 9y^2 - 18y = 11$$
$$4(x^2 - 4x + \underline{\hphantom{xx}}) + 9(y^2 - 2y + \underline{\hphantom{xx}}) = 11$$
$$4(x^2 - 4x + 4) + 9(y^2 - 2y + 1) = 11 + 16 + 9$$
$$4(x - 2)^2 + 9(y - 1)^2 = 36$$
$$\frac{4(x - 2)^2}{36} + \frac{9(y - 1)^2}{36} = 1$$
$$\frac{(x - 2)^2}{9} + \frac{(y - 1)^2}{4} = 1$$

We now have the form

$$\frac{(x - h)^2}{a^2} + \frac{(y - k)^2}{b^2} = 1$$

where $h = 2$, $k = 1$, $a = 3$, $b = 2$. By Table 3, $c^2 = a^2 - b^2 = 5$, so $c = \sqrt{5}$, and we have:

Center: $(h, k) = (2, 1)$
Foci: $(h \pm c, k)$: $\left(2 - \sqrt{5}, 1\right)$ and $\left(2 + \sqrt{5}, 1\right)$
Vertices: $(h \pm a, k)$: $(-1, 1)$ and $(5, 1)$

9 - CHAPTER REVIEW

17. $4x^2 - 16x + 16y + 32 = 0$ is a <u>parabola</u>.

$$4x^2 - 16x = -16y - 32$$
$$4(x^2 - 4x + \underline{\quad}) = -16y - 32 + \underline{\quad}$$
$$4(x^2 - 4x + 4) = -16y - 32 + \overline{16}$$

$$(4)(4) = 16$$
$$4(x - 2)^2 = -16y - 16$$
$$4(x - 2)^2 = -16(y + 1)$$
$$(x - 2)^2 = -4(y + 1)$$

We have: (1) $a = 1$
           (2) $h = 2$
           (3) $k = -1$

and: Vertex: $(2, -1)$
           Focus: $(2, -2)$
           Directrix: $y = 0$

19. $9x^2 + 4y^2 - 18x + 8y = 23$

Both variables are squared, and the squared terms are both positive, so this is an ellipse.

$$9x^2 - 18x + 4y^2 + 8y = 23$$
$$9(x^2 - 2x + \underline{\quad}) + 4(y^2 + 2y + \underline{\quad}) = 23$$
$$9(x^2 - 2x + 1) + 4(y^2 + 2y + 1) = 23 + 9 + 4$$
$$9(x - 1)^2 + 4(y + 1)^2 = 36$$

$$\frac{(x - 1)^2}{4} + \frac{(y + 1)^2}{9} = 1$$

This is in the form:

$$\frac{(x - h)^2}{b^2} + \frac{(y - k)^2}{a^2} = 1$$

where $h = 1$, $k = -1$, $a = 3$, $b = 2$.

From Table 3, $c^2 = a^2 - b^2 = 5$, so $c = \sqrt{5}$, and we have:

           Center: $(h, k) = (1, -1)$
           Foci: $(h, k \pm c)$, or $\left(1, -1 - \sqrt{5}\right)$ and $\left(1, -1 + \sqrt{5},\right)$
           Vertices: $(h, k \pm a)$, or $(1, -4)$ and $(1, 2)$

21.   We are given:

           Type of graph: Parabola
           Focus: $F(-2, 0)$
           Directrix: $x = 2$

From Table 2, we see that to find the equation of a parabola, we need to know the coordinates of the vertex, (h, k), the distance from the vertex to the focus (a), and whether the parabola opens up, down, left, or right.

Here, the directrix, $x = 2$, is a vertical line, so the axis of symmetry is a horizontal line (which passes through the focus, $(-2, 0)$, i.e., $y = 0$.

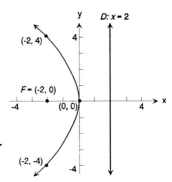

The vertex is on the axis, midway between the focus and the directrix, at $V(0, 0)$. Then we have:

$$a = d(V, F) = 2$$

Finally, the focus is to the <u>left</u> of the vertex, so the parabola opens to the left. By Table 2, the parabola has an equation of the form:

$$(y - k)^2 = -4a(x - h), \text{ or}$$
$$y^2 = -4ax, \text{ since } h = 0, \ k = 0$$
$$y^2 = -8x, \text{ since } a = 2$$

23. We are given:

  Type of graph: Hyperbola
     Center: $C(0, 0)$
     Focus: $F_2(0, 4)$
     Vertex: $V_1(0, -2)$

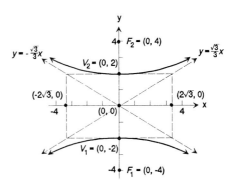

As we see from Table 4, to determine the equation of a hyperbola, we need to know the orientation of the transverse axis, the coordinates of the center, (h, k) and the constants a and b.

We know $h = 0, \ k = 0$
    $c = d(C, F_2) = 4$
    $a = d(C, V_1) = 2$
and $b^2 = c^2 - a^2$
     $= 16 - 4 = 12$
or $b = \sqrt{12} = 2\sqrt{3}$

The center, focus, and vertex all lie on the <u>vertical</u> line $x = 0$, so the transverse axis is parallel to the y-axis. Therefore,

$$\frac{(y - k)^2}{a^2} - \frac{(x - h)^2}{b^2} = 1, \text{ or}$$

$$\frac{y^2}{4} - \frac{x^2}{12} = 1$$

As a further aid in graphing the hyperbola, the asymptotes are:

$$y - k = \pm \frac{a}{b}(x - h), \text{ or}$$

$$y = \pm \frac{1}{\sqrt{3}} x$$

$$y = \pm \frac{\sqrt{3}}{3} x$$

(Lines through $(0, 0)$ with slopes $\dfrac{\sqrt{3}}{3}$ and $\dfrac{-\sqrt{3}}{3}$).

25. We are given:

Type of Graph: Ellipse
        Foci: $F_1(-3, 0)$, $F_2(3, 0)$
      Vertex: $V_2(4, 0)$

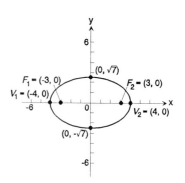

From Table 3, we see that we need the center, (h, k), a(the distance from the center to a vertex), and b, and we need to know whether the major axis is vertical or horizontal.

In this problem, the foci both lie on the <u>horizontal</u> line $y = 0$. The center is the midpoint between F_1 and F_2: $C(0, 0)$, so $h = 0$, $k = 0$.

Also: $c = d(C, F_1) = 3$
           $a = d(C, V_2) = 4$
and $b^2 = a^2 - c^2 = 7$, so
           $b = \sqrt{7}$

Therefore, the equation is of the form:

$$\frac{(x - h)^2}{a^2} + \frac{(y - k)^2}{b^2} = 1, \text{ or}$$

$$\frac{x^2}{16} + \frac{y^2}{7} = 1$$

27. We are given:

Type of graph: Parabola
       Vertex: $V(2, -3)$
        Focus: $F(2, -4)$

From Table 2, we need to know the vertex, $V(h, k)$; the distance, a, from the vertex to the focus; and which direction the parabola opens.

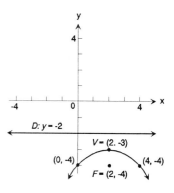

Since the focus is <u>below</u> the vertex, the parabola opens <u>down</u>. Also, $a = d(V, F) = 1$.

Therefore, the equation will be of the form:

$(x - h)^2 = -4a(y - k)$
or $(x - 2)^2 = -4(y + 3)$

29. We are given:

Type of graph: Hyperbola
Center: $C(-2, -3)$
Focus: $F_1(-4, -3)$
Vertex: $V_1(-3, -3)$

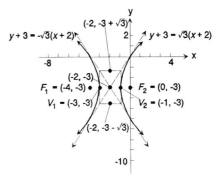

The transverse axis is the <u>horizontal</u> line $y = -3$, and:

$a = d(C, V_1) = 1$
$c = d(C, F_1) = 2$
$b^2 = c^2 - a^2 = 3$, so

$b = \sqrt{3}$

Foci: $(h \pm c, k)$: $(-4, -3)$ and $(0, -3)$
Vertices: $(h \pm a, k)$: $(-3, -3)$ and $(-1, -3)$

Asymptotes: $y - k = \pm \dfrac{b}{a}(x - h)$ or

$y + 3 = \pm\sqrt{3}(x + 2)$

The equation is:

$$\frac{(x - h)^2}{a^2} - \frac{(y - k)^2}{b^2} = 1, \text{ or}$$

$$\frac{(x + 2)^2}{1} - \frac{(y + 3)^2}{3} = 1$$

31. We are given:

Type of graph: Ellipse
Foci: $F_1(-4, 2)$ and
$F_2(-4, 8)$
Vertex: $V_2(-4, 10)$

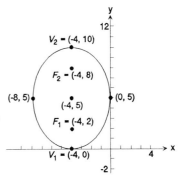

As we see in Table 3, we need to know the center, (h, k); the distance, a, from the center to either vertex; b, the distance from the center to either endpoint of the minor axis; and whether the major axis is vertical or horizontal. The center is the midpoint between F_1 and F_2: $C(-4, 5)$.

Then $a = d(C, V_2) = 5$, and $c = d(C, F_1) = 3$.

Therefore, $b^2 = a^2 - c^2 = 16$, so that $b = 4$.

Finally, F_1, F_2 and V_2 all lie on the <u>vertical</u> line $x = -4$, so the major axis is parallel to the y-axis. By Table 3, the equation of the ellipse is

$$\frac{(x - h)^2}{b^2} + \frac{(y - k)^2}{a^2} = 1, \text{ or}$$

$$\frac{(x + 4)^2}{16} + \frac{(y - 5)^2}{25} = 1$$

33. We are given:

Center: $C(-1, 2)$
$a = 3$
$c = 4$

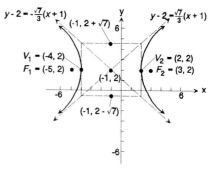

Transverse axis parallel to the x-axis.

Since the conic section has a transverse axis, it must be a hyperbola. For hyperbolas, $b^2 = c^2 - a^2$, so $b^2 = 7$, or $b = \sqrt{7}$.

From Table 4, the equation is of the form:

$$\frac{(x - h)^2}{a^2} - \frac{(y - k)^2}{b^2} = 1$$

where (h, k) are the coordinates of the center. Therefore, we have:

$$\frac{(x + 1)^2}{9} - \frac{(y - 2)^2}{7} = 1$$

As an aid to graphing, the vertices are at $(h \pm a, k)$, or $(-4, 2)$ and $(2, 2)$, and the asymptotes are:

$$y - 2 = \pm \frac{\sqrt{7}}{3}(x + 1)$$

(Lines through $(-1, 2)$ with slopes $\frac{\sqrt{7}}{3} \approx .88$ and $-\frac{\sqrt{7}}{3} \approx -.88$)

35. We are given:

Vertices: $V_1(0, 1)$
and $V_2(6, 1)$
Asymptote: $3y + 2x - 9 = 0$

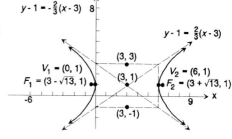

Since this conic section has asymptotes, it must be a hyperbola. The center is midway between V_1 and V_2: $C(3, 1)$, and the transverse axis is the <u>horizontal</u> line $y = 1$. Also, $a = d(C, V_1) = 3$.
But we need to know b to determine the equation. From Table 4, the asymptotes are:

$$y - k = \pm \frac{b}{a}(x - h), \text{ or}$$

$$y - 1 = \pm \frac{b}{a}(x - 3)$$

9 ANALYTIC GEOMETRY

We must put the given asymptote into this form:

$$3y + 2x - 9 = 0$$

$$3y = -2x + 9$$

$$y = -\frac{2}{3}x + 3$$

$$y - 1 = -\frac{2}{3}x + 2$$

$$y - 1 = -\frac{2}{3}x + \frac{2 \cdot 3}{3}$$

$$y - 1 = -\frac{2}{3}(x - 3)$$

Therefore, $-\dfrac{b}{a} = -\dfrac{2}{3}$

$$3b = 2a$$

$$3b = 6 \quad \text{(since } a = 3)$$

$$b = 2$$

The equation of the hyperbola is:

$$\frac{(x - 3)^2}{9} - \frac{(y - 1)^2}{4} = 1$$

37. $y^2 + 4x + 3y - 8 = 0$

Here $A = 0$ and $C = 1$ so that $AC = 0$. The equation is a parabola.

39. $x^2 + 2y^2 + 4x - 8y + 2 = 0$

Here $A = 1$ and $C = 2$ so that $AC = 2$. The equation is a ellipse.

41. $9x^2 - 12xy + 4y^2 + 8x + 12y = 0$

Here $A = 9$ and $B = -12$, and $C = 4$ so that $B^2 - 4AC = 144 - 144 = 0$. The equation is a parabola.

43. $4x^2 + 10xy + 4y^2 - 9 = 0$

Here $A = 4$ and $B = 10$, and $C = 4$ so that $B^2 - 4AC = 100 - 64 = 36$. The equation is a hyperbola.

45. $x^2 - 2xy + 3y^2 + 2x + 4y - 1 = 0$

Here $A = 1$ and $B = -2$, and $C = 3$ so that $B^2 - 4AC = 4 - 12 = -8$. The equation is an ellipse.

47. $A = 2$, $B = 5$, $C = 2$, $\cot 2\theta = 0$ so that $\theta = 45°$

$$x = \frac{\sqrt{2}}{2}(x' - y'); \quad y = \frac{\sqrt{2}}{2}(x' + y')$$

$$2\left[\frac{\sqrt{2}}{2}(x' - y')\right]^2 + 5\left[\frac{\sqrt{2}}{2}(x' - y')\right]\left[\frac{\sqrt{2}}{2}(x' + y')\right] + 2\left[\frac{\sqrt{2}}{2}(x' + y')\right]^2 - \frac{9}{2} = 0$$

$$2 \cdot \frac{1}{2}\left(x'^2 - 2x'y'^2 + y'^2\right) + 5 \cdot \frac{1}{2}\left(x'^2 - y'^2\right) + 2 \cdot \frac{1}{2}\left(x'^2 + 2x'y' + y'^2\right) = \frac{9}{2}$$

$$\frac{9}{2}x'^2 - \frac{1}{2}y'^2 = \frac{9}{2}$$

$$x'^2 - \frac{-y'^2}{9} = 1$$

Hyperbola; center at origin; transverse
axis is the x'-axis; Vertices at $(\pm 1, 0)$.

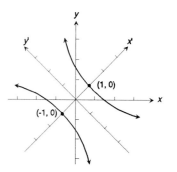

49. $A = 6$, $B = 4$, $C = 9$, $\cot 2\theta = \frac{-3}{4}$; $\cos 2\theta = \frac{-3}{5}$

$$\sin \theta = \sqrt{\frac{1 + \frac{3}{5}}{2}} = \frac{2}{\sqrt{5}} = \frac{2\sqrt{5}}{5}; \quad \cos \theta = \sqrt{\frac{1 - \frac{3}{5}}{2}} = \frac{1}{\sqrt{5}} = \frac{\sqrt{5}}{5}$$

$\theta = 63°$

$$x = \frac{\sqrt{5}}{5}(x' - 2y'), \quad y = \frac{\sqrt{5}}{5}(2x' + y')$$

$$6\left[\frac{\sqrt{5}}{5}(x' - 2y')\right]^2 + 4\left[\frac{\sqrt{5}}{5}(x' - 2y')\right]\left[\frac{\sqrt{5}}{5}(2x' + y')\right] + 9\left[\frac{\sqrt{5}}{5}(2x' + y')\right]^2$$
$$= 20$$

$$\frac{6}{5}\left(x'^2 - 4x'y' + 4y'^2\right) + \frac{4}{5}\left(2x'^2 - 3x'y' - 2y'^2\right)$$
$$+ \frac{9}{5}\left(4x'^2 + 4x'y' + y'^2\right) = 20$$

$$10x'^2 + 5y'^2 = 20$$

$$\frac{x'^2}{2} + \frac{y'^2}{4} = 1$$

Ellipse; center at origin; major axis is
the y'-axis; Vertices at $(0, \pm 2)$

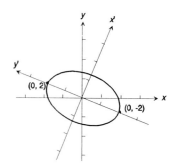

51. $A = 4$, $B = -12$, $C = 9$, $\cot 2\theta = \dfrac{5}{12}$; $\cos 2\theta = \dfrac{5}{13}$

$$\sin \theta = \sqrt{\dfrac{1 - \dfrac{5}{13}}{2}} = \dfrac{2}{\sqrt{13}} = \dfrac{2\sqrt{13}}{13}; \quad \cos \theta = \sqrt{\dfrac{1 + \dfrac{5}{13}}{2}} = \dfrac{3}{\sqrt{13}} = \dfrac{3\sqrt{13}}{13}$$

$\theta \approx 34°$

$$x = \dfrac{3\sqrt{13}}{13}x' - \dfrac{2\sqrt{13}}{13}y' = \dfrac{\sqrt{13}}{13}(3x' - 2y')$$

$$y = \dfrac{2\sqrt{13}}{13}x' + \dfrac{3\sqrt{13}}{13}y' = \dfrac{\sqrt{13}}{13}(2x' + 3y')$$

$$4\left[\dfrac{\sqrt{13}}{13}(3x' - 2y')\right]^2 - 12\left[\dfrac{\sqrt{13}}{13}(3x' - 2y')\right]\left[\dfrac{\sqrt{13}}{13}(2x' + 3y')\right]$$

$$+ 9\left[\dfrac{\sqrt{13}}{13}(2x' + 3y')\right]^2 - 12\left[\dfrac{\sqrt{13}}{13}(3x' - 2y')\right]$$

$$+ 8\left[\dfrac{\sqrt{13}}{13}(2x' + 3y')\right] = 0$$

$$\dfrac{4}{13}\left(9x'^2 - 12x'y'^2 + 4y'^2\right) - \dfrac{12}{13}\left(6x'^2 + 5x'y' - 6y'^2\right)$$

$$+ \dfrac{9}{13}\left(4x'^2 + 12x'y' + 9y'^2\right) + 52\dfrac{\sqrt{13}}{13}x' = 0$$

$$13y'^2 + 4\sqrt{13}\,x' = 0 = 0$$

$$y'^2 = \dfrac{-4\sqrt{13}}{13}x'$$

Parabola; vertex at the origin; focus on

the x'-axis at $\left(\dfrac{-\sqrt{13}}{13},\ 0\right)$.

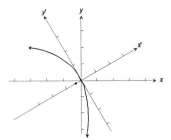

53. $r = \dfrac{4}{1 - \cos \theta}$

$ep = 4$, $e = 1$, $p = 4$

Parabola; directrix is perpendicular to the polar axis 4 units to the left of the pole.

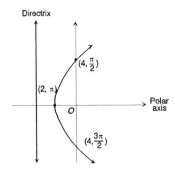

55.

$$r = \frac{6}{2 - \sin\theta} = \frac{3}{1 - \frac{1}{2}\sin\theta}$$

$ep = 3$, $e = \frac{1}{2}$, $p = 6$

Ellipse; directrix parallel to the polar axis 6 units below the pole. Vertices are at $\left(6, \frac{\pi}{2}\right)$ and $\left(2, \frac{3\pi}{2}\right)$.

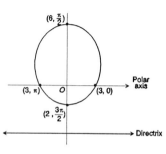

57.

$$r = \frac{8}{4 + 8\cos\theta} = \frac{2}{1 + 2\cos\theta}$$

$ep = 2$, $e = 2$, $p = 1$

Hyperbola, directrix is perpendicular to the polar axis 1 unit to the right of the pole.

Vertices are at $\left(\frac{2}{3}, 0\right)$ and $(-2, \pi)$.

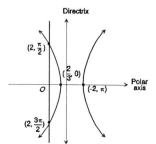

59.

$$r = \frac{4}{1 - \cos\theta}$$

$r(1 - \cos\theta) = 4$

$r - r\cos\theta = 4$

$r = 4 + r\cos\theta$

$r^2 = (4 + r\cos\theta)^2$

$x^2 + y^2 = (4 + x)^2$

$x^2 + y^2 = 16 + 8x + x^2$

$y^2 - 8x - 16 = 0$

61.

$$r = \frac{8}{4 + 8\cos\theta}$$

$r(4 + 8\cos\theta) = 8$

$4r = 8 - 8r\cos\theta$

$r = 2(1 - r\cos\theta)$

$r^2 = 4(1 - r\cos\theta)^2$

$x^2 + y^2 = 4(1 - x)^2$

$x^2 + y^2 = 4(1 - 2x + x^2)$

$3x^2 - y^2 - 8x + 4 = 0$

63.

$x = 4t - 2$,

$y = 1 - t$

$x = 4(1 - y) - 2$

$x = 4 - 4y - 2$

$x + 4y = 2$

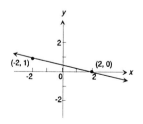

65.

$x = 3\sin t$, $y = 4\cos t + 2$

$\frac{x}{3} = \sin t$; $\frac{y - 2}{4} = \cos t$

$\sin^2 t + \cos^2 t = 1$

$\frac{x^2}{9} + \frac{(y - 2)^2}{16} = 1$

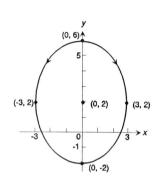

67. $x = \sec^2 t$, $y = \tan^2 t$
$1 + \tan^2 t = \sec^2 t$
$1 + y = x$

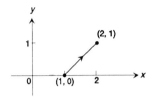

69. We start with the ellipse and determine its foci and vertices:

$$4x^2 + 9y^2 = 36$$
$$\frac{x^2}{9} + \frac{y^2}{4} = 1$$

This is an ellipse centered at $(0, 0)$. The larger denominator is associated with the x^2 term, so we know that the major axis is horizontal, and $a^2 = 9$, $b^2 = 4$, and $c^2 = a^2 - b^2 = 5$, so $c = \sqrt{5}$.

From Table 3:

Vertices of the ellipse: $(h \pm a, k)$, or $(-3, 0)$ and $(3, 0)$

Foci of the ellipse: $(h \pm c, k)$, or $\left(-\sqrt{5}, 0\right)$ and $\left(\sqrt{5}, 0\right)$

Therefore,

Foci of the hyperbola: $F_1(-3, 0)$ and $F_2(3, 0)$

Vertices of the hyperbola: $V_1\left(-\sqrt{5}, 0\right)$ and $V_2\left(\sqrt{5}, 0\right)$

The center of the hyperbola is midway between F_1 and F_2: $C(0, 0)$, and the transverse axis is the <u>horizontal</u> line, $y = 0$.

Finally,

$a = d(C, V_1) = \sqrt{5}$
$c = d(C, F_1) = 3$, and
$b^2 = c^2 - a^2 = 9 - 5 = 4$, so
$b = 2$

By Table 3, the equation of the <u>hyperbola</u> is:

$$\frac{(x - h)^2}{a^2} - \frac{(y - k)^2}{b^2} = 1, \text{ or}$$

$$\frac{x^2}{5} - \frac{y^2}{4} = 1$$

71.   Let (x, y) be any point in this collection of points.

The distance from (x, y), to $(3, 0) = \sqrt{(x - 3)^2 + y^2}$

The distance (x, y) to the line $x = \frac{16}{3}$ is $\left| x - \frac{16}{3} \right|$

Therefore, we have

$$\sqrt{(x - 3)^2 + y^2} = \frac{3}{4}\left| x - \frac{16}{3} \right|$$

$$(x - 3)^2 + y^2 = \frac{9}{16}\left(x - \frac{16}{3} \right)^2 \quad \text{square both sides}$$

$$x^2 - 6x + 9 + y^2 = \frac{9}{16}\left(x^2 - \frac{32}{3}x + \frac{256}{9} \right)$$

$$16x^2 - 96x + 144 + 16y^2 = 9x^2 - 96x + 256$$

$$7x^2 + 16y^2 = 122$$

$$\frac{7x^2}{112} + \frac{16y^2}{112} = 1$$

$$\frac{x^2}{16} + \frac{y^2}{7} = 1$$

Thus the set of points is an ellipse.

9 ANALYTIC GEOMETRY

CHAPTER
10

SYSTEMS OF EQUATIONS AND INEQUALITIES

≡ EXERCISE 10.1 SYSTEMS OF LINEAR EQUATIONS: SUBSTITUTION; ELIMINATION

1. $\left.\begin{array}{l} 2x - y = 5 \\ 5x + 2y = 8 \end{array}\right\}$ $x = 2,\ y = -1$ $\begin{array}{ll} 2(2) - (-1) = 5 & 5 = 5 \\ 5(2) + 2(-1) = 8 & 8 = 8 \end{array}$

Therefore, $x = 2$, $y = -1$ is a solution to the system of equations, since they satisfy <u>both</u> equations.

3. $\left.\begin{array}{l} 3x - 4y = 4 \\ x - 3y = \frac{1}{2} \end{array}\right\}$ $x = 2,\ y = \frac{1}{2}$ $\begin{array}{ll} 3(2) - 4\left(\frac{1}{2}\right) = 4 & 4 = 4 \\[2mm] 2 - 3\left(\frac{1}{2}\right) = \frac{1}{2} & \frac{1}{2} = \frac{1}{2} \end{array}$

Each equation is satisfied, so $x = 2$, $y = \frac{1}{2}$ is a solution of the system.

5. $\left.\begin{array}{l} x^2 - y^2 = 5 \\ xy = 2 \end{array}\right\}$ $x = 2,\ y = 1$ $\begin{array}{ll} 2^2 - 1^2 = 3 & 3 = 3 \\ (2)(1) = 2 & 2 = 2 \end{array}$

7. $\left.\begin{array}{l} \dfrac{x}{1 + x} + 3y = 6 \\[2mm] x + 9y^2 = 36 \end{array}\right\}$ $x = 0,\ y = 2$ $\begin{array}{ll} \dfrac{0}{1 + 0} + 3(2) = 6 & 6 = 6 \\[2mm] 0 + 9(2)^2 = 36 & 36 = 36 \end{array}$

9. $\left.\begin{array}{l} 3x + 3y + 2z = 4 \\ x - y - z = 0 \\ 2y - 3z = -8 \end{array}\right\}$ $x = 1,\ y = -1,\ z = 2$ $\begin{array}{ll} 3(1) + 3(-1) + 2(2) = 4 & 4 = 4 \\ 1 - (-1) - 2 = 0 & 0 = 0 \\ 2(-1) - 3(2) = -8 & -8 = -8 \end{array}$

Therefore, $x = 1$, $y = 2$, $z = 2$ is a solution to the system.

11. $\begin{cases} x + y = 8 & (1) \\ x - y = 4 & (2) \end{cases}$

We can use substitution.

Step 1: We can easily solve either equation for x or y. Let us begin by solving equation (1) for x:

$$x + y = 8$$
$$x = -y + 8$$

Step 2: Now substitute $x = -y + 8$ into (2):

$$x - y = 4$$
$$-y + 8 - y = 4$$
$$-2y + 8 = 4$$

Step 3: Solve the last equation for y:

$$-2y + 8 = 4$$
$$-2y = -4$$
$$y = 2$$

Step 4: We can now find x by using the fact that $x = -y + 8$ (Step 1) and $y = 2$ (Step 3):

$$x = -y + 8$$
$$x = -(2) + 8, \text{ since } y = 2$$
$$x = 6$$

The solution of the system is $x = 6$, $y = 2$.

13. $\begin{cases} 5x - y = 13 & (1) \\ 2x + 3y = 12 & (2) \end{cases}$

We use elimination.

$\begin{cases} 15x - 3y = 39 & (1) \quad \text{Multiply both sides by 3.} \\ 2x + 3y = 12 & (2) \end{cases}$

$\begin{cases} 15x - 3y = 39 & (1) \\ 17x = 51 & (2) \quad \text{Replace (2) by (1) + (2)} \end{cases}$

$\begin{cases} 15x - 3y = 39 & (1) \\ x = 3 & (2) \end{cases}$

$\begin{cases} 45 - 3y = 39 & (1) \quad \text{Back-substitute; } x = 3 \\ x = 3 & (2) \end{cases}$

$\begin{cases} -3y = -6 & (1) \\ x = 3 & (2) \end{cases}$

$\begin{cases} y = 2 & (1) \\ x = 3 & (2) \end{cases}$

The solution is $x = 3$, $y = 2$.

15.
$$\begin{cases} 3x \quad\quad = 24 \quad (1) \\ \ x + 2y = 0 \quad\ (2) \end{cases}$$

We use substitution.

Step 1: Since equation (1) contains only one variable, it is best to start there:

$$3x = 24$$
$$x = 8$$

Step 2: Substitute $x = 8$ into (2):

$$x + 2y = 0$$
$$8 + 2y = 0$$

Step 3: Solve for y:

$$8 + 2y = 0$$
$$2y = -8$$
$$y = -4$$

We already know x, so the solution of the system is $x = 8$, $y = -4$.

17.
$$\begin{cases} 3x - 6y = 24 \quad (1) \\ 5x + 4y = 12 \quad (2) \end{cases}$$

We use substitution.

Step 1: It is harder to decide which equation to solve for which variable. Since the coefficient of x in equation (1) divides all other constants in equation (1), I choose to solve for x, in order to avoid fractions for as long as possible:

$$3x - 6y = 24$$
$$3x = 6y + 24$$
$$x = 2y + 8$$

Step 2: Substitute $x = 2y + 8$ into (2):

$$5x + 4y = 12$$
$$5(2y + 8) + 4y = 12$$
$$10y + 40 + 4y = 12$$
$$14y + 40 = 12$$

Step 3: Solve for y:

$$14y + 40 = 12$$
$$14y = -28$$
$$y = -2$$

Step 4: Determine x:

$$x = 2y + 8 \quad\quad \text{(Step 1)}$$
$$x = 2(-2) + 8 \quad (y = -2)$$
$$x = 4$$

The solution is $x = 4$, $y = -2$.

19. $\begin{cases} 2x + y = 1 & (1) \\ 4x + 2y = 6 & (2) \end{cases}$

We use substitution.

Step 1: It is easiest to solve equation (1) for y:

$$2x + y = 1$$
$$y = -2x + 1$$

Step 2: Substitute $y = -2x + 1$ into (2):

$$4x + 2y = 6$$
$$4x + 2(-2x + 1) = 6$$
$$4x - 4x + 2 = 6$$
$$0x = 4$$

This has no solution, so the system is inconsistent.

21. $\begin{cases} 2x - 4y = -2 & (1) \\ 3x + 2y = 3 & (2) \end{cases}$

We use elimination.

$$\begin{cases} 2x - 4y = -2 & (1) \\ 6x + 4y = 6 & (2) \quad \text{Multiply both sides by 2.} \end{cases}$$

$$\begin{cases} 2x - 4y = -2 & (1) \\ 8x \quad\quad = 4 & (2) \quad \text{Replace (2) by (1) + (2).} \end{cases}$$

$$\begin{cases} 2x - 4y = -2 & (1) \\ \quad\quad x = \dfrac{1}{2} & (2) \end{cases}$$

$$\begin{cases} 1 - 4y = -2 & (1) \quad \text{Back-substitute; } x = \dfrac{1}{2}. \\ \quad\quad x = \dfrac{1}{2} & (2) \end{cases}$$

$$\begin{cases} -4y = -3 & (1) \\ \quad x = \dfrac{1}{2} & (2) \end{cases}$$

$$\begin{cases} y = \dfrac{3}{4} & (1) \\ x = \dfrac{1}{2} & (2) \end{cases}$$

The solution is $x = \dfrac{1}{2}$, $y = \dfrac{3}{4}$.

23.
$$\begin{cases} x + 2y = 4 & (1) \\ 2x + 4y = 8 & (2) \end{cases}$$

We use substitution.

Step 1: Solve for x in (1):

$$x + 2y = 4$$
$$x = -2y + 4$$

Step 2: Substitute $x = -2y + 4$ into (2):

$$2x + 4y = 8$$
$$2(-2y + 4) + 4y = 8$$
$$-4y + 8 + 4y = 8$$
$$0y = 0$$

This is an identity (i.e., true for <u>any</u> value of y), so <u>any</u> value of y is a solution, and we let $x = 4 - 2y$, y any real number.

25.
$$\begin{cases} 2x - 3y = -1 & (1) \\ 10x + 10y = 5 & (2) \end{cases}$$

We use elimination.

$$\begin{cases} -10x + 15y = 5 & (1) \quad \text{Multiply both sides by } -5. \\ 10x + 10y = 5 & (2) \end{cases}$$

$$\begin{cases} -10x + 15y = 5 & (1) \\ 25y = 10 & (2) \quad \text{Replace (2) by (1) + (2).} \end{cases}$$

$$\begin{cases} -10x + 15y = 5 & (1) \\ y = \dfrac{2}{5} & (2) \end{cases}$$

$$\begin{cases} -10x + 15\left(\dfrac{2}{5}\right) = 5 & (1) \quad \text{Back-substitute; } y = \dfrac{2}{5}. \\ y = \dfrac{2}{5} & (2) \end{cases}$$

$$\begin{cases} -10x + 6 = 5 & (1) \\ y = \dfrac{2}{5} & (2) \end{cases}$$

$$\begin{cases} x = \dfrac{1}{10} & (1) \\ y = \dfrac{2}{5} & (2) \end{cases}$$

The solution is $x = \dfrac{1}{10}$, $y = \dfrac{2}{5}$.

27. $\begin{cases} 2x + 3y = 6 & (1) \\ x - y = \dfrac{1}{2} & (2) \end{cases}$

We use substitution.

<u>Step 1</u>: Solve for (2) for x:

$$x - y = \frac{1}{2}$$
$$x = y + \frac{1}{2}$$

<u>Step 2</u>: Substitute $x = y + \dfrac{1}{2}$ into (1):

$$2x + 3y = 6$$
$$2\left(y + \frac{1}{2}\right) + 3y = 6$$
$$2y + 1 + 3y = 6$$
$$5y + 1 = 6$$

<u>Step 3</u>: Solve for y:

$$5y + 1 = 6$$
$$5y = 5$$
$$y = 1$$

<u>Step 4</u>: Determine x:

$$x = y + \frac{1}{2} \quad \text{(Step 1)}$$
$$x = 1 + \frac{1}{2} \quad (y = 1)$$
$$x = \frac{3}{2}$$

The solution is $x = \dfrac{3}{2}$, $y = 1$.

29. $\begin{cases} 2x + 3y = 5 & (1) \\ 4x + 6y = 10 & (2) \end{cases}$

We use elimination.

$\begin{cases} 2x + 3y = 5 & (1) \\ -2x - 3y = -5 & (2) \quad \text{Multiply both sides by } -\dfrac{1}{2}. \end{cases}$

$\begin{cases} 2x + 3y = 5 & (1) \\ \quad\quad 0 = 0 & (2) \quad \text{Replace (2) by (1) + (2).} \end{cases}$

We can write the solution as:

$$x = -\frac{3}{2}y + \frac{5}{2}, \text{ where } y \text{ is any real number,}$$

or $\quad y = -\dfrac{2}{3}x + \dfrac{5}{3}$, where x is any real number.

10 SYSTEMS OF EQUATIONS AND INEQUALITIES

31. $\begin{cases} 3x - 5y = 3 & (1) \\ 15x + 5y = 21 & (2) \end{cases}$

We use elimination.

$\begin{cases} 3x - 5y = 3 & (1) \\ 18x = 24 & (2) \end{cases}$ Replace (2) by (1) + (2).

$\begin{cases} 3x - 5y = 3 & (1) \\ x = \dfrac{4}{3} & (2) \end{cases}$

$\begin{cases} 4 - 5y = 3 & (1) \\ x = \dfrac{4}{3} & (2) \end{cases}$ Back-substitution; $x = \dfrac{4}{3}$.

$\begin{cases} 5y = 1 & (1) \\ x = \dfrac{4}{3} & (2) \end{cases}$

$\begin{cases} y = \dfrac{1}{5} & (1) \\ x = \dfrac{4}{3} & (2) \end{cases}$

The solution is $x = \dfrac{4}{3}$, $y = \dfrac{1}{5}$.

33. $\begin{cases} x - y = 6 & (1) \\ 2x - 3z = 16 & (2) \\ 2y + z = 4 & (3) \end{cases}$

We use substitution.

Step 1: We can solve for x in equation (1):
$$x - y = 6$$
$$x = y + 6$$

Step 2: Substitute this expression for x into equations (2) and (3):

$\begin{cases} 2x - 3z = 16 & (2) \\ 2y + z = 4 & (3) \end{cases}$

$\begin{cases} 2(y + 6) - 3z = 16 & (2) \\ 2y + z = 4 & (3) \end{cases}$

$\begin{cases} 2y - 3z + 12 = 16 & (2) \\ 2y + z = 4 & (3) \end{cases}$

$\begin{cases} 2y - 3z + = 4 & (2) \\ 2y + z = 4 & (3) \end{cases}$

Now we must return to Step 1 and solve (2) or (3) for y or z. It is easiest to solve for z in equation (3):

$$2y + z = 4$$
$$z = 4 - 2y$$

Now substitute $z = 4 - 2y$ into (2):

$$2y - 3z = 4$$
$$2y - 3(4 - 2y) = 4$$
$$2y - 12 + 6y = 4$$
$$8y = 16$$

Step 3: $\quad y = 2$

Step 4: From $z = 4 - 2y$, we have:

$$z = 4 - 2(2)$$
$$z = 0,$$

and from $\quad x = y + 6 \quad$ (Step 1)

we have $\quad x = 2 + 6$
$$x = 8$$

The solution is: $x = 8$, $y = 2$, and $z = 0$.

35. $\begin{cases} x - 2y + 3z = 7 & (1) \\ 2x + y + z = 4 & (2) \\ -3x + 2y - 2z = -10 & (3) \end{cases}$

We use substitution.

Step 1A: Solve equation (1) for x:

$$x - 2y + 3z = 7$$
$$x = 7 + 2y - 3z$$

Step 2A: Substitute $x = 7 + 2y - 3z$ into (2) and (3):

$$\begin{cases} 2x + y + z = 4 & (2) \\ -3x + 2y - 2z = -10 & (3) \end{cases}$$

$$\begin{cases} 2(7 + 2y - 3z) + y + z = 4 & (2) \\ -3(7 + 2y - 3z) + 2y - 2z = -10 & (3) \end{cases}$$

$$\begin{cases} 14 + 4y - 6z + y + z = 4 & (2) \\ -21 - 6y + 9z + 2y - 2z = -10 & (3) \end{cases}$$

$$\begin{cases} 5y - 5z = -10 & (2) \\ -4y + 7z = 11 & (3) \end{cases}$$

Step 1B: We must now solve for y or z in (2) or (3). If we work with (2), we can avoid fractions on this step:

$$5y - 5z = -10 \qquad (2)$$
$$-5z = -5y - 10$$
$$z = y + 2$$

Substitute this into (3):

$$-4y + 7z = 11$$
$$-4y + 7(y + 2) = 11$$
$$-4y + 7y + 14 = 11$$
$$3y = -3$$

Step 3: $y = -1$

Step 4: Determine z:

$$z = y + 2 \qquad \text{(Step 1B)}$$
$$z = -1 + 2 \qquad (y = -1)$$
$$z = 1$$

Determine x:

$$x = 7 + 2y - 3z \qquad \text{(Step 1A)}$$
$$x = 7 + 2(-1) - 3(1) \qquad (y = -1),\ z = 1$$
$$x = 2$$

The solution is $x = 2$, $y = -1$, $z = 1$.

37.
$$\begin{cases} x - y - z = 1 & (1) \\ 2x + 3y + z = 2 & (2) \\ 3x + 2y = 0 & (3) \end{cases}$$

We use elimination.

Since z is already missing from equation (3), we try to eliminate it from either (1) or (2).

$$\begin{cases} x - y - z = 1 & (1) \\ 3x + 2y = 3 & (2) \quad \text{Replace (2) by (1) + (2).} \\ 3x + 2y = 0 & (3) \end{cases}$$

$$\begin{cases} x - y - z = 1 & (1) \\ -3x - 2y = -3 & (2) \quad \text{Multiply both sides by -1.} \\ 3x + 2y = 0 & (3) \end{cases}$$

$$\begin{cases} x - y - z = 1 & (1) \\ -3x - 2y = -3 & (2) \\ 0x + 0y = -3 & (3) \quad \text{Replace (3) by (2) + (3).} \end{cases}$$

(3) has no solution, so the original system is inconsistent.

39.
$$\begin{cases} x - y - z = 1 & (1) \\ -x + 2y - 3z = -4 & (2) \\ 3x - 2y - 7z = 0 & (3) \end{cases}$$

We use substitution.

Step 1A: Solve for x in (1):

$$x - y - z = 1$$
$$x = 1 + y + z$$

<u>Step 2A:</u>   Substitute into (2) and (3):

$$\begin{cases} -x + 2y - 3z = -4 & (2) \\ 3x - 2y - 7z = 0 & (3) \end{cases}$$

$$\begin{cases} -(1 + y + z) + 2y - 3z = -4 & (2) \\ 3(1 + y + z) - 2y - 7z = 0 & (3) \end{cases}$$

$$\begin{cases} y - 4z = -3 & (2) \\ y - 4z = -3 & (3) \end{cases}$$

<u>Step 1B:</u>   Solve (2) for y:

$$y - 4z = -3$$
$$y = -3 + 4z$$

<u>Step 2B:</u>   Substitute this into (3):

$$y - 4z = -3$$
$$-3 + 4z - 4z = -3$$
$$0 \cdot z = 0$$

This equation is satisfied by all values of z, so z can be any real number.

Then, from Step 1B, $y = 4z - 3$, and from Step 1A,

$$x = 1 + y + z$$
$$x = 1 + (4z - 3) + z$$
$$x = 5z - 2$$

Thus, we can write the solution as:

$x = 5z - 2$; $y = 4z - 3$; where z is <u>any</u> real number.

We could also express our answer in terms of the following:

$x = \dfrac{5}{4}y + \dfrac{7}{4}$; $z = \dfrac{1}{4}y + \dfrac{3}{4}$, where y is any real number.

Also, the solution is:

$y = \dfrac{4}{5}x - \dfrac{7}{5}$; $z = \dfrac{1}{5}x + \dfrac{2}{5}$, where x is any real number.

41. $\begin{cases} 2x - 2y + 3z = 6 & (1) \\ 4x - 3y + 2z = 0 & (2) \\ -2x + 3y - 7z = 1 & (3) \end{cases}$

We use elimination.

We can eliminate x from equation (3) by adding (1) and (3):

$\begin{cases} 2x - 2y + 3z = 6 & (1) \\ 4x - 3y + 2z = 0 & (2) \\ y - 4z = 7 & (3) \quad \text{Replace (3) by (1) + 3.} \end{cases}$

526   10 SYSTEMS OF EQUATIONS AND INEQUALITIES

We need to eliminate the <u>same</u> variable, x, from either (1) or (2):

$$\begin{cases} -4x + 4y - 6z = -12 & (1) \quad \text{Multiply both sides by } -2. \\ 4x - 3y + 2z = 0 & (2) \\ y - 4z = 7 & (3) \end{cases}$$

$$\begin{cases} -4x + 4y - 6z = -12 & (1) \\ y - 4z = -12 & (2) \quad \text{Replace (2) by (1) + (2).} \\ y - 4z = 7 & (3) \end{cases}$$

We can see that (2) and (3) are contradictory: if we subtract (3) from (2), we obtain $0 = -19$. Thus, the system is inconsistent.

43. $$\begin{cases} x + y - z = 6 & (1) \\ 3x - 2y + z = -5 & (2) \\ x + 3y - 2z = 14 & (3) \end{cases}$$

We use substitution.

<u>Step 1A</u>: I choose to solve equation (2) for z:

$$3x - 2y + z = -5$$
$$z = -5 - 3x + 2y$$

<u>Step 2A</u>: Now substitute this expression for z into (1) and (3):

$$\begin{cases} x + y - z = 6 & (1) \\ x + 3y - 2z = 14 & (3) \end{cases}$$

$$\begin{cases} x + y - (-5 - 3x + 2y) = 6 & (1) \\ x + 3y - 2(-5 - 3x + 2y) = 14 & (3) \end{cases}$$

$$\begin{cases} 4x - y = 1 & (1) \\ 7x - y = 4 & (3) \end{cases}$$

<u>Step 1B</u>: Now I solve for y in (1). (Solving equation (3) for y would be equally easy.)

$$4x - y = 1$$
$$-y = 1 - 4x$$
$$y = -1 + 4x$$

<u>Step 2B</u>: Substitute this into (3):

$$7x - y = 4$$
$$7x - (-1 + 4x) = 4$$
$$3x + 1 = 4$$

<u>Step 3</u>: $$x = 1$$

<u>Step 4</u>: We now find y and z:

$$y = -1 + 4x \qquad \text{(Step 1B)}$$
$$y = -1 + 4(1) \quad (x = 1)$$
$$y = 3$$

Also, $$z = -5 - 3x + 2y \qquad \text{(Step 1A)}$$
$$z = -5 - 3(1) + 2(3) \quad (x = 1, \ y = 3)$$
$$z = -2$$

So, the solution is: $x = 1, \ y = 3, \ z = -2$.

45.
$$\begin{cases} x + 2y - z = -3 & (1) \\ 2x - 4y + z = -7 & (2) \\ -2x + 2y - 3z = 4 & (3) \end{cases}$$

We can eliminate z by adding (1) and (2):

$$\begin{cases} x + 2y - z = -3 & (1) \\ 3x - 2y = -10 & (2) \quad \text{Replace (2) by (1) + (2).} \\ -2x + 2y - 3z = 4 & (3) \end{cases}$$

Now eliminate z from (1) or (3):

$$\begin{cases} -3x - 6y + 3z = 9 & (1) \quad \text{Multiply both sides by } -3. \\ 3x - 2y = -10 & (2) \\ -2x + 2y - 3z = 4 & (3) \end{cases}$$

$$\begin{cases} -3x - 6y + 3z = 9 & (1) \\ 3x - 2y = -10 & (2) \\ -5x - 4y = 13 & (3) \quad \text{Replace (3) by (1) + (3).} \end{cases}$$

$$\begin{cases} -3x - 6y + 3z = 9 & (1) \\ -6x + 4y = 20 & (2) \quad \text{Multiply both sides by } -2. \\ -5x - 4y = 13 & (3) \end{cases}$$

$$\begin{cases} -3x - 6y + 3z = 9 & (1) \\ -6x + 4y = 20 & (2) \\ -11x = 33 & (3) \quad \text{Replace (3) by (2) + (3).} \end{cases}$$

$$\begin{cases} -3x - 6y + 3z = 9 & (1) \\ -6x + 4y = 20 & (2) \\ x = -3 & (3) \end{cases}$$

$$\begin{cases} 9 - 6y + 3z = 9 & (1) \quad \text{Back-substitute; } x = -3. \\ 18 + 4y = 20 & (2) \quad \text{Back-substitute; } x = -3. \\ x = -3 & (3) \end{cases}$$

$$\begin{cases} -6y + 3z = 0 & (1) \\ y = \dfrac{1}{2} & (2) \\ x = -3 & (3) \end{cases}$$

$$\begin{cases} -3 + 3z = 0 & (1) \quad \text{Back-substitute; } y = \dfrac{1}{2}. \\ y = \dfrac{1}{2} & (2) \\ x = -3 & (3) \end{cases}$$

$$\begin{cases} z = 1 & (1) \\ y = \dfrac{1}{2} & (2) \\ x = -3 & (3) \end{cases}$$

The solution is $x = -3$, $y = \dfrac{1}{2}$, $z = 1$.

10 SYSTEMS OF EQUATIONS AND INEQUALITIES

47.
$$\begin{cases} \dfrac{1}{x} + \dfrac{1}{y} = 8 \quad (1) \\[2mm] \dfrac{3}{x} - \dfrac{5}{y} = 0 \quad (2) \end{cases}$$

$$\begin{cases} \dfrac{1}{x} + \dfrac{1}{y} = 8 \quad (1) \\[2mm] 3\left(\dfrac{1}{x}\right) - 5\left(\dfrac{1}{y}\right) = 0 \quad (2) \end{cases}$$

Now let $u = \dfrac{1}{x}$, $v = \dfrac{1}{y}$:

$$\begin{cases} u + v = 8 \quad (1) \\ 3u - 5v = 0 \quad (2) \end{cases}$$

Step 1: Solve (1) for v:

$$u + v = 8$$
$$v = 8 - u$$

Step 2: Substitute into (2):

$$3u - 5v = 0$$
$$3u - 5(8 - u) = 0$$
$$3u - 40 + 5u = 0$$

Step 3: Solve for u:

$$8u = 40$$
$$u = 5$$

Step 4:
$$v = 8 - u \; \text{(Step 1)}$$
$$v = 8 - 5 \; (u = 5)$$
$$v = 3$$

So $u = 5$, $v = 3$.

But we are supposed to find x and y.

We have:
$$u = \frac{1}{x}, \; v = \frac{1}{y}$$
$$5 = \frac{1}{x}, \; 3 = \frac{1}{y}$$
$$x = \frac{1}{5}, \; y = \frac{1}{3}$$

49.
$$\begin{cases} y = \sqrt{2}\,x - 20\sqrt{7} \\ y = -0.1x + 20 \end{cases}$$

Using the ZOOM and TRACE features on the graphing calculator, the point of intersection is $x = 48.3$, $y = 15.3$.

51.
$$\begin{cases} \sqrt{2}\,x + \sqrt{3}\,y + \sqrt{6} = 0 \\ \sqrt{3}\,x - \sqrt{2}\,y + 60 = 0 \end{cases}$$

Using the ZOOM and TRACE features on the graphing calculator, the point of intersection is $x = -22.5$, $y = 14.9$.

53.
$$\begin{cases} \sqrt{3}\,x + \sqrt{2}\,y = \sqrt{0.3} \\ 100x - 95y = 20 \end{cases}$$

Using the ZOOM and TRACE features on the graphing calculator, the point of intersection is $x = 0.08$, $y = 0.3$.

In Problems 55-64, start by giving variable names (x, y, etc.) to the unknowns. Then translate each statement about the unknowns into an equation involving the variables. (The number of equations and the number of variables must be equal.) Solve the equations for the unknowns. Finally, be sure you have answered the original question.

55. Let the two numbers be x and y.
$$\begin{cases} x + y = 81 & \text{(Their sum is 81.)} \\ 2x - 3y = 62 & \text{(Twice one minus three times the other is 62.)} \end{cases}$$

Solve the first equation for x:
$$x + y = 81$$
$$x = 81 - y$$

Substitute this into the second equation:
$$2x - 3y = 62$$
$$2(81 - y) - 3y = 62$$
$$-5y = -100$$
$$y = 20$$

Now find x: We know $x = 81 - y$, and $y = 20$,
$$\text{so } x = 81 - 20$$
$$x = 61$$

The two numbers are 20 and 61.

57. Denote the width and length of the rectangle by w and ℓ. Then:

(1) $2w + 2\ell = 90$ (the perimeter is 90)
(2) $\ell = 2w$ (length is twice the width)

Substitute $\ell = 2w$ into (1):
$$2w + 2\ell = 90$$
$$2w + 2(2w) = 90$$
$$6w = 90$$
$$w = 15$$

Then $\ell = 2w$, or $\ell = 30$.

The room is 15 feet by 30 feet.

59. Let x = cost of one cheeseburger, in cents
y = cost of a shake, in cents

Then:　$4x + 2y = 790$　(1)
$2y = x + 15$

From (2), we have:$y = \dfrac{1}{2}x + \dfrac{15}{2}$

Substitute this into equation (1):
$$4x + 2y = 790$$
$$4x + 2\left(\dfrac{1}{2}x + \dfrac{15}{2}\right) = 790$$
$$4x + x + 15 = 790$$
$$5x = 775$$
$$x = 155$$

Then, since $y = \dfrac{1}{2}x + \dfrac{15}{2}$, we have
$$y = \dfrac{155}{2} + \dfrac{15}{2}$$
$$y = 85$$

Therefore, a cheeseburger costs \$1.55, and a shake costs 85 cents.

61. Here, the unknown quantities are speeds.

Let x = average windspeed
y = average air speed of the Piper

both in miles per hour.

Going <u>with</u> the wind, the plane has a groundspeed of $x + y$.　Then:

(rate) · (time) = distance
$$(x + y)(3) = 600, \text{ or}$$
$$x + y = 200 \qquad (1)$$

<u>Against</u> the wind, the speed of the plane will be $(y - x)$.

(rate) · (time) = distance
$$(y - x)(4) = 600, \text{ or}$$
$$y - x = 150 \qquad (2)$$

We have:　$\begin{cases} x + y = 200 & (1) \\ y - x = 150 & (2) \end{cases}$

Solve (2) for y:　　$y - x = 150$
$$y = 150 + x$$

Then from (1):　　　$x + y = 200$
$$x + (150 + x) = 200$$
$$2x = 50$$
$$x = 25$$

Finally, $y = 150 + x = 175$

Thus, the wind speed is $x = 25$ mph, and the air speed of the plane is $y = 175$ mph.

63. Here, we really have only one unknown:

x = Number of pounds of cashews to use

Since we are using 30 pounds of peanuts, the mixture will contain $30 + x$ pounds.

The basic formula is:

Revenue = (Number of pounds) · (Price per pound)

Revenue from peanuts alone = $(30)(1.50)$
= 45 dollars
Revenue from cashews alone = $x(5)$
= $5x$ dollars
Revenue from mixture = $(30 + x) \cdot 3$
= $90 + 3x$ dollars

So we want:

$$90 + 3x = 45 + 5x$$
$$45 = 2x$$
$$x = \frac{45}{2} = 22\frac{1}{2}$$

The manager should use 22.5 pounds of cashews.

65. Let x = Number of 2-person work stations,
and y = Number of 3-person work stations.

Then:

$2x + 3y = 38$ (The lab can be used by 38 students at one time.)
$x + y = 16$ (The total number of work stations is 16.)

Now solve for x and y:

$$\begin{cases} 2x + 3y = 38 & (1) \\ x + y = 16 & (2) \end{cases}$$

$$\begin{cases} 2x + 3y = 38 & (1) \\ -2x - 2y = -32 & (2) \quad \text{Multiply both sides by -2.} \end{cases}$$

$$\begin{cases} 2x + 3y = 38 & (1) \\ y = 6 & (2) \quad \text{Replace(2) by (1) + (2).} \end{cases}$$

$$\begin{cases} 2x + 18 = 38 & (1) \quad \text{Back-substitute; } y = 6. \\ y = 6 & (2) \end{cases}$$

$$\begin{cases} x = 10 & (1) \\ y = 6 & (2) \end{cases}$$

Therefore, the lab contains $x = 10$ work stations for 2 students, and $y = 6$ work stations for 3 students.

67. In order to determine the size of the refund, we must determine
the cost-per-package for bacon and eggs. We will use the general
principle:

Total cost for an Item = (Cost per package) · (Number of Packages)

Let x = Cost-per-package of bacon
and y = Cost-per-carton of eggs

Now translate the sentences into equations:

(1) Three packages of bacon and two cartons of eggs cost $7.45:
$3x + 2y = 7.45$ (1)

(2) Two packages of bacon and three cartons of eggs cost $6.45:
$2x + 3y = 6.45$ (2)

Solve for x and y:

$$\begin{cases} 3x + 2y = 7.45 & (1) \\ 2x + 3y = 6.45 & (2) \end{cases}$$

$$\begin{cases} 6x + 4y = 14.90 & (1) \quad \text{Multiply both sides by 2.} \\ -6x - 9y = -19.35 & (2) \quad \text{Multiply both sides by -3.} \end{cases}$$

$$\begin{cases} 6x + 4y = 14.90 & (1) \\ \quad\; - 5y = -4.45 & (2) \quad \text{Replace (2) by (1) + (2).} \end{cases}$$

$$\begin{cases} 6x + 4y = 14.90 & (1) \\ \quad\quad\; y = .89 & (2) \end{cases}$$

$$\begin{cases} 6x + 3.56 = 14.90 & (1) \quad \text{Back-substitute; } y = .89. \\ \quad\quad\quad y = .89 & (2) \end{cases}$$

$$\begin{cases} 6x = 11.34 & (1) \\ y = .89 & (2) \end{cases}$$

$$\begin{cases} x = 1.89 & (1) \\ y = .89 & (2) \end{cases}$$

We now know that bacon costs $1.89 per package, and eggs cost $.89
per carton. But the question is, how much money will be refunded
on two packages of bacon and two cartons of eggs?

Refund = 2(1.89) + 2(.89)
     = $5.56

69. Let the numbers be x, y, and z, from smallest to largest $(x \le y \le z)$.

Then we have:

$$\begin{cases} x + y + z = 48 & (1) \\ y + z = 3x & (2) \\ x + y = z + 6 & (3) \end{cases}$$

$$\begin{cases} x + y + z = 48 & (1) \\ -3x + y + z = 0 & (2) \\ x + y - z = 6 & (3) \end{cases}$$

We can eliminate z twice by adding (3) to (1) and (3) to (2):

$$\begin{cases} 2x + 2y \quad\; = 54 & (1) \quad \text{Replace (1) by (1) + (3).} \\ -2x + 2y \quad\; = 6 & (2) \quad \text{Replace (2) by (2) + (3).} \\ x + \; y - z = 6 & (3) \end{cases}$$

$$\begin{cases} 4y \quad\;\; = 60 & (1) \quad \text{Replace (1) by (1) + (2).} \\ -2x + 2y \quad\; = 6 & (2) \\ x + \; y - z = 6 & (3) \end{cases}$$

$$\begin{cases} y \quad\;\; = 15 & (1) \\ -2x + 2y \quad\; = 6 & (2) \\ x + \; y - z = 6 & (3) \end{cases}$$

$$\begin{cases} y \quad\;\; = 15 & (1) \\ -2x + 30 \quad\; = 6 & (2) \quad \text{Back-substitute; } y = 15. \\ x + 15 - z = 6 & (3) \quad \text{Back-substitute; } y = 15. \end{cases}$$

$$\begin{cases} y = 15 & (1) \\ -2x \quad\;\; = -24 & (2) \\ x - z = -9 & (3) \end{cases}$$

$$\begin{cases} y = 15 & (1) \\ x = 12 & (2) \\ x - z = -9 & (3) \end{cases}$$

$$\begin{cases} y = 15 & (1) \\ x = 12 & (2) \\ 12 - z = -9 & (3) \quad \text{Back-substitute; } x = 12. \end{cases}$$

$$\begin{cases} y = 15 & (1) \\ x = 12 & (2) \\ z = 21 & (3) \end{cases}$$

The three numbers are: 12, 15, and 21.

10 SYSTEMS OF EQUATIONS AND INEQUALITIES

71. $\begin{cases} I_2 = I_1 + I_3 \\ 5 - 3I_1 - 5I_2 = 0 \\ 10 - 5I_2 - 7I_3 = 0 \end{cases}$

$\begin{cases} I_2 = I_1 + I_3 \\ 5 - 3I_1 - 5(I_1 + I_3) = 0 \\ 10 - 5(I_1 + I_3) - 7I_3 = 0 \end{cases}$

$\begin{cases} I_2 = I_1 + I_3 \\ -8I_1 - 5I_3 + 5 = 0 \\ -5I_1 - 12I_3 + 10 = 0 \end{cases}$

$\begin{cases} I_2 = I_1 + I_3 \\ 40I_1 + 25I_3 = 25 \\ -40I_1 - 96I_3 = -80 \end{cases}$

$-71I_3 = -55$

$I_3 = \dfrac{55}{71}$

$-8I_1 - 5\left(\dfrac{55}{71}\right) + 5 = 0$

$-8I_1 = -5 + \dfrac{275}{71}$

$-8I_1 = \dfrac{-355 + 275}{71}$

$-8I_1 = \dfrac{-80}{71}$

$I_1 = \dfrac{10}{71}$

$I_2 = I_1 + I_3$

$I_2 = \dfrac{10}{71} + \dfrac{55}{71}$

$I_2 = \dfrac{65}{71}$

$I_1 = \dfrac{10}{71}, \; I_2 = \dfrac{65}{71}, \; I_3 = \dfrac{55}{71}$

73. Let x = Number of orchestra seats,
    y = Number of main seats,
    and z = Number of balcony seats.

Then $x + y + z = 500$, since there are a total of 500 seats.

If all seats are sold, the revenue is \$17,100:

$$50x + 35y + 25z = 17,100$$

Finally, if we sell only half of the orchestra seats, revenue is \$14,600:

$$50\left(\frac{1}{2}x\right) + 35y + 25z = 14,600$$

$$\begin{cases} x + y + z = 500 & (1) \\ 50x + 35y + 25z = 17,100 & (2) \\ 25x + 35y + 25z = 14,600 & (3) \end{cases}$$

$$\begin{cases} x + y + z = 500 & (1) \\ 10x + 7y + 5z = 3420 & (2) \quad \text{Divide both sides by 5.} \\ -5x - 7y - 5z = -2920 & (3) \quad \text{Divide both sides by -5.} \end{cases}$$

$$\begin{cases} x + y + z = 500 & (1) \\ 10x + 7y + 5z = 3420 & (2) \\ 5x = 500 & (3) \quad \text{Replace (3) by (2) + (3).} \end{cases}$$

$$\begin{cases} x + y + z = 500 & (1) \\ 10x + 7y + 5z = 3420 & (2) \\ x = 100 & (3) \end{cases}$$

$$\begin{cases} 100 + y + z = 500 & (1) \quad \text{Back-substitute; x = 100.} \\ 1000 + 7y + 5z = 3420 & (2) \quad \text{Back-substitute; x = 100.} \\ x = 100 & (3) \end{cases}$$

$$\begin{cases} y + z = 400 & (1) \\ 7y + 5z = 2420 & (2) \\ x = 100 & (3) \end{cases}$$

$$\begin{cases} -5y - 5z = -2000 & (1) \quad \text{Multiply both sides by -5.} \\ 7y + 5z = 2420 & (2) \\ x = 100 & (3) \end{cases}$$

$$\begin{cases} 2y = 420 & (1) \quad \text{Replace (1) by (1) + (2).} \\ 7y + 5z = 2420 & (2) \\ x = 100 & (3) \end{cases}$$

$$\begin{cases} y = 210 & (1) \\ 7y + 5z = 2420 & (2) \\ x = 100 & (3) \end{cases}$$

$$\begin{cases} y = 210 & (1) \\ 1470 + 5z = 2420 & (2) \quad \text{Back-substitute; } y = 210. \\ x = 100 & (3) \end{cases}$$

$$\begin{cases} y = 210 & (1) \\ 5z = 950 & (2) \\ x = 100 & (3) \end{cases}$$

$$\begin{cases} y = 210 & (1) \\ z = 190 & (2) \\ x = 100 & (3) \end{cases}$$

Thus, there are: $x = 100$ orchestra seats,
$y = 210$ main seats, and
$z = 190$ balcony seats.

75. We have $y = x^2 + bx + c$.

If this passes through $(1, 2)$, then

$$y = x^2 + bx + c$$
$$2 = 1^2 + b(1) + c \quad (x = 1, \ y = 2)$$
$$2 = 1 + b + c$$
$$b + c = 1 \qquad (1)$$

If $(-1, 3)$ lies on the graph, then

$$y = x^2 + bx + c$$
$$3 = (-1)^2 + b(-1) + c$$
$$3 = 1 - b + c$$
$$-b + c = 2 \qquad (2)$$

Let's solve (2) for c: $\quad -b + c = 2$
$$c = 2 + b$$

Substitute this into (1): $\qquad b + c = 1$
$$b + (2 + b) = 1$$
$$2b = -1$$
$$b = -\frac{1}{2}$$

Now find c: $\quad c = 2 + b$
$$c = 2 + \left(-\frac{1}{2}\right)$$
$$c = \frac{3}{2}$$

The solution is $b = -\frac{1}{2}$, $c = \frac{3}{2}$.

77.   We have $y = ax^2 + bx + c$.

From $(-1, 4)$: $4 = a(-1)^2 + b(-1) + c$
                     $4 = a - b + c$ (1)

From $(2, 3)$: $3 = a(2)^2 + b(2) + c$
                     $3 = 4a + 2b + c$ (2)

From $(0, 1)$: $1 = a(0)^2 + b(0) + c$
                     $c = 1$ (3)

Substitute $c = 1$ into (1) and (2):

$$\begin{cases} a - b + c = 4 & (1) \\ 4a + 2b + c = 3 & (2) \end{cases}$$

$$\begin{cases} a - b + 1 = 4 & (1) \\ 4a + 2b + 1 = 3 & (2) \end{cases}$$

$$\begin{cases} a - b = 3 & (1) \\ 4a + 2b = 2 & (2) \end{cases}$$

Now solve for (1) for a:

$$a - b = 3$$
$$a = 3 + b$$

Then, from (2):

$$4a + 2b = 2$$
$$4(3 + b) + 2b = 2$$
$$6b + 12 = 2$$
$$6b = -10$$
$$b = -\frac{10}{6}$$

or $$b = -\frac{5}{3}$$

Finally, $$a = 3 + b$$
$$a = 3 - \frac{5}{3}$$
$$a = \frac{4}{3}$$

The solution is $a = \frac{4}{3}$, $b = -\frac{5}{3}$, $c = 1$.

79. Solve:

$$\begin{cases} y = m_1x + b_1 & (1) \\ y = m_2x + b_2 & (2) \end{cases}$$

From (1), $y = m_1x + b_1$. Substitute this into (2):

$$y = m_2x + b_2 \qquad (2)$$
$$m_1x + b_1 = m_2x + b_2$$
$$m_1x - m_2x = b_2 - b_1$$
$$(m_1 - m_2)x = b_2 - b_1$$
$$x = \frac{b_2 - b_1}{m_1 - m_2} = \frac{b_1 - b_2}{m_2 - m_1} \qquad \text{(Note } m_1 - m_2 \neq 0\text{)}$$

Then

$$y = m_1x + b_1$$
$$y = m_1\left(\frac{b_2 - b_1}{m_1 - m_2}\right) + b_1$$
$$y = \frac{m_1b_2 - m_1b_1}{m_1 - m_2} + \frac{b_1(m_1 - m_2)}{m_1 - m_2}, \quad \text{common denominator)}$$
$$y = \frac{m_1b_2 - m_1b_1 + m_1b_1 - m_2b_1}{m_1 - m_2}$$
$$y = \frac{m_1b_2 - m_2b_1}{m_1 - m_2} = \frac{m_2b_1 - m_1b_2}{m_2 - m_1}$$

81. We have:

$$\begin{cases} y = m_1x + b_1 & (1) \\ y = m_2x + b_2 & (2) \end{cases}$$

Equation (1) is already solved for y, so we can substitute this into (2):

$$y = m_2x + b_2 \qquad (2)$$
$$m_1x + b_1 = m_2x + b_2$$
$$m_1x - m_2x = b_2 - b_1$$
$$mx - mx = b - b$$
$$0 \cdot x = 0$$

This is solved by every value of x. So the solution is:

$y = mx + b$, where x can be any real number.

≡ EXERCISE 10.2 SYSTEMS OF LINEAR EQUATIONS: MATRICES

1. $\begin{cases} x - 3y = 5 \\ 4x + y = 6 \end{cases}$ becomes $\begin{bmatrix} 1 & -3 & | & 5 \\ 4 & 1 & | & 6 \end{bmatrix}$

3. $\begin{cases} 2x + 3y - 6 = 0 \\ 4x - 6y + 2 = 0 \end{cases}$

First, put the constants on the right-hand side:

$\begin{cases} 2x + 3y = 6 \\ 4x - 6y = -2 \end{cases}$

This can be represented as:

$\begin{bmatrix} 2 & 3 & | & 6 \\ 4 & -6 & | & -2 \end{bmatrix}$

5. $\begin{cases} 0.01x - 0.03y = 0.06 \\ 0.13x + 0.10y = 0.20 \end{cases}$ becomes $\begin{bmatrix} 0.01 & -0.03 & | & 0.06 \\ 0.13 & 0.10 & | & 0.20 \end{bmatrix}$

7. $\begin{cases} x - y + z = 10 \\ 3x + 2y = 5 \\ x + y + 2z = 2 \end{cases}$ becomes $\begin{bmatrix} 1 & -1 & 1 & | & 10 \\ 3 & 2 & 0 & | & 5 \\ 1 & 1 & 2 & | & 2 \end{bmatrix}$

9. $\begin{cases} x + y - z = 2 \\ 3x - 2y = 2 \end{cases}$ becomes $\begin{bmatrix} 1 & 1 & -1 & | & 2 \\ 3 & -2 & 0 & | & 2 \end{bmatrix}$

11. We would need to use $R_1 = -1r_2 + r_1$.

13. We would need to use $R_1 = -4r_2 + r_1$.

15. $\begin{bmatrix} 1 & 2 & 3 & | & 0 \\ 2 & 4 & 3 & | & 3 \\ -3 & 2 & 1 & | & -2 \end{bmatrix} \rightarrow \begin{bmatrix} 1 & 2 & 3 & | & 0 \\ 0 & 0 & -3 & | & 3 \\ -3 & 2 & 1 & | & -2 \end{bmatrix}$
$$\uparrow$$
$$R_2 = -2r_1 + r_2$$

Note that row one is not affected by this operation, only row two.

17. $\begin{bmatrix} 1 & 2 & 3 & | & 0 \\ 2 & 4 & 3 & | & 3 \\ -3 & 2 & 1 & | & -2 \end{bmatrix} \rightarrow \begin{bmatrix} 1 & 2 & 3 & | & 0 \\ 8 & 0 & 1 & | & 7 \\ -3 & 2 & 1 & | & -2 \end{bmatrix}$
$$\uparrow$$
$$R_2 = -2r_3 + r_2$$

19. $\begin{bmatrix} 1 & 2 & 3 & | & 0 \\ 2 & 4 & 3 & | & 3 \\ -3 & 2 & 1 & | & -2 \end{bmatrix} \rightarrow \begin{bmatrix} 1 & 2 & 3 & | & 0 \\ 2 & 4 & 3 & | & 3 \\ -1 & 6 & 4 & | & 1 \end{bmatrix}$

$$\uparrow$$
$$R_3 = r_2 + r_3$$

21. $\begin{cases} x = 2 \\ y = 3 \end{cases}$ consistent $x = 2,\ y = 3$

23. $\begin{cases} x = 1 \\ y = 2 \\ 0 = 3 \end{cases}$ inconsistent

25. $\begin{cases} x + 2z = -1 \\ y - 4z = -2 \\ 0 = 0 \end{cases}$ consistent $x = -1 - 2z,\ y = -2 + 4z,$ z any real number

27. $\begin{cases} x_1 = 1 \\ x_2 + x_4 = 2 \\ x_3 + 2x_4 = 3 \end{cases}$ consistent $x_1 = 1,\ x_2 = 2 - x_4,\ x_3 = 3 - 2x_4,$ x_4 any real number.

29. $\begin{cases} x_1 + 4x_4 = 2 \\ x_2 + x_3 + 3x_4 = 3 \\ 0 = 0 \end{cases}$ consistent $x_1 = 2 - 4x_4,\ x_2 = 3 - x_3 - 3x_4,$ $x_3,\ x_4$ any real numbers.

31. $\begin{cases} x + y = 8 \\ x - y = 4 \end{cases}$ can be represented as: $\begin{bmatrix} 1 & 1 & | & 8 \\ 1 & -1 & | & 4 \end{bmatrix}$

Since we already have a 1 in the first row, first column, we proceed to obtain 0's below it:

$\begin{bmatrix} 1 & 1 & | & 8 \\ 1 & -1 & | & 4 \end{bmatrix} \rightarrow \begin{bmatrix} 1 & 1 & | & 8 \\ 0 & -2 & | & -4 \end{bmatrix} \rightarrow \begin{bmatrix} 1 & 1 & | & 8 \\ 0 & 1 & | & 2 \end{bmatrix}$

$$\uparrow \qquad\qquad \uparrow$$
$$R_2 = -1r_1 + r_2 \qquad R_2 = -\frac{1}{2}r_2$$

That means:

$\begin{cases} x + y = 8 \\ y = 2 \end{cases}$

$\begin{cases} x + 2 = 8 \\ y = 2 \end{cases}$ Back-substitute; $y = 2$.

$\begin{cases} x = 6 \\ y = 2 \end{cases}$

The solution is $x = 6,\ y = 2$.

10.2 SYSTEMS OF LINEAR EQUATIONS: MATRICES

33. $\begin{cases} x - 5y = -13 \\ 3x + 2y = 12 \end{cases}$

$$\begin{bmatrix} 1 & -5 & | & -13 \\ 3 & 2 & | & 12 \end{bmatrix} \rightarrow \begin{bmatrix} 1 & -5 & | & -13 \\ 0 & 17 & | & 51 \end{bmatrix} \rightarrow \begin{bmatrix} 1 & -5 & | & -13 \\ 0 & 1 & | & 3 \end{bmatrix}$$

$y = 3 \qquad x - 5y = -13$
$\qquad\qquad\qquad x = 5(3) - 13$
$\qquad\qquad\qquad x = 2$

$x = 2, \; y = 3$

35. $\begin{cases} 3x \qquad = 24 \\ x + 2y = 0 \end{cases}$ becomes: $\begin{bmatrix} 3 & 0 & | & 24 \\ 1 & 2 & | & 0 \end{bmatrix} \rightarrow \begin{bmatrix} 1 & 2 & | & 0 \\ 3 & 0 & | & 24 \end{bmatrix}$

$\uparrow$
Interchange r_1 and r_2
(to get a 1 in row 1, column 1).

$$\rightarrow \begin{bmatrix} 1 & 2 & | & 0 \\ 0 & -6 & | & 24 \end{bmatrix}$$
$\uparrow$
$R_2 = -3r_1 + r_2$
(to get a zero below the 1)

$$\rightarrow \begin{bmatrix} 1 & 2 & | & 0 \\ 0 & 1 & | & -4 \end{bmatrix}$$
$\uparrow$
$R_2 = -\dfrac{1}{6}r_2$

Therefore, we have:

$\begin{cases} x + 2y = 0 \\ \qquad y = -4 \end{cases}$

$\begin{cases} x - 8 = 0 \\ \qquad y = -4 \end{cases}$ Back-substitute

$\begin{cases} x = 8 \\ y = -4 \end{cases}$

The solution is $x = 8, \; y = -4$.

37. $\begin{cases} 3x - 6y = 24 \\ 5x + 4y = 12 \end{cases}$ becomes: $\begin{bmatrix} 3 & -6 & | & 24 \\ 5 & 4 & | & 12 \end{bmatrix} \rightarrow \begin{bmatrix} 1 & -2 & | & 8 \\ 5 & 4 & | & 12 \end{bmatrix}$

$\uparrow$
$R_1 = \dfrac{1}{3}r_1$

$$\rightarrow \begin{bmatrix} 1 & -2 & | & 8 \\ 0 & 14 & | & -28 \end{bmatrix}$$
$\uparrow$
$R_2 = -5r_1 + r_2$

10 SYSTEMS OF EQUATIONS AND INEQUALITIES

$$\rightarrow \begin{bmatrix} 1 & -2 & | & 8 \\ 0 & 1 & | & -2 \end{bmatrix}$$

$$\uparrow$$

$$R_2 = \frac{1}{14} r_2$$

$$\rightarrow \begin{bmatrix} 1 & 0 & | & 4 \\ 0 & 1 & | & -2 \end{bmatrix}$$

$$\uparrow$$

$$R_1 = 2r_2 + r_1$$

The solution is $x = 4$, $y = -2$.

39. $\begin{cases} 2x + y = 1 \\ 4x + 2y = 6 \end{cases}$ becomes: $\begin{bmatrix} 2 & 1 & | & 1 \\ 4 & 2 & | & 6 \end{bmatrix} \rightarrow \begin{bmatrix} 1 & \frac{1}{2} & | & \frac{1}{2} \\ 4 & 2 & | & 6 \end{bmatrix}$

$$\uparrow$$

$$R_1 = \frac{1}{2} r_1$$

$$\rightarrow \begin{bmatrix} 1 & \frac{1}{2} & | & \frac{1}{2} \\ 0 & 0 & | & 4 \end{bmatrix}$$

$$\uparrow$$

$$R_2 = -4r_1 + r_2$$

The system is inconsistent.

41. $\begin{cases} 2x - 4y = -2 \\ 3x + 2y = 3 \end{cases}$ becomes: $\begin{bmatrix} 2 & -4 & | & -2 \\ 3 & 2 & | & 3 \end{bmatrix} \rightarrow \begin{bmatrix} 1 & -2 & | & -1 \\ 3 & 2 & | & 3 \end{bmatrix}$

$$\uparrow$$

$$R_1 = \frac{1}{2} r_1$$

$$\rightarrow \begin{bmatrix} 1 & -2 & | & -1 \\ 0 & 8 & | & 6 \end{bmatrix}$$

$$\uparrow$$

$$R_2 = -3r_1 + r_2$$

$$\rightarrow \begin{bmatrix} 1 & -2 & | & -1 \\ 0 & 1 & | & \frac{3}{4} \end{bmatrix}$$

$$\uparrow$$

$$R_2 = \frac{1}{8} r_2$$

$$\rightarrow \begin{bmatrix} 1 & 0 & | & \frac{1}{2} \\ 0 & 1 & | & \frac{3}{4} \end{bmatrix}$$

$$\uparrow$$

$$R_1 = 2r_2 + r_1$$

The solution is $x = \frac{1}{2}$, $y = \frac{3}{4}$.

43. $\begin{cases} x + 2y = 4 \\ 2x + 4y = 8 \end{cases}$ becomes: $\begin{bmatrix} 1 & 2 & | & 4 \\ 2 & 4 & | & 8 \end{bmatrix} \rightarrow \begin{bmatrix} 1 & 2 & | & 4 \\ 0 & 0 & | & 0 \end{bmatrix}$

$$\uparrow$$
$$R_2 = -2r_1 + r_2$$

Therefore, the system is equivalent to the single equation:

$$x + 2y = 4$$

(a) In terms of y,

$$x = -2y + 4$$

where y can be any real number.

(b) In terms of x,

$$2y = -x + 4,$$

or $y = -\dfrac{1}{2}x + 2,$

where x can be any real number.

45. $\begin{cases} 2x - 3y = -1 \\ 10x + 10y = 5 \end{cases}$ becomes: $\begin{bmatrix} 2 & -3 & | & -1 \\ 10 & 10 & | & 5 \end{bmatrix} \rightarrow \begin{bmatrix} 2 & -3 & | & -1 \\ 1 & 1 & | & \frac{1}{2} \end{bmatrix}$

$$\uparrow$$
$$R_2 = \frac{1}{10}r_2$$

$$\rightarrow \begin{bmatrix} 1 & 1 & | & \frac{1}{2} \\ 2 & -3 & | & -1 \end{bmatrix}$$

$$\uparrow$$

Interchange r_1 and r_2. This puts a 1 in row 1, column 1, and only introduces one fraction.

$$\rightarrow \begin{bmatrix} 1 & 1 & | & \frac{1}{2} \\ 0 & -5 & | & -2 \end{bmatrix}$$

$$\uparrow$$
$$R_2 = -2r_1 + r_2$$

$$\rightarrow \begin{bmatrix} 1 & 1 & | & \frac{1}{2} \\ 0 & 1 & | & \frac{2}{5} \end{bmatrix}$$

$$\uparrow$$
$$R_2 = -\frac{1}{5}r_2$$

$$\rightarrow \begin{bmatrix} 1 & 0 & | & -\frac{2}{5} + \frac{1}{2} \\ 0 & 1 & | & \frac{2}{5} \end{bmatrix}$$

$$\uparrow$$
$$R_1 = -1r_2 + r_1$$

Therefore, the solution is $x = -\dfrac{2}{5} + \dfrac{1}{2}$, or $x = \dfrac{1}{10}$, and $y = \dfrac{2}{5}$.

10 SYSTEMS OF EQUATIONS AND INEQUALITIES

47. $\begin{cases} 2x + 3y = 6 \\ x - y = \frac{1}{2} \end{cases}$ becomes: $\begin{bmatrix} 2 & 3 & | & 6 \\ 1 & -1 & | & \frac{1}{2} \end{bmatrix} \rightarrow \begin{bmatrix} 1 & -1 & | & \frac{1}{2} \\ 2 & 3 & | & 6 \end{bmatrix}$

$\uparrow$

Interchange r_1 and r_2

$\rightarrow \begin{bmatrix} 1 & -1 & | & \frac{1}{2} \\ 0 & 5 & | & 5 \end{bmatrix}$

$\uparrow$

$R_2 = -2r_1 + r_2$

$\rightarrow \begin{bmatrix} 1 & -1 & | & \frac{1}{2} \\ 0 & 1 & | & 1 \end{bmatrix}$

$\uparrow$

$R_2 = \frac{1}{5}r_2$

$\rightarrow \begin{bmatrix} 1 & 0 & | & \frac{3}{2} \\ 0 & 1 & | & 1 \end{bmatrix}$

$\uparrow$

$R_1 = r_2 + r_1$

The solution is $x = \frac{3}{2}$, $y = 1$.

49. $\begin{cases} 2x + 3y = 5 \\ 4x + 6y = 10 \end{cases}$ becomes: $\begin{bmatrix} 2 & 3 & | & 5 \\ 4 & 6 & | & 10 \end{bmatrix} \rightarrow \begin{bmatrix} 1 & \frac{3}{2} & | & \frac{5}{2} \\ 4 & 6 & | & 10 \end{bmatrix}$

$\uparrow$

$R_1 = \frac{1}{2}r_1$

$\rightarrow \begin{bmatrix} 1 & \frac{3}{2} & | & \frac{5}{2} \\ 0 & 0 & | & 0 \end{bmatrix}$

$\uparrow$

$R_2 = -4r_1 + r_2$

Thus, the system is equivalent to the single equation:

$$x + \frac{3}{2}y = \frac{5}{2}$$

We can write the solution as

$$x = -\frac{3}{2}y + \frac{5}{2};$$

where y can be any real number.

10.2 SYSTEMS OF LINEAR EQUATIONS: MATRICES

51. $\begin{cases} 3x - 5y = 3 \\ 15x + 5y = 21 \end{cases}$ becomes: $\begin{bmatrix} 3 & -5 & | & 3 \\ 15 & 5 & | & 21 \end{bmatrix} \rightarrow \begin{bmatrix} 1 & -\frac{5}{3} & | & 1 \\ 15 & 5 & | & 21 \end{bmatrix}$

$$\uparrow$$
$$R_1 = \tfrac{1}{3} r_1$$

$$\rightarrow \begin{bmatrix} 1 & -\frac{5}{3} & | & 1 \\ 0 & 30 & | & 6 \end{bmatrix}$$

$$\uparrow$$
$$R_2 = -15 r_1 + r_2$$

$$\rightarrow \begin{bmatrix} 1 & -\frac{5}{3} & | & 1 \\ 0 & 1 & | & \frac{1}{5} \end{bmatrix}$$

$$\uparrow$$
$$R_2 = \tfrac{1}{30} r_2$$

Thus, we have:

$$\begin{cases} x - \dfrac{5}{3} y = 1 \\ y = \dfrac{1}{5} \end{cases}$$

$$\begin{cases} x - \dfrac{5}{3}\left(\dfrac{1}{5}\right) = 1 \quad \text{(Back-substitution)} \\ y = \dfrac{1}{5} \end{cases}$$

$$\begin{cases} x = \dfrac{4}{3} \\ y = \dfrac{1}{5} \end{cases}$$

The solution is $x = \dfrac{4}{3}$, $y = \dfrac{1}{5}$

10 SYSTEMS OF EQUATIONS AND INEQUALITIES

53. $\begin{cases} x - y = 6 \\ 2x - 3z = 16 \\ 2y + z = 4 \end{cases}$ becomes: $\begin{bmatrix} 1 & -1 & 0 & | & 6 \\ 2 & 0 & -3 & | & 16 \\ 0 & 2 & 1 & | & 4 \end{bmatrix}$ (We already have a 1 in row 1, column 1.)

$\rightarrow \begin{bmatrix} 1 & -1 & 0 & | & 6 \\ 0 & 2 & -3 & | & 4 \\ 0 & 2 & 1 & | & 4 \end{bmatrix}$ (Use the 1 to get 0's below it.)

$\uparrow$
$R_2 = -2r_1 + r_2$

$\rightarrow \begin{bmatrix} 1 & -1 & 0 & | & 6 \\ 0 & 1 & -\dfrac{3}{2} & | & 2 \\ 0 & 2 & 1 & | & 4 \end{bmatrix}$ (We need a 1 in row 2, column 2.)

$\uparrow$
$R_2 = \dfrac{1}{2}r_2$

$\rightarrow \begin{bmatrix} 1 & -1 & 0 & | & 6 \\ 0 & 1 & -\dfrac{3}{2} & | & 2 \\ 0 & 0 & 4 & | & 0 \end{bmatrix}$ (Obtain a 0 below the 1 in row 2.)

$\uparrow$
$R_3 = -2r_2 + r_3$

$\rightarrow \begin{bmatrix} 1 & -1 & 0 & | & 6 \\ 0 & 1 & -\dfrac{3}{2} & | & 2 \\ 0 & 0 & 1 & | & 0 \end{bmatrix}$

$\uparrow$
$R_3 = \dfrac{1}{4}r_3$

Thus, we have:

$\begin{cases} x - y = 6 \\ y - \dfrac{3}{2}z = 2 \\ z = 0 \end{cases}$

$\begin{cases} x - y = 6 \\ y = 2 \\ z = 0 \end{cases}$ (Back-substitution; z = 0)

$\begin{cases} x - 2 = 6 \\ y = 2 \\ z = 0 \end{cases}$

The solution is $x = 8$, $y = 2$, $z = 0$.

55.
$$\begin{cases} x - 2y + 3z = 7 \\ 2x + y + z = 4 \\ -3x + 2y - 2z = -10 \end{cases} \quad \text{becomes:} \quad \begin{bmatrix} 1 & -2 & 3 & | & 7 \\ 2 & 1 & 1 & | & 4 \\ -3 & 2 & -2 & | & -10 \end{bmatrix}$$

$$\rightarrow \begin{bmatrix} 1 & -2 & 3 & | & 7 \\ 0 & 5 & -5 & | & -10 \\ 0 & -4 & 7 & | & 11 \end{bmatrix} \quad \text{(Use the 1 in row 1 to get 0's below it.)}$$
$\uparrow$
$R_2 = -2r_1 + r_2$
$R_3 = 3r_1 + r_3$

To obtain a 1 in row 2, column 2, we can either multiply row 2 by $\frac{1}{5}$, or add row 3 to row 2:

$$\rightarrow \begin{bmatrix} 1 & -2 & 3 & | & 7 \\ 0 & 1 & -1 & | & -2 \\ 0 & -4 & 7 & | & 11 \end{bmatrix}$$
$\uparrow$
$R_2 = \frac{1}{5}r_2$

$$\rightarrow \begin{bmatrix} 1 & 0 & 1 & | & 3 \\ 0 & 1 & -1 & | & -2 \\ 0 & 0 & 3 & | & 3 \end{bmatrix}$$
$\uparrow$
$R_1 = 2r_2 + r_1$
$R_3 = 4r_2 + r_3$

The zero <u>above</u> the 1 in row 2 will eliminate the need to do back-substitution at the end of the problem.

$$\rightarrow \begin{bmatrix} 1 & 0 & 1 & | & 3 \\ 0 & 1 & -1 & | & -2 \\ 0 & 0 & 1 & | & 1 \end{bmatrix}$$
$\uparrow$
$R_3 = \frac{1}{3}r_3$

$$\rightarrow \begin{bmatrix} 1 & 0 & 0 & | & 2 \\ 0 & 1 & 0 & | & -1 \\ 0 & 0 & 1 & | & 1 \end{bmatrix}$$
$\uparrow$
$R_1 = -1r_3 + r_1$
$R_2 = r_3 + r_2$

The solution is $x = 2$, $y = -1$, $z = 1$.

10 SYSTEMS OF EQUATIONS AND INEQUALITIES

57. $\begin{cases} 2x - 2y - 2z = 2 \\ 2x + 3y + z = 2 \\ 3x + 2y \quad\quad = 0 \end{cases}$ becomes: $\begin{bmatrix} 2 & -2 & -2 & | & 2 \\ 2 & 3 & 1 & | & 2 \\ 3 & 2 & 0 & | & 0 \end{bmatrix}$

$\rightarrow \begin{bmatrix} 2 & -2 & -2 & | & 2 \\ 0 & 5 & 3 & | & 0 \\ 1 & 4 & 2 & | & -2 \end{bmatrix}$

$\uparrow$

$R_2 = -r_1 + r_2$
$R_3 = -r_1 + r_3$

$\rightarrow \begin{bmatrix} 1 & 4 & 2 & | & -2 \\ 0 & 5 & 3 & | & 0 \\ 2 & -2 & -2 & | & 2 \end{bmatrix}$

$\uparrow$

$R_1 \leftrightarrow R_3$

$\rightarrow \begin{bmatrix} 1 & 4 & 2 & | & -2 \\ 0 & 5 & 3 & | & 0 \\ 0 & -10 & -6 & | & 6 \end{bmatrix}$

$\uparrow$

$R_3 = -2r_1 + r_3$

$\rightarrow \begin{bmatrix} 1 & 4 & 2 & | & -2 \\ 0 & 1 & \frac{3}{5} & | & 0 \\ 0 & -10 & -6 & | & 6 \end{bmatrix}$

$\uparrow$

$R_2 = \frac{1}{5}r_2$

$\rightarrow \begin{bmatrix} 1 & 0 & \frac{22}{5} & | & -2 \\ 0 & 5 & \frac{3}{5} & | & 0 \\ 0 & 0 & 0 & | & 6 \end{bmatrix}$

$\uparrow$

$R_1 = -4r_2 + r_1$
$R_3 = 10r_2 + r_3$

No solution. Inconsistent.

59. $\begin{cases} -x + y + z = -1 \\ -x + 2y - 3z = -4 \\ 3x - 2y - 7z = 0 \end{cases}$ becomes: $\begin{bmatrix} -1 & 1 & 1 & | & -1 \\ -1 & 2 & -3 & | & -4 \\ 3 & -2 & -7 & | & 0 \end{bmatrix}$

$$\rightarrow \begin{bmatrix} 1 & -1 & -1 & | & 1 \\ -1 & 2 & -3 & | & -4 \\ 3 & -2 & -7 & | & 0 \end{bmatrix}$$
$\uparrow$
$R_1 = -r_1$

$$\rightarrow \begin{bmatrix} 1 & -1 & -1 & | & 1 \\ 0 & 1 & -4 & | & -3 \\ 0 & 1 & -4 & | & -3 \end{bmatrix}$$
$\uparrow$
$R_2 = r_1 + r_2$
$R_3 = -3r_1 + r_3$

$$\rightarrow \begin{bmatrix} 1 & 0 & -5 & | & -2 \\ 0 & 1 & -4 & | & -3 \\ 0 & 0 & 0 & | & 0 \end{bmatrix}$$
$\uparrow$
$R_1 = r_2 + r_1$
$R_3 = -r_2 + r_3$

$x = 5z - 2; \ y = 4z - 3$ where z is any real number.

61. $\begin{cases} 2x - 2y + 3z = 6 \\ 4x - 3y + 2z = 0 \\ -2x + 3y - 7z = 1 \end{cases}$ becomes: $\begin{bmatrix} 2 & -2 & 3 & | & 6 \\ 4 & -3 & 2 & | & 0 \\ -2 & 3 & -7 & | & 1 \end{bmatrix}$

$$\rightarrow \begin{bmatrix} 1 & -1 & \frac{3}{2} & | & 3 \\ 4 & -3 & 2 & | & 0 \\ -2 & 3 & -7 & | & 1 \end{bmatrix}$$
$\uparrow$
$R_1 = \frac{1}{2}r_1$

$$\rightarrow \begin{bmatrix} 1 & -1 & \frac{3}{2} & | & 3 \\ 0 & 1 & -4 & | & -12 \\ 0 & 1 & -4 & | & 7 \end{bmatrix} \quad \text{Trouble!}$$
$\uparrow$
$R_2 = -4r_1 + r_2$
$R_3 = 2r_1 + r_3$

$$\rightarrow \begin{bmatrix} 1 & -1 & \frac{3}{2} & | & 3 \\ 0 & 1 & -4 & | & -12 \\ 0 & 0 & 0 & | & 19 \end{bmatrix}$$
$\uparrow$
$R_3 = -1r_2 + r_3$

The system is inconsistent.

10 SYSTEMS OF EQUATIONS AND INEQUALITIES

63.
$$\begin{cases} x + y - z = 6 \\ 3x - 2y + z = -5 \\ x + 3y - 2z = 14 \end{cases}$$
becomes:
$$\begin{bmatrix} 1 & 1 & -1 & | & 6 \\ 3 & -2 & 1 & | & -5 \\ 1 & 3 & -2 & | & 14 \end{bmatrix}$$

$$\rightarrow \begin{bmatrix} 1 & 1 & -1 & | & 6 \\ 0 & -5 & 4 & | & -23 \\ 0 & 2 & -1 & | & 8 \end{bmatrix}$$ We now need a 1 in row 2,
column 2, and there $\underline{is}$ a
way to avoid fractions.

↑
$R_2 = -3r_1 + r_2$
$R_3 = -1r_1 + r_3$

$$\rightarrow \begin{bmatrix} 1 & 1 & -1 & | & 6 \\ 0 & -1 & 2 & | & -7 \\ 0 & 2 & -1 & | & 8 \end{bmatrix}$$

↑
$R_2 = 2r_3 + r_2$

$$\rightarrow \begin{bmatrix} 1 & 1 & -1 & | & 6 \\ 0 & 1 & -2 & | & 7 \\ 0 & 2 & -1 & | & 8 \end{bmatrix}$$

↑
$R_2 = -1r_2$

$$\rightarrow \begin{bmatrix} 1 & 0 & 1 & | & -1 \\ 0 & 1 & -2 & | & 7 \\ 0 & 0 & 3 & | & -6 \end{bmatrix}$$

↑
$R_1 = -1r_2 + r_1$
$R_3 = -2r_2 + r_3$

$$\rightarrow \begin{bmatrix} 1 & 0 & 1 & | & -1 \\ 0 & 1 & -2 & | & 7 \\ 0 & 0 & 1 & | & -2 \end{bmatrix}$$

↑
$R_3 = \frac{1}{3}r_3$

$$\rightarrow \begin{bmatrix} 1 & 0 & 0 & | & 1 \\ 0 & 1 & 0 & | & 3 \\ 0 & 0 & 1 & | & -2 \end{bmatrix}$$ I prefer $\underline{reduced}$ $\underline{echelon}$ $\underline{form}$ to
back-substitution, in most cases.

↑
$R_1 = -1r_3 + r_1$
$R_2 = 2r_3 + r_2$

The solution is $x = 1$, $y = 3$, $z = -2$.

65. $\begin{cases} x + 2y - z = -3 \\ 2x - 4y + z = -7 \\ -2x + 2y - 3z = 4 \end{cases}$ becomes: $\begin{bmatrix} 1 & 2 & -1 & | & -3 \\ 2 & -4 & 1 & | & -7 \\ -2 & 2 & -3 & | & 4 \end{bmatrix}$

$\rightarrow \begin{bmatrix} 1 & 2 & -1 & | & -3 \\ 0 & -8 & 3 & | & -1 \\ 0 & 6 & -5 & | & -2 \end{bmatrix}$

$\uparrow$
$R_2 = -2r_1 + r_2$
$R_3 = 2r_1 + r_3$

$\rightarrow \begin{bmatrix} 1 & 2 & -1 & | & -3 \\ 0 & -2 & -2 & | & -3 \\ 0 & 6 & -5 & | & -2 \end{bmatrix}$ This will make the fractions in row 2 easier to work with.

$\uparrow$
$R_2 = r_3 + r_2$

$\rightarrow \begin{bmatrix} 1 & 2 & -1 & | & -3 \\ 0 & 1 & 1 & | & \frac{3}{2} \\ 0 & 6 & -5 & | & -2 \end{bmatrix}$

$\uparrow$
$R_2 = -\frac{1}{2}r_2$

$\rightarrow \begin{bmatrix} 1 & 0 & -3 & | & -6 \\ 0 & 1 & 1 & | & \frac{3}{2} \\ 0 & 0 & -11 & | & -11 \end{bmatrix}$

$\uparrow$
$R_1 = -2r_2 + r_1$
$R_3 = -6r_2 + r_3$

$\rightarrow \begin{bmatrix} 1 & 0 & -3 & | & -6 \\ 0 & 1 & 1 & | & \frac{3}{2} \\ 0 & 0 & 1 & | & 1 \end{bmatrix}$

$\uparrow$
$R_3 = -\frac{1}{11}r_3$

$\rightarrow \begin{bmatrix} 1 & 0 & 0 & | & -3 \\ 0 & 1 & 0 & | & \frac{1}{2} \\ 0 & 0 & 1 & | & 1 \end{bmatrix}$

$\uparrow$
$R_1 = 3r_3 + r_1$
$R_2 = -1r_3 + r_2$

The solution is $x = -3$, $y = \frac{1}{2}$, $z = 1$.

67.
$$\begin{cases} 3x + y - z = \dfrac{2}{3} \\ 2x - y + z = 1 \\ 4x + 2y = \dfrac{8}{3} \end{cases} \quad \text{becomes:} \quad \begin{bmatrix} 3 & 1 & -1 & \bigm| & \dfrac{2}{3} \\ 2 & -1 & 1 & \bigm| & 1 \\ 4 & 2 & 0 & \bigm| & \dfrac{8}{3} \end{bmatrix}$$

$$\rightarrow \begin{bmatrix} 1 & 2 & -2 & \bigm| & -\dfrac{1}{3} \\ 2 & -1 & 1 & \bigm| & 1 \\ 4 & 2 & 0 & \bigm| & \dfrac{8}{3} \end{bmatrix}$$

$\uparrow$

$R_1 = -1r_2 + r_1$

$$\rightarrow \begin{bmatrix} 1 & 2 & -2 & \bigm| & -\dfrac{1}{3} \\ 0 & -5 & 5 & \bigm| & \dfrac{5}{3} \\ 0 & -6 & 8 & \bigm| & \dfrac{12}{3} \end{bmatrix}$$

$\uparrow$

$R_2 = -2r_1 + r_2$
$R_3 = -4r_1 + r_3$

$$\rightarrow \begin{bmatrix} 1 & 2 & -2 & \bigm| & -\dfrac{1}{3} \\ 0 & 1 & -1 & \bigm| & -\dfrac{1}{3} \\ 0 & -6 & 8 & \bigm| & 4 \end{bmatrix}$$

$\uparrow$

$R_2 = -\dfrac{1}{5}r_2$

$$\rightarrow \begin{bmatrix} 1 & 0 & 0 & \bigm| & \dfrac{1}{3} \\ 0 & 1 & -1 & \bigm| & -\dfrac{1}{3} \\ 0 & 0 & 2 & \bigm| & 2 \end{bmatrix}$$

$\uparrow$

$R_1 = -2r_2 + r_1$
$R_3 = 6r_2 + r_3$

$$\rightarrow \begin{bmatrix} 1 & 0 & 0 & \bigm| & \dfrac{1}{3} \\ 0 & 1 & -1 & \bigm| & -\dfrac{1}{3} \\ 0 & 0 & 1 & \bigm| & 1 \end{bmatrix}$$

$\uparrow$

$R_3 = \dfrac{1}{2}r_3$

10.2 SYSTEMS OF LINEAR EQUATIONS: MATRICES

$$\rightarrow \begin{bmatrix} 1 & 0 & 0 & | & \frac{1}{3} \\ 0 & 1 & 0 & | & \frac{2}{3} \\ 0 & 0 & 1 & | & 1 \end{bmatrix}$$

$\uparrow$

$R_2 = r_3 + r_2$

The solution is $x = \frac{1}{3}$, $y = \frac{2}{3}$, $z = 1$.

69. $\begin{cases} x + y + z + w = 4 \\ 2x - y + z = 0 \\ 3x + 2y + z - w = 6 \\ x - 2y - 2z + 2w = -1 \end{cases}$ becomes: $\begin{bmatrix} 1 & 1 & 1 & 1 & | & 4 \\ 2 & -1 & 1 & 0 & | & 0 \\ 3 & 2 & 1 & -1 & | & 6 \\ 1 & -2 & -2 & 2 & | & -1 \end{bmatrix}$

$$\rightarrow \begin{bmatrix} 1 & 1 & 1 & 1 & | & 4 \\ 0 & -3 & -1 & -2 & | & -8 \\ 0 & -1 & -2 & -4 & | & -6 \\ 0 & -3 & -3 & 1 & | & -5 \end{bmatrix}$$

$\uparrow$

$R_2 = -2r_1 + r_2$
$R_3 = -3r_1 + r_3$
$R_4 = -1r_1 + r_4$

$$\rightarrow \begin{bmatrix} 1 & 1 & 1 & 1 & | & 4 \\ 0 & -1 & -2 & -4 & | & -6 \\ 0 & -3 & -1 & -2 & | & -8 \\ 0 & -3 & -3 & 1 & | & -5 \end{bmatrix}$$

$\uparrow$

Interchange r_2 and r_3

$$\rightarrow \begin{bmatrix} 1 & 1 & 1 & 1 & | & 4 \\ 0 & 1 & 2 & 4 & | & 6 \\ 0 & -3 & -1 & -2 & | & -8 \\ 0 & -3 & -3 & 1 & | & -5 \end{bmatrix}$$

$\uparrow$

$R_2 = -1r_2$

$$\rightarrow \begin{bmatrix} 1 & 0 & -1 & -3 & | & -2 \\ 0 & 1 & 2 & 4 & | & 6 \\ 0 & 0 & 5 & 10 & | & 10 \\ 0 & 0 & 3 & 13 & | & 13 \end{bmatrix}$$

$\uparrow$

$R_1 = -1r_2 + r_1$
$R_3 = 3r_2 + r_3$
$R_4 = 3r_2 + r_4$

$$\rightarrow \begin{bmatrix} 1 & 0 & -1 & -3 & | & -2 \\ 0 & 1 & 2 & 4 & | & 6 \\ 0 & 0 & 1 & 2 & | & 2 \\ 0 & 0 & 3 & 13 & | & 13 \end{bmatrix}$$

$\uparrow$

$R_3 = \frac{1}{5}r_3$

10 SYSTEMS OF EQUATIONS AND INEQUALITIES

$$\rightarrow \begin{bmatrix} 1 & 0 & 0 & -1 & | & 0 \\ 0 & 1 & 0 & 0 & | & 2 \\ 0 & 0 & 1 & 2 & | & 2 \\ 0 & 0 & 0 & 7 & | & 7 \end{bmatrix}$$

↑

$R_1 = r_3 + r_1$
$R_2 = -2r_3 + r_2$
$R_4 = -3r_3 + r_4$

$$\rightarrow \begin{bmatrix} 1 & 0 & 0 & -1 & | & 0 \\ 0 & 1 & 0 & 0 & | & 2 \\ 0 & 0 & 1 & 2 & | & 2 \\ 0 & 0 & 0 & 1 & | & 1 \end{bmatrix}$$

↑

$R_4 = \dfrac{1}{7}r_4$

$$\rightarrow \begin{bmatrix} 1 & 0 & 0 & 0 & | & 0 \\ 0 & 1 & 0 & 0 & | & 2 \\ 0 & 0 & 1 & 0 & | & 0 \\ 0 & 0 & 0 & 1 & | & 1 \end{bmatrix}$$

↑

$R_1 = r_4 + r_1$
$R_3 = -2r_4 + r_3$

The solution is $x = 1$, $y = 2$, $z = 0$, $w = 1$.

71. $\begin{cases} x + 2y + z = 1 \\ 2x - y + 2z = 2 \\ 3x + y + 3z = 3 \end{cases}$ becomes: $\begin{bmatrix} 1 & 2 & 1 & | & 1 \\ 2 & -1 & 2 & | & 2 \\ 3 & 1 & 3 & | & 3 \end{bmatrix}$

$$\rightarrow \begin{bmatrix} 1 & 2 & 1 & | & 1 \\ 0 & -5 & 0 & | & 0 \\ 0 & -5 & 0 & | & 0 \end{bmatrix}$$

↑

$R_2 = -2r_1 + r_2$
$R_3 = -3r_1 + r_3$

The system is equivalent to two equations:

$\begin{cases} x + 2y + z = 1 \quad (1) \\ \quad\;\; -5y \quad\quad = 0 \quad (2) \end{cases}$

From (2) we have: $y = 0$, and back-substitution into (1) yields:

$x + z = 1$

We can write the solution as:

$y = 0$; $x = -z + 1$; where z is any real number,

or

$y = 0$; $z = -x + 1$; where x is any real number.

73. $\begin{cases} x - y + z = 5 \\ 3x + 2y - 2z = 0 \end{cases}$ becomes: $\begin{bmatrix} 1 & -1 & 1 & | & 5 \\ 3 & 2 & -2 & | & 0 \end{bmatrix} \rightarrow \begin{bmatrix} 1 & -1 & 1 & | & 5 \\ 0 & 5 & -5 & | & -15 \end{bmatrix}$

$\uparrow$
$R_2 = -3r_1 + r_2$

$\rightarrow \begin{bmatrix} 1 & -1 & 1 & | & 5 \\ 0 & 1 & -1 & | & -3 \end{bmatrix}$

$\uparrow$
$R_2 = \frac{1}{5}r_2$

$\rightarrow \begin{bmatrix} 1 & 0 & 0 & | & 2 \\ 0 & 1 & -1 & | & -3 \end{bmatrix}$

$\uparrow$
$R_1 = r_2 + r_1$

This represents the system:

$\begin{cases} x = 2 \\ y - z = -3 \end{cases}$

The solution is:

$x = 2; \; y = z - 3;$ where z is any real number,

or

$x = 2; \; z = y + 3;$ where y is any real number.

75. $\begin{cases} 2x + 3y - z = 3 \\ x - y - z = 0 \\ -x + y + z = 0 \\ x + y + 3z = 5 \end{cases}$ becomes: $\begin{bmatrix} 2 & 3 & -1 & | & 3 \\ 1 & -1 & -1 & | & 0 \\ -1 & 1 & 1 & | & 0 \\ 1 & 1 & 3 & | & 5 \end{bmatrix} \rightarrow \begin{bmatrix} 1 & -1 & -1 & | & 0 \\ 2 & 3 & -1 & | & 3 \\ -1 & 1 & 1 & | & 0 \\ 1 & 1 & 3 & | & 5 \end{bmatrix}$

$\uparrow$
Interchange rows one and two

$\rightarrow \begin{bmatrix} 1 & -1 & -1 & | & 0 \\ 0 & 5 & 1 & | & 3 \\ 0 & 0 & 0 & | & 0 \\ 0 & 2 & 4 & | & 5 \end{bmatrix}$

$\uparrow$
$R_2 = -2r_1 + r_2$
$R_3 = r_1 + r_3$
$R_4 = -1r_1 + r_4$

$\rightarrow \begin{bmatrix} 1 & -1 & -1 & | & 0 \\ 0 & 5 & 1 & | & 3 \\ 0 & 2 & 4 & | & 5 \\ 0 & 0 & 0 & | & 0 \end{bmatrix}$

$\uparrow$
Interchange r_3 and r_4

10 SYSTEMS OF EQUATIONS AND INEQUALITIES

$$\rightarrow \begin{bmatrix} 1 & -1 & -1 & | & 0 \\ 0 & 1 & -7 & | & -7 \\ 0 & 2 & 4 & | & 5 \\ 0 & 0 & 0 & | & 0 \end{bmatrix}$$

↑
$$R_2 = -2r_3 + r_2$$

$$\rightarrow \begin{bmatrix} 1 & 0 & -8 & | & -7 \\ 0 & 1 & -7 & | & -7 \\ 0 & 1 & 18 & | & 19 \\ 0 & 0 & 0 & | & 0 \end{bmatrix}$$

↑
$$R_1 = r_2 + r_1$$
$$R_3 = -2r_2 + r_3$$

$$\rightarrow \begin{bmatrix} 1 & 0 & -8 & | & -7 \\ 0 & 1 & -7 & | & -7 \\ 0 & 0 & 1 & | & \frac{19}{18} \\ 0 & 0 & 0 & | & 0 \end{bmatrix}$$

↑
$$R_3 = \frac{1}{18}r_3$$

Because of the unusual fraction, I will do back-substitution. We have:

$$\begin{cases} x - 8z = -7 \\ y - 7z = -7 \\ \qquad z = \frac{19}{18} \end{cases}$$

$$\begin{cases} x - 8\left(\frac{19}{18}\right) = -7 \\ y - 7\left(\frac{19}{18}\right) = -7 \\ \qquad z = \frac{19}{18} \end{cases}$$

$$\begin{cases} x = \dfrac{-7 \cdot 18 + 8 \cdot 19}{18} = \dfrac{26}{18} = \dfrac{13}{9} \\ y = \dfrac{-7 \cdot 18 + 7 \cdot 19}{18} = \dfrac{7}{18} \\ z = \dfrac{19}{18} \end{cases}$$

Thus, the solution is $x = \dfrac{13}{9}$, $y = \dfrac{7}{18}$, $z = \dfrac{19}{18}$.

77.
$$\begin{cases} 4x + y + z - w = 4 \\ x - y + 2z + 3w = 3 \end{cases}$$

$$\begin{bmatrix} 4 & 1 & 1 & -1 & | & 4 \\ 1 & -1 & 2 & 3 & | & 3 \end{bmatrix} \rightarrow \begin{bmatrix} 1 & -1 & 2 & 3 & | & 3 \\ 4 & 1 & 1 & -1 & | & 4 \end{bmatrix}$$
$$\uparrow$$
Interchange rows

$$\rightarrow \begin{bmatrix} 1 & -1 & 2 & 3 & | & 3 \\ 0 & 5 & -7 & -13 & | & -8 \end{bmatrix}$$
$$\uparrow$$
$$R_2 = -4r_1 + r_2$$

This is equivalent to the system

$$\begin{cases} x - y + 2z - 3w = 3 \quad (1) \\ \quad\quad 5y - 7z - 13w = -8 \quad (2) \end{cases}$$

From (2),

$$5y = 7z + 13w - 8$$

$$y = \frac{7}{5}z + \frac{13}{5}w - \frac{8}{5}$$

Then from (1),

$$x = y - 2z - 3w + 3$$

$$\text{or} \quad x = \left(\frac{7}{5}z + \frac{13}{5}w - \frac{8}{5}\right) - 2z - 3w + 3$$

$$x = -\frac{3}{5}z - \frac{2}{5}w + \frac{7}{5}$$

The solution is

$$x = -\frac{3}{5}z - \frac{2}{5}w + \frac{7}{5}$$

$$y = \frac{7}{5}z + \frac{13}{5}w - \frac{8}{5}$$

where z and w are any real numbers.

79. We have $y = ax^2 + bx + c$, and each of the three points must satisfy this equation.

(1, 2): $2 = a + b + c$
(-2, -7): $-7 = 4a - 2b + c$
(-2, -3): $-3 = 4a + 2b + c$

We have three equations in three unknowns which can be represented by:

$$\left[\begin{array}{ccc|c} 1 & 1 & 1 & 2 \\ 4 & -2 & 1 & -7 \\ 4 & 2 & 1 & -3 \end{array}\right] \rightarrow \left[\begin{array}{ccc|c} 1 & 1 & 1 & 2 \\ 0 & -6 & -3 & -15 \\ 0 & -2 & -3 & -11 \end{array}\right]$$

$$\begin{array}{l} \uparrow \\ R_2 = -4r_1 + r_2 \\ R_3 = -4r_1 + r_3 \end{array}$$

$$\rightarrow \left[\begin{array}{ccc|c} 1 & 1 & 1 & 2 \\ 0 & 1 & \frac{1}{2} & \frac{5}{2} \\ 0 & -2 & -3 & -11 \end{array}\right]$$

$$\begin{array}{l} \uparrow \\ R_2 = -\frac{1}{6}r_2 \end{array}$$

$$\rightarrow \left[\begin{array}{ccc|c} 1 & 0 & \frac{1}{2} & -\frac{1}{2} \\ 0 & 1 & \frac{1}{2} & \frac{5}{2} \\ 0 & 0 & -2 & -6 \end{array}\right]$$

$$\begin{array}{l} \uparrow \\ R_1 = -1r_2 + r_1 \\ R_3 = 2r_2 + r_3 \end{array}$$

$$\rightarrow \left[\begin{array}{ccc|c} 1 & 0 & \frac{1}{2} & -\frac{1}{2} \\ 0 & 1 & \frac{1}{2} & \frac{5}{2} \\ 0 & 0 & 1 & 3 \end{array}\right]$$

$$\begin{array}{l} \uparrow \\ R_3 = -\frac{1}{2}r_3 \end{array}$$

$$\rightarrow \left[\begin{array}{ccc|c} 1 & 0 & 0 & -2 \\ 0 & 1 & 0 & 1 \\ 0 & 0 & 1 & 3 \end{array}\right]$$

$$\begin{array}{l} \uparrow \\ R_1 = -\frac{1}{2}r_3 + r_1 \\ R_2 = -\frac{1}{2}r_3 + r_2 \end{array}$$

The solution is $a = -2$, $b = 1$, $c = 3$, so the parabola is

$$y = -2x^2 + x + 3$$

You can verify that each of the given points satisfies this equation.

10.2 SYSTEMS OF LINEAR EQUATIONS: MATRICES   559

81. $f(x) = ax^3 + bx^2 + cx + d$

$f(-3) = -112$ implies $-27a + 9b - 3c + d = -112$
$f(-1) = -2$ implies $-a + b - c + d = -2$
$f(1) = 4$ implies $a + b + c + d = 4$

and

$f(2) = 13$ implies $-8a + 4b + 2c + d = 13$

Thus, we want to find the solution to a system of four equations in four unknowns.

$$\begin{bmatrix} -27 & 9 & -3 & 1 & | & -112 \\ -1 & 1 & -1 & 1 & | & -2 \\ 1 & 1 & 1 & 1 & | & 4 \\ 8 & 4 & 2 & 1 & | & 13 \end{bmatrix} \rightarrow \begin{bmatrix} 1 & 1 & 1 & 1 & | & 4 \\ -1 & 1 & -1 & 1 & | & -2 \\ -27 & 9 & -3 & 1 & | & -112 \\ 8 & 4 & 2 & 1 & | & 13 \end{bmatrix}$$

↑
Interchange r_3 and r_1

$$\rightarrow \begin{bmatrix} 1 & 1 & 1 & 1 & | & 4 \\ 0 & 2 & 0 & 2 & | & 2 \\ 0 & 36 & 24 & 28 & | & -4 \\ 0 & -4 & -6 & -7 & | & -19 \end{bmatrix}$$

↑
$R_2 = r_1 + r_2$
$R_3 = 27r_1 + r_3$
$R_4 = -8r_1 + r_4$

$$\rightarrow \begin{bmatrix} 1 & 1 & 1 & 1 & | & 4 \\ 0 & 1 & 0 & 1 & | & 1 \\ 0 & 9 & 6 & 7 & | & -1 \\ 0 & -4 & -6 & -7 & | & -19 \end{bmatrix}$$

↑
$R_2 = \frac{1}{2}r_2$

$R_3 = \frac{1}{4}r_3$

$$\rightarrow \begin{bmatrix} 1 & 0 & 1 & 0 & | & 3 \\ 0 & 1 & 0 & 1 & | & 1 \\ 0 & 0 & 6 & -2 & | & -10 \\ 0 & 0 & -6 & -3 & | & -15 \end{bmatrix}$$
Now can we get a 1 in row 3 column 3, <u>and</u> avoid fractions?

↑
$R_1 = -1r_2 + r_1$
$R_3 = -9r_2 + r_3$
$R_4 = 4r_2 + r_4$

$$\rightarrow \begin{bmatrix} 1 & 0 & 1 & 0 & | & 3 \\ 0 & 1 & 0 & 1 & | & 1 \\ 0 & 0 & 3 & -1 & | & -5 \\ 0 & 0 & -2 & -1 & | & -5 \end{bmatrix}$$

↑
$R_3 = \frac{1}{2}r_3$

$R_4 = \frac{1}{3}r_4$

$$\rightarrow \begin{bmatrix} 1 & 0 & 1 & 0 & | & 3 \\ 0 & 1 & 0 & 1 & | & 1 \\ 0 & 0 & 1 & -2 & | & -10 \\ 0 & 0 & -2 & -1 & | & -5 \end{bmatrix}$$

$\uparrow$
$R_3 = r_4 + r_3$

$$\rightarrow \begin{bmatrix} 1 & 0 & 0 & 2 & | & 13 \\ 0 & 1 & 0 & 1 & | & 1 \\ 0 & 0 & 1 & -2 & | & -10 \\ 0 & 0 & 0 & -5 & | & -25 \end{bmatrix}$$

$\uparrow$
$R_1 = -1r_3 + r_1$
$R_4 = 2r_3 + r_4$

$$\rightarrow \begin{bmatrix} 1 & 0 & 0 & 2 & | & 13 \\ 0 & 1 & 0 & 1 & | & 1 \\ 0 & 0 & 1 & -2 & | & -10 \\ 0 & 0 & 0 & 1 & | & 5 \end{bmatrix}$$

$\uparrow$
$R_4 = -\dfrac{1}{5}r_4$

$$\rightarrow \begin{bmatrix} 1 & 0 & 0 & 0 & | & 3 \\ 0 & 1 & 0 & 0 & | & -4 \\ 0 & 0 & 1 & 0 & | & 0 \\ 0 & 0 & 0 & 1 & | & 5 \end{bmatrix}$$

$\uparrow$
$R_1 = -2r_4 + r_1$
$R_2 = -1r_4 + r_2$
$R_3 = 2r_4 + r_3$

So we have: $a = 3$, $b = -4$, $c = 0$, $d = 5$.

The function is:

$$f(x) = 3x^3 - 4x^2 + 5$$

83.   Let x, y, and z represent the number of liters of 15%, 25% and 50% solutions which will be mixed.   Then,

$$x + y + z = 100 \quad (1)$$

Also, in x liters of 15% solution, there will be .15x liters of H_2SO_4, y liters of 25% solution contain .25y liters of H_2SO_4, and the z liters contain .50z liters of H_2SO_4.   Meanwhile, our final 100 liter mixture is 40% H_2SO_4, so it contains .40(100) = 40 liters of H_2SO_4.

Thus, .15x + .25y + .50z = 40   (2)

We have 2 equations in three unknowns:

$$\begin{bmatrix} 1 & 1 & 1 & | & 100 \\ 0.15 & 0.25 & 0.50 & | & 40 \end{bmatrix} \rightarrow \begin{bmatrix} 1 & 1 & 1 & | & 100 \\ 0 & 0.10 & 0.35 & | & 25 \end{bmatrix}$$

$$\uparrow$$
$$R_2 = -.15r_1 + r_2$$

$$\rightarrow \begin{bmatrix} 1 & 1 & 1 & | & 100 \\ 0 & 1 & 3.5 & | & 250 \end{bmatrix}$$

$$\uparrow$$
$$R_2 = 10r_2$$

$$\rightarrow \begin{bmatrix} 1 & 0 & -2.5 & | & -150 \\ 0 & 1 & 3.5 & | & 250 \end{bmatrix}$$

$$\uparrow$$
$$R_1 = -1r_2 + r_1$$

This gives

$$\begin{cases} x - 2.5z = -150 \\ y + 3.5z = 250 \end{cases}$$

so $x = 2.5z - 150$
    $y = -3.5z + 250$

where z can be any real number.

But, we require $x \geq 0$, $y \geq 0$, and $z \geq 0$.

Since $x \geq 0$, we have:

$$2.5z - 150 \geq 0$$
$$2.5 \geq 150$$
$$z \geq 60$$

Also, $y \geq 0$ implies

$$-3.5z + 250 \geq 0$$
$$-3.5 \geq -250$$
$$z \leq 71.43$$

Some possible solutions are given below:

z (50%)	$x = 2.5z - 150$ (15%)	$y = -3.5z + 250$ (25%)	40%
60	0	40	100
64	10	26	100
68	20	12	100
70	25	5	100

85.　x = price of hamburger, y = price of fries, z = price of colas

$$\begin{cases} 8x + 6y + 6z = 26.10 \\ 10x + 6y + 8z = 31.60 \end{cases}$$

$$\begin{bmatrix} 8 & 6 & 6 & | & 26.10 \\ 10 & 6 & 8 & | & 31.60 \end{bmatrix}$$

$$\begin{bmatrix} 4 & 3 & 3 & | & 13.05 \\ 5 & 3 & 4 & | & 15.80 \end{bmatrix}$$

$$\begin{bmatrix} 4 & 3 & 3 & | & 13.05 \\ 1 & 0 & 1 & | & 12.75 \end{bmatrix}$$

$$\begin{bmatrix} 1 & 0 & 1 & | & 2.75 \\ 4 & 3 & 3 & | & 13.05 \end{bmatrix}$$

$$\begin{bmatrix} 1 & 0 & 1 & | & 2.75 \\ 0 & 3 & -1 & | & 2.05 \end{bmatrix}$$

$$\begin{bmatrix} 1 & 0 & 1 & | & 2.75 \\ 0 & 1 & -\dfrac{1}{3} & | & \dfrac{2.05}{3} \end{bmatrix}$$

$x = 2.75 - z$, z any real number

$y = \dfrac{2.05}{3} + \dfrac{1}{3}z$, z any real number

$y = 0.68 + \dfrac{1}{3}z$, z any real number

x	\$2.15	\$2.00	\$1.85
y	\$0.88	\$0.93	\$0.98
z	\$0.60	\$0.75	\$0.90

87.　Let x = amount in Treasury bills, y = amount in corporate bonds, z = amount in junk bonds

$$\begin{cases} x + y + z = 25000 \\ .07x + .09y + .11z = 2000 \end{cases}$$

$$\begin{bmatrix} 1 & 1 & 1 & | & 25000 \\ .07 & .09 & .11 & | & 2000 \end{bmatrix}$$

$$\begin{bmatrix} 1 & 1 & 1 & | & 25,000 \\ 7 & 9 & 11 & | & 200,000 \end{bmatrix}$$

$$\begin{bmatrix} 1 & 1 & 1 & | & 25,000 \\ 0 & 2 & 4 & | & 25,000 \end{bmatrix}$$

$$\begin{bmatrix} 1 & 1 & 1 & | & 25,000 \\ 0 & 1 & 2 & | & 12,500 \end{bmatrix}$$

$$\begin{bmatrix} 1 & 0 & -1 & | & 12,500 \\ 0 & 1 & 2 & | & 12,500 \end{bmatrix}$$

$x = 12,500 + z$, $y = 12,500 - 2z$, z any real number

x	\$12,500	\$14,500	\$16,500
y	\$12,500	\$ 8,500	\$ 4,500
z	0	\$ 2,000	\$ 4,000

89. (a)

$$\begin{bmatrix} 1 & 1 & 1 & | & 25,000 \\ .07 & .09 & .11 & | & 1,500 \end{bmatrix}$$

$$\begin{bmatrix} 1 & 1 & 1 & | & 25,000 \\ 7 & 9 & 11 & | & 150,000 \end{bmatrix}$$

$$\begin{bmatrix} 1 & 1 & 1 & | & 25,000 \\ 0 & 2 & 4 & | & -25,000 \end{bmatrix}$$

$$\begin{bmatrix} 1 & 1 & 1 & | & 25,000 \\ 0 & 1 & 2 & | & -12,500 \end{bmatrix}$$

$$\begin{bmatrix} 1 & 0 & -1 & | & 37,500 \\ 0 & 1 & 2 & | & -12,500 \end{bmatrix}$$

$x = 37,500 + z$, $y = -12,500 - 2z$, z any real number

(b)

$$\begin{bmatrix} 1 & 1 & 1 & | & 25,000 \\ .07 & .09 & .11 & | & 2,500 \end{bmatrix}$$

$$\begin{bmatrix} 1 & 1 & 1 & | & 25,000 \\ 7 & 9 & 11 & | & 250,000 \end{bmatrix}$$

$$\begin{bmatrix} 1 & 1 & 1 & | & 25,000 \\ 0 & 2 & 4 & | & 75,000 \end{bmatrix}$$

$$\begin{bmatrix} 1 & 1 & 1 & | & 25,000 \\ 0 & 1 & 2 & | & 37,500 \end{bmatrix}$$

$$\begin{bmatrix} 1 & 0 & -1 & | & -12,500 \\ 0 & 1 & 2 & | & 37,500 \end{bmatrix}$$

$x = -12,500 + z$, $y = 37,500 - 2z$, z any real number

(c) The \$1500 requirement is exceeded even if all the money is invested in Treasury bills; the \$2500 requirement leaves little flexibility and requires a large amount to be invested in junk bonds.

$$91. \quad \begin{cases} I_1 + I_2 = I_3 \\ 16 - 8 - 9I_3 - 3I_1 = 0 \\ 16 - 4 - 9I_3 - 9I_2 = 0 \\ 8 - 4 - 9I_2 + 3I_1 = 0 \end{cases}$$

$$I_1 + I_2 - I_3 = 0$$
$$-3I_1 \quad\quad -9I_3 = -8$$
$$-9I_2 - 9I_3 = -12$$
$$3I_1 - 9I_2 \quad\quad = -4$$

$$\left[\begin{array}{ccc|c} 1 & 1 & -1 & 0 \\ -3 & 0 & -9 & -8 \\ 0 & -9 & -9 & -12 \\ 3 & -9 & 0 & -4 \end{array} \right]$$

$$\left[\begin{array}{ccc|c} 1 & 1 & -1 & 0 \\ 0 & 3 & -12 & -8 \\ 0 & -9 & -9 & -12 \\ 0 & -12 & 3 & -4 \end{array} \right]$$

$$\left[\begin{array}{ccc|c} 1 & 0 & 3 & \frac{8}{3} \\ 0 & 1 & -4 & -\frac{8}{3} \\ 0 & 0 & -45 & -36 \\ 0 & 0 & -45 & -36 \end{array} \right]$$

$$\left[\begin{array}{ccc|c} 1 & 0 & 3 & \frac{8}{3} \\ 0 & 1 & -4 & -\frac{8}{3} \\ 0 & 0 & 1 & \frac{36}{45} \\ 0 & 0 & 0 & 0 \end{array} \right]$$

$$I_3 = \frac{36}{45} = \frac{4}{5}$$

$$I_2 = \frac{-8}{3} + 4\left(\frac{4}{5}\right)$$
$$= \frac{-40 + 48}{15}$$
$$= \frac{8}{15}$$

$$I_1 = \frac{8}{3} - 3\left(\frac{4}{5}\right)$$
$$= \frac{40 - 36}{15} = \frac{4}{15}$$

$$I_1 = \frac{4}{15}, \; I_2 = \frac{8}{15}, \; I_3 = \frac{4}{5}$$

93. $\begin{cases} a_1x + b_1y = c_1 \\ a_2x + b_2y = c_2 \end{cases}$ becomes:

$$\begin{bmatrix} a_1 & b_1 & \bigm| & c_1 \\ a_2 & b_2 & \bigm| & c_2 \end{bmatrix}$$ If $a_1 \neq 0$, we can divide row one by a_1 to obtain a 1 in the top left corner:

$$\rightarrow \begin{bmatrix} 1 & \dfrac{b_1}{a_1} & \bigm| & \dfrac{c_1}{a_1} \\ a_2 & b_2 & \bigm| & c_2 \end{bmatrix}$$

$\uparrow$
$R_1 = \dfrac{1}{a_1} r_1$, provided $a_1 \neq 0$

Our next move would depend on whether a_2 is zero or not. If $a_2 \neq 0$, we continue:

$$\rightarrow \begin{bmatrix} 1 & \dfrac{b_1}{a_1} & \Bigm| & \dfrac{c_1}{a_1} \\ 0 & \dfrac{-a_2b_1}{a_1} + b_2 & \Bigm| & \dfrac{-a_2c_1}{a_1} + c_2 \end{bmatrix}$$

$\uparrow$
$R_2 = -a_2r_1 + r_2$

$$\rightarrow \begin{bmatrix} 1 & \dfrac{b_1}{a_1} & \Bigm| & \dfrac{c_1}{a_1} \\ 0 & \dfrac{a_1b_2 - a_2b_1}{a_1} & \Bigm| & \dfrac{a_1c_2 - a_2c_1}{a_1} \end{bmatrix}$$

$\uparrow$
Simplifying

Now recall that $D = a_1b_2 - a_2b_1$ so we have:

$$\begin{cases} x + \dfrac{b_1}{a_1}y = \dfrac{c_1}{a_1} & (1) \\ \dfrac{D}{a_1}y = \dfrac{a_1c_2 - a_2c_1}{a_1} & (2) \end{cases}$$

Solve for (2) for y: $y = \dfrac{a_1c_2 - a_2c_1}{D}$

Then use back-substitution to find x

$$x + \dfrac{b_1}{a_1}\left(\dfrac{a_1c_2 - a_2c_1}{D}\right) = \dfrac{c_1}{a_1}$$

$$x = \dfrac{c_1}{a_1} - \dfrac{b_1(a_1c_2 - a_2c_1)}{a_1D}$$

We now need to get a common denominator (a_1D) to simplify x:

$$x = \frac{c_1D}{a_1D} - \frac{b_1a_1c_2 - b_1a_2c_1}{a_1D}$$

$$= \frac{c_1(a_1b_2 - a_2b_1) - b_1a_1c_2 + b_1a_2c_1}{a_1D}$$

$$= \frac{c_1a_1b_2 - c_1a_2b_1 - b_1a_1c_2 + b_1a_2c_1}{a_1D}$$

$$= \frac{c_1a_1b_2 - b_1a_1c_2}{a_1D}$$

$$= \frac{c_1b_2 - b_1c_2}{D}$$

and our solution is:

$$x = \frac{1}{D}(c_1b_2 - b_1c_2)$$

$$y = \frac{1}{D}(a_1c_2 - a_2c_1), \text{ provided } a_1 \neq 0, \ a_2 \neq 0, \text{ as desired.}$$

But what if a_2 <u>is</u> zero?

Then we have:
$$\left[\begin{array}{cc|c} 1 & \dfrac{b_1}{a_1} & \dfrac{c_1}{a_1} \\ 0 & b_2 & c_2 \end{array}\right]$$

Also, $D = a_1b_2 - a_2b_1$
$ = a_1b_2 \quad$ (since $a_2 = 0$)

Therefore, b_2 <u>cannot</u> be 0, and we continue:

$$\rightarrow \left[\begin{array}{cc|c} 1 & \dfrac{b_1}{a_1} & \dfrac{c_1}{a_1} \\ 0 & 1 & \dfrac{c_2}{b_2} \end{array}\right]$$

$\uparrow$

$$R_2 = \frac{1}{b_2}r_2$$

$$\rightarrow \left[\begin{array}{cc|c} 1 & 0 & -\dfrac{b_1c_2}{a_1b_2} + \dfrac{c_1}{a_1} \\ 0 & 1 & \dfrac{c_2}{b_2} \end{array}\right]$$

$\uparrow$

$$R_1 = \frac{-b_1}{a_1}r_2 + r_1$$

$$\rightarrow \left[\begin{array}{cc|c} 1 & 0 & \dfrac{b_2c_1 - b_1c_2}{a_1b_2} \\ 0 & 1 & \dfrac{c_2}{b_2} \end{array}\right] \qquad \text{Simplifying}$$

The solution is:

$$x = \frac{b_2c_1 - b_1c_2}{a_1b_2} = \frac{1}{D}(b_2c_1 - b_1c_2)$$

$$y = \frac{c_2}{b_2} = \frac{a_1c_2}{a_1b_2} = \frac{1}{D}(a_1c_2)$$

$(D = a_1b_2$ if $a_2 = 0)$

This takes care of the case $a_1 \ne 0$, $a_2 = 0$.

Finally, what if $a_1 = 0$? Then

$D = a_1b_2 - a_2b_1 = -a_2b_1$, so $a_2 \ne 0$, $b_1 \ne 0$, and we want to show:

$$x = \frac{1}{D}(c_1b_2 - c_2b_1) = \frac{1}{-a_2b_1}(c_1b_2 - c_2b_1)$$

$$y = \frac{1}{D}(a_1c_2 - a_2c_1) = \frac{1}{-a_2b_1}(-a_2c_1) = \frac{c_1}{b_1}$$

Since $a_1 = 0$, we start with:

$$\left[\begin{array}{cc|c} 0 & b_1 & c_1 \\ a_2 & b_2 & c_2 \end{array}\right] \rightarrow \left[\begin{array}{cc|c} a_2 & b_2 & c_2 \\ 0 & b_1 & c_1 \end{array}\right]$$

$\uparrow$
Interchange rows

$$\rightarrow \left[\begin{array}{cc|c} 1 & \dfrac{b_2}{a_2} & \dfrac{c_2}{a_2} \\[2ex] 0 & 1 & \dfrac{c_1}{b_1} \end{array}\right]$$

$\uparrow$

$$R_1 = \frac{1}{a_2}r_1 \quad (\text{since } a_2 \ne 0)$$

$$R_2 = \frac{1}{b_1}r_2 \quad (\text{since } b_1 \ne 0)$$

Therefore, $y = \dfrac{c_1}{b_1}$, as desired, and by back-substitution,

$$x + \frac{b_2}{a_2}y = \frac{c_2}{a_2}$$

$$x + \frac{b_2}{a_2}\left(\frac{c_1}{b_1}\right) = \frac{c_2}{a_2}$$

$$x = \frac{c_2}{a_2} - \frac{c_1b_2}{a_2b_1}$$

$$x = \frac{c_2b_1 - c_1b_2}{a_2b_1}$$

or

$$x = \frac{-1}{a_2b_1}(c_1b_2 - c_2b_1)$$

as desired.

10 SYSTEMS OF EQUATIONS AND INEQUALITIES

☰ EXERCISE 10.3 SYSTEMS OF LINEAR EQUATIONS: DETERMINANTS

1. $\begin{vmatrix} 3 & 1 \\ 4 & 2 \end{vmatrix} = (3)(2) - (4)(1) = 2$

3. $\begin{vmatrix} 6 & 4 \\ -1 & 3 \end{vmatrix} = (6)(3) - (-1)(4) = 18 + 4 = 22$

5. $\begin{vmatrix} -3 & -1 \\ 4 & 2 \end{vmatrix} = (-3)(2) - (4)(-1) = -2$

7. $\begin{vmatrix} 3 & 4 & 2 \\ 1 & -1 & 5 \\ 1 & 2 & -2 \end{vmatrix} = 3 \begin{vmatrix} -1 & 5 \\ 2 & -2 \end{vmatrix} - 4 \begin{vmatrix} 1 & 5 \\ 1 & -2 \end{vmatrix} + 2 \begin{vmatrix} 1 & -1 \\ 1 & 2 \end{vmatrix}$

$= 3[(-1)(-2) - (2)(5)] - 4[(1)(-2) - (1)(5)]$
$\quad + 2[(1)(2) - (1)(-1)]$
$= 3[2 - 10] - 4[-2 - 5] + 2[2 + 1]$
$= (3)(-8) - (4)(-7) + (2)(3)$
$= -24 + 28 + 6$
$= 10$

9. $\begin{vmatrix} 4 & -1 & 2 \\ 6 & -1 & 0 \\ 1 & -3 & 4 \end{vmatrix} = 4 \begin{vmatrix} -1 & 0 \\ -3 & 4 \end{vmatrix} - (-1) \begin{vmatrix} 6 & 0 \\ 1 & 4 \end{vmatrix} + 2 \begin{vmatrix} 6 & -1 \\ 1 & -3 \end{vmatrix}$

$= 4[(-1)(4) - (-3)(0)] - (-1)[(6)(4) - (1)(0)]$
$\quad + 2[(6)(-3) - (1)(-1)]$
$= 4[-4] - (-1)[24] + 2[-18 + 1]$
$= -16 + 24 - 34$
$= -26$

11. $\begin{cases} x + y = 8 \\ x - y = 4 \end{cases}$

Here, $D = \begin{vmatrix} 1 & 1 \\ 1 & -1 \end{vmatrix} = -1 - 1 = -2$

Since $D \neq 0$, we proceed to find D_x and D_y.

To obtain D_x, replace the first column in D by the constants on the right-hand-side of the original system of equations:

$D_x = \begin{vmatrix} 8 & 1 \\ 4 & -1 \end{vmatrix} = -8 - 4 = -12$

To obtain D_y, replace the second column in D by the constants:

$D_y = \begin{vmatrix} 1 & 8 \\ 1 & 4 \end{vmatrix} = 4 - 8 = -4$

Then by Cramer's Rule,

$$x = \frac{D_x}{D} = \frac{-12}{-2} = 6$$

$$y = \frac{D_y}{D} = \frac{-4}{-2} = 2$$

13. $\begin{cases} 5x - y = 13 \\ 2x + 3y = 12 \end{cases}$

Here, $D = \begin{vmatrix} 5 & -1 \\ 2 & 3 \end{vmatrix} = 15 - (-2) = 17$

Since $D \neq 0$, we find D_x and D_y:

To obtain D_x, replace the first column in D by the constants:

$$D_x = \begin{vmatrix} 13 & -1 \\ 12 & 3 \end{vmatrix} = 39 - (-12) = 51$$

Similarly,

$$D_y = \begin{vmatrix} 5 & 13 \\ 2 & 12 \end{vmatrix} = 60 - 26 = 34$$

Then by Cramer's Rule,

$$x = \frac{D_x}{D} = \frac{51}{17} = 3$$

$$\text{and } y = \frac{D_y}{D} = \frac{34}{17} = 2$$

15. $\begin{cases} 3x \quad\;\; = 24 \\ x + 2y = 0 \end{cases}$

This is <u>easily</u> solved by inspection, but, to use Cramer's Rule:

$$D = \begin{vmatrix} 3 & 0 \\ 1 & 2 \end{vmatrix} = 6$$

$$D_x = \begin{vmatrix} 24 & 0 \\ 0 & 2 \end{vmatrix} = 48$$

$$\text{and } D_y = \begin{vmatrix} 3 & 24 \\ 1 & 0 \end{vmatrix} = -24$$

so that $x = \dfrac{D_x}{D} = \dfrac{48}{6} = 8$

and $y = \dfrac{D_y}{D} = \dfrac{-24}{6} = -4$

17. $\begin{cases} 3x - 6y = 24 \\ 5x + 4y = 12 \end{cases}$

Here, $D = \begin{vmatrix} 3 & -6 \\ 5 & 4 \end{vmatrix} = 12 - (-30) = 42$

$D_x = \begin{vmatrix} 24 & -6 \\ 12 & 4 \end{vmatrix} = 96 - (-72) = 168$

and $D_y = \begin{vmatrix} 3 & 24 \\ 5 & 12 \end{vmatrix} = 36 - 120 = -84$

Therefore,

$$x = \frac{D_x}{D} = \frac{168}{42} = 4$$

and $$y = \frac{D_y}{D} = \frac{-84}{42} = -2$$

19. $\begin{cases} 3x - 2y = 4 \\ 6x - 4y = 0 \end{cases}$

$D = \begin{vmatrix} 3 & -2 \\ 6 & -4 \end{vmatrix} = -12 - (-12) = 0$

Since $D = 0$, we cannot use Cramer's Rule. It is not applicable.

21. $\begin{cases} 3x - 4y = -2 \\ 3x + 2y = 3 \end{cases}$

Here, $D = \begin{vmatrix} 2 & -4 \\ 3 & 2 \end{vmatrix} = 4 - (-12) = 16$

$D_x = \begin{vmatrix} -2 & -4 \\ 3 & 2 \end{vmatrix} = -4 - (-12) = 8$

and $D_y = \begin{vmatrix} 2 & -2 \\ 3 & 3 \end{vmatrix} = 6 - (-6) = 12$

By Cramer's Rule,

$$x = \frac{D_x}{D} = \frac{8}{16} = \frac{1}{2}$$

and $$y = \frac{D_y}{D} = \frac{12}{16} = \frac{3}{4}$$

10.3 SYSTEMS OF LINEAR EQUATIONS: DETERMINANTS   571

23. $\begin{cases} 2x - 3y = -1 \\ 10x + 10y = 5 \end{cases}$

Here, $D = \begin{vmatrix} 2 & -3 \\ 10 & 10 \end{vmatrix} = 20 - (-30) = 50$

$D_x = \begin{vmatrix} -1 & -3 \\ 5 & 10 \end{vmatrix} = -10 - (-15) = 5$

and $D_y = \begin{vmatrix} 2 & -1 \\ 10 & 5 \end{vmatrix} = 10 - (-10) = 20$

By Cramer's Rule,

$$x = \frac{D_x}{D} = \frac{5}{50} = \frac{1}{10}$$

and $$y = \frac{D_y}{D} = \frac{20}{50} = \frac{2}{5}$$

25. $\begin{cases} 2x + 3y = 6 \\ x - y = \dfrac{1}{2} \end{cases}$

Here, $D = \begin{vmatrix} 2 & 3 \\ 1 & -1 \end{vmatrix} = -2 - 3 = -5$

$D_x = \begin{vmatrix} 6 & 3 \\ \frac{1}{2} & -1 \end{vmatrix} = -6 - \frac{3}{2} = -\frac{15}{2}$

and $D_y = \begin{vmatrix} 2 & 6 \\ 1 & \frac{1}{2} \end{vmatrix} = 1 - 6 = -5$

By Cramer's Rule,

$$x = \frac{D_x}{D} = \frac{-\frac{15}{2}}{-5} = \frac{3}{2}$$

and $$y = \frac{D_y}{D} = \frac{-5}{-5} = 1$$

27. $\begin{cases} 3x - 5y = 3 \\ 15x + 5y = 21 \end{cases}$

Here, $D = \begin{vmatrix} 3 & -5 \\ 15 & 5 \end{vmatrix} = 15 - (-5)(15) = 90$

$D_x = \begin{vmatrix} 3 & -5 \\ 21 & 5 \end{vmatrix} = 15 - (-105) = 120$

and $D_y = \begin{vmatrix} 3 & 3 \\ 15 & 21 \end{vmatrix} = (3)(21) - (3)(15) = (3)(21 - 15) = 18$

By Cramer's Rule,

$$x = \frac{D_x}{D} = \frac{120}{90} = \frac{12}{9} = \frac{4}{3}$$

and $\quad y = \frac{D_y}{D} = \frac{18}{90} = \frac{1}{5}$

29. $\begin{cases} x + y - z = 6 \\ 3x - 2y + z = -5 \\ x + 3y - 2z = 14 \end{cases}$

Here,

$$D = \begin{vmatrix} 1 & 1 & -1 \\ 3 & -2 & 1 \\ 1 & 3 & -2 \end{vmatrix} = 1 \begin{vmatrix} -2 & 1 \\ 3 & -2 \end{vmatrix} - 1 \begin{vmatrix} 3 & 1 \\ 1 & -2 \end{vmatrix} + (-1) \begin{vmatrix} 3 & -2 \\ 1 & 3 \end{vmatrix}$$

$$= 1(4 - 3) - (1)(-6 - 1) + (-1)(9 - (-2))$$
$$= 1 - (-7) + (-11)$$
$$= -3$$

To obtain D_x, replace the first column in D by the column of constants:

$$D_x = \begin{vmatrix} 6 & 1 & -1 \\ -5 & -2 & 1 \\ 14 & 3 & -2 \end{vmatrix} = 6 \begin{vmatrix} -2 & 1 \\ 3 & -2 \end{vmatrix} - 1 \begin{vmatrix} -5 & 1 \\ 14 & -2 \end{vmatrix} + (-1) \begin{vmatrix} -5 & -2 \\ 14 & 3 \end{vmatrix}$$

$$= 6(4 - 3) - 1(10 - 14) + (-1)(-15 - (-28))$$
$$= 6 - (-4) + (-13)$$
$$= -3$$

Similarly,

$$D_y = \begin{vmatrix} 1 & 6 & -1 \\ 3 & -5 & 1 \\ 1 & 14 & -2 \end{vmatrix} = 1 \begin{vmatrix} -5 & 1 \\ 14 & -2 \end{vmatrix} - 6 \begin{vmatrix} 3 & 1 \\ 1 & -2 \end{vmatrix} + (-1) \begin{vmatrix} 3 & -5 \\ 1 & 14 \end{vmatrix}$$

$$= 1(10 - 14) - 6(-6 - 1) + (-1)(42 - (-5))$$
$$= -4 - (-42) + (-47)$$
$$= -9$$

Finally,

$$D_z = \begin{vmatrix} 1 & 1 & 6 \\ 3 & -2 & -5 \\ 1 & 3 & 14 \end{vmatrix} = 1 \begin{vmatrix} -2 & -5 \\ 3 & 14 \end{vmatrix} - 1 \begin{vmatrix} 3 & -5 \\ 1 & 14 \end{vmatrix} + 6 \begin{vmatrix} 3 & -2 \\ 1 & 3 \end{vmatrix}$$

$$= 1(-28 - (-15)) - 1(42 - (-5)) + 6(9 - (-2))$$
$$= -13 - 47 + 66$$
$$= 6$$

$$x = \frac{D_x}{D} = \frac{-3}{-3} = 1$$

$$y = \frac{D_y}{D} = \frac{-9}{-3} = 3$$

and $\quad z = \frac{D_z}{D} = \frac{6}{-3} = -2$

31. $\begin{cases} x + 2y - z = -3 \\ 2x - 4y + z = -7 \\ -2x + 2y - 3z = 4 \end{cases}$

$$D = \begin{vmatrix} 1 & 2 & -1 \\ 2 & -4 & 1 \\ -2 & 2 & -3 \end{vmatrix} = 1\begin{vmatrix} -4 & 1 \\ 2 & -3 \end{vmatrix} - 2\begin{vmatrix} 2 & 1 \\ -2 & -3 \end{vmatrix} + (-1)\begin{vmatrix} 2 & -4 \\ -2 & 2 \end{vmatrix}$$

$$= 1(12 - 2) - 2(-6 - (-2)) + (-1)(4 - 8)$$
$$= 10 - (-8) + 4$$
$$= 22$$

$$D_x = \begin{vmatrix} -3 & 2 & -1 \\ -7 & -4 & 1 \\ 4 & 2 & -3 \end{vmatrix} = -3\begin{vmatrix} -4 & 1 \\ 2 & -3 \end{vmatrix} - 2\begin{vmatrix} -7 & 1 \\ 4 & -3 \end{vmatrix} + (-1)\begin{vmatrix} -7 & -4 \\ 4 & 2 \end{vmatrix}$$

$$= -3(12 - 2) - 2(21 - 4) + (-1)(-14 - (-16))$$
$$= -30 - 34 + (-2)$$
$$= -66$$

$$D_y = \begin{vmatrix} 1 & -3 & -1 \\ 2 & -7 & 1 \\ -2 & 4 & -3 \end{vmatrix} = 1\begin{vmatrix} -7 & 1 \\ 4 & -3 \end{vmatrix} - (-3)\begin{vmatrix} 2 & 1 \\ -2 & -3 \end{vmatrix} + (-1)\begin{vmatrix} 2 & -7 \\ -2 & 4 \end{vmatrix}$$

$$= 1(21 - 4) - (-3)(-6 - (-2)) + (-1)(8 - 14)$$
$$= 17 - 12 + 6$$
$$= 11$$

and

$$D_z = \begin{vmatrix} 1 & 2 & -3 \\ 2 & -4 & -7 \\ -2 & 2 & 4 \end{vmatrix} = 1\begin{vmatrix} -4 & -7 \\ 2 & 4 \end{vmatrix} - 2\begin{vmatrix} 2 & -7 \\ -2 & 4 \end{vmatrix} + (-3)\begin{vmatrix} 2 & -4 \\ -2 & 2 \end{vmatrix}$$

$$= 1(-16 - (-14)) - 2(8 - 14) + (-3)(4 - 8)$$
$$= -2 - (-12) + 12$$
$$= 22$$

By Cramer's Rule,

$$x = \frac{D_x}{D} = \frac{-66}{22} = -3$$

$$y = \frac{D_y}{D} = \frac{11}{22} = \frac{1}{2}$$

and $\quad z = \dfrac{D_z}{D} = \dfrac{22}{22} = 1$

33. $\begin{cases} x - 2y + 3z = 1 \\ 3x + y - 2z = 0 \\ 2x - 4y + 6z = 2 \end{cases}$

$$D = \begin{vmatrix} 1 & -2 & 3 \\ 3 & 1 & -2 \\ 2 & -4 & 6 \end{vmatrix} = 1\begin{vmatrix} 1 & -2 \\ -4 & 6 \end{vmatrix} - (-2)\begin{vmatrix} 3 & -2 \\ 2 & 6 \end{vmatrix} + 3\begin{vmatrix} 3 & 1 \\ 2 & -4 \end{vmatrix}$$

$$= 1(-2) - (-2)(22) + 3(-14)$$
$$= 0$$

Since $D = 0$, Cramer's Rule cannot be applied.

574   10 SYSTEMS OF EQUATIONS AND INEQUALITIES

35.
$$\begin{cases} x + 2y - z = 0 \\ 2x - 4y + z = 0 \\ -2x + 2y - 3z = 0 \end{cases}$$

$$D = \begin{vmatrix} 1 & 2 & -1 \\ 2 & -4 & 1 \\ -2 & 2 & -3 \end{vmatrix} = 1 \begin{vmatrix} -4 & 1 \\ 2 & -3 \end{vmatrix} - 2 \begin{vmatrix} 2 & 1 \\ -2 & -3 \end{vmatrix} + (-1) \begin{vmatrix} 2 & -4 \\ -2 & 2 \end{vmatrix}$$
$$= 1(10) - 2(-4) + (-1)(-4)$$
$$= 22$$

$$D_x = \begin{vmatrix} 0 & 2 & -1 \\ 0 & -4 & 1 \\ 0 & 2 & -3 \end{vmatrix} = 0 \begin{vmatrix} -4 & 1 \\ 2 & -3 \end{vmatrix} - 2 \begin{vmatrix} 0 & -1 \\ 0 & -3 \end{vmatrix} + (-1) \begin{vmatrix} 0 & -4 \\ 0 & -3 \end{vmatrix}$$
$$= 0(12 - 2) - 2(0 - 0) + (-1)(0 - 0)$$
$$= 0$$

(We could have used (12) in the text, which states that if any row or column contains only 0's, the value of the determinant is 0.)

$$D_y = \begin{vmatrix} 1 & 0 & -1 \\ 2 & 0 & 1 \\ -2 & 0 & -3 \end{vmatrix} = 0$$

Similarly, $D_z = 0$

Therefore,

$$x = \frac{D_x}{D} = \frac{0}{22} = 0$$

and $y = 0, z = 0$

37.
$$\begin{cases} x - 2y + 3z = 0 \\ 3x + y - 2z = 0 \\ 2x - 4y + 6z = 0 \end{cases}$$

$$D = \begin{vmatrix} 1 & -2 & 3 \\ 3 & 1 & -2 \\ 2 & -4 & 6 \end{vmatrix} = 1 \begin{vmatrix} 1 & -2 \\ -4 & 6 \end{vmatrix} - (-2) \begin{vmatrix} 3 & -2 \\ 2 & 6 \end{vmatrix} + 3 \begin{vmatrix} 3 & 1 \\ 2 & -4 \end{vmatrix}$$
$$= 1(6 - 8) - (-2)(18 - (-4)) + 3(-12 - 2)$$
$$= -2 - (-44) + (-42)$$
$$= 0$$

Since $D = 0$, Cramer's Rule cannot be applied.

39.
$$\begin{cases} \dfrac{1}{x} + \dfrac{1}{y} = 8 \\ \dfrac{3}{x} - \dfrac{5}{y} = 0 \end{cases}$$

$$\begin{cases} \dfrac{1}{x} + \dfrac{1}{y} = 8 \\ 3\left(\dfrac{1}{x}\right) - 5\left(\dfrac{1}{y}\right) = 0 \end{cases}$$

Let $u = \dfrac{1}{x}$ and $v = \dfrac{1}{y}$.   Then we have:

$$u + v = 8$$
$$3u - 5v = 0. \quad \text{Now we solve for } u \text{ and } v:$$

Here, $D = \begin{vmatrix} 1 & 1 \\ 3 & -5 \end{vmatrix} = -5 - 3 = -8$

$$D_u = \begin{vmatrix} 8 & 1 \\ 0 & -5 \end{vmatrix} = -40 - 0 = -40$$

and $D_v = \begin{vmatrix} 1 & 8 \\ 3 & 0 \end{vmatrix} = 0 - 24 = -24$

Therefore, by Cramer's Rule,

$$u = \frac{D_u}{D} = \frac{-40}{-8} = 5$$

and $v = \dfrac{D_v}{D} = \dfrac{-24}{-8} = 3$

But, we are supposed to find x and y.

Since $u = \dfrac{1}{x}$, we have

$$5 = \frac{1}{x}$$
$$5x = 1$$
$$x = \frac{1}{5}$$

Also, $v = \dfrac{1}{y}$

$$3 = \frac{1}{y}$$
$$3y = 1$$
$$y = \frac{1}{3}$$

10 SYSTEMS OF EQUATIONS AND INEQUALITIES

41. Since $\begin{vmatrix} x & x \\ 4 & 3 \end{vmatrix} = 3x - 4x = -x$, we have

$$-x = 5$$
$$x = -5$$

43. $\begin{vmatrix} x & 1 & 1 \\ 4 & 3 & 2 \\ -1 & 2 & 5 \end{vmatrix} = x \begin{vmatrix} 3 & 2 \\ 2 & 5 \end{vmatrix} - 1 \begin{vmatrix} 4 & 2 \\ -1 & 5 \end{vmatrix} + 1 \begin{vmatrix} 4 & 3 \\ -1 & 2 \end{vmatrix}$

$$= x(15 - 4) - 1(20 - (-2)) + 1(8 - (-3))$$
$$= 11x - 22 + 11$$
$$= 11x - 11$$

so we have

$$11x - 11 = 2$$
$$11x = 13$$
$$x = \frac{13}{11}$$

45. $\begin{vmatrix} x & 2 & 3 \\ 1 & x & 0 \\ 6 & 1 & -2 \end{vmatrix} = x \begin{vmatrix} x & 0 \\ 1 & -2 \end{vmatrix} - 2 \begin{vmatrix} 1 & 0 \\ 6 & -2 \end{vmatrix} + 3 \begin{vmatrix} 1 & x \\ 6 & 1 \end{vmatrix}$

$$= x(-2x - 0) - 2(-2 - 0) + 3(1 - 6x)$$
$$= -2x^2 + 4 + 3 - 18x$$
$$= -2x^2 - 18x + 7$$

so we have

$$-2x^2 - 18x + 7 = 7$$
$$-2x^2 - 18x = 0$$
$$-2x(x + 9) = 0$$

so $x = 0$ or $x = -9$

47. Let $D = \begin{vmatrix} x & y & z \\ u & v & w \\ 1 & 2 & 3 \end{vmatrix} = 4$

Then, $\begin{vmatrix} 1 & 2 & 3 \\ u & v & w \\ x & y & z \end{vmatrix} = -4$, because the value of a determinant changes sign if any two rows are interchanged.

49. We try to use row operations to put

$\begin{vmatrix} x & y & z \\ -3 & -6 & -9 \\ u & v & w \end{vmatrix}$ into the form $\begin{vmatrix} x & y & z \\ u & v & w \\ 1 & 2 & 3 \end{vmatrix}$

10.3 SYSTEMS OF LINEAR EQUATIONS: DETERMINANTS   577

Since we know that the value of the determinant on the right is 4.

$$\begin{vmatrix} x & y & z \\ -3 & -6 & -9 \\ u & v & w \end{vmatrix} = -3 \begin{vmatrix} x & y & z \\ 1 & 2 & 3 \\ u & v & w \end{vmatrix} \text{ by (14)}$$

$$= (-3)(-1) \begin{vmatrix} x & y & z \\ u & v & w \\ 1 & 2 & 3 \end{vmatrix} \text{ by (11)}$$

Therefore, $\begin{vmatrix} x & y & z \\ -3 & -6 & -9 \\ u & v & w \end{vmatrix} = (-3)(-1)(4) = 12$

51. Let $D = \begin{vmatrix} x & y & z \\ u & v & w \\ 1 & 2 & 3 \end{vmatrix} = 4$

Now,

$$\begin{vmatrix} 1 & 2 & 3 \\ x-3 & y-6 & z-9 \\ 2u & 2v & 2w \end{vmatrix} = 2 \begin{vmatrix} 1 & 2 & 3 \\ x-3 & y-6 & z-9 \\ u & v & w \end{vmatrix} \text{ by (14)}$$

$$= 2(-1) \begin{vmatrix} x-3 & y-6 & z-9 \\ 1 & 2 & 3 \\ u & v & w \end{vmatrix} \text{ by (11)}$$

$$= 2(-1)(-1) \begin{vmatrix} x-3 & y-6 & z-9 \\ u & v & w \\ 1 & 2 & 3 \end{vmatrix} \text{ by (11)}$$

Note that in this last determinant, row one can be obtained from D by the operation

$$R_1 = -3r_3 + r_1$$

By (15), that operation leaves the value of the determinant unchanged, so that

$$\begin{vmatrix} x-3 & y-6 & z-9 \\ u & v & w \\ 1 & 2 & 3 \end{vmatrix} = \begin{vmatrix} x & y & z \\ u & v & w \\ 1 & 2 & 3 \end{vmatrix} = 4$$

Therefore,

$$\begin{vmatrix} 1 & 2 & 3 \\ x-3 & y-6 & z-9 \\ 2u & 2v & 2w \end{vmatrix} = (2)(-1)(-1)(4) = 8$$

10 SYSTEMS OF EQUATIONS AND INEQUALITIES

53. Let $D = \begin{vmatrix} x & y & z \\ u & v & w \\ 1 & 2 & 3 \end{vmatrix} = 4$

$$\begin{vmatrix} 1 & 2 & 3 \\ 2x & 2y & 2z \\ u-1 & v-2 & w-3 \end{vmatrix} = 2 \begin{vmatrix} 1 & 2 & 3 \\ x & y & z \\ u-1 & v-2 & w-3 \end{vmatrix} \text{ by (14)}$$

$$= 2(-1) \begin{vmatrix} x & y & z \\ 1 & 2 & 3 \\ u-1 & v-2 & w-3 \end{vmatrix} \text{ by (11)}$$

$$= 2(-1)(-1) \begin{vmatrix} x & y & z \\ u-1 & v-2 & w-3 \\ 1 & 2 & 3 \end{vmatrix}$$

Note, in this last determinant, row two can be obtained from D by the operation

$$R_2 = -1r_3 + r_2$$

which leaves the value of the determinant unchanged by (15). In other words,

$$\begin{vmatrix} x & y & z \\ u-1 & v-2 & w-3 \\ 1 & 2 & 3 \end{vmatrix} = \begin{vmatrix} x & y & z \\ u & v & w \\ 1 & 2 & 3 \end{vmatrix} = 4$$

and

$$\begin{vmatrix} 1 & 2 & 3 \\ 2x & 2y & 2z \\ u-1 & v-2 & w-3 \end{vmatrix} = 2(-1)(-1)4 = 8$$

55. $\begin{vmatrix} x^2 & x & 1 \\ y^2 & y & 1 \\ z^2 & z & 1 \end{vmatrix} = x^2 \begin{vmatrix} y & 1 \\ z & 1 \end{vmatrix} - x \begin{vmatrix} y^2 & 1 \\ z^2 & 1 \end{vmatrix} + 1 \begin{vmatrix} y^2 & y \\ z^2 & z \end{vmatrix}$

$$\begin{aligned}
&= x^2(y - z) - x(y^2 - z^2) + 1(y^2z - yz^2) \\
&= x^2(y - z) - x(y - z)(y + z) + yz(y - z) \\
&= (y - z)[x^2 - x(y + z) + yz] \\
&= (y - z)[x^2 - xy - xz + yz] \\
&= (y - z)[x(x - y) - z(x + y)] \\
&= (y - z)(x - y)(x - z)
\end{aligned}$$

as desired.

57. Generally, below is a 3 × 3 determinant.

$$\begin{vmatrix} a_{13} & a_{12} & a_{11} \\ a_{23} & a_{22} & a_{21} \\ a_{33} & a_{32} & a_{31} \end{vmatrix} = a_{13}(a_{22}a_{31} - a_{32}a_{21}) - a_{12}(a_{23}a_{31} - a_{33}a_{21})$$
$$+ a_{11}(a_{23}a_{32} - a_{33}a_{22})$$
$$= a_{13}a_{22}a_{31} - a_{13}a_{32}a_{21} - a_{12}a_{23}a_{31} + a_{12}a_{33}a_{21}$$
$$+ a_{11}a_{23}a_{32} - a_{11}a_{33}a_{22}$$
$$= a_{11}a_{22}a_{33} + a_{11}a_{32}a_{23} + a_{12}a_{21}a_{33} - a_{12}a_{31}a_{23}$$
$$- a_{13}a_{21}a_{32} + a_{13}a_{31}a_{22}$$
$$= [a_{11}(a_{22}a_{33} - a_{32}a_{23}) - a_{12}(a_{21}a_{33} - a_{31}a_{23})$$
$$+ a_{13}(a_{21}a_{32} - a_{31}a_{22})]$$

$$= -\begin{bmatrix} a_{11} & a_{12} & a_{13} \\ a_{21} & a_{22} & a_{23} \\ a_{31} & a_{32} & a_{33} \end{bmatrix}$$

As an example, let $A = \begin{vmatrix} 1 & 3 & 2 \\ -1 & 4 & -3 \\ 2 & 1 & 6 \end{vmatrix}$

Then, $A = 1\begin{vmatrix} 4 & -3 \\ 1 & 6 \end{vmatrix} - 3\begin{vmatrix} -1 & -3 \\ 2 & 6 \end{vmatrix} + 2\begin{vmatrix} -1 & 4 \\ 2 & 1 \end{vmatrix}$

$$= 1(24 - (-3)) - 3(-6 - (-6)) + 2(-1 - 8)$$
$$= 27 - 0 - 18 = 9$$

Interchange columns one and three:

$$B = \begin{vmatrix} 2 & 3 & 1 \\ -3 & 4 & -1 \\ 6 & 1 & 2 \end{vmatrix} = 2\begin{vmatrix} 4 & -1 \\ 1 & 2 \end{vmatrix} - 3\begin{vmatrix} -3 & -1 \\ 6 & 2 \end{vmatrix} + 1\begin{vmatrix} -3 & 4 \\ 6 & 1 \end{vmatrix}$$
$$= 2(8 - (-1)) - 3(-6 - (-6)) + 1(-3 - 24)$$
$$= 18 - 0 - 27 = -9$$

Therefore, $B = -9 = (-1)9 = (-1)A$

59. To give a proof of a theorem, we cannot simply show an example. Instead, let D represent <u>any</u> 3 by 3 determinant in which the entries in column one equal those in column three. Then D will be of the form

$$D = \begin{vmatrix} a & d & a \\ b & e & b \\ c & f & c \end{vmatrix}$$

where a, b, c, d, e, f can be <u>any</u> real numbers.

Then, $D = \begin{vmatrix} a & d & a \\ b & e & b \\ c & f & c \end{vmatrix} = a\begin{vmatrix} e & b \\ f & c \end{vmatrix} - d\begin{vmatrix} b & b \\ c & c \end{vmatrix} + a\begin{vmatrix} b & e \\ c & f \end{vmatrix}$

$$= a(ce - bf) - d(bc - bc) + a(bf - ce)$$
$$= a(ce - bf) - \qquad 0 \qquad + a(bf - ce)$$
$$= ace - abf + abf - ace$$
$$= 0$$

10 SYSTEMS OF EQUATIONS AND INEQUALITIES

≡ HISTORICAL PROBLEM

1.   We wish to solve

$$\begin{cases} x^2 + y^2 = 100 & (1) \\ \quad x = \frac{3}{4}y & (2) \end{cases}$$

This is readily done by substitution, since we know from (2) that $x = \frac{3}{4}y$. Substituting this expression for x into (1) produces:

$$x^2 + y^2 = 100$$

$$\left(\frac{3}{4}y\right)^2 + y^2 = 100$$

$$\frac{9}{16}y^2 + y^2 = 100$$

$$\frac{25}{16}y^2 = 100$$

$$y^2 = \frac{16}{25}(100)$$

$$y = \pm\sqrt{\frac{16(100)}{25}}$$

$$y = \pm\frac{(4)(10)}{5}, \quad \text{or} \quad y = \pm 8$$

Now determine x from (2):

$$x = \frac{3}{4}y$$

If $y = -8$, $x = \frac{3}{4}(-8) = -6$

If $y = 8$, $x = \frac{3}{4}(8) = 6$

Thus, we have two solutions:

$x = -6$ and $y = -8$;
$x = 6$ and $y = 8$

Each solution checks, as you can verify.

≡ EXERCISES

1. $\begin{cases} x + 2y + 3 = 0 & (1) \\ x^2 + y^2 = 5 & (2) \end{cases}$

(a) First we graph (1):

$$x + 2y + 3 = 0$$

Since both x and y appear only to the first power, this represents a straight line. An easy way to graph a line is to find the x-intercept and y-intercept, and then draw a straight line.

For the x-intercept, set $y = 0$:

$$x + 3 = 0$$
$$x = -3$$

For the y-intercept, set $x = 0$:

$$2y + 3 = 0$$
$$2y = -3$$
$$y = -\frac{3}{2}$$

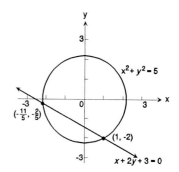

(b) We now graph equation (2):

$$x^2 + y^2 = 5$$

This is the equation of a circle, centered at $(0, 0)$, with $r^2 = 5$, so that $r = \sqrt{5} \approx 2.24$.

(c) Now solve the system (i.e., find the points of intersection). We cannot use elimination, since both x and y are <u>squared</u> in one equation, but not in the other. Instead, we use substitution, solving first for x or y in (1):

$$x + 2y + 3 = 0 \qquad (1)$$
$$x = -2y - 3$$

(It is easier to solve for x than y). Now substitute this expression into (2):

$$x^2 + y^2 = 5$$
$$(-2y - 3)^2 + y^2 = 5$$
$$4y^2 + 12y + 9 + y^2 = 5 \quad \text{This is a quadratic equation.}$$
$$5y^2 + 12y + 4 = 0$$
$$(5y + 2)(y + 2) = 0$$
$$5y + 2 = 0 \quad \text{or} \quad y + 2 = 0$$
$$5y = -2 \qquad\qquad y = -2$$
$$y = \frac{-2}{5}$$

10 SYSTEMS OF EQUATIONS AND INEQUALITIES

Now $x = 2y - 3$, so we have:

$$x = -2\left(-\frac{2}{5}\right) - 3 \quad \text{or} \quad x = -2(-2) - 3$$

$$x = \frac{4}{5} - 3 \qquad\qquad x = 1$$

$$x = -\frac{11}{5}$$

We have two possible solutions:

$$x = -\frac{11}{5}, \; y = -\frac{2}{5}$$

and $x = 1$, $y = -2$

Both solutions check, as you can verify.

3. $\begin{cases} x^2 + y^2 = 4 & (1) \\ y^2 - x = 4 & (2) \end{cases}$

(a) To graph (1):

$$x^2 + y^2 = 4$$

Notice that it is a circle of radius 2, centered at (0, 0).

(b) Now graph (2):

$$y^2 - x = 4$$
$$-x = -y^2 + 4$$
$$x = y^2 - 4$$

This may be graphed by finding points and connecting them with a smooth curve:

y	$x = y^2 - 4$	Point (x, y)
-2	0	(0, -2)
-1	-3	(-3, -1)
0	-4	(-4, 0)
1	-3	(-3, 1)
2	0	(0, 2)

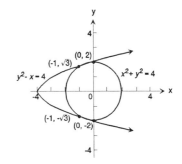

(c) Now solve the system:

$\begin{cases} x^2 + y^2 = 4 & (1) \\ y^2 - x = 4 & (2) \end{cases}$

$\begin{cases} x^2 + y^2 = 4 & (1) \\ -y^2 + x = -4 & (2) \quad \text{Multiply by -1.} \end{cases}$

$\begin{cases} x^2 + y^2 = 4 & (1) \\ x^2 + x = 0 & (2) \quad \text{Replace (2) by (1) + (2).} \end{cases}$

Now solve (2) for x:

$$x^2 + x = 0 \quad (2)$$
$$x(x + 1) = 0$$
$$x = 0 \quad \text{or} \quad x + 1 = 0$$
$$x = -1$$

Now use (1) to find y:

If $x = 0$, then
$$x^2 + y^2 = 4 \quad (1)$$
$$0 + y^2 = 4$$
$$y^2 = 4$$
$$y = \pm 2$$

If $x = -1$, then
$$x^2 + y^2 = 4 \quad (1)$$
$$1 + y^2 = 4$$
$$y^2 = 3$$
$$y = \pm\sqrt{3}$$

We have four possible solutions:

$x = 0$ and $y = -2$;
$x = 0$ and $y = 2$;
$x = -1$ and $y = -\sqrt{3}$;
$x = -1$ and $y = \sqrt{3}$

Let's check these:

For $x = 0$, $y = \pm 2$:

$$0^2 + (\pm 2)^2 = 0 + 4 = 4 \quad (1)$$
$$(\pm 2)^2 - 0 = 4 - 0 = 4 \quad (2)$$

For $x = -1$, $y = \pm\sqrt{3}$:

$$(-1)^2 + (\pm\sqrt{3})^2 = 1 + 3 = 4 \quad (1)$$
$$(\pm\sqrt{3})^2 - (-1) = 3 + 1 = 4 \quad (2)$$

Thus, all four possibilities check.

5.
$$\begin{cases} x^2 + y^2 = 36 & (1) \\ x + y = 8 & (2) \end{cases}$$

(a) Graph (1):
$$x^2 + y^2 = 36$$

This is a circle of radius 6, centered at $(0, 0)$.

(b) Now graph the line $x + y = 8(2)$. First find the intercepts and then draw a straight line through them:

For the x-intercept, set $y = 0$:

$$x + y = 8$$
$$x + 0 = 8$$
$$x = 8$$

For the y-intercept, set $x = 0$:

$$x + y = 8$$
$$y = 8$$

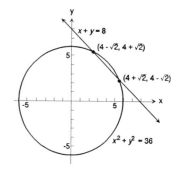

10 SYSTEMS OF EQUATIONS AND INEQUALITIES

(c) Now solve the system:

$$\begin{cases} x^2 + y^2 = 36 & (1) \\ \quad x + y = 8 & (2) \end{cases}$$

Let's use substitution, solving for y in (2):

$$x + y = 8 \qquad (2)$$
$$y = -x + 8$$

Now substitute this expression into (1):

$$x^2 + y^2 = 36 \qquad (1)$$
$$x^2 + (-x + 8)^2 = 36$$
$$x^2 + x^2 - 16x + 64 = 36 \qquad \text{This is a quadratic.}$$
$$2x^2 - 16x + 28 = 0$$

Use the quadratic formula:

$$x = \frac{-b \pm \sqrt{b^2 - 4ac}}{2a}$$

$$x = \frac{16 \pm \sqrt{256 - 4(56)}}{4} = \frac{16 \pm \sqrt{32}}{4}$$

$$= \frac{16 \pm 4\sqrt{2}}{4}$$

or $x = 4 + \sqrt{2}$ and $x = 4 - \sqrt{2}$

Now find y:

If $x = 4 + \sqrt{2}$, then:

$$y = -x + 8 \qquad \text{(from (2))}$$
$$y = -(4 + \sqrt{2}) + 8$$
$$y = 4 - \sqrt{2}$$

If $x = 4 - \sqrt{2}$, then:

$$y = -x + 8 \qquad \text{(from (2))}$$
$$y = -(4 - \sqrt{2}) + 8$$
$$y = 4 + \sqrt{2}$$

We have two <u>possible</u> solutions:

$$x = 4 + \sqrt{2} \text{ and } y = 4 - \sqrt{2}$$
$$\text{and } x = 4 - \sqrt{2} \text{ and } y = 4 + \sqrt{2}$$

Now to check:

For $x = 4 + \sqrt{2}$, $y = 4 - \sqrt{2}$:

$$\begin{cases} \left(4 + \sqrt{2}\right)^2 + \left(4 - \sqrt{2}\right)^2 = 16 + 8\sqrt{2} + 2 + 16 - 8\sqrt{2} + 2 = 36 & (1) \\ \left(4 + \sqrt{2}\right) + \left(4 - \sqrt{2}\right) = 8 & (2) \end{cases}$$

For $x = 4 - \sqrt{2}$ and $y = 4 + \sqrt{2}$:

$$\begin{cases} \left(4 - \sqrt{2}\right)^2 + \left(4 + \sqrt{2}\right)^2 = 16 - 8\sqrt{2} + 2 + 16 + 8\sqrt{2} + 2 = 36 & (1) \\ \left(4 - \sqrt{2}\right) + \left(4 + \sqrt{2}\right) = 8 & (2) \end{cases}$$

Both solutions check.

10.4 SYSTEMS OF NONLINEAR EQUATIONS

7. $\begin{cases} xy = 4 & (1) \\ x^2 + y^2 = 8 & (2) \end{cases}$

(a) To graph (1):
$$xy = 4$$
$$y = \frac{4}{x}$$

Make a table:

x	$y = \dfrac{4}{x}$	Point (x, y)
0	Undef.	no y-intercept
-2	-2	$(-2, -2)$
-1	-4	$(-1, -4)$
1	4	$(1, 4)$
2	2	$(2, 2)$

(b) For (2), $x^2 + y^2 = 8$ is a circle centered at $(0, 0)$ with radius $r = \sqrt{8} \approx 2.83$.

(c) Solve the system:
$$\begin{cases} xy = 4 & (1) \\ x^2 + y^2 = 8 & (2) \end{cases}$$

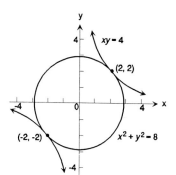

Solve (1) for y:
$$xy = 4 \qquad (1)$$
$$y = \frac{4}{x}$$

Substitute this into (2):
$$x^2 + y^2 = 8$$
$$x^2 + \left(\frac{4}{x}\right)^2 = 8$$
$$x^2 + \frac{16}{x^2} = 8$$
$$x^4 + 16 = 8x^2 \qquad \text{Multiply both sides by } x^2.$$
$$x^4 - 8x^2 + 16 = 0 \qquad \text{This is a disguised quadratic.}$$

Let $u = x^2$ (so that $u^2 = x^4$):
$$u^2 - 8u + 16 = 0$$
$$(u - 4)(u - 4) = 0$$

Therefore, $u = 4$, or
$$x^2 = 4$$
$$x = \pm 2$$

If $x = 2$, then:
$$y = \frac{4}{x} \qquad \text{(from (1))}$$
$$y = \frac{4}{x} = 2$$

If $x = -2$, then:
$$y = \frac{4}{x}$$
$$y = -2$$

We have two possible solutions:

$x = 2$ and $y = 2$;
$x = -2$ and $y = -2$

both of which check.

9. $\begin{cases} x^2 + y^2 = 4 & (1) \\ \quad y = x^2 - 9 & (2) \end{cases}$

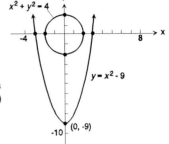

(a) To graph (1), $x^2 + y^2 = 4$, we have a circle of radius 2, centered at $(0, 0)$.

(b) Now graph the parabola $y = x^2 - 9$ with vertex $(0, 9)$ and x-intercepts, $(3, 0)$ and $(-3, 0)$.

(c) Now solve the system:

$\begin{cases} x^2 + y^2 = 4 & (1) \\ \quad y = x^2 - 9 & (2) \end{cases}$

Let's substitute expression (2) into (1):
$$x^2 + (x^2 - 9)^2 = 4$$
$$x^2 + x^4 - 18x^2 + 81 = 4$$
$$x^4 - 17x^2 + 77 = 0 \text{ This is quadratic.}$$

Use the quadratic formula:
$$x^2 = \frac{17 \pm \sqrt{289 - 4(77)}}{2}$$
$$x^2 = \frac{17 \pm \sqrt{289 - 308}}{2}$$
$$= \frac{17 \pm \sqrt{-19}}{2}$$

No real solution. Inconsistent.

11. $\begin{cases} y = x^2 - 4 & (1) \\ y = 6x - 13 & (2) \end{cases}$

(a) Graph (1):
$$y = x^2 - 4$$

This is a parabola with y-intercept (vertex), $(0, -4)$ and x-intercepts, $(-2, 0)$, $(2, 0)$.

(b) Now graph the line $y = 6x - 13(2)$.
First find the intercepts and then
draw a straight line through them.
For the x-intercept, set $y = 0$:

$$0 = 6x - 13$$
$$13 = 6x$$
$$\frac{13}{6} = x$$

For the y-intercept, set $x = 0$:

$$y = 6(0) - 13$$
$$y = -13$$

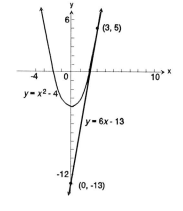

(c) Now solve the system:

$$y = x^2 - 4 \quad (1)$$
$$y = 6x - 13 \quad (2)$$

Substituting equation (2) into equation (1),

$$6x - 13 = x^2 - 4$$
$$x^2 - 6x + 9 = 0$$
$$(x - 3)^2 = 0$$
$$x = 3$$

Now, back-substitute this result into (2),

$$y = 6(3) - 13$$
$$y = 5$$

We have <u>one</u> solution:

$$x = 3 \text{ and } y = 5$$

Now to check:

$$\begin{cases} 5 = (3)^2 - 4 = 5 & (1) \\ 5 = 6(3) - 13 = 5 & (2) \end{cases}$$

The solution checks.

13. $\begin{cases} 2x^2 + y^2 = 18 & (1) \\ xy = 4 & (2) \end{cases}$

Solve equation (2) for y:

$$xy = 4$$
$$y = \frac{4}{x}$$

Substitute this into (1):

$$2x^2 + y^2 = 18$$
$$2x^2 + \left(\frac{4}{x}\right)^2 = 18$$
$$2x^2 + \frac{16}{x^2} = 18$$
$$2x^4 + 16 = 18x^2 \quad \text{Multiply both sides by } x^2.$$
$$x^4 + 8 = 9x^2 \quad \text{Divide all terms by 2.}$$
$$x^4 - 9x^2 + 8 = 0 \quad \text{This is a disguised quadratic.}$$

10 SYSTEMS OF EQUATIONS AND INEQUALITIES

Let $u = x^2$:
$$u^2 - 9u + 8 = 0$$
$$(u - 1)(u - 8) = 0$$

Therefore, $u = 1$ or $u = 8$, so that
$$x^2 = 1 \text{ or } x^2 = 8, \text{ and}$$
$$x = -1, \ 1, \ -\sqrt{8}, \ \sqrt{8}$$

If $x = -1$, then $y = \dfrac{4}{x} = -4$

If $x = 1$, then $y = \dfrac{4}{x} = 4$

If $x = -\sqrt{8} = -2\sqrt{2}$, then $y = \dfrac{4}{x} = \dfrac{4}{-2\sqrt{2}} = -\sqrt{2}$

If $x = \sqrt{8} = 2\sqrt{2}$, then $y = \sqrt{2}$

We have four <u>possible</u> solutions:

$x = -1$ and $y = -4$;
$x = 1$ and $y = 4$
$x = -2\sqrt{2}$ and $y = -\sqrt{2}$
and $x = 2\sqrt{2}$ and $y = \sqrt{2}$

All four check, as you may verify.

15. $\begin{cases} 3x - y = 1 & (1) \\ x^2 + 4y^2 = 17 & (2) \end{cases}$

Solve for y in equation (1):
$$3x - y = 1$$
$$-y = -3x + 1$$
$$y = 3x - 1$$

Substitute this expression for y into (2):
$$x^2 + 4y^2 = 17$$
$$x^2 + 4(3x - 1)^2 = 17$$
$$x^2 + 4(9x^2 - 6x + 1) = 17$$
$$x^2 + 36x^2 - 24x + 4 = 17$$
$$37x^2 - 24x - 13 = 0$$
$$(37x + 13)(x - 1) = 0$$

So we have $x = \dfrac{-13}{37}$ or $x = 1$.

If $x = \dfrac{-13}{37}$, then:

$y = 3x - 1$ from (1)

$y = 3\left(\dfrac{-13}{37}\right) - 1$

$y = \dfrac{-39}{37} - \dfrac{37}{37} = \dfrac{-76}{37}$

If $x = 1$, then:

$y = 3x - 1$
$y = 2$

10.4 SYSTEMS OF NONLINEAR EQUATIONS   589

We have two possible solutions:

$$x = \frac{-13}{37} \text{ and } y = \frac{-76}{37}$$
$$\text{and } x = 1 \text{ and } y = 2$$

Let's check these:

For $x = \frac{-13}{37}$, $y = \frac{-76}{37}$:

$$\begin{cases} 3\left(\frac{-13}{37}\right) - \left(\frac{-76}{37}\right) = \frac{-39}{37} + \frac{76}{37} = \frac{37}{37} = 1 & (1) \\ \left(\frac{-13}{37}\right)^2 + 4\left(\frac{-76}{37}\right)^2 = \frac{169}{(37)^2} + \frac{4(5776)}{(37)^2} = \frac{23273}{(37)^2} = 17 & (2) \end{cases}$$

For $x = 1$, $y = 2$:

$$3(1) - 2 = 1 \qquad (1)$$
$$(1)^2 + 4(2)^2 = 17 \qquad (2)$$

Both solutions check:

17. $$\begin{cases} x + y + 1 = 0 & (1) \\ x^2 + y^2 + 6y - x = 7 & (2) \end{cases}$$

We could add (1) and (2) and eliminate the x-terms, but we would still have an x^2-term. Instead, let's solve (1) for x:

$$x + y + 1 = 0$$
$$x = -y - 1$$

Now substitute this expression for x into (2):

$$x^2 + y^2 + 6y - x = 7$$
$$(-y - 1)^2 + y^2 + 6y - (-y - 1) = 7$$
$$y^2 + 2y + 1 + y^2 + 6y + y + 1 = 7 \quad \text{This is a quadratic.}$$
$$2y^2 + 9y - 5 = 0$$
$$(2y - 1)(y + 5) = 0$$

So we have $y = \frac{1}{2}$ or $y = -5$.

If $y = \frac{1}{2}$, then:

$$x = -y - 1 \qquad \text{from (1)}$$
$$x = -\frac{1}{2} - 1 = \frac{-3}{2}$$

If $y = -5$, then:

$$x = -y - 1$$
$$x = 4$$

We have two possible solutions:

$$x = \frac{-3}{2} \text{ and } y = \frac{1}{2};$$
$$x = 4 \text{ and } y = -5$$

You can verify that both of these solutions check in the original system of equations.

19. $\begin{cases} 4x^2 - 3xy + 9y^2 = 15 & (1) \\ 2x + 3y = 5 & (2) \end{cases}$

Solve for x in (2):

$$2x + 3y = 5$$
$$2x = -3y + 5$$
$$x = \frac{-3}{2}y + \frac{5}{2}$$

Substitute this expression into (1):

$$4x^2 - 3xy + 9y^2 = 15$$

$$4\left(\frac{-3}{2}y + \frac{5}{2}\right)^2 - 3\left(\frac{-3}{2}y + \frac{5}{2}\right)y + 9y^2 = 15$$

$$4\left(\frac{9}{4}y^2 - \frac{15}{2}y + \frac{25}{4}\right) - 3\left(\frac{-3}{2}y^2 + \frac{5}{2}y\right) + 9y^2 = 15$$

$$9y^2 - 30y + 25 + \frac{9}{2}y^2 - \frac{15}{2}y + 9y^2 = 15$$

$$18y^2 - 60y + 50 + 9y^2 - 15y + 18y^2 = 30$$

$$45y^2 - 75y + 20 = 0$$

$$9y^2 - 15y + 4 = 0$$

$$(3y - 1)(3y - 4) = 0$$

So $y = \frac{1}{3}$ or $y = \frac{4}{3}$

If $y = \frac{1}{3}$, then:

$$x = \frac{-3}{2}y + \frac{5}{2} \qquad \text{from (2)}$$

$$x = \frac{-3}{2}\left(\frac{1}{3}\right) + \frac{5}{2} = 2$$

If $y = \frac{4}{3}$, then:

$$x = \frac{-3}{2}y + \frac{5}{2}$$

$$x = \frac{-3}{2}\left(\frac{4}{3}\right) + \frac{5}{2} = \frac{1}{2}$$

The two possible solutions are:

$$x = 2 \text{ and } y = \frac{1}{3};$$

$$x = \frac{1}{2} \text{ and } y = \frac{4}{3}$$

Both of these solutions check.

21. $\begin{cases} x^2 - 4y^2 + 7 = 0 & (1) \\ 3x^2 + y^2 - 31 = 0 & (2) \end{cases}$

Here we can use the method of elimination:

$\begin{cases} x^2 - 4y^2 + 7 = 0 & (1) \\ 12x^2 + 4y^2 - 124 = 0 & (2) \quad \text{Multiply both sides by 4.} \end{cases}$

$$\begin{cases} x^2 - 4y^2 + 7 = 0 & (1) \\ 13x^2 - 117 = 0 & (2) \end{cases} \quad \text{Replace (2) by (1) + (2).}$$

Now solve (2) for x:

$$13x^2 - 117 = 0$$
$$13x^2 = 117$$
$$x^2 = 9$$

so, $x = 3$ or $x = -3$

If $x = 3$, then:

$$x^2 - 4y^2 + 7 = 0 \quad (1)$$
$$9 - 4y^2 + 7 = 0$$
$$-4y^2 = -16$$
$$y^2 = 4$$
$$y = \pm2$$

If $x = -3$, then:

$$x^2 - 4y^2 + 7 = 0$$
$$9 - 4y^2 + 7 = 0$$
$$-4y^2 = -16$$
$$y^2 = 4$$
$$y = \pm2$$

We have four possible solutions:

$x = 3, \ y = 2;$
$x = 3, \ y = -2;$
$x = -3, \ y = 2;$
$x = -3, \ y = -2$

We can check all four at once. If $x = \pm3$, and $y = \pm2$:

$$\begin{cases} x^2 - 4y^2 + 7 = 9 - 16 + 7 = 0 & (1) \\ 3x^2 + y^2 - 31 = 27 + 4 - 31 = 0 & (2) \end{cases}$$

All four solutions check.

23. $\begin{cases} 7x^2 - 3y^2 + 5 = 0 & (1) \\ 3x^2 + 5y^2 - 12 = 0 & (2) \end{cases}$

Let's solve (2) for $\underline{x^2}$:

$$3x^2 + 5y^2 - 12 = 0$$
$$3x^2 = -5y^2 + 12$$
$$x^2 = \frac{-5}{3}y^2 + 4$$

Substitute this expression for x^2 into (1):

$$7x^2 - 3y^2 + 5 = 0$$
$$7\left(\frac{-5}{3}y^2 + 4\right) - 3y^2 + 5 = 0$$
$$\frac{-35}{3}y^2 + 28 - 3y^2 + 5 = 0$$
$$-35y^2 + 84 - 9y^2 + 15 = 0 \qquad \text{Multiply by 3.}$$
$$-44y^2 = -99$$
$$y^2 = \frac{99}{44} = \frac{9}{4}$$
$$y = \pm\frac{3}{2}$$

If $y = \frac{3}{2}$, then:

$$x^2 = \frac{-5}{3}y^2 + 4 \qquad \text{from (2)}$$
$$x^2 = \frac{-5}{3}\left(\frac{9}{4}\right) + 4$$
$$x^2 = \frac{1}{4}$$
$$x = \pm\frac{1}{2}$$

If $y = \frac{-3}{2}$, then:

$$x^2 = \frac{-5}{3}y^2 + 4$$
$$x^2 = \frac{-5}{3}\left(\frac{9}{4}\right) + 4, \text{ as above}$$
$$x = \pm\frac{1}{2}$$

We have four solutions:

$$x = \frac{1}{2},\ y = \frac{3}{2};\ x = -\frac{1}{2},\ y = \frac{3}{2}$$
$$x = \frac{1}{2},\ y = \frac{-3}{2};\ x = \frac{-1}{2},\ y = \frac{-3}{2}$$

(All of them check.)

25. $\begin{cases} x^2 + 2xy = 10 & (1) \\ 3x^2 - xy = 2 & (2) \end{cases}$

$\begin{cases} x^2 + 2xy = 10 & (1) \\ 6x^2 - 2xy = 4 & (2) \quad \text{Multiply by 2.} \end{cases}$

$\begin{cases} x^2 + 2xy = 10 & (1) \\ 7x^2 = 14 & (2) \quad \text{Replace (2) by (1) + (2).} \end{cases}$

From (2): $\quad x^2 = 2$

$$x = \pm\sqrt{2}$$

If $x = \sqrt{2}$, then:

$$x^2 + 2xy = 10 \qquad (1)$$
$$2 + 2\sqrt{2}\,y = 10$$
$$2\sqrt{2}\,y = 8$$
$$y = \frac{8}{2\sqrt{2}} = 2\sqrt{2}$$

If $x = -\sqrt{2}$, then:

$$x^2 + 2xy = 10$$
$$2 - 2\sqrt{2}\,y = 10$$
$$-2\sqrt{2}\,y = 8$$
$$y = \frac{8}{-2\sqrt{2}} = -2\sqrt{2}$$

The two possible solutions are:

$$x = \sqrt{2},\ y = 2\sqrt{2} \text{ and } x = -\sqrt{2},\ y = -2\sqrt{2}$$

Both solutions check:

27. $$\begin{cases} 2x^2 + y^2 = 2 & (1) \\ x^2 - 2y^2 + 8 = 0 & (2) \end{cases}$$

$$\begin{cases} 4x^2 + 2y^2 = 4 & (1) \quad \text{Multiply equation (1) by 2.} \\ x^2 - 2y^2 = -8 & (2) \end{cases}$$

Adding equation (1) and (2),

$$5x^2 = -4$$
$$x^2 = \frac{-4}{5}$$

No solution. The system is inconsistent.

29. $$\begin{cases} x^2 + 2y^2 = 16 & (1) \\ 4x^2 - y^2 = 24 & (2) \end{cases}$$

$$\begin{cases} x^2 + 2y^2 = 16 & (1) \\ 8x^2 - 2y^2 = 48 & (2) \quad \text{Multiply by 2.} \end{cases}$$

$$\begin{cases} x^2 + 2y^2 = 16 & (1) \\ 9x^2 = 64 & (2) \quad \text{Replace(2) by (1) + (2).} \end{cases}$$

From (2):

$$x^2 = \frac{64}{9}$$
$$x = \pm\frac{8}{3}$$

If $x = \pm\frac{8}{3}$, then:

$$x^2 + 2y^2 = 16 \qquad (1)$$
$$\frac{64}{9} + 2y^2 = 16$$
$$2y^2 = \frac{144}{9} - \frac{64}{9} = \frac{80}{9}$$
$$y^2 = \frac{40}{9}$$

and $y = \pm\sqrt{\dfrac{40}{9}} = \pm\dfrac{-2\sqrt{10}}{3}$

We have four solutions, all of which check:

$$x = \frac{8}{3},\ y = \frac{2\sqrt{10}}{3};\ x = \frac{8}{3},\ y = \frac{-2\sqrt{10}}{3};$$

$$x = \frac{-8}{3},\ y = \frac{2\sqrt{10}}{3};\ x = \frac{-8}{3},\ y = \frac{-2\sqrt{10}}{3}$$

31.
$$\begin{cases} \dfrac{5}{x^2} - \dfrac{2}{y^2} + 3 = 0 & (1) \\[3mm] \dfrac{3}{x^2} + \dfrac{1}{y^2} - 7 = 0 & (2) \end{cases}$$

$$\begin{cases} \dfrac{5}{x^2} - \dfrac{2}{y^2} + 3 = 0 & (1) \\[3mm] \dfrac{6}{x^2} + \dfrac{2}{y^2} - 14 = 0 & (2) \quad \text{Multiply both sides by 2.} \end{cases}$$

$$\begin{cases} \dfrac{5}{x^2} - \dfrac{2}{y^2} + 3 = 0 & (1) \\[3mm] \dfrac{11}{x^2} - 11 = 0 & (2) \quad \text{Replace (2) by (1) + (2).} \end{cases}$$

Now, from (2):

$$\frac{11}{x^2} = 11$$
$$11 = 11x^2$$
$$x^2 = 1$$
$$x = \pm 1$$

If $x = 1$ or $x = -1$, we have:

$$\frac{5}{x^2} - \frac{2}{y^2} + 3 = 0 \quad (9((1)$$
$$\frac{5}{1} - \frac{2}{y^2} + 3 = 0 \quad (\text{since } x = \pm 1)$$
$$8 = \frac{2}{y^2}$$
$$8y^2 = 2$$
$$y^2 = \frac{1}{4}$$
$$y = \pm \frac{1}{2}$$

We have four solutions, all of which can be checked:

$$x = 1,\ y = \frac{1}{2};\ x = 1,\ y = \frac{-1}{2};$$

$$x = -1,\ y = \frac{1}{2};\ x = -1,\ y = \frac{-1}{2}$$

33.

$$\begin{cases} \dfrac{1}{x^4} + \dfrac{6}{y^4} = 6 & (1) \\[2mm] \dfrac{2}{x^4} - \dfrac{2}{y^4} = 19 & (2) \end{cases}$$

Let's make a substitution: let $u = \dfrac{1}{x^4}$ and $v = \dfrac{1}{y^4}$. Then we have:

$$\begin{cases} u + 6v = 6 & (1) \\ 2u - 2v = 19 & (2) \end{cases}$$

$$\begin{cases} u + 6v = 6 & (1) \\ 6u - 6v = 57 & (2) \quad \text{Multiply each term by 3.} \end{cases}$$

$$\begin{cases} u + 6v = 6 & (1) \\ 7u = 63 & (2) \quad \text{Replace (2) by (1) + (2).} \end{cases}$$

From (2):
$$7u = 63$$
$$u = 9$$

Then, from (1),
$$u + 6v = 6$$
$$9 + 6v = 6$$
$$6v = -3$$
$$v = \frac{-1}{2}$$

Recall that $u = \dfrac{1}{x^4}$ and $v = \dfrac{1}{y^4}$. Thus,

$$\frac{1}{y^4} = v = \frac{-1}{2}$$
$$2 = -y^4$$
$$y^4 = -2$$

This is impossible, since an even power can never be negative, so we have no real solution. The system is inconsistent.

35.

$$\begin{cases} x^2 - 3xy + 2y^2 = 0 & (1) \\ x^2 + xy = 6 & (2) \end{cases}$$

Since both x and y appear in two terms of equation (1), we cannot use elimination. We can either solve for y in (2), or factor (1). Let's try the latter approach:

$$x^2 - 3xy + 2y^2 = 0 \qquad (1)$$
$$(x - y)(x - 2y) = 0$$

Therefore, either $x = y$ or $x = 2y$.

If $x = y$, then from (2):
$$x^2 + xy = 6$$
$$y^2 + y^2 = 6 \quad (x = y)$$
$$y^2 = 3$$
$$y = \pm\sqrt{3}$$

We know $x = y$, so we have two possible solutions: $x = \sqrt{3}$, $y = \sqrt{3}$ and $x = -\sqrt{3}$, $y = -\sqrt{3}$.

If $x = 2y$, then

$$x^2 + xy = 6 \qquad (2)$$
$$4y^2 + 2y^2 = 6 \qquad (x = 2y)$$
$$y^2 = 1$$
$$y = \pm 1$$

Then, from $x = 2y$, we have two more possible solutions:

$$x = 2, \ y = 1 \text{ and } x = -2, \ y = -1$$

You can check that all four solutions are valid.

37. $\begin{cases} xy - x^2 = -3 & (1) \\ 3xy - 4y^2 = 2 & (2) \end{cases}$

Solve equation (1) for y:

$$xy - x^2 = -3 \qquad (1)$$
$$xy = x^2 - 3$$
$$y = \frac{x^2 - 3}{x}, \text{ provided } x \neq 0.$$

We will check the possibility $x = 0$ later.

Substitute $y = \frac{x^2 - 3}{x}$ into (2):

$$3xy - 4y^2 = 2 \qquad (2)$$
$$3x\left(\frac{x^2 - 3}{x}\right) - 4\left(\frac{x^2 - 3}{x}\right)^2 = 2$$
$$3x^2 - 9 - \frac{4(x^2 - 3)^2}{x^2} = 2$$
$$3x^4 - 9x^2 - 4(x^4 - 6x^2 + 9) = 2x^2 \quad \text{Multiply by } x^2.$$
$$3x^4 - 9x^2 - 4x^4 + 24x^2 - 36 - 2x^2 = 0$$
$$-x^4 + 13x^2 - 36 = 0$$
$$x^4 - 13x^2 + 36 = 0$$
$$(x^2 - 4)(x^2 - 9) = 0$$
$$x^2 = 4 \quad \text{or} \quad x^2 = 9$$

$$x = \pm 2 \quad \text{or} \quad x = \pm 3$$

If $x = 2$, then $\qquad y = \frac{x^2 - 3}{x}$

or $\qquad\qquad\quad y = \frac{4 - 3}{2} = \frac{1}{2}$

If $x = -2$, then $\qquad y = \frac{x^2 - 3}{x}$

or $\qquad\qquad\quad y = \frac{4 - 3}{-2} = \frac{-1}{2}$

If $x = 3$, then $\quad y = \dfrac{x^2 - 3}{x}$

$\quad\quad$ or $\quad\quad\quad\quad y = \dfrac{9 - 3}{3} = 2$

If $x = -3$, then $\quad y = \dfrac{x^2 - 3}{x}$

$\quad\quad$ or $\quad\quad\quad\quad y = \dfrac{9 - 3}{-3} = -2$

This gives us four possible solutions. What about the possibility $x = 0$? From (1) we would have $0 = -3$. Thus, there are no solutions for which $x = 0$. Each of the four solutions above checks:

$$x = 2, \ y = \frac{1}{2}; \ x = -2, \ y = \frac{-1}{2},$$
$$x = 3, \ y = 2; \ x = -3, \ y = -2$$

39. $\begin{cases} x^3 - y^3 = 26 & (1) \\ x - y = 2 & (2) \end{cases}$

Solve (2) for x:

$$x - y = 2$$
$$x = y + 2$$

and substitute this into (1):

$$x^3 - y^3 = 26$$
$$(y + 2)^3 - y^3 = 26$$
$$y^3 + 6y^2 + 12y + 8 - y^3 = 26$$
$$6y^2 + 12y - 18 = 0$$
$$y^2 + 2y - 3 = 0$$
$$(y + 3)(y - 1) = 0$$

so $y = -3$ or $y = 1$.

If $y = -3$, then $x = y + 2 = -1$

If $y = 1$, then $x = y + 2 = 3$.

The two solutions (both of which check) are $x = -1$, $y = -3$, and $x = 3$, $y = 1$.

41. $\begin{cases} y^2 + y + x^2 - x - 2 = 0 & (1) \\ y + 1 + \dfrac{x - 2}{y} = 0 & (2) \end{cases}$

$\begin{cases} y^2 + y + x^2 - x - 2 = 0 & (1) \\ -y^2 - y - x + 2 = 0 & (2) \quad \text{Multiply both sides by } -y. \end{cases}$

$\begin{cases} y^2 + y + x^2 - x - 2 = 0 & (1) \\ x^2 - 2x = 0 & (2) \quad \text{Replace (2) by (1) + (2).} \end{cases}$

From (2): $\quad x - 2x = 0$
$$x(x - 2) = 0$$
$$x = 0 \quad \text{or} \quad x = 2$$

Now use (1):

If $x = 0$: $y^2 + y + x^2 - x - 2 = 0$ (1)
$$y^2 + y - 2 = 0$$
$$(y + 2)(y - 1) = 0$$
and $y = -2$ or $y = 1$

If $x = 2$: $y^2 + y + x^2 - x - 2 = 0$ (1)
$$y^2 + y + 4 - 2 - 2 = 0$$
$$y^2 + y = 0$$
$$y(y + 1) = 0$$
and $y = 0$ or $y = -1$

Thus, we have four possible solutions:

$x = 0$, $y = -2$; $x = 0$, $y = 1$;
$x = 2$, $y = 0$; and $x = 2$, $y = -1$

We see from equation (2) that y cannot be zero. That eliminates the third solution above. The others can be checked:

$x = 0$, $y = -2$
$x = 0$, $y = 1$
$x = 2$, $y = -1$

43. $\begin{cases} \log_x y = 3 \\ \log_x (4y) = 5 \end{cases}$

$\begin{cases} y = x^3 \\ 4y = x^5 \end{cases}$

$$4x^3 = x^5$$
$$x^5 - 4x^3 = 0$$
$$x^3(x^2 - 4) = 0$$
$$x = 0, \ x = -2, \ x = 2$$

0 and -2 are extraneous (the base of a logarithm must be positive). Thus, $x = 2$ and $y = 2^3 = 8$.

45. Let x and y be the two numbers. Then we have:

$\begin{cases} x + y = 8 & (1) \\ x^2 + y^2 = 36 & (2) \end{cases}$

Solve (1) for y: $x + y = 8$
$$y = 8 - x$$

Substitute this into (2):

$$x^2 + y^2 = 36 \quad (2)$$
$$x^2 + (8 - x)^2 = 36$$
$$x^2 + 64 - 16x + x^2 = 36$$
$$2x^2 - 16x + 28 = 0$$
$$x^2 - 8x + 14 = 0$$

Here we must use the quadratic formula:

$$x = \frac{8 \pm \sqrt{64 - 4(14)}}{2} = \frac{8 \pm \sqrt{8}}{2} = 4 \pm \sqrt{2}$$

If $x = 4 + \sqrt{2}$, then $y = 8 - x = 4 - \sqrt{2}$

10.4 SYSTEMS OF NONLINEAR EQUATIONS

599

If $x = 4 - \sqrt{2}$, then $y = 8 - x = 4 + \sqrt{2}$

Thus, either way, the two numbers are $4 + \sqrt{2}$ and $4 - \sqrt{2}$.
Obviously, their sum is 8. Check the sum of squares:

$$(4 + \sqrt{2})^2 + (4 - \sqrt{2})^2 = 16 + 8\sqrt{2} + 2 + 16 - 8\sqrt{2} + 2$$
$$= 16 + 2 + 16 + 2$$
$$= 36 .$$

so the solution checks.

47. Let x and y be the two numbers. Then we have:

$$\begin{cases} xy = 7 & (1) \\ x^2 + y^2 = 50 & (2) \end{cases}$$

Solve for y in (1): $y = \dfrac{7}{x}$

Then from (2):
$$x^2 + y^2 = 50$$
$$x^2 + \left(\dfrac{7}{x}\right)^2 = 50$$
$$x^2 + \dfrac{49}{x^2} = 50$$
$$x^4 + 49 = 50x^2 \quad \text{Multiply by } x^2.$$
$$x^4 - 50x^2 + 49 = 0$$
$$(x^2 - 49)(x^2 - 1) = 0$$
$$x^2 = 49 \quad \text{or} \quad x^2 = 1$$
$$x = \pm 7 \quad \text{or} \quad x = \pm 1$$

Recall that $y = \dfrac{7}{x}$.

If $x = 7$, then $y = 1$

If $x = -7$, then $y = -1$

If $x = 1$, then $y = 7$

If $x = -1$, then $y = -7$

Thus, we have <u>two</u> pairs of numbers: 7 and 1, or -7 and -1. Both
pairs solve the original problem.

49. Let x and y be the two numbers. Then:

$$\begin{cases} x - y = xy & (1) \\ \dfrac{1}{x} + \dfrac{1}{y} = 5 & (2) \end{cases}$$

$$\begin{cases} x - y = xy & (1) \\ y + x = 5xy & (2) \end{cases} \quad \text{Multiply by } xy.$$

$$\begin{cases} x - y = xy & (1) \\ 2x = 6xy & (2) \end{cases} \quad \text{Replace (2) by (1) + (2).}$$

From (2) we have:
$$2x - 6xy = 0$$
$$x - 3xy = 0$$
$$x(1 - 3y) = 0$$

10 SYSTEMS OF EQUATIONS AND INEQUALITIES

So either $x = 0$ or $y = \frac{1}{3}$. But from the original equation (2), we see that $x = 0$ is impossible. Thus, $y = \frac{1}{3}$.

From (1):

$$x - y = xy \quad (1)$$

$$x - \frac{1}{3} = \frac{1}{3}x \quad \left(y = \frac{1}{3}\right)$$

$$\frac{2}{3}x = \frac{1}{3}$$

$$x = \frac{1}{2}$$

Thus, the two numbers are $x = \frac{1}{2}$ and $y = \frac{1}{3}$.

51. $\begin{cases} \dfrac{a}{b} = \dfrac{2}{3} \\ a + b = 10 \end{cases}$

$$\frac{10 - b}{b} = \frac{2}{3}$$
$$3(10 - b) = 2b$$
$$30 = 5b$$
$$b = 6$$
$$a = 4$$

$a + b = 10, \quad b - a = 2$

Ratio of $a + b$ to $a - b$ is $\dfrac{10}{2} = 5$.

53. $\begin{cases} y = x^2 + 1 \\ y = x + 1 \end{cases}$

$$x^2 + 1 = x + 1$$
$$x^2 - x = 0$$
$$x(x - 1) = 0$$

$$x = 0, \quad x = 1$$

$$x = 0, \quad y = 1; \quad x = 1, \quad y = 2$$

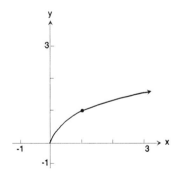

55.
$$\begin{cases} y = \sqrt{36 - x^2} \\ x = 8 - x \end{cases}$$

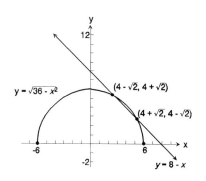

$$\sqrt{36 - x^2} = 8 - x$$
$$36 - x^2 = 64 - 16x + x^2$$

$$2x^2 - 16x + 28 = 0$$
$$x^2 - 8x + 14 = 0$$
$$x = \frac{8 \pm \sqrt{64 - 56}}{2}$$
$$x = \frac{8 \pm 2\sqrt{2}}{2}$$
$$x = 4 \pm \sqrt{2}$$
$$x = 8 - x$$
$$x = 8 - (4 \pm \sqrt{2})$$
$$y = 4 \mp \sqrt{2}$$
$$\left(4 + \sqrt{2},\ 4 - \sqrt{2}\right) \text{ and } \left(4 - \sqrt{2},\ 4 + \sqrt{2}\right)$$

57.
$$\begin{cases} y = \sqrt{x} \\ y = 2 - x \end{cases}$$

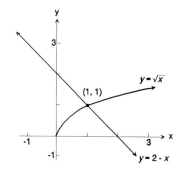

$$\sqrt{x} = 2 - x$$
$$x = 4 - 4x + x^2$$
$$x^2 - 5x + 4 = 0$$
$$(x - 1)(x - 4) = 0$$

If $x = 1$, then $y = 1$.

$$(1,\ 1)$$

59.
$$\begin{cases} x = 2y \\ x = y^2 - 2y \end{cases}$$

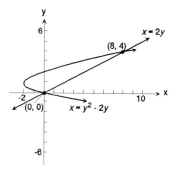

$$2y = y^2 - 2y$$
$$y^2 - 4y = 0$$
$$y(y - 4) = 0$$
$$y = 0,\ y = 4$$

If $y = 0$, $x = 0$; if $y = 4$, $x = 8$

$$(0,\ 0) \text{ and } (8,\ 4)$$

10 SYSTEMS OF EQUATIONS AND INEQUALITIES

61. ℓ = length, w = width

$$\begin{cases} 2\ell + 2w = 16 & (1) \quad \text{(Perimeter)} \\ \ell w = 15 & (2) \quad \text{(Area)} \end{cases}$$

$2\ell + 2\left(\dfrac{15}{\ell}\right) = 16$ Substituting $w = \dfrac{15}{\ell}$ from (2) into equation (1).

$2\ell^2 - 16\ell + 30 = 0$ Multiply by ℓ.

$2(\ell^2 - 8\ell + 15) = 0$

$2(\ell - 5)(\ell - 3) = 0$

$\ell = 5 \qquad\qquad \ell = 3$

$w = \dfrac{15}{\ell} = 3 \quad w = \dfrac{15}{3} = 5$

The dimensions are 3 inches × 5 inches, or 5 inches × 3 inches.

63. Let the piece of cardboard have width x, and length y. Then its area is $A = xy$. From the drawing, we see that the width of the box will be $w = x - 4$, its length will be $\ell = y - 4$, and its height will be $h = 2$. Therefore, its volume is

$$V = \ell \cdot w \cdot h = 2(x - 4)(y - 4)$$

We have:

$$\begin{cases} xy = 216 & (1) \\ 2(x - 4)(y - 4) = 224 & (2) \end{cases}$$

Solve (1) for y: $y = \dfrac{216}{x}$

Then from (2):

$$2(x - 4)(y - 4) = 224$$

$$(2x - 8)\left(\dfrac{216}{x} - 4\right) = 224$$

$$432 - 8x - \dfrac{1728}{x} + 32 = 224$$

$$432x - 8x^2 - 1728 + 32x = 224x \quad \text{Multiply by } x.$$

$$-8x^2 + 240x - 1728 = 0$$

$$x^2 - 30x + 216 = 0 \qquad \text{Divide by } -8.$$

With some trial and error, this can be factored:

$$(x - 12)(x - 18) = 0$$

so $x = 12$ or $x = 18$

If $x = 12$, then $y = \dfrac{216}{x} = 18$

If $x = 18$, then $y = \dfrac{216}{x} = 12$

Thus, the cardboard should be 12 centimeters by 18 centimeters.

65. Let x be the length of the two equal sides (an isosceles triangle has two sides equal); let y be the base, and let h be the altitude drawn to the base. We want to determine y, and we are given:

$$h = 3$$
$$\text{and} \quad x + x + y = 18$$

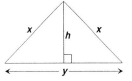

From the Pythagorean Theorem, applied to <u>half</u> of the triangle, we have:

$$\left(\frac{y}{2}\right)^2 + h^2 = x^2$$

$$\frac{y^2}{4} + 9 = x^2 \quad (h = 3)$$

$$y^2 + 36 = 4x^2$$

This gives two equations in x and y:

$$\begin{cases} 2x + y = 18 & (1) \\ y^2 + 36 = 4x^2 & (2) \end{cases}$$

We are asked to find y, so let's eliminate x:

From (1): $2x + y = 18$

$$x = \frac{-y}{2} + 9$$

Then from (2):

$$y^2 + 36 = 4\left(\frac{-y}{2} + 9\right)^2$$

$$y^2 + 36 = 4\left(\frac{y^2}{4} - \frac{18}{2}y + 81\right)$$

$$y^2 + 36 = y^2 - 36y + 324$$

$$36y = 288$$

$$y = 8$$

Thus, the base of the triangle is $y = 8$ cm.

67. Let v_1, v_2, v_3 = speeds of runners 1, 2, 3
     t_1, t_2, t_3 = times of runners 1, 2, 3

$$\begin{cases} 5280 = v_1 t_1 \\ 5270 = v_2 t_1 \\ 5260 = v_3 t_1 \\ 5280 = v_2 t_2 \end{cases}$$

Distance between 2nd runner and 3rd runner after t_2 seconds is:

$$5280 - v_3 t_2 = 5280 - \left(\sqrt{3}\, t_1\right)\left(\frac{t_2}{t_1}\right)$$

$$= 5280 - 5260\left(\frac{5280}{5270}\right)$$

$$= 10.02 \text{ feet}$$

69. $$M^2 - 4(2M - 4) = 0$$
     $$M^2 - 8M + 16 = 0$$
     $$(M - 4)^2 = 0$$
     $$M = 4$$

Now, using the point-slope equation with slope 4 and the point, $(2, 4)$,

$$y - 4 = 4(x - 2)$$
$$y = 4x - 8 + 4$$
$$y = 4x - 4$$

71. Refer to Problem 69 for the method to be used. We want the system

$$\begin{cases} y = x^2 + 2 & (1) \\ y = mx + b & (2) \end{cases}$$

to have one solution. Substitute $y = mx + b$ into the first equation:

$$mx + b = x^2 + 2$$
$$-x^2 + mx + b - 2 = 0$$
$$x^2 - mx + 2 - b = 0 \quad \text{(a quadratic)}$$

Here $A = 1$, $B = -m$ and $C = 2 - b$. The quadratic formula will produce just one solution provided $B^2 - 4AC = 0$, i.e.,

$$(-m)^2 - 4(1)(2 - b) = 0$$
$$m^2 - 8 + 4b = 0 \quad (1)$$

We also want $(1, 3)$ to lie on the line $y = mx + b$, so that $3 = m(1) + b$. We have two equations:

$$\begin{cases} m^2 - 8 + 4b = 0 & (1) \\ 3 = m + b & (2) \end{cases}$$

From (2), $b = 3 - m$, so, by (1):

$$m^2 - 8 + 4b = 0 \quad (1)$$
$$m^2 - 8 + 4(3 - m) = 0$$
$$m^2 - 8 + 12 - 4m = 0$$
$$m^2 - 4m + 4 = 0$$
$$(m - 2)(m - 2) = 0$$

so that $m = 2$. Then $b = 3 - m - 1$, and the equation of the tangent line is: $y = mx + b$, or

$$y = 2x + 1$$

73. Refer to Problems 69 and 71 to see the method used. The system

$$\begin{cases} 2x^2 + 3y^2 = 14 & (1) \\ y = mx + b & (2) \end{cases}$$

is to have just one solution. Substitute $y = mx + b$ into the first equation:

$$2x^2 + 3y^2 = 14$$
$$2x^2 + 3(mx + b)^2 = 14$$
$$2x^2 + 3(m^2x^2 + 2mbx + b^2) = 14$$
$$2x^2 + 3m^2x^2 + 6mbx + 3b^2 - 14) = 0$$
$$(2 + 3m^2)x^2 + (6mb)x + (3b^2 - 14) = 0 \text{ (a quadratic)}$$

Here $A = 2 + 3m^2$, $B = 6mb$ and $C = 3b^2 - 14$.

There will be one solution to the quadratic if $B^2 - 4AC = 0$, i.e.,

$$(6mb)^2 - 4(2 + 3m^2)(3b^2 - 14) = 0$$
$$36m^2b^2 - 4(6b^2 - 28 + 9m^2b^2 - 42m^2) = 0$$
$$36m^2b^2 - 24b^2 + 112 - 36m^2b^2 + 168m^2 = 0$$
$$-24b^2 + 112 + 168m^2 = 0$$
$$3b^2 - 14 - 21m^2 = 0 \text{ (1) Divide by } -8.$$

10.4 SYSTEMS OF NONLINEAR EQUATIONS

605

We also want $(1, 2)$ to lie on the line $y = mx + b$, i.e, we need

$\qquad 2 = m + b \qquad (2)$

We have two equations to solve:

$$\begin{cases} 3b^2 - 14 - 21m^2 = 0 & (1) \\ \qquad\qquad 2 = m + b & (2) \end{cases}$$

From (2), $b = 2 - m$, and by (1),

$$\begin{aligned} 3b^2 - 14 - 21m^2 &= 0 \\ 3(2 - m)^2 - 14 - 21m^2 &= 0 \\ 3(4 - 4m + m^2) - 14 - 21m^2 &= 0 \\ 12 - 12m + 3m^2 - 14 - 21m^2 &= 0 \\ -18m^2 - 12m - 2 &= 0 \\ 9m^2 + 6m + 1 &= 0 \\ (3m + 1)(3m + 1) &= 0 \end{aligned}$$

so $m = \dfrac{-1}{3}$ and $b = 2 - m = \dfrac{7}{3}$, and the equation of the tangent line is

$$y = mx + b, \text{ or}$$
$$y = \frac{-1}{3}x + \frac{7}{3}$$

75. We want the system

$$\begin{cases} x^2 - y^2 = 3 \\ \qquad y = mx + b \end{cases}$$

to have just <u>one</u> solution.

Substitute $y = mx + b$ into the first equation:

$$\begin{aligned} x^2 - y^2 &= 3 \\ x^2 - (mx + b)^2 &= 3 \\ x^2 - (m^2x^2 + 2mbx - b^2) &= 3 \\ x^2 - m^2x^2 + 2mbx - b^2 &= 3 \\ (1 - m^2)x^2 + (-2mb)x + (-b^2 - 3) &= 0 \end{aligned}$$

Let $A = 1 - m^2$, $B = -2mb$, $C = -b^2 - 3$

We want $B^2 - 4AC = 0$, or

$$\begin{aligned} (-2mb)^2 - 4(1 - m^2)(-b^2 - 3) &= 0 \\ 4m^2b^2 - 4(-b^2 - 3 + m^2b^2 + 3m^2) &= 0 \\ 4b^2 + 12 - 12m^2 &= 0 \\ b^2 + 3 - 3m^2 &= 0 \qquad (1) \end{aligned}$$

We want $(2, 1)$ to lie on the line $y = mx + b$:

$\qquad 1 = 2m + b \qquad (2)$

This gives us the system:

$$\begin{cases} b^2 + 3 - 3m^2 = 0 & (1) \\ \qquad\quad 1 = 2m + b & (2) \end{cases}$$

From (2): $b = 1 - 2m$, and we have

$$b^2 + 3 - 3m^2 = 0 \qquad (1)$$
$$(1 - 2m)^2 + 3 - 3m^2 = 0$$
$$1 - 4m + 4m^2 + 3 - 3m^2 = 0$$
$$m^2 - 4m + 4 = 0$$
$$(m - 2)(m - 2) = 0$$

so $m = 2$, $b = 1 - 2m = -3$, and the tangent line is

$$y = mx + b, \text{ or}$$
$$y = 2x - 3$$

77. Let ℓ and w be the length and width of a rectangle. Then the area, A, and perimeter, P, are given by:

$$\begin{cases} A = \ell w & (1) \\ P = 2\ell + 2w & (2) \end{cases}$$

We solve for ℓ and w, treating A and P as constants:

From (2), $2\ell = P - 2w$

$$\ell = \frac{P}{2} - w$$

Then from (1),

$$\ell = A \qquad (1)$$
$$\left(\frac{P}{2} - 3\right)w = A$$
$$\frac{P}{2}w - w^2 = A$$
$$-w^2 + \frac{P}{2}w - A = 0$$
$$w^2 - \frac{P}{2}w + A = 0$$

This is a quadratic, in w, with $a = 1$, $b = \frac{-P}{2}$ and $c = A$.

Then,

$$w = \frac{-b \pm \sqrt{b^2 - 4ac}}{2a}$$

$$= \frac{\frac{P}{2} \pm \sqrt{\frac{P^2}{4} - 4A}}{2}$$

$$= \frac{\frac{P}{2} \pm \sqrt{\frac{P^2 - 16A}{4}}}{2}$$

$$= \frac{\frac{P}{2} - \frac{\sqrt{P^2 - 16A}}{2}}{2}$$

or

$$w = \frac{P \pm \sqrt{P^2 - 16A}}{4} \qquad \text{Multiplying top and bottom by 2.}$$

Recall that $\ell = \frac{P}{2} - w$.

If $w = \dfrac{P \pm \sqrt{P^2 - 16A}}{4}$, then we have

$$\ell = \dfrac{P}{2} - \dfrac{P}{4} + \dfrac{\sqrt{P^2 - 16A}}{4}$$

or $\ell = \dfrac{P}{4} - \dfrac{\sqrt{P^2 - 16A}}{4} = \dfrac{P - \sqrt{P^2 - 16A}}{4}$

This gives a length <u>smaller</u> than the width, so we reject this solution.

If $w = \dfrac{P - \sqrt{P^2 - 16A}}{4}$, then we obtain

$$\ell = \dfrac{P + \sqrt{P^2 - 16A}}{4},$$

and these are the proper formulas.

79. $\begin{cases} x + 2y = 0 & (1) \\ (x - 1)^2 + (y - 1)^2 = 5 & (2) \end{cases}$

The line $x + 2y = 0$ can be rewritten as

$$y = \dfrac{-1}{2}x,$$

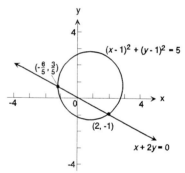

so we see it is a line with slope $m = \dfrac{-1}{2}$ which passes through $(0, 0)$.

Also, $(x - 1)^2 + (y - 1)^2 = 5$ is a circle with center $C(1, 1)$ and radius $r = \sqrt{5} \approx 2.24$.

To find the points of intersection, substitute $y = \dfrac{-1}{2}x$ (from (1)) into (2):

$$(x - 1)^2 + (y - 1)^2 = 5 \quad (2)$$
$$(x - 1)^2 + \left(\dfrac{-1}{2}x - 1\right)^2 = 5$$
$$x^2 - 2x + 1 + \dfrac{1}{4}x^2 + x + 1 = 5$$
$$\dfrac{5}{4}x^2 - x - 3 = 0$$
$$5x^2 - 4x - 12 = 0 \quad \text{Multiply by 4.}$$
$$(5x + 6)(x - 2) = 0$$
$$\text{so } x = \dfrac{-6}{5}, \text{ or } x = 2$$

If $x = \dfrac{-6}{5}$, then $y = \dfrac{-1}{2}x = \dfrac{3}{5}$

If $x = 2$, then $y = \dfrac{-1}{2}x = -1$

Both solutions check, so we have two points of intersection:

$$\left(\dfrac{-6}{5}, \dfrac{3}{5}\right) \text{ and } (2, -1)$$

81. $\begin{cases} (x - 1)^2 + (y + 2)^2 = 4 & (1) \\ y^2 + 4y - x + 1 = 0 & (2) \end{cases}$

Equation (1) is a circle with center at $C(1, -2)$ and radius $r = 2$. To graph the parabola, (2), let's complete the square and then plot points:

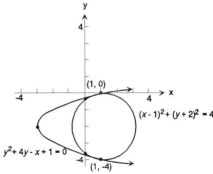

$y^2 + 4y + \underline{} = x - 1 + \underline{}$
$y^2 + 4y + 4 = x - 1 + 4$
$(y + 2)^2 = x + 3$
$ x = (y + 2)^2 - 3$

y	$x = (y + 2)^2 - 3$	(x, y)
-4	1	$(1, -4)$
-3	-2	$(-2, -3)$
-2	-3	$(-3, -2)$
-1	-2	$(-2, -1)$
0	1	$(1, 0)$

To solve the system, let's use the standard form we obtained for the parabola:

$\begin{cases} (x - 1)^2 + (y + 2)^2 = 4 & (1) \\ (y + 2)^2 - 3 = x & (2) \end{cases}$

$\begin{cases} (x - 1)^2 + 3 = 4 - x & (1) \quad \text{Replace (1) by (1) - (2)} \\ (y + 2)^2 - 3 = x & (2) \end{cases}$

From (1): $\quad (x - 1)^2 + 3 = 4 - x$
$x - 2x + 1 + 3 = 4 - x$
$x^2 - x = 0$
$x(x - 1) = 0$
$x = 0 \text{ or } x = 1$

If $x = 0$, then by (2): $(y + 2)^2 - 3 = x$
$(y + 2)^2 - 3 = 0$
$(y + 2)^2 = 3$
$y + 2 = \pm\sqrt{3}$
$y = -2 \pm \sqrt{3}$

If $x = 1$, then: $\quad (y + 2)^2 - 3 = 1$
$(y + 2)^2 = 4$
$y + 2 = \pm 2$
$y = -2 \pm 2$

i.e., $y = 0$ or $y = -4$

We have <u>four</u> points of intersection:

$(0, -2 - \sqrt{3})$, $(0, -2 + \sqrt{3})$, $(1, 0)$, and $(1, -4)$.

83. $\begin{cases} y = \dfrac{4}{x - 3} & (1) \\ x^2 - 6x + y^2 + 1 = 0 & (2) \end{cases}$

To graph (1), plot points:

x	$y = \dfrac{4}{x - 3}$	(x, y)
6	$\dfrac{4}{3}$	$\left(6, \dfrac{4}{3}\right)$
4	$\dfrac{4}{1}$	$(4, 4)$
3	Undefined	
2	$\dfrac{4}{-1}$	$(2, -4)$
0	$\dfrac{4}{-3}$	$\left(0, -\dfrac{4}{3}\right)$

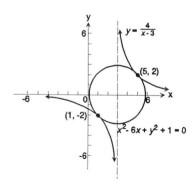

To graph (2), complete the square:

$$x^2 - 6x + \underline{} + y^2 = -1 + \underline{}$$
$$x^2 - 6x + 9 + y^2 = -1 + 9$$
$$(x - 3)^2 + y^2 = 8$$

This is a circle with center at $C(3, 0)$ and radius $r = \sqrt{8} \approx 2.83$. Now solve the system:

$$\begin{cases} y = \dfrac{4}{x - 3} & (1) \\ x^2 - 6x + y^2 + 1 = 0 & (2) \end{cases}$$

We could substitute $y = \dfrac{4}{x - 3}$ into (2), but then we obtain a complicated denominator, $(x - 3)^2 = x^2 - 6x + 9$.

Instead, let's solve (1) for x:

$$y(x - 3) = 4$$
$$x - 3 = \dfrac{4}{y}$$
$$x = \dfrac{4}{y} + 3$$

and substitute this into (2):

$$x^2 - 6x + y^2 + 1 = 0$$
$$\left(\dfrac{4}{y} + 3\right)^2 - 6\left(\dfrac{4}{y} + 3\right) + y^2 + 1 = 0$$
$$\dfrac{16}{y^2} + \dfrac{24}{y} + 9 - \dfrac{24}{y} - 18 + y^2 + 1 = 0$$
$$\dfrac{16}{y^2} + y^2 - 8 = 0$$
$$16 + y^4 - 8y^2 = 0 \quad \text{Multiply by } y^2.$$
$$y^4 - 8y^2 + 16 = 0$$
$$(y^2 - 4)(y^2 - 4) = 0$$

　　　　10 SYSTEMS OF EQUATIONS AND INEQUALITIES

so $y^2 = 4$
$y = \pm 2$

If $y = 2$, then $x = \dfrac{4}{y} + 3 = 5$

If $y = -2$, then $x = \dfrac{4}{y} + 3 = 1$

We have two points of intersection:

$(5, 2)$ and $(1, -2)$

85. Solve for r_1 and r_2:

$$\begin{cases} r_1 + r_2 = \dfrac{-b}{a} & (1) \\[2mm] r_1 r_2 = \dfrac{c}{a} & (2) \end{cases}$$

From (1),

$$r_1 = -r_2 - \frac{b}{a}, \text{ and}$$

$$r_1 r_2 = \frac{c}{a} \quad (2)$$

$$\left(-r_2 - \frac{b}{a}\right) r_2 = \frac{c}{a}$$

$$-r_2^2 - \frac{b}{a} r_2 - \frac{c}{a} = 0$$

$$a r_2^2 + b r_2 + c = 0 \quad \text{Multiply by } -a.$$

By the quadratic formula,

$$r_2 = \frac{-b \pm \sqrt{b^2 - 4ac}}{2a}$$

Then,

$$r_1 = -r_2 - \frac{b}{a}$$

$$= -\left(\frac{-b \pm \sqrt{b^2 - 4ac}}{2a}\right) - \frac{2b}{2a}$$

$$= \frac{b \mp \sqrt{b^2 - 4ac} - 2b}{2a}$$

$$= \frac{-b \mp \sqrt{b^2 - 4ac}}{2a}$$

Thus, we have one pair of numbers:

$$\frac{-b + \sqrt{b^2 - 4ac}}{2a} \text{ and } \frac{-b - \sqrt{b^2 - 4ac}}{2a}$$

≡ EXERCISE 10.5 SYSTEMS OF LINEAR INEQUALITIES

1. $x \geq 0$:

    Step 1: Graph the line $x = 0$ (a vertical line through the origin, i.e., the y-axis). Use a solid line, since the inequality is nonstrict.

    Step 2: Choose a test point not on the line, say, the point (2, 0), which is to the right of the line.

    Step 3: For this point we have $x = 2 \geq 0$, so that the inequality is satisfied.

    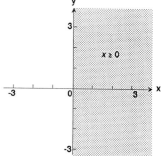

    Step 4: Therefore, we shade to the right of the line $x = 0$.

3. $x + y \geq 4$:

    Step 1: Graph $x + y = 4$, using a solid line (for a nonstrict inequality). The x-intercept is 4, and the y-intercept is 4.

    Step 2: Choose a test point say, (0, 0), which is below the line.

    Step 3: For this point (0, 0), we have:

    $$x + y = 0 + 0 = 0 < 4$$

    so the inequality is not satisfied.

    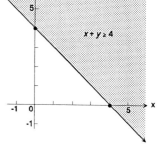

    Step 4: Therefore, we shade the region <u>above</u> the line $x + y = 4$.

5. $2x + y \geq 6$:

    Step 1: Graph $2x + y = 6$, using a <u>solid</u> line. The x-intercept is $x = 3$, and the y-intercept is $y = 6$.

    Step 2: Choose a test point, say, (0, 0), which is below the line.

    Step 3: For this point (0, 0), we have:

    $$2x + y = 0 < 6$$

    so the point does <u>not</u> satisfy the inequality.

    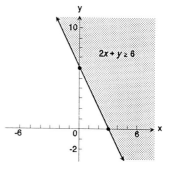

    Step 4: Therefore, we shade the region <u>above</u> the line $2x + y = 6$.

7. $4x - y \geq 4$:

    Step 1: Graph $4x - y = 4$, using a solid line. The x-intercept is $x = 1$, and the y-intercept is $y = -4$.

    Step 2: Choose a test point, say, $(0, 0)$, which is above the line.

    Step 3: For this point $(0, 0)$, we have:

$$4x - y = 0 < 4$$

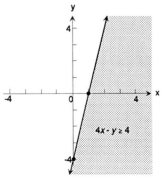

    so the point does <u>not</u> satisfy the inequality.

    Step 4: Therefore, we shade the region <u>below</u> the line $4x - y = 4$.

9. $3x - 2y \geq -6$:

    Step 1: Graph $3x - 2y = -6$, using a solid line. The x-intercept is $x = -2$, and the y-intercept is $y = 3$.

    Step 2: Choose a test point: $(0, 0)$ is <u>below</u> the line.

    Step 3: At $(0, 0)$, we have:

$$3x - 2y = 0 > -6$$

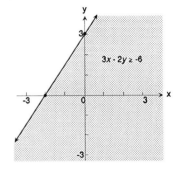

    so the point <u>does</u> satisfy the inequality.

    Step 4: Therefore, we shade the region below the line $3x - 2y = -6$.

11. $3x + 4y \geq 0$:

    Step 1: Graph $3x + 4y = 0$, using a solid line. This is a line through $(0, 0)$ with a slope of $\frac{-3}{4}$ (since we can rewrite the equation as $y = \frac{-3}{4}x$).

    Step 2: We cannot use $(0, 0)$ as a test point since it is <u>on</u> the line. Let's try $(1, 0)$, which is above the line.

    Step 3: At the point $(1, 0)$, we have:

$$3x + 4y = 3 > 0$$

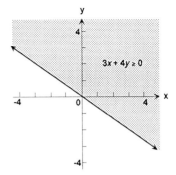

    so the point <u>does</u> satisfy the inequality.

    Step 4: We shade the entire region <u>above</u> the line $3x + 4y = 0$.

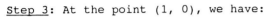

10.5 SYSTEMS OF LINEAR INEQUALITIES   613

13. $\begin{cases} x + y \le 2 & (1) \\ 2x + y \ge 4 & (2) \end{cases}$

(a) $x + y \le 2$:

Step 1: Graph $x + y = 2$ with a solid line. The x-intercept is $x = 2$, and the y-intercept is $y = 2$.

Step 2: We will use $(0, 0)$, below the line, as a test point.

Step 3: At $(0, 0)$, $x + y = 0 < 2$, so the inequality is satisfied.

Step 4: Shade the region below the line $x + y = 2$.

(b) $2x + y \ge 4$:

Step 1: Graph $2x + y = 4$ with a solid line. The x-intercept is $x = 2$, and the y-intercept is $y = 4$.

Step 2: Use $(0, 0)$, below the line, as a test point.

Step 3: At $(0, 0)$, $2x + y = 0 < 4$, so the inequality is <u>not</u> satisfied.

Step 4: Shade the region above the line.

(c) Where the two shaded regions overlap is the graph of the system.

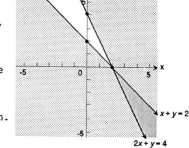

15. $\begin{cases} 2x - y \le 4 & (1) \\ 3x + 2y \ge -6 & (2) \end{cases}$

(a) $2x - y \le 4$:

Step 1: Graph $2x - y = 4$ with a solid line. The x-intercept is $x = 2$, and the y-intercept is $y = -4$.

Step 2: We will use $(0, 0)$ as a test point (above the line).

Step 3: At $(0, 0)$, $2x - y = 0 < 4$.

Step 4: Shade the region above the line $2x - y = 4$.

(b) $3x + 2y \ge -6$:

Step 1: Graph $3x + 2y = -6$ with a solid line. The x-intercept is $x = -2$, and the y-intercept is $y = -3$.

Step 2: Use $(0, 0)$, above the line.

Step 3: At $(0, 0)$, $3x + 2y = 0 > -6$.

Step 4: Shade the region above the line.

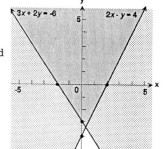

17. $\begin{cases} 2x - 3y \le 0 & (1) \\ 3x + 2y \le 6 & (2) \end{cases}$

(a) $2x - 3y \le 0$:

Step 1: Graph $2x - 3y = 0$, a line through $(0, 0)$ with slope $= \dfrac{2}{3}$.

Step 2: Let's use $(1, 0)$ below the line.

Step 3: At $(1, 0)$, $2x - 3y = 2 > 0$, so the point does not satisfy the inequality.

Step 4: Shade the region <u>above</u> the line $2x - 3y = 0$.

(b) $3x + 2y \le 6$:

Step 1: Graph $3x + 2y = 6$ with a solid line. The x-intercept is $x = 2$, and the y-intercept is $y = 3$.

Step 2: We can use $(0, 0)$, below the line.

Step 3: At $(0, 0)$, $3x + 2y = 0 < 6$.

Step 4: Shade the region below the line $3x + 2y = 6$.

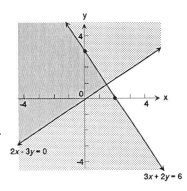

19. $\begin{cases} x - 2y \le 6 & (1) \\ 2x - 4y \ge 0 & (2) \end{cases}$

(a) $x - 2y \le 6$:

Step 1: Graph $x - 2y = 6$, using a solid line. The x-intercept is $x = 6$; the y-intercept is $y = -3$.

Step 2: Use $(0, 0)$ above the line.

Step 3: At $(0, 0)$, $x - 2y = 0 < 6$.

Step 4: Shade the region <u>above</u> the line $x - 2y = 6$.

(b) $2x - 4y \ge 0$:

Step 1: Graph $2x - 4y = 0$ with a solid line. This is a line through $(0, 0)$ with slope $= \dfrac{1}{2}$.

Step 2: Let's use $(1, 0)$ below the line.

Step 3: At $(1, 0)$, $2x - 4y = 2 > 0$.

Step 4: Shade the region <u>below</u> $2x - 4y = 0$.

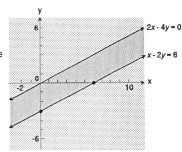

10.5 SYSTEMS OF LINEAR INEQUALITIES

21. $\begin{cases} 2x + y \geq -2 & (1) \\ 2x + y \geq 2 & (2) \end{cases}$

(a) $2x + y \geq -2$:

    <u>Step 1</u>: Graph $2x + y = -2$, with a solid line. The x-intercept is $x = -1$; the y-intercept is $y = -2$.

    <u>Step 2</u>: Use $(0, 0)$ above the line.

    <u>Step 3</u>: At $(0, 0)$, $2x + y = 0 > -2$, so the inequality is satisfied.

    <u>Step 4</u>: Shade the region above the line $2x + y = -2$.

(b) $2x + y \geq 2$:

    <u>Step 1</u>: Graph $2x + y = 2$ with a solid line. The x-intercept is $x = 1$; the y-intercept is $y = 2$.

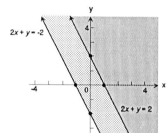

    <u>Step 2</u>: Use $(0, 0)$ <u>below</u> the line.

    <u>Step 3</u>: At $(0, 0)$, $2x + y = 0 < 2$, so $(0, 0)$ does <u>not</u> satisfy the inequality.

    <u>Step 4</u>: Therefore, shade the region <u>above</u> the line $2x + y = 2$.

23. $\begin{cases} 2x + 3y \geq 6 & (1) \\ 2x + 3y \leq 0 & (2) \end{cases}$

This system has no solution, because $2x + 3y$ cannot be greater than 6 <u>and</u> less than 0 at the same time.

(a) $2x + 3y \geq 6$:

    <u>Step 1</u>: Graph $2x + 3y = 6$, with a solid line. The x-intercept is $x = 3$; the y-intercept is $y = 2$.

    <u>Step 2</u>: Use $(0, 0)$ below the line as a test point.

    <u>Step 3</u>: At $(0, 0)$, $2x + 3y = 0 < 6$, so the inequality is <u>not</u> satisfied.

    <u>Step 4</u>: Shade the region <u>above</u> the line $2x + 3y = 6$.

(b) $2x + 3y \leq 0$:

    <u>Step 1</u>: Graph $2x + 3y = 0$ with a solid line. This is a line through $(0, 0)$ with slope $= \dfrac{-2}{3}$.

10 SYSTEMS OF EQUATIONS AND INEQUALITIES

Step 2: Use (1, 0) above the line
as a test point.

Step 3: At (1, 0), $2x + 3y = 2 > 0$.

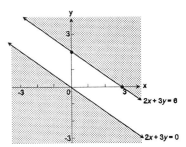

Step 4: Therefore, we shade the
region <u>below</u> the line
$2x + 3y = 0$.

(c) Notice that the two shaded regions
do <u>not</u> overlap. Thus, the system
of inequalities has no solution.

25.
$$\begin{cases} x \geq 0 & (1) \\ y \geq 0 & (2) \\ 2x + y \leq 6 & (3) \\ x + 2y \leq 6 & (4) \end{cases}$$

(a) $x \geq 0$; $y \geq 0$:

These two inequalities require that our shaded region must be
restricted to quadrant I.

(b) $2x + y \leq 6$:

Step 1: Graph $2x + y = 6$ with a solid line. The x-intercept
is $x = 3$; the y-intercept is $y = 6$.

Step 2: Use (0, 0) below the line as a test point.

Step 3: At (0, 0), $2x + y = 0 < 6$, so the inequality is
satisfied.

Step 4: Shade the region below the line $2x + y = 6$.

(c) $x + 2y \leq 6$:

Step 1: Graph $x + 2y = 6$ with a solid line. The x-intercept
is $x = 6$; the y-intercept is $y = 3$.

Step 2: Let's use (0, 0) <u>below</u> the
line as a test point.

Step 3: At (0, 0), $x + 2y = 0 < 6$.

Step 4: Therefore, shade the region
<u>below</u> the line $x + 2y = 6$.

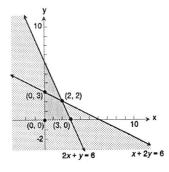

(d) The graph is bounded.

(e) To list the vertices, consult the
graph. We see that there are four
vertices:

(1) Intersection of x-axis and y-axis: (0, 0).
(2) Intersection of $x + 2y = 6$ and y-axis: (0, 3).

(3) Intersection of $x + 2y = 6$ and $2x + y = 6$. To find this point of intersection, it will be necessary to solve a system of equations:

$$\begin{cases} x + 2y = 6 & (1) \\ 2x + y = 6 & (2) \end{cases}$$

$$\begin{cases} -2x - 4y = -12 & (1) \quad \text{Multiply by } -2. \\ 2x + y = 6 & (2) \end{cases}$$

$$\begin{cases} -3y = -6 & (1) \quad \text{Replace (1) by (1) + (2).} \\ 2x + y = 6 & (2) \end{cases}$$

$$\begin{cases} y = 2 & (1) \quad \text{Divide by } -3. \\ 2x + y = 6 & (2) \end{cases}$$

$$\begin{cases} y = 2 & (1) \\ 2x + 2 = 6 & (2) \quad \text{Back-substitution: } y = 2. \end{cases}$$

$$\begin{cases} y = 2 & (1) \\ x = 2 & (2) \end{cases}$$

Thus, the point of intersection is $(2, 2)$.

(4) The intersection of $2x + y = 6$ and the x-axis: $(3, 0)$

So, the four vertices are: $(0, 0)$, $(0, 3)$, $(2, 2)$, and $(3, 0)$.

27. $$\begin{cases} x \geq 0 & (1) \\ y \geq 0 & (2) \\ x + y \geq 2 & (3) \\ 2x + y \geq 4 & (4) \end{cases}$$

(a) $x \geq 0$; $y \geq 0$:

These inequalities require that the final shaded area be in quadrant I.

(b) $x + y \geq 2$:

Step 1: Graph $x + y = 2$ with a solid line. The x-intercept is $x = 2$; the y-intercept is $y = 2$.

Step 2: Use $(0, 0)$ below the line as a test point.

Step 3: At $(0, 0)$, $x + y = 0 < 2$, so $(0, 0)$ does not satisfy the inequality.

Step 4: Shade the region above the line $x + y = 2$.

(c) $2x + y \geq 4$:

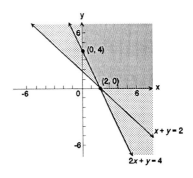

Step 1: Graph $2x + y = 4$. The x-intercept is 2; the y-intercept is 4.

Step 2: Use $(0, 0)$ <u>below</u> the line as a test point.

Step 3: At $(0, 0)$, $2x + y = 0 < 4$.

Step 4: Therefore, we shade the region <u>above</u> the line $2x + y = 4$.

(d) The graph is <u>unbounded</u> since it extends forever toward the upper right.

(e) We only have two vertices:

(1) The intersection of $2x + y = 4$ and the y-axis: $(0, 4)$.
(2) The intersection of either $x + y = 2$ or $2x + y = 4$ and the x-axis: $(2, 0)$.

29.
$$\begin{cases} x \geq 0 & (1) \\ y \geq 0 & (2) \\ x + y \geq 2 & (3) \\ 2x + 3y \leq 12 & (4) \\ 3x + y \leq 12 & (5) \end{cases}$$

(a) $x \geq 0$; $y \geq 0$:

Our graph will lie in quadrant I.

(b) $x + y \geq 2$:

Step 1: Graph $x + y = 2$. The x-intercept is 2; the y-intercept is 2.

Step 2: Use $(0, 0)$ below the line

Step 3: At $(0, 0)$, $x + y = 0 < 2$.

Step 4: Therefore, we shade the region <u>above</u> the line $x + y = 2$.

(c) $2x + 3y \leq 12$:

Step 1: Graph $2x + 3y = 12$. The x-intercept is $= 6$; the y-intercept is 4.

Step 2: Use $(0, 0)$ below the line.

Step 3: At $(0, 0)$, $2x + 3y = 0 < 12$.

Step 4: Therefore, shade <u>below</u> the line $2x + 3y = 12$.

10.5 SYSTEMS OF LINEAR INEQUALITIES 619

(d) $3x + y \leq 12$:

Step 1: Graph $3x + y = 12$. The x-intercept is $x = 4$; the y-intercept is 12.

Step 2: Use $(0, 0)$ <u>below</u> the line as a test point.

Step 3: At $(0, 0)$, $3x + y = 0 < 12$.

Step 4: Shade the region <u>below</u> the line $3x + y = 12$.

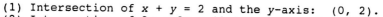

(e) We see that the graph is bounded.

(f) We have five vertices:

   (1) Intersection of $x + y = 2$ and the y-axis: $(0, 2)$.
   (2) Intersection of $2x + 3y = 12$ and the y-axis: $(0, 4)$.
   (3) Intersection of $2x + 3y = 12$ and $3x + y = 12$:

$$\begin{cases} 2x + 3y = 12 & (1) \\ 3x + y = 12 & (2) \end{cases}$$

$$\begin{cases} 2x + 3y = 12 & (1) \\ -9x - 3y = -36 & (2) \quad \text{Multiply by } -3. \end{cases}$$

$$\begin{cases} 2x + 3y = 12 & (1) \\ -7x = -24 & (2) \quad \text{Replace (2) by (1) + (2).} \end{cases}$$

$$\begin{cases} 2x + 3y = 12 & (1) \\ x = \dfrac{24}{7} & (2) \quad \text{Divide by } -7. \end{cases}$$

$$\begin{cases} \dfrac{48}{7} + 3y = 12 & (1) \quad \text{Back-substitution: } x = \dfrac{24}{7} \\ x = \dfrac{24}{7} & (2) \end{cases}$$

$$\begin{cases} 3y = \dfrac{84}{7} - \dfrac{48}{7} & (1) \\ x = \dfrac{24}{7} & (2) \end{cases}$$

$$\begin{cases} y = \dfrac{12}{7} & (1) \\ x = \dfrac{24}{7} & (2) \end{cases}$$

The point of intersection is $\left(\dfrac{24}{7}, \dfrac{12}{7} \right)$

   (4) The intersection of $3x + y = 12$ and the x-axis: $(4, 0)$
   (5) The intersection of $x + y = 2$ and the x-axis: $(2, 0)$

                    10 SYSTEMS OF EQUATIONS AND INEQUALITIES

31.
$$\begin{cases} x \geq 0 & (1) \\ y \geq 0 & (2) \\ x + y \geq 2 & (3) \\ x + y \leq 8 & (4) \\ 2x + y \leq 10 & (5) \end{cases}$$

(a) $x \geq 0$; $y \geq 0$:

This places our final graph in quadrant I.

(b) $x + y \geq 2$:

Step 1: Graph $x + y = 2$. The x-intercept is 2; the y-intercept is 2.

Step 2: Use $(0, 0)$ below the line, as a test point.

Step 3: At $(0, 0)$, $x + y = 0 < 2$, so the inequality is not satisfied.

Step 4: Therefore, shade above the line $x + y = 2$.

(c) $x + y \leq 8$:

Step 1: Graph $x + y = 8$. The x-intercept is 8; the y-intercept is 8.

Step 2: Use $(0, 0)$ below the line.

Step 3: At $(0, 0)$, $x + y = 0 < 8$.

Step 4: Therefore, shade below the line $x + y = 8$.

(d) $2x + y \leq 10$:

Step 1: Graph $2x + y = 10$. The x-intercept is $x = 5$; the y-intercept is $y = 10$.

Step 2: Use $(0, 0)$ below the line as a test point.

Step 3: At $(0, 0)$, $2x + y = 0 < 10$.

Step 4: Shade the region below the line $2x + y = 10$.

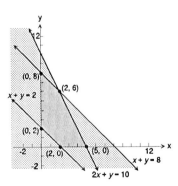

(e) The graph is bounded.

(f) We have five vertices:

    (1) Intersection of $x + y = 2$ and the y-axis: (0, 2).
    (2) Intersection of $x + y = 8$ and the y-axis: (0, 8).
    (3) Intersection of $x + y = 8$ and $2x + y = 10$:

$$\begin{cases} x + y = 8 & (1) \\ 2x + y = 10 & (2) \end{cases}$$

$$\begin{cases} x + y = 8 & (1) \\ -2x - y = -10 & (2) \end{cases}$$

$$\begin{cases} x + y = 8 & (1) \\ -x = -2 & (2) \end{cases} \quad \text{Replace (2) by (1) + (2).}$$

$$\begin{cases} x + y = 8 & (1) \\ x = 2 & (2) \end{cases}$$

$$\begin{cases} y = 6 & (1) \quad \text{Back-substitution: } x = 2. \\ x = 2 & (2) \end{cases}$$

   The vertex is (2, 6).

    (4) The intersection of $2x + y = 10$ and the x-axis: (5, 0)
    (5) The intersection of $x + y = 2$ and the x-axis: (2, 0)

33.
$$\begin{cases} x \geq 0 & (1) \\ y \geq 0 & (2) \\ x + 2y \geq 1 & (3) \\ x + 2y \leq 10 & (4) \end{cases}$$

(a) $x \geq 0$; $y \geq 0$:

Our graph will lie in quadrant I.

(b) $x + 2y \geq 1$:

   <u>Step 1</u>: Graph $x + 2y = 1$. The x-intercept is 1; the
             y-intercept is $\frac{1}{2}$.

   <u>Step 2</u>: Use (0, 0) below the line as a test point.

   <u>Step 3</u>: At (0, 0), $x + 2y = 0 < 1$.

   <u>Step 4</u>: Therefore, we shade the region <u>above</u> the line
            $x + 2y = 1$.

(c) $x + 2y \leq 10$:

   <u>Step 1</u>: Graph $x + 2y = 10$. The
            x-intercept is $= 10$; the
            y-intercept is 5.

   <u>Step 2</u>: Use (0, 0) below the line.

   <u>Step 3</u>: At (0, 0), $x + 2y = 0 < 10$.

   <u>Step 4</u>: Therefore, we shade the
           region <u>below</u> the line
           $x + 2y = 10$.

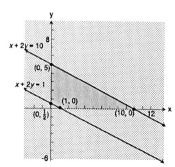

10 SYSTEMS OF EQUATIONS AND INEQUALITIES

(d) We see that the region is bounded.

(e) From the graph, there are four vertices:

    (1) The intersection of $x + 2y = 1$ and the y-axis: $\left(0, \frac{1}{2}\right)$.

    (2) The intersection of $x + 2y = 10$ and the y-axis: $(0, 5)$.
    (3) The intersection of $x + 2y = 10$ and the x-axis: $(10, 0)$.
    (4) The intersection of $x + 2y = 1$ and the x-axis: $(1, 0)$

35. $\begin{cases} x \leq 4 \\ x + y \leq 6 \\ x \geq 0, \ y \geq 0 \end{cases}$

37. $\begin{cases} x \leq 20 \\ y \geq 15 \\ x + y \leq 50 \\ x \leq y \\ x \geq 0 \end{cases}$

39. (a) x denotes the number of packages of the economy blend.
    y denotes the number of packages of the superior blend.

    $4x + 8y \leq 75(16)$ denotes the equation which states that the blends cannot exceed 75 pounds of A grade coffee or $(76)(16) = 1200$ ounces.

    $12x + 8y \leq 120(16)$ denotes the equation which states that the blends cannot exceed 120 pounds of B grade coffee or $(120)(16) = 1920$ ounces.

We also must denote that the number of packages of both blends must be non-negative.

The following equations are simplified:

    from $4x + 8y \leq 1200$
       to $x + 2y \leq 300$

and from $12x + 8y \leq 1920$
       to $3x + 2y \leq 480$

We have the following system: $\begin{aligned} x &\geq 0, \ y \geq 0 \\ x + 2y &\leq 300 \\ 3x + 2y &\leq 480 \end{aligned}$

(b)

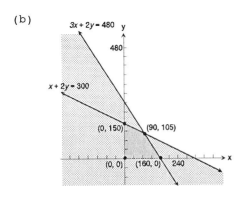

The vertices are $(0, 0)$, $(0, 150)$, $(90, 105)$, and $(160, 0)$.

1. $\begin{cases} 2x - y = 5 & (1) \\ 5x + 2y = 8 & (2) \end{cases}$

Step 1: It is easiest to solve (1) for y:

$$2x - y = 5$$
$$-y = 5 - 2x$$
$$y = -5 + 2x$$

Step 2: Substitute $y = -5 + 2x$ into (2):

$$5x + 2y = 8$$
$$5x + 2(-5 + 2x) = 8$$
$$5x - 10 + 4x = 8$$
$$9x = 18$$

Step 3: $\qquad\qquad x = 2$

Step 4: Determine y:

$$y = -5 + 2x \quad (\text{Step 1})$$
$$y = -5 + 2(2) \quad (x = 2)$$
$$y = -1$$

The solution is $x = 2$, $y = -1$.

3. $\begin{cases} 3x - 4y = 4 & (1) \\ x - 3y = \dfrac{1}{2} & (2) \end{cases}$

Step 1: Let's solve for x in (2):

$$x - 3y = \frac{1}{2}$$
$$x = \frac{1}{2} + 3y$$

Step 2: Substitute $x = \dfrac{1}{2} + 3y$ into (1):

$$3x - 4y = 4$$
$$3\left(\frac{1}{2} + 3y\right) - 4y = 4$$
$$\frac{3}{2} + 9y - 4y = 4$$
$$3 + 18y - 8y = 8 \quad \text{Multiply by 2.}$$
$$10y = 5$$

Step 3: $y = \dfrac{5}{10}$

$\quad\quad\quad y = \dfrac{1}{2}$

Step 4: Determine x:

$$x = \dfrac{1}{2} + 3y \quad\quad \text{(Step 1)}$$

$$x = \dfrac{1}{2} + 3\left(\dfrac{1}{2}\right) \quad \left(y = \dfrac{1}{2}\right)$$

$$x = 2$$

The solution is $x = 2$, $y = \dfrac{1}{2}$.

5. $\quad \begin{cases} x - 2y - 4 = 0 & (1) \\ 3x + 2y - 4 = 0 & (2) \end{cases}$

or

$\quad \begin{cases} x - 2y = 4 & (1) \\ 3x + 2y = 4 & (2) \end{cases}$

Step 1: It is easiest to solve for x in (1):

$$x - 2y = 4$$
$$x = 4 + 2y$$

Step 2: Substitute $x = 4 + 2y$ into (2):

$$3x + 2y = 4$$
$$3(4 + 2y) + 2y = 4$$
$$8y = -8$$

Step 3: $\quad\quad\quad\quad\quad\quad y = -1$

Step 4: Determine x: $\quad\quad x = 4 + 2y \quad\quad\quad \text{(Step 1)}$

$$x = 4 + 2(-1)$$
$$x = 2$$

The solution is $x = 2$, $y = -1$.

7. $\quad \begin{cases} y = 2x - 5 & (1) \\ x = 3y + 4 & (2) \end{cases}$

We can substitute $y = 2x - 5$ into (2):

$$x = 3y + 4$$
$$x = 3(2x - 5) + 4$$
$$x = 6x - 11$$
$$-5x = -11$$
$$x = \dfrac{11}{5}$$

Then we have: $\quad\quad y = 2x - 5$

$$y = \dfrac{22}{5} - 5$$

$$y = \dfrac{-3}{5}$$

The solution is $x = \dfrac{11}{5}$, $y = \dfrac{-3}{5}$.

9.
$$\begin{cases} x - y + 4 = 0 & (1) \\ \dfrac{1}{2}x + \dfrac{1}{6}y + \dfrac{2}{5} = 0 & (2) \end{cases}$$

or

$$\begin{cases} x - y = 4 & (1) \\ 15x + 5y = -12 & (2) \quad \text{Multiply by 30.} \end{cases}$$

Step 1: Solve equation (1) for x:

$$x - y = -4$$
$$x = -4 + y$$

Step 2: Substitute $x = -4 + y$ into (2):

$$15x + 5y = -12$$
$$15(-4 + y) + 5y = -12$$
$$-60 + 20y = -12$$
$$20y = 48$$
$$y = \frac{48}{20}$$

Step 3: $\qquad\qquad\qquad\qquad y = \dfrac{12}{5}$

Step 4: Determine x: $\qquad x = -4 + y \qquad\qquad$ (Step 1)

$$x = -4 + \frac{12}{5}$$
$$x = \frac{-20}{5} + \frac{12}{5}$$
$$x = \frac{-8}{5}$$

The solution is $x = \dfrac{-8}{5}$, $y = \dfrac{12}{5}$.

11.
$$\begin{cases} x - 2y - 8 = 0 & (1) \\ 2x + 2y - 10 = 0 & (2) \end{cases}$$

$$\begin{cases} x - 2y - 8 = 0 & (1) \\ 3x \quad\quad - 18 = 0 & (2) \quad \text{Replace (2) by (1) + (2).} \end{cases}$$

$$\begin{cases} x - 2y - 8 = 0 & (1) \\ \qquad\qquad x = 6 & (2) \end{cases}$$

$$\begin{cases} 6 - 2y - 8 = 0 & (1) \quad \text{Back-substitute; } x = 6. \\ \qquad\qquad x = 6 & (2) \end{cases}$$

$$\begin{cases} -2y = 2 & (1) \\ \quad x = 6 & (2) \end{cases}$$

$$\begin{cases} y = -1 & (1) \\ x = 6 & (2) \end{cases}$$

The solution is $x = 6$; $y = -1$.

10 SYSTEMS OF EQUATIONS AND INEQUALITIES

13. $\begin{cases} y - 2x = 11 & (1) \\ 2y - 3x = 18 & (2) \end{cases}$

$\begin{cases} -2y + 4x = -22 & (1) \quad \text{Multiply both sides by -2.} \\ 2y - 3x = 18 & (2) \end{cases}$

$\begin{cases} -2y + 4x = -22 & (1) \\ x = -4 & (2) \quad \text{Replace (2) by (1) + (2).} \end{cases}$

$\begin{cases} -2y - 16 = -22 & (1) \quad \text{Back-substitute; } x = -4. \\ x = -4 & (2) \end{cases}$

$\begin{cases} -2y = -6 & (1) \\ x = -4 & (2) \end{cases}$

$\begin{cases} y = 3 & (1) \\ x = -4 & (2) \end{cases}$

The solution is $x = -4$; $y = 3$.

15. $\begin{cases} 2x + 3y - 13 = 0 & (1) \\ 3x - 2y = 0 & (2) \end{cases}$

$\begin{cases} 6x + 9y - 39 = 0 & (1) \quad \text{Multiply both sides by 3.} \\ -6x + 4y = 0 & (2) \quad \text{Multiply both sides by -2.} \end{cases}$

$\begin{cases} 6x + 9y - 39 = 0 & (1) \\ 13y - 39 = 0 & (2) \quad \text{Replace (2) by (1) + (2).} \end{cases}$

$\begin{cases} 6x + 9y - 39 = 0 & (1) \\ y = 3 & (2) \end{cases}$

$\begin{cases} 6x + 27 - 39 = 0 & (1) \quad \text{Back-substitute; } y = 3. \\ y = 3 & (2) \end{cases}$

$\begin{cases} x = 2 & (1) \\ y = 3 & (2) \end{cases}$

The solution is $x = 2$, $y = 3$.

17. $\begin{cases} 3x - 2y = 8 & (1) \\ x - \dfrac{2}{3}y = 12 & (2) \end{cases}$

$\begin{cases} 3x - 2y = 8 & (1) \\ -3x + 2y = -36 & (2) \quad \text{Multiply both sides by -3.} \end{cases}$

$\begin{cases} 3x - 2y = 8 & (1) \\ 0x + 0y = -28 & (2) \quad \text{Replace (2) by (1) + (2).} \end{cases}$

The system is inconsistent.

10 - CHAPTER REVIEW

19.
$$\begin{cases} x + 2y - z = 6 & (1) \\ 2x - y + 3z = -13 & (2) \\ 3x - 2y + 3z = -16 & (3) \end{cases}$$

We can eliminate the variable y by adding (1) and (3):

$$\begin{cases} x + 2y - z = 6 & (1) \\ 2x - y + 3z = -13 & (2) \\ 4x\qquad + 2z = -10 & (3) \quad \text{Replace (3) by (1) + (3).} \end{cases}$$

Now we must eliminate the same variable, y, from either (1) or (2):

$$\begin{cases} x + 2y - z = 6 & (1) \\ 4x - 2y + 6z = -26 & (2) \quad \text{Multiply both sides by 2.} \\ 4x\qquad + 2z = -10 & (3) \end{cases}$$

$$\begin{cases} x + 2y - z = 6 & (1) \\ 5x\qquad + 5z = -20 & (2) \quad \text{Replace (2) by (1) + (2).} \\ 4x\qquad + 2z = -10 & (3) \end{cases}$$

Now attack (2) and (3):

$$\begin{cases} x + 2y - z = 6 & (1) \\ 20x\qquad + 20z = -80 & (2) \quad \text{Multiply both sides by 4.} \\ -20x\qquad - 10z = 50 & (3) \quad \text{Mutliply both sides by -5.} \end{cases}$$

$$\begin{cases} x + 2y - z = 6 & (1) \\ 20x\qquad + 20z = -80 & (2) \\ \qquad\qquad 10z = -30 & (3) \quad \text{Replace (3) by (2) + (3).} \end{cases}$$

$$\begin{cases} x + 2y - z = 6 & (1) \\ 20x\qquad + 20z = -80 & (2) \\ \qquad\qquad z = -3 & (3) \end{cases}$$

$$\begin{cases} x + 2y + 3 = 6 & (1) \quad \text{Back-substitute: } z = -3 \\ 20x\qquad - 60 = -80 & (2) \quad \text{Back-substitute; } z = -3 \\ \qquad\qquad z = -3 & (3) \end{cases}$$

$$\begin{cases} x + 2y = 3 & (1) \\ \qquad x = -1 & (2) \\ \qquad z = -3 & (3) \end{cases}$$

$$\begin{cases} -1 + 2y = 3 & (1) \quad \text{Back-substitute; } x = -1 \\ \qquad x = -1 & (2) \\ \qquad z = -3 & (3) \end{cases}$$

$$\begin{cases} y = 2 & (1) \\ x = -1 & (2) \\ z = -3 & (3) \end{cases}$$

The solution is $x = -1$, $y = 2$, $z = -3$.

10 SYSTEMS OF EQUATIONS AND INEQUALITIES

21. $\begin{cases} 3x - 2y = 1 \\ 10x + 10y = 5 \end{cases}$ becomes: $\begin{bmatrix} 3 & -2 & | & 1 \\ 10 & 10 & | & 5 \end{bmatrix}$ In two steps, we can get a 1 in row 1, column 1, <u>and</u> avoid fractions:

$\rightarrow \begin{bmatrix} 3 & -2 & | & 1 \\ 1 & 16 & | & 2 \end{bmatrix}$

↑
$R_2 = -3r_1 + r_2$

$\rightarrow \begin{bmatrix} 1 & 16 & | & 2 \\ 3 & -2 & | & 1 \end{bmatrix}$

↑
Interchange rows

$\rightarrow \begin{bmatrix} 1 & 16 & | & 2 \\ 0 & -50 & | & -5 \end{bmatrix}$

↑
$R_2 = -3r_1 + r_2$

$\rightarrow \begin{bmatrix} 1 & 16 & | & 2 \\ 0 & 1 & | & \frac{1}{10} \end{bmatrix}$

↑
$R_2 = -\frac{1}{50}r_2$

$\rightarrow \begin{bmatrix} 1 & 0 & | & \frac{20}{10} - \frac{16}{10} \\ 0 & 1 & | & \frac{1}{10} \end{bmatrix}$

↑
$R_1 = -16r_2 + r_1$

The solution is $x = \frac{2}{5}$, $y = \frac{1}{10}$.

23. $\begin{cases} 5x + 6y - 3z = 6 \\ 4x - 7y - 2z = -3 \\ 3x + y - 7z = 1 \end{cases}$ becomes: $\begin{bmatrix} 5 & 6 & -3 & | & 6 \\ 4 & -7 & -2 & | & -3 \\ 3 & 1 & -7 & | & 1 \end{bmatrix}$

$\rightarrow \begin{bmatrix} 1 & 13 & -1 & | & 9 \\ 4 & -7 & -2 & | & -3 \\ 3 & 1 & -7 & | & 1 \end{bmatrix}$

↑
$R_1 = -1r_2 + r_1$

$\rightarrow \begin{bmatrix} 1 & 13 & -1 & | & 9 \\ 0 & -59 & 2 & | & -39 \\ -1 & -38 & -4 & | & -26 \end{bmatrix}$ Let's get some smaller numbers:

↑
$R_2 = -4r_1 + r_2$
$R_3 = -3r_1 + r_3$

$$\rightarrow \begin{bmatrix} 1 & 13 & -1 & | & 9 \\ 0 & -59 & 2 & | & -39 \\ 0 & 19 & 2 & | & 13 \end{bmatrix}$$

↑
$R_3 = -\dfrac{1}{2} r_3$

$$\rightarrow \begin{bmatrix} 1 & 13 & -1 & | & 9 \\ 0 & -2 & 8 & | & 0 \\ 0 & 19 & 2 & | & 13 \end{bmatrix}$$

↑
$R_2 = 3r_3 + r_2$

$$\rightarrow \begin{bmatrix} 1 & 13 & -1 & | & 9 \\ 0 & 1 & -4 & | & 0 \\ 0 & 19 & 2 & | & 13 \end{bmatrix}$$

↑
$R_2 = -\dfrac{1}{2} r_2$

We got a 1 in row 2, column 2, <u>and</u> avoided fractions, by taking a couple of extra steps!

$$\rightarrow \begin{bmatrix} 1 & 0 & 51 & | & 9 \\ 0 & 1 & -4 & | & 0 \\ 0 & 0 & 78 & | & 13 \end{bmatrix}$$

↑
$R_1 = -13r_2 + r_1$
$R_3 = -19r_2 + r_3$

$$\rightarrow \begin{bmatrix} 1 & 0 & 51 & | & 9 \\ 0 & 1 & -4 & | & 0 \\ 0 & 0 & 1 & | & \frac{13}{78} = \frac{1}{6} \end{bmatrix}$$

↑
$R_3 = \dfrac{1}{78} r_3$

$$\rightarrow \begin{bmatrix} 1 & 0 & 0 & | & \frac{-51}{6} + \frac{54}{6} \\ 0 & 1 & 0 & | & \frac{4}{6} \\ 0 & 0 & 1 & | & \frac{1}{6} \end{bmatrix}$$

↑
$R_1 = -51r_3 + r_1$
$R_2 = 4r_3 + r_2$

The solution is $x = \dfrac{1}{2}$, $y = \dfrac{2}{3}$, $z = \dfrac{1}{6}$.

10 SYSTEMS OF EQUATIONS AND INEQUALITIES

25. $\begin{cases} x \qquad -2z = 1 \\ 2x + 3y \qquad = -3 \\ 4x - 3y - 4z = 3 \end{cases}$ becomes: $\begin{bmatrix} 1 & 0 & -2 & | & 1 \\ 2 & 3 & 0 & | & -3 \\ 4 & -3 & -4 & | & 3 \end{bmatrix}$

$\rightarrow \begin{bmatrix} 1 & 0 & -2 & | & 1 \\ 0 & 3 & 4 & | & -5 \\ 0 & -3 & 4 & | & -1 \end{bmatrix}$

↑

$R_2 = -2r_1 + r_2$
$R_3 = -4r_1 + r_3$

$\rightarrow \begin{bmatrix} 1 & 0 & -2 & | & 1 \\ 0 & 1 & \frac{4}{3} & | & -\frac{5}{3} \\ 0 & -3 & 4 & | & -1 \end{bmatrix}$

↑

$R_2 = \frac{1}{3}r_2$

$\rightarrow \begin{bmatrix} 1 & 0 & -2 & | & 1 \\ 0 & 1 & \frac{4}{3} & | & -\frac{5}{3} \\ 0 & 0 & 8 & | & -6 \end{bmatrix}$

↑

$R_3 = 3r_2 + r_3$

$\rightarrow \begin{bmatrix} 1 & 0 & -2 & | & 1 \\ 0 & 1 & \frac{4}{3} & | & -\frac{5}{3} \\ 0 & 0 & 1 & | & -\frac{3}{4} \end{bmatrix}$

↑

$R_3 = -\frac{1}{8}r_3$

$\rightarrow \begin{bmatrix} 1 & 0 & 0 & | & -\frac{1}{2} \\ 0 & 1 & 0 & | & -\frac{2}{3} \\ 0 & 0 & 1 & | & -\frac{3}{4} \end{bmatrix}$

↑

$R_1 = 2r_3 + r_1$
$R_2 = -\frac{4}{3}r_3 + r_2$

The solution is $x = -\frac{1}{2}$, $y = -\frac{2}{3}$, $z = -\frac{3}{4}$.

10 - CHAPTER REVIEW

27. $\begin{cases} x - y + z = 0 \\ x - y - 5z = 6 \\ 2x - 2y + z = 1 \end{cases}$ (Get the constants on the right-hand-side.)

$$\begin{bmatrix} 1 & -1 & 1 & | & 0 \\ 1 & -1 & -5 & | & 6 \\ 2 & -2 & 1 & | & 1 \end{bmatrix}$$

$$\rightarrow \begin{bmatrix} 1 & -1 & 1 & | & 0 \\ 0 & 0 & -6 & | & 6 \\ 0 & 0 & -1 & | & 1 \end{bmatrix}$$
$\uparrow$
$R_2 = -1r_1 + r_2$
$R_3 = -2r_1 + r_3$

It is impossible to obtain a 1 in row 2 column 2 (we cannot use row 1, since that would mess up the 0's we have in column 1).

So we focus on row 2, column 3:

$$\rightarrow \begin{bmatrix} 1 & -1 & 1 & | & 0 \\ 0 & 0 & 1 & | & -1 \\ 0 & 0 & -1 & | & 1 \end{bmatrix}$$
$\uparrow$
$R_2 = -\dfrac{1}{6}r_2$

$$\rightarrow \begin{bmatrix} 1 & -1 & 0 & | & 1 \\ 0 & 0 & 1 & | & -1 \\ 0 & 0 & 0 & | & 0 \end{bmatrix}$$
$\uparrow$
$R_1 = -r_2 + r_1$
$R_3 = r_2 + r_3$

Thus, we have: $x - y = 1$
$\qquad\qquad\qquad\quad z = -1$

We can write the solution as:

$z = -1$, $x = y + 1$, where y can be any real number,
or
$z = -1$, $y = x - 1$, where x can be any real number.

29.
$$\begin{cases} x - y - z - t = 16 \\ 2x + y + z + 2t = 3 \\ x - 2y - 2z - 3t = 0 \\ 3x - 4y + z + 5t = -3 \end{cases}$$
becomes:
$$\begin{bmatrix} 1 & -1 & -1 & -1 & | & 1 \\ 2 & 1 & 1 & 2 & | & 3 \\ 1 & -2 & -2 & -3 & | & 0 \\ 3 & -4 & 1 & 5 & | & -3 \end{bmatrix}$$

$$\rightarrow \begin{bmatrix} 1 & -1 & -1 & -1 & | & 1 \\ 0 & 3 & 3 & 4 & | & 1 \\ 0 & -1 & -1 & -2 & | & -1 \\ 0 & -1 & 4 & 8 & | & -6 \end{bmatrix}$$
$\uparrow$
$R_2 = -2r_1 + r_2$
$R_3 = -1r_2 + r_3$
$R_4 = -3r_1 + r_4$

$$\rightarrow \begin{bmatrix} 1 & -1 & -1 & -1 & | & 1 \\ 0 & 1 & 1 & 0 & | & -1 \\ 0 & -1 & -1 & -2 & | & -1 \\ 0 & -1 & 4 & 8 & | & -6 \end{bmatrix}$$
$\uparrow$
$R_2 = 2r_3 + r_2$

$$\rightarrow \begin{bmatrix} 1 & 0 & 0 & -1 & | & 0 \\ 0 & 1 & 1 & 0 & | & -1 \\ 0 & 0 & 0 & -2 & | & -2 \\ 0 & 0 & 5 & 8 & | & -7 \end{bmatrix}$$ Now we need a 1 in row 3, column 3.
$\uparrow$
$R_1 = r_2 + r_1$
$R_3 = r_2 + r_3$
$R_4 = r_2 + r_4$

$$\rightarrow \begin{bmatrix} 1 & 0 & 0 & -1 & | & 0 \\ 0 & 1 & 1 & 0 & | & -1 \\ 0 & 0 & 5 & 8 & | & -7 \\ 0 & 0 & 0 & -2 & | & -2 \end{bmatrix}$$
$\uparrow$
Interchange r_3 and r_4

$$\rightarrow \begin{bmatrix} 1 & 0 & 0 & -1 & | & 0 \\ 0 & 1 & 1 & 0 & | & -1 \\ 0 & 0 & 1 & \frac{8}{5} & | & -\frac{7}{5} \\ 0 & 0 & 0 & 1 & | & 1 \end{bmatrix}$$
$\uparrow$
$R_3 = \frac{1}{5} r_3$
$R_4 = -\frac{1}{2} r_4$

$$\rightarrow \begin{bmatrix} 1 & 0 & 0 & -1 & | & 0 \\ 0 & 1 & 0 & -\dfrac{8}{5} & | & \dfrac{2}{5} \\ 0 & 0 & 1 & \dfrac{8}{5} & | & -\dfrac{7}{5} \\ 0 & 0 & 0 & 1 & | & 1 \end{bmatrix}$$

↑

$R_2 = -1r_3 + r_2$

$$\rightarrow \begin{bmatrix} 1 & 0 & 0 & 0 & | & 1 \\ 0 & 1 & 0 & 0 & | & 2 \\ 0 & 0 & 1 & 0 & | & -3 \\ 0 & 0 & 0 & 1 & | & 1 \end{bmatrix}$$

↑

$R_1 = r_4 + r_1$

$R_2 = \dfrac{8}{5}r_4 + r_2$

$R_3 = -\dfrac{8}{5}r_4 + r_3$

The solution is: $x = 1$, $y = 2$, $z = -3$, $t = 1$

31. $\begin{vmatrix} 3 & 4 \\ 1 & 3 \end{vmatrix} = (3)(3) - (1)(4) = 9 - 4 = 5$

33. $\begin{vmatrix} 1 & 4 & 0 \\ -1 & 2 & 6 \\ 4 & 1 & 3 \end{vmatrix} = 1 \begin{vmatrix} 2 & 6 \\ 1 & 3 \end{vmatrix} - 4 \begin{vmatrix} -1 & 6 \\ 4 & 3 \end{vmatrix} + 0 \begin{vmatrix} -1 & 2 \\ 4 & 1 \end{vmatrix}$

$= 1(6 - 6) - 4(-3 - 24) + 0$
$= 0 - 4(-27) + 0$
$= 108$

35. $\begin{vmatrix} 2 & 1 & -3 \\ 5 & 0 & 1 \\ 2 & 6 & 0 \end{vmatrix} = 2 \begin{vmatrix} 0 & 1 \\ 6 & 0 \end{vmatrix} - 1 \begin{vmatrix} 5 & 1 \\ 2 & 0 \end{vmatrix} + (-3) \begin{vmatrix} 5 & 0 \\ 2 & 6 \end{vmatrix}$

$= 2(0 - 6) - 1(0 - 2) + (-3)(30 - 0)$
$= -12 + 2 - 90$
$= -100$

37. $\begin{cases} x - 2y = 4 \\ 3x + 2y = 4 \end{cases}$

Here,

$$D = \begin{vmatrix} 1 & -2 \\ 3 & 2 \end{vmatrix} = 2 - (-6) = 8$$

To obtain D_x, replace the first column in D by the column of constants.

$$D_x = \begin{vmatrix} 4 & -2 \\ 4 & 2 \end{vmatrix} = 8 - (-8) = 16$$

To obtain D_y, replace the second column of D by the column of constants.

$$D_y = \begin{vmatrix} 1 & 4 \\ 3 & 4 \end{vmatrix} = 4 - 12 = -8$$

Then, $x = \dfrac{D_x}{D} = \dfrac{16}{8} = 2$

and $y = \dfrac{D_y}{D} = \dfrac{-8}{8} = -1$

39. $\begin{cases} 2x + 3y - 13 = 0 \\ 3x - 2y \quad\;\; = 0 \end{cases}$

$\begin{cases} 2x + 3y = 13 \\ 3x - 2y = 0 \end{cases}$

Here, $D = \begin{vmatrix} 2 & 3 \\ 3 & -2 \end{vmatrix} = -4 - 9 = -13$

$$D_x = \begin{vmatrix} 13 & 3 \\ 0 & -2 \end{vmatrix} = -26 - 0 = -26$$

$$D_y = \begin{vmatrix} 2 & 13 \\ 3 & 0 \end{vmatrix} = 0 - 39 = -39$$

By Cramer's Rule,

$$x = \dfrac{D_x}{D} = \dfrac{-26}{-13} = 2$$

and $y = \dfrac{D_y}{D} = \dfrac{-39}{-13} = 3$

41. $\begin{cases} x + 2y - \;\;z = 6 \\ 2x - \;\;y + 3z = -13 \\ 3x - 2y + 3z = -16 \end{cases}$

$$D = \begin{vmatrix} 1 & 2 & -1 \\ 2 & -1 & 3 \\ 3 & -2 & 3 \end{vmatrix} = 1 \begin{vmatrix} -1 & 3 \\ -2 & 3 \end{vmatrix} - 2 \begin{vmatrix} 2 & 3 \\ 3 & 3 \end{vmatrix} + (-1) \begin{vmatrix} 2 & -1 \\ 3 & -2 \end{vmatrix}$$

$$= 1(-3 + 6) - 2(6 - 9) + (-1)(-4 + 3)$$
$$= 3 + 6 + 1$$
$$= 10$$

$$D_x = \begin{vmatrix} 6 & 2 & -1 \\ -13 & -1 & 3 \\ -16 & -2 & 3 \end{vmatrix} = 6 \begin{vmatrix} -1 & 3 \\ -2 & 3 \end{vmatrix} - 2 \begin{vmatrix} -13 & 3 \\ -16 & 3 \end{vmatrix} + (-1) \begin{vmatrix} -13 & -1 \\ -16 & -2 \end{vmatrix}$$

$$= 6(-3 + 6) - 2(-39 + 48) + (-1)(26 - 16)$$
$$= 18 - 18 - 10$$
$$= -10$$

10 - CHAPTER REVIEW

$$D_y = \begin{vmatrix} 1 & 6 & -1 \\ 2 & -13 & 3 \\ 3 & -16 & 3 \end{vmatrix} = 1 \begin{vmatrix} -13 & 3 \\ -16 & 3 \end{vmatrix} - 6 \begin{vmatrix} 2 & 3 \\ 3 & 3 \end{vmatrix} + (-1) \begin{vmatrix} 2 & -13 \\ 3 & -16 \end{vmatrix}$$

$$= 1(-39 + 48) - 6(6 - 9) + (-1)(-32 + 39)$$
$$= 9 + 18 - 7$$
$$= 20$$

$$D_z = \begin{vmatrix} 1 & 2 & 6 \\ 2 & -1 & -13 \\ 3 & -2 & -16 \end{vmatrix} = 1 \begin{vmatrix} -1 & -13 \\ -2 & -16 \end{vmatrix} - 2 \begin{vmatrix} 2 & -13 \\ 3 & -16 \end{vmatrix} + 6 \begin{vmatrix} 2 & -1 \\ 3 & -2 \end{vmatrix}$$

$$= 1(16 - 26) - 2(-32 + 39) + 6(-4 + 3)$$
$$= -10 - 14 - 6$$
$$= -30$$

By Cramer's Rule,

$$x = \frac{D_x}{D} = \frac{-10}{10} = -1$$

$$y = \frac{D_y}{D} = \frac{20}{10} = 2$$

and $y = \dfrac{D_z}{D} = \dfrac{-30}{10} = -3$

43.
$$\begin{cases} 2x + y + 3 = 0 & (1) \\ x^2 + y^2 = 5 & (2) \end{cases}$$

We solve (1) for y:

$$2x + y + 3 = 0$$
$$y = -2x - 3$$

Substitute this expression for y into (2):

$$x^2 + y^2 = 5$$
$$x^2 + (-2x - 3)^2 = 5$$
$$x^2 + 4x^2 + 12x + 9 = 5$$
$$5x^2 + 12x + 4 = 0$$
$$(5x + 2)(x + 2) = 0$$
$$\text{so } x = \frac{-2}{5} \quad \text{or} \quad x = -2$$

If $x = \dfrac{-2}{5}$, then, from (1),

$$y = -2x - 3$$
$$y = \frac{4}{5} - 3 = \frac{-11}{5}$$

If $x = -2$, then $y = -2x - 3 = 1$

Thus, we have two possible solutions:

$$x = \frac{-2}{5}, \ y = \frac{-11}{5} \text{ and } x = -2, \ y = 1$$

Both of these check.

45.

$$\begin{cases} 2xy + y^2 = 10 & (1) \\ 3y^2 - xy = 2 & (2) \end{cases}$$

$$\begin{cases} 2xy + y^2 = 10 & (1) \\ -2xy + 6y^2 = 4 & (2) \quad \text{Multiply by 2.} \end{cases}$$

$$\begin{cases} 2xy + y^2 = 10 & (1) \\ 7y^2 = 14 & (2) \quad \text{Replace (2) by (1) + (2).} \end{cases}$$

From (2), $y^2 = 2$

$$y = \pm\sqrt{2}$$

If $y = \sqrt{2}$, then:

$$2xy + y^2 = 10 \quad (1)$$

$$2\sqrt{2}x + 2 = 10$$

$$2\sqrt{2}x = 8$$

$$x = \frac{8}{2\sqrt{2}} = 2\sqrt{2}$$

If $y = -\sqrt{2}$, then:

$$-2\sqrt{2}x + 2 = 10$$

$$-2\sqrt{2}x = 8$$

$$x = -2\sqrt{2}$$

We have two possible solutions:

$$x = 2\sqrt{2}, \ y = \sqrt{2}; \text{ and } x = -2\sqrt{2}, \ y = -\sqrt{2}$$

and they both check.

47.

$$\begin{cases} x^2 + y^2 = 6y & (1) \\ x^2 = 3y & (2) \end{cases}$$

From (2), $y = \frac{1}{3}x^2$

Then, by (1):

$$x^2 + y^2 = 6y$$

$$x^2 + \left(\frac{1}{3}x^2\right)^2 = 6\left(\frac{1}{3}x^2\right)$$

$$x^2 + \frac{x^4}{9} = 2x^2$$

$$9x^2 + x^4 = 18x^2 \quad \text{Multiply by 9.}$$

$$x^4 - 9x^2 = 0$$

$$x^2(x^2 - 9) = 0$$

$$x^2(x - 3)(x + 3) = 0$$

Thus, $x = 0$, $x = 3$, or $x = -3$

If $x = 0$, then $y = \dfrac{x^2}{3} = 0$

If $x = 3$, then $y = \dfrac{x^2}{3} = 3$

If $x = -3$, then $y = \dfrac{x^2}{3} = 3$

All of these solutions check:

$$x = 0, \ y = 0; \ x = 3, \ y = 3; \text{ and } x = -3, \ y = 3$$

49. $\begin{cases} 3x^2 + 4xy + 5y^2 = 8 & (1) \\ x^2 + 3xy + 2y^2 = 0 & (2) \end{cases}$

Equation (2) can be factored:

$x^2 + 3xy + 2y^2 = 0$
$(x + y)(x + 2y) = 0$

So $x = -y$ or $x = -2y$

From (1), if $x = -y$, then

$$\begin{aligned} 3(-y)^2 + 4(-y)y + 5y^2 &= 8 \\ 3y^2 - 4y^2 + 5y^2 &= 8 \\ 4y^2 &= 8 \\ y^2 &= 2 \\ y &= \pm\sqrt{2} \end{aligned}$$

Since $x = -y$, if $y = \sqrt{2}$, $x = -\sqrt{2}$, and if $y = -\sqrt{2}$, $x = \sqrt{2}$

This gives two possible solutions:

$x = \sqrt{2}$, $y = -\sqrt{2}$; $x = -\sqrt{2}$, $y = \sqrt{2}$, both of which check.

On the other hand, if $x = -2y$, then from (1):

$$\begin{aligned} 3(-2y)^2 + 4(-2y)y + 5y^2 &= 8 \\ 12y^2 - 8y^2 + 5y^2 &= 8 \\ 9y^2 &= 8 \\ y^2 &= \frac{8}{9} \\ y &= \pm\sqrt{\frac{8}{9}} = \pm\frac{2\sqrt{2}}{3} \end{aligned}$$

10 SYSTEMS OF EQUATIONS AND INEQUALITIES

Then, since $x = -2y$, we have:

If $y = \dfrac{2\sqrt{2}}{3}$, then $x = \dfrac{-4\sqrt{2}}{3}$

If $y = \dfrac{-2\sqrt{2}}{3}$, then $x = \dfrac{4\sqrt{2}}{3}$

Let's check these two possible solutions:

For $x = \dfrac{4\sqrt{2}}{3}$, $y = \dfrac{-2\sqrt{2}}{3}$

$$\begin{cases} 3x^2 + 4xy + 5y^2 = 3\left(\dfrac{32}{9}\right) + 4\left(\dfrac{-16}{9}\right) + 5\left(\dfrac{8}{9}\right) = \dfrac{72}{9} = 8 \\ x^2 + 3xy + 2y^2 = \left(\dfrac{32}{9}\right) + 3\left(\dfrac{-16}{9}\right) + 2\left(\dfrac{8}{9}\right) = \dfrac{0}{9} = 0 \end{cases}$$

The solution $x = \dfrac{-4\sqrt{2}}{3}$, $y = \dfrac{2\sqrt{2}}{3}$ also checks.

Thus, we have four solutions:

$$x = \sqrt{2}, \ y = -\sqrt{2}; \ x = -\sqrt{2}, \ y = \sqrt{2}$$

$$x = \dfrac{-4\sqrt{2}}{3}, \ y = \dfrac{-2\sqrt{2}}{3}; \ \text{and } x = \dfrac{-4\sqrt{2}}{3}, \ y = \dfrac{2\sqrt{2}}{3}$$

51. $\begin{cases} x^2 - 3x + y^2 + y = -2 \quad (1) \\ \dfrac{x^2 - x}{y} + y + 1 = 0 \quad (2) \end{cases}$

$\begin{cases} x^2 - 3x + y^2 + y + 2 = 0 \quad (1) \\ x^2 - x + y^2 + y = 0 \quad (2) \ \text{Multiply by y.} \end{cases}$

We can now eliminate y:

$\begin{cases} x^2 - 3x + y^2 + y + 2 = 0 \quad (1) \\ -x^2 + x - y^2 - y = 0 \quad (2) \ \text{Multiply by -1.} \end{cases}$

$\begin{cases} x^2 - 3x + y^2 + y + 2 = 0 \quad (1) \\ -2x + 2 = 0 \quad (2) \ \text{Replace (2) by (1) + (2).} \end{cases}$

From (2): $x = 1$. Then from (1):

$$x^2 - 3x + y^2 + y + 2 = 0$$
$$1 - 3 + y^2 + y + 2 = 0$$
$$y^2 + y = 0$$
$$y(y + 1) = 0$$

So, $y = 0$ or $y = -1$.

Thus, we have two possible solutions:

$$x = 1, \ y = 0 \text{ or } x = 1, \ y = -1.$$

From the original equation (2), we see that y <u>cannot</u> be 0. But the other solution checks. Thus, we have just one solution:

$$x = 1, \ y = -1.$$

53. $\begin{cases} -2x + y \le 2 & (1) \\ x + y \ge 2 & (2) \end{cases}$

(a) $-2x + y \le 2$:

Step 1: Graph $-2x + y = 2$ with a solid line (since the inequality is nonstrict). The x-intercept is $x = -1$; the y-intercept is $y = 2$.

Step 2: Let's use $(0, 0)$, which lies <u>below</u> the line, as a test point.

Step 3: At $(0, 0)$, $-2x + y = 0 \le 2$, so the inequality is <u>satisfied</u>.

Step 4: Therefore, we shade the region <u>below</u> the line $-2x + y = 2$.

(b) $x + y \ge 2$:

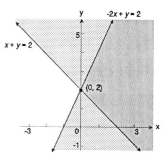

Step 1: Graph $x + y = 2$ with a solid line. The x-intercept is 2; the y-intercept is 2.

Step 2: Use $(0, 0)$ <u>below</u> the line as a test point.

Step 3: At $(0, 0)$, $x + y = 0 < 2$, so the inequality is <u>not</u> satisfied.

Step 4: Shade the region <u>above</u> the line $x + y = 2$.

(c) We see that the overlapping shaded region is unbounded.

(d) There is only one vertex, the intersection of $-2x + y = 2$ and $x + y = 2$:

$$\begin{cases} -2x + y = 2 & (1) \\ x + y = 2 & (2) \end{cases}$$

$$\begin{cases} -2x + y = 2 & (1) \\ -x - y = -2 & (2) \end{cases} \quad \text{Multiply by } -1.$$

$$\begin{cases} -2x + y = 2 & (1) \\ -3x = 0 & (2) \end{cases} \quad \text{Replace (2) by (1) + (2).}$$

$$\begin{cases} -2x + y = 2 & (1) \\ x = 0 & (2) \end{cases}$$

$$\begin{cases} y = 2 & (1) \\ x = 0 & (2) \end{cases}$$

The vertex is $(0, 2)$.

55. $$\begin{cases} x \geq 0 \\ y \geq 0 \\ x + y \leq 4 \\ 2x + 3y \leq 6 \end{cases}$$

(a) $x \geq 0$; $y \geq 0$:

These inequalities require that our graph be located in quadrant I.

(b) $x + y \leq 4$:

Step 1: Graph $x + y = 4$. The x-intercept is $x = 4$; the y-intercept is $y = 4$.

Step 2: Use $(0, 0)$ below the line as a test point.

Step 3: At $(0, 0)$, $x + y = 0 < 4$.

Step 4: Therefore, we shade below the line $x + y = 4$.

(c) $2x + 3y \leq 6$:

Step 1: Graph $2x + 3y = 6$. The x-intercept is 3; the y-intercept is 2.

Step 2: Use $(0, 0)$ below the line as a test point.

Step 3: At $(0, 0)$, $2x + 3y = 0 < 6$.

Step 4: Shade below the line $2x + 3y = 6$.

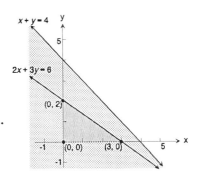

(d) We see that the graph is bounded.

(e) There are <u>three</u> vertices:

  (1) The intersection of $x = 0$ and $y = 0$: $(0, 0)$.
  (2) The intersection of $2x + 3y = 6$ and the y-axis: $(0, 2)$.
  (3) The intersection of $2x + 3y = 6$ and the x-axis: $(3, 0)$.

57.
$$\begin{cases} x \geq 0 & (1) \\ y \geq 0 & (2) \\ 2x + y \leq 8 & (3) \\ x + 2y \geq 2 & (4) \end{cases}$$

(a) $x \geq 0$; $y \geq 0$:

The graph will lie in quadrant I.

(b) $2x + y \leq 8$:

<u>Step 1</u>: Graph $2x + y = 8$. The x-intercept is $x = 4$; the y-intercept is 8.

<u>Step 2</u>: Use $(0, 0)$ below the line as a test point.

<u>Step 3</u>: At $(0, 0)$, $2x + y = 0 < 8$.

<u>Step 4</u>: Therefore, we shade <u>below</u> the line $2x + y = 8$.

(c) $x + 2y \geq 2$:

<u>Step 1</u>: Graph $x + 2y = 2$. The x-intercept is 2; the y-intercept is 1.

<u>Step 2</u>: Use $(0, 0)$ below the line as a test point.

<u>Step 3</u>: At $(0, 0)$, $x + 2y = 0 < 2$, so the inequality is <u>not</u> satisfied.

<u>Step 4</u>: Therefore, we shade <u>above</u> the line $x + 2y = 2$.

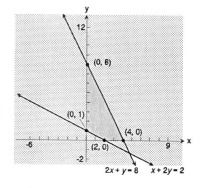

(d) The graph is bounded.

(e) There are <u>four</u> vertices, all are either x-intercepts or y-intercepts.

  $(0, 1)$, $(0, 8)$, $(4, 0)$, and $(2, 0)$.

59. We are given the system:

$$\begin{cases} 2x + 5y = 5 & (1) \\ 4x + 10y = A & (2) \end{cases}$$

and we want to find a value of A such that the system will have infinitely many solutions.

Solve by elimination:

$$\begin{cases} 2x + 5y = 5 & (1) \\ 4x + 10y = A & (2) \end{cases}$$

$$\begin{cases} -4x - 10y = -10 & (1) \quad \text{Multiply by -2.} \\ \;\;4x + 10y = A & (2) \end{cases}$$

$$\begin{cases} -4x - 10y = -10 & (1) \\ \qquad\qquad 0 = A - 10 & (2) \quad \text{Replace (2) by (1) + (2).} \end{cases}$$

If $A - 10 = 0$, the system will have infinitely many solutions of the form

$$-4x - 10y = -10$$
$$-4x = 10y - 10$$
$$x = \frac{-5}{2}y + \frac{5}{2}$$

where y is any real number.

Therefore, we need $A = 10$.

61. We are given $y = ax^2 + bx + c$. If $(0, 1)$ satisfies this equation, then:

$$1 = 0a + 0b + c,$$
$$\text{or} \quad 1 = c$$

From $(1, 0)$, we have:

$$0 = a + b + c$$

$$\text{or} \qquad a + b = -c$$
$$\qquad\quad a + b = -1 \quad \text{since } c = 1$$

From $(-2, 1)$,

$$1 = 4a - 2b + c$$
$$4a - 2b = 1 - c$$
$$4a - 2b = 0 \qquad \text{since } c = 1$$

We now have two equations in the two unknowns a and b.

$$\begin{cases} \;\;a + b = -1 & (1) \\ 4a - 2b = 0 & (2) \end{cases}$$

$$\begin{cases} 2a - 2b = -2 & (1) \quad \text{Multiply by 2.} \\ 4a - 2b = 0 & (2) \end{cases}$$

$$\begin{cases} 2a + 2b = -2 & (1) \\ \qquad 6a = -2 & (2) \quad \text{Replace (2) by (1) + (2).} \end{cases}$$

From (2): $\quad a = \dfrac{-1}{3}$

Then from (1),

$$2a + 2b = -2$$

$$2\left(\frac{-1}{3}\right) + 2b = -2$$

$$2b = -2 + \frac{2}{3}$$

$$b = -1 + \frac{1}{3}$$

$$b = -\frac{2}{3}$$

Therefore, $y = ax^2 + bx + c = -\frac{1}{3}x^2 - \frac{2}{3}x + 1.$

63. Let x = the amount Katy receives.
     y = the amount Mike receives.
     z = the amount Danny receives.
and w = the amount Colleen receives.

We are told: $x + y + z + w = 45$, and

$$y = 2x$$

$$w = x$$

$$z = \frac{x}{2}$$

Therefore, by substitution:

$$x + y + z + w = 45$$

$$x + 2x + \frac{x}{2} + x = 45$$

$$2x + 4x + x + 2x = 90 \quad \text{Multiply by 2.}$$

$$9x = 90$$

$$x = 10$$

Katy receives $x = 10$ dollars.
Mike receives $y = 2x = 20$ dollars.
Danny receives $z = \frac{x}{2} = 5$ dollars.
Colleen receives $w = x = 10$ dollars.

65. Let x = airspeed of the plane (i.e, the speed of the plane if it
          was flying in still air)
     y = speed of jet-stream
     d = distance from Chicago to Ft. Lauderdale.

We use the basic formula:

rate × time = distance

We first need to determine if the jet-stream flows from Chicago to Ft. Lauderdale or vice-versa. The trip from Chicago to Ft. Lauderdale takes less time than the return trip, so the jet-stream flows from Chicago to Ft. Lauderdale.

From Chicago to Ft. Lauderdale, we have:

rate: $x + y$ (going <u>with</u> the jet-stream)

time: 2 hrs, 30 min = $2\frac{1}{2} = \frac{5}{2}$ hr.

distance: d

Therefore,

$$(x + y)\left(\frac{5}{2}\right) = d$$

or

$5x + 5y = 2d$ (1) Multiply by 2.

From Ft. Lauderdale to Chicago we have:

rate: $x - y$ (going <u>against</u> jet-stream)

time: 2 hrs. 50 min = $2\frac{5}{6} = \frac{17}{6}$ hr.

distance: d

Therefore,

$$(x - y)\left(\frac{17}{6}\right) = d$$

or

$17x - 17y = 6d$ (1) Multiply by 6.

We have two equations:

$$\begin{cases} 5x + 5y = 2d & (1) \\ 17x - 17y = 6d & (2) \end{cases}$$

But we know $x = 475$ (airspeed of plane), so we have:

$$\begin{cases} 2375 + 5y = 2d & (1) \\ 8075 - 17y = 6d & (2) \end{cases}$$

$$\begin{cases} -7125 - 15y = -6d & (1) \quad \text{Multiply by } -3. \\ 8075 - 17y = 6d & (2) \end{cases}$$

$$\begin{cases} -7125 - 15y = -6d & (1) \\ 950 - 32y = 0 & (2) \quad \text{Replace (2) by (1) + (2).} \end{cases}$$

From (2): $-32y = -950$

$$y = \frac{-950}{-32} = 29.69$$

Therefore, the speed of the jet-stream is 29.69 mph.

10 - CHAPTER REVIEW   645

67. Let x = # of hours for Katy to do the job alone,
    y = # of hours for Mike to do the job alone,
    z = # of hours for Danny to do the job alone.

We have:

	Fraction of the job done in one hour
Katy	$\frac{1}{x}$
Mike	$\frac{1}{y}$
Danny	$\frac{1}{z}$
Katy & Mike	$\frac{1}{x} + \frac{1}{y}$
Mike & Danny	$\frac{1}{y} + \frac{1}{z}$
Danny & Katy	$\frac{1}{x} + \frac{1}{z}$

But we are told that, working together, Katy and Mike can do the job in 1 hour and 20 minutes $\left(1\frac{2}{6} = \frac{8}{6} = \frac{4}{3} \text{ hours}\right)$.

Therefore, together, they complete $\dfrac{1}{\left(\frac{4}{3}\right)} = \dfrac{3}{4}$ of the job per hour:

$$\frac{1}{x} + \frac{1}{y} = \frac{3}{4} \quad (1)$$

Similarly, Mike and Danny do the job in 1 hour and 36 minutes

$$\left(1\frac{36}{60} = 1\frac{6}{10} = \frac{16}{10} = \frac{8}{5} \text{ hr.}\right)$$

Therefore,

$$\frac{1}{y} + \frac{1}{z} = \frac{1}{\left(\frac{8}{5}\right)} = \frac{5}{8} \quad (2)$$

Finally, Danny and Katy take 2 hours 40 minutes

$$\left(2\frac{4}{6} = \frac{16}{6} = \frac{8}{3} \text{ hr.}\right) \text{ to do the job:}$$

$$\frac{1}{x} + \frac{1}{z} = \frac{1}{\left(\frac{8}{3}\right)} = \frac{3}{8}$$

10 SYSTEMS OF EQUATIONS AND INEQUALITIES

We have:

$$\begin{cases} \dfrac{1}{x} + \dfrac{1}{y} = \dfrac{3}{4} & (1) \\[2mm] \dfrac{1}{y} + \dfrac{1}{z} = \dfrac{5}{8} & (2) \\[2mm] \dfrac{1}{x} + \dfrac{1}{z} = \dfrac{3}{8} & (3) \end{cases}$$

Let $u = \dfrac{1}{x}$, $v = \dfrac{1}{y}$, $w = \dfrac{1}{z}$:

$$\begin{cases} u + v = \dfrac{3}{4} & (1) \\[2mm] v + w = \dfrac{5}{8} & (2) \\[2mm] u + w = \dfrac{3}{8} & (3) \end{cases}$$

$$\begin{cases} u + v = \dfrac{3}{4} & (1) \\[2mm] v + w = \dfrac{5}{8} & (2) \\[2mm] -v + w = -\dfrac{3}{8} & (3) \quad \text{Replace (3) by (3) } - \text{ (1).} \end{cases}$$

$$\begin{cases} u + v = \dfrac{3}{4} & (1) \\[2mm] v + w = \dfrac{5}{8} & (2) \\[2mm] 2w = \dfrac{1}{4} & (3) \quad \text{Replace (3) by (3) } + \text{ (2).} \end{cases}$$

$$\begin{cases} u + v = \dfrac{3}{4} & (1) \\[2mm] v + w = \dfrac{5}{8} & (2) \\[2mm] w = \dfrac{1}{8} & (3) \quad \text{Divide by 2.} \end{cases}$$

$$\begin{cases} u + v = \dfrac{3}{4} & (1) \\[2mm] v + \dfrac{1}{8} = \dfrac{5}{8} & (2) \quad \text{Back-substitution; } w = \dfrac{1}{8}. \\[2mm] w = \dfrac{1}{8} & (3) \end{cases}$$

$$\begin{cases} u + v = \dfrac{3}{4} & (1) \\[2mm] \phantom{u + {}} v = \dfrac{1}{2} & (2) \\[2mm] \phantom{u + v = {}} w = \dfrac{1}{8} & (3) \end{cases}$$

$$\begin{cases} u = \dfrac{3}{4} & (1) \quad \text{Back-substitution; } v = \dfrac{1}{2} \\[2mm] v = \dfrac{1}{2} & (2) \\[2mm] w = \dfrac{1}{8} & (3) \end{cases}$$

Finally, $\dfrac{1}{x} = u = \dfrac{1}{4}$, so $x = 4$; $\dfrac{1}{y} = v = \dfrac{1}{2}$, so $y = 2$;

and $\dfrac{1}{z} = w = \dfrac{1}{8}$, so $z = 8$.

Katy can do the job in $x = 4$ hours;
Mike can do it in $y = 2$ hours;
Danny can do it in $z = 8$ hours.

69.　Let w = width of the plot
　　　　ℓ = length of the plot
　　　　d = length of the diagonal
　　　　p = perimeter of the rectangle.

Now w, ℓ, d form a right triangle with hypotenuse = d, so we have

$$w^2 + \ell^2 = d^2 \qquad (1)$$
Also　　$2w + 2\ell = p \qquad (2)$

So we have two equations:

$$\begin{cases} w^2 + \ell^2 = (26)^2 & (1) \\[2mm] 2w + 2\ell = 68 & (2) \end{cases}$$

$$\begin{cases} w^2 + \ell^2 = 676 & (1) \\[2mm] w + \ell = 34 & (2) \quad \text{Divide by 2.} \end{cases}$$

Solve for ℓ in (2):

$$w + \ell = 34$$
$$\ell = 34 - w$$

Substitute this into (1):

$$w^2 + \ell^2 = 676$$
$$w^2 + (34 - w)^2 = 676$$
$$w^2 + 1156 - 68w + w^2 = 676$$
$$2w^2 - 68w + 480 = 0$$
$$w^2 - 34w + 240 = 0$$
$$(w - 24)(w - 10) = 0$$

so $\quad w = 24$ or $w = 10$

Finally, from (2), $\ell = 34 - w$, so:

If $w = 24$, then $\ell = 10$;
If $w = 10$, then $\ell = 24$

Thus, the rectangle is 10 feet by 24 feet.

71. Let x and y be the legs of the triangle. We have:

$$\begin{cases} x + y + 6 = 14 & (1) \quad \text{The perimeter.} \\ x^2 + y^2 = 36 & (2) \quad \text{Pythagorean Theorem.} \end{cases}$$

From (1): $x + y = 8$
$$y = -x + 8$$

Then, by (2):

$$x^2 + y^2 = 36$$
$$x^2 + (-x + 8)^2 = 36$$
$$x^2 + x^2 - 16x + 64 = 36$$
$$2x^2 - 16x + 28 = 0$$
$$x^2 - 8x + 14 = 0$$

$$x = \frac{8 \pm \sqrt{64 - 4(14)}}{2}$$

so $\qquad = \frac{8 \pm \sqrt{8}}{2}$

or

$$x = 4 \pm \sqrt{2}$$

If $x = 4 + \sqrt{2}$, then $y = 8 - x = 4 - \sqrt{2}$

If $x = 4 - \sqrt{2}$, then $y = 8 - x = 4 + \sqrt{2}$

In either case, the lengths of the legs are $4 + \sqrt{2}$ inches and $4 - \sqrt{2}$ inches.

C H A P T E R
11

INDUCTION; SEQUENCES

≡ EXERCISE 11.1 MATHEMATICAL INDUCTION

1. (I) $n = 1$: $2 \cdot 1 = 2$ and $1(1 + 1) = 2$

    (II)   If $2 + 4 + 6 + \dots + 2k = k(k + 1)$,
    then $2 + 4 + 6 + \dots + 2k + 2(k + 1)$
    $= [2 + 4 + 6 + \dots + 2k] + 2(k + 1)$
    $= k(k + 1) + 2(k + 1)$
    $= k^2 + k + 2k + 2$
    $= k^2 + 3k + 2$
    $= (k + 1)(k + 2)$

3. (I) $n = 1$: $1 + 2 = 3$ and $\frac{1}{2}(1)(1 + 5) = \frac{1}{2}(6) = 3$

    (II)   If $3 + 4 + 5 + \dots + (k + 2) = \frac{1}{2}k(k + 5)$,

    then $3 + 4 + 5 + \dots + (k + 2) + [(k + 1) + 2]$
    $= [3 + 4 + 5 + \dots + (k + 2)] + (k + 3)$
    $= \frac{1}{2}k(k + 5) + k + 3$
    $= \frac{1}{2}k^2 + \frac{5}{2}k + k + 3$
    $= \frac{1}{2}k^2 + \frac{7}{2}k + 3$
    $= \frac{1}{2}(k^2 + 7k + 6)$
    $= \frac{1}{2}(k + 1)(k + 6)$

5.　(I)　$n = 1$:　$3 \cdot 1 - 1 = 2$ and $\frac{1}{2}(1)[3(1) + 1] = \frac{1}{2}(4) = 2$

(II)　If　$2 + 5 + 8 + \ldots + (3k - 1) = \frac{1}{2}k(3k + 1)$,

then　$2 + 5 + 8 + \ldots + (3k - 1) + [3(k + 1) - 1]$

$$= [2 + 5 + 8 + \ldots + (3k - 1)] + 3k + 2$$

$$= \frac{1}{2}k(3k + 1) + (3k + 2)$$

$$= \frac{3}{2}k^2 + \frac{1}{2}k + 3k + 2$$

$$= \frac{3}{2}k^2 + \frac{7}{2}k + 2$$

$$= \frac{1}{2}(3k^2 + 7k + 4)$$

$$= \frac{1}{2}(k + 1)(3k + 4)$$

7.　(I)　$n = 1$:　$2^{1-1} = 1$ and $2^1 - 1 = 1$

(II)　If
then
$1 + 2 + 2^2 + \ldots + 2^{k-1} = 2^k - 1$,
$1 + 2 + 2^2 + \ldots + 2^{k-1} + 2^{(k+1)} - 1$
$= (1 + 2 + 2^2 + \ldots + 2^{k-1}) + 2^k$
$= 2k - 1 + 2^k$
$= 2(2^k) - 1$
$= 2^{k+1} - 1$

9.　(I)　$n = 1$:　$4^{1-1} = 1$ and $\frac{1}{3}(4^1 - 1) = \frac{1}{3}(3) = 1$

(II)　If　$1 + 4 + 4^2 + \ldots + 4^{k-1} = \frac{1}{3}(4^k - 1)$,

then　$1 + 4 + 4^2 + \ldots + 4^{k-1} + 4^{(k+1)-1}$

$$= (1 + 4 + 4^2 + \ldots + 4^{(k-1)}) + 4^k$$

$$= \frac{1}{3}(4^k - 1) + 4^k$$

$$= \frac{1}{3}[4^k - 1 + 3(4^k)]$$

$$= \frac{1}{3}[4(4^k) - 1]$$

$$= \frac{1}{3}(4^{k+1} - 1)$$

11.1　MATHEMATICAL INDUCTION

651

11. (I) $n = 1$: $\dfrac{1}{1 \cdot 2} = \dfrac{1}{2}$ and $\dfrac{1}{1 + 1} = \dfrac{1}{2}$

(II)   If $\dfrac{1}{1 \cdot 2} + \dfrac{1}{2 \cdot 3} + \dfrac{1}{3 \cdot 4} + \ldots + \dfrac{1}{k(k + 1)} = \dfrac{k}{k + 1}$,

then $\dfrac{1}{1 \cdot 2} + \dfrac{1}{2 \cdot 3} + \dfrac{1}{3 \cdot 4} + \ldots + \dfrac{1}{k(k + 1)}$

$$+ \dfrac{1}{(k + 1)[(k + 1) + 1]}$$

$$= \left[\dfrac{1}{1 \cdot 2} + \dfrac{1}{2 \cdot 3} + \dfrac{1}{3 \cdot 4} + \ldots + \dfrac{1}{k(k + 1)} \right] + \dfrac{1}{(k + 1)(k + 2)}$$

$$= \dfrac{k}{k + 1} + \dfrac{1}{(k + 1)(k + 2)}$$

$$= \dfrac{k(k + 2) + 1}{(k + 1)(k + 2)}$$

$$= \dfrac{k^2 + 2k + 1}{(k + 1)(k + 2)}$$

$$= \dfrac{(k + 1)(k + 1)}{(k + 1)(k + 2)}$$

$$= \dfrac{k + 1}{k + 2}$$

13. (I) $n = 1$: $1^2 = 1$ and $\dfrac{1}{6} \cdot 1 \cdot 2 \cdot 3 = 1$

(II)   If $1^2 + 2^2 + 3^2 + \ldots + k^2 = \dfrac{1}{6} k(k + 1)(2k + 1)$,

then $1^2 + 2^2 + 3^2 + \ldots + k^2 + (k + 1)^2$

$$= (1^2 + 2^2 + 3^2 + \ldots + k^2) + (k + 1)^2$$

$$= \dfrac{1}{6} k(k + 1)(2k + 1) + (k + 1)^2$$

$$= \dfrac{1}{6} k(2k^2 + 3k + 1) + (k^2 + 2k + 1)$$

$$= \dfrac{1}{3} k^3 + \dfrac{1}{2} k^2 + \dfrac{1}{6} k + k^2 + 2k + 1$$

$$= \dfrac{1}{3} k^3 + \dfrac{3}{2} k^2 + \dfrac{13}{6} k + 1$$

$$= \dfrac{1}{6} (2k^3 + 9k^2 + 13k + 6)$$

$$= \dfrac{1}{6} (k + 1)(2k^2 + 7k + 6)$$

$$= \dfrac{1}{6} (k + 1)(k + 2)(2k + 3)$$

15.   (I) $n = 1$: $5 - 1 = 4$ and $\frac{1}{2}(9 - 1) = \frac{1}{2} \cdot 8 = 4$

     (II)   If $4 + 3 + 2 + \ldots + (5 - k) = \frac{1}{2}k(9 - k)$,

            then $4 + 3 + 2 + \ldots + (5 - k) = [5 - (k + 1)]$

$$= [4 + 3 + 2 + \ldots + (5 - k)] + [5 - (k + 1)]$$

$$= \frac{1}{2}k(9 - k) + 4 - k$$

$$= \frac{9}{2}k - \frac{1}{2}k^2 + 4 - k$$

$$= -\frac{1}{2}k^2 + \frac{7}{2}k + 4$$

$$= \frac{1}{2}(-k^2 + 7k + 8)$$

$$= \frac{1}{2}(k + 1)(8 - k)$$

$$= \frac{1}{2}(k + 1)[9 - (k + 1)]$$

17.   (I) $n = 1$: $1 \cdot (1 + 1) = 2$ and $\frac{1}{3} \cdot 1 \cdot 2 \cdot 3 = 2$

     (II)   If $1 \cdot 2 + 2 \cdot 3 + 3 \cdot 4 + \ldots + k(k + 1)$
$$= \frac{1}{3}k(k + 1)(k + 2),$$

            then $1 \cdot 2 + 2 \cdot 3 + 3 \cdot 4 + \ldots + k(k + 1)$
$$+ (k + 1)(k + 1 + 1)$$

$$= [1 \cdot 2 + 2 \cdot 3 + 3 \cdot 4 + \ldots + k(k + 1)]$$
$$+ (k + 1)(k + 2)$$

$$= \frac{1}{3}k(k + 1)(k + 2) + (k + 1)(k + 2)$$

$$= (k + 1)(k + 2)\left[\frac{1}{3}k + 1\right]$$

$$= (k + 1)(k + 2)\frac{1}{3}(k + 3)$$

$$= \frac{1}{3}(k + 1)(k + 2)(k + 3)$$

11.1 MATHEMATICAL INDUCTION   653

19.   (I) $n = 1$: $1^2 + 1 = 2$ is divisible by 2.

      (II)   If $k^2 + k$ is divisible by 2,

            then $(k + 1)^2 + (k + 1)$

$$= k^2 + 2k + 1 + k + 1$$
$$= (k^2 + k) + (2k + 2)$$

            Since $k^2 + k$ is divisible by 2 and $2k + 2$ is divisible by 2, then $(k + 1)^2 + k + 1$ is divisible by 2.

21.   (I) $n = 1$: $1^2 - 1 + 2 = 2$ is divisible by 2.

      (II)   If $k^2 - k + 2$ is divisible by 2,

            then $(k + 1)^2 - (k + 1) + 2$

$$= k^2 + 2k + 1 - k - 1 + 2$$
$$= (k^2 - k + 2) + 2k + 1 - 1$$
$$= (k^2 - k + 2) + 2k$$

            Since $k^2 - k + 2$ is divisible by 2 and $2k$ is divisible by 2, then $(k + 1)^2 - (k + 1) + 2$ is divisible by 2.

23.   (I) $n = 1$: If $x > 1$, then $x^1 = x > 1$.

      (II)   Assume, for any natural number k, that if $x > 1$, then $x^k > 1$. Show that if $x^k > 1$, then $x^{k+1} > 1$:

$$x^{k+1} = x^k \cdot x > 1 \cdot x = x > 1$$
$$\uparrow$$
$$x^k > 1$$

25.   (I) $n = 1$: $a - b$ is a factor of $a' - b' = a - b$.

      (II)   If $a - b$ is a factor of $a^k - b^k$, show that $a - b$ is a factor of $a^{k+1} - b^{k+1} = a(a^k - b^k) + b^k(a - b)$.

            Since $a - b$ is a factor of $a^k - b^k$ and $a - b$ is a factor of $a - b$, then $a - b$ is a factor of $a^{k+1} - b^{k+1}$.

27. $n = 1$: $1^2 - 1 + 41 = 41$ is a prime number.
      $n = 41$: $41^2 - 41 + 41 = 1681 = 41^2$ is not prime.

29. (I) $n = 1$: $ar^{1-1} = a \cdot 1 = a$ and $a \cdot \dfrac{1 - r'}{1 - r} = a$ because $r \neq 1$.

(II) If $a + ar + ar + ar^2 + \dots + ar^{k-1} = a \cdot \left(\dfrac{1 - r^k}{1 - r} \right)$

then, $a + ar + ar^2 + \dots + ar^{k-1} + ar^{(k+1)-1}$

$= (a + ar + ar^2 + \dots + ar^{k-1}) + ar^k$

$= a \cdot \left(\dfrac{1 - r^k}{1 - r} \right) + ar^k$

$= \dfrac{a(1 - r^k) + ar^k(1 - r)}{1 - r}$

$= \dfrac{a - ar^k + ar^k - ar^{k+1}}{1 - r}$

$= a \cdot \left(\dfrac{1 - r^{k+1}}{1 - r} \right)$

31. (I) $n = 3$: The sum of the angles of a triangle is $(3 - 2) \cdot 180° = 180°$.

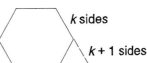
k sides

k + 1 sides

(II) Assume that for any integer k the sum of the angles of a convex polygon of k sides is $(k - 2) \cdot 180°$. A convex polygon of $k + 1$ sides consists of a convex polygon k sides plus a triangle. See the illustration. The sum of the angles is $(k - 2)180° + 180° = (k - 1)180°$. Since Conditions I and II have been met, the result follows.

≡ EXERCISE 11.2 THE BINOMIAL THEOREM

1. $9! = 9 \cdot 8 \cdot 7 \cdot 6 \cdot 5 \cdot 4 \cdot 3 \cdot 2 \cdot 1 = 362,880$

3. $\dfrac{8!5!}{6!6!} = \dfrac{8 \cdot 7 \cdot 6! \cdot 5!}{6! \cdot 6 \cdot 5!} = \dfrac{8 \cdot 7}{6} = \dfrac{28}{3}$

5. $\dbinom{5}{3} = \dfrac{5!}{3!2!} = \dfrac{5 \cdot 4}{2} = 10$

7. $\dbinom{7}{5} = \dfrac{7!}{5!2!} = \dfrac{7 \cdot 6}{2} = 21$

9. $\dbinom{50}{49} = \dfrac{50!}{49!1!} = 50$

11. $\dbinom{1000}{1000} = 1$

13. $\dbinom{55}{13} = 1.451 \times 10^{12}$

15. $\dbinom{47}{25} = 1.483 \times 10^{13}$

17. $(x + 1)^5 = \dbinom{5}{0}x^5 + \dbinom{5}{1}x^4 + \dbinom{5}{2}x^3 + \dbinom{5}{3}x^2 + \dbinom{5}{4}x + \dbinom{5}{5}$

$= x^5 + 5x^4 + 10x^3 + 10x^2 + 5x + 1$

19. $(x - 2)^6 = \binom{6}{0}x^6 + \binom{6}{1}(-2)x^5 + \binom{6}{2}(-2)^2x^4 + \binom{6}{3}(-2)^3x^3 + \binom{6}{4}(-2)^4x^2$

$\qquad + \binom{6}{5}(-2)^5x + \binom{6}{6}(-2)^6$

$\qquad = x^6 + 6 \cdot -2x^5 + 15 \cdot 4x^4 + 20 \cdot -8x^3 + 15 \cdot 16x^2$

$\qquad + 6 \cdot -32x + 64$

$\qquad = x^6 - 12x^5 + 60x^4 - 160x^3 + 240x^2 - 192x + 64$

21. $(3x + 1)^4 = \binom{4}{0}(3x)^4 + \binom{4}{1}(3x)^3 + \binom{4}{2}(3x)^2 + \binom{4}{3}(3x) + \binom{4}{4}$

$\qquad = 81x^4 + 4 \cdot 27x^3 + 6 \cdot 9x^2 + 4 \cdot 3x + 1$

$\qquad = 81x^4 + 108x^3 + 54x^2 + 12x + 1$

23. $(x^2 + y^2)^5 = \binom{5}{0}(x^2)^5 + \binom{5}{1}(y^2)(x^2)^4 + \binom{5}{2}(y^2)^2(x^2)^3 + \binom{5}{3}(y^2)^3(x^2)^2$

$\qquad + \binom{5}{4}(y^2)^4(x^2) + \binom{5}{5}(y^2)^5$

$\qquad = x^{10} + 5y^2x^8 + 10y^4x^6 + 10y^6x^4 + 5y^8x^2 + y^{10}$

25. $\left(\sqrt{x} + \sqrt{2}\right)^6 = \binom{6}{0}\left(\sqrt{x}\right)^6 + \binom{6}{1}\sqrt{2}\left(\sqrt{x}\right)^5 + \binom{6}{2}\left(\sqrt{2}\right)^2\left(\sqrt{x}\right)^4 + \binom{6}{3}\left(\sqrt{2}\right)^3\left(\sqrt{x}\right)^3$

$\qquad + \binom{6}{4}\left(\sqrt{2}\right)^4\left(\sqrt{x}\right)^2 + \binom{6}{5}\left(\sqrt{2}\right)^5\left(\sqrt{x}\right) + \binom{6}{6}\left(\sqrt{2}\right)^6$

$\qquad = x^3 + 6\sqrt{2}\left(\sqrt{x}\right)^5 + 15\left(\sqrt{2}\right)^2\left(\sqrt{x}\right)^4 + 20\left(\sqrt{2}\right)^3\left(\sqrt{x}\right)^3$

$\qquad + 15\left(\sqrt{2}\right)^4\left(\sqrt{x}\right)^2 + 6\left(\sqrt{2}\right)^5\sqrt{x} + \left(\sqrt{2}\right)^6$

$\qquad = x^3 + 6\sqrt{2}x^{5/2} + 30x^2 + 40\sqrt{2}x^{3/2} + 60x + 24\sqrt{2}x^{1/2} + 8$

27. $(ax + by)^5 = \binom{5}{0}(ax)^5 + \binom{5}{1}(by)(ax)^4 + \binom{5}{2}(by)^2(ax)^3 + \binom{5}{3}(by)^3(ax)^2$

$\qquad + \binom{5}{4}(by)^4(ax) + \binom{5}{5}(by)^5$

$\qquad = (ax)^5 + 5by(ax)^4 + 10(by)^2(ax)^3 + 10(by)^3(ax)^2$

$\qquad + 5(by)^4(ax) + (by)^5$

29. $n = 10$, $a = 3$, $x = x$, and $j = 6$

$\binom{10}{10 - 6}3^{10-6}x^6 = \binom{10}{4}3^4 \cdot x^6 = \frac{10!}{4!6!} \cdot 81 \cdot x^6 = \frac{10 \cdot 9 \cdot 8 \cdot 7}{4 \cdot 3 \cdot 2} \cdot 81x^6$

$\qquad\qquad\qquad\qquad\qquad\qquad\qquad\qquad\qquad = 17,010x^6$

The coefficient of x^6 is 17,010.

31. $n = 12$, $a = -1$, $x = 2x$, and $j = 8$

$$\binom{12}{12-8}(-1)^{12-8}(2x)^8 = \binom{12}{4}(-1)^4 \cdot 256 \cdot x^8 = \frac{12!}{4!8!} \cdot 256 \cdot x^8$$

$$= 126{,}720x^8$$

The coefficient of x^8 is $126{,}720$.

33. $n = 9$, $a = 3$, $x = 2x$, and $j = 7$

$$\binom{9}{9-7}(3)^{9-7}(2x)^7 = \binom{9}{2}3^2 \cdot 128 \cdot x^7 = \frac{9!}{2!7!} \cdot 9 \cdot 128 \cdot x^7 = 41{,}472x^7$$

The coefficient of x^7 is $41{,}472$.

35. The fifth term contains x^3. $n = 7$, $a = 3$, $x = x$, and $j = 3$

$$\binom{7}{7-3}(3)^{7-3}x^3 = \binom{7}{4}3^4 \cdot x^3 = \frac{7!}{4!3!} \cdot 81 \cdot x^3 = 2835x^3$$

37. The third term contains x^7. $n = 9$, $a = -2$, $x = 3x$, and $j = 7$

$$\binom{9}{9-7}(-2)^{9-7}(3x)^7 = \binom{9}{2}(-2)^2(3x)^7 = \frac{9!}{2!7!} \cdot 4 \cdot 2187 \cdot x^7 = 314{,}928x^7$$

39. $\left(x^2 + \dfrac{1}{x}\right)^{12}$

Constant term $= \binom{12}{12-j}\left(\dfrac{1}{x}\right)^{12-j}(x^2)^j$ where $2j = 12 - j$ or $j = 4$

Constant term $= \binom{12}{8}\left(\dfrac{1}{x}\right)^8(x^2)^4 = \binom{12}{8} = \dfrac{12!}{8!4!} = \dfrac{12 \cdot 11 \cdot 10 \cdot 9}{4 \cdot 3 \cdot 2} = 495$

41. $\left(x - \dfrac{2}{\sqrt{x}}\right)^{10}$

x^4 term $= \binom{10}{10-j}\left(\dfrac{-2}{\sqrt{10}}\right)^{10-j}x^j$ where $j - \dfrac{10-j}{2} = 4$ or $\dfrac{3}{2}j = 9$ or $j = 6$

x^4 term $= \binom{10}{4}\left(\dfrac{-2}{\sqrt{10}}\right)^4 x^6 = 16\binom{10}{6}\left(\dfrac{x^6}{x^2}\right) = \dfrac{16 \cdot 10 \cdot 9 \cdot 8 \cdot 7}{4 \cdot 3 \cdot 2} = 3360$

43. $(1.001)^5 = (1 + 10^{-3})^5 = 1 + \binom{5}{1} \cdot 10^{-3} + \binom{5}{2} \cdot (10^{-3})^2$

$$+ \binom{5}{3}(10^{-3})^3 + \ldots$$

$$= 1 + .005 + 10(.000001) + 10(.000000001) + \ldots$$
$$= 1 + .005 + .00001 + \ldots$$
$$\approx 1.00501 \text{ correct to five decimal places}$$

45. $\binom{n}{n} = 1$

$$\dfrac{n!}{n!(n-n)!} = \dfrac{n!}{n!0!} = \dfrac{n!}{n! \cdot 1} = \dfrac{n!}{n!} = 1$$

11.2 THE BINOMIAL THEOREM

47. $\quad \dbinom{n}{0} + \dbinom{n}{1} + \dots + \dbinom{n}{n} = 2^n$

$\qquad \dbinom{n}{0} + \dbinom{n}{1} + \dots + \dbinom{n}{n} = (1 + 1)^n$

$$= \dbinom{n}{0}1^n + \dbinom{n}{1}(1)^1(1)^{n-1} + \dbinom{n}{2}(1)^2(1)^{n-2} + \dots$$

$$+ \dbinom{n}{n}(1)^n(1)^{n-n}$$

$$= \dbinom{n}{0} + \dbinom{n}{1} + \dots + \dbinom{n}{n}$$

49. $\quad 12! = 4.790016 \times 10^8$

$\qquad 20! = 2.432902 \times 10^{18}$

$\qquad 25! = 1.551121 \times 10^{25}$

$\qquad 12! \approx \sqrt{2 \cdot 12\pi}\left(\dfrac{12}{e}\right)^{12}\left(1 + \dfrac{1}{12 \cdot 12 - 1}\right)$

$\qquad 12! \approx \sqrt{24\pi}\,(54782414.5)(1.006993007)$

$\qquad 12! \approx (8.683215055)(54782414.5)(1.006993007)$

$\qquad 12! \approx 479013972.2$

$\qquad 12! \approx 4.790139722 \times 10^8$

$\qquad 20! \approx \sqrt{2 \cdot 20\pi}\left(\dfrac{20}{e}\right)^{20}\left(1 + \dfrac{1}{12 \cdot 20 - 1}\right)$

$\qquad 20! \approx \sqrt{40\pi}\,(2.1612762 \times 10^{17})(1.0041841)$

$\qquad 20! \approx (11.20998243)(2.1612762 \times 10^{17})(1.0041841)$

$\qquad 20! \approx 2.432924 \times 10^{18}$

$\qquad 25! \approx \sqrt{2 \cdot 25\pi}\left(\dfrac{25}{e}\right)^{25}\left(1 + \dfrac{1}{12 \cdot 25 - 1}\right)$

$\qquad 25! \approx \sqrt{50\pi}\,(1.2334972 \times 10^{24})(1.003344481)$

$\qquad 25! \approx (12.53314137)(1.2334972 \times 10^{24})(1.003344481)$

$\qquad 25! \approx 1.5511299 \times 10^{25}$

≡ EXERCISE 11.3 SEQUENCES

1. $\quad a_1 = 1$
$\qquad a_2 = 2$
$\qquad a_3 = 3$
$\qquad a_4 = 4$
$\qquad a_5 = 5$

3. $\quad a_1 = \dfrac{1}{1 + 1} = \dfrac{1}{2}$

$\qquad a_2 = \dfrac{2}{2 + 1} = \dfrac{2}{3}$

$\qquad a_3 = \dfrac{3}{3 + 1} = \dfrac{3}{4}$

$\qquad a_4 = \dfrac{4}{4 + 1} = \dfrac{4}{5}$

$\qquad a_5 = \dfrac{5}{5 + 1} = \dfrac{5}{6}$

5. $a_1 = (1)(1^2) = 1$
    $a_2 = (-1)(2^2) = -4$
    $a_3 = (1)(3^2) = 9$
    $a_4 = (-1)(4^2) = -16$
    $a_5 = (1)(5^2) = 25$

7. $a_1 = \dfrac{2}{4} = \dfrac{1}{2}$

    $a_2 = \dfrac{4}{10} = \dfrac{2}{5}$

    $a_3 = \dfrac{8}{28} = \dfrac{2}{7}$

    $a_4 = \dfrac{16}{82} = \dfrac{8}{41}$

    $a_5 = \dfrac{32}{244} = \dfrac{8}{61}$

9. $a_1 = \dfrac{-1}{(2)(3)} = \dfrac{-1}{6}$

    $a_2 = \dfrac{1}{(3)(4)} = \dfrac{1}{12}$

    $a_3 = \dfrac{-1}{(4)(5)} = \dfrac{-1}{20}$

    $a_4 = \dfrac{1}{(5)(6)} = \dfrac{1}{30}$

    $a_5 = \dfrac{-1}{(6)(7)} = \dfrac{-1}{42}$

11. $a_1 = \dfrac{1}{e}$

    $a_2 = \dfrac{2}{e^2}$

    $a_3 = \dfrac{3}{e^3}$

    $a_4 = \dfrac{4}{e^4}$

    $a_5 = \dfrac{5}{e^5}$

13. $\dfrac{n}{n+1}$

15. $\dfrac{1}{2^{n-1}}$

17. $(-1)^{n+1}$

19. $(-1)^{n+1}n$

21. $a_1 = 1 = 1$
    $a_2 = 2 + 1 = 3$
    $a_3 = 2 + 3 = 5$
    $a_4 = 2 + 5 = 7$
    $a_5 = 2 + 7 = 9$

23. $a_1 = -2$
    $a_2 = 1 + -2 = -1$
    $a_3 = 2 + -1 = 1$
    $a_4 = 3 + 1 = 4$
    $a_5 = 4 + 4 = 8$

25. $a_1 = 5$
    $a_2 = 2 \cdot 5 = 10$
    $a_3 = 2 \cdot 10 = 20$
    $a_4 = 2 \cdot 20 = 40$
    $a_5 = 2 \cdot 40 = 80$

27. $a_1 = 3$

    $a_2 = \dfrac{3}{1} = 3$

    $a_3 = \dfrac{3}{2}$

    $a_4 = \dfrac{\frac{3}{2}}{3} = \dfrac{1}{2}$

    $a_5 = \dfrac{\frac{1}{2}}{4} = \dfrac{1}{8}$

11.3 SEQUENCES

29. $a_1 = 1$
$a_2 = 2$
$a_3 = 1 \cdot 2 = 2$
$a_4 = 2 \cdot 2 = 4$
$a_5 = 2 \cdot 4 = 8$

31. $a_1 = A$
$a_2 = A + d$
$a_3 = A + 2d$
$a_4 = A + 3d$
$a_5 = A + 4d$

33. $a_1 = \sqrt{2}$

$a_2 = \sqrt{2 + \sqrt{2}}$

$a_3 = \sqrt{2 + \sqrt{2 + \sqrt{2}}}$

$a_4 = \sqrt{2 + \sqrt{2 + \sqrt{2 + \sqrt{2}}}}$

$a_5 = \sqrt{2 + \sqrt{2 + \sqrt{2 + \sqrt{2 + \sqrt{2}}}}}$

35. $\displaystyle\sum_{k=1}^{n} (k + 1) = 2 + 3 + 4 + \ldots + (n + 1)$

37. $\displaystyle\sum_{k=1}^{n} \frac{k^2}{2} = \frac{1}{2} + 2 + \frac{9}{2} + \ldots + \frac{n^2}{2}$

39. $\displaystyle\sum_{k=0}^{n} \frac{1}{3^k} = 1 + \frac{1}{3} + \frac{1}{9} + \ldots + \frac{1}{3^n}$

41. $\displaystyle\sum_{k=0}^{n-1} \frac{1}{3^{k+1}} = \frac{1}{3} + \frac{1}{9} + \frac{1}{27} + \ldots + \frac{1}{3^n}$

43. $\displaystyle\sum_{k=2}^{n} (-1)^k \ln k = \ln 2 - \ln 3 + \ln 4 = \ldots + (-1)^n \ln n$

45. $1 + 2 + 3 + \ldots + n = \displaystyle\sum_{k=1}^{n} k$

47. $\dfrac{1}{2} + \dfrac{2}{3} + \dfrac{3}{4} + \ldots + \dfrac{n}{n+1} = \displaystyle\sum_{k=1}^{n} \dfrac{k}{k+1}$

49. $1 - \dfrac{1}{3} + \dfrac{1}{9} - \dfrac{1}{27} + \ldots + (-1)^n \dfrac{1}{3^n} = \displaystyle\sum_{k=0}^{n} (-1)^k \left(\dfrac{1}{3^k}\right)$

51. $3 + \dfrac{3^2}{2} + \dfrac{3^3}{3} + \ldots + \dfrac{3^n}{n} = \displaystyle\sum_{k=1}^{n} \dfrac{3k}{k}$

11 INDUCTION; SEQUENCES

53. $$a + (a + d) + (a + 2d) + \ldots + (a + nd) = \sum_{k=1}^{n} a + (k - 1)d$$

$$= \sum_{k=0}^{n} (a + kd)$$

55. $a_1 = 1,\ a_2 = 1,\ a_3 = 2,\ a_4 = 3,\ a_5 = 5,\ a_6 = 8,\ a_7 = 13,\ a_8 = 21,$

$a_{n+2} = a_{n+1} + a_n$

$a_8 = a_6 + a_7$

$a_8 = 8 + 13$

$a_8 = 21$

59. $1,\ 1,\ 2,\ 3,\ 5,\ 8,\ 13$

The Fibonacci sequence.

≡ EXERCISE 11.4 ARITHMETIC AND GEOMETRIC SEQUENCES; GEOMETRIC SERIES

1. $d = a_{n+1} - a_n$
$d = (n + 6) - (n + 5)$
$d = n + 6 - n - 5$
$d = 1$

$a_1 = 6$
$a_2 = 7$
$a_3 = 8$
$a_4 = 9$

3. $d = [2(n + 1) - 5] - (2n - 5)$
$d = 2n + 2 - 5 - 2n + 5$
$d = 2$

$a_1 = -3$
$a_2 = -1$
$a_3 = 1$
$a_4 = 3$

5. $d = [6 - 2(n + 1)] - (6 - 2n)$
$d = 6 - 2n - 2 - 6 + 2n$
$d = -2$

$a_1 = 4$
$a_2 = 2$
$a_3 = 0$
$a_4 = -2$

7. $d = \left[\frac{1}{2} - \frac{1}{3}(n + 1)\right] - \frac{1}{2} - \frac{1}{3}n$

$d = \frac{1}{2} - \frac{1}{3}n - \frac{1}{3} - \frac{1}{2} + \frac{1}{3}n$

$d = -\frac{1}{3}$

$a_1 = \frac{1}{6}$

$a_2 = -\frac{1}{6}$

$a_3 = -\frac{1}{2}$

$a_4 = -\frac{5}{6}$

9.
$$d = (\ln 3^{n+1}) - (\ln 3^n)$$
$$d = (n + 1)\ln 3 - n(\ln 3)$$
$$d = \ln 3(n + 1 - n)$$
$$d = \ln 3$$

$$a_1 = \ln 3$$
$$a_2 = 2 \ln 3$$
$$a_3 = 3 \ln 3$$
$$a_4 = 4 \ln 3$$

11.
$$r = \frac{2^{n+1}}{2^n}$$

$$r = 2^{n+1-n}$$

$$r = 2$$

$$a_1 = 2$$
$$a_2 = 4$$
$$a_3 = 8$$
$$a_4 = 16$$

13.
$$r = \frac{-3\left(\frac{1}{2}\right)^{n+1}}{-3\left(\frac{1}{2}\right)^n}$$

$$r = \left(\frac{1}{2}\right)^{n+1-n}$$

$$r = \frac{1}{2}$$

$$a_1 = -\frac{3}{2}$$

$$a_2 = -\frac{3}{4}$$

$$a_3 = -\frac{3}{8}$$

$$a_4 = -\frac{3}{16}$$

15.
$$r = \frac{\frac{2^{(n+1)-1}}{4}}{\frac{2^{n-1}}{4}}$$

$$r = \frac{2^n}{2^{n-1}}$$

$$r = 2^{n-(n-1)}$$

$$r = 2$$

$$a_1 = \frac{1}{4}$$

$$a_2 = \frac{1}{2}$$

$$a_3 = 1$$

$$a_4 = 2$$

17.
$$r = \frac{2^{\frac{n+1}{3}}}{2^{\frac{n}{3}}}$$

$$r = 2^{\frac{n+1}{3} - \frac{n}{3}}$$

$$r = 2^{1/3}$$

$$a_1 = 2^{1/3}$$

$$a_2 = 2^{2/3}$$

$$a_3 = 2$$

$$a_4 = 2^{4/3}$$

19.
$$r = \frac{\frac{3^{(n+1)-1}}{2^{n+1}}}{\frac{3^{n-1}}{2^n}}$$

$$r = \frac{3^n}{2\left(3^{n-1}\right)}$$

$$r = \frac{1}{2}3^{n-(n-1)}$$

$$r = \frac{3}{2}$$

$$a_1 = \frac{1}{2}$$

$$a_2 = \frac{3}{4}$$

$$a_3 = \frac{9}{8}$$

$$a_4 = \frac{27}{16}$$

21. Arithmetic

$d = [(n + 1) + 2] - (n + 2)$
$d = n + 3 - n - 2$
$d = n + 3 - n - 2$
$d = 1$

23. Neither

25. Arithmetic

$d = \left[3 - \dfrac{2}{3}(n + 1)\right] - \left(3 - \dfrac{2}{3}n\right)$

$d = 3 - \dfrac{2}{3}n - \dfrac{2}{3} - 3 + \dfrac{2}{3}n$

$d = \dfrac{-2}{3}$

27. Neither

29. Geometric

$r = \dfrac{\left(\dfrac{2}{3}\right)^{n+1}}{\left(\dfrac{2}{3}\right)^{n}}$

$r = \dfrac{2^{\,n+1-n}}{3}$

$r = \dfrac{2}{3}$

31. Geometric

$r = \dfrac{-\left(2^{(n+1)-1}\right)}{-\left(2^{n-1}\right)}$

$r = 2^{n-(n-1)}$

$r = 2$

33. Geometric

$r = \dfrac{3^{\frac{n+1}{2}}}{3^{\frac{n}{2}}}$

$r = 3^{\frac{n+1}{2} - \frac{n}{2}}$

$r = 3$

$r = 3^{1/2}$

35. $a_n = a_1 + (n - 1)d$
$a_5 = 1 + 4(2)$
$a_5 = 1 + 8$
$a_5 = 9$
$a_n = 1 + (n - 1)2$
$a_n = 1 + 2n - 2$
$a_n = 2n - 1$

37. $a_5 = 5 + 4(-3) = -7$
$a_n = 5 + (n - 1) - 3$
$a_n = 5 - 3n + 3$
$a_n = 8 - 3n$

39. $a_5 = 0 + 4\left(\dfrac{1}{2}\right) = 2$

$a_n = 0 + (n - 1)\dfrac{1}{2}$

$a_n = \dfrac{1}{2}n - \dfrac{1}{2}$

$a_n = \dfrac{1}{2}(n - 1)$

41.
$$a_5 = \sqrt{2} + 4\left(\sqrt{2}\right)$$
$$a_5 = \sqrt{2} + 4\sqrt{2}$$
$$a_5 = \sqrt{2}(1 + 4)$$
$$a_5 = 5\sqrt{2}$$
$$a_n = \sqrt{2} + (n - 1)\sqrt{2}$$
$$a_n = \sqrt{2}(1 + n - 1)$$
$$a_n = n\sqrt{2} = \sqrt{2}\,n$$

43.
$$a_n = ar^{n-1}$$
$$a_5 = 2^4(1) = 2^4 = 16$$
$$a_n = 2^{n-1}(1)$$
$$a_n = 2^{n-1}$$

45.
$$a_5 = -1^4(5) = 1(5) = 5$$
$$a_n = (-1)^{n-1}(5)$$

47.
$$a_5 = \left(\frac{1}{2}\right)^4(0) = 0$$
$$a_n = \left(\frac{1}{2}\right)^{n-1}(0)$$
$$a_n = 0$$

49.
$$a_5 = \sqrt{2}^4\left(\sqrt{2}\right) = \sqrt{2}^5 = 4\sqrt{2}$$
$$a_n = \sqrt{2}^{n-1}\sqrt{2}$$
$$a_n = \left(\sqrt{2}\right)^n$$

51. The first term of the arithmetic sequence is $a = 2$, and the common difference is 2. The nth term obeys the formula:
$$a_n = 2 + (n - 1)2$$
Hence, the 12th term is
$$a_{12} = 2 + (11)2$$
$$a_{12} = 2 + 22$$
$$a_{12} = 24$$

53.
$$a = 1, \quad d = -3$$
$$a_n = 1 + (n - 1) - 3$$
$$a_n = 1 - 3n + 3$$
$$a_n = 4 - 3n$$
$$a_{10} = 1 + 9(-3)$$
$$a_{10} = 1 - 27 = -26$$

Also, $\quad a_{10} = 4 - 3(10)$
$$a_{10} = 4 - 30$$
$$a_{10} = -26$$

55.
$$a = a, \quad d = b$$
$$a_n = a + (n - 1)b$$
$$a_8 = a + 7b$$

57. The first term of this geometric sequence is $a = 1$, and the common ratio is $\frac{1}{2}$. $\quad \dfrac{\frac{1}{2}}{1} = \frac{1}{2}.$ The nth term obeys the formula:
$$a_n = \left(\frac{1}{2}\right)^{n-1}(1) = \left(\frac{1}{2}\right)^{n-1}$$
$$a_7 = \left(\frac{1}{2}\right)^8 = \frac{1}{64}$$

59. $a = 1,\quad r = -1$

$a_n = (-1)^{n-1}(1)$

$a_n = (-1)^{n-1}$

$a_9 = (-1)^8 = 1$

61. $a = 0.4,\quad r = 0.1$

$a_n = (0.1)^{n-1}(0.4)$

$a_8 = (0.1)^7(0.4)$

$a_8 = (0.0000001)(0.4)$

$a_8 = 0.00000004$

63. $a_8 = a + 7d = 8$

$a_{20} = a + 19d = 44$

$\phantom{a_{20} = a + 19}-12d = -36$

$\phantom{a_{20} = a + 19d}d = 3$

$a = 8 - 7d$

$a = 8 - 7(3)$

$a = 8 - 21$

$a = -13$

65. $a_9 = a + 8d = -5$

$a_{15} = a + 14d = 31$

$\phantom{a_{15} = a + 1}-6d = -36$

$\phantom{a_{15} = a + 14}d = 6$

$a = -5 - 8d$

$a = -5 - 8(6)$

$a = -53$

67. $a_{15} = a + 14d = 0$

$a_{40} = a + 39d = -50$

$\phantom{a_{40} = a + 3}-25d = 50$

$\phantom{a_{40} = a + 39}d = -2$

$a = -14d$

$a = -14(-2)$

$a = 28$

69. $a_{15} = a + 13d = -1$

$a_{18} = a + 17d = -9$

$\phantom{a_{18} = a + 1}-4d = 8$

$\phantom{a_{18} = a + 17}d = -2$

$a = -1 - 13d$

$a = -1 - 13(-2)$

$a = 25$

71. The sequence $\{1 + 5n\}$ is an arithmetic sequence with first term $a = 6$ and nth-term $(1 + 5n)$. To find the sum S_n we use Formula (4).

$$S_n = \frac{n}{2}[a + a_n] = \frac{n}{2}[6 + (1 + 5n)] = \frac{n}{2}(5n + 7)$$

73. The sequence $\dfrac{2^{n-1}}{4}$ is a geometric sequence with $a = \dfrac{1}{4}$ and $r = 2$.

$$S_n = \sum_{k=1}^{n} a_k = a_1 + a_2 + \ldots + a_n = a\left(\frac{1 - r^n}{1 - r}\right)$$

In this sequence,

$$S_n = \sum_{k=1}^{n} \frac{2^{k-1}}{4} = \frac{1}{4} + \frac{2}{4} + \frac{2^2}{4} + \frac{2^3}{4} + \ldots + \frac{2^{n-1}}{4} = \frac{1}{4}\left[\frac{1 - 2^n}{1 - 2}\right]$$

$$= -\frac{1}{4}(1 - 2^n)$$

75. Geometric sequence: $a = \dfrac{2}{3},\quad r = \dfrac{2}{3}$

$$S_n = \sum_{k=1}^{n} \left(\frac{2}{3}\right)^k = \frac{2}{3} + \left(\frac{2}{3}\right)^2 + \left(\frac{2}{3}\right)^3 + \ldots + \left(\frac{2}{3}\right)^n$$

$$= \frac{2}{3}\left[\frac{1 - \left(\frac{2}{3}\right)^n}{1 - \left(\frac{2}{3}\right)}\right] = \frac{2}{3}\left[\frac{1 - \left(\frac{2}{3}\right)^n}{\frac{1}{3}}\right] = 2\left[1 - \left(\frac{2}{3}\right)^n\right]$$

77. Arithmetic sequence: $a = 2$, $a_n = 2n$

$$S_n = \frac{n}{2}[2 + 2n] = n(1 + n)$$

79. Geometric sequence: $a = -1$, $r = 2$

$$S_n = \sum_{k=1}^{n} -\left(2^{k-1}\right) = -1 - 2 - 4 - 8 - \ldots - \left(2^{n-1}\right)$$

$$= -1\left(\frac{1 - 2^n}{1 - 2}\right) = 1 - 2^n$$

81. This geometric series has first term $a = 1$ and common ratio $r = \frac{1}{3}$. Since $|r| < 1$, its sum is

$$1 + \frac{1}{3} + \frac{1}{9} + \ldots = \frac{1}{1 - \frac{1}{3}} = \frac{3}{2}$$

83. This geometric series has first terms $a = 8$ and common ratio $r = \frac{1}{2}$. Since $|r| < 1$, its sum is

$$8 + 4 + 2 + \ldots = \frac{8}{1 - \frac{1}{2}} = 16$$

85. This geometric series has first terms $a = 2$ and common ratio $r = \frac{-1}{4}$. Since $|r| < 1$, its sum is

$$2 + \frac{1}{2} + \frac{1}{8} - \frac{1}{32} + \ldots = \frac{2}{1 + \frac{1}{4}} = \frac{8}{5}$$

87. This geometric series has first terms $a = 3$ and common ratio $r = \frac{1}{4}$. Since $|r| < 1$, its sum is

$$\sum_{k=1}^{\infty} 3\left(\frac{1}{4}\right)^{k-1} = \frac{3}{1 - \frac{1}{4}} = 4$$

89. This geometric series has first term $a = 6$ and common ratio $r = \frac{-2}{3}$. Since $|r| < 1$, its sum is

$$\sum_{k=1}^{\infty} 6\left(\frac{-2}{3}\right)^{k-1} = \frac{6}{1 + \frac{2}{3}} = \frac{18}{5}$$

91. The difference of successive terms must be a constant.

$$(2x + 1) - (x + 3) = x - 2$$
$$(5x + 2) - (2x + 1) = 3x + 1$$

Thus, $x - 2 = 3x + 1$
$$-3 = 2x$$
$$x = \frac{-3}{2}$$

93. $a = \frac{3}{4}(20)$, geometric sequence

$a = 15$

$r = \frac{3}{4}$

$a_3 = \left(\frac{3}{4}\right)^2 (15) = \frac{9}{16} \cdot 15 = \frac{135}{16}$ ft.

$a_n = \left(\frac{3}{4}\right)^{n-1} \cdot 15 = \left(\frac{3}{4}\right)^n \left(\frac{3}{4}\right)^{-1} (15) = \left(\frac{3}{4}\right)^n \left(\frac{4}{3} \cdot 15\right)$

$a_n = \left(\frac{3}{4}\right)^n \cdot 20$ ft.

After striking the ground the nth time, the height h is

$$h = \left(\frac{3}{4}\right)^n \cdot 20$$

If $h = 6$, $6 = \left(\frac{3}{4}\right)^n \cdot 20$

$$\left(\frac{3}{4}\right)^n = \frac{6}{20}$$

$$n = \log_{\frac{3}{4}} \frac{6}{20} = \frac{\ln \frac{6}{20}}{\ln \frac{3}{4}} = 4.185$$

After the 5th time, its height is less than 6 inches.

95.

Begin	After 1 year:	After 2 years:
18,000	$18000 + .05(18000)$	$18000(1 + .05)$ $+ 18000(1 + .05)(.05)$ $= 18000(1 + .05)^2$

After 4 years, the salary is $18,000(1 + .05)^4 = \$21,879.11$

97. Note: The ball actually bounces
straight upward and downward in a
vertical position.

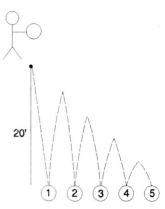

$$S_5 = 20 + \sum_{k=1}^{4} 2\left(\frac{3}{4}\right)^k \cdot 20 = 20 + 15(2) + \frac{45}{4}(2) + \frac{135}{16}(2) + \frac{405}{64}(2)$$

$$= 20 + 30 + 23 + 17 + 12$$

$$= 102 \text{ feet}$$

$$= 20 + 30\frac{1 - \left(\frac{3}{4}\right)^4}{1 - \frac{3}{4}}$$

$$= 20 + 30\frac{175}{256} \cdot 4$$

$$= 102 \text{ feet}$$

99. Geometric sequence, $a = 1$, $r = 2$

$$S_{64} = \sum_{k=1}^{64} 2^{k-1} = 1 + 2 + 4 + \ldots + 2^{63}$$

$$= 1\left[\frac{1 - 2^{64}}{1 - 2}\right] = -(1 - 2^{64}) = -1 + 2^{64}$$

$$= 1.845 \times 10^{19}$$

101. The shaded region is

$$\frac{1}{4} + \frac{1}{4}\left(\frac{1}{4}\right) + \frac{1}{4}\left(\frac{1}{4}\right)\left(\frac{1}{4}\right) + \ldots = \frac{1}{4}\left(1 + \frac{1}{4} + \frac{1}{4^2} + \frac{1}{4^3} + \ldots\right)$$

$$= \frac{1}{4}\left(\frac{1}{1 - \frac{1}{4}}\right) = \frac{1}{4}\left(\frac{1}{3/4}\right) = \frac{1}{3}$$

1.   (I) $n = 1$: $3 \cdot 1 = 3$ and $\dfrac{3 \cdot 1}{2}(2) = 3$

  (II)   If $3 + 6 + 9 + \ldots + 3k = \dfrac{3k}{2}(k + 1)$,

  then $3 + 6 + 9 + \ldots + 3k + 3(k + 1)$

  $= (3 + 6 + 9 + \ldots + 3k) + (3k + 3)$

  $= \dfrac{3k^2}{2}(k + 1) + (3k + 3)$

  $= \dfrac{3k^2}{2} + \dfrac{3k}{2} + 3k + 3$

  $= \dfrac{3k^2}{2} + \dfrac{3k}{2} + \dfrac{6k}{2} + \dfrac{6}{2}$

  $= \dfrac{3k^2}{2} + \dfrac{9k}{2} + \dfrac{6}{2}$

  $= \dfrac{3}{2}(k^2 + 3k + 2)$

  $= \dfrac{3}{2}(k + 1)(k + 2)$

3.   (I) $n = 1$: $2 \cdot 3^{1-1} = 2$ and $3^1 - 1 = 2$

  (II)   If $2 + 6 + 18 + \ldots + 2 \cdot 3^{k-1} = 3^k - 1$,

  then $2 + 6 + 18 + \ldots + 2 \cdot 3^{k-1} + 2 \cdot 3^{(k+1)-1}$

  $= (2 + 6 + 18 + \ldots + 2 \cdot 3^{k-1}) + 2 \cdot 3^k$

  $= 3^k - 1 + 2 \cdot 3^k$

  $= 3^k(1 + 2) - 1$

  $= 3 \cdot 3^k - 1$

  $= 3^{k+1} - 1$

5.   (I) $n = 1$: $1^2 = 1$ and $\frac{1}{2}(6 - 3 - 1) = \frac{1}{2}(2) = 1$

       (II)   If $1^2 + 4^2 + 7^2 + \ldots + (3k - 2)^2 = \frac{1}{2}k(6k^2 - 3k - 1)$

             then $1^2 + 4^2 + 7^2 + \ldots + (3k - 2)^2 + [3(k + 1) - 2]^2$

$$= [1^2 + 4^2 + 7^2 + \ldots + (3k - 2)^2] + (3k + 1)^2$$

$$= \frac{1}{2}k(6k^2 - 3k - 1) + (3k + 1)^2$$

$$= \frac{1}{2}(6k^3 + 15k^2 + 11k + 2)$$

$$= \frac{1}{2}(k + 1)\left[6k^2 + 12k - 3k + 6 - 3 - 1\right]$$

$$= \frac{1}{2}(k + 1)\left[6k^2 + 12k + 6 - 3k - 3 - 1\right]$$

$$= \frac{1}{2}(k + 1)\left[6(k^2 + 2k + 1) - 3(k + 1) - 1\right]$$

$$= \frac{1}{2}(k + 1)\left[6(k + 1)^2 - 3(k + 1) - 1\right]$$

7. $5! = 5 \cdot 4 \cdot 3 \cdot 2 \cdot 1 = 120$

9. $\binom{5}{2} = \frac{5!}{2!\,3!} = \frac{5 \cdot 4}{2} = 10$

11. $(x + 2)^4 = \binom{4}{0}x^4 + \binom{4}{1}2 \cdot x^3 + \binom{4}{2}2^2x^2 + \binom{4}{3}2^3 \cdot x + \binom{4}{4}2^4$

$$= x^4 + 4 \cdot 2x^3 + 6 \cdot 4x^2 + 4 \cdot 8x + 16$$

$$= x^4 + 8x^3 + 24x^2 + 32x + 16$$

13. $(2x + 3)^5 = \binom{5}{0}(2x)^5 + \binom{5}{1}3(2x)^4 + \binom{5}{2}3^2(2x)^3 + \binom{5}{3}3^3(2x^2)$

$$+ \binom{5}{4}3^4(2x) + \binom{5}{5}3^5$$

$$= 32x^5 + 5 \cdot 3 \cdot 16x^4 + 10 \cdot 9 \cdot 8x^3 + 10 \cdot 27 \cdot 4x^2$$

$$+ 5 \cdot 81 \cdot 2x + 243$$

$$= 32x^5 + 240x^4 + 720x^3 + 1080x^2 + 810x + 243$$

15. $n = 9$, $a = 2$, $x = x$, and $j = 7$

$$\binom{9}{9 - 7}2^{9-7}x^7 = \binom{9}{2} \cdot 2^2 x^7$$

$$= \frac{9!}{2!7!} \cdot 4 \cdot x^7$$

$$= \frac{9 \cdot 8}{2} \cdot 4 \cdot x^7$$

$$= 144x^7$$

The coefficient of x^7 is 144.

17. $n = 7$, $a = 1$, $x = 2x$, and $j = 2$

$$\binom{7}{7 - 2}1^{7-2}(2x)^2 = \binom{7}{5}1^5 \cdot 4x^2$$

$$= \frac{7!}{5!2!} \cdot 1 \cdot 4 \cdot x^2$$

$$= \frac{7 \cdot 6}{2} \cdot 1 \cdot 4x^2$$

$$= 84x^2$$

The coefficient of x^2 is 84.

19. $a_1 = (-1)^1\dfrac{1 + 1}{1 + 2} = \dfrac{-2}{3}$

$a_2 = (-1)^2\dfrac{2 + 1}{2 + 2} = \dfrac{3}{4}$

$a_3 = (-1)^3\dfrac{3 + 1}{3 + 2} = \dfrac{-4}{5}$

$a_4 = (-1)^4\dfrac{4 + 1}{4 + 2} = \dfrac{5}{6}$

$a_5 = (-1)^5\dfrac{5 + 1}{5 + 2} = \dfrac{-6}{7}$

21. $a_1 = \dfrac{2^1}{1^2} = 2$

$a_2 = \dfrac{2^2}{2^2} = 1$

$a_3 = \dfrac{2^3}{3^2} = \dfrac{8}{9}$

$a_4 = \dfrac{2^4}{4^2} = 1$

$a_5 = \dfrac{2^5}{5^2} = \dfrac{32}{25}$

23. $a_1 = 3$

$a_2 = \dfrac{2}{3}(3) = 2$

$a_3 = \dfrac{2}{3}(2) = \dfrac{4}{3}$

$a_4 = \dfrac{2}{3}\left(\dfrac{4}{3}\right) = \dfrac{8}{9}$

$a_5 = \dfrac{2}{3}\left(\dfrac{8}{9}\right) = \dfrac{16}{27}$

25. $a_1 = 2$

$a_2 = 2 - 2 = 0$
$a_3 = 2 - 0 = 2$
$a_4 = 2 - 2 = 0$
$a_7 = 2 - 0 = 2$

27. Arithmetic

$$a_{n+1} - a_n = d$$
$$n + 4 - (n + 3) = d = 1$$
$$a = 4, \quad a_n = (n + 3)$$
$$S_n = \frac{n}{2}[4 + (n + 3)] = \frac{n}{2}(n + 7)$$

29. Neither

31. Geometric

$$r = \frac{a_{n+1}}{a_n}$$
$$r = \frac{64}{8} = 8$$
$$S_r = \sum_{k=1}^{n} 2^{3k} = 8 + 64 + 512 + \ldots + 2^{3n}$$
$$= 8\left(\frac{1 - 8^n}{1 - 8}\right)$$
$$= \frac{-8}{7}(1 - 8^n)$$
$$= \frac{-8}{7} + \frac{8^{n+1}}{7}$$
$$= \frac{8}{7}(8^n - 1)$$

33. Arithmetic

$$d = 4 - 0$$
$$d = 4$$
$$a = 0, \quad a_n = 4(n - 1)$$
$$S_n = \frac{n}{2}[0 + 4(n-1)]$$
$$= \frac{n}{2}[4(n - 1)]$$
$$= 2n(n - 1)$$

35. Geometric

$$r = \frac{\frac{3}{2}}{3} = \frac{1}{2}$$
$$S_n = \sum_{k=1}^{n} 3\left(\frac{1}{2}\right)^{k-1} = 3 + \frac{3}{2} + \frac{3}{4} + \ldots + 3\left(\frac{1}{2}\right)^{n-1}$$
$$= 3\frac{\left(1 - \left(\frac{1}{2}\right)^n\right)}{1 - \frac{1}{2}} = 3\frac{\left(1 - \left(\frac{1}{2}\right)^n\right)}{\frac{1}{2}}$$
$$= 6\left[1 - \left(\frac{1}{2}\right)^n\right]$$

37. Neither

39. Arithmetic: $d = 4, a = 3$

$$a_7 = 3 + 6(4)$$
$$a_7 = 27$$

41. Geometric: $r = \frac{1}{10}, \ a_1 = 1$

$$a_{11} = \left(\frac{1}{10}\right)^{10}(1)$$
$$a_{11} = \left(\frac{1}{10}\right)^{10}$$

43. Arithmetic: $d = \sqrt{2}$, $a = \sqrt{2}$

$$a_9 = \sqrt{2} + 8\sqrt{2}$$
$$a_9 = 9\sqrt{2}$$

11 INDUCTION; SEQUENCES

45. $a_7 = a + 6d = 31$

$a_{20} = a + 19d = 96$
$\phantom{a_{20} = a + 1}13d = 65$
$\phantom{a_{20} = a + 19d}d = 5$

$a = 31 - 6d$
$a = 31 - 6(5)$
$a = 1$

$a_n = 1 + (n - 1)5$
$a_n = 1 + 5n - 5$
$a_n = 5n - 4$

47. $a_{10} = a + 9d = 0$

$a_{18} = a + 17d = 8$
$\phantom{a_{18} = a + 1}8d = 8$
$\phantom{a_{18} = a + 17}d = 1$

$a = -9d$
$a = -9$

$a_n = -9 + (n - 1)1$
$a_n = -9 + n - 1$
$a_n = n - 10$

49. $a = 3$, $r = \dfrac{1}{3}$

$$3 + 1 + \frac{1}{3} + \frac{1}{9} + \ldots = \frac{3}{1 - \dfrac{1}{3}} = \frac{9}{2}$$

51. $a = 2$, $r = \dfrac{-1}{2}$

$$2 - 1 + \frac{1}{2} - \frac{1}{4} + \ldots = \frac{2}{1 + \dfrac{1}{2}} = \frac{4}{3}$$

53. $a = 4$, $r = \dfrac{1}{2}$

$$\sum_{k=1}^{\infty} 4\left(\frac{1}{2}\right)^{k-1} = \frac{4}{1 - \dfrac{1}{2}} = 8$$

CHAPTER
12

MISCELLANEOUS TOPICS

≡ EXERCISE 12.1 MATRIX ALGEBRA

≡ HISTORICAL PROBLEM

1. (a) Using the correspondence $a + bi \longleftrightarrow \begin{bmatrix} a & b \\ -b & a \end{bmatrix}$,

we have: $2 - 5i \longleftrightarrow \begin{bmatrix} 2 & -5 \\ 5 & 2 \end{bmatrix}$

$1 + 3i \longleftrightarrow \begin{bmatrix} 1 & 3 \\ -3 & 1 \end{bmatrix}$

(b) $\begin{bmatrix} 2 & -5 \\ 5 & 2 \end{bmatrix} \begin{bmatrix} 1 & 3 \\ -3 & 1 \end{bmatrix} = \begin{bmatrix} 17 & 1 \\ -1 & 17 \end{bmatrix}$

(c) Now $\begin{bmatrix} 17 & 1 \\ -1 & 17 \end{bmatrix} \longleftrightarrow 17 + i$

(d) On the other hand, we have

$$(2 - 5i)(1 + 3i) = 2 - 15i^2 + 6i - 5i$$
$$= 17 + i$$

≡ EXERCISES

In Problems 1 - 16, we are using: $A = \begin{bmatrix} 0 & 3 & -5 \\ 1 & 2 & 6 \end{bmatrix}$; $B = \begin{bmatrix} 4 & 1 & 0 \\ -2 & 3 & -2 \end{bmatrix}$; $C = \begin{bmatrix} 4 & 1 \\ 6 & 2 \\ -2 & 3 \end{bmatrix}$

1. $A + B = \begin{bmatrix} 0 & 3 & -5 \\ 1 & 2 & 6 \end{bmatrix} + \begin{bmatrix} 4 & 1 & 0 \\ -2 & 3 & -2 \end{bmatrix}$

$= \begin{bmatrix} 0 + 4 & 3 + 1 & -5 + 0 \\ 1 + (-2) & 2 + 3 & 6 + (-2) \end{bmatrix} = \begin{bmatrix} 4 & 4 & -5 \\ -1 & 5 & 4 \end{bmatrix}$

3. $4A = 4\begin{bmatrix} 0 & 3 & -5 \\ 1 & 2 & 6 \end{bmatrix} = \begin{bmatrix} 4 \cdot 0 & 4 \cdot 3 & 4(-5) \\ 4 \cdot 1 & 4 \cdot 2 & 4 \cdot 6 \end{bmatrix}$

$$= \begin{bmatrix} 0 & 12 & -20 \\ 4 & 8 & 24 \end{bmatrix}$$

5. $3A - 2B = \begin{bmatrix} 0 & 9 & -15 \\ 3 & 6 & 18 \end{bmatrix} - \begin{bmatrix} 8 & 2 & 0 \\ -4 & 6 & -4 \end{bmatrix} = \begin{bmatrix} -8 & 7 & -15 \\ 7 & 0 & 22 \end{bmatrix}$

7. $AC = \begin{bmatrix} 0 & 3 & -5 \\ 1 & 2 & 6 \end{bmatrix} \begin{bmatrix} 4 & 1 \\ 6 & 2 \\ -2 & 3 \end{bmatrix}$

$$= \begin{bmatrix} 0 \cdot 4 + 3 \cdot 6 + (-5)(-2) & 0 \cdot 1 + 3 \cdot 2 + (-5)3 \\ 1 \cdot 4 + 2 \cdot 6 + 6(-2) & 1 \cdot 1 + 2 \cdot 2 + 6 \cdot 3 \end{bmatrix}$$

$$= \begin{bmatrix} 28 & -9 \\ 4 & 23 \end{bmatrix}$$

9. $CA = \begin{bmatrix} 4 & 1 \\ 6 & 2 \\ -2 & 3 \end{bmatrix} \begin{bmatrix} 0 & 3 & -5 \\ 1 & 2 & 6 \end{bmatrix}$

$$= \begin{bmatrix} 4 \cdot 0 + 1 \cdot 1 & 4 \cdot 3 + 1 \cdot 2 & 4(-5) + 1 \cdot 6 \\ 6 \cdot 0 + 2 \cdot 1 & 6 \cdot 3 + 2 \cdot 2 & 6(-5) + 2 \cdot 6 \\ (-2) \cdot 0 + 3 \cdot 1 & (-2) \cdot 3 + 3 \cdot 2 & (-2)(-5) + 3 \cdot 6 \end{bmatrix}$$

$$= \begin{bmatrix} 1 & 14 & -14 \\ 2 & 22 & -18 \\ 3 & 0 & 28 \end{bmatrix}$$

11. $C(A + B) = \begin{bmatrix} 4 & 1 \\ 6 & 2 \\ -2 & 3 \end{bmatrix} \begin{bmatrix} 4 & 4 & -5 \\ -1 & 5 & 4 \end{bmatrix} = \begin{bmatrix} 15 & 21 & -16 \\ 22 & 34 & -22 \\ -11 & 7 & 22 \end{bmatrix}$

13. $AC - 3I_2 = \begin{bmatrix} 0 & 3 & -5 \\ 1 & 2 & 6 \end{bmatrix} \begin{bmatrix} 4 & 1 \\ 6 & 2 \\ -2 & 3 \end{bmatrix} - 3\underbrace{\begin{bmatrix} 1 & 0 \\ 0 & 1 \end{bmatrix}}_{I_2}$

$$= \begin{bmatrix} 28 & -9 \\ 4 & 23 \end{bmatrix} - \begin{bmatrix} 3 & 0 \\ 0 & 3 \end{bmatrix} = \begin{bmatrix} 25 & -9 \\ 4 & 20 \end{bmatrix}$$

12.1 MATRIX ALGEBRA

15. $CA - CB = \begin{bmatrix} 4 & 1 \\ 6 & 2 \\ -2 & 3 \end{bmatrix} \begin{bmatrix} 0 & 3 & -5 \\ 1 & 2 & 6 \end{bmatrix} - \begin{bmatrix} 4 & 1 \\ 6 & 2 \\ -2 & 3 \end{bmatrix} \begin{bmatrix} 4 & 1 & 0 \\ -2 & 3 & -2 \end{bmatrix}$

$= \begin{bmatrix} 1 & 14 & -14 \\ 2 & 22 & -18 \\ 3 & 0 & 28 \end{bmatrix} - \begin{bmatrix} 14 & 7 & -2 \\ 20 & 12 & -4 \\ -14 & 7 & -6 \end{bmatrix} = \begin{bmatrix} -13 & 7 & -12 \\ -18 & 10 & -14 \\ 17 & -7 & 34 \end{bmatrix}$

17. $\begin{bmatrix} 2 & -2 \\ 1 & 0 \end{bmatrix} \begin{bmatrix} 2 & 1 & 4 & 6 \\ 3 & -1 & 3 & 2 \end{bmatrix}$

$= \begin{bmatrix} 2 \cdot 2 + (-2)3 & 2 \cdot 1 + (-2)(-1) & 2 \cdot 4 + (-2)3 & 2 \cdot 6 + (-2)2 \\ 1 \cdot 2 + 0 \cdot 3 & 1 \cdot 1 + 0(-1) & 1 \cdot 4 + 0 \cdot 3 & 1 \cdot 6 + 0 \cdot 2 \end{bmatrix}$

$= \begin{bmatrix} -2 & 4 & 2 & 8 \\ 2 & 1 & 4 & 6 \end{bmatrix}$

19. $\begin{bmatrix} 1 & 0 & 1 \\ 2 & 4 & 1 \\ 3 & 6 & 1 \end{bmatrix} \begin{bmatrix} 1 & 3 \\ 6 & 2 \\ 8 & -1 \end{bmatrix} = \begin{bmatrix} 9 & 2 \\ 34 & 13 \\ 47 & 20 \end{bmatrix}$

21. $A = \begin{bmatrix} 2 & 1 \\ 1 & 1 \end{bmatrix}$ <u>Step 1</u>: Form $\begin{bmatrix} 2 & 1 & | & 1 & 0 \\ 1 & 1 & | & 0 & 1 \end{bmatrix} = [A \,|\, I_2]$

<u>Step 2</u>: $\begin{bmatrix} 2 & 1 & | & 1 & 0 \\ 1 & 1 & | & 0 & 1 \end{bmatrix} \rightarrow \begin{bmatrix} 1 & 1 & | & 0 & 1 \\ 2 & 1 & | & 1 & 0 \end{bmatrix}$
↑
Interchange rows one and two

$\rightarrow \begin{bmatrix} 1 & 1 & | & 0 & 1 \\ 0 & -1 & | & 0 & -2 \end{bmatrix}$
↑
$R_2 = -2r_1 + r_2$

$\rightarrow \begin{bmatrix} 1 & 0 & | & 1 & -1 \\ 0 & -1 & | & 1 & -2 \end{bmatrix}$
↑
$R_2 = r_2 + r_1$

$\rightarrow \begin{bmatrix} 1 & 0 & | & 1 & -1 \\ 0 & 1 & | & -1 & 2 \end{bmatrix}$
↑
$R_2 = (-1)r_2$

<u>Step 3</u>: We have now achieved the form $[I_2 \,|\, A^{-1}]$,

so $A^{-1} = \begin{bmatrix} 1 & -1 \\ -1 & 2 \end{bmatrix}$.

Check: $A \cdot A^{-1} = \begin{bmatrix} 2 & 1 \\ 1 & 1 \end{bmatrix} \begin{bmatrix} 1 & -1 \\ -1 & 2 \end{bmatrix} = \begin{bmatrix} 1 & 0 \\ 0 & 1 \end{bmatrix} = I_2!$

12 MISCELLANEOUS TOPICS

23.　$A = \begin{bmatrix} 6 & 5 \\ 2 & 2 \end{bmatrix}$

Step 1: $[A \mid I_2] = \begin{bmatrix} 6 & 5 & 1 & 0 \\ 2 & 2 & 0 & 1 \end{bmatrix}$

Step 2: $\begin{bmatrix} 6 & 5 & 1 & 0 \\ 2 & 2 & 0 & 1 \end{bmatrix} \rightarrow \begin{bmatrix} 2 & 2 & 0 & 1 \\ 6 & 5 & 1 & 0 \end{bmatrix}$
↑

Interchange rows

$\rightarrow \begin{bmatrix} 2 & 2 & 0 & 1 \\ 6 & -1 & 1 & -3 \end{bmatrix}$
↑

$R_2 = -3r_1 + r_2$

$\rightarrow \begin{bmatrix} 2 & 0 & 2 & -5 \\ 0 & -1 & 1 & -3 \end{bmatrix}$
↑

$R_1 = 2r_2 + R_1$

$\rightarrow \begin{bmatrix} 1 & 0 & 1 & -\dfrac{5}{2} \\ 0 & 1 & -1 & 3 \end{bmatrix}$
↑

$R_1 = \dfrac{1}{2}r_1$ and $R_2 = (-1)r_2$

Step 3: We have:　$A^{-1} = \begin{bmatrix} 1 & -\dfrac{5}{2} \\ -1 & 3 \end{bmatrix}$

25.　$A = \begin{bmatrix} 2 & 1 \\ a & a \end{bmatrix}$, where $a \neq 0$.

$\begin{bmatrix} 2 & 1 & 1 & 0 \\ a & a & 0 & 1 \end{bmatrix} \rightarrow \begin{bmatrix} 1 & \dfrac{1}{2} & \dfrac{1}{2} & 0 \\ a & a & 0 & 1 \end{bmatrix}$
↑

$R_1 = \dfrac{1}{2}r_1$

$\rightarrow \begin{bmatrix} 1 & \dfrac{1}{2} & \dfrac{1}{2} & 0 \\ 0 & \dfrac{1}{2}a & -\dfrac{1}{2}a & 1 \end{bmatrix}$
↑

$R_2 = -ar_1 + r_2$

$\rightarrow \begin{bmatrix} 1 & \dfrac{1}{2} & \dfrac{1}{2} & 0 \\ 0 & 1 & -1 & \dfrac{2}{a} \end{bmatrix}$
↑

$R_2 = \left(\dfrac{2}{a}\right)r_2$

12.1　MATRIX ALGEBRA

$$\rightarrow \begin{bmatrix} 1 & 0 & 1 & -\dfrac{1}{a} \\ 0 & 1 & -1 & \dfrac{2}{a} \end{bmatrix}$$

↑

$$R_1 = -\frac{1}{2}r_2 + r_1$$

Therefore, $A^{-1} = \begin{bmatrix} 1 & -\dfrac{1}{a} \\ -1 & \dfrac{2}{a} \end{bmatrix}$

27. $A = \begin{bmatrix} 1 & -1 & 1 \\ 0 & -2 & 1 \\ -2 & -3 & 0 \end{bmatrix}$

$$\begin{bmatrix} 1 & -1 & 1 & | & 1 & 0 & 0 \\ 0 & -2 & 1 & | & 0 & 1 & 0 \\ -2 & -3 & 0 & | & 0 & 0 & 1 \end{bmatrix} \rightarrow \begin{bmatrix} 1 & -1 & 1 & | & 1 & 0 & 0 \\ 0 & -2 & 1 & | & 0 & 1 & 0 \\ 0 & -5 & 2 & | & 2 & 0 & 1 \end{bmatrix}$$

↑

$$R_3 = 2r_1 + r_3$$

$$\rightarrow \begin{bmatrix} 1 & -1 & 1 & | & 1 & 0 & 0 \\ 0 & 1 & -\dfrac{1}{2} & | & 0 & -\dfrac{1}{2} & 0 \\ 0 & -5 & 2 & | & 2 & 0 & 1 \end{bmatrix}$$

↑

$$R_2 = -\frac{1}{2}r_2$$

$$\rightarrow \begin{bmatrix} 1 & 0 & \dfrac{1}{2} & | & 1 & -\dfrac{1}{2} & 0 \\ 0 & 1 & -\dfrac{1}{2} & | & 0 & -\dfrac{1}{2} & 0 \\ 0 & 0 & -\dfrac{1}{2} & | & 2 & -\dfrac{5}{2} & 1 \end{bmatrix}$$

↑

$$R_1 = r_2 + r_1$$
$$R_3 = 5r_2 + r_3$$

$$\rightarrow \begin{bmatrix} 1 & 0 & 0 & | & 3 & -3 & 1 \\ 0 & 1 & 0 & | & -2 & 2 & -1 \\ 0 & 0 & -\dfrac{1}{2} & | & 2 & -\dfrac{5}{2} & 1 \end{bmatrix}$$

↑

$$R_1 = r_3 + r_1$$
$$R_2 = (-1)r_3 + r_2$$

$$\rightarrow \begin{bmatrix} 1 & 0 & 0 & | & 3 & -3 & 1 \\ 0 & 1 & 0 & | & -2 & 2 & -1 \\ 0 & 0 & 1 & | & -4 & 5 & -2 \end{bmatrix}$$

↑

$$R_3 = -2r_3$$

12 MISCELLANEOUS TOPICS

We have $A^{-1} = \begin{bmatrix} 3 & -3 & 1 \\ -2 & 2 & -1 \\ -4 & 5 & -2 \end{bmatrix}$

29. $A = \begin{bmatrix} 1 & 1 & 1 \\ 3 & 2 & -1 \\ 3 & 1 & 2 \end{bmatrix}$

Then,

$$\left[\begin{array}{ccc|ccc} 1 & 1 & 1 & 1 & 0 & 0 \\ 3 & 2 & -1 & 0 & 1 & 0 \\ 3 & 1 & 2 & 0 & 0 & 1 \end{array}\right] \rightarrow \left[\begin{array}{ccc|ccc} 1 & 1 & 1 & 1 & 0 & 0 \\ 0 & -1 & -4 & -3 & 1 & 0 \\ 0 & -2 & -1 & -3 & 0 & 1 \end{array}\right]$$

$\uparrow$

$R_2 = -3r_1 + r_2$
$R_3 = -3r_1 + r_3$

$$\rightarrow \left[\begin{array}{ccc|ccc} 1 & 1 & 1 & 1 & 0 & 0 \\ 0 & 1 & 4 & 3 & -1 & 0 \\ 0 & -2 & -1 & -3 & 0 & 1 \end{array}\right]$$

$\uparrow$

$R_2 = (-1)r_2$

$$\rightarrow \left[\begin{array}{ccc|ccc} 1 & 0 & -3 & -2 & 1 & 0 \\ 0 & 1 & 4 & 3 & -1 & 0 \\ 0 & 0 & 7 & -3 & -2 & 1 \end{array}\right]$$

$\uparrow$

$R_1 = (-1)r_2 + r_1$
$R_3 = 2r_2 + r_3$
$\uparrow$

$$\rightarrow \left[\begin{array}{ccc|ccc} 1 & 0 & -3 & -2 & 1 & 0 \\ 0 & 1 & 4 & 3 & -1 & 0 \\ 0 & 0 & 1 & \frac{3}{7} & -\frac{2}{7} & \frac{1}{7} \end{array}\right]$$

$\uparrow$

$R_3 = \left(\frac{1}{7}\right)r_3$

$$\rightarrow \left[\begin{array}{ccc|ccc} 1 & 0 & 0 & -\frac{5}{7} & \frac{1}{7} & \frac{3}{7} \\ 0 & 1 & 0 & \frac{9}{7} & \frac{1}{7} & -\frac{4}{7} \\ 0 & 0 & 1 & \frac{3}{7} & -\frac{2}{7} & \frac{1}{7} \end{array}\right]$$

$\uparrow$

$R_1 = 3r_3 + r_1$
$R_2 = -4r_3 + r_2$

12.1 MATRIX ALGEBRA

679

Thus, $A^{-1} = \begin{bmatrix} -\dfrac{5}{7} & \dfrac{1}{7} & \dfrac{3}{7} \\[2mm] \dfrac{9}{7} & \dfrac{1}{7} & -\dfrac{4}{7} \\[2mm] \dfrac{3}{7} & -\dfrac{2}{7} & \dfrac{1}{7} \end{bmatrix} = \dfrac{1}{7}\begin{bmatrix} -5 & 1 & 3 \\ 9 & 1 & -4 \\ 3 & -2 & 1 \end{bmatrix}$

31. Let $A = \begin{bmatrix} 2 & 1 \\ 1 & 1 \end{bmatrix}$, $X = \begin{bmatrix} x \\ y \end{bmatrix}$, $B = \begin{bmatrix} 8 \\ 5 \end{bmatrix}$.

Then $2x + y = 8$

$\qquad x + y = 5$

can be written compactly as $A \cdot X = B$.

From Problem 21, $A^{-1} = \begin{bmatrix} 1 & -1 \\ -1 & 2 \end{bmatrix}$, and

$X = A^{-1}B = \begin{bmatrix} 1 & -1 \\ -1 & 2 \end{bmatrix}\begin{bmatrix} 8 \\ 5 \end{bmatrix} = \begin{bmatrix} 3 \\ 2 \end{bmatrix}$, or

in other words, $x = 3$ and $y = 2$.

33. Here, $A = \begin{bmatrix} 2 & 1 \\ 1 & 1 \end{bmatrix}$, $X = \begin{bmatrix} x \\ y \end{bmatrix}$, $B = \begin{bmatrix} 0 \\ 5 \end{bmatrix}$.

From Problem 21, $A^{-1} = \begin{bmatrix} 1 & -1 \\ -1 & 2 \end{bmatrix}$, so

$X = A^{-1}B = \begin{bmatrix} 1 & -1 \\ -1 & 2 \end{bmatrix}\begin{bmatrix} 0 \\ 5 \end{bmatrix} = \begin{bmatrix} -5 \\ 10 \end{bmatrix}$

or $x = -5$ and $y = 10$.

35. $A = \begin{bmatrix} 6 & 5 \\ 2 & 2 \end{bmatrix}$, $X = \begin{bmatrix} x \\ y \end{bmatrix}$, $B = \begin{bmatrix} 7 \\ 2 \end{bmatrix}$.

From Problem 23, $A^{-1} = \begin{bmatrix} 1 & -\dfrac{5}{2} \\[2mm] -1 & 3 \end{bmatrix}$, so

$X = A^{-1}B = \begin{bmatrix} 1 & -\dfrac{5}{2} \\[2mm] -1 & 3 \end{bmatrix}\begin{bmatrix} 7 \\ 2 \end{bmatrix} = \begin{bmatrix} 2 \\ -1 \end{bmatrix}$

or $x = 2$ and $y = -1$.

37. $A = \begin{bmatrix} 6 & 5 \\ 2 & 2 \end{bmatrix}$, $X = \begin{bmatrix} x \\ y \end{bmatrix}$, $B = \begin{bmatrix} 13 \\ 5 \end{bmatrix}$.

From Problem 23, $A^{-1} = \begin{bmatrix} 1 & -\dfrac{5}{2} \\[2mm] -1 & 3 \end{bmatrix}$, so

12 MISCELLANEOUS TOPICS

$$X = A^{-1}B = \begin{bmatrix} 1 & -\dfrac{5}{2} \\ -1 & 3 \end{bmatrix} \begin{bmatrix} 13 \\ 5 \end{bmatrix} = \begin{bmatrix} \dfrac{1}{2} \\ 2 \end{bmatrix},$$

or $x = \dfrac{1}{2}$ and $y = 2$.

39. $A = \begin{bmatrix} 2 & 1 \\ a & a \end{bmatrix}$, $X = \begin{bmatrix} x \\ y \end{bmatrix}$, $B = \begin{bmatrix} -3 \\ -a \end{bmatrix}$, where $a \neq 0$.

From Problem 25, $A^{-1} = \begin{bmatrix} 1 & -\dfrac{1}{a} \\ -1 & \dfrac{2}{a} \end{bmatrix}$, so

$$X = A^{-1}B = \begin{bmatrix} 1 & -\dfrac{1}{a} \\ -1 & \dfrac{2}{a} \end{bmatrix} \begin{bmatrix} -3 \\ -a \end{bmatrix} = \begin{bmatrix} -2 \\ 1 \end{bmatrix},$$

or $x = -2$ and $y = 1$.

41. $A = \begin{bmatrix} 2 & 1 \\ a & a \end{bmatrix}$, $X = \begin{bmatrix} x \\ y \end{bmatrix}$, $B = \begin{bmatrix} \dfrac{7}{a} \\ 5 \end{bmatrix}$.

From Problem 25, $A^{-1} = \begin{bmatrix} 1 & -\dfrac{1}{a} \\ -1 & \dfrac{2}{a} \end{bmatrix}$, so

$$X = A^{-1}B = \begin{bmatrix} 1 & -\dfrac{1}{a} \\ -1 & \dfrac{2}{a} \end{bmatrix} \begin{bmatrix} \dfrac{7}{a} \\ 5 \end{bmatrix} = \begin{bmatrix} \dfrac{2}{a} \\ \dfrac{3}{a} \end{bmatrix},$$

or $x = \dfrac{2}{a}$ and $y = \dfrac{3}{a}$.

43. $A = \begin{bmatrix} 1 & -1 & 1 \\ 0 & -2 & 1 \\ -2 & -3 & 0 \end{bmatrix}$, $X = \begin{bmatrix} x \\ y \\ z \end{bmatrix}$, $B = \begin{bmatrix} 0 \\ -1 \\ -5 \end{bmatrix}$.

By Problem 27, $A^{-1} = \begin{bmatrix} 3 & -3 & 1 \\ -2 & 2 & -1 \\ -4 & 5 & -2 \end{bmatrix}$, so

$$X = A^{-1}B = \begin{bmatrix} 3 & -3 & 1 \\ -2 & 2 & -1 \\ -4 & 5 & -2 \end{bmatrix} \begin{bmatrix} 0 \\ -1 \\ -5 \end{bmatrix} = \begin{bmatrix} -2 \\ 3 \\ 5 \end{bmatrix},$$

or $x = -2$, $y = 3$, and $z = 5$.

12.1 MATRIX ALGEBRA

45. $A = \begin{bmatrix} 1 & -1 & 1 \\ 0 & -2 & 1 \\ -2 & -3 & 0 \end{bmatrix}$, $X = \begin{bmatrix} x \\ y \\ z \end{bmatrix}$, $B = \begin{bmatrix} 2 \\ 2 \\ \frac{1}{2} \end{bmatrix}$.

By Problem 27, $A^{-1} = \begin{bmatrix} 3 & -3 & 1 \\ -2 & 2 & -1 \\ -4 & 5 & -2 \end{bmatrix}$, so

$$X = A^{-1}B = \begin{bmatrix} 3 & -3 & 1 \\ -2 & 2 & -1 \\ -4 & 5 & -2 \end{bmatrix} \begin{bmatrix} 2 \\ 2 \\ \frac{1}{2} \end{bmatrix} = \begin{bmatrix} \frac{1}{2} \\ -\frac{1}{2} \\ 1 \end{bmatrix},$$

or $x = \frac{1}{2}$, $y = -\frac{1}{2}$, and $z = 1$.

47. $A = \begin{bmatrix} 1 & 1 & 1 \\ 3 & 2 & -1 \\ 3 & 1 & 2 \end{bmatrix}$, $X = \begin{bmatrix} x \\ y \\ z \end{bmatrix}$, $B = \begin{bmatrix} 9 \\ 8 \\ 1 \end{bmatrix}$.

By Problem 29, $A^{-1} = \frac{1}{7}\begin{bmatrix} -5 & 1 & 3 \\ 9 & 1 & -4 \\ 3 & -2 & 1 \end{bmatrix}$, so

$$X = A^{-1}B = \frac{1}{7}\begin{bmatrix} -5 & 1 & 3 \\ 9 & 1 & -4 \\ 3 & -2 & 1 \end{bmatrix} \begin{bmatrix} 9 \\ 8 \\ 1 \end{bmatrix} = \frac{1}{7}\begin{bmatrix} -34 \\ 85 \\ 12 \end{bmatrix} = \begin{bmatrix} -\frac{34}{7} \\ \frac{85}{7} \\ \frac{12}{7} \end{bmatrix}$$

or $x = -\frac{34}{7}$, $y = \frac{85}{7}$, $z = \frac{12}{7}$

49. $A = \begin{bmatrix} 1 & 1 & 1 \\ 3 & 2 & -1 \\ 3 & 1 & 2 \end{bmatrix}$, $X = \begin{bmatrix} x \\ y \\ z \end{bmatrix}$, $B = \begin{bmatrix} 2 \\ 7 \\ 3 \\ \frac{10}{3} \end{bmatrix}$.

By Problem 29, $A^{-1} = \frac{1}{7}\begin{bmatrix} -5 & 1 & 3 \\ 9 & 1 & -4 \\ 3 & -2 & 1 \end{bmatrix}$, so

$$X = A^{-1}B = \frac{1}{7}\begin{bmatrix} -5 & 1 & 3 \\ 9 & 1 & -4 \\ 3 & -2 & 1 \end{bmatrix} \begin{bmatrix} 2 \\ 7 \\ 3 \\ \frac{10}{3} \end{bmatrix} = \frac{1}{7}\begin{bmatrix} \frac{7}{3} \\ 7 \\ \frac{14}{3} \end{bmatrix} = \begin{bmatrix} \frac{1}{3} \\ 1 \\ \frac{2}{3} \end{bmatrix},$$

or $x = \frac{1}{3}$, $y = 1$, $z = \frac{2}{3}$

12 MISCELLANEOUS TOPICS

51. $A = \begin{bmatrix} 4 & 2 \\ 2 & 1 \end{bmatrix}$.

We start with $[A \mid I_2]$ and put it into reduced echelon form. If I_2 does **not** appear to the left of the vertical bar, then A has no inverse.

$$[A \mid I_2] = \begin{bmatrix} 4 & 2 & 1 & 0 \\ 2 & 1 & 0 & 1 \end{bmatrix} \rightarrow \begin{bmatrix} 4 & 2 & 1 & 0 \\ 0 & 0 & -\frac{1}{2} & 1 \end{bmatrix}$$

$\uparrow$

$$R_2 = -\frac{1}{2}r_1 + r_2$$

$$\rightarrow \begin{bmatrix} 1 & \frac{1}{2} & \frac{1}{4} & 0 \\ 0 & 0 & -\frac{1}{2} & 1 \end{bmatrix}$$

$\uparrow$

$$R_1 = -\frac{1}{4}r_1$$

This is in reduced echelon form, but the identity matrix I_2 does not appear on the left. Thus, A has no inverse.

53. $$[A \mid I_2] = \begin{bmatrix} 15 & 3 & 1 & 0 \\ 10 & 2 & 0 & 1 \end{bmatrix} \rightarrow \begin{bmatrix} 1 & \frac{1}{5} & \frac{1}{15} & 0 \\ 10 & 2 & 0 & 1 \end{bmatrix}$$

$\uparrow$

$$R_1 = \frac{1}{15}r_1$$

$$\rightarrow \begin{bmatrix} 1 & \frac{1}{5} & \frac{1}{15} & 0 \\ 0 & 0 & -\frac{2}{3} & 1 \end{bmatrix}$$

$\uparrow$

$$R_2 = -10r_1 + r_2$$

We cannot obtain I_2 on the left, so A has no inverse.

55. $$\begin{bmatrix} -3 & 1 & -1 & 1 & 0 & 0 \\ 1 & -4 & -7 & 0 & 1 & 0 \\ 1 & 2 & 5 & 0 & 0 & 1 \end{bmatrix} \rightarrow \begin{bmatrix} 1 & 2 & 5 & 0 & 0 & 1 \\ 1 & -4 & -7 & 0 & 1 & 0 \\ -3 & 1 & -1 & 1 & 0 & 0 \end{bmatrix}$$

$\uparrow$

Interchange rows one and three

$$\rightarrow \begin{bmatrix} 1 & 2 & 5 & 0 & 0 & 1 \\ 0 & -6 & -12 & 0 & 1 & -1 \\ 0 & 7 & 14 & 1 & 0 & 3 \end{bmatrix}$$

$\uparrow$

$$R_2 = -r_1 + r_2$$
$$R_3 = 3r_1 + r_3$$

12.1 MATRIX ALGEBRA

$$\rightarrow \begin{bmatrix} 1 & 2 & 5 & 0 & 0 & 1 \\ 0 & 1 & 2 & 0 & -\frac{1}{6} & \frac{1}{6} \\ 0 & 1 & 2 & \frac{1}{7} & 0 & \frac{3}{7} \end{bmatrix}$$

$\uparrow$

$R_2 = -\frac{1}{6}r_2$

$R_3 = \frac{1}{7}r_3$

$$\rightarrow \begin{bmatrix} 1 & 2 & 5 & 0 & 0 & 1 \\ 0 & 1 & 2 & 0 & -\frac{1}{6} & \frac{1}{6} \\ 0 & 0 & 0 & \frac{1}{7} & \frac{1}{6} & \frac{11}{42} \end{bmatrix}$$

$\uparrow$

$R_3 = -r_2 + r_3$

The left side is not the identity matrix, I_3, so A has no inverse.

57. (a) Since we want to use a 2 by 3 matrix to represent the data, we can let rows represent stainless steel and aluminum, respectively, while the columns can represent 10-gallon, 5-gallon, and 1-gallon containers:

	PRODUCTION		
	10 g.	5 g.	1 g.
Stainless steel	500	350	400
Aluminum	700	500	850

$\begin{bmatrix} 500 & 350 & 400 \\ 700 & 500 & 850 \end{bmatrix}$

This could have been represented by a 3 by 2 matrix, by letting rows represent size of container, and columns represent type of material.

$$\begin{bmatrix} 500 & 700 \\ 350 & 500 \\ 400 & 850 \end{bmatrix}$$

(b) Now we are given the following information:

	Pounds of Material
10-gal.	15
5-gal.	8
1-gal.	3

$\begin{bmatrix} 15 \\ 8 \\ 3 \end{bmatrix}$

(c) From (a) and (b):

$$\begin{bmatrix} 500 & 350 & 400 \\ 700 & 500 & 850 \end{bmatrix} \begin{bmatrix} 15 \\ 8 \\ 3 \end{bmatrix} = \begin{bmatrix} 11,500 \\ 17,050 \end{bmatrix}$$

The first row represents stainless steel containers (11,500 pounds), and the second row represents aluminum (17,050 pounds).

(d)

	Stainless Steel	Aluminum
Cost per pound	0.10	0.05

$\left.\right\}$ [0.10 0.05]

(e) $[0.10 \ 0.05] \begin{bmatrix} 11,500 \\ 17,050 \end{bmatrix} = [2002.50]$

i.e., total cost of material = \$2002.50.

Note: Since the first entry in (c) was for stainless steel, the first entry, a_{11}, in (d) had to be for stainless steel.

59.   Let $A = \begin{bmatrix} a & b \\ c & d \end{bmatrix}$.

We are assuming that $D = ad - bc \neq 0$, and we wish to show that A has no inverse. We start, as usual, with

$$[A \mid I_2] = \begin{bmatrix} a & b & | & 1 & 0 \\ c & d & | & 0 & 1 \end{bmatrix}.$$

Our first step in putting this into echelon form would be to multiply row one by $\frac{1}{a}$, to get a 1 in the first row, first column. But that will be impossible if $a = 0$. Thus, we need to consider two cases:

Case 1: If $a \neq 0$, proceed as usual:

$$\begin{bmatrix} a & b & | & 1 & 0 \\ c & d & | & 0 & 1 \end{bmatrix} \rightarrow \begin{bmatrix} 1 & \frac{b}{a} & | & \frac{1}{a} & 0 \\ c & d & | & 0 & 1 \end{bmatrix}$$

$$\uparrow$$
$$R_1 = \frac{1}{a} r_1$$

$$\rightarrow \begin{bmatrix} 1 & \frac{b}{a} & | & \frac{1}{a} & 0 \\ 0 & d - \frac{cb}{a} & | & -\frac{c}{a} & 1 \end{bmatrix}$$

$$\uparrow$$
$$R_2 = -cr_1 + r_2$$

12.1 MATRIX ALGEBRA

$$= \begin{bmatrix} 1 & \dfrac{b}{a} & \Bigg| & \dfrac{1}{a} & 0 \\ 0 & \dfrac{ad-bc}{a} & \Bigg| & -\dfrac{c}{a} & 1 \end{bmatrix} \quad \left(\text{since } d - \dfrac{cb}{a} = \dfrac{ad-bc}{a} \right)$$

$$\rightarrow \begin{bmatrix} 1 & \dfrac{b}{a} & \Bigg| & \dfrac{1}{a} & 0 \\ 0 & 1 & \Bigg| & \dfrac{-c}{ad-bc} & \dfrac{a}{ad-bc} \end{bmatrix}$$

↑

$$R_2 = \left(\dfrac{a}{ad-bc} \right) r_2$$

↑

$$\rightarrow \begin{bmatrix} 1 & 0 & \Bigg| & \dfrac{1}{a} + \dfrac{bc}{a(ad-bc)} & \dfrac{-b}{ad-bc} \\ 0 & 1 & \Bigg| & \dfrac{-c}{ad-bc} & \dfrac{a}{ad-bc} \end{bmatrix}$$

↑

$$R_1 = \left(\dfrac{-b}{a} \right) r_2 + r_1$$

Now some algebra:

$$\dfrac{1}{a} + \dfrac{bc}{a(ad-bc)} = \dfrac{ad-bc}{a(ad-bc)} + \dfrac{bc}{a(ad-bc)}$$

$$= \dfrac{ad}{a(ad-bc)}$$

$$= \dfrac{d}{ad-bc}$$

Thus, our inverse is:

$$A^{-1} = \begin{bmatrix} \dfrac{d}{ad-bc} & \dfrac{-b}{ad-bc} \\ \dfrac{-c}{ad-bc} & \dfrac{a}{ad-bc} \end{bmatrix},$$

$$= \dfrac{1}{D} \begin{bmatrix} d & -b \\ -c & a \end{bmatrix}$$

since $D = ad - bc$.

Case 2: But what if $a = 0$? Then, $ad - bc \neq 0$, so we know $b \neq 0$ and $c \neq 0$. Also, we will have:

$$[A \,|\, I_2] = \begin{bmatrix} 0 & b & | & 1 & 0 \\ c & d & | & 0 & 1 \end{bmatrix}$$

So $\begin{bmatrix} 0 & b & | & 1 & 0 \\ c & d & | & 0 & 1 \end{bmatrix} \rightarrow \begin{bmatrix} c & d & | & 0 & 1 \\ 0 & b & | & 1 & 0 \end{bmatrix}$

↑

Interchange rows

$$\rightarrow \left[\begin{array}{cc|cc} 1 & \dfrac{d}{c} & 0 & \dfrac{1}{c} \\ 0 & b & 1 & 0 \end{array}\right]$$

$\uparrow$

$R_1 = \dfrac{1}{c}r_1$ (since $c \neq 0$)

$$= \left[\begin{array}{cc|cc} 1 & \dfrac{d}{c} & 0 & \dfrac{1}{c} \\ 0 & 1 & \dfrac{1}{b} & 0 \end{array}\right]$$

$\uparrow$

$R_2 = \dfrac{1}{b}r_2$ (since $b \neq 0$)

$$\rightarrow \left[\begin{array}{cc|cc} 1 & 0 & \dfrac{-d}{bc} & \dfrac{1}{c} \\ 0 & 1 & \dfrac{1}{b} & 0 \end{array}\right]$$

$\uparrow$

$R_1 = \left(-\dfrac{d}{c}\right)r_2 + r_1$

Therefore, $A^{-1} = \left[\begin{array}{cc} \dfrac{-d}{bc} & \dfrac{1}{c} \\ \dfrac{1}{b} & 0 \end{array}\right]$.

That looks odd, but if we get a common denominator, we have:

$$A^{-1} = \left[\begin{array}{cc} \dfrac{-d}{bc} & \dfrac{b}{bc} \\ \dfrac{c}{bc} & \dfrac{0}{bc} \end{array}\right] = \dfrac{1}{-bc}\left[\begin{array}{cc} d & -b \\ -c & 0 \end{array}\right] = \dfrac{1}{D}\left[\begin{array}{cc} d & -b \\ -c & a \end{array}\right],$$

since $a = 0$, and $D = ad - bc = -bc$!

≡ EXERCISE 12.2 LINEAR PROGRAMMING

In Problems 1-6, we have the same set of vertices: (0, 3), (0, 6), (5, 6), (5, 2), and (4, 0).

1. $z = x + y$

VERTEX	VALUE OF OBJECTIVE FUNCTION ($z = x + y$)
(0, 3)	$z = 0 + 3 = 3$
(0, 6)	$z = 0 + 6 = 6$
(5, 6)	$z = 5 + 6 = 11$
(5, 2)	$z = 5 + 2 = 7$
(4, 0)	$z = 4 + 0 = 4$

The maximum value is 11, at (5, 6), and the minimum value is 3, at (0, 3).

3.

VERTEX	VALUE OF $z = x + 10y$
(0, 3)	$z = 0 + 10 \cdot 3 = 30$
(0, 6)	$z = 0 + 10 \cdot 6 = 60$
(5, 6)	$z = 5 + 10 \cdot 6 = 65$ ← Maximum value
(5, 2)	$z = 5 + 10 \cdot 2 = 25$
(4, 0)	$z = 4 + 10 \cdot 0 = 4$ ← Minimum value

5.

VERTEX	VALUE OF OBJECTIVE FUNCTION ($z = 5x + 7y$)
(0, 3)	$z = 5 \cdot 0 + 7 \cdot 3 = 21$
(0, 6)	$z = 5 \cdot 0 + 7 \cdot 6 = 42$
(5, 6)	$z = 5 \cdot 5 + 7 \cdot 6 = 67$ ← Maximum value
(5, 2)	$z = 5 \cdot 5 + 7 \cdot 2 = 39$
(4, 0)	$z = 5 \cdot 4 + 7 \cdot 0 = 20$ ← Minimum value

7. Maximize $z = 2x + y$

Subject to $x \geq 0$, $y \geq 0$, $x + y \leq 6$, $x + y \geq 1$

(a) We have two lines:
$y = -x + 6$ and $y = -x + 1$.

There is no point of intersection, since they are parallel lines. (Their slopes are both -1).

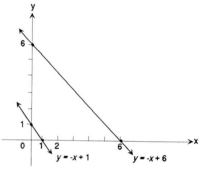

(b) Now evaluate $z = 2x + y$ at each vertex:

VERTEX	VALUE OF $z = 2x + y$
(0, 1)	$z = 2 \cdot 0 + 1 = 1$
(0, 6)	$z = 2 \cdot 0 + 6 = 6$
(6, 0)	$z = 2 \cdot 6 + 0 = 12$
(1, 0)	$z = 2 \cdot 1 + 0 = 2$

The maximum possible value for z is 12, at the point (6, 0).

9. Minimize $z = 2x + 5y$

Subject to $x \geq 0$, $y \geq 0$, $x + y \geq 2$, $x \leq 5$, $y \leq 3$.

(a) Graph the constraints:

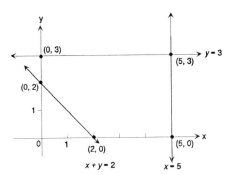

VERTEX	VALUE OF $z = 2x + 5y$
(0, 2)	$z = 10$
(0, 3)	$z = 15$
(5, 3)	$z = 25$
(5, 0)	$z = 10$
(2, 0)	$z = 4$

The minimum value of z is 4, at the point $(2, 0)$.

11. $z = 3x + 5y$

Subject to $x \geq 0$, $y \geq 0$, $x + y \geq 2$, $2x + 3y \leq 12$, $3x + 2y \leq 12$.

We have three lines:

(1) $y = -x + 2$

(2) $y = -\frac{2}{3}x + 4$

(3) $y = -\frac{3}{2}x + 6$

Let's find the points of intersection:

(1) and (2):
$$-x + 2 = -\frac{2}{3}x + 4$$
$$-3x + 6 = -2x + 12$$
$$-x = 6$$
$$x = -6 \text{ and } y = -x + 2$$
$$y = 8$$

But, $(-6, 8)$ is not in the feasible region $(x \geq 0)$.

(1) and (3):
$$-x + 2 = -\frac{3}{2}x + 6$$
$$-2x + 4 = -3x + 12$$
$$x = 8$$
and $\quad y = -x + 2 = -6$
But $(8, -6)$ is not in the feasible region.

(2) and (3):
$$-\frac{2}{3}x + 4 = -\frac{3}{2}x + 6$$
$$-4x + 24 = -9x + 36$$
$$5x = 12$$
$$x = \frac{12}{5}$$
and
$$y = -\frac{3}{2}x + 6$$
$$y = -\frac{3}{2}\left(\frac{12}{5}\right) + 6 = \frac{12}{5}$$

This gives the point $\left(\dfrac{12}{5},\ \dfrac{12}{5}\right)$

We now graph the constraints:

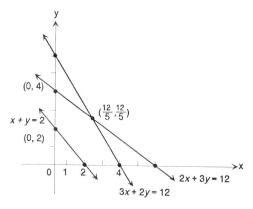

VERTEX	VALUE OF $z = 3x + 5y$
$(0,\ 2)$	$z = 10$
$(0,\ 4)$	$z = 20$
$\left(\dfrac{12}{5},\ \dfrac{12}{5}\right)$	$z = \dfrac{96}{5} = 19.2$
$(4,\ 0)$	$z = 12$
$(2,\ 0)$	$z = 6$

The maximum value of z is 20, at the point $(0,\ 4)$.

13. Minimize $z = 5x + 4y$

Subject to $x \geq 0$, $y \geq 0$, $x + y \geq 2$, $2x + 3y \leq 12$, $3x + y \leq 12$.

We have three lines to graph:

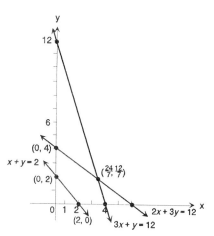

(1) $y = -x + 2$; y-intercept = 2; x-intercept = 2

(2) $y = -\dfrac{2}{3}x + 4$; y-intercept = 4; x-intercept = 6

(3) $y = -3x + 12$; y-intercept = 12; x-intercept = 4

The only intersection we need is:

(2) and (3): $-\dfrac{2}{3}x + 4 = -3x + 12$

$$-2x + 12 = -9x + 36$$
$$7x = 24$$
$$x = \dfrac{24}{7}$$

and $y = -3x + 12$; $y = \dfrac{12}{7}$

$$\left(\dfrac{24}{7},\ \dfrac{12}{7}\right)$$

VERTEX	VALUE OF $z = 5x + 4y$
$(0, 2)$	$z = 8$
$(0, 4)$	$z = 16$
$\left(\dfrac{24}{7}, \dfrac{12}{7}\right)$	$z = \dfrac{168}{7} = 24$
$(4, 0)$	$z = 20$
$(2, 0)$	$z = 10$

The minimum value of z is 8, at the point $(0, 2)$.

15.   Maximize $z = 5x + 2y$

Subject to $x \geq 0$, $y \geq 0$,
$x + y \leq 10$, $2x + y \geq 10$,
$x + 2y \geq 10$.

We have:

(1) $y = -x + 10$;
     y-intercept $= 10$;
     x-intercept $= 10$

(2) $y = -2x + 10$;
     y-intercept $= 10$;
     x-intercept $= 5$

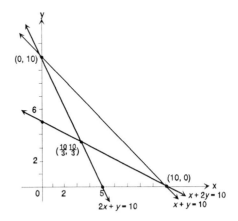

(3) $y = -\dfrac{1}{2}x + 5$;
     y-intercept $= 5$;
     x-intercept $= 10$

To find the intersection of (2) and (3):

$$-2x + 10 = -\frac{1}{2}x + 5$$
$$-4x + 20 = -x + 10$$
$$-3x = -10$$
$$x = \frac{10}{3}$$

and $y = -2x + 10$
$$y = -2\left(\frac{10}{3}\right) + 10$$
$$y = \frac{10}{3}; \left(\frac{10}{3}, \frac{10}{3}\right)$$

VERTEX	VALUE OF $z = 5x + 2y$
$(0, 10)$	$z = 20$
$(10, 0)$	$z = 50$
$\left(\dfrac{10}{3}, \dfrac{10}{3}\right)$	$z = \dfrac{70}{3} = 23\dfrac{1}{3}$

The maximum value of z is 50, at the point $(10, 0)$.

12.2 LINEAR PROGRAMMING

17. As in Example 3, the low-grade mixture contains 4 ounces of Colombian coffee and 12 ounces of special-blend, and the high-grade mixture contains 8 ounces of Colombian and 8 ounces of special-blend, per one-pound bag. We have 100 pounds (1600 ounces) of Colombian and 120 pounds (1920 ounces) of special-blend.

Again, let x = Number of packages of the low-grade mixture
y = Number of packages of the high-grade mixture

Now, the objective function (profit), is:

$P = 0.40x + 0.30y$

Our constraints are $x \geq 0$, $y \geq 0$,

$4x + 8y \leq 1600$ (Colombian coffee)
$12x + 8y \leq 1920$ (Special-blend).

These are the same constraints as in Example 3 (see Figure 4 in the text). Thus, we have the same vertices, but a different objective function:

VERTEX	VALUE OF $P = 0.40x + 0.30y$
(0, 0)	$P = 0$
(0, 200)	$P = 0.3(200) = 60$
(40, 180)	$P = 0.4(40) + 0.3(180) = 70$
(160, 0)	$P = 0.40(160) = 64$

Thus, the maximum profit is $70, when

$x = $ 40 bags of low-grade mixture
$y = $ 180 bags of high-grade mixture

19. Let x = Number of Downhill skis produced
y = Number of Cross-country skis produced

We want to maximize profit (which is $70 per Downhill ski, $50 per Cross-Country ski). Thus, total profit is:

$P = 70x + 50y$

This is our objective function. Now we need to determine our constraints:

$x \geq 0$, $y \geq 0$ Nonnegative constraints

We only have 40 hours manufacturing time available:

$2x + y \leq 40$ Manufacturing time

Also, $x + y \leq 32$ Finishing time constraint

So we have two lines to graph:

(1) $y = -2x + 40$;
    y-intercept = 40;
    x-intercept = 20
(2) $y = -x + 32$;
    y-intercept = 32;
    x-intercept = 32

For the point of inter- section of (1) and (2):

$$-2x + 40 = -x + 32$$
$$-x = -8$$
$$x = 8$$

and
$$y = -x + 32$$
$$y = 24; \ (8, \ 24)$$

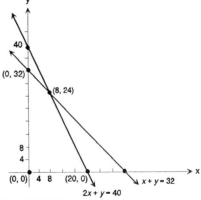

VERTEX	VALUE OF PROFIT: $P = 70x + 50y$
(0, 0)	$P = 0$
(0, 32)	$P = 50 \cdot 32 = 1600$
(8, 24)	$P = 70 \cdot 8 + 50 \cdot 24 = 1760$
(20, 0)	$P = 70 \cdot 20 = 1400$

The maximum profit is \$1,760, obtained when

    $x = 8$ Downhill skis
    $y = 24$ Cross-country skis

are produced.

21.   Let x = Number of acres of soybeans
        y = Number of acres of wheat

The objective function is the profit:

$$P = 180x + 100y$$

(\$180 per acre of soybeans; \$100 per acre of wheat)

For constraints, we have:

    $x \geq 0, \ y \geq 0$ Nonnegative constraints
    $60x + 30y \leq 1800$ Preparation Costs
and $3x + y \leq 120$ Workdays

We graph the two lines:

(1) $y = -2x + 60$;
  y-intercept = 60;
  x-intercept = 30

(2) $y = -\frac{3}{4}x + 30$;
  y-intercept = 30;
  x-intercept = 40

The point of intersection of (1) and (2):

$$-2x + 60 = -\frac{3}{4}x + 30$$
$$-8x + 240 = -3x + 120$$
$$-5x = -120$$
$$x = 24$$
$$y = -2x + 60 = 12;$$
$$(24, 12)$$

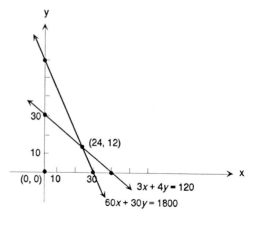

VERTEX	VALUE OF PROFIT: $P = 180x + 100y$
(0, 0)	$P = 0$
(0, 30)	$P = 3000$
(24, 12)	$P = 4320 + 1200 = 5520$
(30, 0)	$P = 5400$

The maximum profit is $5,520, from $x = 24$ acres of soybeans and $y = 12$ acres of wheat.

23. We can make the following table:

	Certificate of Deposit	Treasury Bills	Total
Annual Yield	9% = 0.09	7% = 0.07	
Amount Invested	≤12,000	≥8,000	≤20,000

Let x = Amount invested in the Certificate of Deposit
  y = Amount invested in Treasury bills

We want to maximize the return, R:

$$R = 0.09x + 0.07y$$

Our constraints are:

$x \geq 0$, $y \geq 0$	Nonnegative constraints
$x + y \leq 20,000$	Total to be invested
$x \leq 12,000$	Certificate constraint
$y \geq 8,000$	Treasury bill constraint
$y \geq x$	Amount in Treasury bills must equal or exceed amount in Certificate of Deposit

We graph two lines:

    (1) $y = -x + 20,000$;
       y-intercept $= 20,000$
       x-intercept $= 20,000$
    (2) $y = x$;
       y-intercept $= 0$
       x-intercept $= 0$

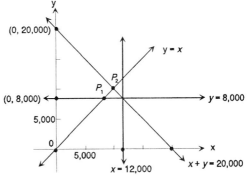

The two points of intersection, P_1, P_2, are:

$P_1(8000, 8000)$,
$P_2(10000, 10000)$

VERTEX	VALUE OF RETURN $R = 0.09x + 0.07y$
(0, 8000)	$R = 0.07(8000) = 560$
(0, 20000)	$R = 0.07(20,000) = 1,400$
(10000, 10000)	$R = 0.09(10,000) + 0.07(10,000) = 1,600$
(8000, 8000)	$R = 0.09(8000) + 0.07(8000) = 1,280$

The maximum profit is \$1,600 obtained from investing \$10,000 in the Certificate and \$10,000 in Treasury bills.

25.   Make a table:

	Racing	Figure	Total Time Available
Fabrication	6 hrs	4 hrs	120 hrs
Finishing	1 hr	2 hr	40 hrs

Let x = Number of racing skates
    y = Number of figure skates

The Profit, P, is given by:

    $P = 10x + 12y$

Our constraints are:

    $x \geq 0, y \geq 0$ Nonnegative constraints
    $6x + 4y \leq 120$ Fabrication time constraint
    $x + 2y \leq 40$ Finishing time constraint

We have the two lines:

    (1) $y = -\frac{3}{2}x + 30$; y-intercept $= 30$;
       x-intercept $= 20$
    (2) $y = -\frac{1}{2}x + 20$;
       y-intercept $= 20$;
       x-intercept $= 40$

The point of intersection of (1) and (2):

$$-\frac{3}{2}x + 30 = -\frac{1}{2}x + 20$$

$$-3x + 60 = -x + 40$$

$$-2x = -20$$

$$x = 10$$

and

$$y = -\frac{1}{2}x + 20$$

$$y = 15; \quad (10, 15)$$

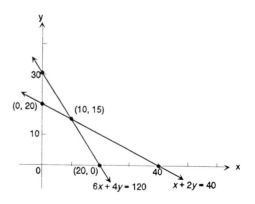

VERTEX	VALUE OF PROFIT: $P = 10x + 12y$
(0, 0)	$P = 0$
(0, 20)	$P = 240$
(10, 15)	$P = 280$
(20, 0)	$P = 200$

The maximum profit is \$280, obtained by manufacturing

and $x = 10$ racing skates
       $y = 15$ figure skates

each day.

27.   Let x = Number of first class seats
           y = Number of coach seats

The constraints are:

(1) $8 \le x \le 16$
(2) $80 \le y \le 120$
(3) $\dfrac{x}{y} \le \dfrac{1}{12}$ Ratio of first class to coach cannot exceed $\dfrac{1}{12}$.

Common sense solves this one: Let $y = 120$. Then by (3), $\dfrac{x}{120} \le \dfrac{1}{12}$, or $x \le 10$, so we take $x = 10$, since that is allowed by (1). Thus, the largest number of seat we can have is 120 coach and 10 first class.

29.

	UNITS PER OUNCE	
	Supplement A	Supplement B
Vitamin I	5	25
Vitamin II	25	10
Vitamin III	10	10
Vitamin IV	35	20

12 MISCELLANEOUS TOPICS

Let x = Number of ounces of Supplement A per 100 ounces of feed
y = Number of ounces of Supplement B per 100 ounces of feed

Our cost will be:

$C = 0.06x + 0.08y,$

per 100 ounces of feed.

Our constraints are:

$x \geq 0, \; y \geq 0$
$5x + 25y \geq 50$ Vitamin I requirement
$25x + 10y \geq 90$ Vitamin II
$10x + 10y \geq 60$ Vitamin III
$35x + 20y \geq 100$ Vitamin IV

We have four lines:

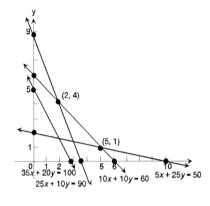

(1) $y = -\frac{1}{5}x + 2$; y-intercept
= 2;
x-intercept = 10
(2) $y = -\frac{5}{2}x + 9$; y-intercept
= 9;
x-intercept = 3.6
(3) $y = -x + 6$; y-intercept
= 6;
x-intercept = 6
(4) $y = -\frac{7}{4}x + 5$; y-intercept
= 5;
x-intercept = $\frac{20}{7} \approx 2.86$

We see from the graph that we need two points of intersection:

(2) and (3):
$$-\frac{5}{2}x + 9 = -x + 6$$
$$-5x + 18 = -2x + 12$$
$$-3x = -6$$
$$x = 2$$
and $y = -x + 6 = 4$; (2, 4)

(1) and (3):
$$-\frac{1}{5}x + 2 = -x + 6$$
$$-x + 10 = -5x + 30$$
$$4x = 20$$
$$x = 5$$
and $y = -x + 6 = 1$; (5, 1)

Notice that feasible region is <u>unbounded</u>; i.e., it extends infinitely far in both the x- and y- directions. For example, $x = 1000$, $y = 10,000$ is a <u>feasible</u> solution (but it certainly would <u>not</u> minimize our cost!).

12.2 LINEAR PROGRAMMING   697

VERTEX	VALUE OF COST: $C = 0.06x + 0.08y$
(0, 9)	$C = .72$
(2, 4)	$C = .44$
(5, 1)	$C = .38$
(10, 0)	$C = .60$

The total cost will be a minimum ($0.38 per 100 ounces of feed) if the farmer uses 5 ounces of Supplement A and 1 ounce of Supplement B.

≡ EXERCISE 12.3 PARTIAL FRACTION DECOMPOSITION

1. The expression $\dfrac{x}{x^2 - 1}$ is proper, since the degree of the numerator is less than that of the denominator.

3. $\dfrac{x^2 + 5}{x^2 - 4}$ is improper, so we do the division:

$$
\begin{array}{r}
1 \quad \leftarrow \text{Quotient} \\
x^2 - 4\overline{)x^2 + 0x + 5} \\
\underline{x^2 \qquad - 4} \\
9 \quad \leftarrow \text{Remainder}
\end{array}
$$

Then, $\dfrac{x^2 + 5}{x^2 - 4} = 1 + \dfrac{9}{x^2 - 4}$.

5. $\dfrac{5x^3 + 2x - 1}{x^2 - 4}$ is improper, so we must do long division:

$$
\begin{array}{r}
5x \\
x^2 - 4\overline{)5x^3 + 0x^2 + 2x - 1} \\
\underline{5x^3 \qquad\quad -20x} \\
22x - 1
\end{array}
$$

Thus, $\dfrac{5x^3 + 2x - 1}{x^2 - 4} = 5x + \dfrac{22x - 1}{x^2 - 4}$

7. Here, $\dfrac{x(x - 1)}{(x + 4)(x - 3)} = \dfrac{x^2 - x}{x^2 + x - 12}$ is improper:

$$
\begin{array}{r}
1 \\
x^2 + x - 12\overline{)x^2 - x + 0} \\
\underline{x^2 + x - 12} \\
-2x + 12
\end{array}
$$

Hence, $\dfrac{x(x - 1)}{(x + 4)(x - 3)} = 1 + \dfrac{-2x + 12}{x^2 + x - 12} = 1 + \dfrac{-2(x - 6)}{(x + 4)(x - 3)}$

12 MISCELLANEOUS TOPICS

In Problems 9-38, we will use the following steps:

> *Step 1: Perform long division, if necessary, to obtain a proper*
> *fraction, $\frac{p(x)}{q(x)}$, and put it in lowest terms.*
> *Step 2: Factor the denominator, $q(x)$, completely into linear and*
> *irreducible quadratic factors, and identify which of the*
> *four cases in the text apply.*
> *Step 3: Write out the partial fraction expansion, based on the*
> *factors of $q(x)$.*
> *Step 4: Solve for the unknown coefficients in Step 3.*
> *Step 5: Write the final decomposition.*

9. $\dfrac{4}{x(x-1)}$

Step 1: Already proper.

Step 2: Done: $q(x) = x(x-1)$.
This is Case 1 (only non-repeated linear factors).

Step 3: $\dfrac{4}{x(x-1)} = \dfrac{A}{x} + \dfrac{B}{x-1}$

Step 4: Multiply both sides by $x(x-1)$:

$4 = A(x-1) + Bx$

Let $x = 1$:   then $4 = A(0) + B$, or $B = 4$
Let $x = 0$:   then $4 = A(-1) + B \cdot 0$, or $A = -4$

Step 5: Therefore, $\dfrac{4}{x(x-1)} = \dfrac{-4}{x} + \dfrac{4}{x-1}$

11. $\dfrac{1}{x(x^2 + 1)}$

Step 1: Already proper.

Step 2: $q(x) = x(\underbrace{x^2 + 1}_{})$

cannot be factored

This is Case 1 (nonrepeated linear) and Case 3
(nonrepeated irreducible quadratic).

Step 3: $\dfrac{1}{x(x^2 + 1)} = \dfrac{A}{x} + \dfrac{Bx + C}{x^2 + 1}$

(Remember, irreducible quadratic factors in the
denominator require <u>first degree</u> numerators in the
decomposition.)

12.3 PARTIAL FRACTION DECOMPOSITION   699

Multiply by $x(x^2 + 1)$:

$$1 = A(x^2 + 1) + (Bx + C)x$$

Let $x = 0$: then $1 = A$.

We still have two unknowns, B and C, so we need two equations:

Let $-x = 1$: $1 = A(1 + 1) + (B \cdot 1 + C) \cdot 1$
   or $1 = 2A + B + C$
   or $B + C = -1$

Let $x = -1$: $1 = 2A + (-B + C)(-1)$
   or $1 = 2 + B - C$ (since $A = 1$)
   or $B - C = -1$ (since $A = 1$)

We have $B + C = -1$
            $\underline{B - C = -1}$
   Add: $2B = -2$
                   $B = -1$

Then, from $B + C = -1$, we get:

$$-1 + C = -1$$
$$C = 0$$

Step 5: $\dfrac{1}{x(x^2 + 1)} = \dfrac{1}{x} + \dfrac{-x}{x^2 + 1}$

13. $\dfrac{1}{(x - 1)(x - 2)}$

Step 1: This is proper.

Step 2: $q(x) = (x - 1)(x - 2)$. Case 1 only.

Step 3: $\dfrac{x}{(x - 1)(x - 2)} = \dfrac{A}{x - 1} + \dfrac{B}{x - 2}$

Step 4: Multiply by $(x - 1)(x - 2)$:

$$x = A(x - 2) + B(x - 1)$$

Let $x = 2$: $2 = B$

Let $x = 1$: $1 = -A$, or $A = -1$

Step 5: $\dfrac{x}{(x - 1)(x - 2)} = \dfrac{-1}{x - 1} + \dfrac{2}{x - 2}$

15. $$\frac{x^2}{(x - 1)^2(x + 1)}$$

Step 1: This is proper.

Step 2: $q(x) = (x - 1)^2(x + 1)$.

This involves both Case 1 and Case 2 (a repeated linear factor).

Step 3: For the factor $(x - 1)^2$, we need two terms, $\frac{A}{x - 1} + \frac{B}{(x - 1)^2}$, since it is a linear factor raised to the <u>second</u> power.

$$\frac{x^2}{(x - 1)^2(x + 1)} = \frac{A}{x - 1} + \frac{B}{(x - 1)^2} + \frac{C}{x + 1}$$

Step 4: Multiply by $(x - 1)^2(x + 1)$:

$$x^2 = A(x - 1)(x + 1) + B(x + 1) + C(x - 1)^2$$

Since we have three unknowns, we need three equations, so we will choose three values of x starting with the <u>zeros</u> of $q(x)$:

Let $x = 1$: $1 = 2B$, or $B = \frac{1}{2}$

Let $x = -1$: $1 = 4C$, or $C = \frac{1}{4}$

Now pick a third value for x:

Let $x = 0$: $0 = -A + B + C$

$0 = -A + \frac{1}{2} + \frac{1}{4}$, since we know B and C.

$A = \frac{3}{4}$

Step 5: $$\frac{x^2}{(x - 1)^2(x + 1)} = \frac{\frac{3}{4}}{x - 1} + \frac{\frac{1}{2}}{(x - 1)^2} + \frac{\frac{1}{4}}{x + 1}$$

17. $$\frac{1}{x^3 - 8}$$

Step 1: This is proper.

Step 2: We need to do some factoring:

$$q(x) = (x - 2)(\underbrace{x^2 + 2x + 4})$$

can this be factored?

12.3 PARTIAL FRACTION DECOMPOSITION   701

To determine if $x^2 + 2x + 4$ can be factored or if it is irreducible, check its discriminant:

$$b^2 - 4ac = 4 - 4(4) = 4 - 16 = -12 < 0$$

Therefore, $x^2 + 2x + 4$ has no real zeros; i.e., it is irreducible over the reals.

We have Case 1 and Case 3.

<u>Step 3</u>: $\dfrac{1}{x^3 - 8} = \dfrac{1}{(x - 2)(x^2 + 2x + 4)} = \dfrac{A}{x - 2} + \dfrac{Bx + C}{x^2 + 2x + 4}$

<u>Step 4</u>: Multiply both sides by $x^3 - 8 = (x - 2)(x^2 + 2x + 4)$:

$$1 = A(x^2 + 2x + 4) + (Bx + C)(x - 2) \tag{1}$$

We only have one zero for $q(x)$, $x = 2$:

Let $x = 2$: $1 = 12A$, or $A = \dfrac{1}{12}$

But we need three equations. Another approach is to expand (1) and collect like terms:

$$1 = Ax^2 + 2Ax + 4A + Bx^2 - 2Bx + Cx - 2C$$
or
$$1 = (A + B)x^2 + (2A - 2B + C)x + (4A - 2C)$$

There is no x^2 on the left-hand side, so we must have $A + B = 0$. Also, there is no x-term on the left, so

$$2A - 2B + C = 0$$

Finally, the constant terms must be equal:

$$1 = 4A - 2C$$

But, we already know $A = \dfrac{1}{12}$, so from $A + B = 0$, we have $B = -\dfrac{1}{12}$, and from $1 = 4A - 2C$ we have:

$$2C = 4A - 1$$
$$2C = \dfrac{4}{12} - 1$$
$$2C = \dfrac{-8}{12}$$
$$C = \dfrac{-4}{12} = \dfrac{-1}{3}$$

<u>Step 5</u>: $\dfrac{1}{x^3 - 8} = \dfrac{\frac{1}{12}}{x - 2} + \dfrac{\frac{-1}{12}x - \frac{1}{3}}{x^2 + 2x + 4} = \dfrac{\frac{1}{12}}{x - 2} + \dfrac{\frac{-1}{12}(x + 4)}{x^2 + 2x + 4}$

12 MISCELLANEOUS TOPICS

19. $\dfrac{x^2}{(x - 1)^2(x + 1)^2}$

<u>Step 1</u>: This is proper.

<u>Step 2</u>: $q(x) = (x - 1)^2(x + 1)^2$

This is case 2.

<u>Step 3</u>: $\dfrac{x^2}{(x - 1)^2(x + 1)^2} = \dfrac{A}{x - 1} + \dfrac{B}{(x - 1)^2} + \dfrac{C}{x + 1} + \dfrac{D}{(x + 1)^2}$

<u>Step 4</u>: Multiply by $(x - 1)^2(x + 1)^2$:

$$x^2 = A(x - 1)(x + 1)^2 + B(x + 1)^2 + C(x - 1)^2(x + 1) + D(x - 1)^2$$

Let $x = 1$: $\quad 1 = 4B, \quad B = \dfrac{1}{4}$

Let $x = -1$: $1 = 4D, \quad D = \dfrac{1}{4}$

Let $x = 0$: $\quad 0 = -A + B + C + D$

$\qquad$ or $\quad A - C = \dfrac{1}{2}$

Let $x = 2$: $\quad 4 = 9A + 9B + 3C + D$

$\qquad$ or $\quad -9A - 3C = -\dfrac{3}{2}$

$\qquad$ or $\quad 3A + C = \dfrac{1}{2}$

We have two equations in two unknowns:

$$A - C = \dfrac{1}{2}$$

$$3A + C = \dfrac{1}{2}$$

Add: $\quad \overline{4A \qquad = 1}$

$$A = \dfrac{1}{4}$$

and $\quad A - C = \dfrac{1}{2}$

$$\dfrac{1}{4} - C = \dfrac{1}{2}$$

$$C = -\dfrac{1}{4}$$

<u>Step 5</u>: $\dfrac{x^2}{(x - 1)^2(x + 1)^2} = \dfrac{\frac{1}{4}}{x - 1} + \dfrac{\frac{1}{4}}{(x - 1)^2} + \dfrac{\frac{-1}{4}}{x + 1} + \dfrac{\frac{1}{4}}{(x + 1)^2}$

21. $\dfrac{x-3}{(x+2)(x+1)^2}$

Steps 1 and 2: Proper, with $q(x) = (x+2)(x+1)^2$, Case 1 and Case 2.

Step 3: $\dfrac{x-3}{(x+2)(x+1)^2} = \dfrac{A}{x+2} + \dfrac{B}{(x+1)} + \dfrac{C}{(x+1)^2}$

Step 4: Clear of fractions:

$$x - 3 = A(x+1)^2 + B(x+2)(x+1) + C(x+2)$$

Let $x = -1$: $-4 = C$
Let $x = -2$: $-5 = A$
Let $x = 0$: $\quad -3 = A + 2B + 2C$
$\qquad\qquad$ or $\quad -3 = -5 + 2B - 8$
$\qquad\qquad\qquad 10 = 2B$
$\qquad\qquad\qquad B = 5$

Step 5: $\dfrac{x-3}{(x+2)(x+1)^2} = \dfrac{-5}{x+2} + \dfrac{5}{x+1} + \dfrac{-4}{(x+1)^2}$

23. $\dfrac{x+4}{x^2(x^2+4)}$

Step 1: This is proper.

Step 2: $q(x) = x^2(x^2+4)$.

Be careful to note the difference between x^2 and $x^2 + 4$:

The factor x^2 is a linear factor (x), repeated:

$x^2 = x \cdot x$, so it is Case 2.

The factor $x^2 + 4$ is a quadratic that cannot be factored as a product of linear factors (Case 3).

Step 3: $\dfrac{x+4}{x^2(x^2+4)} = \dfrac{A}{x} + \dfrac{B}{x^2} + \dfrac{Cx+D}{x^2+4}$

Step 4: $x + 4 = Ax(x^2+4) + B(x^2+4) + (Cx+D)x^2.$ $\qquad$ (1)

There is only one zero, $x = 0$:

Let $x = 0$: $\quad 4 = 4B$, or $B = 1$.

Let's expand equation (1) and combine like terms:

$$x + 4 = Ax^3 + 4Ax + Bx^2 + 4B + Cx^3 + Dx^2$$
$$= (A+C)x^3 = (B+D)x^2 + (4A)x + 4B$$

Now equate coefficients of like powers of x on the left and right:

For x^3: $0 = A + C$
For x^2: $0 = B + D$
For x^1: $1 = 4A$, or $A = \dfrac{1}{4}$
For x^0: $4 = 4B$, or $B = 1$

Then: $A + C = 0 \qquad B + D = 0$

$\dfrac{1}{4} + C = 0 \qquad 1 + D = 0$

$C = \dfrac{-1}{4} \qquad D = -1$

Step 5: $\dfrac{x + 4}{x^2(x^2 + 4)} = \dfrac{\frac{1}{4}}{x} + \dfrac{1}{x^2} + \dfrac{-\frac{1}{4}x - 1}{x^2 + 4} = \dfrac{\frac{1}{4}}{x} + \dfrac{1}{x^2} - \dfrac{\frac{1}{4}(x + 4)}{x^2 + 4}$

25. $\dfrac{x^2 + 2x + 3}{(x + 1)(x^2 + 2x + 4)}$

Step 1: This is proper.

Step 3: $q(x) = (x + 1)(x^2 + 2x + 4)$

$\qquad\qquad b^2 - 4ac = 4 - 16 = -12 < 0$

We have Case 1 and Case 3.

Step 3: $\dfrac{x^2 + 2x + 3}{(x + 1)(x^2 + 2x + 4)} = \dfrac{A}{x + 1} = \dfrac{Bx + C}{x^2 + 2x + 4}$

Step 4: $x^2 + 2x + 3 = A(x^2 + 2x + 4) + (Bx + C)(x + 1)$

Let $x = -1$: $2 = 3A$; $A = \dfrac{2}{3}$

Let $x = 0$: $3 = 4A + C$; $C = 3 - 4A = 3 - \dfrac{8}{3} = \dfrac{1}{3}$

Let $x = 1$: $6 = 7A + 2B + 2C$

$2B = 6 - 7A - 2C$

$2B = 6 - \dfrac{14}{3} - \dfrac{2}{3} = \dfrac{2}{3}$; $B = \dfrac{1}{3}$

Step 5: $\dfrac{x^2 + 2x + 3}{(x + 1)(x^2 + 2x + 4)} = \dfrac{\frac{2}{3}}{x + 1} + \dfrac{\frac{1}{3}(x + 1)}{x^2 + 2x + 4}$

27. $\dfrac{x}{(3x - 2)(2x + 1)}$

Step 1: This is proper.

Step 2: $q(x) = (3x - 2)(2x + 1)$, Case 1.

12.3 PARTIAL FRACTION DECOMPOSITION

Step 3: $\dfrac{x}{(3x - 2)(2x + 1)} = \dfrac{A}{3x - 2} + \dfrac{B}{2x + 1}$

Step 4: $x = A(2x + 1) + B(3x - 2)$

The two zeros are $x = -\dfrac{1}{2}$ and $x = \dfrac{2}{3}$

Let $x = -\dfrac{1}{2}$: $-\dfrac{1}{2} = B\left(-\dfrac{7}{2}\right)$

$1 = 7B$

$B = \dfrac{1}{7}$

Let $x = \dfrac{2}{3}$: $\dfrac{2}{3} = A\left(\dfrac{7}{3}\right)$

$2 = 7A$

$A = \dfrac{2}{7}$

Step 5: $\dfrac{x}{(3x - 2)(2x + 1)} = \dfrac{\frac{2}{7}}{3x - 2} + \dfrac{\frac{1}{7}}{2x + 1}$

29. $\dfrac{x}{x^2 + 2x - 3}$

Step 1: Already proper.

Step 2: $q(x) = x^2 + 2x - 3 = (x + 3)(x - 1)$, Case 1.

Step 3: $\dfrac{x}{x^2 + 2x - 3} = \dfrac{A}{x + 3} + \dfrac{B}{x - 1}$

Step 4: $x = A(x - 1) + B(x + 3)$

Let $x = 1$: $1 = 4B$, $B = \dfrac{1}{4}$

Let $x = -3$: $-3 = -4A$, $A = \dfrac{3}{4}$

Step 5: $\dfrac{x}{(x^2 + 2x - 3)} = \dfrac{\frac{3}{4}}{x + 3} + \dfrac{\frac{1}{4}}{x - 1}$

31. $\dfrac{x^2 + 2x + 3}{(x^2 + 4)^2}$

Steps 1 and 2: This is proper, and falls under Case 4 (repeated irreducible quadratic factor).

Step 3: $\dfrac{x^2 + 2x + 3}{(x^2 + 4)^2} = \dfrac{Ax + B}{x^2 + 4} + \dfrac{Cx + D}{(x^2 + 4)^2}$

12 MISCELLANEOUS TOPICS

Step 4: Clear of fractions:

$$x^2 + 2x + 3 = (Ax + B)(x^2 + 4) + (Cx + D)$$

We have no zeros of $q(x)$, so expand the right-hand side:

$$x^2 + 2x + 3 = Ax^3 + Bx^2 + 4Ax + 4B + Cx + D$$
$$x^2 + 2x + 3 = Ax^3 + Bx^2 + (4A + C)x + (4B + D)$$

x^3: $0 = A$

x^2: $1 = B$

x: $2 = 4A + C$, $C = 2$ (since $A = 0$)

Constant: $3 = 4B + D$, $D = 3 - 4B = -1$

Step 5: $\dfrac{x^2 + 2x + 3}{(x^2 + 4)^2} = \dfrac{1}{x^2 + 4} + \dfrac{2x - 1}{(x^2 + 4)^2}$

33. $\dfrac{7x + 3}{x^3 - 2x^2 - 3x}$

Step 1: This is proper.

Step 2: $q(x) = x^3 - 2x^2 - 3x = x(x^2 - 2x - 3)$
$$= x(x - 3)(x + 1)$$

This is Case 1.

Step 3: $\dfrac{7x + 3}{x^3 - 2x^2 - 3x)} = \dfrac{A}{x} = \dfrac{B}{x - 3} + \dfrac{C}{x + 1}$

Step 4: Multiply both sides by $x^3 - 2x^2 - 3x = x(x - 3)(x + 1)$:

$$7x + 3 = A(x - 3)(x + 1) + Bx(x + 1) + Cx(x - 3)$$

Let $x = 0$: $3 = -3A$; $A = -1$

Let $x = 3$: $24 = 12B$, $B = 2$

Let $x = -1$: $-4 = 4C$, $C = -1$

Step 5: $\dfrac{7x + 3}{x^3 - 2x^2 - 3x} = \dfrac{-1}{x} + \dfrac{2}{x - 3} + \dfrac{-1}{x + 1}$

35. $\dfrac{x^2}{x^3 - 4x^2 + 5x - 2}$

Step 1: This is proper.

Step 2: Try to find a zero by synthetic division:

```
1)1  -4   5   -2          -1)1  -4   5   -2
      1  -5   10                -1   3   -2
   1  -5  10  -12            1  -3   2   |0|
```

$$x^2 - 3x + 2$$

Thus, $x - 1$ is a factor:

$$q(x) = x^3 - 4x^2 + 5x - 2 = (x - 1)(x^2 - 3x + 2)$$
$$= (x - 1)(x - 1)(x - 2)$$
$$= (x - 1)^2(x - 2)$$

This involves Case 1 and Case 2.

<u>Step 3</u>: $\dfrac{x^2}{x^3 - 4x^2 + 5x - 2} = \dfrac{A}{x - 1} + \dfrac{B}{(x - 1)^2} + \dfrac{C}{x - 2}$

<u>Step 4</u>: Multiply by $x^3 - 4x^2 + 5x - 2 = (x - 1)^2(x - 2)$:

$$x^2 = A(x - 1)(x - 2) + B(x - 2) + C(x - 1)^2$$

Let $x = 1$: $1 = -B$, $B = -1$
Let $x = 2$: $4 = C$
Let $x = 0$: $0 = 2A - 2B + C$
                     $2A = 2B - C$
                     $2A = -2 - 4$
                      $A = -3$

<u>Step 5</u>: $\dfrac{x^2}{x^3 - 4x^2 + 5x - 2} = \dfrac{-3}{x - 1} + \dfrac{-1}{(x - 1)^2} + \dfrac{4}{x - 2}$

37. $\dfrac{x^3}{(x^2 + 16)^3}$

<u>Step 1 and 2</u>:   Proper, Case 4

<u>Step 3</u>: $\dfrac{x^3}{(x^2 + 16)^3} = \dfrac{Ax + B}{x^2 + 16} + \dfrac{Cx + D}{(x^2 + 16)^2} + \dfrac{Ex + F}{(x^2 + 16)^3}$

<u>Step 4</u>: $x^3 = (Ax + B)(x^2 + 16)^2 + (Cx + D)(x^2 + 16) + Ex + F$ or

$$x^3 = (Ax + B)(x^4 + 32x^2 + 256) + Cx^3 + Dx^2 + 16Cx + 16D$$
$$+ Ex + F$$

$$x^3 = Ax^5 + 32Ax^3 + 256Ax + Bx^4 + 32Bx^2 + 256B + Cx^3$$

$$+ Dx^2 + 16Cx + 16D + Ex + F$$

$$x^3 = Ax^5 + Bx^4 + (32A + C)x^3 + (32B + D)x^2$$

$$+ (256A + 16C + E)x + (256B + 16D + F)$$

Now we equate coefficients of like powers of x to obtain six equations in six unknowns.

x^5: $0 = A$
x^4: $0 = B$
x^3: $1 = 32A + C$, $C = 1$
x^2: $0 = 32B + D$, $D = 0$
x^1: $0 = 256A + 16C + E$, $E = -16C = -16$
x^0: $0 = 256B + 16D + F$, $F = 0$

Step 5: $\dfrac{x^3}{(x^2 + 16)^3} = \dfrac{x}{(x^2 + 16)^2} + \dfrac{-16x}{(x^2 + 16)^3}$

39. $\dfrac{4}{2x^2 - 5x - 3}$

Step 1: This is proper.

Step 2: $q(x) = 2x^2 - 5x - 3 = (2x + 1)(x - 3)$, Case 1.

Step 3: $\dfrac{4}{2x^2 - 5x - 3} = \dfrac{A}{2x + 1} + \dfrac{B}{x - 3}$

Step 4: $4 = A(x - 3) + B(2x + 1)$

The two zeros are $x = 3$ and $x = \dfrac{-1}{2}$

Let $x = 3$: $\quad 4 = 7B$

$\qquad\qquad\quad B = \dfrac{4}{7}$

Let $x = \dfrac{-1}{2}$: $\quad 4 = \dfrac{-7}{2}A$

$\qquad\qquad\qquad A = \dfrac{-8}{7}$

Step 5: $\dfrac{4}{2x^2 - 5x - 3} = \dfrac{-\dfrac{8}{7}}{2x + 1} + \dfrac{\dfrac{4}{7}}{x - 3}$

41. $\dfrac{2x + 3}{x^4 - 9x^2}$

Step 1: This is proper.

Step 2: $q(x) = x^4 - 9x^2 = x^2(x - 3)(x + 3)$, Case 1 and 2.

Step 3: $\dfrac{2x + 3}{x^4 - 9x^2} = \dfrac{A}{x} + \dfrac{B}{x^2} + \dfrac{C}{x - 3} + \dfrac{D}{x + 3}$

Step 4: $2x + 3 = Ax(x - 3)(x + 3) + B(x - 3)(x + 3) + Cx^2(x - 3)$
$\qquad\qquad\qquad + Dx^2(x - 3)$

Let $x = 0$: $\quad 3 = B(-9)$

$\qquad\qquad\qquad B = \dfrac{-1}{3}$

Let $x = 3$: $\quad 9 = C(9)(6)$

$\qquad\qquad\qquad C = \dfrac{1}{6}$

Let $x = -3$: $-3 = D(9)(-6)$

$\qquad\qquad\qquad D = \dfrac{1}{18}$

12.3 PARTIAL FRACTION DECOMPOSITION

$$2x + 3 = Ax(x - 3)(x + 3) - \frac{1}{3}(x - 3)(x + 3) + \frac{1}{6}x^2(x + 3)$$
$$+ \frac{1}{18}x^2(x - 3)$$

Let $x = 1$: $5 = A(-2)(4) - \frac{1}{3}(-2)(4) + \frac{1}{6}(4) + \frac{1}{18}(-2)$

$$5 = A(-8) + \frac{8}{3} + \frac{2}{3} - \frac{1}{9}$$

$$\frac{45}{9} - \frac{30}{9} + \frac{1}{9} = A(-8)$$

$$\frac{16}{9} = A(-8)$$

$$A = \frac{-2}{9}$$

Step 5: $\dfrac{2x + 3}{x^4 - 9x^2} = \dfrac{-\frac{2}{9}}{x} - \dfrac{\frac{1}{3}}{x^2} + \dfrac{\frac{1}{6}}{x - 3} + \dfrac{\frac{1}{18}}{x + 3}$

≡ EXERCISE 12.4 VECTORS

Problems 1-8 use the diagram at right.

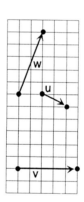

1.   **v + w**:

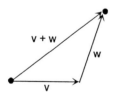

3.   **3v**:

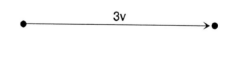

12 MISCELLANEOUS TOPICS

5. **v - w:**   7. **3v + u - 2w:**

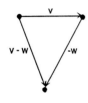

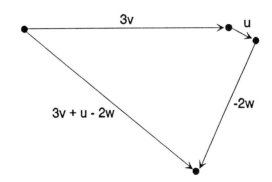

In Problems 9-16, refer
to the figure at right.

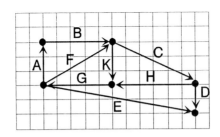

9.   We see that $A + B = F$, so $x + B = F$ implies $x = A$.

11. From the figure, $F + C + D = E$. Therefore, $C = -F + E - D$.

13. We see that $H + G + E = D$, so that $E = -G - H + D$.

15. Notice that $A + B + K + G = 0$, so x must be 0.

17. If $\|v\| = 4$, then $\|3v\| = |3| \|v\| = (3)(4) = 12$.

For Problems 19-26, use the fact that if $P = (x_1, y_1)$ and $Q = (x_2, y_2)$,
then, $\overrightarrow{PQ}$ represents the vector $v = (x_2 - x_1)i + (y_2 - y_1)j$, where i and j
are unit vectors in the positive x-direction and positive y-direction
respectively.

19. $P = (0, 0)$, $Q = (3, 4)$
      Then, $v = (3 - 0)i + (4 - 0)j = 3i + 4j$

21. $P = (3, 2)$, $Q = (5, 6)$
      Then, $v = (5 - 3)i + (6 - 2)j = 2i + 4j$

23. $P = (-2, -1)$, $Q = (6, -2)$
      Then, $v = (6 - (-2))i + (-2 - (-1))j = 8i - j$

12.4 VECTORS   711

25. $P = (1, 0)$, $Q = (0, 1)$
    Then, $\mathbf{v} = -\mathbf{i} + \mathbf{j}$

27. If $\mathbf{v} = 3\mathbf{i} - 4\mathbf{j}$, then $\|\mathbf{v}\| = \sqrt{3^2 + (-4)^2} = 5$

29. For $\mathbf{v} = \mathbf{i} - \mathbf{j}$, $\|\mathbf{v}\| = \sqrt{1^2 + (-1)^2} = \sqrt{2}$

31. For $\mathbf{v} = -2\mathbf{i} + 3\mathbf{j}$, $\|\mathbf{v}\| = \sqrt{4 + 9} = \sqrt{13}$

33. $\mathbf{v} = 3\mathbf{i} - 5\mathbf{j}$ and $\mathbf{w} = -2\mathbf{i} + 3\mathbf{j}$

    So $2\mathbf{v} + 3\mathbf{w} = (6\mathbf{i} - 10\mathbf{j}) + (6\mathbf{i} + 9\mathbf{j})$
    $$= -\mathbf{j}$$

35. For $\mathbf{v} = 3\mathbf{i} - 5\mathbf{j}$, $\mathbf{w} = -2\mathbf{i} + 3\mathbf{j}$, we have

    $$\|\mathbf{v} - \mathbf{w}\| = \|5\mathbf{i} - 8\mathbf{j}\| = \sqrt{25 + 64} = \sqrt{89}$$

37. For $\mathbf{v} = 3\mathbf{i} - 5\mathbf{j}$ and $\mathbf{w} = -2\mathbf{i} + 3\mathbf{j}$,

    $$\|\mathbf{v}\| - \|\mathbf{w}\| = \sqrt{9 + 25} - \sqrt{4 + 9}$$
    $$= \sqrt{34} - \sqrt{13}$$

39. For $\mathbf{v} = 5\mathbf{i}$, $\|\mathbf{v}\| = \sqrt{25} = 5$, and a unit vector in the same direction as $\mathbf{v}$ is:

    $$\mathbf{u} = \frac{1}{\|\mathbf{v}\|} \mathbf{v} = \frac{1}{5}\mathbf{v} = \frac{1}{5}(5\mathbf{i}) = \mathbf{i}$$

41. For $\mathbf{v} = 3\mathbf{i} - 4\mathbf{j}$, $\|\mathbf{v}\| = \sqrt{9 + 16} = 5$, and a unit vector in the same direction as $\mathbf{v}$ is:

    $$\mathbf{u} = \frac{1}{\|\mathbf{v}\|} \mathbf{v}$$
    $$= \frac{1}{5}(3\mathbf{i} - 4\mathbf{j})$$
    $$= \frac{3}{5}\mathbf{i} - \frac{4}{5}\mathbf{j}$$

    Notice: $\|\mathbf{u}\| = \sqrt{\frac{9}{25} + \frac{16}{25}} = \sqrt{\frac{25}{25}} = \sqrt{1} = 1$,
    so $\mathbf{u}$ is indeed a unit vector.

43. For $\mathbf{v} = \mathbf{i} - \mathbf{j}$, $\|\mathbf{v}\| = \sqrt{1 + 1} = \sqrt{2}$,
and a unit vector in the same direction as $\mathbf{v}$ is:

$$\begin{aligned} \mathbf{u} &= \frac{1}{\|\mathbf{v}\|}\,\mathbf{v} \\ &= \frac{1}{\sqrt{2}}(\mathbf{i} - \mathbf{j}) \\ &= \frac{\sqrt{2}}{2}\mathbf{i} - \frac{\sqrt{2}}{2}\mathbf{j} \end{aligned}$$

45. Let $\mathbf{v} = a\mathbf{i} + b\mathbf{j}$.

We want $\|\mathbf{v}\| = 4$ and $a = 2b$.

Now $\|\mathbf{v}\| = \sqrt{a^2 + b^2} = \sqrt{(2b)^2 + b^2} = \sqrt{4b^2 + b^2} = \sqrt{5b^2}$

We want

$$\begin{aligned} \sqrt{5b^2} &= 4 \\ 5b^2 &= 16 \\ b^2 &= \frac{16}{3} \\ b &= \sqrt{\frac{16}{5}} = \frac{4}{\sqrt{5}} = \frac{\pm 4\sqrt{5}}{5} \end{aligned}$$

Then, $a = 2b = \dfrac{\pm 8\sqrt{5}}{5}$

and $\mathbf{v} = \dfrac{\pm 8\sqrt{5}}{5}\mathbf{i} + \dfrac{4\sqrt{5}}{5}\mathbf{j}$ or $\mathbf{v} = \dfrac{-8\sqrt{5}}{5}\mathbf{i} - \dfrac{4\sqrt{5}}{5}\mathbf{j}$

47. Given: $\mathbf{v} = 2\mathbf{i} - \mathbf{j}$
$\mathbf{w} = x\mathbf{i} + 3\mathbf{j}$
$\|\mathbf{v} + \mathbf{w}\| = 5$,

solve for x.

We have: $\mathbf{v} + \mathbf{w} = (2 + x)\mathbf{i} + 2\mathbf{j}$, so that

$$\begin{aligned} \|\mathbf{v} + \mathbf{w}\| &= \sqrt{(2 + x)^2 + 4} \\ &= \sqrt{4 + 4x + x^2 + 4} \\ &= \sqrt{x^2 + 4x + 8} \end{aligned}$$

12.4 VECTORS

Therefore, $\|\mathbf{v} + \mathbf{w}\| = 5$ implies

$$\sqrt{x^2 + 4x + 8} = 5$$

$$\sqrt{x^2 + 4x + 8} = 25$$

$$\sqrt{x^2 + 4x - 17} = 0$$

so by the quadratic formula,

$$x = \frac{-4 \pm \sqrt{16 - 4(-17)}}{2}$$

$$= \frac{-4 \pm \sqrt{84}}{2}$$

$$= \frac{-4 \pm 2\sqrt{21}}{2}$$

or $\quad x = -2 + \sqrt{21} \approx 2.58 \quad$ or $\quad x = -2 - \sqrt{21} \approx -6.58$

$$\left\{ -2 + \sqrt{21}, \ -2 - \sqrt{21} \right\}$$

49. Let: $\mathbf{v}_a$ = Velocity of aircraft relative to the air
$\mathbf{v}_w$ = Wind velocity
$\mathbf{v}_g$ = Velocity of aircraft relative to ground

Then $\mathbf{v}_g = \mathbf{v}_a + \mathbf{v}_w$

So, we need to find expressions for $\mathbf{v}_a$ and $\mathbf{v}_w$

(a) A unit vector in the easterly direction is $\mathbf{i}$. Therefore, $\mathbf{v}_a = 500\mathbf{i}$.

(b) A vector in the northwesterly direction is $-\mathbf{i} + \mathbf{j}$, with length $\|-\mathbf{i} + \mathbf{j}\| = \sqrt{2}$. Therefore, a <u>unit</u> vector in the northwesterly direction is:

$$\mathbf{u} = \frac{1}{\sqrt{2}} (-\mathbf{i} + \mathbf{j}) = \frac{-\sqrt{2}}{2}\mathbf{i} + \frac{\sqrt{2}}{2}\mathbf{j}$$

From this, $\mathbf{v}_w = 60\mathbf{u} = -30\sqrt{2}\,\mathbf{i} + 30\sqrt{2}\,\mathbf{j}$

(c) Now, $\mathbf{v}_g = \mathbf{v}_a + \mathbf{v}_w = (500 - -30\sqrt{2})\mathbf{i} + 30\sqrt{2}\mathbf{j}$

The <u>speed</u> of the aircraft is $\|\mathbf{v}_g\|$:

$$\|\mathbf{v}_g\| = \sqrt{(500 - 30\sqrt{2})^2 + (30\sqrt{2})^2}$$

or $\|\mathbf{v}_g\| = \sqrt{500^2 - 2(30)(500)\sqrt{2} + 2(30)^2 + 2(30)^2}$

$$= \sqrt{250,000 - 30,000\sqrt{2} + 3,600}$$

$$= \sqrt{253,600 - 42,426.4}$$

$$= \sqrt{211,173.6}$$

$$= 460 \text{ kilometers per hour}$$

51. Refer to the figure and the notation used in the solution to Problem 49 (above).

We will use the basic equation

$$\mathbf{v}_g = \mathbf{v}_a + \mathbf{v}_w \qquad (1)$$

If there was no wind, the plane's velocity would be simply $\mathbf{v}_a$. Here we are given information about the velocity of the airplane relative to the ground, $\mathbf{v}_g$, and the speed of the wind, $\mathbf{v}_w$, and we want to find $\mathbf{v}_a$, the plane's velocity in still air. By (1),

$$\mathbf{v}_a = \mathbf{v}_g - \mathbf{v}_w$$

(a) What is $\mathbf{v}_g$? We are told that, relative to the ground, the plane has a northwesterly direction and a speed of 250 mph.

From the solution to Problem 49, a <u>unit</u> vector in the northwesterly direction is

$$\mathbf{u} = \frac{-\sqrt{2}}{2}\mathbf{i} + \frac{\sqrt{2}}{2}\mathbf{j}$$

Therefore, $\mathbf{v}_g = 250(\mathbf{u})$
$$= -125\sqrt{2}\mathbf{i} + 125\sqrt{2}\mathbf{j}$$

(b) What is $\mathbf{v}_w$? An easterly wind is a wind <u>from</u> the east, so $\mathbf{v}_w = -50\mathbf{i}$.

12.4 VECTORS

715

(c) Therefore, $\mathbf{v}_a = \mathbf{v}_g - \mathbf{v}_w$

$$= \left(-125\sqrt{2} + 50\right)\mathbf{i} + 125\sqrt{2}\,\mathbf{j},$$

and the speed <u>would have</u> been

$$\|\mathbf{v}_a\| = \sqrt{\left(-125\sqrt{2} + 50\right)^2 + \left(125\sqrt{2}\right)^2}$$

$$= \sqrt{31250 - 12500\sqrt{2} + 2500 + 31250}$$

$$\approx 217.5 \text{ or } 218 \text{ mph}$$

53.   Referring to the figure and using the grid,

$$\mathbf{F}_1 = -3\mathbf{i}$$
$$\mathbf{F}_2 = -\mathbf{i} + 4\mathbf{j}$$
$$\mathbf{F}_3 = 4\mathbf{i} - 2\mathbf{j}$$
$$\mathbf{F}_4 = -4\mathbf{j}$$

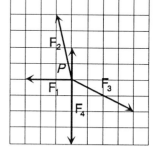

We will add an unknown force, $\mathbf{X} = a\mathbf{i} + b\mathbf{j}$. The object at P will not move provided the sum of all the forces acting upon it is $\mathbf{0}$; i.e., we want

$$\mathbf{F}_1 + \mathbf{F}_2 + \mathbf{F}_3 + \mathbf{F}_3 + \mathbf{X} = \mathbf{0}$$

or, $-3\mathbf{i} + (-\mathbf{i} + 4\mathbf{j}) + (4\mathbf{i} - 2\mathbf{j}) + (-4\mathbf{j}) + (a\mathbf{i} + b\mathbf{j}) = \mathbf{0}$

or, $-2\mathbf{j} + (a\mathbf{i} + b\mathbf{j}) = \mathbf{0}$
$a\mathbf{i} + (-2 + b)\mathbf{j} = \mathbf{0}$

This means $a = 0$
$-2 + b = 0, \ b = 2$

Therefore, $\mathbf{X} = 2\mathbf{j}$

≡ EXERCISE 12.5 THE DOT PRODUCT

1. $\mathbf{v} = \mathbf{i} - \mathbf{j}, \ \mathbf{w} = \mathbf{i} + \mathbf{j}$

(a) $\mathbf{v} \cdot \mathbf{w} = 1 \cdot 1 + (-1) \cdot 1 = 1 - 1 = 0$

(b) $\cos \theta = \dfrac{\mathbf{v} \cdot \mathbf{w}}{\|\mathbf{v}\|\,\|\mathbf{w}\|} = \dfrac{0}{\|\mathbf{v}\|\,\|\mathbf{w}\|} = 0$

(Note: If $\cos \theta = 0$, then $\theta = 90°$, so $\mathbf{v}$ and $\mathbf{w}$ are perpendicular.)

3. $v = 2i + j$, $w = i + 2j$

   (a) $v \cdot w = 2 + 2 = 4$

   (b) $\|v\| = \sqrt{4 + 1} = \sqrt{5}$
       $\|w\| = \sqrt{1 + 4} = \sqrt{5}$

       $\cos \theta = \dfrac{v \cdot w}{\|v\| \, \|w\|} = \dfrac{4}{\sqrt{5}\sqrt{5}} = \dfrac{4}{5}$

5. $v = \sqrt{3}\,i - j$, $w = i + j$

   (a) $v \cdot w = \sqrt{3} - 1$

   (b) $\|v\| = \sqrt{3 + 1} = 2$
       $\|w\| = \sqrt{1 + 1} = \sqrt{2}$

       $\cos \theta = \dfrac{\sqrt{3} - 1}{2\sqrt{2}} = \dfrac{\sqrt{6} - \sqrt{2}}{4}$

7. $v = 3i + 4j$, $w = 4i + 3j$

   (a) $v \cdot w = 12 + 12 = 24$

   (b) $\|v\| = \sqrt{9 + 16} = 5$
       $\|w\| = \sqrt{16 + 9} = 5$

       $\cos \theta = \dfrac{24}{5 \cdot 5} = \dfrac{24}{25}$

9. $v = 4i$, $w = j$

   (a) $v \cdot w = 4 \cdot 0 + 0 \cdot 1 = 0$

       (Because $v = 4i = 4i + 0j$, and $w = 0i + 1j$)

   (b) $\cos \theta = \dfrac{v \cdot w}{\|v\| \, \|w\|} = 0$

11.   Given: $v = ai - j$, $w = 2i + 3j$

   We want $\theta = \dfrac{\pi}{2}$. Then $\cos \theta = \cos \dfrac{\pi}{2} = 0$.

   But $\cos \theta = \dfrac{v \cdot w}{\|v\| \, \|w\|}$ only if $v \cdot w = 0$.

   Therefore, we want $v \cdot w = 2a - 3 = 0$,

       or $2a = 3$
       $a = \dfrac{3}{2}$

12.5 THE DOT PRODUCT

13. $v = 2i - 3j, \quad w = i - j$

$$v_1 = \text{proj}_w v = \frac{2 + 3}{\left(\sqrt{2}\right)^2}(i - j) = \frac{5}{2}(i - j)$$

$$v_2 = v - v_1 = (2i - 3j) - \frac{5}{2}(i - j)$$

$$= \frac{-1}{2}i - \frac{1}{2}j$$

15. $v = i - j, \quad w = i + 2j$

$$v_1 = \text{proj}_w v = \frac{1 - 2}{\left(\sqrt{5}\right)^2}(i + 2j) = \frac{-1}{5}(i + 2j)$$

$$v_2 = v - v_1 = (i - j) + \frac{1}{5}(i + 2j)$$

$$= i + \frac{1}{5}i - j + \frac{2}{5}j$$

$$= \frac{6}{5}i - \frac{3}{5}j$$

17. $v = 3i + j, \quad w = -2i - j$

$$v_1 = \text{proj}_w v = \frac{-7}{\left(\sqrt{5}\right)^2}(-2i - j) = \frac{7}{5}(2i + j)$$

$$v_2 = v - v_1 = (3i + j) - \frac{7}{5}(2i + j)$$

$$= 3i - \frac{14}{5}i + j - \frac{7}{5}j$$

$$= \frac{1}{5}i - \frac{2}{5}j$$

19.   Let

v_a = velocity of plane in still air
v_w = velocity of the wind
v_g = velocity of the plane relative to the ground

Then $v_g = v_a + v_w$

(a) To derive an expression for v_a, we first need to find a unit vector in the direction southwest. The vector $-i - j$ points in the desired direction, so a unit vector in a southwesterly direction is

$$u = \frac{1}{\|-i - j\|}(-i - j) = \frac{\sqrt{2}}{2}i - \frac{\sqrt{2}}{2}j$$

Hence, $v_a = 550\left(\frac{-\sqrt{2}}{2}i - \frac{\sqrt{2}}{2}j\right)$

or $v_a = -275\sqrt{2}\,i - 275\sqrt{2}\,j$

(b) The wind is <u>from</u> the west, so

$$\mathbf{v}_w = 80\mathbf{i}$$

(c) We know $\mathbf{v}_g = \mathbf{v}_a + \mathbf{v}_w$

so $\mathbf{v}_g = (-275\sqrt{2}\,\mathbf{i} - 275\sqrt{2}\,\mathbf{j} + 80\mathbf{i}$

or $\mathbf{v}_g = (80 - 275\sqrt{2}\,)\mathbf{i} - 275\sqrt{2}\,\mathbf{j}$

Note that both components of $\mathbf{v}_g$ are negative, so $\mathbf{v}_g$ points into the third quadrant, somewhere between W and S.

(d) The speed of the jet, relative to the earth, is

$$\|\mathbf{v}_g\| = \sqrt{\left(80 - 275\sqrt{2}\right)^2 + \left(275\sqrt{2}\right)^2}$$

$$= \sqrt{6400 - 44000\sqrt{2} + 151250 + 151250}$$

$$= \sqrt{246674.6}$$

$$\approx 496.7 \text{ miles per hour}$$

(e) To find the direction of $\mathbf{v}_g$, pick a simple vector in a known direction such as $-\mathbf{i}$ which points due west, and then determine the angle between $\mathbf{v}_g$ and $-\mathbf{i}$

We use the formula

$$\cos\,\theta = \frac{\mathbf{v}_g \cdot (-\mathbf{i})}{\|\mathbf{v}_g\|\,\|-\mathbf{i}\|}$$

Now $\mathbf{v}_g = (80 - 275\sqrt{2}\,)\mathbf{i} - 275\sqrt{2}\,\mathbf{j}$

so $\mathbf{v}_g \cdot (-\mathbf{i}) = \left(80 - 275\sqrt{2}\right)(-1) + \left(-275\sqrt{2}\right)(0)$

$$= 275\sqrt{2} - 80$$

$$\approx 308.91$$

In part (d), we found $\|\mathbf{v}_g\| \approx 496.7$

Finally $\|-\mathbf{i}\| = 1$.

Therefore, $\cos\,\theta \approx \dfrac{308.91}{496.7} = .6219$

Then $\theta - \cos^{-1}(.6219) \approx 51.5°$

The angle between $\mathbf{v}_g$ and due west is 51.5°; i.e., the plane's direction is 51.5° south of west.

21. For convenience, let the positive x-axis point downstream, so that the velocity of the current, $\mathbf{v}_c$, is:

$$\mathbf{v}_c = 3\mathbf{i}$$

Let $\mathbf{v}_w$ = velocity of the boat in the water.
and $\mathbf{v}_g$ = velocity of the boat relative to the land.

Then $\mathbf{v}_g = \mathbf{v}_w + \mathbf{v}_c$

Here is what we are given.

(1) $\mathbf{v}_c = 3\mathbf{i}$

(2) The <u>speed</u> of the boat in water is given:

$$\|\mathbf{v}_w\| = 20,$$

but the <u>direction</u> of $\mathbf{v}_w$ is what we want to find.

(3) We know the <u>direction</u> of the boat relative to the land, $\mathbf{j}$ (directly <u>across</u> the stream), but we do not know $\|\mathbf{v}_g\|$.

To get started, let $\mathbf{v}_w = a\mathbf{i} + b\mathbf{j}$

By (2), $\|\mathbf{v}_w\| = \sqrt{a^2 + b^2} = 20$,

so that $a^2 + b^2 = 400$.

We know the direction of $\mathbf{v}_g$, so we let

$$\mathbf{v}_g = k\mathbf{j}$$

Since $\mathbf{v}_g = \mathbf{v}_w + \mathbf{v}_c$, we have

$$k\mathbf{j} = a\mathbf{i} + b\mathbf{j} + 3\mathbf{i}$$
$$k\mathbf{j} = (a + 3)\mathbf{i} + b\mathbf{j}$$

So, $a + 3 = 0$ and $b = k$
$\qquad\quad a = -3$

Now, $\qquad a^2 + b^2 = 400$
$\qquad\qquad\quad 9 + b^2 = 400$
$\qquad\qquad\qquad\;\; b^2 = 391$
$\qquad\qquad\qquad\;\; b \approx 19.77$
and $\qquad\qquad\;\; k = b \approx 19.77$

We now have

$$\mathbf{v}_w = -3\mathbf{i} + 19.77\mathbf{j}$$
and $\mathbf{v}_g = 19.77\mathbf{j}$

Since the $\mathbf{i}$ component of $\mathbf{v}_w$ is negative, the boat is headed slightly <u>upstream</u>.

12 MISCELLANEOUS TOPICS

Let's find the angle between $\mathbf{v}_w$ and $\mathbf{j}$:

$$\cos\theta = \frac{\mathbf{v}_w \cdot \mathbf{j}}{\|\mathbf{v}_w\|\,\|\mathbf{j}\|}$$

$$= \frac{19.77}{20}, \text{ since by (2), } \|\mathbf{v}_w\| = 20$$

$$\approx .9885$$

So $\theta \approx \cos^{-1}(.9885) \approx 8.7°$.

Thus, the heading of the boat needs to be 8.7° upstream from the line directly across the stream.

Finally, the speed of the boat, relative to the land is $\|\mathbf{v}_g\| = \|19.77\mathbf{j}\| = 19.77$, and the time needed to cross the stream is given by

$$\text{time} = \frac{\text{distance}}{\text{rate}} = \frac{0.5 \text{ kilometer}}{19.77 \text{ kilometer/hr}} \approx .0253 \text{ hr} \approx 1.5 \text{ min.}$$

23. $\mathbf{v}_m = 2(\cos\theta\mathbf{i} + \sin\theta\mathbf{j})$
$\mathbf{v}_a = k\mathbf{j}$
$\mathbf{v}_R = \mathbf{i}$
$\mathbf{v}_m + \mathbf{v}_R = \mathbf{v}_a$

$k^2 + 1 = 4$

$k = \sqrt{3}$
$2\cos\theta\mathbf{i} + 2\sin\theta\mathbf{j} + \mathbf{i} = k\mathbf{j}$

$2\cos\theta + 1 = 0, \ 2\sin\theta = \sqrt{3}$

$$\cos\theta = \frac{-1}{2} \quad \sin\theta = \frac{\sqrt{3}}{2}$$

$$\theta = 120°$$

The swimmer should head at an angle of 60° to the shore.

$$\text{time} = \frac{\text{distance}}{\text{velocity}} = \frac{\frac{1}{2} \text{ km}}{\sqrt{3} \text{ km/hour}} = .289 \text{ hours} = 17.32 \text{ minutes}$$

25. $\mathbf{F} = 3\dfrac{(2\mathbf{i} + \mathbf{j})}{\sqrt{5}}$ $\overrightarrow{\mathbf{AB}} = 2\mathbf{j}$

$$W = \mathbf{F} \cdot \overrightarrow{\mathbf{AB}} = \left(\frac{6\mathbf{i}}{\sqrt{5}} + \frac{3}{\sqrt{5}}\mathbf{j}\right) \cdot 2\mathbf{j} = \frac{6}{\sqrt{5}}\mathbf{j} = \sqrt{\left(\frac{6}{\sqrt{5}}\right)^2} = \frac{6}{\sqrt{5}} \approx 2.68 \text{ ft-lb}$$

27. $\mathbf{F} = 20\cos(30\mathbf{i} - \sin 30\mathbf{j})$

$\overrightarrow{\mathbf{AB}} = 100\mathbf{i}$

$$W = \mathbf{F} \cdot \overrightarrow{\mathbf{AB}} = 2000 \cdot \frac{\sqrt{3}}{2} = 1732 \text{ ft.-lb.}$$

29. Let $\mathbf{u} = a_1\mathbf{i} + b_1\mathbf{j}$
     $\mathbf{v} = a_2\mathbf{i} + b_2\mathbf{j}$
     $\mathbf{w} = a_3\mathbf{i} + b_3\mathbf{j}$

Then,

$$\begin{aligned}
\mathbf{u} \cdot (\mathbf{v} + \mathbf{w}) &= (a_1\mathbf{i} + b_1\mathbf{j}) \cdot [a_2\mathbf{i} + b_2\mathbf{j} + a_3\mathbf{i} + b_3\mathbf{j}] \\
&= (a_1\mathbf{i} + b_1\mathbf{j}) \cdot [(a_2 + a_3)\mathbf{i} + (b_2 + b_3)\mathbf{j}] \\
&= a_1(a_2 + a_3) + b_1(b_2 + b_3)
\end{aligned}$$

On the other hand,

$$\begin{aligned}
\mathbf{u} \cdot \mathbf{v} + \mathbf{u} \cdot \mathbf{w} &= (a_1\mathbf{i} + b_1\mathbf{j}) \cdot (a_2\mathbf{i} + b_2\mathbf{j}) + (a_1\mathbf{i} + b_1\mathbf{j} \cdot (a_3\mathbf{i} + b_3\mathbf{j}) \\
&= a_1a_2 + b_1b_2 + a_1a_3 + b_1b_3 \\
&= a_1a_2 + a_1a_3 + b_1b_2 + b_1b_3 \\
&= a_1(a_2 + a_3) + b_1(b_2 + b_3)
\end{aligned}$$

Thus, $\mathbf{u} \cdot (\mathbf{v} + \mathbf{w}) = \mathbf{u} \cdot \mathbf{v} + \mathbf{u} \cdot \mathbf{w}$

31. Let $\mathbf{v} = a\mathbf{i} + b\mathbf{j}$

Since $\mathbf{v}$ is a unit vector, we know:

$$\|\mathbf{v}\| = \sqrt{a^2 + b^2} = 1$$

or $a^2 + b^2 = 1$

If α is the angle between $\mathbf{v}$ and $\mathbf{i}$,

then $\cos \alpha = \dfrac{\mathbf{v} \cdot \mathbf{i}}{\|\mathbf{v}\| \, \|\mathbf{i}\|}$

or $\cos \alpha = \dfrac{(a\mathbf{i} + b\mathbf{j}) \cdot (\mathbf{i})}{1 \cdot 1}$

$\cos \alpha = a$

Then, since $a^2 + b^2 = 1$, we have

$$\begin{aligned}
\cos^2 \alpha + b^2 &= 1 \\
b^2 &= 1 - \cos^2 \alpha \\
b^2 &= \sin^2 \alpha \\
b &= \sin \alpha
\end{aligned}$$

Hence $\mathbf{v} = \cos \alpha \mathbf{i} + \sin \alpha \mathbf{j}$

33. Let $\mathbf{v} = a\mathbf{i} + b\mathbf{j}$

$$\text{proj}_\mathbf{i}\mathbf{v} = \frac{\mathbf{v} \cdot \mathbf{i}}{\|\mathbf{i}\|^2}\mathbf{i} = \mathbf{v}(\mathbf{v} \cdot \mathbf{i})\mathbf{i}$$

$\mathbf{v} \cdot \mathbf{i} = a$, $\mathbf{v} \cdot \mathbf{j} = b$, so $\mathbf{v} = (\mathbf{v} \cdot \mathbf{i}) + (\mathbf{v} \cdot \mathbf{j})\mathbf{j}$

35. $(\mathbf{v} - \alpha\mathbf{w})\mathbf{w} = \mathbf{v} \cdot \mathbf{w} - \alpha\mathbf{w} \cdot \mathbf{w} = \mathbf{v} \cdot \mathbf{w} - \alpha\|\mathbf{w}\|^2$

$$= \mathbf{v} \cdot \mathbf{w} - \frac{\mathbf{v} \cdot \mathbf{w}}{\|\mathbf{w}\|^2}\|\mathbf{w}\|^2 = 0$$

37. If **F** is orthagonal to $\overrightarrow{\textbf{AB}}$, then because

$$\text{work} = \textbf{F} \cdot \overrightarrow{\textbf{AB}} = (\textbf{i} + \textbf{j})(\textbf{i} - \textbf{j}) = 1 - 1 = 0$$

☰ 12 - CHAPTER REVIEW

1. $A + C = \begin{bmatrix} 1 & 0 \\ 2 & 4 \\ -1 & 2 \end{bmatrix} + \begin{bmatrix} 3 & -4 \\ 1 & 5 \\ 5 & -2 \end{bmatrix} = \begin{bmatrix} 4 & -4 \\ 3 & 9 \\ 4 & 0 \end{bmatrix}$

3. $6A = 6\begin{bmatrix} 1 & 0 \\ 2 & 4 \\ -1 & 2 \end{bmatrix} = \begin{bmatrix} 6 & 0 \\ 12 & 24 \\ -6 & 12 \end{bmatrix}$

5. $AB = \begin{bmatrix} 1 & 0 \\ 2 & 4 \\ -1 & 2 \end{bmatrix}\begin{bmatrix} 4 & -3 & 0 \\ 1 & 1 & -2 \end{bmatrix} = \begin{bmatrix} 4+0 & -3+0 & 0+0 \\ 8+4 & -6+4 & 0+(-8) \\ -4+2 & 3+2 & 0+(-4) \end{bmatrix}$

$$= \begin{bmatrix} 4 & -3 & 0 \\ 12 & -2 & -8 \\ -2 & 5 & -4 \end{bmatrix}$$

7. $CB = \begin{bmatrix} 3 & -4 \\ 1 & 5 \\ 5 & -2 \end{bmatrix}\begin{bmatrix} 4 & -3 & 0 \\ 1 & 1 & -2 \end{bmatrix} = \begin{bmatrix} 8 & -13 & 8 \\ 9 & 2 & -10 \\ 18 & -17 & 4 \end{bmatrix}$

9. $A = \begin{bmatrix} 4 & 6 \\ 1 & 3 \end{bmatrix}$. We start with $[A|I_2]$, and proceed to put it into reduced echelon form.

$$[A|I_2] = \begin{bmatrix} 4 & 6 & | & 1 & 0 \\ 1 & 3 & | & 0 & 1 \end{bmatrix} \rightarrow \begin{bmatrix} 1 & 3 & | & 0 & 1 \\ 4 & 6 & | & 1 & 0 \end{bmatrix}$$
↑
Interchange rows

$$\rightarrow \begin{bmatrix} 1 & 3 & | & 0 & 1 \\ 0 & -6 & | & 1 & -4 \end{bmatrix}$$
↑
$R_2 = -4r_1 + r_2$

$$\rightarrow \begin{bmatrix} 1 & 0 & | & \frac{1}{2} & -1 \\ 1 & -6 & | & 1 & -4 \end{bmatrix}$$
↑
$R_1 = \frac{1}{2}r_2 + r_1$

$$\rightarrow \begin{bmatrix} 1 & 0 & \Big| & \frac{1}{2} & -1 \\[2mm] 0 & 1 & \Big| & -\frac{1}{6} & \frac{2}{3} \end{bmatrix}$$

$$\uparrow$$

$$R_2 = -\frac{1}{6}r_2$$

Therefore, $A^{-1} = \begin{bmatrix} \frac{1}{2} & -1 \\[2mm] -\frac{1}{6} & \frac{2}{3} \end{bmatrix}$

11. $A = \begin{bmatrix} 1 & 3 & 3 \\ 1 & 2 & 1 \\ 1 & -1 & 2 \end{bmatrix}$

$$\begin{bmatrix} 1 & 3 & 3 & | & 1 & 0 & 0 \\ 1 & 2 & 1 & | & 0 & 1 & 0 \\ 1 & -1 & 2 & | & 0 & 0 & 1 \end{bmatrix} \rightarrow \begin{bmatrix} 1 & 3 & 3 & | & 1 & 0 & 0 \\ 0 & -1 & -2 & | & -1 & 1 & 0 \\ 0 & -4 & -1 & | & -1 & 0 & 1 \end{bmatrix}$$

$$\uparrow$$
$$R_2 = -1r_1 + r_2$$
$$R_3 = -1r_1 + r_3$$

$$\rightarrow \begin{bmatrix} 1 & 3 & 3 & | & 1 & 0 & 0 \\ 0 & 1 & 2 & | & 1 & -1 & 0 \\ 0 & -4 & -1 & | & -1 & 0 & 1 \end{bmatrix}$$

$$\uparrow$$
$$R_2 = -1r_2$$

$$\rightarrow \begin{bmatrix} 1 & 0 & -3 & | & -2 & 3 & 0 \\ 0 & 1 & 2 & | & 1 & -1 & 0 \\ 0 & 0 & 7 & | & 3 & -4 & 1 \end{bmatrix}$$

$$\uparrow$$
$$R_1 = -3r_2 + r_1$$
$$R_3 = 4r_2 + r_3$$
$$\uparrow$$

$$\rightarrow \begin{bmatrix} 1 & 0 & -3 & | & -2 & 3 & 0 \\ 0 & 1 & 2 & | & 1 & -1 & 0 \\ 0 & 0 & 1 & | & \frac{3}{7} & -\frac{4}{7} & \frac{1}{7} \end{bmatrix}$$

$$\uparrow$$
$$R_3 = \frac{1}{7}r_3$$

$$\rightarrow \begin{bmatrix} 1 & 0 & 0 & | & -\frac{5}{7} & \frac{9}{7} & \frac{3}{7} \\[2mm] 0 & 1 & 0 & | & \frac{1}{7} & \frac{1}{7} & -\frac{2}{7} \\[2mm] 0 & 0 & 1 & | & \frac{3}{7} & -\frac{4}{7} & \frac{1}{7} \end{bmatrix}$$

$$\uparrow$$
$$R_1 = 3r_3 + r_1$$
$$R_2 = -2r_3 + r_2$$

12 MISCELLANEOUS TOPICS

Therefore, $A^{-1} = \begin{bmatrix} -\frac{5}{7} & \frac{9}{7} & \frac{3}{7} \\ \frac{1}{7} & \frac{1}{7} & -\frac{2}{7} \\ \frac{3}{7} & -\frac{4}{7} & \frac{1}{7} \end{bmatrix} = \frac{1}{7}\begin{bmatrix} -5 & 9 & 3 \\ 1 & 1 & -2 \\ 3 & -4 & 1 \end{bmatrix}$

13. $A = \begin{bmatrix} 4 & -8 \\ -1 & 2 \end{bmatrix}$

$\left[\begin{array}{cc|cc} 4 & -8 & 1 & 0 \\ -1 & 2 & 0 & 1 \end{array}\right] \rightarrow \left[\begin{array}{cc|cc} 1 & -2 & \frac{1}{4} & 0 \\ -1 & 2 & 0 & 1 \end{array}\right]$

$\uparrow$
$R_1 = \frac{1}{4}r_1$

$\rightarrow \left[\begin{array}{cc|cc} 1 & -2 & \frac{1}{4} & 0 \\ 0 & 0 & \frac{1}{4} & 1 \end{array}\right]$

$\uparrow$
$R_2 = r_1 + r_2$

This did not take the form $[I_2 | B]$, so A has no inverse; i.e., A is singular.

15. Maximize $z = 3x + 4y$

Subject to $x \geq 0$, $y \geq 0$, $3x + 2y \geq 6$, $x + y \leq 8$.

We graph the two lines:

(1) $y = -\frac{3}{2}x + 3$; y-intercept = 3; x-intercept = 2;

(2) $y = -x + 8$; y-intercept = 8; x-intercept = 8.

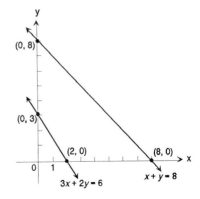

VERTEX	VALUE OF $z = 3x + 4y$
(0, 3)	$z = 12$
(0, 8)	$z = 32$
(8, 0)	$z = 24$
(2, 0)	$z = 6$

The maximum value of z is 32, at the point (0, 8).

17. Minimize $z = 3x + 5y$

Subject to $x \geq 0$, $y \geq 0$,
$x + y \geq 1$, $3x + 2y \leq 12$,
$x + 3y \leq 12$.

We graph the three lines:

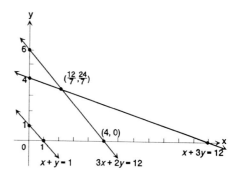

(1) $y = -x + 1$;
   y-intercept $= 1$;
   x-intercept $= 1$

(2) $y = -\frac{3}{2}x + 6$;

   y-intercept $= 6$;
   x-intercept $= 4$

(3) $y = -\frac{1}{3}x + 4$;

   y-intercept $= 4$;
   x-intercept $= 12$

We need the point of intersection of (2) and (3):

$$-\frac{3}{2}x + 6 = -\frac{1}{3}x + 4$$

$$-9x + 36 = -2x + 24$$

$$-7x = -12$$

$$x = \frac{12}{7}$$

and $y = -\frac{1}{3}x + 4 = -\frac{4}{7} + \frac{28}{7} = \frac{24}{7}$; $\left(\frac{12}{7}, \frac{24}{7}\right)$

VERTEX	VALUE OF $z = 3x + 5y$
$(0, 1)$	$z = 5$
$(0, 4)$	$z = 20$
$\left(\frac{12}{7}, \frac{24}{7}\right)$	$z = \frac{36}{7} + \frac{120}{7} = \frac{156}{7} \approx 22.3$
$(4, 0)$	$z = 12$
$(1, 0)$	$z = 3$

The minimum value of z is 3, at $(1, 0)$.

12 MISCELLANEOUS TOPICS

19. Maximize $z = 5x + 4y$

Subject to $x \geq 0$, $y \geq 0$,
$x + 2y \geq 2$, $3x + 4y \leq 12$, $y \geq x$

We have three lines:

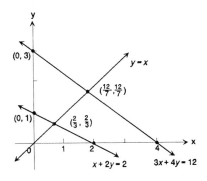

(1) $y = -\dfrac{1}{2}x + 1$;

  y-intercept $= 1$;
  x-intercept $= 2$

(2) $y = -\dfrac{3}{4}x + 3$;

  y-intercept $= 3$;
  x-intercept $= 4$

(3) $y = x$

We need the following points of intersection:

(1) and (3): $x = -\dfrac{1}{2}x + 1$

  $2x = -x + 2$

  $3x = 2$

  $x = \dfrac{2}{3}$

  and $y = x$

  $y = \dfrac{2}{3}$

  $\left(\dfrac{2}{3}, \ \dfrac{2}{3}\right)$

(2) and (3): $x = -\dfrac{3}{4}x + 3$

  $4x = -3x + 12$

  $7x = 12$

  $x = \dfrac{12}{7}$

  and $y = x$

  $y = \dfrac{12}{7}$

  $\left(\dfrac{12}{7}, \ \dfrac{12}{7}\right)$

VERTEX	VALUE OF $z = 5x + 4y$
$(0, 1)$	$z = 4$
$(0, 3)$	$z = 12$
$\left(\dfrac{12}{7}, \dfrac{12}{7}\right)$	$z = \dfrac{108}{7}$
$\left(\dfrac{2}{3}, \dfrac{2}{3}\right)$	$z = \dfrac{18}{3} = 6$

The maximum value of z is $\dfrac{108}{7}$, when $x = \dfrac{12}{7}$, and $y = \dfrac{12}{7}$.

For Problems 21-30, we will use the five-step method from Section 11.3.

21. $\dfrac{6}{x(x - 4)}$

Step 1: This is a proper fraction.

Step 2: $q(x) = x(x - 4)$

This is Case 1 (nonrepeated linear factors).

Step 3: $\dfrac{6}{x(x - 4)} = \dfrac{A}{x} + \dfrac{B}{x - 4}$

Step 4: Multiply both sides by $x(x - 4)$:

$$6 = A(x - 4) + Bx$$

Let $x = 0$: $6 = -4A$, or $A = -\dfrac{6}{4} = -\dfrac{3}{2}$

Let $x = 4$: $6 = 4B$, or $B = \dfrac{6}{4} = \dfrac{3}{2}$

Step 5: $\dfrac{6}{x(x - 4)} = \dfrac{-\dfrac{3}{2}}{x} + \dfrac{\dfrac{3}{2}}{x - 4}$

23. $\dfrac{x - 4}{x^2(x - 1)}$

Step 1: This is a proper fraction.

Step 2: $q(x) = x^2(x - 1)$

Note that x^2 is a <u>repeated linear</u> factor: $x^2 = (x)(x)$.

Step 3: $\dfrac{x - 4}{x^2(x - 1)} = \dfrac{A}{x} + \dfrac{B}{x^2} + \dfrac{C}{x - 1}$

12 MISCELLANEOUS TOPICS

Step 4: Multiply by $x^2(x - 1)$:

$$x - 4 = Ax(x - 1) + B(x - 1) + Cx^2$$

Let $x = 0$: $-4 = -B$, or $B = 4$
Let $x = 1$: $-3 = C$

Now choose any value of x, say

$x = 2$: $-2 = 2A + B + 4C$
$-2 = 2A + 4 - 12$
$2A = -2 - 4 + 12$
$A = 3$

Step 5: $\dfrac{x - 4}{x^2(x - 1)} = \dfrac{3}{x} + \dfrac{4}{x^2} - \dfrac{3}{x - 1}$

25. $\dfrac{x}{(x^2 + 9)(x + 1)}$

Step 1: This is proper.

Step 2: $q(x) = (x^2 + 9)(x + 1)$
$\underbrace{\qquad}$
cannot be factored

This is Case 1 and Case 3.

Step 3: $\dfrac{x}{(x^2 + 9)(x + 1)} = \dfrac{Ax + B}{x^2 + 9} + \dfrac{C}{x + 1}$

Step 4: $x = (Ax + B)(x + 1) + C(x^2 + 9)$ (1)

Now $q(x)$ only has one zero, $x = -1$:

Let $x = -1$: $-1 = 10C = -\dfrac{1}{10}$

To find the coefficients A and B, let's expand (1):

$$x = Ax^2 + Ax + Bx + B + Cx^2 + 9C$$
$$x = (A + C)x^2 + (A + B)x + (B + 9C)$$

Now equate coefficients of like powers of x:

For x^2: $0 = A + C$
For x: $1 = A + B$
For the constant: $0 = B + 9C$

But we know $C = -\dfrac{1}{10}$, so

$0 = A + C$
$0 = A - \dfrac{1}{10}$
$A = \dfrac{1}{10}$

$$\text{and } 0 = B + 9C$$

$$0 = B - \frac{9}{10}$$

$$B = \frac{9}{10}$$

Step 5: $\dfrac{x}{(x^2 + 9)(x + 1)} = \dfrac{\frac{1}{10}x + \frac{9}{10}}{x^2 + 9} = \dfrac{-\frac{1}{10}}{x + 1}$

27. $\dfrac{x^3}{\left(x^2 + 4\right)^2}$

Step 1: This is a proper fraction.

Step 2: $q(x) = (x^2 + 4)^2$

This is Case 4 (a <u>repeated</u> irreducible quadratic factor).

Step 3: $\dfrac{x^3}{\left(x^2 + 4\right)^2} = \dfrac{Ax + B}{x^2 + 4} + \dfrac{Cx + D}{\left(x^2 + 4\right)^2}$

Step 4: Clear of fractions:

$$x^3 = (Ax + B)(x^2 + 4) + Cx + D$$

Since $q(x)$ has <u>no</u> real zeros, we choose to expand the right-hand-side and equate coefficients of like powers of x:

$$x^3 = Ax^3 + Bx^2 + 4Ax + 4B + Cx + D$$
$$x^3 = Ax^3 + Bx^2 + (4A + C)x + (4B + D)$$

Coefficient of x^3: $1 = A$
Coefficient of x^2: $0 = B$
Coefficient of x: $0 = 4A + C$
Constant term: $0 = 4B + D$

Since $A = 1$, we have:

$$0 = 4A + C$$
$$0 = 4 + C$$
$$C = -4$$

Since $B = 0$, we have:

$$0 = 4B + D$$
$$0 = 0 + D$$
$$D = 0$$

Step 5: $\dfrac{x^3}{\left(x^2 + 4\right)^2} = \dfrac{x}{x^2 + 4} + \dfrac{-4x}{(x^2 + 4)^2}$

12 MISCELLANEOUS TOPICS

29. $$\frac{x^2}{(x^2 + 1)(x^2 - 1)}$$

Step 1: This fraction is proper.

Step 2: $q(x) = (x^2 + 1)(x^2 - 1)$
$= (x^2 + 1)(x + 1)(x - 1)$

This is Case 1 and Case 3.

Step 3: $$\frac{x^2}{(x^2 + 1)(x^2 - 1)} = \frac{Ax + B}{x^2 + 1} + \frac{C}{x + 1} + \frac{D}{x - 1}$$

Step 4: $x^2 = (Ax + B)(x + 1)(x - 1) + C(x^2 + 1)(x - 1)$
$+ D(x^2 + 1)(x + 1)$

Let $x = 1$: $1 = 4D$ or $D = \frac{1}{4}$

Let $x = -1$: $1 = 4C$ or $C = -\frac{1}{4}$

Now we need two more equations:

Let $x = 0$: $0 = -B - C + D$
$0 = -B + \frac{1}{4} + \frac{1}{4}$

$B = \frac{1}{2}$

Finally, let $x = 2$:

$4 = (2A + B)(3) + C(5) + D(15)$

$4 = 6A + 3B + 5C + 15D$

$4 = 6A + 3\left(\frac{1}{2}\right) + 5\left(-\frac{1}{4}\right) + 15\left(\frac{1}{4}\right)$

$4 = 6A + \frac{6}{4} - \frac{5}{4} + \frac{15}{4}$

$4 = 6A + \frac{16}{4}$

$0 = 6A$

$A = 0$

Step 5: $$\frac{x^2}{(x^2 + 1)(x^2 - 1)} = \frac{\frac{1}{2}}{x^2 + 1} + \frac{-\frac{1}{4}}{x + 1} + \frac{\frac{1}{4}}{x - 1}$$

12 - CHAPTER REVIEW

731

31. $P = (1, -2)$ and $Q = (3, -6)$, so if **v** is represented by the directed line segment $\overrightarrow{PQ}$, then

$$\begin{aligned} \mathbf{v} &= (3 - 1)\mathbf{i} + (-6 - (-2))\mathbf{j} \\ &= 2\mathbf{i} - 4\mathbf{j} \end{aligned}$$

and

$$\|\mathbf{v}\| = \sqrt{2^2 + (-4)^2} = \sqrt{20} = 2\sqrt{5}$$

33. For $P = (0, -2)$, $Q = (-1, 1)$, we have:

$$\begin{aligned} \mathbf{v} &= (-1 - 0)\mathbf{i} + (1 - (-2))\mathbf{j} \\ &= \mathbf{i} + 3\mathbf{j} \end{aligned}$$

and

$$\|\mathbf{v}\| = \sqrt{(-1)^2 + 3^2} = \sqrt{10}$$

35.

$$\begin{aligned} 4\mathbf{v} - 3\mathbf{w} &= 4(-2\mathbf{i} + \mathbf{j}) - 3(4\mathbf{i} - 3\mathbf{j}) \\ &= -8\mathbf{i} + 4\mathbf{j} - 12\mathbf{i} + 9\mathbf{j} \\ &= -20\mathbf{i} + 13\mathbf{j} \end{aligned}$$

37. $\|\mathbf{v}\| = \|-2\mathbf{i} + \mathbf{j}\| = \sqrt{(-2)^2 + 1^2} = \sqrt{5}$

39. $\|\mathbf{v}\| + \|\mathbf{w}\| = \|-2\mathbf{i} + \mathbf{j}\| = \|4\mathbf{i} - 3\mathbf{j}\|$

$$= \sqrt{4 + 1} + \sqrt{16 + 9}$$

$$= \sqrt{5} + 5 \approx 7.24$$

41. From Problem 37, $\|\mathbf{v}\| = \sqrt{5}$. Then a unit vector having the same direction as **v** is:

$$\begin{aligned} \mathbf{u} &= \frac{1}{\|\mathbf{v}\|} \mathbf{v} \\ &= \frac{1}{\sqrt{5}} (-2\mathbf{i} + \mathbf{j}) \\ &= \frac{\sqrt{5}}{5} (-2\mathbf{i} + \mathbf{j}) \end{aligned}$$

or $\mathbf{u} = \dfrac{-2\sqrt{5}}{5}\mathbf{i} + \dfrac{\sqrt{5}}{5}\mathbf{j}$

Note:

$$\begin{aligned} \|\mathbf{u}\| &= \sqrt{\left(\frac{-2\sqrt{5}}{5}\right)^2 + \left(\frac{\sqrt{5}}{5}\right)^2} \\ &= \sqrt{\frac{4(5)}{25} + \frac{5}{25}} \\ &= \sqrt{\frac{25}{25}} = 1 \end{aligned}$$

12 MISCELLANEOUS TOPICS

43. $\mathbf{v} = -2\mathbf{i} + \mathbf{j}$, $\mathbf{w} = 4\mathbf{i} - 3\mathbf{j}$

$\mathbf{v} \cdot \mathbf{w} = (-2)(4) + (1)(-3) = -11$

$$\cos \theta = \frac{\mathbf{v} \cdot \mathbf{w}}{\|\mathbf{v}\| \, \|\mathbf{w}\|} = \frac{-11}{\sqrt{5} \cdot \sqrt{25}} = \frac{-11\sqrt{5}}{25}$$

45. $\mathbf{v} = \mathbf{i} - 3\mathbf{j}$, $\mathbf{w} = -\mathbf{i} + \mathbf{j}$

$\mathbf{v} \cdot \mathbf{w} = (1)(-1) + (-3)(1) = -4$

$$\cos \theta = \frac{\mathbf{v} \cdot \mathbf{w}}{\|\mathbf{v}\| \, \|\mathbf{w}\|} = \frac{-4}{\sqrt{10} \cdot \sqrt{2}} = \frac{-4}{2\sqrt{5}} = \frac{-2\sqrt{5}}{5}$$

47. $\mathbf{v} = 2\mathbf{i} + 3\mathbf{j}$, $\mathbf{w} = 3\mathbf{i} + \mathbf{j}$

$$\text{proj}_{\mathbf{w}}\mathbf{v} = \frac{6 + 3}{\left(\sqrt{10}\right)^2}(3\mathbf{i} + \mathbf{j}) = \frac{9}{10}(3\mathbf{i} + \mathbf{j})$$

49. For $\mathbf{v} = 3\mathbf{i} - 4\mathbf{j}$ and $\mathbf{w} = 12\mathbf{i} - 5\mathbf{j}$, we have:

$\mathbf{v} \cdot \mathbf{w} = (3\mathbf{i} - 4\mathbf{j}) \cdot (12\mathbf{i} - 5\mathbf{j})$

$\qquad = (3)(12) + (-4)(-5) = 56$

$\|\mathbf{v}\| = \sqrt{9 + 16} = 5$

and $\|\mathbf{w}\| = \sqrt{144 + 25} = 13$

Therefore, if θ is the angle between $\mathbf{v}$ and $\mathbf{w}$,

$$\cos \theta = \frac{\mathbf{v} \cdot \mathbf{w}}{\|\mathbf{v}\| \, \|\mathbf{w}\|} = \frac{56}{5 \cdot 13}$$

or $\cos \theta = .86154$.

From this $\theta = \cos^{-1}(.86154) \approx 30.5°$.

51. Let $\mathbf{v}_c$ = velocity of the current

$\qquad \mathbf{v}_s$ = velocity of the swimmer relative to the water

and $\mathbf{v}_\ell$ = velocity of the swimmer relative to the land

Then $\mathbf{v}_\ell = \mathbf{v}_c + \mathbf{v}_s$

We are given <u>two</u> of the three vectors;

$\qquad \mathbf{v}_c = 2\mathbf{i}$, if we let $\mathbf{i}$ point directly downstream,

and $\mathbf{v}_s = 5\mathbf{j}$, since the swimmer swims perpendicular to the current, at 5 miles per hour.

Then $\mathbf{v}_\ell = \mathbf{v}_c + \mathbf{v}_s$

$\qquad = 2\mathbf{i} + 5\mathbf{j}$

The swimmer's <u>actual</u> speed is:

$\|\mathbf{v}_\ell\| = \sqrt{4 + 25}$

$\qquad = \sqrt{29}$

$\qquad \approx 5.39$ miles per hour

12 - CHAPTER REVIEW

53.   Let x = Number of gasoline engines produced per week, and
        y = Number of diesel engines produced per week

Then the objective function, or <u>cost</u>, is given by:

$$C = 450x + 550y \qquad \text{(in dollars)}$$

Our constraints are:

(1) $x \geq 0$, $y \geq 0$
(2) $x \geq 20$ (must deliver at least 20 gasoline engines)
(3) $y \geq 15$ (must deliver at least 15 diesel engines)
(4) $x \leq 60$ (factory cannot make more than 60 gasoline engines)
(5) $x \leq 40$ (factory cannot make more than 40 diesel engines)
(6) $x + y \geq 50$ (must produce a total of 50 engines)

Equations (2) - (5) bound a rectangle with vertices (20, 15), (20, 40), (60, 40), and (60, 15).

We will need two points of intersection:

(6) and (2): $\begin{cases} x + y = 50 \\ x \quad = 20 \end{cases}$

Then $y = 30$

The point (20, 30) <u>is</u> in the feasible region.

(6) and (3): $\begin{cases} x + y = 50 \\ x \quad = 15 \end{cases}$

Then $x = 35$

The point (35, 15) is also in the feasible region.

VERTEX	VALUE OF COST: $C = 450x + 550y$
(20, 30)	$C = 25{,}500$
(20, 40)	$C = 31{,}000$
(60, 40)	$C = 49{,}000$
(60, 15)	$C = 35{,}250$
(35, 15)	$C = 24{,}000$

The minimum possible cost is $24,000, obtained by producing $x = 35$ gasoline engines, and $y = 15$ diesel engines.

Since the factory is <u>obligated</u> to deliver 20 gasoline engines and 15 diesel engines, their excess capacity is 15 gasoline engines (and no excess diesel engines).

55. $15(\cos \theta i + \sin \theta j) + 5i = 2j$
    $a^2 = (15)^2 - (5)^2$
    $a^2 = 225 - 25$
    $a^2 = 200$

    $a = \sqrt{200}$

$15 \cos \theta i + 5i + 15 \sin \theta j = \sqrt{200}$

So, $15 \cos \theta + 5 = 0$
    $\cos \theta = \dfrac{-1}{3}$

    and $\sin \theta = \dfrac{\sqrt{200}}{15} = \dfrac{10\sqrt{2}}{15} = \dfrac{2\sqrt{2}}{3}$
    $\theta = 109.5°$

The person should head at an angle of 70.5° to the shore.

APPENDIX: ALGEBRA REVIEW

≡ EXERCISE A.1 POLYNOMIALS AND RATIONAL EXPRESSIONS

1. $(10x^5 - 8x^2) + (3x^3 - 2x^2 + 6)$
   $10x^5 + 3x^3 + (-8x^2 - 2x^2) + 6$
   $10x^5 + 3x^3 - 10x^2 + 6$

3. $(x + a)^2 - x^2 = x^2 + 2xa + a^2 - x^2$ by (3a)
   $= (x^2 - x^2) + 2xa + a^2$
   $= 2ax + a^2$

5. $(x + 8)(2x + 1) = 2x^2 + 17x + 8$ by (4b)

7. $(x^2 + x - 1)(x^2 - x + 1)$
   $= x^2(x^2 - x + 1) + x(x^2 - x + 1) - 1(x^2 - x + 1)$
   $\qquad\qquad\qquad\qquad\qquad$ by distributive law
   $= x^2 \cdot x^2 + x^2 \cdot (-x) + x^2 \cdot 1 + x \cdot x^2 + x \cdot (-x) + x \cdot 1$
   $\qquad - 1 \cdot x^2 - 1 \cdot (-x) - 1 \cdot 1$ $\qquad$ by distributive law
   $= x^4 - x^3 + x^2 + x^3 - x^2 + x - x^2 + x - 1$
   $= x^4 - x^2 + 2x - 1$

9. $(x + 1)^3 - (x - 1)^3 = (x^3 + 3x^2 + 3x + 1) - (x^3 - 3x^2 + 3x - 1)$
   $\qquad\qquad\qquad\qquad\qquad\qquad$ by (5a) and (5b)
   $= x^3 + 3x^2 + 3x + 1 - x^3 + 3x^2 - 3x + 1$
   $= 6x^2 + 2$

11.
$$
\begin{array}{r}
4x^2 - 3x + 1 \\
\hline
x)\overline{\,4x^3 - 3x^2 + x + 1\,} \\
4x^3 \\
\hline
-3x^2 \\
-3x^2 \\
\hline
x \\
x \\
\hline
1
\end{array}
$$

The quotient is $4x^2 - 3x + 1$; the remainder is 1.

Check: $x(4x^2 - 3x + 1) + 1 = 4x^3 - 3x + x + 1$

13.

$$x + 2 \overline{)\begin{array}{l} 4x^2 - 11x + 23 \\ 4x^3 - 3x^2 + x + 1 \end{array}}$$

$$\begin{array}{r}
4x^2 - 11x + 23 \\
x + 2 \,\overline{)\,4x^3 - 3x^2 + x + 1} \\
\underline{4x^3 + 8x^2} \\
-11x^2 + x \\
\underline{-11x^2 - 22x} \\
23x + 1 \\
\underline{23x + 46} \\
-45
\end{array}$$

The quotient is $4x^2 - 11x + 23$; the remainder is -45.

Check: $(x + 2)(4x^2 - 11x + 23) + (-45)$

$\qquad = 4x^3 - 11x^2 + 23x + 8x^2 - 22x + 46 + (-45)$

$\qquad = 4x^3 - 3x^2 + x + 1$

15.

$$\begin{array}{r}
4x^2 + 13x + 53 \\
x - 4 \,\overline{)\,4x^3 - 3x^2 + x + 1} \\
\underline{4x^3 - 16x^2} \\
13x^2 + x \\
\underline{13x^2 - 52x} \\
53x + 1 \\
\underline{53x - 212} \\
213
\end{array}$$

The quotient is $4x^2 + 13x + 53$; the remainder is 213.

Check: $(x - 4)(4x^2 + 13x + 53) + 213$

$\qquad = 4x^3 + 13x^2 + 53x - 16x^2 - 52x - 212 + 213$

$\qquad = 4x^3 - 3x^2 + x + 1$

17.

$$\begin{array}{r}
4x - 3 \\
x^2 \,\overline{)\,4x^3 - 3x^2 + x + 1} \\
\underline{4x^3} \\
-3x^2 \\
\underline{-3x^2} \\
x + 1
\end{array}$$

The quotient is $4x - 3$; the remainder is $x + 1$.

Check: $x^2(4x - 3) + (x + 1) = 4x^3 - 3x^2 + x + 1$

19.

$$\begin{array}{r}
4x - 3 \\
x^2 + 2 \,\overline{)\,4x^3 - 3x^2 + x + 1} \\
\underline{4x^3 + 8x} \\
-3x^2 - 7x + 1 \\
\underline{-3x^2 - 6} \\
-7x + 7
\end{array}$$

The quotient is $4x - 3$; the remainder is $-7x + 7$.

Check: $(x^2 + 2)(4x - 3) + (-7x + 7) = 4x^3 - 3x + 8x - 6 - 7x + 7$

$\qquad\qquad\qquad\qquad\qquad\qquad\qquad = 4x^3 - 3x^2 + x + 1$

21.

$$x^2 - 2x - 15 = (x\quad)(x\quad) = \begin{cases} (x\quad 1)(x\quad 15) \\ (x\quad 3)(x\quad 5) \end{cases}$$

$$(\text{signs must alternate}) = \begin{cases} (x + 1)(x - 15) \\ (x - 1)(x + 15) \\ (x + 3)(x - 5) \\ (x - 3)(x + 5) \end{cases}$$

$$= \begin{cases} x^2 - 14x - 15 \\ x^2 + 14x - 15 \\ x^2 - 2x - 15 \\ x^2 + 2x - 15 \end{cases}$$

Thus, $x^2 - 2x - 15 = (x - 5)(x + 3)$

23. $\quad ax^2 - 4a^2x - 45a^3 = a(x^2 - 4ax - 45a^2)$

$\quad$ where $x^2 - 4ax - 45a^2 = (x\quad a)(x\quad a)$

$$= \begin{cases} (x\quad a)(x\quad 45a) \\ (x\quad 3a)(x\quad 15a) \\ (x\quad 5a)(x\quad 9a) \end{cases}$$

$$(\text{signs must alternate}) = \begin{cases} (x + a)(x - 45a) \\ (x - a)(x + 45a) \\ (x + 3a)(x - 15a) \\ (x - 3a)(x + 15a) \\ (x + 5a)(x - 9a) \\ (x - 5a)(x + 9a) \end{cases}$$

$$= \begin{cases} x^2 - 44a - 45a^2 \\ x^2 + 44a - 45a^2 \\ x^2 - 12a - 45a^2 \\ x^2 + 12a - 45a^2 \\ x^2 - 4a - 45a^2 \\ x^2 + 4a - 45a^2 \end{cases}$$

Thus, $x^2 - 4ax - 45a^2 = (x + 5a)(x - 9a)$
$\quad$ and $ax^2 - 4a^2x - 45a^3 = a(x - 9a)(x + 5a)$

25. $\quad x^2 - 27 = x^3 - 3^3 = (x - 3)(x^2 + 3x + 9)$

27. $\quad 3x^2 + 4x + 1 = (3x + 1)(x + 1)$

29. $\quad x^7 - x^5 = x^5(x^2 - 1) = x^5(x - 1)(x + 1)$

31. $\quad \dfrac{3x - 6}{5x} \cdot \dfrac{x^2 - x - 6}{x^2 - 4} = \dfrac{3(x - 2)}{5x} \cdot \dfrac{(x - 3)(x + 2)}{(x + 2)(x - 2)} = \dfrac{3(x - 3)}{5x}$

APPENDIX: ALGEBRA REVIEW

33. $\dfrac{4x^2 - 1}{x^2 - 16} \cdot \dfrac{x^2 - 4x}{2x + 1} = \dfrac{(2x + 1)(2x - 1)}{(x + 4)(x - 4)} \cdot \dfrac{x(x - 4)}{(2x + 1)} = \dfrac{x(2x - 1)}{x + 4}$

35. $\dfrac{x}{x^2 - 7x + 6} - \dfrac{x}{x^2 - 2x - 24} = \dfrac{x}{(x - 1)(x - 6)} - \dfrac{x}{(x + 4)(x - 6)}$

$$= \dfrac{x(x + 4) - x(x - 1)}{(x - 1)(x + 4)(x - 6)}$$

$$= \dfrac{x^2 + 4x - (x^2 - x)}{(x - 1)(x + 4)(x - 6)}$$

$$= \dfrac{5x}{(x - 6)(x - 1)(x + 4)}$$

37. $\dfrac{4}{x^2 - 4} - \dfrac{2}{x^2 + x - 6} = \dfrac{4}{(x + 2)(x - 2)} - \dfrac{2}{(x + 3)(x - 2)}$

$$= \dfrac{4(x + 3) - 2(x + 2)}{(x + 2)(x - 2)(x + 3)}$$

$$= \dfrac{4x + 12 - (2x + 4)}{(x + 2)(x - 2)(x + 3)}$$

$$= \dfrac{2x + 8}{(x + 2)(x - 2)(x + 3)}$$

$$= \dfrac{2(x + 4)}{(x - 2)(x + 2)(x + 3)}$$

39. $\dfrac{x - \dfrac{1}{x}}{x + \dfrac{1}{x}} = \dfrac{\dfrac{x^2 - 1}{x}}{\dfrac{x^2 + 1}{x}} = \dfrac{x^2 - 1}{x} \cdot \dfrac{x}{x^2 + 1} = \dfrac{x^2 - 1}{x^2 + 1} = \dfrac{(x - 1)(x + 1)}{x^2 + 1}$

41. $\dfrac{3 - \dfrac{x^2}{x + 1}}{1 + \dfrac{x}{x^2 - 1}} = \dfrac{\dfrac{3(x + 1)}{x + 1} - \dfrac{x^2}{x + 1}}{\dfrac{x^2 - 1}{x^2 + 1} + \dfrac{x}{x^2 - 1}} = \dfrac{\dfrac{3x + 3 - x^2}{x + 1}}{\dfrac{x^2 - 1 + x}{x^2 - 1}} = \dfrac{\dfrac{-x^2 + 3x + 3}{x + 1}}{\dfrac{x^2 + x - 1}{(x + 1)(x - 1)}}$

$$= \dfrac{-x^2 + 3x + 3}{x + 1} \cdot \dfrac{(x + 1)(x - 1)}{x^2 + x - 1} = \dfrac{(x - 1)(-x^2 + 3x + 3)}{x^2 + x - 1}$$

≡ EXERCISE A.2 RADICALS; RATIONAL EXPRESSIONS

1. $\sqrt{8} = \sqrt{4 \cdot 2} = \sqrt{4}\sqrt{2} = 2\sqrt{2}$

3. $\sqrt[3]{16x^4} = \sqrt[3]{8x^3 \cdot 2x} = \sqrt[3]{8x^3}\sqrt[3]{2x} = 2x\sqrt[3]{2x}$

5. $\sqrt[3]{\sqrt{x^6}} = \sqrt[3]{\sqrt{(x^3)^2}} = \sqrt[3]{x^3} = x$

7. $\sqrt{\dfrac{32x^3}{9x}} = \sqrt{\dfrac{16x^2}{9} \cdot 2} = \dfrac{\sqrt{16x^2}}{9}\sqrt{2} = \dfrac{4}{3}x\sqrt{2}$

9. $\sqrt[4]{x^{12}y^8} = \sqrt[4]{(x^3)^4(y^2)^4} = x^3y^2$

11. $\sqrt[4]{\dfrac{x^9y^7}{xy^3}} = \sqrt[4]{x^8y^4} = \sqrt[4]{(x^2)^4y^4} = x^2y$

13. $\sqrt{36\,x} = \sqrt{36}\,\sqrt{x} = 6\sqrt{x}$

15. $\sqrt{3x^2}\,\sqrt{12x} = \sqrt{36x^3} = \sqrt{36x^2}\,\sqrt{x} = 6x\sqrt{x}$

17. $\left(\sqrt{5}\,\sqrt[3]{9}\right)^2 = \sqrt{5^2}\,\sqrt[3]{9^2} = 5\sqrt[3]{81} = 5\sqrt[3]{3^4} = 5\sqrt[3]{3^3 \cdot 3} = 15\sqrt[3]{3}$

19. $\sqrt{\dfrac{2x-3}{2x^4+3x^3}}\,\sqrt{\dfrac{x}{4x^2-9}} = \sqrt{\dfrac{x(2x-3)}{(2x^4+3x^3)(4x^2-9)}}$

$$= \sqrt{\dfrac{x(2x-3)}{x^3(2x+3)(2x+3)(2x-3)}}$$

$$= \sqrt{\dfrac{1}{x^2(2x+3)^2}} = \dfrac{\sqrt{1}}{\sqrt{x^2(2x+3)^2}}$$

$$= \dfrac{1}{x(2x+3)}$$

21. $\left(3\sqrt{6}\right)\left(2\sqrt{2}\right) = 6\sqrt{12}$
$$= 6\sqrt{4 \cdot 3}$$
$$= 6\sqrt{4}\sqrt{3}$$
$$= 12\sqrt{3}$$

23. $\left(\sqrt{3}+3\right)\left(\sqrt{3}-1\right) = \left(\sqrt{3}\right)^2 + 2\sqrt{3} - 3$
$$= 3 + 2\sqrt{3} - 3$$
$$= 2\sqrt{3}$$

25. $\left(\sqrt{x}-1\right)^2 = \left(\sqrt{x}\right)^2 - 2\sqrt{x} + 1$
$$= x - 2\sqrt{x} + 1$$

27. $\dfrac{1}{\sqrt{2}} = \dfrac{1}{\sqrt{2}} \cdot \dfrac{\sqrt{2}}{\sqrt{2}} = \dfrac{\sqrt{2}}{\left(\sqrt{2}\right)^2} = \dfrac{\sqrt{2}}{2}$

29. $\dfrac{-\sqrt{3}}{\sqrt{5}} = \dfrac{-\sqrt{3}}{\sqrt{5}} \cdot \dfrac{\sqrt{5}}{\sqrt{5}} = \dfrac{-\sqrt{3}\sqrt{5}}{\left(\sqrt{5}\right)^2} = \dfrac{-\sqrt{15}}{5}$

31. $\dfrac{\sqrt{3}}{5-\sqrt{2}} = \dfrac{\sqrt{3}}{5-\sqrt{2}} \cdot \dfrac{5+\sqrt{2}}{5+\sqrt{2}} = \dfrac{\sqrt{3}\left(5+\sqrt{2}\right)}{5^2-\left(\sqrt{2}\right)^2} = \dfrac{\sqrt{3}\left(5+\sqrt{2}\right)}{25-2} = \dfrac{\sqrt{3}\left(5+\sqrt{2}\right)}{23}$

33. $\dfrac{2 - \sqrt{5}}{2 + 3\sqrt{5}} = \dfrac{2 - \sqrt{5}}{2 + 3\sqrt{5}} \cdot \dfrac{2 - 3\sqrt{5}}{2 - 3\sqrt{5}} = \dfrac{\left(2 - \sqrt{5}\right)\left(2 - 3\sqrt{5}\right)}{2^2 - \left(3\sqrt{5}\right)^2} = \dfrac{4 - 8\sqrt{5} + 3\left(\sqrt{5}\right)^2}{4 - 9 \cdot 5}$

$$= \dfrac{4 - 8\sqrt{5} + 15}{4 - 45} = \dfrac{19 - 8\sqrt{5}}{-41} = \dfrac{-19 + 8\sqrt{5}}{41}$$

35. $\dfrac{\sqrt{x + h} - \sqrt{x}}{\sqrt{x + h} + \sqrt{h}} = \dfrac{\sqrt{x + h} - \sqrt{x}}{\sqrt{x + h} + \sqrt{x}} \cdot \dfrac{\sqrt{x + h} - \sqrt{x}}{\sqrt{x + h} - \sqrt{x}} = \dfrac{\left(\sqrt{x + h} - \sqrt{x}\right)^2}{\left(\sqrt{x + h}\right)^2 - \left(\sqrt{x}\right)^2}$

$$= \dfrac{\left(\sqrt{x + h}\right)^2 - 2\sqrt{x + h}\sqrt{x} + \left(\sqrt{x}\right)^2}{x + h - x}$$

$$= \dfrac{x + h - 2\sqrt{x(x + h)} + x}{h} = \dfrac{2x + h - 2\sqrt{x(x + h)}}{h}$$

37. $\sqrt{2t - 1} = 1$

$\left(\sqrt{2t - 1}\right)^2 = 1^2$

$2t - 1 = 1$

$t = 1$

Check: $\sqrt{2(1) - 1} = \sqrt{1} = 1$

39. $\sqrt{15 - 2x} = x$

$\left(\sqrt{15 - 2x}\right) = x^2$

$15 - 2x = x^2$

$-x^2 - 2x + 15 = 0$

$x^2 + 2x - 15 = 0$

$(x + 5)(x - 3) = 0$

$x = -5 \quad \text{or} \quad x = 3$

Check 3: $\sqrt{15 - 2(3)} \overset{?}{=} 3$

$\sqrt{9} \overset{?}{=} 3$

$3 = 3$

Check -5: $\sqrt{15 - 2(-5)} \neq -5$

Does not check.

The solution set is $x = 3$.

41. $8^{2/3} = \left(\sqrt[3]{8}\right)^2 = 2^2 = 4$

43. $(-27)^{1/3} = \sqrt[3]{-27} = -3$

45. $16^{3/2} = \left(\sqrt{16}\right)^3 = 4^3 = 64$

47. $9^{-3/2} = \left(\sqrt{9}\right)^{-3} = 3^{-3} = \dfrac{1}{3^3} = \dfrac{1}{27}$

49. $\left(\dfrac{9}{8}\right)^{3/2} = \left(\sqrt{\dfrac{9}{8}}\right)^3 = \left(\dfrac{\sqrt{9}}{\sqrt{8}}\right)^3 = \left(\dfrac{3}{2\sqrt{2}}\right)^3 = \dfrac{3^3}{\left(2\sqrt{2}\right)^3} = \dfrac{27}{8\left(\sqrt{2}\right)^3}$

$$= \dfrac{27}{16\sqrt{2}} \cdot \dfrac{\sqrt{2}}{\sqrt{2}} = \dfrac{27\sqrt{2}}{32}$$

51. $\left(\dfrac{8}{9}\right)^{-3/2} = \left(\sqrt{\dfrac{8}{9}}\right)^{-3} = \left(\dfrac{\sqrt{8}}{\sqrt{9}}\right)^{-3} = \left(\dfrac{2\sqrt{2}}{3}\right)^{-3} = \dfrac{1}{\left(\dfrac{2\sqrt{2}}{3}\right)^3} = \dfrac{1}{\dfrac{\left(2\sqrt{2}\right)^3}{3^3}}$

$$= \dfrac{3^3}{\left(2\sqrt{2}\right)^3} = \dfrac{27}{16\sqrt{2}} \dfrac{\sqrt{2}}{\sqrt{2}} = \dfrac{27\sqrt{2}}{32}$$

53. $x^{3/4}x^{1/3}x^{-1/2} = x^{(3/4+1/3-1/2)} = x^{(9/12)+(4/12)-(6/12)} = x^{7/12}$

55. $(x^3y^6)^{1/3} = x^{3/3}y^{6/3} = xy^2$

57. $(x^2y)^{1/3}(xy^2)^{2/3} = x^{2/3}y^{1/3} \cdot x^{2/3}y^{4/3} = x^{(2/3)+(2/3)}y^{(1/3)+(4/3)}$
$$= y^{4/3}y^{5/3}$$

59. $(16x^2y^{-1/3})^{3/4} = 16^{3/4}x^{6/4}y^{-3/12} = (2^4)^{3/4}x^{3/2}y^{-1/4}$

$$= 2^3\, x^{3/2}\, \dfrac{1}{y^{1/4}} = \dfrac{8x^{3/2}}{y^{1/4}}$$

61. $\dfrac{x}{(1+x)^{1/2}} + 2(1+x)^{1/2} = \dfrac{x + 2(1+x)^{1/2}(1+x)^{1/2}}{(1+x)^{1/2}}$

$$= \dfrac{x + 2(1+x)}{(1+x)^{1/2}} = \dfrac{3x+2}{(1+x)^{1/2}}$$

63. $\dfrac{\sqrt{1+x} - x \cdot \dfrac{1}{2\sqrt{1+x}}}{1+x} = \dfrac{\sqrt{1+x} - \dfrac{x}{2\sqrt{1+x}}}{1+x}$

$$= \dfrac{\dfrac{\sqrt{1+x}\left(2\sqrt{1+x}\right) - x}{2\sqrt{1+x}}}{\dfrac{1+x}{1}}$$

$$= \dfrac{\dfrac{2(1+x) - x}{2(1+x)^{1/2}}}{\dfrac{1+x}{1}}$$

$$= \dfrac{2+x}{2(1+x)^{1/2}(1+x)} = \dfrac{2+x}{2(1+x)^{3/2}}$$

APPENDIX: ALGEBRA REVIEW

65. $\dfrac{(x + 4)^{1/2} - 2x(x + 4)^{-1/2}}{x + 4} = \dfrac{(x + 4)^{1/2} - \dfrac{2x}{(x + 4)^{1/2}}}{x + 4}$

$= \dfrac{\dfrac{(x + 4)(x + 4)^{1/2} - 2x}{(x + 4)^{1/2}}}{x + 4}$

$= \dfrac{\dfrac{x + 4 - 2x}{(x + 4)^{1/2}}}{\dfrac{x + 4}{1}} = \dfrac{4 - x}{(x + 4)^{3/2}}$

67. $(x + 1)^{3/2} + x \cdot \dfrac{3}{2}(x + 1)^{1/2} = (x + 1)^{1/2}\left[(x + 1)^{2/2} + \dfrac{3x}{2}\right]$

$= (x + 1)^{1/2}\left[x + 1 + \dfrac{3}{2}x\right]$

$= (x + 1)^{1/2}\left[\dfrac{3}{2}x + 1\right]$

$= \dfrac{1}{2}(5x + 2)(x + 1)^{1/2}$

69. $6x^{1/2}(x^2 + x) - 8x^{3/2} - 8x^{1/2} = 2x^{1/2}(3(x^2 + x) - 4x^{2/2} - 4)$

$= 2x^{1/2}(3x^2 + 3x - 4x - 4)$

$= 2x^{1/2}(3x^2 - x - 4)$

$= 2x^{1/2}(3x - 4)(x + 1)$

≡ EXERCISE A.3 COMPLETING THE SQUARE; THE QUADRATIC FORMULA

1. $x^2 - 4x +$ _?_   4 should be added to complete the square.

3. $x^2 + \dfrac{1}{2}x +$ _?_ $\dfrac{1}{16}$ should be added to complete the square.

5. $x^2 + \dfrac{2}{3}x +$ _?_ $\dfrac{1}{9}$ should be added to complete the square.

7. $x^2 + 4x - 21 = 0$
        $x^2 + 4x + 4 = 21 + 4$
        $(x + 2)^2 = 25$

            $x + 2 = \pm\sqrt{25}$
            $x + 2 = \pm 5$
                $x = -2 \pm 5$

    $x = -2 + 5 = 3$ or $x = -2 - 5 = -7$

    The solution set is {-7, 3}.

9.
$$x^2 - \frac{1}{2}x = \frac{3}{16}$$

$$x^2 - \frac{1}{2}x + \frac{1}{16} = \frac{3}{16} + \frac{1}{16}$$

$$\left(x - \frac{1}{4}\right)^2 = \frac{1}{4}$$

$$x - \frac{1}{4} = \pm\sqrt{\frac{1}{4}}$$

$$x - \frac{1}{4} = \pm\frac{1}{2}$$

$$x = \frac{1}{4} \pm \frac{1}{2}$$

$$x = \frac{1}{4} + \frac{1}{2} = \frac{3}{4} \quad \text{or} \quad x = \frac{1}{4} - \frac{1}{2} = \frac{-1}{4}$$

The solution set is $\left\{\frac{-1}{4}, \frac{3}{4}\right\}$

11.
$$3x^2 + x - \frac{1}{2} = 0$$

$$x^2 + \frac{1}{3}x = \frac{1}{6}$$

$$x^2 + \frac{1}{3}x + \frac{1}{36} = \frac{1}{6} + \frac{1}{36}$$

$$\left(x + \frac{1}{6}\right)^2 = \frac{7}{36}$$

$$x + \frac{1}{6} = \pm\sqrt{\frac{7}{36}}$$

$$x = \frac{-1}{6} \pm \frac{\sqrt{7}}{6}$$

$$x = \frac{-1 - \sqrt{7}}{6} \quad \text{or} \quad x = \frac{-1 + \sqrt{7}}{6}$$

The solution set is $\left\{\frac{-1 - \sqrt{7}}{6}, \frac{-1 + \sqrt{7}}{6}\right\}$